TOURS SYMPOSIUM ON NUCLEAR PHYSICS V

Previous Proceedings in the Series of Tours Symposia on Nuclear Physics

Year	Volume	Publisher	ISBN
2000	IV	AIP Conf. Proceedings vol. 561	1-56396-996-3
1997	III	AIP Conf. Proceedings vol. 425	1-56396-749-9
1994	II	World Scientific Publishing	981-02-2156-8
1991	I	World Scientific Publishing	981-02-0892-8

Related Titles from AIP Conference Proceedings

701 The Labyrinth in Nuclear Structure: International Conference on The Labyrinth in Nuclear Structure, an EPS Nuclear Physics Divisional Conference
Edited by Angela Bracco and Constantine A. Kalfas, March 2004, 0-7354-0174-8

698 Intersections of Particle and Nuclear Physics, 8th Conference; CIPANP 2003
Edited by Zhoreh Parsa, February 2004, CD-ROM included, 0-7354-0169-1

681 Proton Emitting Nuclei: Second International Symposium; PROCON 2003
Edited by Enrico Maglione and Francesca Soramel, September 2003, 0-7354-0150-0

660 Hadron Physics: Effective Theories of Low Energy QCD; Second International Workshop on Hadron Physics
Edited by A. H. Blin, B. Hiller, A. A. Osiplov, M. C. Ruivo, and E. van Beveren, March 2003, 0-7354-0120-9

656 Frontiers of Nuclear Structure
Edited by Paul Fallon and Rod Clark, March 2003, 0-7354-0116-0

644 Exotic Clustering: 4th Catania Relativistic Ion Studies; CRIS 2002
Edited by Salvatore Costa, Antonio Insolia, and Cristina Tuvè, November 2002, 0-7354-0099-7

638 Mapping the Triangle: International Conference on Nuclear Structure
Edited by Ani Aprahamian, Jolie A. Cizewski, Stuart Pittel, and N. Victor Zamfir, November 2002, 0-7354-0093-8

610 Nuclear Physics in the 21st Century: International Nuclear Physics Conference, INPC 2001
Edited by Eric Norman, Lee Schroeder, and Gordon Wozninak, April 2002, 0-7354-0056-3

594 Hadrons and Nuclei: First International Symposium
Edited by Il-Tong Cheon, Taekeun Choi, Seung-Woo Hong, and Su Houng Lee, November 2001, 0-7354-0037-7

To learn more about these titles, or the AIP Conference Proceedings Series, please visit the webpage **http://proceedings.aip.org/proceedings**

TOURS SYMPOSIUM ON NUCLEAR PHYSICS V

Tours 2003

Tours, France   26-29 August 2003

EDITORS

M. Arnould
ULB, Belgium

M. Lewitowicz
GANIL, France

G. Münzenberg
GSI, Germany

H. Akimune
M. Ohta
H. Utsunomiya
T. Wada
T. Yamagata
Konan University, Japan

SPONSORING ORGANIZATIONS
Konan University, Japan
Conseil Général d'Indre-et-Loire, France
GANIL, France
GSI, Germany
Université François Rabelais, Tours, France
Lycée Collège Konan de Touraine, France

Melville, New York, 2004
AIP CONFERENCE PROCEEDINGS ■ VOLUME 704

Editors:

M. Arnould
Institut d'Astronomie et d'Astrophysique
Université Libre de Bruxelles
Campus Plaine CP 226
B-1050 Brussels
BELGIUM

E-mail: marnould@astro.ulb.ac.be

M. Lewitowicz
GANIL
BP 5027
F-14076 Caen Cedex
FRANCE

E-mail: lewitowicz@ganil.fr

G. Münzenberg
Kernphysik II
Gesellschaft für Schwerionenforschung
64291 Darmstadt
GERMANY

E-mail: G.Muenzenberg@gsi.de

H. Akimune, M. Ohta, H. Utsunomiya,
T. Wada, T. Yamagata

Department of Physics
Konan University
8-9-1 Okamoto
Higashinada
Kobe 658-8501
JAPAN

E-mail: akimune@konan-u.ac.jp
         masaota@konan-u.ac.jp
         hiro@konan-u.ac.jp
         wada@konan-u.ac.jp
         yamagata@center.konan-u.ac.jp

Authorization to photocopy items for internal or personal use, beyond the free copying permitted under the 1978 U.S. Copyright Law (see statement below), is granted by the American Institute of Physics for users registered with the Copyright Clearance Center (CCC) Transactional Reporting Service, provided that the base fee of $22.00 per copy is paid directly to CCC, 222 Rosewood Drive, Danvers, MA 01923. For those organizations that have been granted a photocopy license by CCC, a separate system of payment has been arranged. The fee code for users of the Transactional Reporting Service is: 0-7354-0177-2/04/$22.00.

© 2004 American Institute of Physics

Individual readers of this volume and nonprofit libraries, acting for them, are permitted to make fair use of the material in it, such as copying an article for use in teaching or research. Permission is granted to quote from this volume in scientific work with the customary acknowledgment of the source. To reprint a figure, table, or other excerpt requires the consent of one of the original authors and notification to AIP. Republication or systematic or multiple reproduction of any material in this volume is permitted only under license from AIP. Address inquiries to Office of Rights and Permissions, Suite 1NO1, 2 Huntington Quadrangle, Melville, N.Y. 11747-4502; phone: 516-576-2268; fax: 516-576-2450; e-mail: rights@aip.org.

L.C. Catalog Card No. 2004102081
ISBN 0-7354-0177-2
ISSN 0094-243X
Printed in the United States of America

CONTENTS

Preface xi
Organizing Committee xiii
Schedule xv

SUPERHEAVY ELEMENTS (SHE)

Vacuum, Matter and Antimatter 3
    W. Greiner
Study of Decays of ^{271}110 and ^{272}111 Produced with ^{208}Pb(^{64}Ni,n) and ^{209}Bi(^{64}Ni,n) Reactions 13
    K. Morita, K. Morimoto, D. Kaji, S. Goto, H. Haba, E. Idegichi,
    R. Kanungo, K. Katori, H. Koura, H. Kudo, T. Ohnishi, A. Ozawa,
    J. C. Peter, T. Suda, K. Sueki, I. Tanihata, F. Tokanai, H. Xu,
    A. V. Yeremin, A. Yoneda, A. Yoshida, Y.-L. Zhao, and T. Zheng
Properties of Heavy Nuclei Measured at the GSI SHIP 21
    S. Hofmann
Fusion-Fission Dynamics of Super-Heavy Element Formation and Decay 31
    V. I. Zagrebaev
Properties of Superheavy Nuclei 41
    I. Muntian, O. Parkhomenko, and A. Sobiczewski
Calculation of High-Dimensional Fission-Fusion Potential-Energy Surfaces in the SHE Region 49
    P. Möller, A. J. Sierk, T. Ichikawa, and A. Iwamoto
Ground-State Properties of Heavy and Superheavy Nuclei Predicted by Nuclear Mass Models 60
    H. Koura

FUSION-FISSION DYNAMICS (FFD)

Effect of Closed Shell Structure on Heavy-Ion Fusion Reactions 73
    H. Ikezoe, K.-i Satou, S. Mitsuoka, K. Nishio, K. Tsuruta, S.-C. Jeong, and
    C.-J. Lin
Fusion Dynamics around the Coulomb Barrier 82
    K. Hagino, N. Rowley, T. Ohtsuki, M. Dasgupta, J. O. Newton, and
    D. J. Hinde
Superheavy Element Production, Nucleus-Nucleus Potential and μ-Catalysis 92
    V. Y. Denisov
Shell Stabilization in Compound Nucleus Survival 102
    A. R. Junghans, K.-H. Schmidt, A. M. Heinz, and A. V. Ignatyuk
Multi-Modal Nuclear Fission in the Actinide Nuclei 111
    T. Asano, T. Wada, M. Ohta, T. Ichikawa, S. Yamaji, and H. Nakahara

Effects of Nuclear Structure in the Transport Coefficients of
Large-Scale Collective Motion 120
 F. Ivanyuk
Stochastic Model of the Tilting Mode in Nuclear Fission....................130
 V. A. Drozdov, D. O. Eremenko, O. V. Fotina, S. Y. Platonov, and
 O. A. Yuminov
Tracking Dissipation in Capture Reactions..................................139
 T. Materna, V. Bouchat, V. Kinnard, F. Hanappe, O. Dorvaux, C. Schmitt,
 L. Stuttgé, K. Siwek-Wilczynska, Y. Aritomo, A. Bogatchev,
 E. Prokhorova, and M. Ohta
Dynamics of Fusion-Fission Process in Superheavy Mass Region..............147
 Y. Aritomo, M. Ohta, T. Materna, F. Hanappe, and L. Stuttge

PHYSICS WITH EXOTIC NUCLEI (PEN)

Physics of Nuclei Far From Stability

Two-Proton Radioactivity—A Curiosity of Nature?..........................159
 B. Blank
The Search for Neutral Nuclei 169
 F. M. Marqués Morano
Recent Results from the β-Decay Studies in the ^{100}Sn Region................176
 Z. Janas, L. Batist, A. Blazhev, W. Brüchle, J. Döring, M. Gierlik,
 M. Górska, H. Grawe, T. Faestermann, S. Harissopulos, A. Jungclaus,
 M. Karny, M. Kavatsyuk, O. Kavatsyuk, R. Kirchner, M. La Commara,
 C. Mazzocchi, I. Mukha, C. Plettner, A. Płochocki, E. Roeckl, M. Romoli,
 M. Schädel, R. Schwengner, and J. Żylicz
Transfer Reaction Studies with Exotic Nuclei 185
 W. N. Catford, R. C. Lemmon, C. N. Timis, M. Labiche, L. Caballero, and
 R. Chapman

New Results with RI Beams

Study of ^{19}Na at SPIRAL 195
 F. de Oliveira Santos and the E400S Collaboration
The EXODET Apparatus and Its First Experimental Results : ^{17}F
Scattering by ^{208}Pb Below the Coulomb Barrier 202
 M. Romoli, M. Mazzocco, E. Vardaci, R. Bonetti, A. De Francesco, A. De
 Rosa, M. Di Pietro, T. Glodariu, A. Guglielmetti, G. Inglima, M. La
 Commara, B. Martin, V. Masone, P. Parascandolo, D. Pierroutsakou,
 M. Sandoli, P. Scopel, C. Signorini, F. Soramel, L. Stroe, J. Greene,
 A. Heinz, D. Henderson, C. L. Jiang, E. F. Moore, R. C. Pardo,
 K. E. Rehm, A. Wuosmaa, and J. F. Liang

Subbarrier Fusion in the Systems ^{11,10}Be+^{209}Bi: The
Experimental Data 212
    C. Signorini, A. Yoshida, Y. Watanabe, D. Pierroutsakou, L. Stroe,
    T. Fukuda, M. Mazzocco, N. Fukuda, Y. Mizoi, M. Ishihara, H. Sakurai,
    and F. Soramel

A Non Perturbative Approach to Neutron and Proton Halo Breakup 218
    A. Bonaccorso

Present Status and Future Plan of RIB Facilities

The New Radioactive Ion Beam Facility at GSI 228
    H. Weick

The SPIRAL2 Project at GANIL.. 234
    A. C. C. Villari for the SPIRAL2 Group

Perspective of the RIKEN Radioisotope Beam Factory Project 245
    T. Motobayashi

Tri-Nucleon Cluster Structure in ^{6}He and ^{6}Be........................... 253
    H. Akimune, T. Yamagata, S. Nakayama, M. Fujiwara, K. Fushimi,
    K. Hara, K. Y. Hara, K. Ichihara, K. Kawase, Y. Matsui, K. Nakanishi,
    A. Shiokawa, M. Tanaka, H. Utsunomiya, and M. Yosoi

New Aspects of Clustering Structure of Light Nuclei

Search for Excited α-Cluster Resonances and Their Analogs in A=6
and 7 Nuclei... 261
    T. Yamagata, S. Nakayama, H. Akimune, M. Fujiwara, K. Fushimi,
    M. B. Greenfield, K. Hara, K. Y. Hara, K. Hashimoto, K. Ichihara,
    K. Kawase, M. Kinoshita, Y. Matsui, K. Nakanishi, M. Tanaka,
    H. Utsunomiya, and M. Yosoi

^{6}Li Excitation above the Breakup Threshold in the ^{6}Li+^{208}Pb System
at Coulomb Barrier Energies 273
    M. Mazzocco, P. Scopel, C. Signorini, L. Fortunato, F. Soramel,
    I. J. Thompson, A. Vitturi, M. Barbui, A. Brondi, M. Cinausero, D. Fabris,
    E. Fioretto, G. La Rana, M. Lunardon, R. Moro, A. Ordine, G. Prete,
    V. Rizzi, L. Stroe, M. Trotta, E. Vardaci, and G. Viesti

Resonance Structure of ^{9}Be and ^{10}Be in a Microscopic Cluster Model 283
    K. Arai

Physics of Cluster Structure of Nuclei

Structure of Continuum States in Unstable Nuclei 293
    K. Kato, M. Myo, and K. Ikeda

Clustering in Exotic Nuclei Studies by Transfer Reactions 301
    R. Wolski, Y. M. Tchuvil'sky, S. D. Kurgalin, G. M. Ter-Akopian,
    P. Roussel-Chomaz, L. Giot, and K. Rusek

PHYSICS FOR NUCLEAR TRANSMUTATION (PNT)

Nuclear Waste Incineration by ADS and Main Aspects of the Accelerator Studied within the European PDS-XADS 313
    A. C. Mueller

ADS Network in Japan and in Asia..................................323
    Y. Nagai

NUCLEAR ASTROPHYSICS (NAP)

Nuclear Physics in Space: From Solar Energetic Particles to Ultra-High Energy Cosmic Rays

Nucleosynthesis by Spallation Reactions in the Early Solar System — The Need for Spallation Cross Sections...............................331
    I. Leya

Updated Big-Bang Nucleosynthesis Compared to WMAP Results 341
    A. Coc, E. Vangioni-Flam, P. Descouvemont, A. Adahchour, and C. Angulo

Propagation of Ultra-High-Energy Nuclei in the Extra-Galactic Photon Field 351
    T. Yamamoto and M. Teshima

Radionuclides in the Galaxy: Some Selected Aspects

Radio Nuclides in the Galaxy Seen in Gamma-Rays.......................361
    V. Schönfelder

Radioactivity of the Key Isotope ^{44}Ti in SN 1987A........................369
    Y. Motizuki and S. Kumagai

Nuclear Data for Low-Energy Astrophysics and Other Applications

BRUSLIB: The Brussels Nuclear Library for Astrophysics Applications 375
    S. Goriely

Building Better Optical Model Potentials for Nuclear Astrophysics Applications 385
    E. Bauge and M. Dupuis

Microscopic Dipole Strength Predictions for Neutron Capture Rates..........395
    E. Khan, M. Samyn, and S. Goriely

Fission Barriers from a Microscopic Model 401
    M. Samyn and G. Goriely

Weak Interaction Processes in Stars 408
    I. N. Borzov

Nuclear Data for Low-Energy Astrophysics and Other Applications —
An Addendum 418
    K. Takahashi

Low-Energy Nuclear Reactions for Astrophysics

Cross Section Measurements of Capture Reactions Relevant to the p
Process: Status and Perspectives 422
    S. V. Harissopulos

Experiments with Radioactive Beams in Nuclear Astrophysics:
Evolutions and Perspectives 432
    P. Leleux

Photoreaction Cross Section Measurements for Astrophysics 439
    H. Utsunomiya, S. Goko, K. Y. Hara, H. Akimune, T. Yamagata,
    M. Ohta, H. Ohgaki, H. Toyokawa, T. Hayakawa, T. Shizuma,
    Y.-W. Lui, and P. Mohr

Low-Energy Radioactive-Ion Beam Separator at CNS and Resonance
Scattering Experiments 447
    T. Teranishi, S. Kubono, J. J. He, M. Notani, T. Fukuchi, S. Michimasa,
    S. Shimoura, S. Nishimura, M. Nishimura, Y. Yanagisawa, M. Kurokawa,
    Y. Wakabayashi, N. Hokoiwa, Y. Gono, T. Morikawa, A. Odahara,
    H. Ishiyama, Y. X. Watanabe, T. Hashimoto, T. Ishikawa, M. H. Tanaka,
    H. Miyatake, J. Y. Moon, J. H. Lee, J. C. Kim, C. S. Lee, V. Guimarães,
    R. F. Lihitenthaler, H. Baba, A. Saito, K. Sato, T. Kawamura, S. Kato,
    H. Iwasaki, K. Ue, Y. Satou, and Z. Fülöp

Study of Astrophysical (α,n) and (p,n) Reactions on Light Neutron-
Rich Nuclei by Means of Low-Energy RNB 453
    H. Ishiyama, H. Miyatake, M.-H. Tanaka, Y. Watanabe, N. Yoshikawa,
    S. Jeong, Y. Matsuyama, Y. Fuchi, I. Katayama, T. Nomura, T. Hashimoto,
    T. Ishikawa, K. Nakai, S. K. Das, P. K. Saha, T. Fukuda, K. Nishio,
    S. Mitsuoka, H. Ikezoe, M. Matsuda, S. Ichikawa, T. Furukawa,
    H. Izumi, T. Shimoda, Y. Mizoi, and M. Terasawa

The (n,γ) Cross Sections of Short-Living s-Process Branching Points 463
    K. Sonnabend, A. Mengoni, P. Mohr, T. Rauscher, K. Vogt, and A. Zilges

Neutron Stars and Other Stars

Microscopic Nuclear Equation of State with Three-Body Forces and
Neutron Star Structure 473
    U. Lombardo, G. F. Burgio, H.-J. Schulze, W. Zuo, and X. R. Zhou

Unexpected Goings-On in the Structure of a Neutron Star Crust 483
    A. Bulgac, P.-H. Heenen, P. Magierski, A. Wirzba, and Y. Yu

Nucleosynthesis in Supernovae and the Early Universe 488
    T. Kajino, T. Sasaqui, M. Orito, K. Otsuki, G. J. Mathews, S. Honda,
    W. Aoki, and S. Chiba

POSTER SESSIONS

Neutron Evaporation as a Probe for Dynamical Effects in Heavy Ion Fusion Reactions 501
    Aj. Kumar, A. Kumar, G. Singh, H. Singh, R. P. Singh, R. Kumar, K. S. Golda, S. K. Datta, and I. M. Govil

What Can We Learn About the Fission Process from the Energy Dependence of the Induced Fission Times Obtained by the Crystal Blocking Technique? 507
    V. A. Drozdov, D. O. Eremenko, O. V. Fotina, S. Y. Platonov, and O. A. Yuminov

Energy Dependence of the Shell Corrections Obtained from Analysis of Fission Fragment 513
    V. A. Drozdov, D. O. Eremenko, S. Y. Platonov, O. V. Fotina, and O. A. Yuminov

Kaon Condensation and the Non-Uniform Nuclear Matter 519
    T. Maruyama, T. Tatsumi, D. N. Voskresensky, T. Tanigawa, and S. Chiba

Production of Zero-Energy Radioactive Nuclear Beams through Extraction from the Liquid-Vapour Interface of Superfluid Helium 526
    N. Takahashi, W. X. Huang, P. Dendooven, K. Gloos, J. P. Pekola, and J. Äystö

Photon-Induced Reactions in Stars and in the Laboratory: A Critical Comparison 532
    P. Mohr

Microscopic Calculations of Spontaneous Fission Life-times and Neutron-Induced Fission Cross Sections 540
    P. Demetriou, M. Samyn, and S. Goriely

Measurement of Photo-Destruction Cross Sections for ^{180}Ta^m at SPring-8 546
    S. Goko and H. Utsunomiya

Photodisintegration of Deuterium in the Precision Era of Big Bang Nucleosynthesis 551
    K. Y. Hara, H. Utsunomiya, H. Akimune, S. Goko, K. Kudo, Y.-W. Lui, H. Ohgaki, M. Ohta, Y. Shibata, H. Toyokawa, A. Uritani, and T. Yamagata

On the Excitation Energy for Maximum Cold Fusion Reactions in Superheavy Mass Region 557
    A. Fukushima, T. Wada, M. Ohta, and Y. Aritomo

New ANC Measurement of the Astrophysical S_{17}-Factor from d(^{7}Be,^{8}B)n Reaction 563
    J. J. Das, V. M. Datar, P. Sugathan, N. Madhaven, P. V. Madhusudhana Rao, A. Navin, A. Jhingan, T. Varughese, S. Nath, A. Ray, S. Barua, R. Singh, S. K. Dhiman, R. Shyam, R. G. Kulkarni, A. K. Sinha, and D. L. Sastry

List of Participants 569
Conference Photos 577
Author Index 581

PREFACE

Tours Symposium on Nuclear Physics V (August 26 – 29, 2003) was devoted to 5 topics in nuclear physics: SHE (Super-heavy Elements), FFD (Fusion-fission Dynamics), PEN (Physics with Exotic Nuclei), PNT (Physics for Nuclear Transmutation) and NAP (Nuclear Astrophysics). We had active discussions in every topic that enhance further developments in the future, thanks to all participants. Round table discussions added lively accents to NAP sessions, thanks to moderators. We thank the members of the organizing committee for their efforts in bringing science and people together to Tours, France. It was a great memory for us to celebrate Professor Nomura's (KEK) pioneering role toward synthesis of super-heavy elements in Japan at his retirement age. During the symposium, we had an informal meeting to discuss construction of an extended nuclear database for astrophysics in the guesthouse of Conceil Général d'Indre-et-Loire.

Ten young scientists including Ph.D. students were invited from India, Japan, Germany, Belgium, Italy and Russia based on the financial support of GSI. We are grateful to the continued financial support of GANIL for publication of the proceedings. We enjoyed a spectacular of the traditional band of hunting horns at the Castle of Montpoupon and a gorgeous conference dinner in the Logis Royal de Loches.

Tours2003 was strongly supported by the following organizations: Conceil Général d'Indre-et-Loire (Mr. POMMEREAU, Président; Mr. FORTIER, Président de l'ADT; Mr. BARBE, Directeur Général des Services; Angèle PLOQUIN, Chargée de Mission-ADT), Université Françoir Rabelais, Tours (Prof. LUSSAULT, Président), GANIL (Prof. GOUTTE, Directeur), GSI (Prof. MÜNZENBER), Lycée Collège Konan de Touraine (Mr. TANAKA, Directeur; Mrs. CREOLA, Assistante), and Konan University (Mr. IKEGAMI, Chief Director; Prof. YOSHIZAWA, President; Prof. SUGIMURA, Vice-President).

We look forward to getting together in the sixth Tours Symposium in 2006.

M. Ohta
H. Utsunomiya
Konan University

ORGANIZING COMMITTEE

H. Akimune (*Konan, Japan*)
M. Arnould (*ULB, Belgium*)
D. Goutte (*GANIL, France*)
W. Greiner (*Frankfurt, Germany*)
D. Guerreau (*IN2P3/CNRS, France*)
A. Iwamoto (*JAERI, Japan*)
T. Kajino (*NAO, Japan*)
S. Kubono (*CNS, Japan*)
M. Lewitowicz (*GANIL, France*)
G. Münzenberg (*GSI, Germany*)
T. Nomura (*KEK, Japan*)
Yu. Ts. Oganessian (*FLNR-JINR, Russia*)
M. Ohta (*Konan, Japan*)
C. Rolfs (*Bochum, Germany*)
C. Signorini (*Padova, Italy*)
H. Utsunomiya (*Konan, Japan*)
T. Wada (*Konan, Japan*)
T. Yamagata (*Konan, Japan*)

HOST INSTITUTE

Konan University

SUPPORTED BY

Conseil Général d'Indre-et-Loire
GANIL
GSI
Université François Rabelais, Tours
Lycée Collège Konan de Touraine

SCHEDULE

	Aug. 26 (Tue)		Aug. 27 (Wed)	Aug. 28 (Thu)		Aug. 29 (Fri)
09:00-10:30	SHE1		FFD3	NAP1		NAP6
	Coffee break					
11:00-12:30	SHE2		Poster	NAP2		NAP7
12:30-14:00	Lunch					
14:00-15:30	SHE3		PEN3	PEN4	NAP3	NAP8
	Coffee break					
16:00-17:30	FFD1	PEN1	PNT	PEN5	NAP4	
	Coffee break		Excursion Banquet	Coffee break		
18:00-19:30	FFD2	PEN2		NAP5		

SHE : Superheavy Elements
FFD : Fusion-Fission Dynamics
PEN : Physics with Exotic Nuclei
PNT : Physics for Nuclear Transmutation
NAP : Nuclear Astrophysics

Chairpersons

SHE1	G. Münzenberg (GSI)
SHE2	T. Nomura (KEK)
SHE3	M. Ohta (Konan)
FFD1	V. Zagrebaev (FLNR-JINR)
FFD2	T. Wada (Konan)
FFD3	K. Morita (RIKEN)
PEN1	C. Signorini (Padova)
PEN2	M. Lewitowicz (GANIL)
PEN3	D. Goutte (GANIL)
PEN4	K. Kato (Hokkaido)

PEN5	N. Takahashi (Osaka Gakuin)
PNT	A. Iwamoto (JAERI)
NAP1	T. Kajino (NAO)
NAP2	W. R. Binns (St Louis)
NAP3	J. Knödlseder (Toulouse)
NAP4	M. Arnould (Brussels)
NAP5	K. Takahashi (Heidelberg)
NAP6	Y. Nagai (Osaka)
NAP7	H. Utsunomiya (Konan)
NAP8	A. Coc (Orsay)

SUPERHEAVY ELEMENTS (SHE)

Vacuum, matter and antimatter

Walter Greiner

Institut für Theoretische Physik, Goethe Universität, D–60054 Frankfurt am Main

Abstract. In this talk I first present the vacuum for the e^+-e^- field of QED and show how it is modified for baryons in nuclear environment. Then I discuss the possibility of producing new types of nuclear systems by implanting an antibaryon into ordinary nuclei. The structure of nuclei containing one antiproton or antilambda is investigated within the framework of a relativistic mean-field model. Self-consistent calculations predict an enhanced binding and considerable compression in such systems as compared with normal nuclei. I present arguments that the life time of such nuclei with respect to the antibaryon annihilation might be long enough for their observation.

It is generally accepted that physical vacuum has nontrivial structure. This conclusion was first made by Dirac on the basis of his famous equation for a fermion field which describes simultaneously particles and antiparticles. The Dirac equation in the vacuum has a simple form

$$(i\gamma^\mu \partial_\mu - m)\Psi(x) = 0, \tag{1}$$

where $\gamma^\mu = (\gamma^0, \boldsymbol{\gamma})$ are Dirac matrices, m is the fermion mass and $\Psi(x)$ is a 4-component spinor field. For a plane wave solution $\Psi(x) = e^{-ipx} u_p$ this equation is written as

$$(\hat{p} - m)u_p = 0, \tag{2}$$

where $\hat{p} = \gamma^0 E - \boldsymbol{\gamma}\mathbf{p}$. Multiplying by $(\hat{p} + m)$ and requiring that $u_p \neq 0$ one obtains the equation $E^2 - \mathbf{p}^2 - m^2 = 0$ which has two solutions

$$E^\pm(\mathbf{p}) = \pm\sqrt{\mathbf{p}^2 + m^2}. \tag{3}$$

Here the $+$ sign corresponds to particles with positive energy $E_N(\mathbf{p}) = E^+(\mathbf{p})$, while the $-$ sign corresponds to solutions with negative energy. To ensure stability of the physical vacuum Dirac has assumed that these negative-energy states are occupied forming what is called now the Dirac sea. Then the second solution of eq. (3) receives natural interpretation: it describes holes in the Dirac sea. These holes are identified with antiparticles. Their energies are obviously given by $E_{\overline{N}}(\mathbf{p}) = -E^-(-\mathbf{p}) = \sqrt{\mathbf{p}^2 + m^2}$. Unfortunately, the Dirac sea brings divergent contributions to physical quantities such as energy density, and one should introduce a proper regularization scheme to rid off these divergences. This picture has received numerous confirmations in quantum electrodynamics and other fields.

One of the most fascinating aspects is the structure of the vacuum in QED and its change into charged vacuum states under the influence of strong (supercritical) electric fields [1]. I shortly remind of this phenomenon.

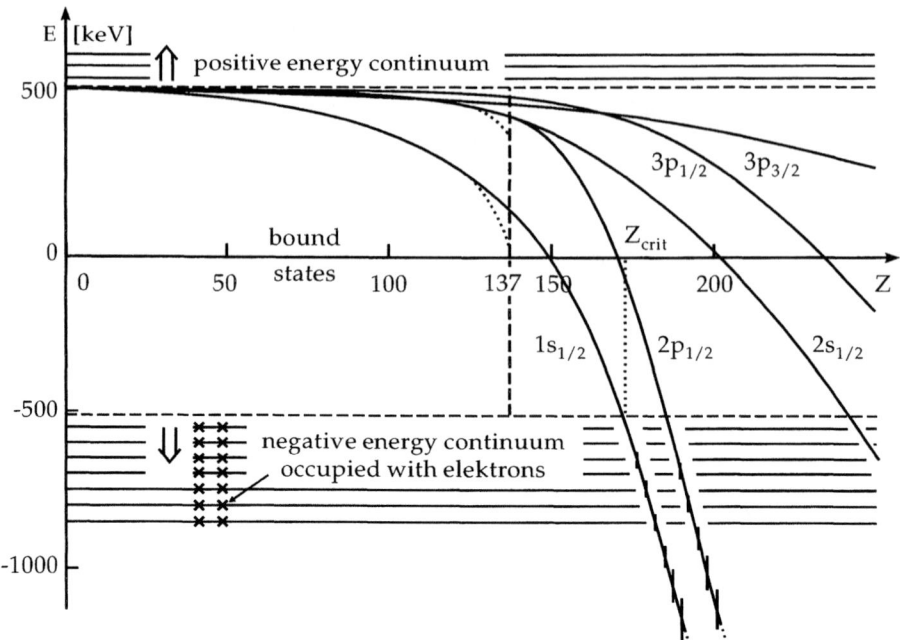

FIGURE 1. Lowest bound states of the Dirac equation for nuclei with charge Z. While the Sommerfeld fine-structure energies (dashed line) for $\xi = 1$ (s states) end at $Z = 137$, the solutions for extended Coulomb potentials (full line) can be traced down to the negative-energy continuum reached at the critical charge Z_{cr} for the $1s$ state. The bound states entering the continuum obtain a spreading witdth as indicated.

Fig. 1 shows the diving of the deeply bound states into the lower energy continuum of the Dirac equation. In the supercritical case the dived state is degenerate with the (occupied) negative electron states. Hence spontaneous $e^+ e^-$ *paircreation* becomes possible, where an electron from the Dirac sea occupies the additional state, leaving a hole in the sea which escapes as a positron while the electron's charge remains near the source. This is a fundamentally new process, whereby the neutral vacuum of QED becomes unstable in supercritical electrical fields. It decays within about $10^{-19}s$ into a charged vacuum. The charged vacuum is now stable due to the Pauli principle, that is the number of emitted particles remains finite. The vacuum is first charged twice because two electrons with opposite spins can occupy the $1s$ shell. After the $2p_{1/2}$ shell has dived beyond $Z_{cr} = 185$, the vacuum is charged four times, etc. This change of the vacuum structure is not a perturbative effect, as are the radiative QED effects (vacuum polarization, self-energy, etc.).

The time-dependence of the energy levels in a supercritical heavy-ion collision is depicted in Fig. 2. An electron (or hole) which was in a certain molecular eigenstate at the beginning of the collision can be transfered with a certain probability into different states by the dynamics of the collision. This can lead to the hole production in an inner shell by excitation of an electron to a higher state and/or hole production by ionization

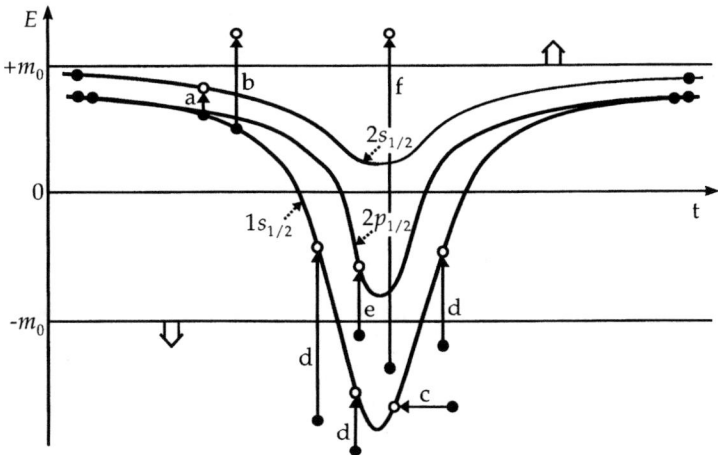

FIGURE 2. Time dependence of the quasi-molecular energy levels in a supercritical heavy ion collision. The arrows denote various excitation processes which lead to the production of holes and positrons.

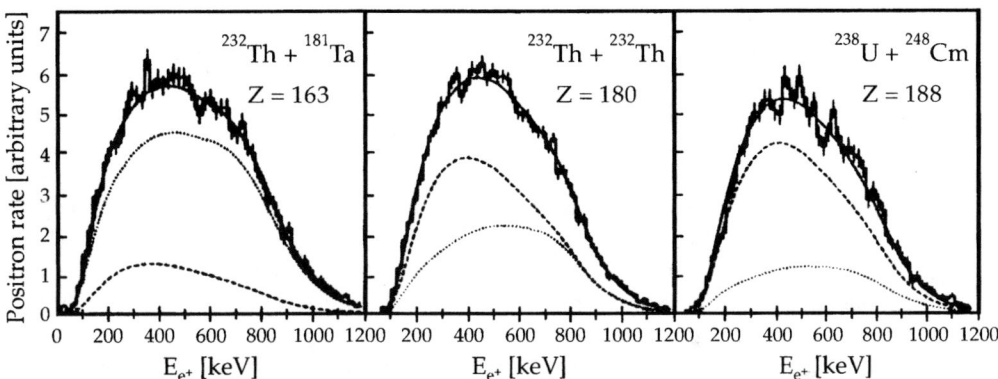

FIGURE 3. Positron energy spectra measured in collisions of Th+Ta, Th+Th, and U+Cm at energies of about 6 MeV per nucleon. The QED predictions (dashed lines) and the experimentally determined background from nuclear pair conversion (dotted lines) add up to the full lines which are in close agreement with experiment.

of an electron to the continuum. Further possibilities are induced positron production by excitation of an electron from the lower continuum to an empty bound level and direct pair production [2].

A comparison of the theoretical predictions and expectations and experimental data is shown in figure 3. Sharp positron peeks can be expected if there were a mechanism in the heavy ion collision leading to a time delay. This may be caused by a pocket in the potential between the two ions. Spontaneous pair production should then be enhanced in supercritical systems. Until now, however, the situation remains inconclusive [2].

It has been noticed already many years ago (see e. g. ref. [3]) that nuclear physics may provide a unique laboratory for investigating the Dirac picture of vacuum. The basis for this is given by relativistic mean-field models which are widely used now for describing nuclear matter and finite nuclei. Within this approach nucleons are described by the Dirac equation coupled to scalar and vector meson fields. Scalar S and vector V potentials generated by these fields modify plane-wave solutions of the Dirac equation as follows

$$E^{\pm}(\mathbf{p}) = V \pm \sqrt{\mathbf{p}^2 + (m-S)^2} . \qquad (4)$$

Again, the $+$ sign corresponds to nucleons with positive energy $E_N(\mathbf{p}) = V + \sqrt{\mathbf{p}^2 + (m-S)^2}$, and the $-$ sign corresponds to antinucleons with energy $E_{\bar{N}}(\mathbf{p}) = -E^-(-\mathbf{p}) = -V + \sqrt{\mathbf{p}^2 + (m-S)^2}$. It is remarkable that changing sign of the vector potential for antinucleons is exactly what is expected from the G-parity transformation of the nucleon potential. As follows from eq. (4), in nuclear environment the spectrum of single-particle states of the Dirac equation is modified in two ways. First, the mass gap between positive- and negative-energy states, $2(m-S)$, is reduced due to the scalar potential and second, all states are shifted upwards due to the vector potential. These changes are illustrated in Fig. 4.

It is well known from nuclear phenomenology that good description of nuclear ground state is achieved with $S \simeq 350$ MeV and $V \simeq 300$ MeV so that the net potential for nucleons is $V - S \simeq -50$ MeV. Using the same values one obtains for antinucleons very a deep potential, $-V - S \simeq -650$ MeV. Such a potential would produce many strongly bound states in the Dirac sea. However, if these states are occupied they are hidden from the direct observation. Only creating a hole in this sea, i.e. inserting a real antibaryon into the nucleus, would produce an observable effect. If this picture is correct one should expect the existence of strongly bound states of antinucleons with nuclei. Below I report on our recent study of antibaryon-doped nuclear systems [4].

Unlike some previous works, we take into account the rearrangement of nuclear structure due to the presence of a real antibaryon. The structure of such systems is calculated using several versions of the relativistic mean–field model (RMF): TM1 [5], NL3 and NL-Z2 [6]. Their parameters were found by fitting binding energies and charge form-factors of spherical nuclei from ^{16}O to ^{208}Pb. The general Lagrangian of the RMF model is written as

$$\begin{aligned}
\mathcal{L} = & \sum_{j=B,\bar{B}} \overline{\psi}_j \left(i\gamma^\mu \partial_\mu - m_j \right) \psi_j \\
& + \frac{1}{2} \partial^\mu \sigma \partial_\mu \sigma - \frac{1}{2} m_\sigma^2 \sigma^2 - \frac{b}{3}\sigma^3 - \frac{c}{4}\sigma^4 \\
& - \frac{1}{4} \omega^{\mu\nu}\omega_{\mu\nu} + \frac{1}{2} m_\omega^2 \omega^\mu \omega_\mu + \frac{d}{4}(\omega^\mu \omega_\mu)^2 \\
& - \frac{1}{4} \vec{\rho}^{\mu\nu}\vec{\rho}_{\mu\nu} + \frac{1}{2} m_\rho^2 \vec{\rho}^{\,\mu}\vec{\rho}_\mu \\
& + \sum_{j=B,\bar{B}} \overline{\psi}_j \left(g_{\sigma j}\sigma + g_{\omega j}\omega^\mu \gamma_\mu + g_{\rho j}\vec{\rho}^{\,\mu}\gamma_\mu \vec{t}_j \right) \psi_j \\
& + \text{Coulomb part}
\end{aligned} \qquad (5)$$

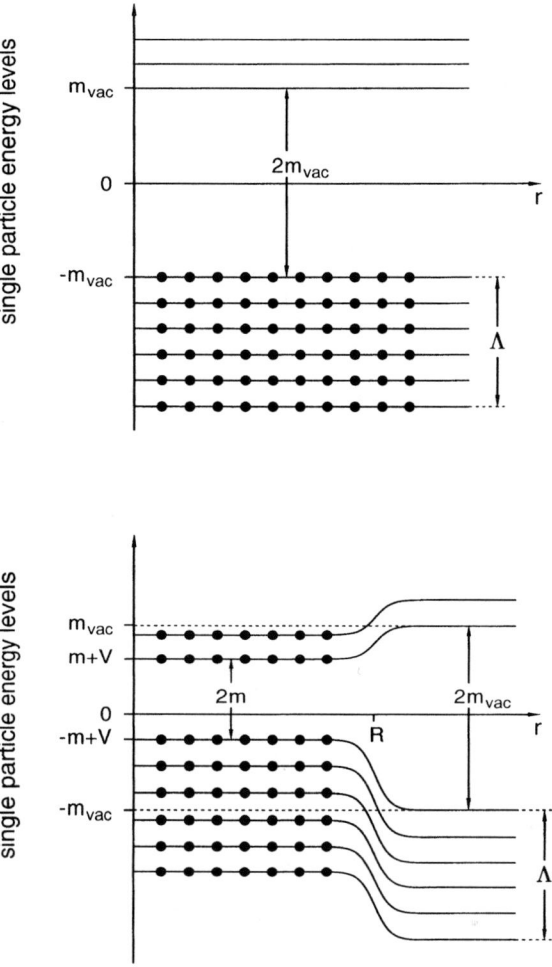

FIGURE 4. Schematic spectrum of Dirac equation in vacuum (upper panel) and in a nucleus of radius R (lower panel). A divergent contribution of negative-energy states is often regularized by introducing a cut-off momentum Λ

Here summation includes valence baryons B, in fact the nucleons forming a nucleus, and valence antibaryons $\bar{B}$ inserted in the nucleus. They are treated as Dirac particles coupled to the scalar-isoscalar (σ), vector-isoscalar (ω) and vector-isovector ($\vec{\rho}$) meson fields. The calculations are carried out within the mean-field approximation where the meson fields are replaced by their expectation values. Also a "no-sea" approximation is used. This implies that all occupied states of the Dirac sea are "integrated out" so that they do not appear explicitly. It is assumed that their effect is taken into account by nonlinear terms in the meson Lagrangian. Most calculations are done with antibaryon coupling constants which are given by the G-parity transformation ($g_{\sigma\bar{N}} = g_{\sigma N}$, $g_{\omega\bar{N}} = -g_{\omega N}$)

and $SU(3)$ flavor symmetry ($g_{\sigma\bar{\Lambda}} = \frac{2}{3} g_{\sigma\bar{N}}$, $g_{\omega\bar{\Lambda}} = \frac{2}{3} g_{\omega\bar{N}}$). In isosymmetric static systems the scalar and vector potentials for nucleons are expressed as $S = g_{\sigma N}\sigma$ and $V = g_{\omega N}\omega^0$.

Following the procedure suggested in Ref. [7] and assuming the axial symmetry of the nuclear system, we solve effective Schrödinger equations for nucleons and an antibaryon together with differential equations for mean meson and Coulomb fields. We explicitly take into account the antibaryon contributions to the scalar and vector densities. It is important that antibaryons give a negative contribution to the vector density, while a positive contribution to the scalar density. This leads to increased attraction and decreased repulsion for surrounding nucleons. To maximize attraction, nucleons move to the center of the nucleus, where the antiproton has its largest occupation probability. This gives rise to a strong local compression of the nucleus and leads to a dramatic rearrangement of its structure.

Results for the ^{16}O nucleus are presented in Fig. 5 which shows 3d plots of nucleon density distributions. The calculations show that inserting an antiproton into the ^{16}O nucleus leads to the increase of central nucleon density by a factor 2–4 depending on the parametrization. Due to a very deep antiproton potential the binding energy of the whole system is increased significantly as compared with 130 MeV for normal ^{16}O. The calculated binding energies of the $\bar{p}-^{16}$O system are 830, 1050 and 1160 MeV for the NL–Z2, NL3 and TM1, respectively. Due to this anomalous binding we call such systems super bound nuclei (SBN). In the case of antilambdas we rescale the coupling constants with a factor 2/3 that leads to the binding energy of 560÷700 MeV for the $\bar{\Lambda}-^{16}$O system.

As a second example, we investigate the effect of a single antiproton inserted into the ^{8}Be nucleus. The normal ^{8}Be nucleus is not spherical, exhibiting a clearly visible 2α structure with the ground state deformation $\beta_2 \simeq 1.20$. As seen in Fig. 6, inserting an antiproton in ^{8}Be results in a much less elongated shape ($\beta_2 \simeq 0.23$) and disappearance of its cluster structure. The binding energy increases from 53 MeV to about 700 MeV. Similar, but weaker effects have been predicted [8] for the K^- bound state in the ^{8}Be nucleus.

The calculations have been performed also with reduced antinucleon coupling constants as compared to the G-parity prescription. We have found that the main conclusions about enhanced binding and considerable compression of $\bar{p}$-doped nuclei remain valid even when coupling constants are reduced by factor 3 or so.

Now I would like to discuss the structural effect of an antiproton in the doubly magic lead nucleus. A contour plot of the sum of proton and neutron densities is shown in figure 7.

In this case we encounter a quite different scenario: again, the complete system is affected, but not in the sense that the whole nucleus shrinks and becomes very dense. Here, a small and localized region of high density develops within the heavy system. Additionally, the lead nucleus deforms itself. This effect is even larger for the case of lead with an $\bar{\alpha}$. The reason for this behaviour can be understood from the properties of the single-particle wavefunctions in the mean-field: In a small region with a deep potential, only states with small angular momenta can be bound deeply. States with higher angular momenta do not have much overlap with the potential. This is exactly what happens here. Basically only the lowest s- and p-states can be bound

Sum of proton and neutron densities for ^{16}O (top),
^{16}O with $\bar{\Lambda}$ (bottom left) and ^{16}O with $\bar{p}$ (bottom right)

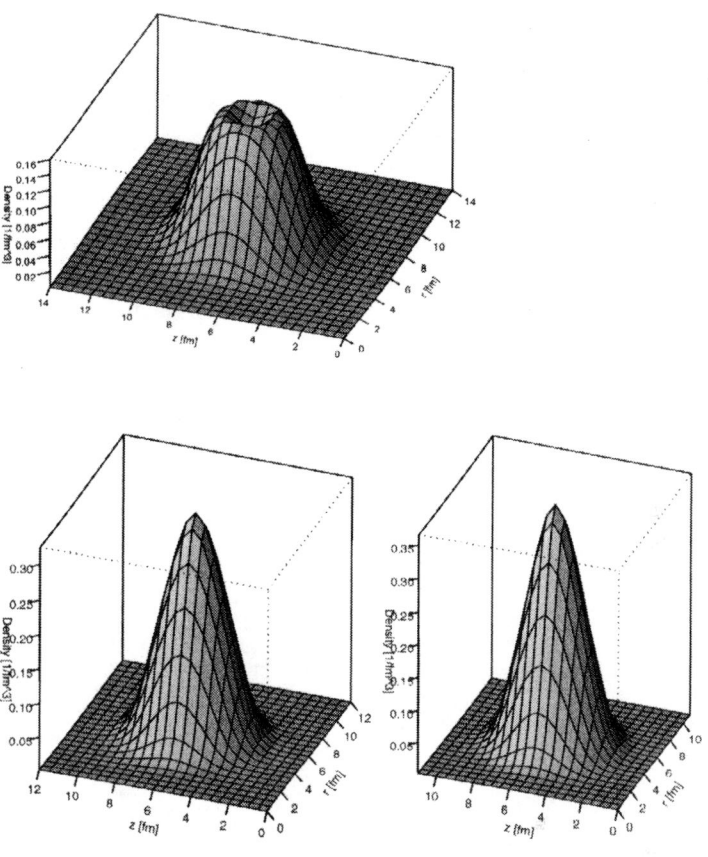

FIGURE 5. Sum of neutron and proton densities for ^{16}O (top), ^{16}O with $\bar{p}$ (bottom right) and ^{16}O with $\bar{\Lambda}$ (bottom left) calculated with the parametrization NL-Z2.

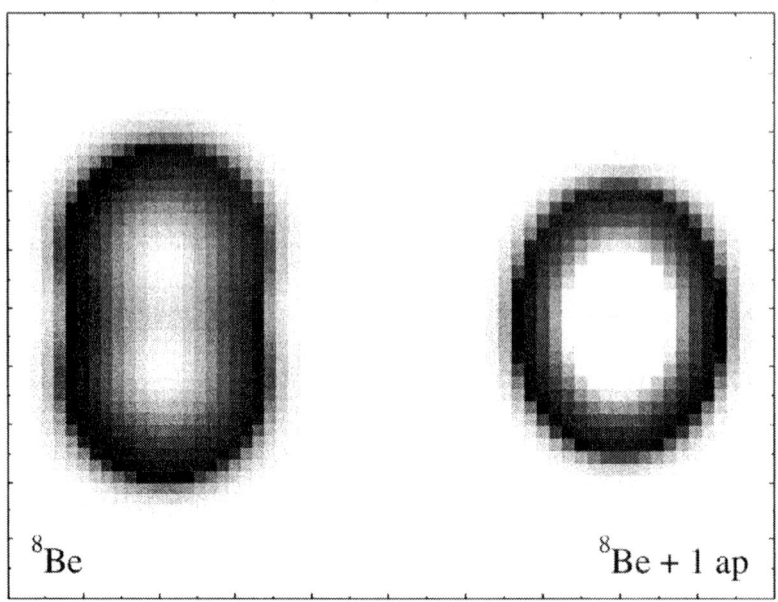

FIGURE 6. Contour plot of nucleon densities for ^{8}Be without (left) and with (right) antiproton calculated with the parametrization NL3.

deeper than for lead without any antiparticles present. Higher lying states do not gain significantly binding or are even lesser bound. The deformation effect probably has two reasons: firstly, a deformation might be energetically favourable to gain some binding for the higher lying states. Secondly, the distortion of the system due to the presence of antiparticles destroys the magicitiy of the system.

The crucial question concerning possible observation of the SBNs is their life time. The main decay channel for such states is the annihilation of antibaryons on surrounding nucleons. The energy available for annihilation of a bound antinucleon equals $Q = 2m_N - B_N - B_{\bar{N}}$, where B_N and $B_{\bar{N}}$ are the corresponding binding energies. In our case this energy is at least by a factor 2 smaller as compared with the vacuum value of $2m_N$. This should lead to a significant suppression of the available phase space and thus to a reduced annihilation rate in medium. We have performed detailed calculations assuming that the annihilation rates into different channels are proportional to the available phase space. All intermediate states with heavy mesons like ρ, ω, η as well as multi-pion channels have been considered. Our conclusion is that decreasing the Q value from 2 GeV to 1 GeV may lead to the reduction of total annihilation rate by factor 20÷30. Then we estimate the SBN life times on the level of 5-25 fm/c which makes their observation feasible. This large margin in the life times is mainly caused by uncertainties in the overlap integral between antinucleon and nucleon scalar densities. Longer life times may be expected for SBNs containing antihyperons. The reason is that instead of pions

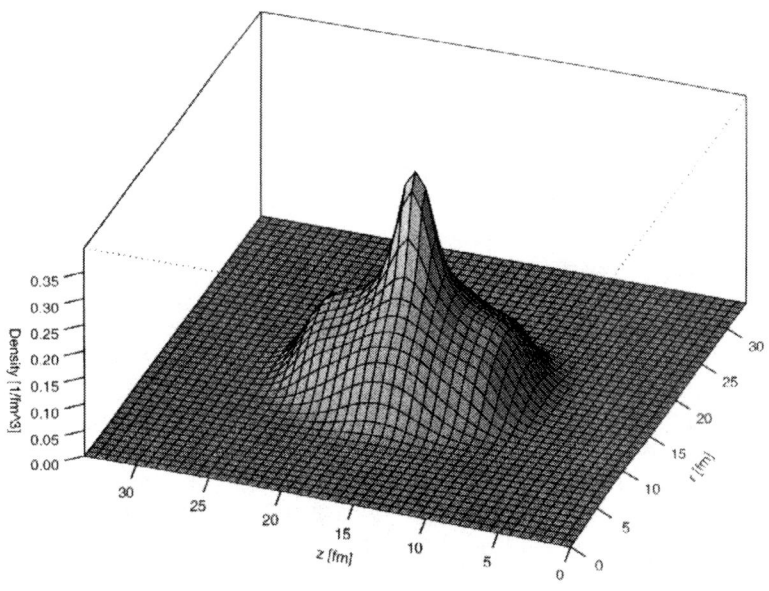

FIGURE 7. Surface plot of the sum of proton and neutron densities for the system ^{208}Pb with $\bar{p}$.

more heavy kaons must be produced in this case. We have also analyzed multi-nucleon annihilation channels (Pontecorvo-like reactions) and have found their contribution to be less than 40% of the single-nucleon annihilation.

We believe that such exotic nuclear states can be produced by using antiproton beams of multi-GeV energy, e.g. at the future GSI facility. It is well known that low-energy antiprotons annihilate on the nuclear periphery (at about 5% of the normal density). Since the annihilation cross section drops significantly with energy, a high-energy antiproton can penetrate deeper into the nuclear interior. Then it can be stopped there in an inelastic collision with a nucleon, e. g. via the reaction $A(\bar{p}, N\pi)_{\bar{p}}A'$, leading to the formation of a $\bar{p}$-doped nucleus. Reactions like $A(\bar{p}, \Lambda)_{\bar{\Lambda}}A'$ can be used to produce a $\bar{\Lambda}$-doped nuclei. Fast nucleons or lambdas can be used for triggering such events. In order to be captured by a target nucleus final antibaryons must be slow in the lab frame. Rough estimates of the SBN formation probability in a central $\bar{p}A$ collision give the values $10^{-5} - 10^{-6}$. With the $\bar{p}$ beam luminocity of $2 \cdot 10^{32}$ cm^{-2}s^{-1} planned at GSI this will correspond to the reaction rate of a few tens of desired events per second.

Several signatures of SBNs can be used for their experimental observation. First, annihilation of a bound antibaryon can proceed via emission of a single photon, pion or kaon with an energy of about 1 GeV (such annihilation channels are forbidden in vacuum). So one may search for relatively sharp lines, with width of $10 \div 40$ MeV, around this energy, emitted isotropically in the SBN rest frame. Another signal may come from explosive disintegration of the compressed nucleus after the antibaryon annihilation. This can be observed by measuring radial collective velocities of nuclear fragments.

It is interesting to look at the antibaryon-nucleus system from somewhat different point of view. An antibaryon implanted into a nucleus acts as an attractor for surrounding nucleons. Due to the uncompensated attractive force these nucleons acquire acceleration towards the center. As the result of this inward collective motion the nucleons pile up producing local compression. If this process would be completely elastic it would generate monopole-like oscillations around the compressed SBN state. The maximum compression is reached when the attractive potential energy becomes equal to the compression energy. Simple estimates show that local baryon densities up to 5 times the normal nuclear density may be obtained in this way. It is most likely that the deconfinment transition will occur at this stage and a high-density cloud containing an antibaryon and a few nucleons will appear in the form of a multi-quark-antiquark cluster. One may speculate that the whole ^{4}He or even ^{16}O nucleus can be transformed into the quark phase by this mechanism. As shown in ref. [9], an admixture of antiquarks to cold quark matter is energetically favorable. The problem of annihilation is now transferred to the quark level. But the argument concerning the reduction of available phase space due to the entrance-channel nuclear effects should work in this case too. Thus one may hope to produce relatively cold droplets of the quark phase by the inertial compression of nuclear matter initiated by an antibaryon.

I am grateful to T. Bürvenich, I.N. Mishustin and L.M. Satarov for fruitful discussions and help in the preparation of this talk.

REFERENCES

1. W. Greiner, B. Müller, J. Rafelksi, *Quantum electrodynamics of strong fields*, Springer Verlag, 2nd edition, December 1985
2. J. Reinhardt and W. Greiner, *Quantum electrodynamics*, Springer Verlag, 3rd edition, February 2003
3. N. Auerbach, A.S. Goldhaber, M.B. Johnson, L.D. Miller, A. Picklesimer, Phys. Lett. **B182** (1986) 221.
4. T. Bürvenich, I.N. Mishustin, L.M. Satarov, H. Stöcker, W. Greiner, Phys. Lett. **B542** (2002) 261.
5. Y. Sugahara and H. Toki, Nucl. Phys. **A579** (1994) 557.
6. M. Bender, K. Rutz, P.–G. Reinhard, J.A. Maruhn, and W. Greiner, Phys. Rev. **C60** (1999) 34304.
7. G. Mao, H. Stöcker, and W. Greiner, Int. J. Mod. Phys. **E8** (1999) 389.
8. Y. Akaishi and T. Yamazaki, Phys. Rev. **C65** (2002) 044005.
9. I.N. Mishustin, L.M. Satarov, H. Stoecker, W. Greiner, Phys. Rev. **C 59** (1999) 3343.

Study of Decays of $^{271}110$ and $^{272}111$ Produced with ^{208}Pb(^{64}Ni,n) and ^{209}Bi(^{64}Ni,n) Reactions

K. Morita[1], K. Morimoto[1], D. Kaji[2,1], S. Goto[3], H. Haba[1], E. Idegichi[2], R. Kanungo[1], K. Katori[1], H. Koura[4,1], H. Kudo[3], T. Ohnishi[1], A. Ozawa[5], J. C. Peter[6], T. Suda[1], K. Sueki[5], I. Tanihata[1], F. Tokanai[7], H. Xu[8], A. V. Yeremin[9], A. Yoneda[1], A. Yoshida[1], Y.-L. Zhao[10], and T. Zheng[11]

[1]*RIKEN (The Institute of Physical and Chemical Research), Wako-shi, Saitama 351-0198, Japan*
[2]*Center for Nuclear Science, University of Tokyo, Wako-shi, Saitama 351-0198, Japan*
[3]*Department of Chemistry, Niigata University, Ikarashi, Niigata 950-2181, Japan*
[4]*Advanced Research Institute for Science and Engineering, Waseda University, Okubo 3-4-1, Shinjuku-ku, Tokyo 169-8555, Japan*
[5]*University of Tsukuba, Tsukuba, Ibaraki 305-8577, Japan*
[6]*Laboratoire de Physique Corpusculaire de Caen, LPC/ISMRA 6, Boulevard du Marechal Juin 14050 CAEN cedex, France*
[7]*Department of Physics, Yamagata University, Shirakawa, Yamagata 990-8560, Japan*
[8]*Institute of Modern Physics, Chinese Academy of Science, Lanzhou 730000, China*
[9]*Flerov Laboratory of Nuclear Reactions, Joint Institute of Nuclear Research, RU-141 980 Dubna, Russia*
[10]*Institute of High Energy Physics, Chinese Academy of Science, Beijing 100039, China*
[11]*School of Physics, Peking University, Beijing 100871, China*

Abstract. Production and decay of an isotope $^{271}110$ of the 110th element were studied. The isotope was produced by ^{208}Pb + ^{64}Ni → $^{271}110$ + n reaction. Fourteen α-decay chains have been assigned to be the decays originating from the isotope $^{271}110$. The excitation function of the production cross section was measured. The results have provided a good confirmation of production and decay of the $^{271}110$ reported by GSI group. The presence of an isomeric state in $^{271}110$ has also been confirmed. Possible isomeric state in ^{267}Hs has also been presented. The production and decay of $^{272}111$ has been also investigated in irradiations of ^{209}Bi targets with ^{64}Ni beam. We have observed 14 α-decay chains in total, that can be assigned, on the basis of their time correlations, to subsequent decays from $^{272}111$.

INTRODUCTION

An attempt to search for heaviest elements is one of the interesting subjects in experimental nuclear physics. New elements whose atomic numbers were greater than 101 were synthesized by heavy-ion induced fusion reactions. To produce heavy system the fusion reaction should proceed by neutron evaporation not by fission process. Because the fissility of a heavy system increases with the increase of the atomic number, the production cross sections of the heavy system tend to decrease in logarithmic manner with the increase of the atomic number. Main difficulty on searching of the heaviest elements comes from the smallness of the production cross sections. For example, the cross section measured for producing $^{277}112$ isotope of the 112th element was reported to be 0.5 pb [1]. The small cross section is limiting further research of nuclei with greater atomic numbers. To overcome the difficulty, more intense primary beams must be used for the production, and a recoil separator with high efficiency and high background reduction must be used. A gas-filled recoil ion separator GARIS [2] was installed in the experimental hall of RILAC facility for studies of heavy elements. As the first attempt, we studied production and decay of an isotope $^{271}110$. The isotopes were produced by ^{208}Pb + ^{64}Ni ¤ $^{271}110$ + n reaction. The present work has confirmed the experimental results by GSI group, reported by S. Hofmann et al., [3-5]. An excitation function of the production cross section was measured.

Then we investigated the production and decay of $^{272}111$ using ^{209}Bi + ^{64}Ni ¤ $^{272}111$ + n reaction. The synthesis of this nuclide was reported by Hofmann et al.[6, 1] using the same reaction with the present work. The present result is the first clear confirmation for the discovery of $^{272}111$ and its α-decay products, ^{264}Bh and ^{268}Mt, reported previously by a GSI group. New information on their half-lives and decay energies as well as the excitation function is presented.

EXPERIMENTAL PROCEDURE

The experimental setups for the studies of $^{271}110$ and $^{272}111$ were almost the same each other except for the targets. Schematic drawing of the experimental setup is shown in Fig.1. Left hand side of the figure (a) shows the plane view of the gas-filled recoil ion separator GARIS and the envelopes. The right hand side of the figure (b) shows a schematic drawing of the arrangement of detectors.

The beam from RILAC irradiated the target after passing through a differential pumping section. Fusion-evaporation reaction products recoiled out of the target in the forward direction were separated from the beam particles by the separator GARIS and focused to the focal plane of GARIS. Then the products were detected by detection system set at the focus.

The ^{64}Ni ion beam was supplied from RILAC. Its typical beam intensity was 5 ~10^{12} s^{-1}. The beam energy was determined by measuring magnetic rigidity of a 90° bending magnet situated in the beamline and by a time-of-flight method. The absolute accuracy of the beam energy was }0.6 MeV.

The beam intensity was monitored by measuring elastically scattered projectiles with a PIN photo diode mounted at 45° with respect to the incident beam direction in a distance of 1.28 m from a target position.

Targets were prepared by vacuum evaporation of metallic lead-208 (isotopically purified up to 98.4%) and bismuth on carbon backing foils of 30 μg/cm^2 thickness. The thickness of the lead-208 targets was about 230 μg/cm2 about and that of bismuth targets was about 280 μg/cm2. The targets were covered by a 10 μg/cm^2 thick carbon to protect the target from sputtering. Targets were mounted on a rotating wheel to prevent the melting of the target material from the heating by the beam irradiation.

The reaction products were separated in-flight from the beam by a gas-filled recoil separator GARIS and were guided into a detector box described later. The details of GARIS is given elsewhere[2]. The separator was filled with helium gas at a pressure of 75 Pa.

In order to separate a gas region of GARIS from a vacuum region of the accelerator without using any foils, a differential pumping system was adopted.

The acceptance of GARIS was 12.2 msr. The value of magnetic rigidity of an analyzer for evaporation residue measurement was set to 2.045 Tm.

The transmission of GARIS depends on effects of reaction kinematics and multiple scattering of reaction product with target material. In the present experiment the transmission was estimated to be 80% for evaporation residues by a simulation taking into account the Coulomb scattering of $^{272}111$ with ^{209}Bi nuclei. The value has an error of }15%.

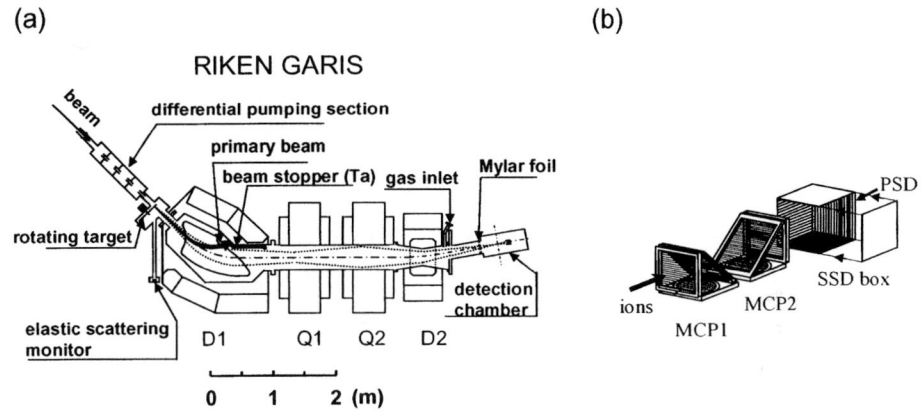

FIGURE 1. (a) Plane view of RIKEN GARIS. The shape of dipole magnets are shown in the figure. Envelopes of evaporation residues are shown by dotted lines. Typical trajectory of primary beam is also shown. (b) Schematic drawing of detection system at the focal plane.

The detection system is separated from the gas-filled region of GARIS by 1 μm Mylar foil.

The detection system consists of two foil detectors with micro-channel plates (MCP) and a silicon semiconductor detector (SSD) box. The foil detectors are made from 0.5 μm Mylar foil covered by layers of 2.7 μg/cm^2 Al and 20 μg/cm^2 CsI to enhance the number of secondary electrons. These detectors were used for two purposes. One is to measure time-of-flight (TOF) of particles coming into the SSD box to obtain information of masses. The other is to distinguish decay signals from the SSD box used in anti-coincidence mode. The signals from the SSD box are treated to be the decay events only when both of the MCPs give no signal. The distance between the two foil detectors was 295 mm and the transmission was 94%. Active area of the MCP is 78 mm in diameter and the typical timing resolution of the detectors was 500 ps in FWHM.

The SSD box consists of five 16-strip silicon detectors. Active area of each detector is 60 -60 mm^2. Each strip is 3.75 mm wide. For one of the detectors, which was used as a stop detector (PSD), each strip is position sensitive in vertical direction. The position resolution is 1 mm in FWHM and the energy resolution is 35 keV for α particles. For the other four detectors (side detectors), each strip is galvanically connected so that these four detectors are only energy sensitive. The side detectors are set upstream with respect to the stop detector in order to detect escaping decay particles with a solid angle of 70% of 2π. The energy resolution obtained by summing the energy-loss signal from the stop detector and the residual energy from the side detector was typically 70 keV for α particles. The singles counting rate of the PSD was about 10 s^{-1} at the typical beam intensity of 5 ~10^{12} s^{-1} and that with no MCP signals was about 1 s^{-1}. By employing a double trigger technique, a lifetime of 5 μs could be measured.

The detection system was periodically checked by measuring evaporation residues from ^{64}Ni on ^{nat}Ce reaction.

RESULTS AND DISCUSSION

^{208}Pb + ^{64}Ni Reaction

We observed totally fourteen decay chains identified to the decays of ^{271}110 from the analysis of generic correlations. Totally 58 α-decays were registered. Thirteen three of 58 decays (57%) were detected only by PSD, and 25 decays (43%) were detected both PSD and SSD. For seven of fourteen chains, 6-fold (ER-α1-α2-α3-α4-α5) correlations were observed where ER denotes the evaporation residue. For four chains, 5-fold correlations were observed. For two chains, 4-fold correlations were observed. For one chain, 2-fold correlation was observed.

Cross Section

We performed the measurement at four beam energies. The energies from the accelerator are listed in TABLE 1. together with the energies at the half-depth of the target, energy spreads in the target, beam dose, average target thickness used, number of observed events, and the corresponding cross sections. Errors in the cross sections in the table are only statistical one in terms of 1σ confidence level. The beam energy giving the maximum cross section was 316 MeV. Corresponding center-of mass energy at the half-depth of the target is 240 MeV. The FWHM value of the excitation function is deduced to be 4 MeV assuming that the shape of the curve is the Gaussian.

TABLE 1. Summary of the excitation function measurements. a: Energy from the accelerator, b: Energy at the half depth of the target. c: width of energy caused by energy loss of the beam by target. d: Errors in the cross section and in the upper limit are given in terms of 1σ confidence level.

E_{in}^a (MeV)	E_c^b (MeV)	ΔE^c (MeV)	Dose (10^{18})	T_{av} (μg/cm²)	number of events	σ^d (pb)
310±0.6	308	±1.3	1.00	230	1	$1.8^{+4.1}_{-1.5}$
313±0.6	311	±1.2	0.95	230	4	$8.0^{+6.0}_{-4.0}$
316±0.6	314	±1.2	1.07	220	9	$16.5^{+7.3}_{-5.5}$
320±0.6	318	±1.1	1.01	210	0	< 3.7

Decays of $^{271}110$

Decay energies and decay times of all α decays in the 14 decay chains observed in ^{208}Pb(^{64}Ni,1n) reaction are plotted in Fig. 2. The measured decay energies of α1s detected just after the implantations of $^{271}110$ into the PSD range from 9.9 to 10.8 MeV shown in the figure. The energy centers of the peak in the spectra were deduced to be 10.45 MeV and 10.73 MeV. The decay energy of $^{271}110$ reported by GSI group [3, 5] was 10.73 MeV. The value agrees well with the higher one of the present result.

Measured decay times of α1 group, which were the time difference between the moment of implantation of the isotopes and the first α emissions, are 7.54, 2.20, 46.0, 87.1, 4.63, 238.7, 1.74, 1.36, 1.31, 0.057, 6.59, 3.80, 1.69, and 0.92 ms in the order of the observation. They could be divided into two groups, one of decay times shorter than 10 ms (11 events), and another of decay times longer than 10 ms (3 events). The mean life of the first group was calculated to be $2.9^{+1.3}_{-0.7}$ ms and that of second was 124^{+169}_{-45} ms with errors of 1σ (68%) confidence level. The mean life times of $^{271}110$ reported by Hofmann et al., [3, 5] were $1.8^{+0.8}_{-0.4}$ ms (11 events) and 65^{+120}_{-26} ms (2 events), respectively. These values obtained in the two independent experiments coincided well both the short decay and long decay groups within 1σ statistical errors. Number ratios of short and long decay time groups, 11/3 by RIKEN group and 11/2 by GSI group, also agrees well.

Based on the comparison of the experimental results of the present study and the one reported by GSI group mentioned above, the production of an isotope $^{271}110$ was confirmed. Furthermore, it became clear that the isotope $^{271}110$ has at least two states

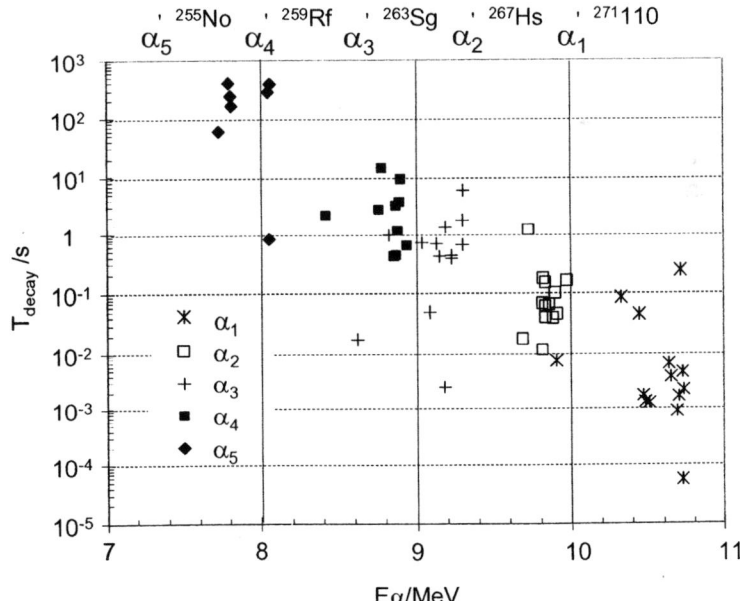

FIGURE 2. Decay energies and decay times of 14 decay chains observed in ^{208}Pb(^{64}Ni,1n) reaction.

with different manner lives. Then we could combine the result to get the values of lifetimes with better counting statistics. The averaged mean life of the shorter one is 2.35 $^{+0.64}_{-0.41}$ ms and that of the longer one is 100 $^{+81}_{-31}$ ms. The values correspond to half-lives ($T_{1/2}$) of 1.63 $^{+0.44}_{-0.29}$ ms, and 69 $^{+56}_{-21}$ ms, respectively.

^{209}Bi + ^{64}Ni Reaction

Cross Section

In the first experimental run, measurements were carried out at the incident ^{64}Ni beam energy of 323 MeV. Three correlated events identified to the decay chains of ^{272}111 isotopes. But during the irradiation the target deterioration was indicated on the beam monitor spectrum as described in section 2. Because the actual beam energy in target is thus not well known in this case, the first three decay chains are excluded from the analysis of excitation function given below.

The irradiations were performed at three different beam energies. We assumed all decay chains from chain 4 to chain 14 belong to ^{272}111 decays. To deduce the cross section, the collection efficiency of GARIS was assumed to be 80%.

The results are summarized in TABLE 2. The attached errors in the cross sections are only statistical ones at 68% confidence level and those in energy were calculated from the energy loss at the target.

At the highest energy, E_c = 323.2 MeV, no event was observed and an upper limit was calculated to be 1.1 pb. From a Gaussian fit assuming the width (FWHM) of 4 MeV which is the same value as in the case of ^{271}110 described in the previous section it was found that the maximum cross section was at E_c = 318.9 MeV which corresponds to the center-of-mass energy of 244.1 MeV.

TABLE 2. Summary of the excitation function measurements. a: Energy from the accelerator, b: Energy at the half depth of the target. d: Errors in the cross section and in the upper limit are given in terms of 1σ confidence level.

E_{in}^a (MeV)	E_c^b (MeV)	Dose (10^{18})	T_{av} (μg/cm^2)	number of events	σ^d (pb)
320.0±0.6	317.5	2.20	252	3	2.6 $^{+2.3}_{-1.5}$
323.0±0.6	320.3	4.49	285	8	2.5 $^{+1.2}_{-0.9}$
326.0±0.6	323.2	2.50	298	0	<1.1

Decay of ^{272}111, ^{268}Mt, and ^{264}Bh

Decay energies and decay times observed in ^{209}Bi(^{64}Ni,1n) reaction are plotted in Fig. 3. A general tendency that the decrease of decay energies, and the increase of decay time, with the order of decay generations (α1-α2-α3-α4-α5) is clearly seen in the figure. We assigned the α1, α2, α3, α4, and α5 to the decay of ^{272}111, ^{268}Mt, ^{264}Bh, ^{260}Db, and ^{256}Lr from the analysis of generic correlations.

We observed 8 decays for α5. Three of 14 decay chains were ended by fission. Two of them were in α3 (^{264}Bh) group and one was in α4 (^{260}Db) group as described below. Three decays were missing in α5 group. Observed decay energies (8.35 – 8.65 MeV) and decay times ($T_{1/2} = 18^{+10}_{-5}$ s) of α5 decay are in good agreement with the literature values for decay of ^{256}Lr (Eα = 8.32 – 8.64 MeV, $T_{1/2} = 28$ s)[7].

We observed 12 decays for α4 group. Eleven decays were detected as alpha decays and one was detected as spontaneous fission decay. Observed decay energies (8.35 – 9.40 MeV) and decay times ($T_{1/2} = 5.7^{+2.3}_{-1.3}$ s) of α4 decay are in agreement with the literature values for decay of ^{260}Db (Eα = 9.04 – 9.12 MeV, $T_{1/2} = 1.5$ s)[7]. ^{260}Db is known to decay by spontaneous fission with branching ratio of 9.6%[7]. Our observation of fission event for α4 group agrees very well with the literature value.

Since the α4 and α5 decays are assigned to ^{260}Db and ^{256}Lr, α1, α2, and α3 decays are unambiguously assigned to the decay of ^{272}111, and ^{268}Mt, and ^{264}Bh.

Decay energies and decay times are summarized in Table 3 together with the values reported by Hofmann et al., in ref. [1]. Errors in the $T_{1/2}$ are 1σ confidence level.

For ^{272}111, (14 α-decays were observed) the $T_{1/2}$ is calculated to $3.8^{+1.4}_{-0.8}$ ms. The present value is more than two times longer than the value reported in ref. [1]. Observed decay energy ranges from 10.20 to 11.56 MeV showing broad distribution, although the 'peak' is located at 11.0 MeV.

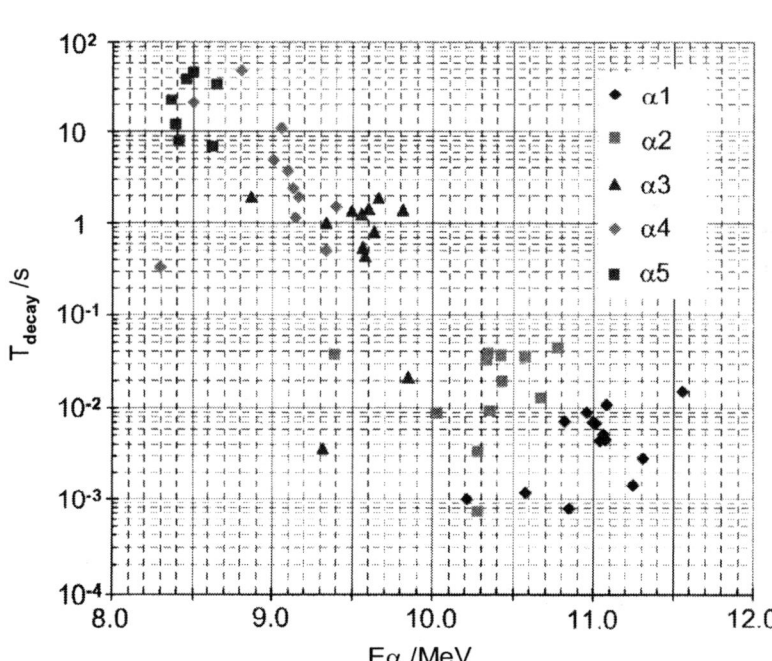

FIGURE 3. Decay energies and decay times of 14 decay chains observed in ^{209}Bi(^{64}Ni,1n) reaction.

For ^{268}Mt, (14 α-decays were observed) the $T_{1/2}$ is calculated to 21^{+8}_{-5} ms. The present value is shorter than the value reported in ref. [1]. Observed decay energy ranges from 9.40 to 10.77 MeV showing broad distribution, although the 'peak' is located at 10.4 MeV.

For ^{264}Bh, (12 α-decays were observed) the $T_{1/2}$ is calculated to $0.9^{+0.3}_{-0.2}$ s. The present value is in good agreement with the value reported in ref. [1]. Observed decay energy ranges from 8.86 to 9.83 MeV showing broad distribution, although the 'peak' is located at 9.7 MeV. We observed 2 fission events in the decay of ^{264}Bh corresponding to a branching ratio of 14%. Decay times for the fission event were 1.0 s and 4.9 s, respectively. In the calculation of $T_{1/2}$ these values are also included. This is the first observation of spontaneous fission decay in Bh isotopes. This fission event may come from EC decay product ^{264}Sg which is unknown nucleus up to now because our setup was not sensitive to distinguish the EC decay.

TABLE 3. Summary of α-decay energies and half-lives of ^{271}111 and its daughter nucleid.

Nucleid	Present Work		from ref. [1]	
	$T_{1/2}$	Eα/MeV	$T_{1/2}$	Eα/MeV
^{271}111	$3.8^{+1.4}_{-0.8}$ ms	10.20 – 11.56	$1.6^{+1.1}_{-0.5}$ ms	10.8 – 11.05
^{268}Mt	21^{+8}_{-5} ms	9.40 – 10.77	42^{+29}_{-12} ms	10.1 – 10.3
^{264}Bh	$0.9^{+0.3}_{-0.2}$ s	8.86 – 9.83	$1.0^{+0.7}_{-0.3}$ s	9.1 – 9.6

ACKNOWLEDGMENTS

Many thanks are due to all accelerator staff members for their excellent operation for long period of time. Authors would like to thank professor T. Chihara and Dr. N. Suzuki for synthesizing material nikkelocene for ion-source. Authors would like to thank greatly professors Y. Yano and M. Ishihara for their continuous support, encouragement, and useful suggestions. We would like to thank also Dr. M. Kase for his all support and arrangement about the beamtime.

REFERENCES

1. S. Hofmann et al., Eur. Phys. A14, 147 (2002).
2. K. Morita et al., Nucl. Instr and Meth. To be published.
3. S. Hofmann, Rep. Prog. Phys. 61 639(1998).
4. S. Hofmann et al., Z. Phys. A350 277(1995).
5. S. Hofmann, J. Nucl. Radiochem. Sci., 4 ,R1 (2003).
6. S. Hofmann et al., Z. Phys. A350 281(1995).
7. R. B. Firestone and V. S. Sherley ed. *Table of Isotopes 8th ed.* (Wiley and Sons, NewYork, 1996)

Properties of heavy nuclei measured at the GSI SHIP

Sigurd Hofmann

*Gesellschaft für Schwerionenforschung (GSI), D–64220 Darmstadt and Physikalisches Institut,
J.W. Goethe-Universität, D–60054 Frankfurt, Germany*

Abstract. The nuclear shell model predicts that the next doubly magic shell-closure beyond ^{208}Pb is at a proton number Z = 114, 120, or 126 and at a neutron number N = 172 or 184. The outstanding aim of experimental investigations is the exploration of this region of spherical 'Super-Heavy Elements' (SHEs). The measured decay data reveal that for the heaviest elements, the dominant decay mode is α emission, not fission. Decay properties as well as reaction cross-sections are compared with results of theoretical investigations. Finally, plans are presented for the further development of the experimental set-up and the application of new techniques. At a higher sensitivity, the exploration of the region of spherical SHEs now becomes feasible, almost forty years after its prediction.

EXPERIMENTAL RESULTS

In this section, recent results are presented dealing with the confirmation of elements 110 to 112. Detailed presentations of the properties of elements 107 to 109 and of earlier results on elements 110 to 112 were given in previous reviews [1, 2, 3].

Element 110, now officially named darmstadtium [4], Ds, was discovered in 1994 using the reaction ^{62}Ni + ^{208}Pb $\rightarrow$ ^{270}Ds* [5]. A total of three decay chains was measured (see also remarks at the end of this section). The main experiment was preceded by a thorough study of the excitation functions for the synthesis of ^{257}Rf and ^{265}Hs in order to determine the optimum beam energy for the production of darmstadtium. The data revealed that the maximum cross-section for the synthesis of hassium was shifted to a lower excitation energy, different from the predictions of reaction theories.

The heavier isotope ^{271}Ds was synthesized with a beam of the more neutron-rich isotope ^{64}Ni [2]. A summary of the measured decay data and a comparison with literature data is given in Fig. 1. The data were fully confirmed by the results of recent experiments performed at RIKEN, Saitama [6] and LBNL, Berkeley [7]. The important result for the further production of elements beyond meitnerium was that the cross-section was enhanced from 2.6 pb to 15 pb by increasing the neutron number of the projectile by two, which gave hope that the cross-sections for production of heavier elements could decrease less steeply with more neutron-rich projectiles. However, this expectation was not proven in the case of element 112.

The even-even nucleus ^{270}Ds was synthesized using the reaction ^{64}Ni + ^{207}Pb [11]. A total of eight α-decay chains was measured during an irradiation time of seven days. Decay data were obtained for the ground-state and a high spin K isomer, for which

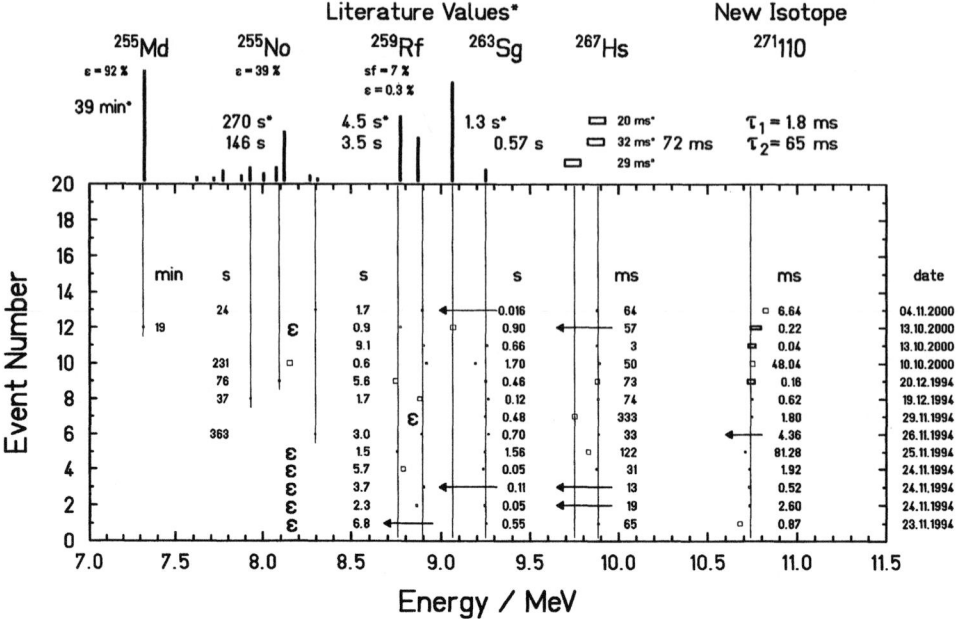

FIGURE 1. Energies and lifetimes of the thirteen α-decay chains resulting from the reaction ^{64}Ni + ^{208}Pb $\rightarrow$ ^{271}Ds + 1n. The isotope ^{271}Ds was identified by comparison of the decay properties of the daughter products with literature data marked by an asterisk (^{267}Hs [8], ^{263}Sg [9], ^{259}Rf [10], ^{255}No [10], ^{255}Md [10]). New decay data of the isotope ^{263}Sg could be deduced. The decay chains are arranged chronologically with the date of production given at the righthand ordinate. The size of the symbols reflects the detector resolution. Arrows mark escaping α particles for which only an energy loss signal, but still time and position information, was obtained.

calculations predict spin and parity 9^-, 10^- or 8^+ [12]. The relevant single particle Nilsson levels are $\nu[613]_{7/2+}$ and $\nu[615]_{9/2+}$ below the Fermi level and $\nu[725]_{11/2-}$ above the Fermi level. Configuration and calculated energy of the excited states are $\{\nu[613]_{7/2+}\,\nu[725]_{11/2-}\}9^-$ at 1.31 MeV, $\{\nu[615]_{9/2+}\,\nu[725]_{11/2-}\}10^-$ at 1.34 MeV, and $\{\nu[613]_{7/2+-}\,\nu[615]_{9/2+-}\}8^+$ at 1.58 MeV.

The new nuclei ^{266}Hs and ^{262}Sg were identified as daughter products after α decay. Spontaneous fission of ^{262}Sg terminates the decay chain. A proposed partial decay scheme of ^{270}Ds is shown in Fig. 2.

Element 111 was synthesized in 1994 using the reaction ^{64}Ni + ^{209}Bi $\rightarrow$ ^{273}111*. A total of three α chains of the isotope ^{272}111 were observed [16]. Another three decay chains were measured in a confirmation experiment in October 2000 [17]. Also these results were confirmed in recent experiments by Morita et al. [6].

Element 112 was investigated at SHIP using the reaction ^{70}Zn + ^{208}Pb $\rightarrow$ ^{278}112* [18]. The irradiation was performed in January-February 1996. Over a period of 24 days, a

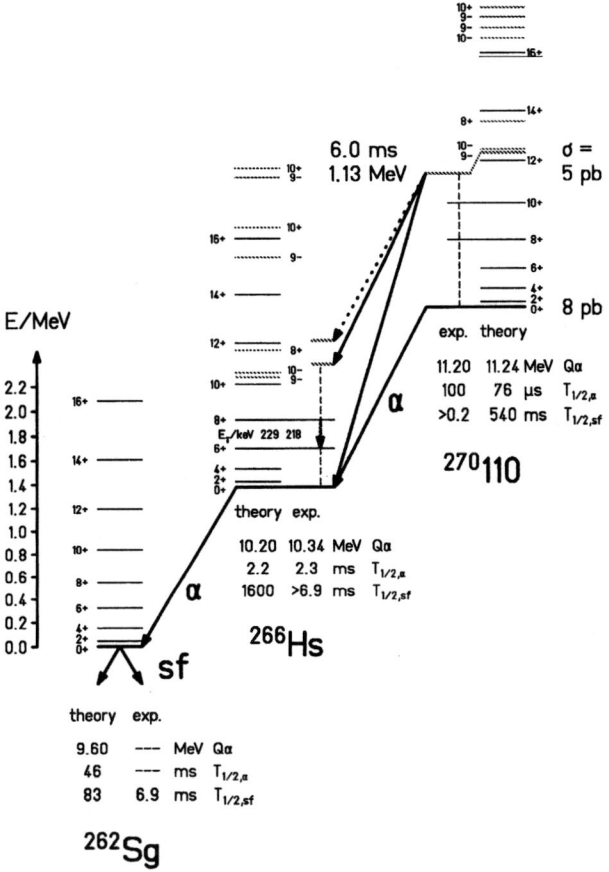

FIGURE 2. Tentative assignment of measured α and γ decay and spontaneous fission data (bold arrows) observed in the reaction ^{64}Ni + ^{207}Pb → ^{271}Ds*. The data were assigned to the ground-state decays of the new isotopes ^{270}Ds, ^{266}Hs, and ^{262}Sg and to a high spin K isomer in ^{270}Ds. The proposed partial level schemes are taken from theoretical studies of Muntian et al. [13] for the rotational levels, of Ćwiok et al. [12] for the K isomers and of Smolanczuk [14] and Smolanczuk et al. [15] for the α energies and spontaneous fission half-lives, respectively. For a detailed discussion see Ref. [11].

total of 3.4×10^{18} projectiles were collected. One α-decay chain, shown in the left side of Fig. 3, was observed resulting in a cross-section of 0.5 pb. The chain was assigned to the one neutron-emission channel. The experiment was repeated in May 2000 aiming to confirm the synthesis of ^{277}112 [17]. During a similar long measuring time, but using slightly higher beam energy, one more decay chain was observed, also shown in Fig. 3. The measured decay pattern of the first four α decays is in agreement with the one

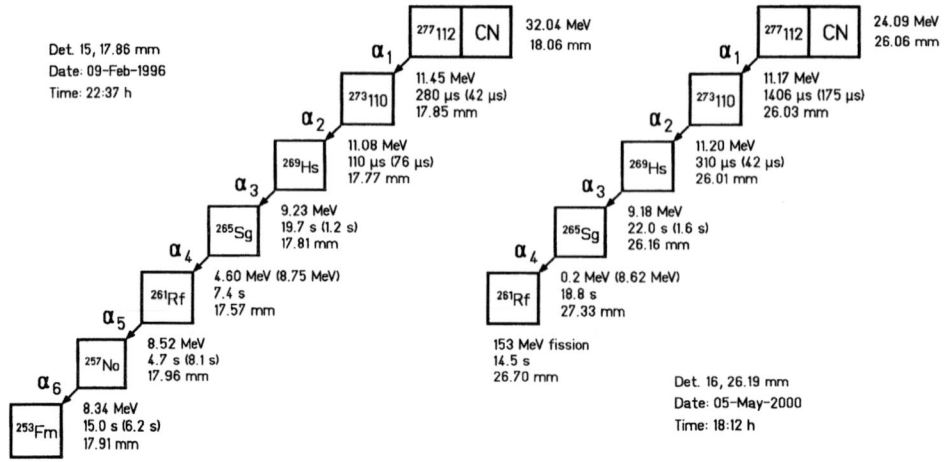

FIGURE 3. Two decay chains measured in experiments at SHIP in the cold fusion reaction ^{70}Zn + ^{208}Pb → ^{278}112*. The chains were assigned to the isotope ^{277}112 produced by evaporation of one neutron from the compound nucleus. The lifetimes given in brackets were calculated using the measured α energies. In the case of escaped α particles the α energies were determined using the measured lifetimes.

observed in the first experiment.

A new result was the occurrence of fission which ended the second decay chain at ^{261}Rf. A spontaneous-fission branch of this nucleus was not yet known, however, it was expected from theoretical calculations. The new results on ^{261}Rf were proven in a recent chemistry experiment [19, 20], in which this isotope was measured as granddaughter in the decay chain of ^{269}Hs.

A reanalysis of all decay chains measured at SHIP since 1994, a total of 34 decay chains was analyzed, revealed that the previously published first decay chain of ^{277}112 [18] (not shown in Fig. 3) and the second of the originally published four chains of ^{269}Ds [5] were spuriously created. Details of the results of the reanalysis are given in Ref. [17].

The excitation function of the reaction ^{54}Cr + ^{208}Pb was studied recently (June 2003, see Sect. 3). At high beam currents of up to 1 pμA the target was continuously monitored using the scattering of 20 keV electrons at the target material [21]. Successfully tested was a PbS target (melting point 1118° C) produced by depositing the target material on a carbon backing which was heated to several 100° C [22]. By heating the backing during evaporation, the formation of a crystalline needle structure of PbS was avoided, which would result in uncontrolled energy loss of the projectiles. Using the 'heated' PbS target, a 1n-excitation function was measured, which was identical to the previously measured one obtained with a metallic Pb target.

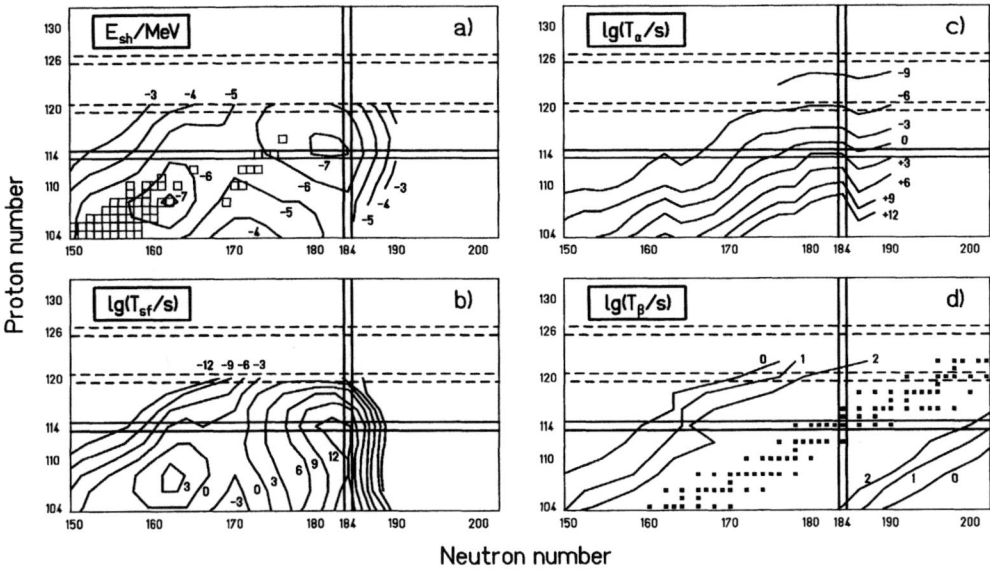

FIGURE 4. Shell-correction energy (a) and partial spontaneous fission, α and β half-lives (b-d). The calculated values in (a)–(c) are taken from Ref. [23, 15] and in (d) from Ref. [28]. The squares in (a) mark the nuclei presently known or under investigation, the filled squares in (d) mark the β stable nuclei.

NUCLEAR STRUCTURE AND DECAY PROPERTIES

The basic step which is necessary for the determination of the stability of SHEs is the calculation of the ground-state binding energy. As a signature for shell effects, we can extract the shell-correction energy by subtracting a smooth macroscopic part (derived from the liquid-drop model) from the measured or calculated total binding energy. In macroscopic-microscopic models the shell-correction energy is of course the essential input value which is calculated directly from the shell model. The shell-correction energy is plotted in Fig. 4a using the data from Ref. [23]. Two equally deep minima are obtained, one at $Z = 108$ and $N = 162$ for deformed nuclei with deformation parameters $\beta_2 \approx 0.22$, $\beta_4 \approx -0.07$ and the other at $Z = 114$ and $N = 184$ for spherical SHEs. Different results are obtained from self-consistent Hartree-Fock-Bogoliubov calculations and relativistic mean-field models [24, 25, 26, 27]. They predict for the spherical nuclei shells at $Z = 114$, 120 or 126 (indicated as dashed lines in Fig. 4) and $N = 184$ or 172, with shell strengths being also a function of the amount of nucleons of the other type.

For the calculation of partial spontaneous fission half-lives the knowledge of ground-state binding energies is not sufficient. It is necessary to determine the fission barrier over a wide range of deformation. The most accurate data were obtained for even-even nuclei using the macroscopic-microscopic model [15, 23]. Partial spontaneous fission half-lives are plotted in Fig. 4b.

Partial α half-lives decrease almost monotonically from 10^{12} s down to 10^{-9} s near $Z = 126$ (Fig. 4c). The valley of β-stable nuclei (marked by black squares in Fig. 4d) passes through $Z = 114$, $N = 184$ [28]. At a distance from the bottom of the valley, the β half-lives decrease gradually down to values of one second.

The interesting question arises, if and how the uncertainty related with the location of the proton and neutron shell closures will change the half-lives of SHEs. Partial α and β half-lives are only insignificantly modified by shell effects, because the decay process occurs between neighboring nuclei. This is different for fission half-lives which are primarily determined by shell effects. However, the uncertainty related with the location of nuclei with the strongest shell-effects and thus longest partial fission half-life at $Z = 114$, 120 or 126 and $N = 172$ or 184, is inconsequential concerning the longest 'total' half-life of SHEs. The regions for SHEs in question are dominated by α decay. And α decay will be modified by only a factor of up to approximately 100, if the double shell closure will not be located at $Z = 114$ and $N = 184$.

The line of reasoning is, however, different concerning the production cross-section. The survival probability of the compound nucleus (CN) is determined among other factors significantly by the fission-barrier. Therefore all present calculations of cross-sections suffer from the uncertainty related with the location and strength of closed shells.

CROSS-SECTIONS, FUSION VALLEYS AND EXCITATION ENERGY

The main features which determine the fusion process of heavy ions are (1) the fusion barrier and related beam energy and excitation energy, (2) the ratio of surface tension versus Coulomb repulsion, which determines the fusion probability and which strongly depends from the degree of asymmetry of the reaction partners (the product $Z_1 Z_2$ at fixed $Z_1 + Z_2$), (3) the impact parameter and related angular momentum, and (4) the ratio of neutron evaporation versus fission probability of the CN. In fusion of SHEs the product $Z_1 Z_2$ reaches extremely large and the fission barrier extremely small values. In addition, the fission barrier is fragile at increasing excitation energy and angular momentum, because it is solely built up from shell effects. For these reasons the fusion of SHEs is drastically hampered, whereas the fusion of lighter elements is advanced through the contracting effect of surface tension.

The effect of Coulomb repulsion on the cross-section starts to act severely for fusion of elements beyond Fm. From there on a continuous decrease of cross-section was measured from microbarns for the synthesis of nobelium down to picobarns for the synthesis of element 112. The data obtained in reactions with ^{208}Pb and ^{209}Bi for the 1n evaporation channel at low excitation energies of about 10–15 MeV (therefore named *cold fusion*) and in reactions with actinide targets for the 4n channel at excitation energies of 35–45 MeV (*hot fusion*) are plotted in Fig. 5. Interesting for further investigation of SHEs are the relatively high cross-sections measured for the synthesis of elements 114 and 116 (4n channel) [29, 30, 31]. In both cases the obtained values of about 0.5 pb are

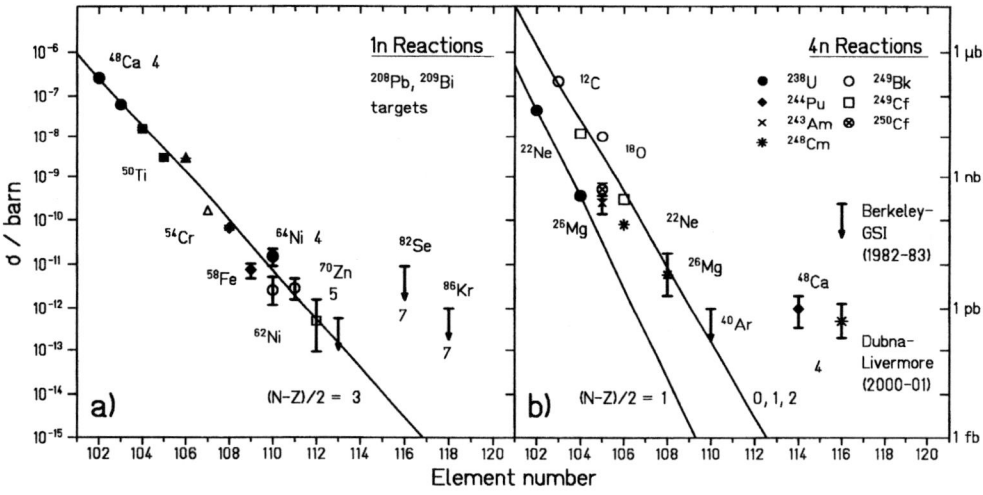

FIGURE 5. Measured cross-sections and cross-section limits for reactions using ^{208}Pb and ^{209}Bi targets and one neutron evaporation (a) and for reactions using actinide targets and four neutron evaporation (b).

considerably larger than expected from the trend set by fusion of the lighter elements. An explanation could be a relatively high and wide fission barrier of the CN, which is created by strong shell effects in the region of spherical SHEs. Note in this context that the experimental sensitivity increased by three orders of magnitude since the 1982-83 search experiments for element 116 using a hot fusion reaction [32].

A number of excitation functions was measured for the synthesis of elements from rutherfordium to darmstadtium using Pb and Bi targets [3]. For the even elements these data are shown in Fig. 6. The figure includes the recently (June 2003) measured excitation function of the reaction ^{54}Cr + ^{208}Pb and an update of the previously obtained data of ^{50}Ti + ^{208}Pb [5, 33].

The maximum evaporation residue cross-section (1n channel) was measured at beam energies well below a fusion barrier calculated in one dimension [34]. At the optimum beam energy projectile and target nuclei are just reaching the contact configuration in a central collision. The relatively simple fusion barrier based on the Bass model [34] is too high and a tunnelling process through this barrier cannot explain the measured cross-section.

Various processes are possible and are discussed in the literature which result in a lowering of the fusion barrier. Among these transfer of nucleons and excitation of vibrational degrees of freedom are the most important [14,35-43]. The theoretical studies are also aimed at reproducing the known cross-section data and further extrapolating the calculations into the region of spherical superheavy nuclei. The measured cross-sections for the formation of ^{257}Rf up to ^{277}112 are reproduced almost within about a factor of 2 by the various models. However, there are significant differences in the cross-section

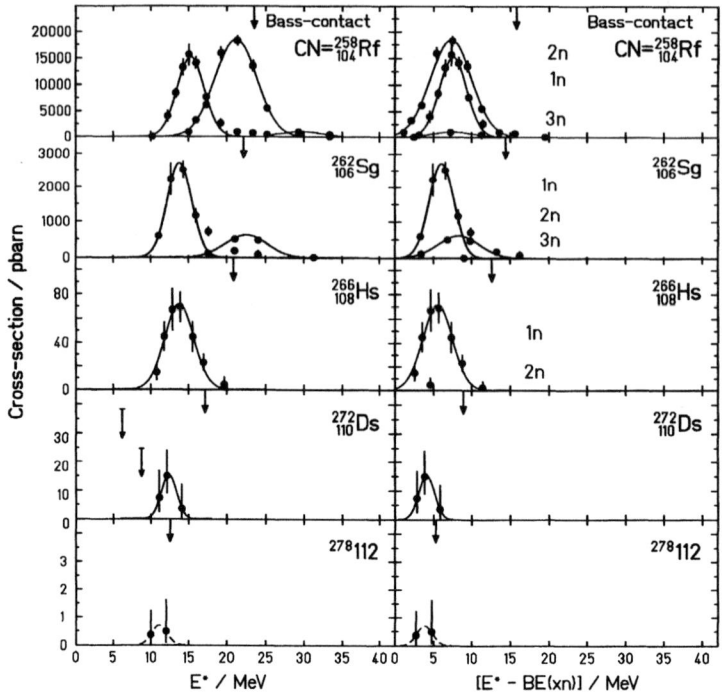

FIGURE 6. Measured even element excitation functions based on ^{208}Pb targets.

values for the synthesis of spherical SHEs at and beyond Z = 114.

In the case of actinide targets, the target nucleus is strongly deformed and the height of the Coulomb barrier is a function of the orientation of the deformation axes. The reaction ^{48}Ca + ^{248}Cm was studied in Dubna [29, 30], and evidence for the 4n channel was obtained at a beam energy resulting in an excitation energy of 30.4 − 35.8 MeV.

It was pointed out in the literature [44] that closed shell nuclei as projectile and target are favorable for fusion of SHEs. The reason is not only a low reaction Q value and thus low excitation energy, but also that fusion of such systems is connected with a minimum of energy dissipation. The fusion path propagates along cold fusion valleys on the potential energy surface, where the reaction partners keep kinetic energy up to the closest possible distance. In this view the difference between 'cold' and 'hot' fusion is not only a result from different values of the excitation energy, but there exists also a qualitative difference. This is on the one side based on a well ordered fusion process along paths of minimum dissipation of energy (cold fusion), and on the other side on a process governed by the formation of a more or less energy equilibrated CN (hot fusion). This qualitative explanation is well in agreement with the results from experimental studies of quasi-fission and compound-nucleus fission [45].

SUMMARY AND OUTLOOK

The experimental work of the last two decades has shown that cross-sections for the synthesis of the heaviest elements decrease almost continuously. However, recent data on the synthesis of element 114 and 116 in Dubna using hot fusion seem to break this trend when the region of spherical superheavy elements is reached.

The progress towards the exploration of the island of spherical SHEs is difficult to predict. Despite the exciting new results, many questions of more general character are still awaiting an answer. New experimental developments will not only make it possible to perform experiments aimed at synthesizing new elements in reasonable measuring times, but will also allow for a number of various other investigations covering reaction physics and spectroscopy.

One can hope that, during the coming years, more data will be measured in order to promote a better understanding of the stability of the heaviest elements and the processes that lead to fusion. A microscopic description of the fusion process will be needed for an effective explanation of all measured phenomena in the case of low dissipative energies. Then, the relationships between fusion probability and stability of the fusion products may also become apparent.

An opportunity for the continuation of experiments in the region of SHEs at decreasing cross-sections afford, among others, further accelerator developments. High current beams and radioactive beams are options for the future. At increased beam currents, values of tens of particle μA's may become accessible, the cross-section level for the performance of experiments can be shifted down into the region of tens of femtobarns, and excitation functions can be measured on the level of tenths of picobarns. High currents, in turn, call for the development of new targets and separator improvements. Radioactive ion beams, not as intense as the ones with stable isotopes, will allow for approaching the closed neutron shell $N = 184$ already at lighter elements. The study of the fusion process using radioactive neutron rich beams will be highly interesting.

The half-lives of spherical SHEs are expected to be relatively long. Based on nuclear models, which are effective predictors of half-lives in the region of the heaviest elements, values from microseconds to years have been calculated for various isotopes. This wide range of half-lives encourages the application of a wide variety of experimental methods in the investigation of SHEs, from the safe identification of short lived isotopes by recoil-separation techniques to atomic physics experiments on trapped ions, and to the investigation of chemical properties of SHEs using long-lived isotopes.

ACKNOWLEDGEMENTS

The recent experiments at SHIP were performed in collaboration with D. Ackermann, F.P. Heßberger, B. Kindler, J. Kojouharova, B. Lommel, R. Mann, G. Münzenberg, S. Reshitko, H.J. Schött (GSI Darmstadt); O.N. Malyshev, A. Popeko, A. Yeremin (JINR Dubna); S. Antalic, P. Cagarda, S. Saro, B. Streicher (University Bratislava); P. Kuusiniemi, M. Leino, J. Uusitalo (University Jyväskylä).

REFERENCES

1. G. Münzenberg, *Rep. Prog. Phys. A* **51**, 57 (1988).
2. S. Hofmann, *Rep. Prog. Phys. A* **61**, 639 (1998).
3. S. Hofmann and G. Münzenberg, *Rev. Mod. Phys.* **72**, 733 (2000).
4. IUPAC 42nd General Assembly, Ottawa, Canada, 9-17 August 2003.
5. S. Hofmann et al., *Z. Phys. A* **350**, 277 (1995).
6. K. Morita et al., contribution to this conference.
7. T.N. Ginter et al., *Phys. Rev. C* **67**, 064609 (2003).
8. Yu.A. Lazarev et al., *Phys. Rev. Lett.* **75**, 1903 (1995).
9. A. Ghiorso et al., *Phys. Rev. Lett.* **33**, 1490 (1974).
10. M.R. Schmorak, *Nuclear Data Sheets* **59**, 507 (1990).
11. S. Hofmann et al., *Eur. Phys. J. A* **10**, 5 (2001).
12. S. Cwiok et al., Phys. Rev. Lett. **83**, 1108 (1999) and private communication.
13. I. Muntian et al., *Phys. Rev. C* **60**, 041302 (1999).
14. R. Smolanczuk, *Phys. Rev. Lett.* **83**, 4705 (1999).
15. R. Smolanczuk et al., *Phys. Rev. C* **52**, 1871 (1995).
16. S. Hofmann et al., *Z. Phys. A* **350**, 281 (1995).
17. S. Hofmann et al., *Eur. Phys. J. A* **14**, 147 (2002).
18. S. Hofmann et al., *Z. Phys. A* **354**, 229 (1996).
19. Ch.E. Düllmann et al., *Nature* **418**, 859 (2002).
20. A. Türler et al., *Eur. Phys. J. A* **17**, 505 (2003).
21. R. Mann et al., to be published.
22. B. Kindler et al., to be published.
23. R. Smolanczuk and A. Sobiczewski, Proc. XV. Nucl. Phys. Divisional Conf. on *Low Energy Nuclear Dynamics*, St.Petersburg, Russia, 1995, p.313, World Scientific, Singapore, 1995.
24. S. Cwiok et al., *Nucl. Phys. A* **611**, 211 (1996).
25. G.A. Lalazissis et al., *Nucl. Phys. A* **608**, 202 (1996).
26. K. Rutz et al., *Phys. Rev. C* **56**, 238 (1997).
27. A.T. Kruppa et al., *Phys. Rev. C* **61**, 034313 (2000).
28. P. Möller et al., *Atomic Data and Nucl. Data Tables* **66**, 131 (1997).
29. Yu.Ts. Oganessian et al., *Phys. Rev. C* **63**, 011301 (2000).
30. Yu.Ts. Oganessian et al., Phys. Atomic Nuclei **64**, 1349 (2001).
31. Yu.Ts. Oganessian et al., *Nature* **400**, 242 (1999).
32. P. Armbruster et al., *Phys. Rev. Lett.* **54**, 406 (1985).
33. S. Hofmann et al., Proc. of the VIII Int. Conference on *Nucleus-Nucleus Collisions*, Moscow, Russia, 2003, to be published.
34. R. Bass, *Nucl. Phys. A* **231**, 45 (1974).
35. Y. Aritomo et al., *Phys. Rev. C* **59**, 796 (1999).
36. E.A. Cherepanov, *Pramana J. Phys.* **53**, 619 (1999).
37. R. Smolanczuk, *Phys. Rev. C* **59**, 2634 (1999).
38. G. Giardina et al., *Eur. Phys. J. A* **8**, 205 (2000).
39. V.Yu. Denisov and S. Hofmann, *Phys. Rev. C* **61**, 034606 (2000).
40. G.G. Adamian et al., *Nucl. Phys. A* **678**, 24 (2000).
41. G.G. Adamian et al., *Phys. Rev. C* **62**, 064303 (2000).
42. R. Smolanczuk, *Phys. Rev. C* **63**, 044607 (2001).
43. V.I. Zagrebaev, in Proc. Int. Workshop on *Fusion Dynamics at the Extremes*, Dubna, Russia, 2000, p. 215, World Scientific, 2001.
44. R.K. Gupta et al., *Z. Phys. A* **283**, 217 (1977).
45. M.G. Itkis et al., in Proc. Int. Workshop on *Fusion Dynamics at the Extremes*, Dubna, Russia, 2000, p. 93, World Scientific, 2001.

Fusion-Fission Dynamics of Super-Heavy Element Formation and Decay

V.I. Zagrebaev

Flerov Laboratory of Nuclear Reaction, JINR, Dubna, 141980, Moscow region, Russia

Abstract. The paper is focused on reaction dynamics of super-heavy nucleus formation and decay at beam energies near the Coulomb barrier. The aim is to review the things we have learned from recent experiments on fusion-fission reactions leading to the formation of compound nuclei with $Z \geq 102$ and from their extensive theoretical analysis. Main attention is paid to the dynamics of formation of very heavy compound nuclei taking place in strong competition with the process of fast fission (quasi-fission). The choice of collective degrees of freedom playing a principal role, finding the multi-dimensional driving potential and the corresponding dynamic equation regulating the whole process are discussed. Theoretical predictions are made for synthesis of SH nuclei up to Z=120 in the asymmetric "hot" fusion reactions basing on use of the heavy transactinide targets.

INTRODUCTION

The interest in the synthesis of super-heavy nuclei has lately grown due to the new experimental results [1, 2, 3] demonstrating a real possibility of producing and investigating the nuclei in the region of the so-called "island of stability". The new reality demands a more substantial theoretical support of these expensive experiments which will allow most optimal choice of fusing nuclei and collision energies as well as a better estimation of the cross sections and unambiguous identification of evaporation residues.

CAPTURE, FUSION, AND EVR FORMATION CROSS SECTIONS

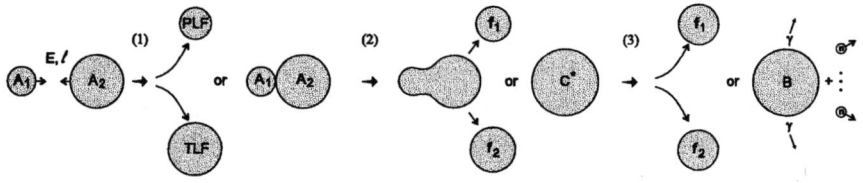

FIGURE 1. Schematic picture of super-heavy nucleus formation.

The process of a cold residual nucleus formation is shown schematically in Fig. 1. A whole process can be divided into three reaction stages. At the first stage, colliding nuclei overcome the Coulomb barrier and approach the point of contact. Quasi-elastic and deep-inelastic reaction channels dominate at this stage leading to formation of projectile-like and target-like fragments (PLF and TLF) in the exit channel. At the second reaction stage the touching nuclei evolve into the configuration of an almost spherical compound mono-nucleus. After dynamic deformation and exchange by several nucleons, two touching heavy nuclei may re-separate into PLF and TLF or may go directly to fission channels without formation of compound nucleus. The later process is usually called quasi-fission. Denote a probability for two touching nuclei to form the compound nucleus as P_{CN}. This probability depends also on initial deformations and orientations of two touching nuclei (see below). At the third reaction stage an excited compound nucleus emits neutrons and γ rays lowering its excitation energy and forming finally a residual nucleus in its ground state. This process takes place in strong competition with a regular fission, and the corresponding survival probability $P_{xn}(l,E^*)$ is usually much less than unity even for a low-excited super-heavy nucleus.

The production cross section of a cold residual nucleus B, which is the product of neutron evaporation and γ emission from an excited compound nucleus C, formed in the fusion process of two heavy nuclei $A_1 + A_2 \to C \to B + xn + N\gamma$ at c.m. energy E close to the Coulomb barrier in the entrance channel, can be decomposed over partial waves and written as follows

$$\sigma_{ER}^{xn}(E) \approx \frac{\pi\hbar^2}{2\mu E} \sum_{l=0}^{\infty} (2l+1) \cdot \int_0^{\infty} f(B) P^{HW}(B,l,E) P_{CN}(B,l,E^*) dB \cdot P_{xn}(l,E^*). \quad (1)$$

Here P^{HW} is the penetration probability of the one-dimensional potential barrier given by the usual Hill-Wheeler formula [4] with the barrier height modified to include a centrifugal term. $f(B)$ is the "barrier distribution function", which takes into account the multi-dimensional character of the realistic barrier given by dynamic deformations of nuclear surfaces and/or different orientations of statically deformed colliding nuclei. Integration over effective barrier B means, in fact, integration over such dynamic deformations and/or different orientations. Semi-empirical [5, 6] and/or channel coupling approaches [7, 8] may be used to calculate rather accurately a penetrability of the multi-dimensional Coulomb barrier and the corresponding capture cross section. The survival probability $P_{xn}(l,E^*)$ of an excited compound nucleus can be calculated within a statistical model [9] rather accurately also, if a realistic integration over neutron evaporation cascade and γ emission is performed and a realistic formula for the level density is used [6]. The most uncertain parameter here is the height of fission barrier of CN. Unfortunately, the fission barriers of super-heavy nuclei calculated within the different approaches differ by several MeV (see, e.g., [10]). The processes of the compound nucleus formation and quasi-fission are the least studied stages of heavy ion fusion reaction. Today there is no consensus for the mechanism of the compound nucleus formation itself, and quite different, sometimes opposite in their physics sense, models are used for its description.

TWO-CORE MODEL

To solve the problem of fusion-fission dynamics one has to answer very principal questions. What are the main degrees of freedom playing most important role at this reaction stage? What is the corresponding driving potential and what is an appropriate equation of motion for description of time evolution of nuclear system at this stage?

Until now quite different degrees of freedom and equations of motion are used for description of the fusion and fission processes. It is justified for not so heavy nuclear systems, when these two reaction stages are well separated in time by a long-lived compound nucleus. For heavy nuclear systems, when quasi-fission channels dominate, the fusion-fission dynamics has to be described as a common inseparable process. At the same time, we have to restrict ourself to a small number of degrees of freedom to be able to solve numerically the corresponding equations of motion. However, any simplified one- or two-dimensional models hardly may be used for adequate description of the complicated reaction dynamics. Elongation of the system (distance between centers of fragments), mass-asymmetry (proton and neutron numbers of two separated nuclei), and deformations of the fragments are quite optimal degrees of freedom for description of fusion-fission process (see Fig. 2), and the two-center shell model [11] seems to be the best for calculation of the corresponding potential energy surface. However, the available numerical codes (being rather complicated) confront with some difficulties for large mass-asymmetry configurations and also for separated nuclei not giving the appropriate values of the Coulomb (fusion) barriers.

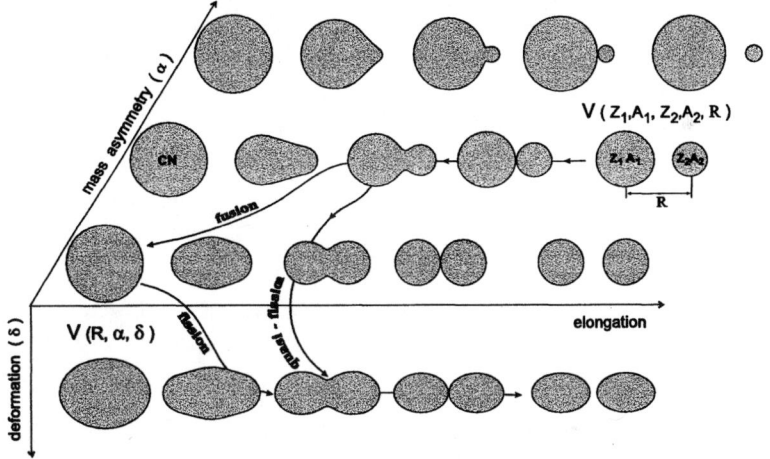

FIGURE 2. Optimal degrees of freedom to describe the fusion-fission dynamics.

In [5, 12] a new approach has been proposed for description of fusion-fission dynamics based on a semi-empirical version of the two-center shell model idea [11]. It is assumed that on a path from the initial configuration of two touching nuclei to the compound nucleus configuration and on a reverse path to the fission channels the nuclear system consists of two cores (z_1, n_1) and (z_2, n_2) surrounded with a certain number of common (shared) nucleons $\Delta A = A_{CN} - a_1 - a_2$ moving in the whole volume occupied by the two cores, see Fig. 3. The processes of compound nucleus formation, fission and quasi-fission take place in the space $(z_1, n_1, \delta_1; z_2, n_2, \delta_2)$, where δ_1 and δ_2 are the dynamic deformations of the cores. The compound nucleus is finally formed when the elongation of the system becomes shorter than a saddle point elongation of CN.

Within the two-core model the fusion-fission driving potential is defined as follows

$$V_{fus-fis}(R; z_1, n_1, \delta_1; z_2, n_2, \delta_2) =$$

$$\tilde{V}_{12}(R,z_1,n_1,\delta_1;z_2,n_2,\delta_2) - [\tilde{B}(a_1) + \tilde{B}(a_2) + \tilde{B}(\Delta A)] + B(A_1^0) + B(A_2^0). \qquad (2)$$

Here $\tilde{B}(a_1) = \tilde{\beta}_1 a_1$, $\tilde{B}(a_2) = \tilde{\beta}_2 a_2$ and $\tilde{B}(\Delta A) = 0.5(\tilde{\beta}_1 + \tilde{\beta}_1)\Delta A$ are the binding energies of the cores and of common nucleons. These quantities depend on the number of shared nucleons. If one defines the measure of the collectivization as $x = \Delta A/\Delta A_{CN}$, then $\tilde{\beta}_{1,2}$ can be roughly approximated as $\tilde{\beta}_{1,2} = \beta_{1,2}^{exp}\varphi(x) + \beta_{CN}^{exp}(1-\varphi(x))$, where $\beta_{1,2}^{exp}$ and β_{CN}^{exp} are specific binding energies of the isolated (free) fragments, which can be derived from the experimental nuclear masses or calculated rather accurately within the macroscopic-microscopic model [13]. Note that $\Delta A < A_{CN}$ and CN formation does not mean $a_{1,2} = 0$ [5]. $\varphi(x)$ is an appropriate monotonous function satisfying the conditions $\varphi(x=0) = 1$, $\varphi(x=1) = 0$. Thus the specific binding energies of the cores $\tilde{\beta}_{1,2}$ approach the specific binding energy of CN with increasing ΔA. All the shell effects enter the total energy (2) by means of $\beta_{1,2}^{exp}$ and β_{CN}^{exp}. The interaction of two fragments $\tilde{V}_{12}$ is defined in usual way at $R \geq R_{cont}$ as a sum of the Coulomb and nuclear potentials (here proximity potential is used). This interaction weakens gradually with increasing the number of shared nucleons ΔA at $R_{CN} < R < R_{cont}$, i.e. with gradual dissolving the two cores in the compound mono-nucleus. The interaction energy transforms to the binding energy of CN. Thus, once the compound nucleus has been formed ($\Delta A = \Delta A_{CN}$), the total energy of the system $V_{fus-fis} = Q_{gg}^{fus} = B(A_1^0) + B(A_2^0) - B(A_{CN})$, as it should be if the energy of two resting at infinity initial nuclei (A_1^0 and A_2^0) is taken as zero.

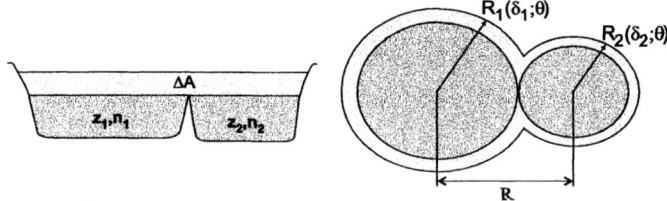

FIGURE 3. Nuclear shape used in the two-core model of fusion-fission dynamics.

The fusion-fission driving potential is shown in Fig. 4 as a function of z_1 and z_2 at $R \leq R_{cont}$ (minimized over n_1,n_2) and as a function of elongation and mass-asymmetry. As can be seen, the shell structure, clearly revealing itself in the contact of two nuclei ($\Delta A = 0$, the diagonal in Fig. 4(b)), is also retained at $R < R_{cont}$ (see the deep minima in the regions of $z_{1,2} \sim 50$ and $z_{1,2} \sim 82$ in Fig. 4b). Following the fission path (dotted curves in Fig. 4b,d) the system overcomes the multi-humped fission barrier (Fig. 4c). It is well-known that the intermediate minima correspond to the shape isomer states. From analysis of the driving potential (see Fig. 4b) we may definitely conclude now that these isomeric states are nothing else but the two-cluster configurations with magic or semi-magic cores [14].

In Fig. 5 the driving potentials calculated within the two-center shell model (version of Ref. [15]) and within the two-core model are compared for the nuclear system formed in collision of ^{48}Ca+^{248}Cm leading to compound nucleus ^{296}116. As can be seen, the results of two calculations are rather close. At the same time, there are several advantages of the proposed approach. The driving potential is derived basing on experimental binding energies of two cores, which means that the "true" shell structure is taken into

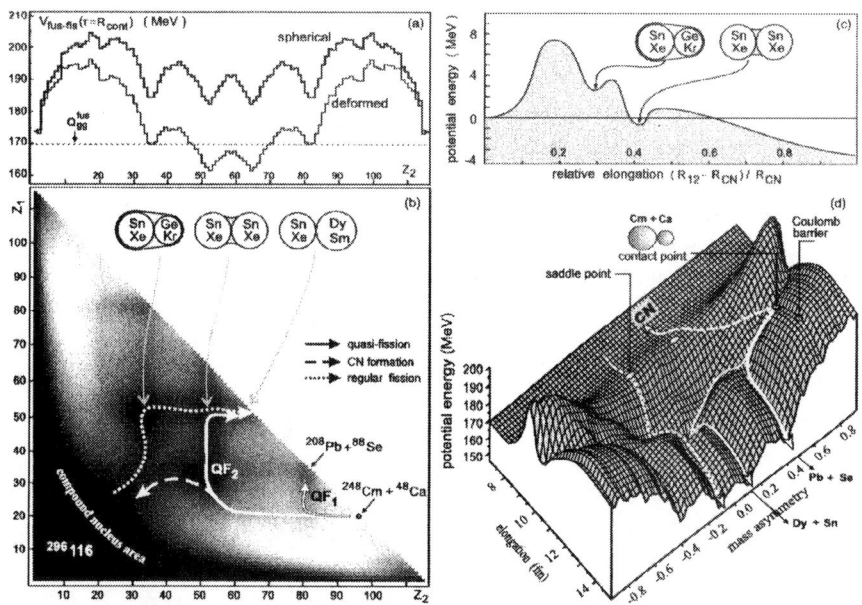

FIGURE 4. Driving potential $V_{fus-fis}$ of the nuclear system consisting of 116 protons and 180 neutrons. (a) Potential energy of two touching nuclei at $a_1 + a_2 = A_{CN}$, $\Delta A = 0$, i.e., along the diagonal of the lower figure. (b) Topographical landscape of the driving potential on the plane (z_1, z_2). The dashed, solid, and dotted curves with arrows show fusion, quasi-fission, and regular fission paths, respectively. (c) Three humped fission barrier calculated along the fission path (dotted curves). (d) Three dimensional plot in the "mass-asymmetry - elongation" space.

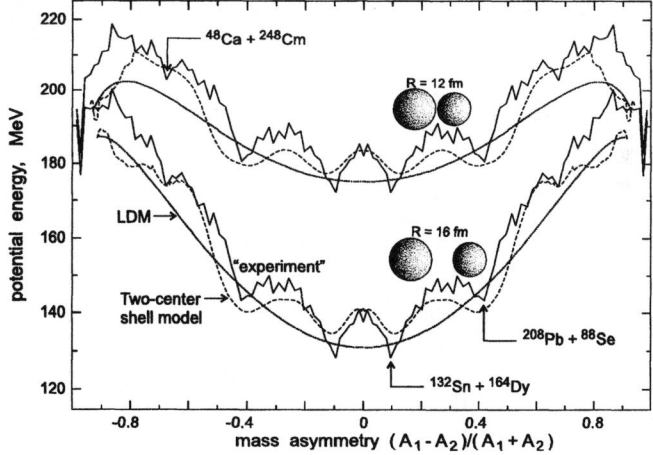

FIGURE 5. Fusion-fission driving potential as a function of mass-asymmetry calculated at two fixed distances between nuclei: in vicinity of the Coulomb barrier, $R = 12$ fm, and for well separated nuclei, $R = 16$ fm. Dotted, dashed and solid curves correspond to the LDM, two-center shell model, and two-core model calculations, respectively.

account and, thus, for well separated nuclei (large values of R) $V_{fus-fis}$ gives *explicit* values of nucleus-nucleus interaction. The fusion-fission driving potential (2) is defined in the whole region $R_{CN} < R < \infty$, it is a continuous function at $R = R_{cont}$, it gives the realistic Coulomb barrier at $R = R_B > R_{cont}$, and may be used for simultaneous description of the whole fusion-fission process. At last, along with using the variables $(z_1,n_1;z_2,n_2)$, one may easily recalculate the driving potential as a function of mass-asymmetry $\alpha = (a_1 - a_2)/(a_1 + a_2)$ and elongation $R_{12} = r_0(a_1^{1/3} + a_2^{1/3})$ (at $R > R_{cont}$, $R_{12} = R = s + R_1 + R_2$, where s is the distance between nuclear surfaces). These variables along with deformations of the fragments δ_1 and δ_2 are commonly used for description of fission process.

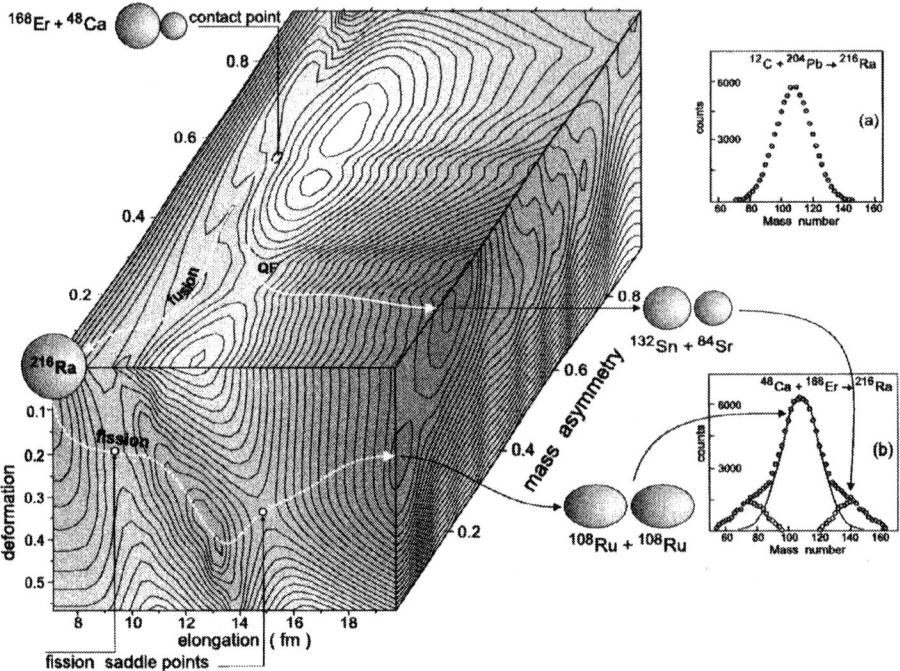

FIGURE 6. Three-dimensional driving potential of formation and fission of ^{216}Ra nucleus. In the insets the fission fragment mass distributions are shown obtained in the ^{12}C+^{204}Pb (a) and ^{48}Ca+^{168}Er (b) fusion-fission reactions [16]. In the last case a contribution from the quasi-fission process can be clearly seen.

Deformations of the fragments were found to be very important both at fusion and fission paths. Deformation energies of the cores $\Delta E_i^{def}(\delta_{i=1,2})$ can be taken into account in (2) by replacement of β_i^{exp} with $\beta_i^{exp} - \Delta E_i^{def}/a_i$ and calculated as follows $\Delta E_i^{def}(\delta_i) = E_i^{LDM}(\delta_i) + W_i(\delta_i) - W_i^0(\delta_i^0)$. Here $E^{LDM}(\delta)$ is the liquid drop model deformation energy and $W(\delta)$ is the shell correction. $W^0(\delta^0)$ is the shell correction to the ground state of a given nucleus, and δ^0 is the quadrupole deformation of the ground state. Thus, using the table of Ref. [13] for the values of $\delta^0(Z_1,A_1)$ and $W^0(Z_1,A_1)$ and knowing the value of the shell correction at zero deformation $W(\delta = 0)$ (if $\delta^0(Z_1,A_1) \neq 0$), one may easily and rather accurately estimate the deformation energy of a given fragment without performing microscopic calculation. It is clear that $\Delta E^{def}(\delta = \delta^0) = 0$ and $W(\delta \to \infty) \to 0$.

In Fig. 6 the calculated three-dimensional driving potential for formation and fission of ^{216}Ra is shown in the space of elongation, mass-asymmetry, and deformation of the fragments (here, for simplicity, the deformations are assumed to be equal to each other: $\delta_1 = \delta_2 = \delta$). In Ref. [16] the two reactions (^{12}C+^{204}Pb and ^{48}Ca+^{168}Er) were studied leading to formation of ^{216}Ra. In the insets of Fig. 6 the corresponding mass distributions of the detected fission fragments are shown. In the case of very asymmetric ^{12}C+^{204}Pb fusion reaction ($\alpha \sim 0.9$) the contact configuration is located behind the Businaro-Gallone maximum, and the nuclei form CN with a probability equal to unity. After formation, CN goes to fission channels along a normal fission path, here it locates preferably at low values of mass-asymmetry and leads to symmetric mass distribution of deformed fission fragments (see Fig. 6). In the case of ^{48}Ca+^{168}Er fusion reaction, from the contact configuration the system may evolve along two different paths, one of them leads to formation of CN, whereas another one leads the system to the quasi-fission channel without formation of CN (see Fig. 6). This quasi-fission valley is caused by the shell effect corresponding to the double-magic ^{132}Sn nucleus. Thus, for the ^{48}Ca+^{168}Er reaction the model predicts a formation of deformed symmetric fission fragments ($\delta_1 \sim \delta_2 \sim 0.3$) and spherical asymmetric quasi-fission fragments, which can be checked in principle by detecting the coincident neutrons and γ-rays.

EQUATIONS OF MOTION ?

The common dynamic equations of motion for the variables (R, α, δ), which would be valid in the whole configuration space and could be applied for description of all the competing processes (deep-inelastic collision, fusion, quasi-fission, and fission, see Fig. 1), are still have to be derived. All these degrees of freedom are quite similar for more or less uniform mono-nucleus, and even the corresponding inertia parameters $(M_R, M_\alpha, M_\delta)$ are very close to each other here. However, for contact configurations and for separated fragments these degrees of freedom are quite different. All them play an important role here. In particular, neutron transfer with positive Q-value (change of α) may significantly increase the sub-barrier penetrability in the entrance channel [17]. Dynamic deformations of the separated fragments are also very important both in the entrance and in exit channels. That is why the common dynamic equations coupling all the degrees of freedom are very desired. Nevertheless, until now the different considerations of the entrance channel and the processes of CN formation, quasi-fission and fission are commonly used. In [18, 19] the Langevin equations were used for description of evolution of the system after contact of two nuclei. In [5] the master equation was proposed and used in [12, 14] to describe approximately the overdamped evolution of the system in the $(z_1, n_1, \delta_1; z_2, n_2, \delta_2)$ space.

Solving the master equation for the distribution function $F(\vec{y} = \{z_1, n_1, \delta_1; z_2, n_2, \delta_2\}; t)$ one may determine the probability of the compound nucleus formation as an integral of the distribution function over the region $R \leq R^{CN}_{saddle}$. Similarly one can define the probabilities of finding the system in the different quasi-fission channels, i.e., the charge and mass distributions of fission fragments measured experimentally. In fact, it is not so easy to perform such realistic calculations due to a large number of the variables. To simplify the problem, we solve the master equation with restricted number of the variables. First, the deformations of the fragments are fixed (not obligatory to zero values). Then the potential energy is minimized over n_1 and n_2 and a two-dimensional driving potential $V_{fus-fis}(z_1, z_2)$ is calculated. Finally the master equation for the distribution function $F(y = \{z_1, z_2\}, t)$

$$\frac{\partial F}{\partial t} = \sum_{y'} \lambda(y, y') F(y', t) - \lambda(y', y) F(y, t) \qquad (3)$$

is solved to determine the value of P_{CN}^0 at a given excitation energy E^*. The macroscopic transition probabilities $\lambda(y, y') \sim exp\{[V_{fus-fis}(y') - V_{fus-fis}(y)]/2T(y)\}$ are used, where $T = \sqrt{[E_{cm} - V_{fus-fis}(y)]/a}$ is the local temperature and a is the level density parameter. Eq. (3) describes an overdamped evolution, when a potential energy of the nuclear system plays a major role. The sum over y' in (3) is extended only to nearest configurations $z_{1,2} \pm 1$ (no fragment transfer). Eq. (3) is solved up to the moment when the total flux comes to the compound nucleus configurations (dark area in Fig. 4b) and/or escapes into the quasi-fission channels, giving us the probability of the compound nucleus formation P_{CN}^0 and the charge distribution of quasi-fission fragments.

Mutual orientation of statically deformed colliding nuclei at their contact configuration is also very important for subsequent evolution of the system. There is a common opinion that the more compact "side-by-side" configuration of two touching nuclei is more favorable for CN formation comparing with the more elongated "nose-to-nose" configuration, which may easily re-separates into the quasi-fission channel (see, for example, [20]). Recently it has been also proved experimentally [21]. This effect is not included in P_{CN}^0 calculated with the simplified master equation (3). To take this effect into account, one may introduce the weight function $g(\theta)$, which turns to zero at $\theta = 0$ (tip orientation, lowest barrier B_1) and is close to unity at $\theta = \pi/2$ (side orientation, highest barrier B_2). Thus the probability of CN formation, bing part of Eq. (1) for the cross section, may be written as $P_{CN}(B(\theta), l, E^*) = P_{CN}^0(l, E^*) \cdot g(\theta)$, and $g(\theta)$ is approximated here by the Fermi function $1/[1 + exp\{(\theta_0 - \theta)/\Delta\theta\}]$, where $0 < \theta_0 < \pi/2$ and $\Delta\theta$ are free parameters. This function simulates somehow a decrease of fusion probability P_{CN} for very elongated "nose-to-nose" configurations of deformed heavy nuclei. Note that in the entrance channel such configurations increase the barrier penetrability due to decrease of the Coulomb barrier. This effect is well-known and can be described very accurately within the semi-empiric [5, 6] and/or CC approaches [7, 8].

CROSS SECTIONS OF SHE FORMATION

In Fig. 7 the calculated capture, fusion, and evaporation residue formation cross sections are shown for the ^{48}Ca induced fusion reactions leading to super-heavy nuclei with $Z = 112 \div 116$. The shell corrections to the ground state energies of super-heavy nuclei proposed by P. Möller et al. [13] were used to estimate the corresponding fission barriers and calculate survival probability $P_{xn}(l,E^*)$. As can be seen, for the "hot" fusion reactions, the EvR cross sections decrease rather slowly with increasing Z and remain at the level of few picobarns. From obtained results one may conclude that the "hot" fusion reactions with the heavy transactinide targets can be successfully used even at existing facilities for a synthesis of super-heavy nuclei with Z up to 120 (^{54}Cr+^{248}Cm or ^{58}Fe+^{244}Pu combinations). The preferable beam energy corresponds to about 40 MeV of CN excitation energy (with maximal yield of 3n and/or 4n evaporation products), i.e., it should be slightly higher than those used in previous experiments [1, 2].

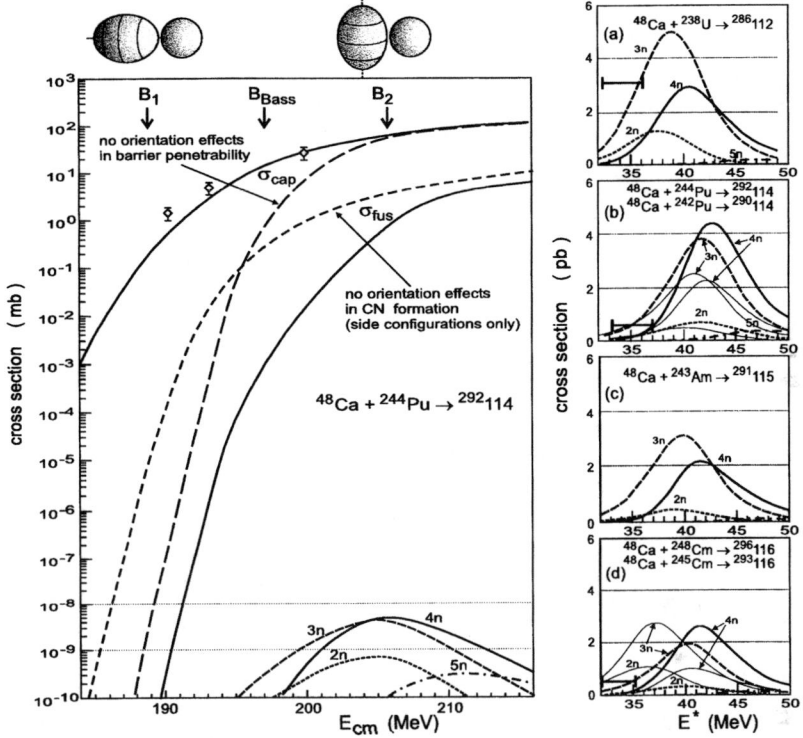

FIGURE 7. Capture, fusion, and evaporation residue formation cross sections in the ^{48}Ca induced fusion reactions. Experimental data for the capture cross sections (left panel) are from [3]. Dashed curves on the left panel are obtained ignoring the orientation effects in the entrance channel (i.e., $f(B) = \delta(B-B_0)$ in (1)) or in CN formation (side orientations only). The dotted, dashed, solid and dash-dotted curves on the right panels (a-d) correspond to 2, 3, 4, and 5 neutron evaporation channels, respectively. The thin curves in (b) and (d) correspond to the fusion reactions with lighter isotopes: ^{242}Pu and ^{245}Cm. Experimental data for production of nuclei with Z=112, 114, and 116 are from [1, 2].

ACKNOWLEDGMENTS

I am indebted to Profs. M.G. Itkis and Yu.Ts. Oganessian for many fruitful discussions of the problem. The work was supported partially by INTAS under grant No. 00-655.

REFERENCES

1. Yu.Ts. Oganessian, A.V. Yeremin, A.G. Popeko, S.L. Bogomolov, G.V. Buklanov, M.L. Chelnokov, V.I. Chepigin, B.N. Gikal, V.A. Gorshkov, G.G. Gulbekian, M.G. Itkis, A.P. Kabachenko, A.Yu. Lavrentev, O.N. Malyshev, J. Rohac, R.N. Sagaidak, S. Hofmann, S. Saro, G. Giardina, K. Morita, Nature **400**, 242 (1999).
2. Yu.Ts. Oganessian, V.K. Utyonkov, Yu.V. Lobanov, F.Sh. Abdulin, A.N. Polyakov, I.V. Shirokovsky, Yu.S. Tsyganov, G.G. Gulbekian, S.L. Bogomolov, B.N. Gikal, A.N. Mezentsev, S. Iliev, V.G. Subbotin, A.M. Sukhov, O.V. Ivanov, G.V. Buklanov, K. Subotic, M.G. Itkis, K.J. Moody, J.F. Wild, N.J. Stoyer, M.A. Stoyer, R.W. Lougheed, Yad.Fiz., **63**, 1769 (2000) [Physics of Atomic Nuclei **63**, 1679 (2000)]; Phys.Rev. C **63**, 011301(R) (2001).
3. M.G. Itkis, Yu.Ts. Oganessian, A.A. Bogatchev, I.M. Itkis, M. Jandel, J. Kliman, G.N. Kniajeva, N.A. Kondratiev, I.V. Korzyukov, E.M. Kozulin, L. Krupa, I.V. Pokrovski, E.V. Prokhorova, B.I. Pustylnik, A.Ya. Rusanov, V.M. Voskresenski, F. Hanappe, B. Benoit, T. Materna, N. Rowley, L. Stuttge, G. Giardina, K.J. Moody, in *Proceedings on Fusion Dynamics at the Extremes*, Dubna, 2000, edited by Yu.Ts. Oganessian and V.I. Zagrebaev, World Scientific, Singapore, 2001, pp.93-109.
4. D.L. Hill, J.A. Wheeler, Phys.Rev. **89**, 1102 (1953).
5. V.I. Zagrebaev, Phys.Rev. C **64**, 034606 (2001).
6. V.I. Zagrebaev, Y. Aritomo, M.G. Itkis, Yu.Ts. Oganessian, and M. Ohta, Phys.Rev. C **65**, 014607 (2002).
7. K. Hagino, N. Rowley, A.T. Kruppa, Comp.Phys.Commun., **123**, 143 (1999).
8. V.I. Zagrebaev and V.V. Samarin, JINR report No. P7-2003-32, Dubna, 2003 (to be published in Yad.Fiz., 2004); CC fusion code of the NRV: http://nrv.jinr.ru/nrv.
9. A.V. Ignatyuk, *Statistical properties of excited atomic nuclei*, Energoatomizdat, Moscow, 1983.
10. M.G. Itkis, Yu.Ts. Oganessian, and V.I. Zagrebaev, Phys.Rev. C **65**, 044602 (2002).
11. U. Mosel, J. Maruhn, and W. Greiner, Phys.Lett. B **34**, 587 (1971); J. Maruhn and W. Greiner, Z.Physik, **251**, 431 (1972).
12. V.I. Zagrebaev, J.Nucl.Radiochem.Sci. 3, No. 1, 13 (2002).
13. P. Möller, J.R. Nix, W.D. Myers, W.J. Swiatecki, At.Data Nucl.Data Tables **59**, 185 (1995).
14. V.I. Zagrebaev, M.G. Itkis, and Yu.Ts. Oganessian, Yad.Fiz., **66**, No.6, 1069 (2003) [Phys.At.Nucl., **66**, No.6, 1033 (2003)].
15. A. Iwamoto, S. Yamaji, S. Suekane, and K. Harada, Prog.Theor.Phys., **55**, 115 (1976).
16. A.Yu. Chizhov, M.G. Itkis, I.M. Itkis, G.N. Kniajeva, E.M. Kozulin, N.A. Kondratiev, I.V. Pokrovski, R.N. Sagaidak, V.M. Voskressenski, A.V. Yeremin, L. Corradi, A. Gadea, A. Latina, A.M. Stefanini, S. Szilner, M. Trotta, A.M. Vinodkumar, S. Beghini, G. Montagnoli, F. Scarlassara, A.Ya. Rusanov, F. Hanappe, O. Dorvaux, N. Rowley, and L. Stuttge, Phys.Rev. C **67**, 011603(R)(2003).
17. V.I. Zagrebaev, Phys.Rev. C **67**, 061601(R) (2003).
18. Y. Aritomo, T. Wada, M. Ohta, Y. Abe, Phys.Rev. C **59**, 796 (1999) (and subsequent papers).
19. Y. Abe, C.W. Shen, G.I. Kosenko, and D. Boilley, Yad.Fiz., **66**, No.6, 1093 (2003).
20. A. Iwamoto, P. Möller, J.R. Nix, H. Sagawa, Nuc.Phys. A **596**, 329 (1996).
21. K. Nishio, H. Ikezoe, S. Mitsuoka, and J. Lu, Phys.Rev. C **62**, 014602 (2000); S. Mitsuoka, H. Ikezoe, K. Nishio, and J. Lu, Phys.Rev. C **62**, 054603 (2000).

Properties of superheavy nuclei

I. Muntian*, O. Parkhomenko* and A. Sobiczewski*

Sołtan Institute for Nuclear Studies, Hoża 69, PL-00-681 Warszawa, Poland

Abstract. The paper concentrates on one property of superheavy nuclei: height of static spontaneous-fission barrier. Results of recent macroscopic-microscopic calculations of this quantity, when axial symmetry of a nucleus is assumed, are illustrated. Effect of non-axiality of a nucleus on it is discussed.

1. INTRODUCTION

One of the basic problems in synthesis of superheavy nuclei is cross section for such a synthesis. There is recently a big activity in calculating this quantity (e.g. [1, 2, 3, 4, 5, 6]). The property of a superheavy nucleus which is needed for such a calculation is the height of fission barrier of a nucleus. This quantity decides on the survival of a synthesized compound nucleus in the process of its cooling.

In the present paper, we concentrate on this property. Results obtained recently within a macroscopic-microscopic model, assuming axial symmetry of a nucleus, are illustrated. Much attention is devoted to discussion of effects of non-axiality of a nucleus on this quantity.

2. FISSION BARRIERS

In this section, we discuss heights of the static spontaneous-fission barriers, B_f^{st}. They are proper to be directly used for calculations of survival of cold or weakly excited compound nuclei, as obtained e.g. in cold (with only one evaporated neutron) fusion reactions. For application of them to more highly excited compound nuclei, they should be reduced (cf. e.g. Ref. [7]), because of a decrease of shell effects with the increase of the excitation energy of a nucleus.

Figure 1 shows the barriers calculated recently within a macroscopic-microscopic model. For nuclei with proton number Z=106-120, the results have been obtained in Ref. [8] and for those with Z=104 in Ref. [9]. The model is described in Ref. [8]. Axial symmetry of a nucleus has been assumed. To put the results together into one figure, only the values calculated for even-even nuclei are shown. As the barriers are totally created by shell effects (no barriers appear for superheavy nuclei, if their shell structure is disregarded, cf. e.g. Refs. [10, 11]), effects of closed shells are observed in the calculated barriers. In particular, one can clearly see the effect (large B_f^{st}) of deformed closed shells at $N = 152$ and also at $N = 162$ in the results for $Z = 104$. The effect of the closed shell at $N = 162$ is also seen for $Z = 110$ and 114, while that of the spherical shell

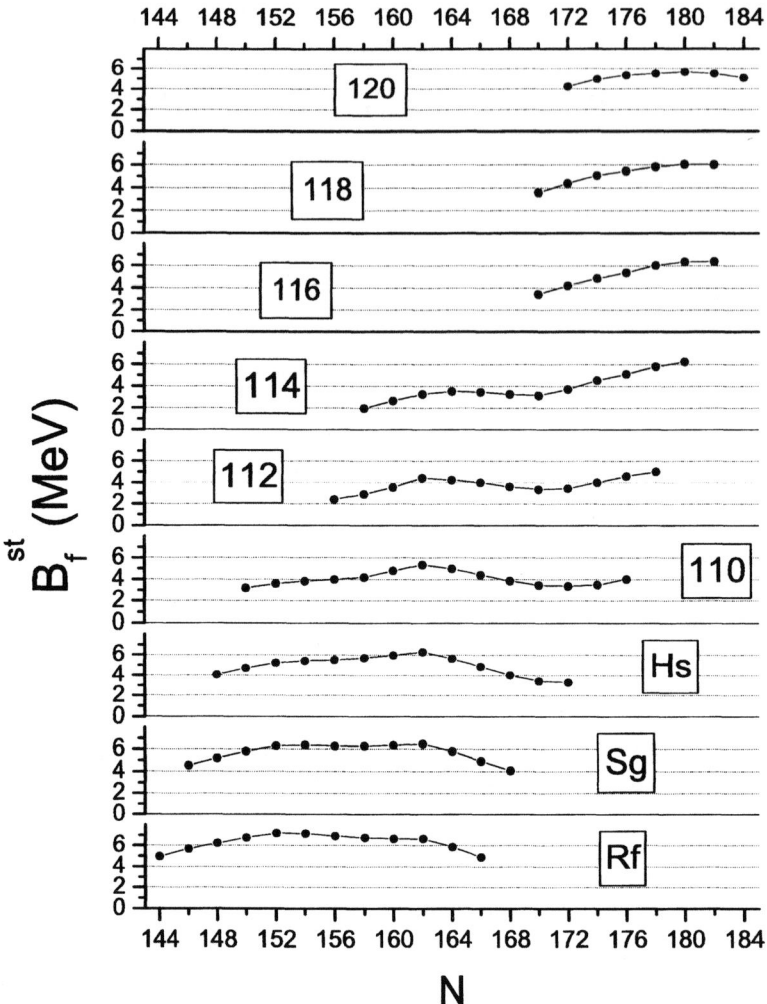

FIGURE 1. Fission barriers calculated for even-even nuclei with proton number Z=104-120.

at $N = 184$ is observed for the elements with $Z = 114$ (increase of B_f^{st} with N approaching 184) and 120. It is also the effect of the shell at $N = 184$, in cooperation with a significant shell correction for protons in the region of $Z = 114 - 126$ (low density of proton single-particle levels in this region, resulting in a large and shallow shell correction for nuclei in this region [12]), that B_f^{st} is still so large (about 6 MeV) for so large atomic number as $Z = 120$.

It is interesting to compare the macroscopic-microscopic results, illustrated in Fig. 1, with results obtained recently within the extended Thomas-Fermi plus Strutinski integral (ETFSI) model [13]. This is done in Fig. 2, where results of both models are shown for superheavy elements with the atomic number Z=104 (Rf), 110 (Ds), 114 and 120. One can see that the difference between the two results is quite large (up to about 3 MeV). Importance of such a large discrepancy is stressed by the fact that already a change of B_f^{st} by only 1 MeV changes the calculated survival probability of a synthesized nucleus by about one order of magnitude or even more [7]. One can also see in Fig. 2 that the dependencies of B_f^{st} on both Z and N, obtained within the two models, are different. In both calculations, axial symmetry of a nucleus has been assumed. The property of the results, which is worth to be noted, and which is common for both models, is a small odd-even effect in the barriers B_f^{st}.

3. EFFECT OF NON-AXIALITY

Results for B_f^{st}, shown and discussed in the previous chapter, have been obtained with the assumption of the axial symmetry of a nucleus. It is known, however, that non-axial shapes may significantly lower the potential energy of a nucleus (e.g. Refs. [14, 15, 16]) and, thus, change B_f^{st}.

To study this change for superheavy nuclei analyzed here, it is sufficient to look at the effect of non-axial shapes on the saddle-point masses. It is because the equilibrium-point masses are stable against these shapes.

Shape of a nucleus is parametrized in our study in the following way

$$R(\theta,\phi) = R_0 \{1 + \beta_2 \left[\cos\gamma Y_{20} + \frac{1}{\sqrt{2}}\sin\gamma(Y_{22}+Y_{2-2})\right]$$
$$+ \beta_4 \left[\frac{1}{12}(7+5\cos 2\gamma)Y_{40} + \frac{1}{2}\sqrt{\frac{5}{6}}\sin 2\gamma(Y_{42}+Y_{4-2})\right.$$
$$+ \left. \frac{1}{12}\sqrt{\frac{35}{2}}(1-\cos 2\gamma)(Y_{44}+Y_{4-4})\right]$$
$$+ \beta_6 Y_{60}\}, \tag{1}$$

where radius of a nucleus $R(\theta,\phi)$ is expressed (in the intrinsic frame of reference) by spherical harmonics $Y_{\lambda\mu}(\theta,\phi)$. Here, besides the quadrupole non-axiality (characterized by the parameter γ), also the hexadecapole non-axiality γ_4 is introduced, according to Ref. [17]. To avoid, however, too large deformation space, only a particular hexadecapole non-axiality is taken (with $\gamma_4 = -2\gamma$ in the notation of Ref. [17]).

The potential energy is analyzed in the 3-dimensional deformation space: $\{\beta_2, \gamma, \beta_4\}$. In each point of the space, the energy is minimized in the β_6 degree of freedom.

Saddle point and the energy at this point are found in the following way. For each analyzed nucleus, we choose two points in the deformation space: one around the equilibrium point of the nucleus, and the other in its fission valley. Then, we connect them by all possible paths passing through dense grid points inside the used space. For each path L, we find the point P_L, which corresponds to the largest energy along this path, $V(P_L)$. Saddle point P_{L_0} is the point, at which the energy $V(P_{L_0})$ is smallest among all the values $V(P_L)$.

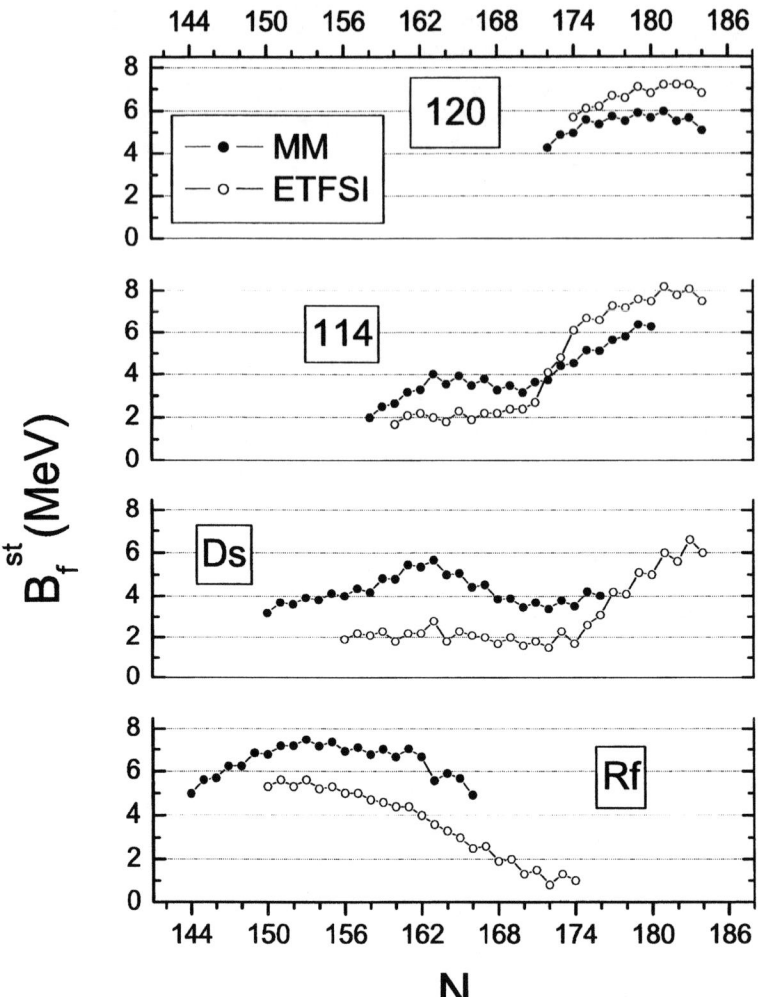

FIGURE 2. Comparison between barriers calculated within a macroscopic-microscopic and the ETFSI models.

The barrier height B_f^{st} is:
$$B_f^{st} = V(P_{L_0}) - E_0, \qquad (2)$$
where the ground-state (corresponding to the equilibrium point P_{eq}) energy E_0 is
$$E_0 = V(P_{eq}) + E_{zp}. \qquad (3)$$
For the zero-point energy E_{zp}, we take 0.7 MeV [18].

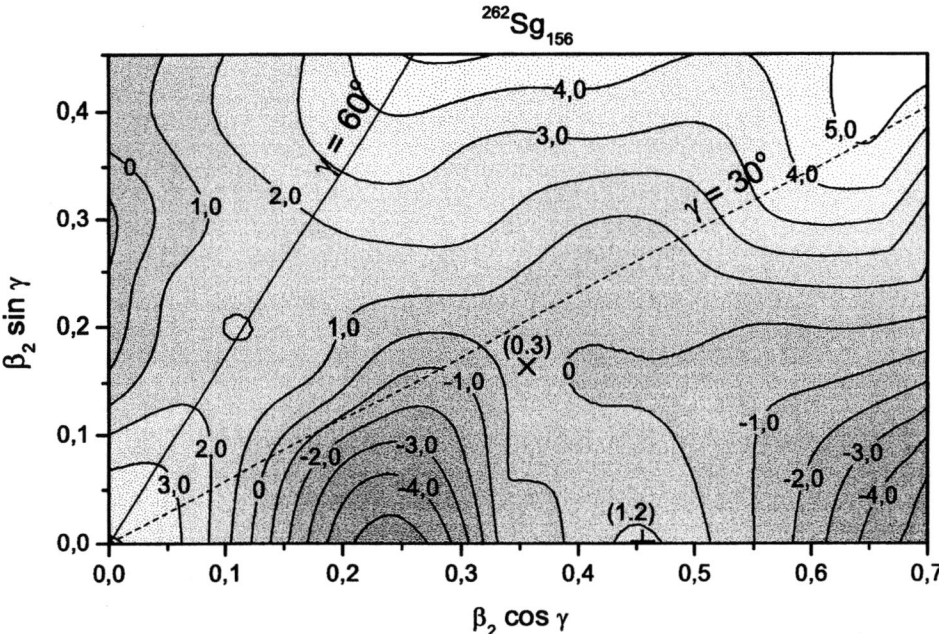

FIGURE 3. Contour map of potential energy calculated for the nucleus ^{262}Sg. Numbers at contour lines specify the value of the energy.

It is worth mentioning that the method of finding saddle point, used by us, avoids problems appearing when saddle point is searched by minimization of energy, at each elongation of a nucleus, in other degrees of freedom. In the latter case, one can easily come to a wrong solution for the saddle point, as discussed e.g. in Ref. [19].

Figure 3 shows contour map of the potential energy of the nucleus ^{262}Sg projected on the plane ($\beta_2 \cos\gamma$, $\beta_2 \sin\gamma$). The line $\gamma=0°$ corresponds to the axially symmetric (with respect to Oz axis) prolate shapes, and the line $\gamma=60°$ is corresponding to the axially symmetric (with respect to the Oy axis) oblate shapes of a nucleus. The line $\gamma=30°$ corresponds to shapes with maximal non-axiality. One can see that the saddle point (denoted by the symbol "+") in the case of axial symmetry has energy 1.2 MeV, while non-axiality shifts the saddle to the point denoted by the symbol "×" and decreases its energy to 0.3 MeV, i.e. by 0.9 MeV. (The energy is normalized so, that its macroscopic part is zero at spherical shape of a nucleus). Thus, the reduction of the saddle-point energy of this nucleus by non-axiality is quite large.

Figure 4 gives contour map of energy of a much heavier nucleus ^{278}112 (which is the compound nucleus obtained at GSI-Darmstadt when synthesizing the nucleus ^{277}112 [20]). One can see that this map is much different from that of Fig. 3. Here, two saddle points are obtained with the same energy (-0.9 MeV) in the case of the axial symmetry. Only the second one (at $\beta_2 \approx 0.52$) is lowered by non-axiality to the energy -1.9 MeV (i.e. by 1.0 MeV). As B_f^{st} is defined by the highest saddle point, it is not decreased by

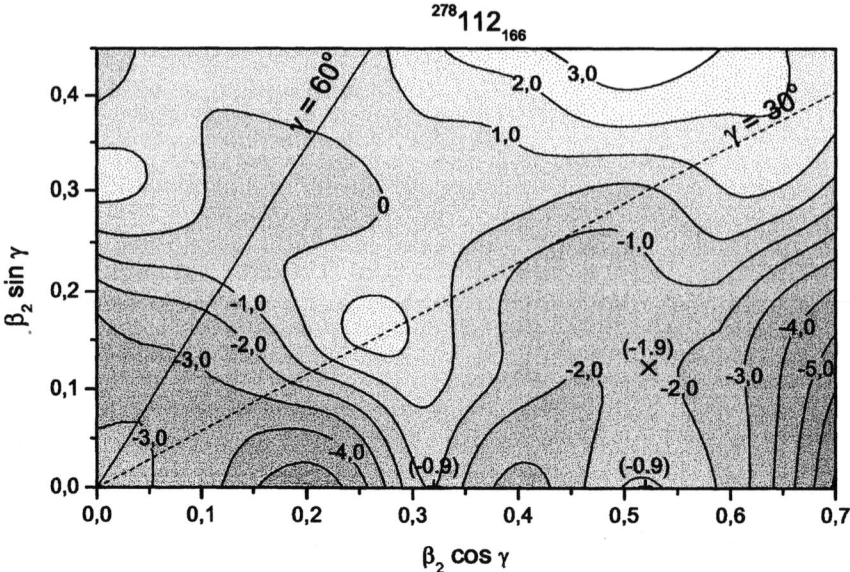

FIGURE 4. Same as in Fig. 3, but for the nucleus $^{278}112$.

non-axial deformations, because the first saddle point (at $\beta_2 \approx 0.32$) remains unchanged.

Figure 5 shows map of energy for the nucleus $^{284}114$. Here, the situation is still somewhat different from both previous ones. For this nucleus, the second saddle point, in the case of axial symmetry, is lower. The decrease of its energy by non-axiality has no influence on the energy of the first (with higher energy) point. Thus, again, non-axiality does not change B_f^{st}. One can also see that this nucleus is not well deformed. It is rather of a transitional type, with a shallow minimum, in its energy, close to spherical shape.

Figure 6 gives map of energy of the nucleus $^{294}116$ (this is the nucleus appearing in the process of cooling of the compound nucleus obtained at JINR-Dubna, when synthesizing the nucleus $^{292}116$ [21]). One can see that non-axiality lowers mass of it at the saddle point by 1.0 MeV.

Concluding, basing on Figs. 3-6, one can say that the potential energy (and, in particular, also the barrier B_f^{st}) of each nucleus is a very individual property of it. It should be carefully (in sufficiently large deformation space) calculated for each nucleus separately, without any averaging.

ACKNOWLEDGMENTS

Support by the Polish State Committee for Scientific Research (KBN), grant no. 2 P03B 039 22, and the Polish-JINR (Dubna) Cooperation Programme is gratefully acknowledged.

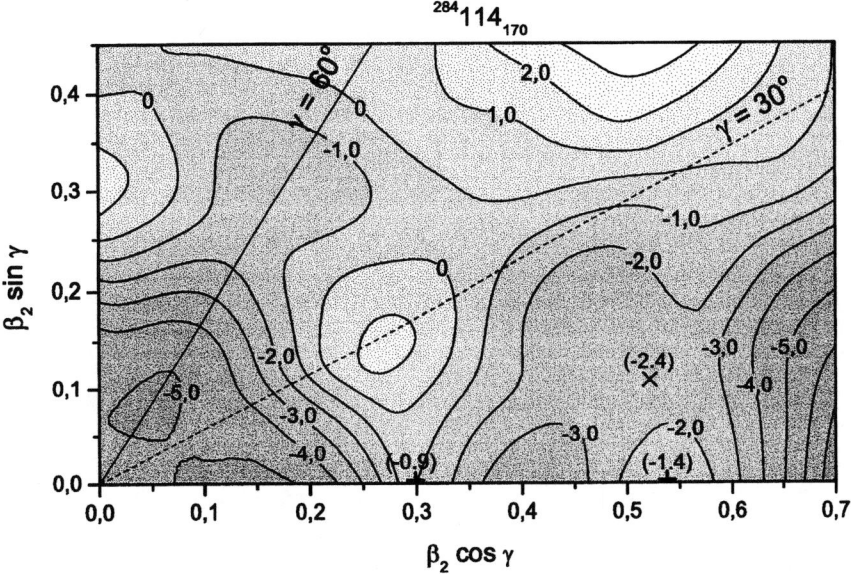

FIGURE 5. Same as in Fig. 3, but for the nucleus $^{284}114$.

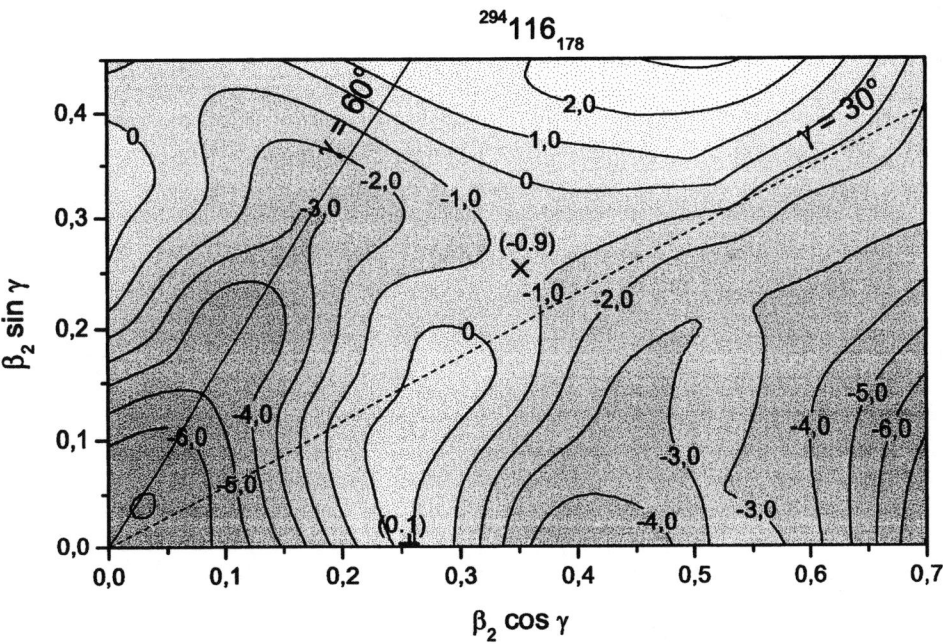

FIGURE 6. Same as in Fig. 3, but for the nucleus $^{294}116$.

REFERENCES

1. V.I. Zagrebaev, Y. Aritomo, M.G. Itkis, Yu.Ts. Oganessian and M. Ohta, *Phys. Rev. C* **65**, 014607 (2001).
2. V.V. Volkov, *Acta Phys. Pol. B* **34**, 1881 (2003).
3. A.S. Zubov, G.G. Adamian, N.V. Antonenko, S.P. Ivanova and W. Scheid, *Acta Phys. Pol. B* **34**, 2083 (2003).
4. G. Giardina, S. Hofmann, A.I. Muminov and A.K. Nasirov, *Eur. Phys. J. A* **8**, 205 (2000).
5. Y. Abe and D. Bouriquet, *Acta Phys. Pol. B* **34**, 1927 (2003).
6. W.J. Świątecki, K. Siwek-Wilczyńska and J. Wilczyński, *Acta Phys. Pol. B* **34**, 2049 (2003).
7. M.G. Itkis, Yu.Ts. Oganessian and V.I. Zagrebaev, *Phys. Rev. C* **65**, 044602 (2002).
8. I. Muntian, Z. Patyk and A. Sobiczewski, *Acta Phys. Pol. B* **34**, 2141 (2003).
9. A. Sobiczewski and I. Muntian, *Nucl. Phys. A*, submitted (2003).
10. Z. Patyk, A. Sobiczewski, P. Armbruster and K.-H. Schmidt, *Nucl. Phys. A* **491**, 267 (1989).
11. A. Sobiczewski, *Usp. Fiz. Nauk* **166**, 943 (1996); *Physics–Uspekhi* **39**, 885 (1996).
12. R. Smolańczuk and A. Sobiczewski, Proc. XV EPS Conf. on nuclear physics: *"Low Energy Nuclear Dynamics"*, St. Petersburg (Russia) 1995, ed. by Yu. Ts. Oganessian, W. von Oertzen and R. Kalpakchieva (World Scientific, Singapore, 1995) p. 313.
13. A. Mamdouh, J.M. Pearson, M. Rayet and F. Tondeur, *Nucl. Phys. A* **679**, 337 (2001).
14. S.E. Larsson, *Phys. Scr.* **8**, 17 (1973).
15. S. Ćwiok and A. Sobiczewski, *Z. Phys. A* **342**, 203 (1992).
16. R.A. Gherghescu, J. Skalski, Z. Patyk and A. Sobiczewski, *Nucl. Phys. A* **651**, 237 (1999).
17. S.G. Rohoziński and A. Sobiczewski, *Acta Phys. Pol. B* **12**, 1001 (1981).
18. R. Smolańczuk, J. Skalski and A. Sobiczewski, *Phys. Rev. C* **52**, 1871 (1995).
19. P. Möller and A. Iwamoto, *Phys. Rev. C* **61**, 047602 (2000).
20. S. Hofmann, V. Ninov, F.P. Hessberger, P. Armbruster, H. Folger, G. Münzenberg, H.J. Schött, A.G. Popeko, A.V. Yeremin, S. Saro, R. Janik and M. Leino, *Z. Phys. A* **354**, 229 (1996).
21. Yu.Ts. Oganessian, V.K. Utyonkov, Yu.V. Lobanov *et al.*, *Phys. Rev. C* **63**, 011301(R) (2000).

Calculation of high-dimensional fission-fusion potential-energy surfaces in the SHE region

Peter Möller*, and Arnold J. Sierk*, Takatoshi Ichikawa† and Akira Iwamoto†

*Theoretical Division, Los Alamos National Laboratory, Los Alamos, New Mexico 87545, USA
†Japan Atomic Energy Research Institute, Tokai-mura, Naka-gun, Ibaraki, 319-1195 Japan

Abstract. We calculate in a macroscopic-microscopic model fission-fusion potential-energy surfaces relevant to the analysis of heavy-ion reactions employed to form heavy-element evaporation residues. We study these multidimensional potential-energy surfaces both inside and outside the touching point.

Inside the point of contact we define the potential on a multi-million-point grid in 5D deformation space where elongation, merging projectile and target spheroidal shapes, neck radius and projectile/target mass asymmetry are independent shape variables. The same deformation space and the corresponding potential-energy surface also describe the shape evolution from the nuclear ground-state to separating fragments in fission, and the fast-fission trajectories in incomplete fusion.

For separated nuclei we study the macroscopic-microscopic potential energy, that is the "collision surface" between a spheroidally deformed target and a spheroidally deformed projectile as a function of three coordinates which are: the relative location of the projectile center-of-mass with respect to the target center-of-mass and the spheroidal deformations of the target and the projectile. We limit our study to the most favorable relative positions of target and projectile, namely that the symmetry axes of the target and projectile are collinear.

INTRODUCTION

It has been a longstanding challenge to understand in detail the collision mechanism in heavy-ion reactions leading to the formation of heavy elements in the vicinity of the predicted superheavy island centered at proton number $Z = 114$ and neutron number $N = 184$. In the discoveries of the elements in the region leading up to the superheavy elements, that is elements in the range $101 \leq Z \leq 106$ the most asymmetric target/projectile combinations available had been used. For example Seaborgium (Sg with 106 protons) was formed in the heavy-ion reaction [1]:

$$^{18}\text{O} + ^{249}\text{Cf} \rightarrow ^{263}\text{Sg} + 4n \qquad (1)$$

However, about 25 years ago it became clear that rather than use the most asymmetric target and projectile combinations to extend the periodic system further, so called cold-fusion reactions were preferable. In a cold-fusion reaction a target with magic or near-magic proton and neutron numbers is used. This double magicity confers extra binding energy to the system. For example relative to a ^{238}U target the extra binding is about 11 MeV and relative to a ^{249}Cf target the extra binding is about 9 MeV. This has no effect on the barrier between the colliding heavy ions relative to infinite separation, but it does lower the barrier relative to the ground-state of the compound system. For collisions "at the Coulomb barrier" this means the compound system is created at a lower excitation energy in a cold-fusion reaction than in other types of heavy-ion reactions. This is thought to favor de-excitation by neutron emission rather than fission.

Here we will identify what other aspects of cold-fusion reactions favor compound nucleus formation, apart from the long-recognized benefit of the lower excitation energy. In particular we will look at the potential energy after touching in a multidimensional deformation space in a model that takes microscopic effects into account, and see how this picture differs from a purely macroscopic picture.

One feature of cold-fusion reactions that is poorly understood is: what is the optimum collision energy for compound nucleus formation? It has been experimentally observed that the optimum energy is 10 or even more MeV below the Coulomb barrier, that is the barrier that is calculated to exist between the colliding nuclei in some simple model. One example of such a simple model is that the colliding ions are assumed to be spherical, and that the force is a purely electrostatic repulsion between the positively charged ions with an attractive, phenomenological "finite-range" nuclear force that starts to become effective when the facing nuclear surfaces have approached each other to within 1.5 fm or so.

Clearly the true barrier must be lower than what is obtained in this oversimplified but until now almost universally used model. We will here calculate a macroscopic-microscopic "collision potential-energy surface" between the approaching ions as a function of three variables. The potential energy after the ions have touched is studied in a five-dimensional macroscopic-microscopic model.

FIVE-DIMENSIONAL COMPOUND-SYSTEM POTENTIAL-ENERGY SURFACES

The five-dimensional macroscopic-microscopic potential-energy surfaces for compound systems reached in cold-fusion heavy-ion reactions are calculated and analyzed using the same techniques as introduced previously in studies of actinide fission. In particular we calculate the potential energy as a function of 5 nuclear-shape coordinates: 15 points each in the neck diameter and left and right fragment deformations, 35 points in the mass asymmetry, and 33 points in the nuclear elongation. This leads to a space of 3898125 grid points. However, as explained elsewhere [2] some grid-point coordinate values do not correspond to physically realizable shapes; therefore the actual number of grid points considered are 3594915. Compared to our previous fission studies we have increased the number of mass-asymmetry grid points from 20 to 35. The extension to such large mass asymmetries means that the expected cold-fusion channel near mass division 208/70 for compound system ^{278}112 is included in our calculated potential-energy surface. The distance between the grid points in the mass-asymmetry coordinate is 2.78 mass units for this compound system. More generally, for a compound nucleus with mass A the grid-point distance in the mass-asymmetry coordinate is $A/100$.

In Figs. 1–3 we show the result of a water-immersion analysis of the calculated 5D deformation spaces for compound systems corresponding to ^{266}Hs, ^{272}Ds, and ^{284}114.

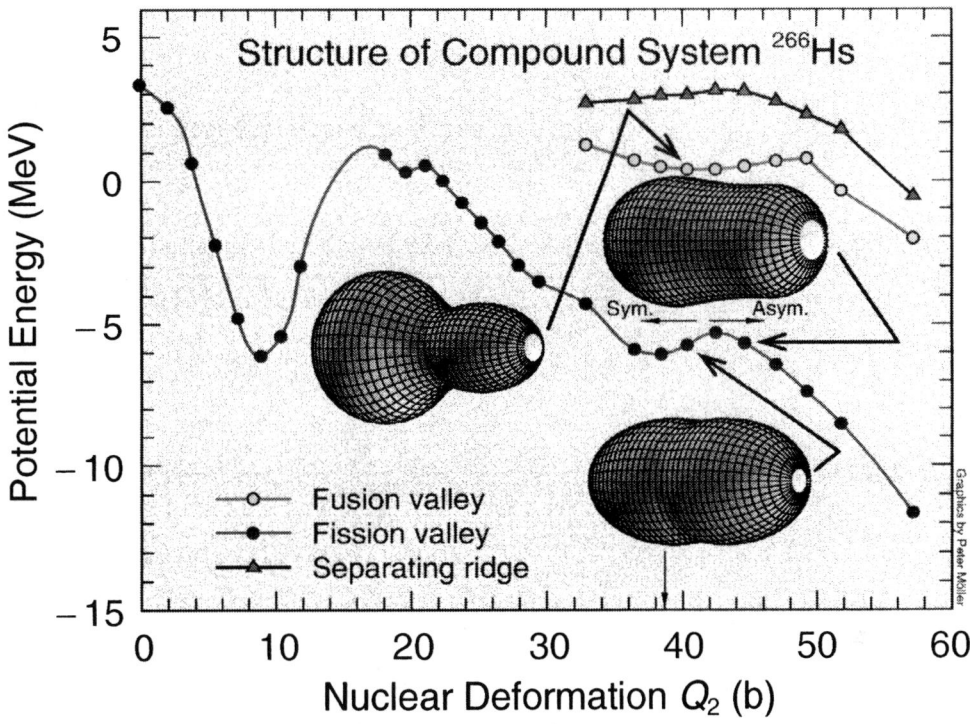

FIGURE 1. Structures in the calculated 5D potential-energy surface of ^{266}Hs. The lower curve corresponds to the fission barrier. For large values of Q_2 there is an additional well-defined valley in the 5D energy surface, which is stabilized with respect to the fission valley by the ridge shown in the top curve. Two of the shapes in the fission valley and one shape in the additional valley are shown. The shapes in the fission valley show that a transition to asymmetric shapes take place around $Q_2 = 43$. The shape in the other valley correspond to $M_H/M_L = 197/70$. The position of the vertical thin arrow on the horizontal axis indicates the value of Q_2 for touching *spherical* target and projectile.

For each system we have identified a fission barrier, which only exists because of large, negative microscopic corrections at the ground-state, which results in a fission barrier that is high enough that these nuclei survive for a sufficiently long time to permit experimental study. For larger deformations all three figures also show a second valley that our water-immersion analysis has identified, as well as the ridge which stabilizes this valley with respect to the fission valley. The shapes in the secondary valley strongly overlap with the target and projectile masses in the incident channel. A little loosely we could say that the colliding projectile and target after touching will slip into this valley as a hand slips into a glove. Of course the energy of this valley is often 10 MeV or more lower than the energy of the fusion collision potential-energy surface saddle discussed below. However this is mainly because the compound shapes we study have developed a substantial neck, which significantly lowers the energy. This appearance of the potential-energy surface is very different from what is obtained in a macroscopic multi-dimensional picture in which these systems just

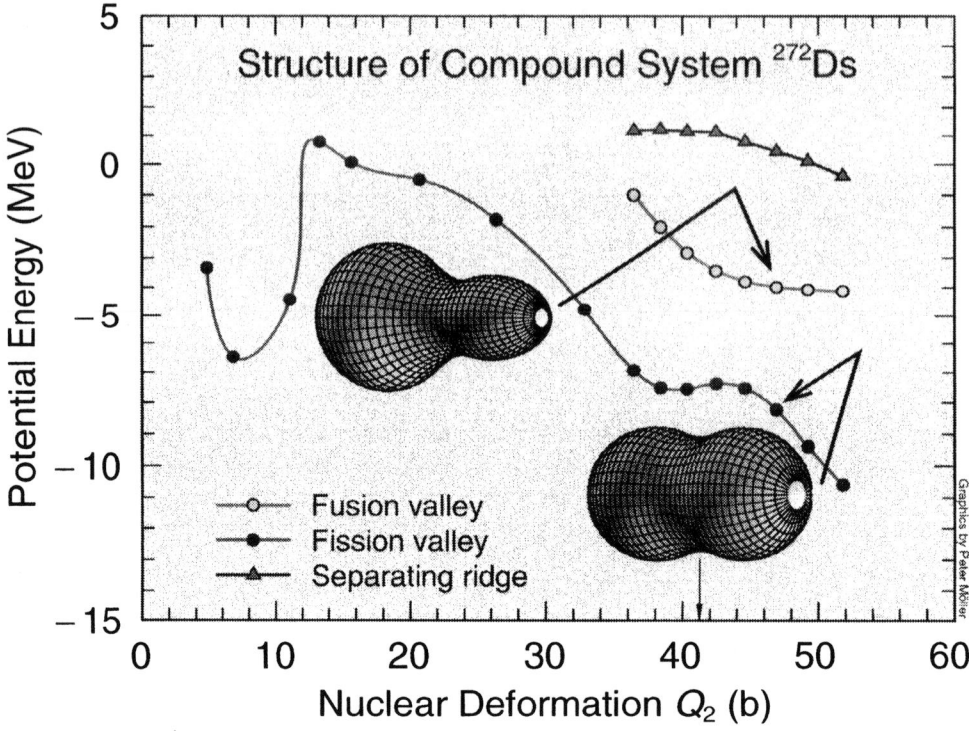

FIGURE 2. Same as Fig. 1 but for ^{272}Ds. The shape in the fission valley is symmetric and the shape in the other valley corresponds to $M_H/M_L = 201/71$.

after touching find themselves on a surface that slopes steeply sideways relative to the incident direction and therefore would immediately deflect systems colliding "at the Coulomb barrier" towards the fission valley and re-separation [3, 4, 5]. Thus, we have identified an additional effect that favors cold fusion: microscopic effects in the compound system potential-energy surface creates a shell-stabilized valley that "fits hand-in-glove" with the entrance-channel configurations. It remains to deduce the effect of how relative differences between systems in the calculated barriers and ridges, and associated dynamical effects, such as damping, affect the evaporation-residue cross sections.

THREE-DIMENSIONAL HEAVY-ION COLLISION POTENTIAL-ENERGY SURFACES

We study here the fusion barrier for some cold-fusion reactions commonly used to reach heavy elements in the region from No to proton number $Z = 114$.

Often the barrier with respect to fusion between two heavy ions is calculated by

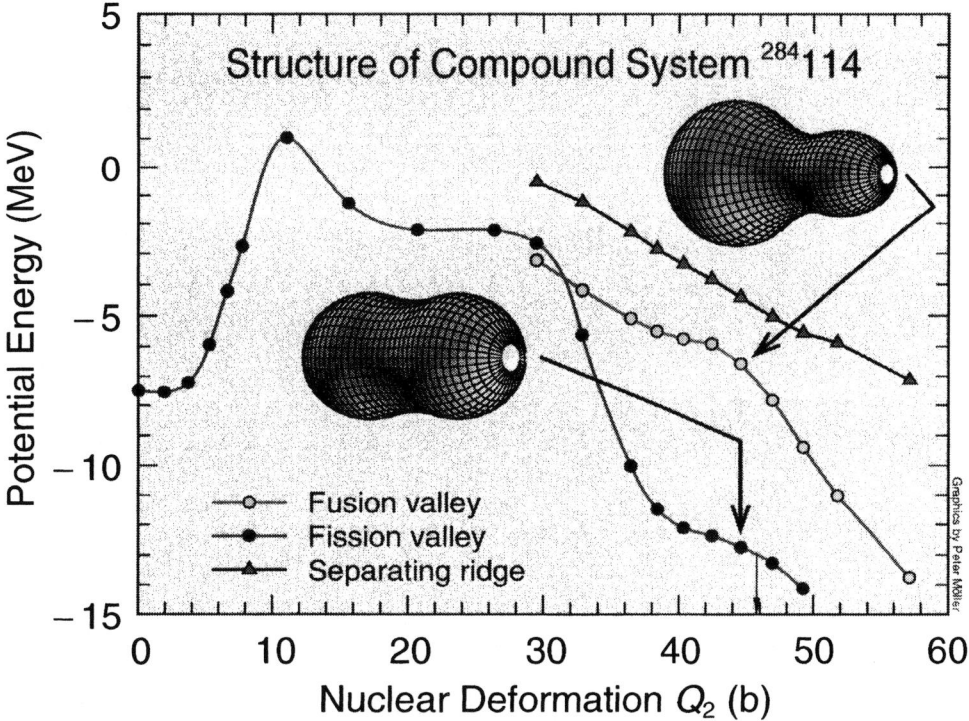

FIGURE 3. Same as Fig. 1 but for $^{284}114$. The shape in the fission valley is symmetric and the shape in the other valley corresponds to $M_H/M_L = 213/71$.

assuming that the colliding heavy ions are spherical in shape and that the energy of the colliding system is the sum of the Coulomb interaction energy between the two ions and a nuclear interaction energy that starts to become important only when the facing surfaces of the colliding ions are within 1.5 fm or so. The nuclear interaction can be given by the proximity-force model [6] or the more general Yukawa-plus-exponential macroscopic model [7]. However, it is well known that the "Coulomb barrier" calculated in such a model is much higher than the optimal energy for forming heavy evaporation residues in heavy-ion collisions. We therefore use a more realistic approach and calculate below the energy of a colliding heavy-ion system as

$$\begin{aligned} E_{P+T}(\varepsilon_{2P},\varepsilon_{2T},x_P,y_P,z_P,\alpha,\beta,\gamma) &= E_P^{\text{self}}(\varepsilon_{2P}) - E_P^{\text{self}}(\varepsilon_{2P}=\text{gs}) \\ &+ E_T^{\text{self}}(\varepsilon_{2T}) - E_T^{\text{self}}(\varepsilon_{2T}=\text{gs}) \\ &+ E_{PT}^{\text{int}}(\varepsilon_{2P},\varepsilon_{2T},x_P,y_P,z_P,\alpha,\beta,\gamma) \end{aligned} \quad (2)$$

Here E_{P+T} is the total energy of the colliding system relative to infinitely separated targets and projectiles in their ground states (gs). The quantity E^{self} is the macroscopic-microscopic potential energy as a function of shape as given by our FRLDM model

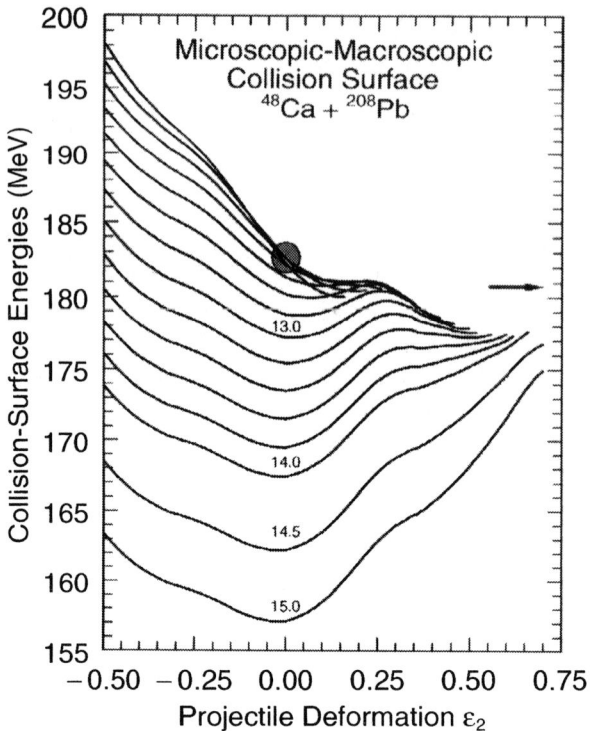

FIGURE 4. Calculated macroscopic-microscopic potential energies for the collision of ^{48}Ca+^{208}Pb. Each of the plotted curves shows the center-of-mass energy of the system versus projectile deformation, for spherical ^{208}Pb and a specific distance between the centers-of-mass of the target and projectile. The distance in fermi, is written above the potential-energy curve. For distances less than 14.0 fm the separation between projectile and target decreases by 0.2 fm for each successive curve. The dot corresponds to the calculated value of the one-dimensional "Coulomb barrier" in a macroscopic model. The arrow indicates the fusion-barrier height obtained by the method corresponding to column 3 in Table 1. It is plotted at an energy that is slightly different from the energy in the table to be "consistent" with the curves in the figure, see text for a discussion.

[8, 2]. Since we give the system energy *relative* to the separated fragments we obviously need to subtract the ground-state self-energies of the target and projectile; thus the second and fourth terms in the right member of Eq. 2 above. The interaction-energy-term calculation is extensively discussed in our Ref. [9]. We assume that the interaction shell-correction energy can be neglected for separated target and projectile. The Cartesian triplet (x_p, y_p, z_p) gives the location of the center of the projectile relative to the center of the target. The Euler angles α, β, and γ specify the orientation of the projectile symmetry axis relative to the target symmetry axis. The energetically most favorable orientation, at least for prolate deformations, for a specific distance between target and projectile, is when the axes of the target and projectile are collinear. To limit the problem to a moderately low-dimensional parameter space we therefore only consider these relative

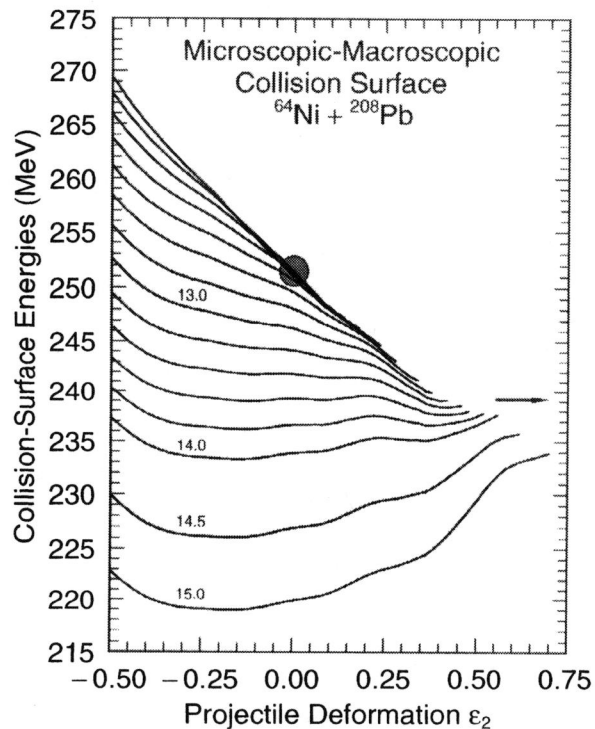

FIGURE 5. Same as figure 4, but for the collision ^{64}Ni+^{208}Pb. Here the difference between the conventional Coulomb barrier and the collision-surface saddle-point height is substantial.

positions of target and projectile and only spheroidal deformations ε_{2T} and ε_{2P} of the target and projectile, respectively. Thus, we have $x_P = 0, y_P = 0, \alpha = 0, \beta = 0$, and $\gamma = 0$, so that the space we investigate is 3-dimensional and is characterized by target and projectile spheroidal deformations and their relative distance z_P. The configuration we consider is also referred to as the polar-parallel configuration [10].

We now argue that if the system loses stability with respect to target and/or projectile deformation as the ions approach each other then the energy at which this occurs defines a more realistic fusion-barrier height. This means that we assume that once the system becomes unstable with respect to deformation, shape changes occur very fast relative to motion in the collision direction. Specifically we determine for a succession of values of z_P the minimum energy with respect to target and projectile deformation and the corresponding target and projectile deformations. If we neglect the microscopic terms in E^{self} we recover the well-known result that the target and projectile are fairly oblate: ε_2 is in the range -0.4 to -0.2 in the situations investigated. However, when the microscopic corrections are included they stabilize the doubly magic ^{208}Pb at spherical or very close to spherical shape; deviations are typically less than 0.03. We can therefore illustrate the important aspects of the collision surface in only two dimensions. In Figs. 4 and

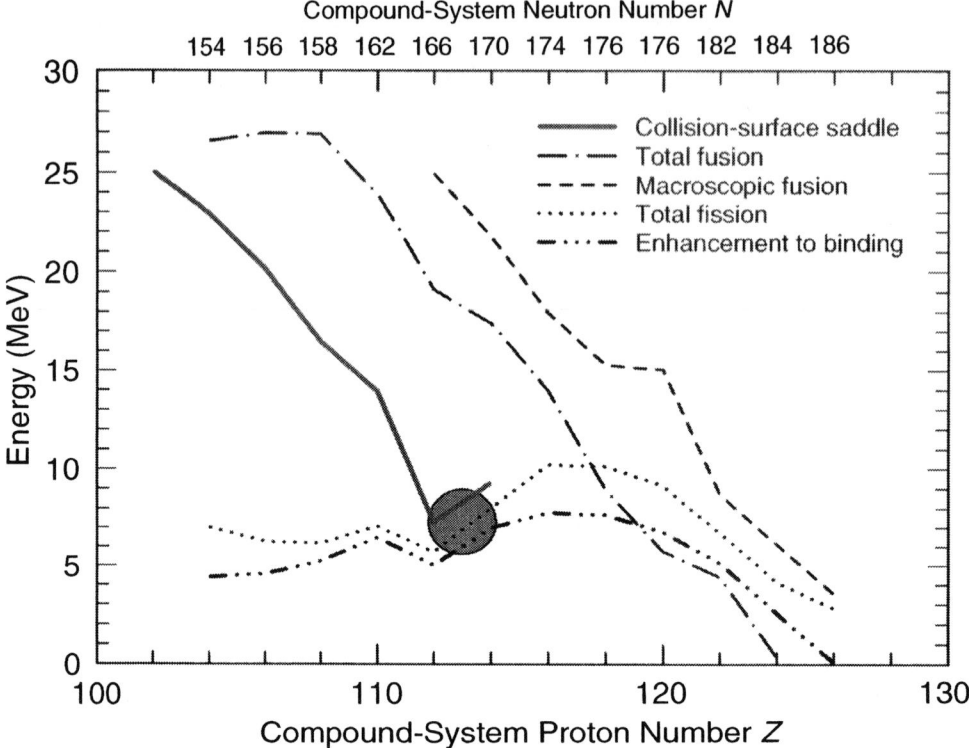

FIGURE 6. Potential-energy-surface barrier and ground-state systematics for cold-fusion reactions leading to compound systems from No to Z = 126. The "Enhancement to binding" curve is exactly the ground-state microscopic correction multiplied by −1. Curves labeled "Total" are calculated with the full macroscopic-microscopic approach. However, both the "Macroscopic fusion" and "Total fusion" results were obtained assuming that target and projectile are spherical during the collision. The "Collision-surface saddle" curve was calculated permitting projectile deformation; otherwise the model is identical to "Total Fusion". See text for further discussions.

5 we show the energy of the colliding system for a spherical target versus projectile deformation for a succession of distances as the projectile approaches the target for two heavy-ion reactions leading to $^{256}_{102}$No and $^{272}_{110}$Ds. We indicate by a big blob the energy corresponding to the "conventional" Coulomb barrier calculated as the maximum in the macroscopic energy between spherical target and projectile. For the light compound system the Coulomb barrier is only 3 MeV higher than the energy at the point where our more realistic collision surface loses stability with respect to projectile deformation. The situation is quite different for the heavy compound system $^{272}_{110}$Ds. Here the collision-surface saddle point is 13 MeV lower than the Coulomb barrier. In Table 1 we present calculations of the collision-surface saddle-point energy for 11 heavy-ion collisions in the second column of numbers. These have been calculated assuming the target is spherical. We have also relaxed this condition but for a ^{208}Pb target the effect on the

TABLE 1. Heavy-ion-reaction fusion barriers calculated in three models. The first model (a) is the conventional Coulomb barrier. In the second model (b) projectile deformation and target and projectile microscopic corrections are included. In the third model (c) the variation in zero-point energy with respect to the projectile spheroidal deformation as the projectile approaches the target is also considered. We have verified that the effect of target deformation is very small, only an MeV or so in most cases, due to the stabilizing effect of the doubly-magic proton and neutron numbers $Z = 82$ and $N = 126$ of the target.

Reaction	1-D Coul. Barr. (a) (MeV)	2-D Saddle (b) (MeV)	2-D Saddle (c) (MeV)
$^{48}_{20}$Ca + $^{208}_{82}$Pb → $^{258}_{102}$No	182.65	179.04	178.70
$^{50}_{22}$Ti + $^{208}_{82}$Pb → $^{258}_{104}$Rf	201.04	194.29	193.87
$^{54}_{24}$Cr + $^{208}_{82}$Pb → $^{262}_{106}$Sg	218.44	209.93	208.60
$^{58}_{26}$Fe + $^{208}_{82}$Pb → $^{266}_{108}$Hs	236.13	224.97	221.96
$^{62}_{28}$Ni + $^{208}_{82}$Pb → $^{270}_{110}$Ds	252.71	244.21	239.04
$^{64}_{28}$Ni + $^{208}_{82}$Pb → $^{272}_{110}$Ds	251.48	238.43	238.58
$^{70}_{30}$Zn + $^{208}_{82}$Pb → ^{278}112	267.14	255.05	252.26
$^{74}_{32}$Ge + $^{208}_{82}$Pb → ^{282}114	284.49	280.07	271.71
$^{76}_{32}$Ge + $^{208}_{82}$Pb → ^{284}114	283.85	275.09	272.40

barrier is usually less than 1 MeV. For comparison we also present, in column one, the conventional Coulomb barrier.

The potential energy curves in Figs. 4 and 5 have all been calculated assuming that target and projectile zero-point energies do not change as the ions approach. However when projectile stability with respect to deformation is lost the projectile zero-point energy vanishes. The correct fusion barrier in this approximation (column 2 in Table 1) is therefore obtained by subtracting the projectile zero-point energy at infinity from the saddles obtained from the energies in this figure. For the case when we calculate the point of instability by taking into account the changing projectile zero-point energy (column 3 in Table 1) we also subtract this zero-point energy and add the appropriate projectile zero-point energy at the current separation. The arrow energies in Figures 4 and 5 are "consistent" with the energy curves, that is the projectile zero-point energies at infinite separation have not been subtracted out. They are therefore located at slightly different energies than the fusion barrier energies listed in Table 1 which are normalized so that they are consistent with experimental energies given in the center-of-mass frame.

In Fig. 5 we observe that the energy curves versus projectile deformation are very flat for a succession of distances before stability with respect to projectile deformation is lost. Therefore the exact energy and deformation where this occurs depends somewhat on the procedure used to find the minimum energy with respect to projectile and target deformation at each separation and on other minor details of the calculations. We therefore consider an improved model for determining the collision-surface saddle point.

We consider the target to be inert, that is, based on the observations above, we keep its shape spherical. For each separation we then calculate the zero-point energy with respect to projectile deformation. That is we calculate the zero-point energies for each of the one-dimensional curves shown in Figs. 4–5. The curves are approximated by parabolas whose minima coincide with the actual function minima and pass through the saddle towards prolate projectile deformations (or when a saddle is non-existent pass through the maxima at the touching point). For the inertia with respect to ε vibrations we use exactly the same model as we have used to calculate the ground-state zero-point energies in our mass model [11] and in fission half-life calculations [12]. When the sum of the zero-point energy and the minimum energy of the curve is equal to the saddle energy towards the right in Figs. 4 and 5 we claim we have found the collision-surface saddle point. We only calculate the zero-point energy for a set of curves at separation distances ... 14.0, 13.8, 13.6, 13.4 ... and then determine by interpolation the point where the zero-point energy coincides with the saddle occurring towards the right on the curves. The collision-surface saddle-point energies determined in this way are given in the third column of numbers in Table 1. We have also calculated the zero-point energies with a WKB method. The collision-surface saddle-point energy obtained by WKB differs from the value obtained in the parabolic approximation by less than 0.1 MeV on the average, despite the non-parabolic appearance of many of the curves in Figs. 4 and 5.

In Fig. 6 we plot the collision-surface saddle relative to the ground-state energy of the compound system. We compare this realistic model of the fusion barrier to two models of the Coulomb barrier. The "Total fusion" takes into account the effect of microscopic corrections on the nuclear masses, whereas "Macroscopic fusion" does not. The "total-fusion" macroscopic-microscopic barrier is considerably lower than the macroscopic barrier, relative to the ground state of the compound system. However, relative to the separated projectile and target configuration the two barriers are equally high. This is the well-known "cold-fusion" effect: The total fusion barrier relative to the ground state of the compound system is lowered due to the extra binding associated with the doubly-magic nucleus ^{208}Pb. Other contributions, which can be positive or negative, arise from the microscopic corrections in the projectile and evaporation residue nuclei. Thus, in a cold fusion reaction "at the Coulomb barrier" the compound system will form at lower excitation energies than in non-cold-fusion reactions leading to the same system. This effect is thought to enhance the evaporation-residue cross section.

We also plot the fission barriers of the compound systems. They are compared to the ground-state microscopic corrections. The "Enhancement to binding" curve shows the ground-state microscopic correction multiplied by -1, to make it more obviously comparable to the fission-barrier height. Since the macroscopic fission barrier is almost non-existent for these systems, the fission-barrier height will be roughly equal to the (negative of the) ground-state microscopic correction [13, 14]. Of interest here is that the calculated fusion-barrier height drops down to about the fission-barrier height at around compound system proton number $Z = 112$. This means that for higher proton number it is roughly the fission-barrier peak that is the highest point on the fusion path. We initially discussed this at the ENAM98 conference, Fig. 3 in that paper [15]. (However, the fusion-barrier calculations there are inaccurate due to incorrect coding of the Wigner shape dependence.) An important consequence is that when the fusion-barrier height drops down to the fission barrier, one must expect a break in the systematics for the optimum energy for evaporation residue formation that is now, for unknown systems typically obtained from the trend of this energy extrapolated from lighter systems.

CONCLUSIONS

Traditionally in dynamical models of colliding heavy ions the potential energy has been described in terms of a macroscopic model. Deformations of target and projectile are sometimes also ignored. We have shown here that for collisions at low energy, that is "at the Coulomb barrier" deformations and microscopic effects contribute to a dramatically different picture of the potential energy.

Inside the touching point we find a substantial "cold-fusion" valley where the shape configuration corresponds closely to the shape of target and projectile just before touching, that is an almost spherical target-like part joined with a deformed projectile part, with a deformation similar to what develops just before touching.

Outside the touching point our more realistic model shows that in collisions leading to very heavy systems, for example $^{272}_{110}$Ds the collision-surface saddle point where stability with respect to projectile deformation is lost is about 13 MeV lower in energy than the conventional Coulomb barrier. This gives a more reasonable correspondence between calculated fusion-barrier heights, which we now equate with the collision-surface saddle-point energy, and the observed most-favorable energies for evaporation residue formation.

Clearly there is a need to go further to understand more completely the mechanism of evaporation-residue formation. Just as shape and microscopic corrections bring major changes to our picture of the static aspects of heavy-ion collisions leading to very heavy systems, there is also a need to consider these in modeling other quantities, for example the inertia and the dissipation mechanisms associated with the collision, and when compound-nucleus formation has occurred the various de-excitation mechanisms need to be described. However, we hope our more realistic modeling of some static aspects of heavy-ion collisions will lead to productive steps also in these directions.

This research is supported by the US DOE.

REFERENCES

1. A. Ghiorso, J. M. Nitschke, J. R. Alonso, C. T. Alonso, M. Nurmia, G. T. Seaborg, E. K. Hulet, R. W. Lougheed, Phys. Rev. Lett. **33** (1974) 1490.
2. P. Möller, D. G. Madland, A. J. Sierk, and A. Iwamoto, Nature **409** (2001) 785.
3. J. R. Nix and A. J. Sierk, Phys. Scr. **10A** (1974) 94.
4. P. Möller and J. R. Nix, Nucl. Phys. **A272** (1976) 502.
5. P. Möller and A. J. Sierk, Nature, **422** (2003) 485.
6. J. Błocki, J. Randrup, W. J. Swiatecki, and C. F. Tsang, Ann. Phys. (N. Y.) **105** (1977) 427.
7. H. J. Krappe, J. R. Nix, and A. J. Sierk, Phys. Rev. C **20** (1979) 992.
8. P. Möller, J. R. Nix, W. D. Myers, and W. J. Swiatecki, Atomic Data Nucl. Data Tables **59** (1995) 185.
9. P. Möller and A. Iwamoto, Nucl. Phys. **A575** (1994) 381.
10. A. Iwamoto, P. Möller, J. R. Nix, and H. Sagawa, Nucl. Phys. **A596** (1996) 329.
11. P. Möller and J. R. Nix, Nucl. Phys. **A361** (1981) 117.
12. P. Möller and J. R. Nix, Phys. Rev. Lett. **37** (1976) 1461.
13. W. J. Swiatecki, Phys. Rev. **100** (1955) 937.
14. Z. Patyk, A. Sobiczewski, P. Armbruster and K.-H. Schmidt, Nucl. Phys. **A491** (1989) 267.
15. P. Möller, Proc. ENAM 98, Exotic Nuclei and Atomic Masses, Bellaire, Michigan, 23–27 June, 1998 (AIP Conference Proceedings, **455** (1998) 698).

Ground-state properties of heavy and superheavy nuclei predicted by nuclear mass models

Hiroyuki Koura[†]

The Institute of Physical and Chemical researches (RIKEN),
Hirosawa 2-1, Wako, Saitama 351-0198, JAPAN
Advanced Research Institute for Science and Engineering, Waseda University,
Okubo 3-4-1, Shinjuku-ku, Tokyo 169-8555, JAPAN

Abstract. Some nuclear mass formulas are reviewed and applied for the heavy and superheavy nuclei. A new mass formula composed of the gross term, the even-odd term, and the shell term is also presented. The new mass formula is a revised version of the spherical basis mass formula published in 2000, that is, the even-odd term is treated more carefully, and a considerable improvement is brought about. The root-mean-square deviation of the new formula from experimental masses is 641 keV for $Z \geq 8$ and $N \geq 8$. Properties on systematics of the neutron-separation energies, alpha-deacy Q-values in the superheavy nuclidic region, and Q-values of the fusion reaction for the heavy nuclei are compared with some mass formulas. Furthermore, a method of estimation for fission half-lives for the superheavy and neutron-rich nuclei is presented. With this estimation, the chart of nuclides for the nuclear decays are drawn, and the prediction of limit of the existence in the nuclidic region is discussed.

INTRODUCTION

Nuclear masses are important quantities to determine the ground-state properties and reactions. Since the Weizsäcker-Bethe nuclear mass formula [1, 2], many mass predictions were presented. At the present time, the main purpose of the study on mass formulas is not only to give more precise mass values of known nuclides but also to predict reliable masses of unknown nuclides, especially the neutron-rich nuclides and the superheavy nuclides. As related to the latter, the mass prediction are required in the study of stability in the superheavy nuclidc region, and in the study of fusion reaction because there are few experimental masses. In this region, furthermore, prediction of dominant decay mode among α-decay, β-decay, and spontaneous fission are also required. Though first two decays can be estimated with Q-values from mass models as input data for each decay model, prediction of spontaneous fission requires, for example, a potential energy surface for the fission and a model of penetration for it.

In this report, we briefly review some mass formulas and outline our mass formula. Properties on systematics of the neutron-separation energies, alpha-deacy Q-values in the superheavy nuclidic region, and Q-values of the fusion reaction for the heavy nuclei are compared with some mass formulas. In our method of obtaining shell energies, fission barriers for the heavy and superheavy nuclei can be estimated. We also estimate the spontaneous-fission half-lives using the potential energy surface calculated by the same method as used for obtaining the shell energies we developed. Furthermore, with this estimation the chart of nuclides for the nuclear decays are drawn, and the prediction of limit of the existence in the nuclidic region is discussed.

MASS FORMULAS

One of the way to reproduce the known nuclear mass values is to use the mass systematics. For examples, the mass formulas by Comay et al. and Jänecke et al. [4] are based on the Garvey-Kelson-like systematics [3], and the root-mean-square (RMS) deviations from then current experimental masses are 350-450 keV. There is also the mass formula by Tachibana et al. [5] (TUYY formula) composed of a gross term with the Coulomb energy treated elaborately, an even-odd term, and an empirical shell term. The first two terms are smooth functions of Z and N without the Wigner term. The RMS deviation of it is about 550 keV. These phenomenological mass formulas are usable to reproduce known masses and unknown ones in the vicinity of known nuclei, however, these cannot be extrapolated to the region of superheavy nuclei where no empirical data are available: the above-mentioned mass formulas can be applied only for $N \leq 157$-160. These formulas may also be difficult to predict the possibility of increasing or decreasing the strength of magicity far from the β-stability line. Moreover, no predictions of the nuclear shapes are available.

On the other hand, some approaches considering the nuclear force and the nuclear deformation have been done based on the models, or kinds of assumptions. The mass formula by Myers et al. [6] is the early study of the liquid-drop model, which is composed of the macroscopic liquid-drop part and the microscopic shell part. (This early version was still in the similar problem to the above-mentioned systematics without the nuclear shapes).

In the last decade, some mass predictions designed for the wide nuclidic regions were presented. We mention three mass predictions. One is the Finite-range droplet model (FRDM, 1995) formula, which is composed of the macroscopic droplet term and microscopic shell term [7]. The macroscopic term is calculated from the finite-range droplet model (FRDM). The shell term is calculated from the folded-Yukawa single-particle potential. Other is the Hartree-Fock plus the BCS-type pairing method with the MSk7 Skyrme force (HFBCS-1, 2001) [8]. The last one is the mass formula by our groups published in 2000 [9], namely KUTY00. This formula is composed of the gross term, the even-odd term, and the shell term, and the first two terms are almost the same as those of TUYY formula. The method of obtaining the shell term is based on the spherical basis, and we solve some problems on the TUYY formula. Furthermore, we recently constructed a new mass formula, we refer to as KTUY03, as a revised version of the KUTY00, that is, the even-odd term is treated more carefully, and a considerable improvement is brought about [10].

These three models can predict the nuclear shapes and apply for superheavy nuclei includes $^{298}[114]_{194}$. The RMS deviations of these three predicted masses are about 600-800 keV in a current status. The differences of properties of these formulas will be discussed in the next section.

PROPERTIES OF RECENT MASS FORMULAS

The root-mean-square (RMS) deviation of masses and separation energies are given in Table 1. The RMS deviation of our present formula, we refer to as KTUY03, is 657.7 keV for 1835 experimental masses [14], which is smaller than that of the KUYT00 mass formula [9], 680.2 keV. In Table 1, we also list the RMS deviations of two other recent mass formulas, FRDM [7] and HFBCS [8]. Among them, our mass formula has the smallest RMS deviation. Although there is not much difference in the RMS deviations among these three mass formulas, there still remain fairly large differences of the estimated masses for some individual nuclides. The RMS deviation of the separation energies of KTUY03 mass formula is listed in Table 1 together with those of the two others. Our RMS deviation is significantly smaller than those of the others.

TABLE 1. RMS deviations of separation energies from experimental data for their mass formulas in keV. The values in the parentheses are the numbers of nuclei.

Mass formula	Mass	neutron		proton	
		S_n	S_{2n}	S_p	S_{2p}
$Z, N \geq 2$	(1835)	(1648)	(1572)	(1592)	(1483)
KTUY03	657.7	361.7	466.0	403.1	542.0
$Z, N \geq 8$	(1768)	(1585)	(1515)	(1527)	(1424)
KTUY03	640.8	319.1	391.9	344.4	465.8
FRDM	678.0	416.7	551.6	409.0	514.2
HFBCS	718.0	464.6	506.1	483.3	529.0

Figure 1 shows the two-neutron separation energies S_{2n} for even-N, the experimental data in (a), our results in (b), ones of FRDM in (c), and HFBCS in (d). We connect the nuclei with the same N by solid lines. In such a figure, magicities are seen as large gaps between two lines. In the panel (a), we see large gaps between $N=8$ and 10 (abbreviated as "at $N=8$"), and at $N=20, 28, 50, 82, 126$ except for the region with very small values of S_{2n}. Similar gaps are seen in the other panels without ones at $N=8$ and $N=20$. On the very neutron-rich region, which corresponds to the region near the $S_{2n}=0$ line, the large gaps of S_{2n} for (b) at $N=20, 28, 50$ decrease, and the gaps at $N=16, 32$ (or 34), 58 become larger compared with the neighboring ones. On the other hand, in the panel (c) the decreasing the magic gaps are not so clear, and unreasonable crossing of the solid lines are shown in the region of the very neutron-rich nuclei. In the panel (d), the similar tendencies on increasing the gaps seem to be seen at $N=16$ and 34, but the gap at $N=20$ and 28 is not so clear compared with those of experimental data and ours, and some unreasonable zigzag lines are seen in the region of heavy and neutron-rich nuclei.

Because there are almost no experimental data in these regions, more experimental information is also required.

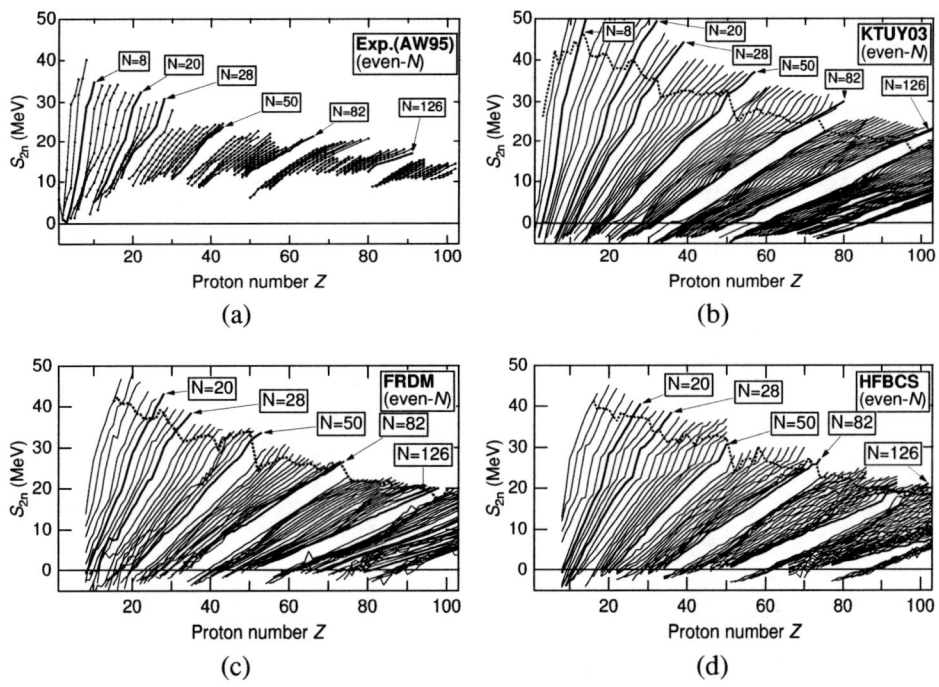

FIGURE 1. Two-neutron separation energies S_{2n} for even-N. The solid lines connent the nuclei with the same N, and dashed line connencts the proton-drip nuclei for fixed N's. (a): for experimental data, (b): for KTUY03, (c): for FRDM, (d): for HFBCS.

APPLICATION FOR SUPERHEAVY NUCLEI

Systematics of α decay Q-values

The main decay mode for the known heaviest elements is α-decay. We compare the calculated α-decay Q-values, with other formulas. our α-decay Q-values present a feature of magicity at $Z = 114$ and at $Z = 126$ as relatively wide gaps between isotope lines, while a similar figure with use of FRDM has a larger gap only at $Z = 114$, and that with use of HFBCS shows no gap. Our Q_α depend on nuclides rather moderately, or regularly, compared with the other two predictions.

We also compare Q_α with the recent experimental data in the vicinity of the nuclide $^{272}110$. Figure 3(a) shows the alpha-decay chains from $^{277}[112]$ predicted by some mass formulas. In these mass formulas, only HFBCS seems to give a similar steep line from $N=161$ to 163 (and from $Z=108$ to 110) to the experimental data, while the other mass formulas could not give such increases, or the KTUY03 lines go straight.

On the other hand, Fig. 3(b) gives the alpha-decay chains from ^{271}Ds ($Z=110$), which is only two-neutron deficient nucleus from ^{273}Ds seen in Fig. 3(a). In this figure we can see the KTUY03 gives as similar smoothness of the alpha-decay chain from ^{271}Ds as

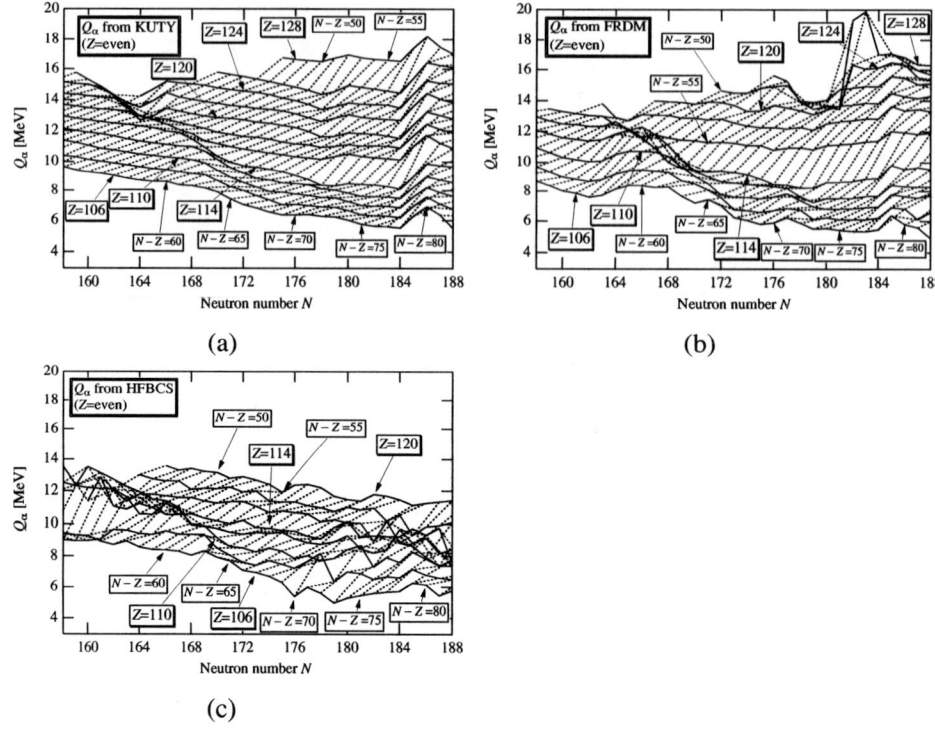

FIGURE 2. Calculated Q_α. (a):KTUY03. (b):FRDM, (c):HFBCS

that of experiment, while only the HFBCS gives a kink at ^{267}[108].

As example of smoothness of Q_α on experient, we also show Q_α of alpha-decay chian in the vicinity of the nuclide ^{234}Bk in Fig. 3(c). Experimental results give the smoothness of Q-values of alpha-decay from ^{234}Cm [20, 21]. In this figure, the KTUY03 formula gives the similar smoothness to the experimental results. On the other hand, both the FRDM and the HFBCS give different fluctuations of the chains between ^{230}Pu and ^{214}Rn.

Q-values of of fusion reaction for heavy nuclei

In the experiments of the fusion reaction for heavy- and superheavy nuclei, determination of beam energy is one of the most important points because of very small cross section of the order of pico barn. Figure 4 shows the peak energies of maximum cross sections subtracted reaction Q-values derived from some mass formulas. This figure shows somewhat large differences of peak energies measured from threshold energies of reactions, however bulk tendencies against mass numbers are similar to each other whether constant or decreasing. Among these reactions, the peak energy of ^{208}Pb(^{64}Ni, n)^{271}110 reaction measured by GSI[19] is systematically smaller than other reactions.

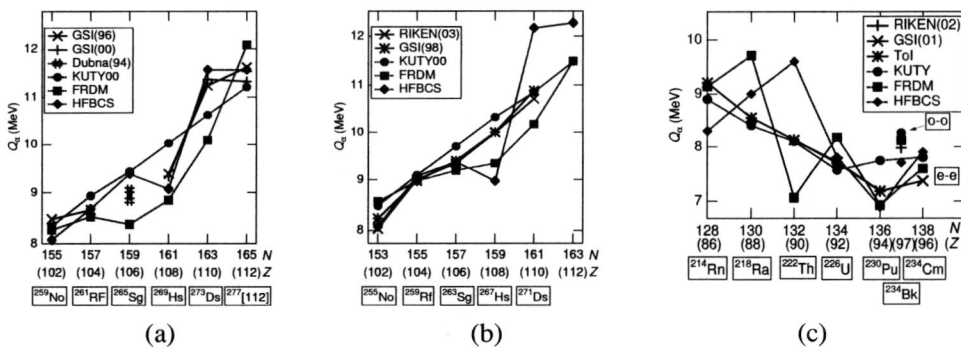

FIGURE 3. Experimental and calculated Q-values of alpha-decay chains. (a): Alpha-decay chains from $^{277}[112]$. (b): Chains from $^{275}[112]$. (c): Chains from ^{234}Cm.

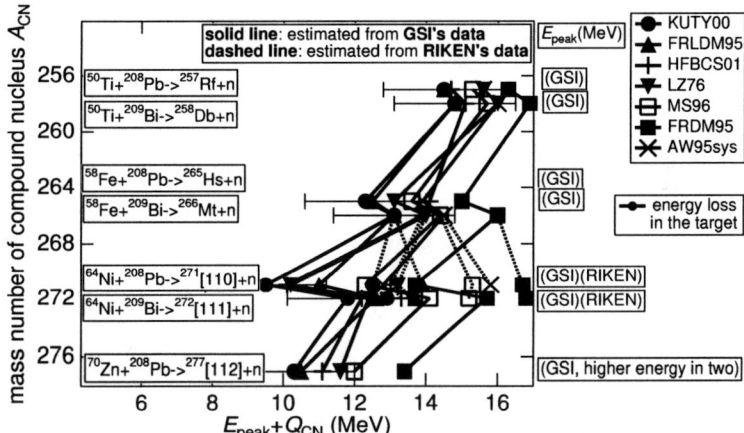

FIGURE 4. Beam energies at maximum cross section. Reaction Q-values are taken from some mass models. All GSI's data are connected by solid lines. As experimental peak energies of $^{271}[110]$ reaction, we take higher energy in two performed in GSI [22]. Some recent experimental data from RIKEN [23, 24] are taken and connected by dashed lines.

On the other hand, experiment of confirmation for this reaction were performed by RIKEN in (2002) [23], and the peak energy of cross section was obtained as higher values than GSI's results. If we take the latter data, the peak energy seem to goes in constant values increasing the mass numbers. More experimental data and theoretical considerations are required.

SPONTANEOUS-FISSION BARRIER-HEIGHTS AND HALF-LIVES

Although our mass formula, KUTY, is concerned only with the equilibrium nuclear shapes, the potential energy surface for spontaneous fission can be calculated by the same method as used for obtaining the shell energies. The fission-barrier height is defined as the highest saddle point from ground-state shell energy towards the prolate shapes. With the method of obtaining the above energy surfaces, we estimate the spontaneous fission half-lives. In this report we limit the α_2, α_4, α_6 deformations in the range $-0.2 < \alpha_2 < 0.5$. We take the one-dimensional WKB method for a virtual particle. That is

$$\log_{10}(T_{SF}) = \log_{10}\left(1 + \exp\left[\frac{2}{\hbar}K\right]\right) + \log_{10}(N_{Coll}) - 0.159 + h\delta_{oddZ} + h\delta_{oddN} - \Delta_{oo}\delta_{oddZ}\delta_{oddN}, \quad (1)$$

with

$$K = \int \sqrt{2k\mu(V(\xi) - E_{gs})}d\xi. \quad (2)$$

Here, μ is a reduced mass of the symmetric fission fragments. The path ξ is described by the differential of the deformation parameters α as

$$d\xi = r_0 A^{1/3} d\alpha. \quad (3)$$

and taken to go the lowest energy towards the prolate shapes. In this study we take the α_2, α_4, α_6 deformations in the range $-0.2 < \alpha_2 < 0.5$.

Among the above parameters, we take $r_0 = 1.2$ fm, and $N_{Coll} = 10^{20.38}$. The value of k in Eq. (2) is adjusted to reproduce the experimental T_{sf}[25] for even-even nuclei. The values of h and Δ_{oo} in Eq. (1) are adjusted for odd-A and odd-odd nuclei after fixed k. The results are $k=6.90$, $h=3.54$, and $\Delta_{oo}=3.0$. Figure 5 shows experimental and calculated $\log_{10}(T_{sf}/(s))$ for even-even nuclei. The root-mean-square deviation from experimental ones is 3.33.

Figure 6 shows the most dominant decay modes calculated from the above estimation. We can find the region of fissioning nuclei is located at three areas: the nuclei near the proton-drip line with $130 \lesssim N \lesssim 150$, the nuclei with $Z \approx 106$ and $N \approx 168$, the nuclei with $N \gtrsim 184$ and $N/Z \lesssim 1.9$.

Existence of the nuclides

With the prediction of the spontaneous-fission half-lives in the superheavy nuclidic region, the limit of the existence of the nuclides is roughly estimated. In the neutron-rich side, the border is estimated as the neutron-drip line defined as S_n and $S_{2n} > 0$. The nuclei outside of them decay by the strong interaction in the order of 10^{20} s.

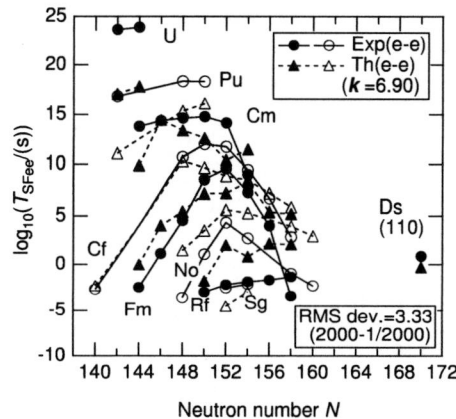

FIGURE 5. Experimental and calculated $\log_{10}(T_{SF})$.

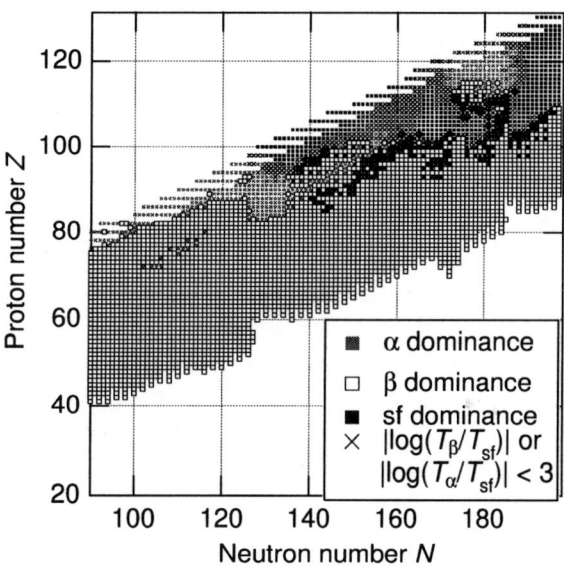

FIGURE 6. Most probable decay modes. Alpha decay: calculated from formula (B) in Ref. [16]. Beta decay: Calculated from the gross theory.[26] Spontaneous fission: This work. The first two are used with Q-values of KUTY's formula.

On the other hand, the proton-drip nuclei outside the nuclei with S_p and $S_{2p}>0$, still have relatively-long half-lives because of the quantum penetrability of protons. In the superheavy nuclidic region, it is possible that the nuclei inside the proton-drip line have very short half-lives because of the spontaneous fission. Though the limit of nuclei for the neutron-deficient side is not so clear compared with the neutron-rich side, if we take

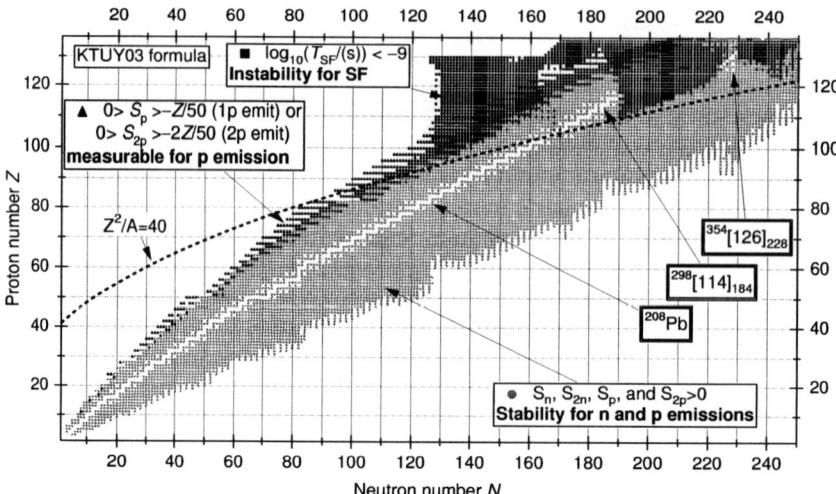

FIGURE 7. The existence of the nuclides. Beta decay: Calculated from the gross theory.[26] Spontaneous fission: This work. The first one is used with Q-values of KUTY's formula. The proton-emissioning nuclei (filled triangles) are estimated from experimental data[27]. The line with Z^2/A is shown for the guide to the eyes.

the lower limit of half-lives, the limit of long-lived nuclides is roughly estimated. For examples, nuclei with $\log_{10}(T_{sf}/(s))<-9$ are shown in Fig. 7. With the use of proton- and neutron-drip lines, the nuclei with $T_{1/2}<10^{-9}$s are estimated. In this figure we can see the well-mentioned "island of stability" around the $^{298}[114]_{184}$ as "peninsula", and another "peninsula" around $^{354}[126]_{228}$ is seen beyond the "island of stability".

By the way, the nuclei $^{354}[126]_{228}$ is found as the β-stable nuclei with relatively long half-lives in our calculation. The α-decay half-life is estimated in the order of 100 years and the spontaneous-fission half-life is estimated in the order of 10 days. This means that total half-life of this nucleus is expected to be long and to decay by spontaneous fission.

ACKNOWLEDGMENTS

The Author thanks Dr. Morita for helpful discussions. The numerical calculations were mainly made with the computer VPP700 at Computer and Information Division, RIKEN.

REFERENCES

1. C.F. von Weizsäcker, Z. Phys. **96** (1935) 431.
2. H.A. Bethe and R.F. Bacher, Rev. Mod. Phys. **8** (1936) 829.

3. G.T. Garvey, *et al.*, Rev. Mod. Phys. **41**, (1969) S1.
4. P.E. Haustein, Special Editor: "*1988-89 Atomic Mass Predictions*", ATOMIC DATA AND NUCLEAR DATA TABLES **39** (1988) 185.
5. T. Tachibana, M. Uno, M. Yamada, S. Yamada, ATOMIC DATA AND NUCLEAR DATA TABLES **39** (1988) 251.
6. W.D. Myers, W.J. Swiatecki, Nucl. Phys. **81** (1966) 1.
7. P. Möller, J.R. Nix, W.D. Myers, J. Swiatecki, ATOMIC DATA AND NUCLEAR DATA TABLES **59**, (1995) 185.
8. S. Goriely, F. Tondeur, J.M. Pearson, ATOMIC DATA AND NUCLEAR DATA TABLES **77** (2001) 311.
9. H. Koura, M. Uno, T. Tachibana and M. Yamada, Nuclear Physics A **674** (2000) 44.
10. H. Koura, T. Tachibana, M. Uno and M. Yamada, in preparation; H. Koura, T. Tachibana, M. Uno and M. Yamada, RIKEN Accelerator Progress Report, **36** (2003), 9.
11. H. Koura, T. Tachibana, M. Uno and M. Yamada, RIKEN Accelerator Progress Report, **36** (2003), 10.
12. H. Koura, M. Uno, T. Tachibana and M. Yamada, Int. Conf. on Exotic Nuclei and Atomic Masses (ENAM98), Bellaire, Michigan, 1998, ed. B.M. Sherrill, D.J. Morrissey, and C.N. Davids, (The American Institute of Physics, 1998) 114.
13. H. Koura and M. Yamada, Nuclear Physics A **671** (2000) 96.
14. G. Audi and A.H. Wapstra, Nuclear Physics A **595** (1995) 409.
15. S. Goriely, F. Tondeur, J.M. Pearson, ATOMIC DATA AND NUCLEAR DATA TABLES **77** (2001) 311.
16. H. Koura, Journal of Nuclear and Radiochemical Sciences, **3** (2001), 201.
17. H. Koura, TOURS Symposium on Nuclear Physics IV (TOURS2000), ed. M. Arnould, et al., (The American Institute of Physics Conference Proceedings 561, 2001) 388.
18. D. Ackermann, TOURS Symposium on Nuclear Physics IV (TOURS2000), ed. M. Arnould, et al., (The American Institute of Physics Conference Proceedings 561, 2001) 323.
19. S. Hofmann, Report on Progress in Physics, **61** (1998) 639.
20. P. Cagarda, S. Antalic, D. Ackermann, F.P. He?berger, S. Hpfmann, B. Kindler, J. Kojouharova, B. Lommel, R. Mann, A.G. Popeko, S. Saro, J. Uusitalo, A.V. Yeremin, GSI Scientific Report 2001 (2002), 15.
21. K. Morimoto, K. Morita, D. Kaji, A. Yoneda, A. Yoshida, T. Suda, E. Ideguchi, T. Ohnishi, Y.-L. Zhao, H. Xu, T. Zheng, H. Haba, H. Kudo, K. Sueki, H. Koura, K. Katori, and, I. Tanihata, RIKEN Accelerator Progress Report, **36** (2003) 89.
22. S. Hofmann, F.P. He?berger1, D. Ackermann, G. Münzenberg, S. Antalic, P. Cagarda, B. Kindler, J. Kojouharova, M. Leino, B. Lommel, R. Mann, A.G. Popeko, S. Reshitko, S. Saro, J. Uusitalo, and A.V. Yeremin, Eur. Phys. J. A **14**, (2002) 147.
23. K. Morita, K. Morimoto, D. Kaji, H. Haba, E. Ideguchi, R. Kanungo, K. Katori, H. Koura, H. Kudo, T. Ohnishi, A. Ozawa, T. Suda, K. Sueki, I. Tanihata, H. Xu, A.V. Yeremin, A. Yoshida, A. Yoneda, Y.-L. Zhao, and T. Zhen, European Physical Journal, submitted for publication.
24. K. Morita, K. Morimoto, D. Kaji, H. Haba, E. Ideguchi, R. Kanungo, K. Katori, H. Koura, H. Kudo, T. Ohnishi, A. Ozawa, T. Suda, K. Sueki, I. Tanihata, H. Xu, A.V. Yeremin, A. Yoshida, A. Yoneda, Y.-L. Zhao, and T. Zhen, Journal of Physical Society of Japan, submitted for publication.
25. Evaluated Nuclear Structure Data File (ENSDF), 2000 August version.
26. T.Tachibana and M. Yamada, Proc. Int. Conf. on exotic nuclei and atomec masses, Arles, 1995, eds. M. de Saint Simon and O. Sorlin (Editions Frontueres, Gif-sur-Yvette, 1995), p.763 , and references therin.
27. R.B. Firestone and V.S Shirley, Table of Isotopes, eighth edition(Wiley, New York, 1996)

FUSION-FISSION DYNAMICS (FFD)

Effect of Closed Shell Structure on Heavy-Ion Fusion Reactions

Hiroshi Ikezoe[*], Ken-ichiro Satou[*,¶], Sinichi Mitsuoka[*], Katsuhisa Nishio[*], Kaoru Tsuruta[*], Sun-Chang Jeong[†], and Cheng-Jian Lin[‡]

[*]*Japan Atomic Energy Research Institute, Tokai, Ibaraki 319-1195, Japan*

[¶]*Institute of Physics and Tandem Accelerator Center, University of Tsukuba, Tsukuba-shi, Ibaraki 305-8577, Japan*

[†]*Institute of Particle and Nuclear Studies, KEK, Tsukuba-shi, Ibaraki 305-0801, Japan*

[‡]*China Institute of Atomic Energy, P.O. Box 275(10), Beijing 102413, P. R. China*

Abstract. The effect of the nuclear shell structure on the heavy-ion fusion reaction was investigated for the reaction systems ^{82}Se + ^{138}Ba, ^{82}Se + ^{134}Ba, ^{16}O + ^{204}Pb, ^{86}Kr + ^{134}Ba, ^{86}Kr + ^{138}Ba, and ^{82}Se + ^{nat}Ce. Evaporation residues for these fusion reactions were measured using a recoil mass separator (JAERI-RMS) near Coulomb barrier region. The measured evaporation residue cross sections for the reactions ^{82}Se + ^{138}Ba and ^{86}Kr + ^{138}Ba were two orders of magnitude and one order of magnitude larger than those for the reaction ^{82}Se + ^{134}Ba and ^{86}Kr + ^{134}Ba, respectively, at the excitation energy region of 10 ~ 30 MeV. The evaporation residue cross sections were compared with those of the other reaction systems that make the same or similar compound nuclei as the present reaction systems. It was found that the evaporation residue cross sections correlate strongly with the sum of the shell energies for both projectile and target nuclei, i.e., the evaporation residue cross sections increase as the sum of the shell energy decreases.

INTRODUCTION

Fusion process between heavy nuclei to produce super-heavy elements has been extensively investigated so far. It is well known that the fusion probability between heavy nuclei depends on the charge product Z_pZ_t of projectile and target, i.e., the Coulomb repulsion at contact. When the charge product is less than 1800, its fusion cross section has been well reproduced by the one-dimensional barrier penetration model. When the charge product is larger than 1800, its fusion cross section is considerably reduced compared with the model calculation. This fact means that interacting two nuclei cannot always fuse to make a compound nucleus even if they have a kinetic energy to overcome the fusion barrier between two nuclei. This is because the saddle point of heavy compound nucleus is more compact than a touching shape of two nuclei. An extra kinetic energy is needed so that the reaction system can reach the saddle point after surmounting the fusion barrier. This extra kinetic energy begins to increase sharply at the charge product 1800 ~ 1900 [1]. Therefore, for a symmetric reaction system where masses of projectile and target are similar each other, it is more difficult to make a compound nucleus compared to an asymmetric reaction system, because of a large charge product for the symmetric reaction system than that of the asymmetric reaction system.

A fusion probability between heavy nuclei at a low excitation energy depends on not only the charge product but also the nuclear structure of projectile and target. It was reported that the number of a valence nucleon outside a major shell changes the fusion probability [1, 2], i.e., the extra kinetic energy increases as the number of the valence nucleon. This effect was incorporated as a small correction in the systematics of the extra kinetic energy [1]. Oganessian et al. [3] reported a large difference in the evaporation residue cross sections between the fusion reactions ^{136}Xe + ^{86}Kr and ^{130}Xe + ^{86}Kr, where the nucleus ^{136}Xe has a closed neutron shell N = 82 and the neutron number of the nucleus ^{130}Xe is 76. They found that the measured evaporation residue cross sections for the fusion reaction ^{136}Xe + ^{86}Kr are almost two to three orders of magnitude larger than those for the fusion reaction ^{130}Xe + ^{86}Kr near the Coulomb barrier region. This result suggests an strong influence of the closed shell on the fusion process. The enhancement of the evaporation residue cross sections between ^{208}Pb and ^{48}Ca is also pointed out in [4]. Recently, We reported the evaporation residue cross sections for the reactions ^{82}Se + ^{138}Ba and ^{82}Se + ^{134}Ba, where the nucleus ^{138}Ba has a closed neutron shell of N = 82. We showed that the measured evaporation residue cross section for the reaction ^{82}Se + ^{138}Ba was two orders of magnitude larger than those for the reaction ^{82}Se + ^{134}Ba [5]. This experimental evidence also suggests that the closed shell structure plays an important role in the heavy-ion fusion process at a low excitation energy.

In this paper, we report the experimental results on the fusion reactions ^{86}Kr + ^{134}Ba, ^{86}Kr + ^{138}Ba and ^{82}Se + $^{nat.}$Ce, where the nucleus ^{86}Kr has the closed neutron shell N = 50. The nucleus $^{nat.}$Ce consists of the isotope ^{140}Ce with the natural abundance of 88.48%, which has the closed neutron shell N = 82. Recently, Hinde et al. [6] showed that the fusion probability for the reaction system ^{16}O + ^{204}Pb, which makes the same compound nucleus ^{220}Th as the reaction system ^{82}Se + ^{138}Ba, is one order of magnitude larger than that of the reaction system ^{82}Se + ^{138}Ba. In order to verify this fact, we also measured the evaporation residues for the reaction ^{16}O + ^{204}Pb. Based on these experimental results, we emphasize that the shell energy of projectile and target is an important quantity to control the heavy-ion fusion process and a large shell energy in negative values gives rise to an enhancement of fusion.

EXPERIMENTAL PROCEDURES

Evaporation residues for the fusion reactions ^{86}Kr + ^{138}Ba, ^{86}Kr + ^{134}Ba and ^{82}Se + $^{nat.}$Ce were measured using ^{86}Kr and ^{82}Se beams from JAERI-tandem booster accelerator. The targets of ^{134}Ba, ^{138}Ba and $^{nat.}$Ce, whose thickness were 0.41, 0.50 and 0.34 mg/cm^2, respectively, were mounted on a rotating frame and rotated at 100 rpm during the beam irradiation to prevent the targets from breaking due to the beam heating. The evaporation residues emitted in the beam direction from a target foil were passed through a thin carbon foil (0.03 mg/cm^2) to reset a charge state of recoiling ions and separated in-flight from the primary beam by the JAERI-recoil mass separator (JAERI-RMS) [7]. The evaporation residues transported through the JAERI-RMS were passed through two micro-channel plate detectors (MCP) and implanted into a double-sided position-sensitive strip detector (DPSD) mounted at the focal position of the JAERI-RMS. The time-of-flight signal from the MCP was used to distinguish the incoming evaporation resides from the subsequent α-decay event. The energies and the positions of the incoming evaporation residues and their subsequent α-particle decays were measured by the DPSD. The typical energy and position resolution were 75 keV (FWHM) and 0.5 mm, respectively.

All evaporation residues produced in the present fusion reactions decay by emitting α-particles. Their α-decay energies and half-lives are known from the literature [8]. The identification of each evaporation residue was made event by event by measuring its subsequent α-decay energy and the time interval between implanted evaporation residue and its decay event. A silicon surface barrier detector was set at 45° with respect to the beam direction to detect the elastic scattering of the beam particles from the targets. The transport efficiency of the evaporation residue through the JAERI-RMS was estimated by the methods written in [9]. The typical transport efficiencies of xn, pxn and αxn channels were 0.37, 0.31 and 0.25, respectively. These elastic scattering yields and the transport efficiency were used to determine the absolute value of the evaporation residue cross sections.

In the measurements of the evaporation residues produced in the reaction ^{16}O + ^{204}Pb, the evaporation residues recoiling out from a target foil were implanted in an aluminum catcher foil which was positioned immediately behind the target foil and its thickness (5 μm) is thick enough to stop the recoiling evaporation residues inside the catcher foil. After the irradiation of 70 to 90 min. by ^{16}O beams, the catcher foil was removed from the target and put on a silicon surface barrier detector to detect subsequent α-decays from implanted evaporation residues. Although the JAERI-RMS was also used to detect the evaporation residues, the absolute cross sections were not obtained in the JAERI-RMS measurements. This was because the transport efficiency for the evaporation residues with a low kinetic energy of about 5 ~ 8 MeV was not accurately estimated and the relative yield of each evaporation residue was obtained.

EXPERIMENTAL RESULTS AND DISCUSSIONS

The obtained evaporation residue cross sections for the reaction system ^{86}Kr + ^{138}Ba are plotted in Fig. 1 as a function of c.m. energy determined in the middle of the target layer. Evaporation channels expected in respective energy regions are indicated in the figure. The evaporation residues corresponding to the $3n$ and $4n$ channels were not measured because the half-lives and the α-decay energies for the parent nuclei ^{220}U and ^{221}U were unknown. The $1n$ channel is clearly seen at about 10 MeV below the Bass barrier 217.2 MeV. This means that the actual fusion barrier is extended to a lower energy than the Bass barrier.

The reduced cross sections $\sigma(ER)/\pi\lambda^2$ obtained from the sum of the detected evaporation residues ($1n$, $2n$, αn, $\alpha 2n$, $\alpha 3n$ and pn channels) were plotted in Fig. 2 together with those for other reaction systems which makes the same or the similar

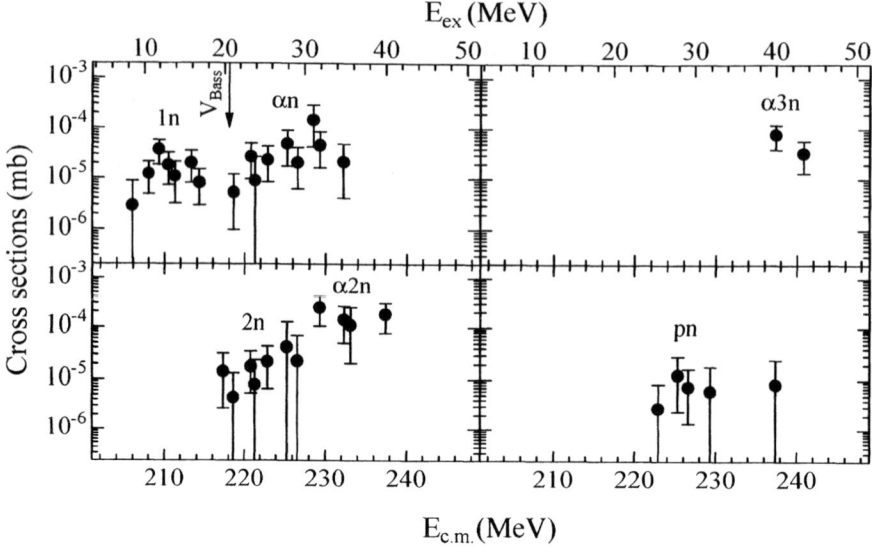

FIGURE 1. Evaporation residue cross sections measured in the reaction ^{86}Kr + ^{138}Ba. The arrow shows the position of the Bass barrier.

compound nuclei $^{222, 226, 228}$U as the present reaction system ^{86}Kr + ^{138}Ba. Here λ is the reduced de Broglie wavelength. The measured evaporation residue cross sections for the reactions ^{86}Kr + ^{134}Ba and ^{82}Se + $^{nat.}$Ce are also plotted for comparison. In the reaction ^{86}Kr + ^{134}Ba, one event corresponding to the αn channel was detected at the excitation energy 29.9 MeV and one event corresponding to the pn channel was detected at each excitation energy of 25.4 MeV and 31.5 MeV. No evaporation residue was detected below the excitation energy of 25 MeV. As for the reaction ^{82}Se + $^{nat.}$Ce, no evaporation residue was observed below the excitation energy of 23 MeV and the upper limits of the evaporation residue cross section were determined.

As shown in Fig. 2, the reduced cross sections for the reaction ^{86}Kr + ^{138}Ba show the largest values among the reaction systems at the excitation energy region less than 35 MeV. Above the excitation energy of 35 MeV, the reduced cross section for the reaction system ^{20}Ne + ^{208}Pb becomes largest. It is noted that the reaction system ^{86}Kr + ^{138}Ba has the largest value of the charge product Z_pZ_t among those of the other reaction systems. The present result suggests that the fusion of heavy nuclei is not always governed only by the single quantity Z_pZ_t. In order to emphasize this point, the reduced evaporation residue cross sections at the Bass barriers are plotted in Fig. 3(b) as a function of the sum of the shell energy for the projectile and target nuclei. The shell energy was obtained from the mass table of [10]. The reaction system for each data point is indicated in the figure. The reduced evaporation residue cross sections may depend on the property of each compound nucleus (excitation energy and mass number). This was simulated using the statistical model code HIVAP [11]. The thin line shows the result of the calculation, where we adopted the statistical model

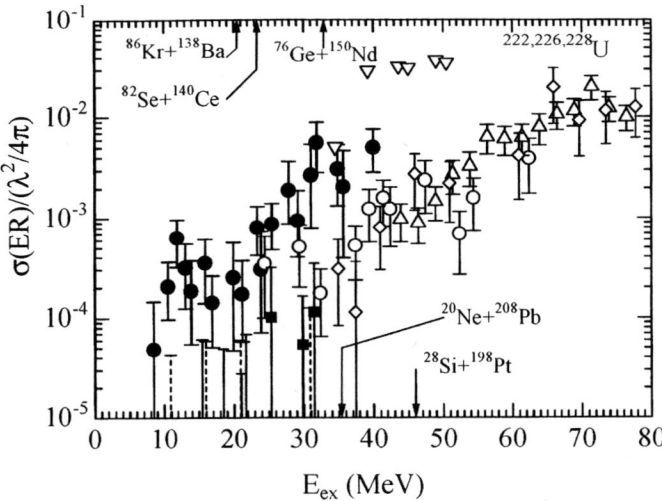

FIGURE 2. Reduced evaporation residue cross sections for various reactions; ^{86}Kr + ^{138}Ba (solid circles), ^{86}Kr + ^{134}Ba (solid squares), ^{82}Se + $^{nat.}$Ce (open circles), ^{76}Ge + ^{150}Nd (diamonds, [16]), ^{28}Si + ^{198}Pt (open triangles, [16]), ^{20}Ne + ^{208}Pb (inverted open triangles, [17]). The Bass barriers for these reaction systems are shown as arrows. The Bass barrier for ^{86}Kr + ^{134}Ba is at the excitation energy of 22.7 MeV. The upper limits for the reactions ^{86}Kr + ^{134}Ba, ^{82}Se + $^{nat.}$Ce and ^{76}Ge + ^{150}Nd are shown as thick lines, thin lines and dashed lines, respectively.

parameters used in [9]. The calculated values for the evaporation residue cross section are not a smooth function of the sum of the shell energy. On the other hand, the measured evaporation residue cross sections increase smoothly as the sum of the shell energy decreases in negative values. This result indicates that the trend shown in Fig.3(b) is not ascribed to the property of each compound nucleus and rather ascribed to each entrance channel (fusion process). Therefore, we conclude that the shell energies of both projectile and target are an important quantity to control the fusion process.

In Fig. 4, we plotted the reduced cross sections $\sigma(\Sigma xn)/\pi\lambda^2$ for the various reaction systems which make the compound nucleus ^{220}Th. Here, $\sigma(\Sigma xn)$ is the sum of the observed xn channels (x = 1 to 7). Our previous data for the reaction ^{82}Se + ^{138}Ba [5] and the present data for the reaction ^{16}O + ^{204}Pb are also included. We see that the reduced cross section for the reaction ^{82}Se + ^{138}Ba shows a peak exactly at the Bass barrier of this reaction system. The cross section decreases below the fusion barrier because of the barrier penetration and decreases also above the fusion barrier because of the competition between a neutron emission and fission in the evaporation process. This result means no fusion barrier shift and a narrow barrier distribution centered at the Bass barrier.

On the other hand, the reaction systems ^{124}Sn + ^{96}Zr and ^{70}Zn + ^{150}Nd show a broad barrier distribution extended from their Bass barrier energies to the energies larger by about 20 MeV than the Bass barriers. These broad barrier distributions are

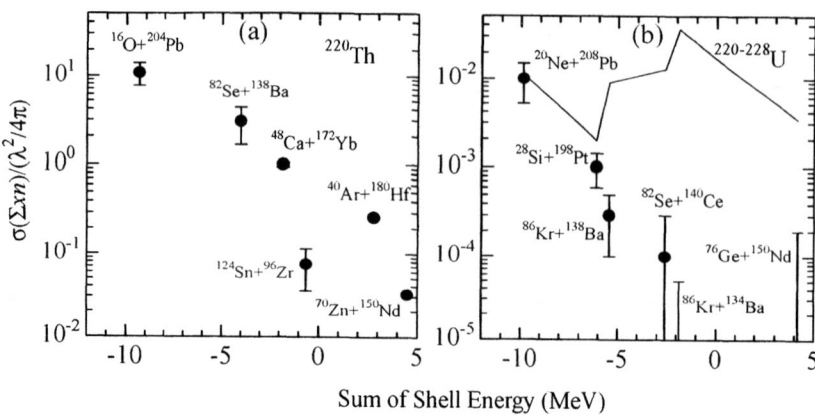

FIGURE 3. Reduced evaporation residue cross sections at the Bass barriers as a function of the sum of the shell energy of projectile and target nuclei for various reactions. The thin line in (b) shows the value calculated by HIVAP [11]. The excitation energies at the Bass barriers for reaction systems listed in (a) are 31.1 MeV (^{16}O + ^{204}Pb), 25.5 MeV (^{82}Se + ^{138}Ba), 36.2 MeV (^{48}Ca + ^{172}Yb), 27.4 MeV (^{124}Sn + ^{96}Zr), 44.8 MeV (^{40}Ar + ^{180}Hf) and 38.0 MeV (^{70}Zn + ^{150}Nd).

commonly observed in the heavy-ion fusion reaction as shown in [1] and well correlated with the extra kinetic energy (the amount of the barrier shift from the Bass barrier) needed for fusion. This systematics where the extra kinetic energy increases as a function of the charge product Z_pZ_t is not true for the reaction ^{82}Se + ^{138}Ba. We expect the extra kinetic energy around 10 ~ 15 MeV for the reaction ^{82}Se + ^{138}Ba from this systematics. This effectively makes the fusion barrier high and considerably decreases the 1n and 2n channels cross-sections at the low excitation energy region less than 30 MeV. There is no barrier shift seen in Fig. 4. On the other hand, the reduced evaporation residue cross section for the reaction ^{82}Se + ^{134}Ba [5] shows the consistent trend as predicted by the systematics [1], i.e., the fusion barrier is shifted to a high energy by the amount of 10 ~ 15 MeV [5, 12].

Our present data for the evaporation residue cross section in the reaction ^{16}O + ^{204}Pb are consistent with and even larger than those of Hinde et al. [6]. If we assume no fusion hindrance for the reaction ^{16}O + ^{204}Pb, the present result shows that the fusion for the reaction ^{82}Se + ^{138}Ba is hindered by the amount of one order of magnitude or more at the energy region above the Bass barrier. In Fig. 3(a), we plotted the reduced cross sections at the Bass barriers for various reaction systems which make the compound nucleus ^{220}Th. The excitation energies for these reaction systems are shown in the figure caption of Fig. 3. The reduced evaporation residue cross sections are not correlated with the excitation energy (the property of the compound nucleus) and rather well correlated with the sum of the shell energy of the target and projectile nuclei (the property of the entrance channel). It is clearly seen that the reduced cross section increases as the sum of the shell energies decreases in negative values, i.e., the target and/or projectile with a closed shell structure makes the fusion probability high. This systematic trend is consistent with that seen in Fig. 3(b).

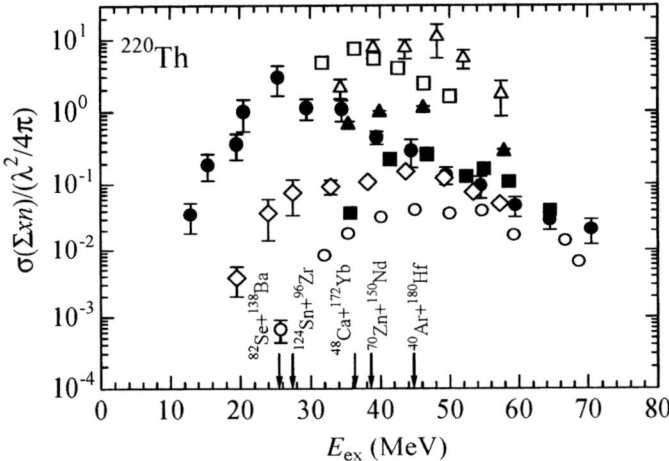

FIGURE 4. Reduced evaporation residues for various reaction systems which make the compound nucleus ^{220}Th; ^{82}Se + ^{138}Ba (solid circles, [5]), ^{48}Ca + ^{172}Yb (solid triangles, [18]), ^{40}Ar + ^{180}Hf (solid squares, [19]), ^{124}Sn + ^{96}Zr (diamonds,[18]), ^{70}Zn + ^{150}Nd (open circles), ^{16}O + ^{204}Pb (open triangles for the present work and open squares for [6]).

Myers and Swiatecki [13] pointed out that the shell energy resists neck growth at the time of contact between projectile and target, and then the projectile nucleus can approaches closely to the target nucleus with a small kinetic energy dissipation. They proposed the sum of the shell energy and the congruence energy for both projectile and target as a resistance factor against the neck growth. The observed trend seen in Fig. 3(a) and 3(b) is essentially unchanged even if the reduced evaporation residue cross section is plotted as a function of the sum of the shell energy and the congruence energy. Therefore, the present data are consistent with this argument.

Möller and Sierk discussed the fusion process on the multi-dimensional microscopic potential energy surface [14]. When heavy nuclei first touch, mainly the Coulomb repulsive force push the system to deformations and eventually the system falls down the fission valley. If the system resists deformations away from the fusion path by the help of the shell energy, the system may proceed to a compound nucleus following the fusion path. This argument qualitatively explains the present data. Oganessian et al. suggests the close relation between the fusion process and the fission process [3]. In the case of the fusion reaction ^{136}Xe + ^{86}Kr, the low excited compound nucleus ^{222}Th can decay into the asymmetric fission components close to the nuclei ^{86}Kr and ^{136}Xe. The compound nucleus ^{220}Th which is produced in the fusion reaction ^{82}Se + ^{138}Ba, may decay into the asymmetric fission components close to the nuclei ^{82}Se and ^{138}Ba [15]. On the other hand, the compound nucleus ^{216}Th, which is formed in the fusion reactions ^{130}Xe + ^{86}Kr and also ^{82}Se + ^{134}Ba, has little chance to decay into such asymmetric fission component [15]. This argument is also related to the shell structure of final fragments. We conclude that the shell energy of the projectile and target plays a major role in the fusion process between heavy nuclei in addition to the Coulomb repulsive force.

SUMMARY AND CONCLUSIONS

Evaporation residues for the reactions ^{82}Se + ^{138}Ba, ^{82}Se + ^{134}Ba, ^{16}O + ^{204}Pb, ^{86}Kr + ^{138}Ba, ^{86}Kr + ^{134}Ba and ^{82}Se + $^{nat.}$Ce were measured to investigate the dependence of fusion reaction on the nuclear shell structure of the colliding nuclei. The evaporation residue cross sections measured in the reaction ^{82}Se + ^{138}Ba and ^{86}Kr + ^{138}Ba were almost two orders of magnitude and one order of magnitude larger near the Coulomb barrier region than those for the reaction ^{82}Se + ^{134}Ba and ^{86}Kr + ^{134}Ba, respectively. These large differences are ascribed to the entrance channel property of the fusion process. The present data were compared with the other reaction systems, which make the same or similar compound nuclei as the present reactions. From these comparisons, we conclude that the hindrance of the fusion for massive reaction systems is strongly affected by the shell structure of colliding partners in addition to the Coulomb repulsion factor $Z_p Z_t$. It is important to realize theoretically the energy dissipation due to the friction after contact by taking into account the shell structure of projectile and target nuclei. Further experimental investigation is needed also to make the relation between fusion and fission clear.

ACKNOWLEDGMENTS

We thank the crew of the JAERI tandem booster facility for the beam operation.

REFERENCES

1. A. B. Quint, W. Reisdorf, K. -H. Schmidt, P. Armbruster, F. P. Hessberger, S. Hofmann, J. Keller, G. Müzenberg, H. Stelzer, H. -G. Clerc, W. Morawek, and C. -C. Sahm, Z. Phys. A **346**, 119 (1993).
2. K. -H. Schmidt, and W. Morawek, Rep. Prog. Phys. **54**, 949 (1991).
3. Yu Ts. Oganessian, *Heavy Elements and Related New Phenomena*, Ed. by W. Greiner and R. K. Gupta (World Scientific, 1999) Vol.1, p. 43
4. Yu Ts. Oganessian, V. K. Utyonkov, Yu. V. Lobanov, F. S. Abdullin, A. N. Polyakov, I. V. Shirokovsky, Yu S. Tsyganov, A. N. Mezentsev, S. Iliev, V. G. Subbotin, A. M. Sukhov, K. Subotic, O. V. Ivanov, A. N. Voinov, V. I. Zagrebaev, K. J. Moody, J. F. Wild, N. J. Stoyer, M. A. Stoyer, and R. W. Lougheed, Phys. Rev. C **64**, 054606 (2001).
5. K. Satou, H. Ikezoe, S. Mitsuoka, K. Nishio, and S. C. Jeong, Phys. Rev. C **65**, 054602 (2002).
6. D. J. Hinde, M. Dasgupta, and A. Mukherjee, Phys. Rev. Lett. **89**, 282701 (2002).
7. H. Ikezoe, Y. Nagame, T. Ikuta, S. Hamada, I. Nishinaka, and T. Ohtsuki, Nucl. Instrum. Methods, A **376**, 420 (1996).
8. R. B. Firestone, in *Table of Isotopes*, Ed. by V. S. Shirley (Wiley, New York, 1996).
9. S. Mitsuoka, H. Ikezoe, K. Nishio, and J. Lu, Phys. Rev. C **62**, 054603 (2000).
10. W.D. Myers and W.J. Świątecki, Nucl. Phys. **A601**, 141 (1996).

11. W. Reisdorf and M Schädel, Z. Phys. A 343, 47 (1992).
12. H. Ikezoe, K. Satou, S. Mitsuoka, K. Nishio, and S.C. Jeong, Phys. Atom. Nucl. 66, 1053 (2003).
13. W. D. Myers and W. J. Światecki, Phys. Rev. C **62**, 044610 (2000).
14. P. Möller and A. J. Sierk, Nature Vol. 422, 485 (2003).
15. K.-H. Schmidt, S. Steinhäuser, C. Böckstiegel, A. Grewe, A. Heinz, A.R. Junghans, J. Benlliure, H.-G. Clerc, M. de Jong, J. Müller, M. Pfützner, and B. Voss, Nucl. Phys. **A665**, 221 (2000).
16. K. Nishio, H. Ikezoe, S. Mitsuoka, and J. Lu, Phys. Rev. C **62**,014602 (2000).
17. R.N. Sagaidak, V.I. Chepigin, A.P. Kabachenko, J. Roháč, Yu. Ts. Oganessian, A.G. Popeko, A.V. Yeremin, Ans. D'Arrigo, G. Fazio, G. Giardina, M. Herman, R. Ruggeri, and R. Sturiale, J. Phys. G: Nucl. Part. Phys. **24** 611 (1998).
18. C.-C. Sahm, H.-G. Clerc, K.-H. Schmidt, W. Reisdorf, P. Arumbruster, F.P. Hessberger, J.G. Keller, G. Münzenberg, and D. Vermeulen, Nucl. Phys. **A441**, 316 (1985).
19. H.-G. Clerc, J.G. keller, C.-C. Sahm, K.-H. Schmidt, H. Schulte, and D. Vermeulen, Nucl. Phys. **A419**, 571 (1984).

Fusion dynamics around the Coulomb barrier

K. Hagino*, N. Rowley†, T. Ohtsuki**, M. Dasgupta‡, J.O. Newton‡ and D.J. Hinde‡

*Yukawa Institute for Theoretical Physics, Kyoto University, Kyoto 606-8502, Japan
†Institut de Recherches Subatomiques, UMR7500, IN2P3-CNRS/Université Louis Pasteur, BP28, F-67037 Strasbourg Cedex 2, France
**Laboratory of Nuclear Science, Tohoku University, Sendai 982-0826, Japan
‡ Department of Nuclear Physics, Research School of Physical Sciences and Engineering, Australian National University, Canberra ACT 0200, Australia

Abstract.
We perform exact coupled-channels calculations, taking into account properly the effects of Coulomb coupling and the finite excitation energy of collective excitations in the colliding nuclei, for three Fm formation reactions, ^{37}Cl + ^{209}Bi, ^{45}Sc + ^{197}Au, and ^{59}Co + ^{181}Ta. For the ^{37}Cl + ^{209}Bi and ^{45}Sc + ^{197}Au reactions, those calculations well reproduce the experimental total fission cross sections, and a part of the extra-push phenomena can be explained in terms of the Coulomb excitations to multi-phonon states. On the other hand, for the heaviest system, the deep-inelastic collisions become much more significant, and the fission cross sections are strongly overestimated. We also discuss the surprisingly large surface diffuseness parameters required to fit recent high-precision fusion data for medium-heavy systems, in connection with the fusion supression observed in massive systems.

INTRODUCTION

The coupled-channels method has been very successful in reproducing experimental cross sections for heavy-ion reactions involving *medium-heavy* nuclei. Particularly, in many systems, it simultaneously reproduces the subbarrier enhancement of fusion cross sections and the shape of the fusion barrier distribution by including a few low-lying collective excitations of colliding nuclei and nucleon transfer channels [1]. It is now a standard theoretical tool to analyse experimental fusion and quasi-elastic cross sections at energies around the Coulomb barrier [2].

However, it has not yet been completely understood to what extent this method works for the fusion of massive systems, where the charge product of the target and projectile nuclei, $Z_P Z_T$, is typically larger than about 1800. For those systems, other reaction processes, such as deep-inelastic collision and quasi-fission, come into play, and the reaction dynamics around the Coulomb barrier becomes much more complex than for medium-heavy systems [3]. In order to calculate fusion cross sections, the competition of fusion with these other processes has to be taken into account properly [4, 5, 6], and the dynamics after the Coulomb barrier is overcome becomes very important. This is the most difficult problem in the fusion of massive nuclei, and there are still large ambiguities in theoretical predictions of fusion cross sections. Because of this reason, one often employs a simplified coupled-channels treatment for the barrier penetration prior to the touching configuration [5, 7], which is essentially based on the constant coupling approximation [8] or a variant [9].

In this contribution, we critically examine the consequences of using such a simplified coupled-channels framework, and point out that the exact treatment for the Coulomb coupling plays an important role in massive systems. We then apply the exact coupled-channels approach to total fission cross sections for three Fm formation reactions [10], where the charge product $Z_P Z_T$ is given by 1411 (^{37}Cl + ^{209}Bi), 1659 (^{45}Sc + ^{197}Au), and 1971 (^{59}Co + ^{181}Ta). We will show that the small hindrance of the reduced cross sections for the second heaviest system compared with the lightest system can be understood in terms of the effect of energy loss due to the Coulomb excitation, while the fusion hindrance for the heaviest system exceeds that effect and an explicit treatment of deep-inelastic collisions is necessary. We also discuss the surface property of the nucleus-nucleus potential, where recent high-precision fusion data for medium-heavy systems systematically show that a surprisingly large diffuseness parameter is required in order to fit them [11, 12, 13, 14]. We argue that this apparant anomaly could originate from the competition between fusion and the deep-inelastic processes which occur in fusion of heavy systems [14].

COUPLED-CHANNELS APPROACH TO FUSION OF MASSIVE SYSTEMS

Effect of Coulomb excitations

A hot compound nucleus formed by a fusion reaction decays either by emitting a few particles (mainly neutrons) and gamma rays, or by fission. The fusion cross section is thus a sum of the evaporation residue and the fusion-fission cross sections. Theoretically, it is computed as

$$\sigma_{\text{fus}}(E) = \frac{\pi}{k^2} \sum_l (2l+1) P_{\text{fus}}(E,l) = \frac{\pi}{k^2} \sum_l (2l+1) T_{\text{bp}}(E,l) \cdot P_{\text{CN}}(E,l), \quad (1)$$

where $T_{\text{bp}}(E,l)$ is the barrier passing probability for the Coulomb barrier while $P_{\text{CN}}(E,l)$ is the probability of compound nucleus formation after barrier penetration.

For medium-heavy systems, the compound nucleus is formed immediately after the Coulomb barrier is overcome, and P_{CN} is essentially 1. This justifies the assumption of strong absorption inside the Coulomb barrier, or equivalently, the incoming wave boundary condition [2]. The coupled-channels approach has been successful here. For massive systems, on the other hand, P_{CN} significantly deviates from 1, due to competition among several reaction processes. The fusion cross sections, therefore, appear to be hindered if one compares fusion cross sections with those obtained by assuming $P_{\text{CN}} = 1$. Recent experimental data clearly indicate the strong competition between compound nucleus formation and quasi-fission [15, 16, 17].

In Eq. (1), the barrier passing probability T_{bp} can in principle be computed by the coupled-channels approach. For this, although exact coupled-channels codes are available [2], a simplified approach has often been taken, even for fusion of massive systems

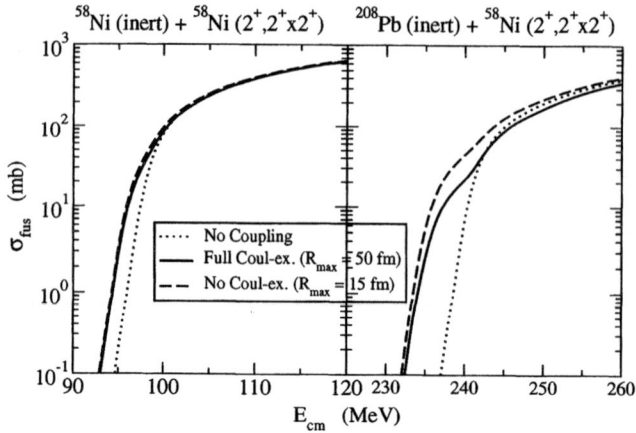

FIGURE 1. Influence of the Coulomb couplings in fusion cross sections for the ^{58}Ni + ^{58}Ni (left panel) and the ^{58}Ni + ^{208}Pb (right panel) systems. The double quadrupole phonon excitation in the projectile nucleus ^{58}Ni is included in the coupled-channels calculations, while the target nucleus is treated as inert. The solid line includes the full effect of the Coulomb excitation, while the dashed line disregards it by matching the numerical wave functions to the asymptotic wave functions at relatively small distance. The fusion cross sections without the coupling are denoted by the dotted line.

[5, 7]. The simplification is achieved by using one or more of the following approximations: i) the linear coupling approximation, where the nuclear coupling potential is assumed to be linear with respect to an excitation operator for intrinsic motions, ii) multi-phonon excitations are neglected, and iii) the eigenchannel approximation, where the intrinsic excitation energies are treated approximately. Since the effective coupling strength is approximately proportional to $Z_P Z_T$ for a given value of deformation parameter [18], the shortcomings of these approximations become more severe for heavier systems.

The validity of the first and the second approximations has been examined in detail in Refs. [18, 19, 20]. We therefore discuss the third point here. The eigenchannel approach is intimately related to the concept of barrier distribution [1, 8, 21]. Although this approach is exact when the intrinsic excitation energy vanishes, it also works well even with a finite excitation energy as long as the coupling potential is localized inside the uncoupled barrier [22]. In realistic cases, however, the coupling potential extends outside the barrier due to the long range Coulomb interaction. If, prior to reaching the barrier, there is appreciable Coulomb excitation to a collective state whose excitation energy is not small, this results in a decrease of the relative energy, leading to the reduction of fusion cross sections and a modification of effective eigen-barriers. Since the eigenchannel approach treats the excitation energy approximately, this effect will be missed in a calculation.

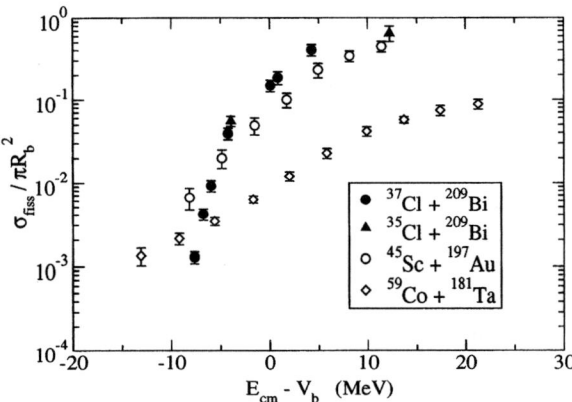

FIGURE 2. Reduced total fission cross sections for four reactions forming Fm, ^{35,37}Cl + ^{209}Bi (the filled circles and triangles), ^{45}Sc + ^{197}Au (the open circles), and ^{59}Co + ^{181}Ta (the open diamonds), as a function of energy measured from the Coulomb barrier height.

In order to demonstrate the effect of Coulomb excitations prior to the barrier, figure 1 compares fusion cross sections for the ^{58}Ni + ^{58}Ni system with those for the ^{58}Ni + ^{208}Pb system. We assume that the target nucleus is inert, and include only the double quadrupole phonon excitation in the projectile ^{58}Ni nucleus. The excitation energy for the single phonon state is 1.45 MeV, and we assume a simple harmonic oscillator coupling for the double phonon excitation. The solid and dashed lines are obtained by integrating the coupled-channels equations from inside the Coulomb barrier to R_{max}=50 fm and 15 fm, respectively. The latter calculation, therefore, effectively disregards the effect of Coulomb excitation outside the Coulomb barrier. For a comparison, we also show the fusion cross sections in the no coupling limit by the dotted line. One clearly sees that the Coulomb coupling considerably alters the fusion cross sections when the target is heavy, while the difference is small for the lighter system. One also notices that the fusion cross sections are even smaller than the no coupling calculations for the heavier system at energies above the Coulomb barrier. Evidently, the Coulomb excitation provides another mechanism of fusion inhibition in massive systems.

Total fission cross sections for Fm formation reactions

Let us now discuss fission cross sections for Z=100 (Fm) formation reactions measured by Ohtsuki *et al.* [10]. The experimental cross sections were obtained by selecting the fission events in the TOF vs ΔE plot, and thus contain both the fusion-fission and the quasi-fission (if any) cross sections. For the heaviest system, the fission events were

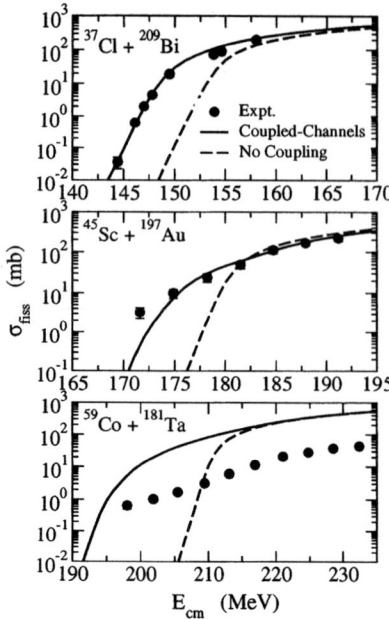

FIGURE 3. Comparison of the coupled-channels calculations with the experimental data for total fission cross sections for the Fm formation reactions. The solid line is the result of the coupled-channels calculations, while the dashed line is obtained without any coupling. The experimental data are denoted by the filled circles.

not well separated from the deep-inelastic collision (DIC) events. For this system, the fission cross sections were estimated by choosing the same mass region of fission fragments as that observed in the ^{35,37}Cl + ^{209}Bi reaction. Therefore, the experimental fission cross sections might be underestimated for the heaviest system if the fission events extend towards the DIC region. Fig. 2 shows the observed reduced fission cross sections as a function of the difference between the center of mass energy and the average barrier energy for four systems, ^{35,37}Cl + ^{209}Bi ($Z_P Z_T$=1411), ^{45}Sc + ^{197}Au ($Z_P Z_T$=1659), and ^{59}Co + ^{181}Ta ($Z_P Z_T$=1971). Notice that the reduced cross sections for the second heaviest system (^{45}Sc + ^{197}Au) are somewhat smaller than those for the lightest system (^{35,37}Cl + ^{209}Bi), and the cross sections for the heaviest system (^{59}Co + ^{181}Ta) are substantially hindered compared with the other two systems.

In order to investigate whether these features can be understood in terms of the Coulomb excitation discussed in the previous subsection, we perform exact coupled-channels calculations. For this purpose, we use an extended version of the computer code CCFULL [2], where the effect of finite ground state spin of colliding nuclei is incorporated within the isocentrifugal approximation. For the ^{37}Cl + ^{209}Bi system, we include three vibrational states, $5/2^+$ (3.09 MeV), $7/2^-$ (3.1 MeV), and $9/2^-$ (4.01 MeV) in ^{37}Cl as well as double octupole vibrations in ^{209}Bi. The deformation parameters are estimated from the experimental B(E2) and B(E3) values. We introduce a single effective channel for the seven octupole states which have a character of $h_{9/2} \otimes {}^{208}$Pb(3$^-$) in ^{209}Bi, and consider a harmonic oscillator coupling for the double phonon state. For the ^{45}Sc + ^{197}Au system, we include 5 quadrupole states which have a character of $f_{7/2} \otimes {}^{44}$Ca(2$^+$) in ^{45}Sc [3/2$^-$ (0.38 MeV), 5/2$^-$ (0.72 MeV), 11/2$^-$ (1.24 MeV), 7/2$^-$ (1.41 MeV), and 9/2$^-$ (1.66 MeV)], and treat the target nucleus ^{197}Au as a classical rotor with $\beta_2 = -0.13$ and $\beta_4 = -0.03$. For ^{45}Sc, since the excitation energies for the quadrupole states are not close to each other, we do not introduce an effective single channel, but treat them exactly. For the ^{59}Co + ^{181}Ta system, we include 5 quadrupole states which have a character of $(f_{7/2})^{-1} \otimes {}^{60}$Ni(2$^+$) in ^{59}Co [3/2$^-$ (1.1 MeV), 9/2$^-$ (1.19 MeV), 11/2$^-$ (1.46 MeV), 5/2$^-$ (1.48 MeV), and 7/2$^-$ (1.74 MeV)], and treat the target nucleus ^{181}Ta as a classical rotor with $\beta_2 = 0.354$ and $\beta_4 = -0.05$. More details of the calculations will be given elsewhere [23].

The results of those calculations are shown in Fig. 3. One finds that the coupled-channels calculations well reproduce the experimental data for the two lightest systems, ^{37}Cl + ^{209}Bi and ^{45}Sc + ^{197}Au. Especially, the reduction of cross sections in the latter system is reproduced nicely. As we discussed in the previous subsection, this small hindrance of cross sections is caused by the strong Coulomb coupling to the collective states outside the Coulomb barrier. In contrast, the coupled-channels calculation considerably overestimates fission cross sections for the heaviest system, ^{59}Co + ^{181}Ta. We will discuss this point in the next subsection.

Role of deep-inelastic collision

In his review article, Reisdorf argued [3] that cross sections for the sum of fusion and other damped reactions may be interpreted as the total barrier passing cross sections. Indeed, he found that the sum of fusion and deep-inelastic collision (DIC) cross sections for the ^{58}Ni + ^{124}Sn reaction could be well reproduced by the standard potential model (see Fig. 4). More recently, Esbensen et al. followed a similar idea and reproduced the experimental cross sections for the sum of fusion and DIC reactions for the same system with the coupled-channels approach [25]. In the semiclassical picture, deep inelastic collisions correspond to those trajectories which overcome the barrier in the entrance channel but eventually escape after appreciable interaction with the target. All of those considerations immediately lead to the idea that the total barrier passing cross sections, which the coupled-channels approach yields, may have to be compared with a sum of fusion, quasi-fission, and DIC, i.e.,

$$\sigma_{bp}(E) = \sigma_{fus}(E) + \sigma_{qf}(E) + \sigma_{DIC}(E). \tag{2}$$

The large reduction of total fission cross sections for the ^{59}Co + ^{181}Ta system, therefore, may be attributed to the competition between fusion, quasi-fission and DIC reactions.

The present framework of the coupled-channels method could be used to obtain inclusive cross sections of fusion and DIC reactions, but it would be difficult to obtain them

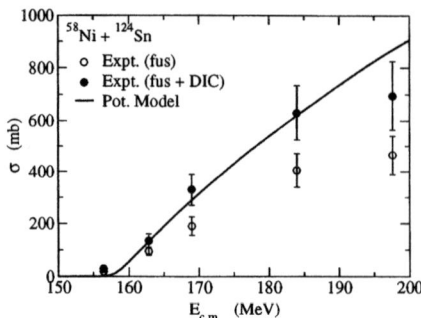

FIGURE 4. Fusion cross sections (the open circles) and the sum of fusion and deep inelastic collision cross sections (the filled circles) for the ^{58}Ni + ^{124}Sn reaction. The solid line is a prediction of the barrier penetration model with a global nucleus-nucleus potential. The experimental data are taken from Ref. [24].

separately. One possible way to isolate fusion from DIC theoretically is to introduce angular momentum truncation in Eq. (1) and exclude higher partial wave contributions, as was done by Zagrebaev et al. [5]. However, a large ambiguity exists in this prescription, since there is no clear and unique way to introduce the angular momentum truncation, especially at energies below the barrier [26]. In recent publications, Abe et al. combined the surface friction model with the Langevin approach in order to take into account the competition between fusion and DIC in the approaching phase for a formation reaction of superheavy elements [4]. This approach may be promising, but is essentially classical. Computation of fusion cross sections with a quantum mechanical model under the influence of DIC process is still an open problem.

SURFACE DIFFUSENESS ANOMALY IN FUSION POTENTIALS

Let us now discuss the second subject, that is the surface property of the nucleus-nucleus potential and its anomaly, recently recognised in fusion reactions for medium-heavy systems. For scattering processes, it seems well accepted that the surface diffuseness parameter a should be around 0.63 fm [27] if the nuclear potential is parametrised by a Woods-Saxon (WS) form. In marked contrast, recent high-precision fusion data suggest that a much larger diffuseness, between 0.8 and 1.4 fm, is needed to fit the data [11, 12]. This is not just for particular systems but seems to be a rather general result [14].

We illustrate this problem in Fig. 5 by comparing experimental data for the ^{16}O + ^{208}Pb fusion reaction with various coupled-channels calculations with the WS potential. We include the double octupole phonon and the single 5^- phonon excitations in ^{208}Pb. At energies well above the barrier, where the fusion cross sections σ_{fus} is relatively in-

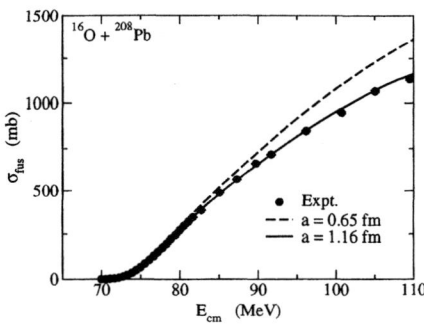

FIGURE 5. Comparison of coupled-channels calculations with the experimental data for the ^{16}O + ^{208}Pb fusion reaction. The double octupole phonon as well as the single 5^- phonon excitations are included. The solid and the dashed lines are obtained by setting the surface diffuseness parameter of Woods-Saxon potential to be a=1.16 and a= 0.65 fm, respectively. The experimental data are taken from Ref. [28].

sensitive to the couplings, a WS potential with a diffuseness a =0.65 fm significantly overestimates fusion cross sections (dashed line). Changing the depth and radius parameters in the WS potential is not helpful, since it merely leads to an energy shift in the calculated fusion cross sections without significantly changing the energy dependence. On the other hand, a potential with a=1.16 fm (full line) fits the data well. A similar problem also exists at energies below the barrier [12, 13], but we do not discuss it in this contribution.

Up to now, several possible reasons for this anomaly have been considered. These include the departure of the nuclear potential from the WS form [11, 12], dissipation effects [12], and unrecognised systematic errors in experimental data [14], but none of them is conclusive yet. We would like to propose here another possible effect, that is, the competition between fusion and deep inelastic collision [14]. This is motivated by an apparent similarity between fusion inhibition in massive systems discussed in the previous section (see fig. 4) and the overestimate of fusion cross sections shown in fig. 5. To this end, let us introduce a suppression factor S which is defined as a ratio between the experimental fusion cross sections and the prediction of the potential model with the standard value for the diffuseness parameter, $a \approx 0.63$ fm. Provided that S is independent of bombarding energy, it can be stated that the fusion cross sections are hindered by a factor S for whatever reason.

The values of S obtained for a large number of systems are shown in fig. 6 [14]. Also shown are two points (large filled stars) at Z_1Z_2 = 1400 derived from the data for the ^{58}Ni + ^{112,124}Sn reactions at energies around the fusion barrier. For these systems, as we indicate in Fig. 4, a very significant contribution from deep-inelastic scattering has been observed experimentally even at energies below the barrier [24]. In this case, the value of S has been taken as σ_{fus} divided by the sum of σ_{fus} and σ_{DIC}. These two points do not

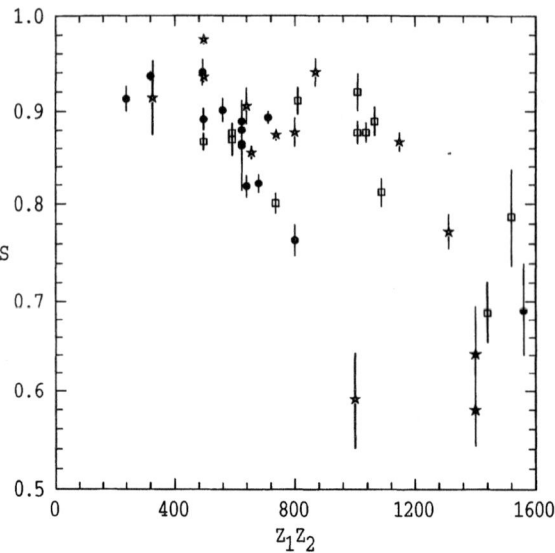

FIGURE 6. Calculated suppression factors S for fusion cross sections with respect to the prediction of potential model with $a \approx 0.63$ fm. The large filled stars at $Z_1 Z_2 = 1400$ refer to the reactions ^{58}Ni + ^{112,124}Sn, where the deep-inelastic collisions have been observed experimentally.

seem inconsistent with the other points with large $Z_1 Z_2$ in Fig. 6, which are derived from the fusion data only. Strongly damped reactions similar to DIC are reported to occur in lighter systems, for example, ^{32}S + ^{64}Ni [29] ($Z_1 Z_2 = 448$) at energies well above the barrier. It would be interesting to see if they also occur at energies closer to the barrier. If they do, the values of S would be reduced below unity and might result in at least a partial explanation of the experimental values for a being much larger than the standard value, 0.63 fm.

SUMMARY

Extensive efforts have been made both experimentally and theoretically to understand fusion of massive systems, especially for synthesis of superheavy elements (SHE), but the reaction mechanism has not yet been completely understood. Here, we have performed coupled-channels calculations and pointed out that the exact treatment of Coulomb excitation becomes important for massive systems. We have applied the coupled-channels framework to three Fm formation reactions, and have shown that Coulomb excitation indeed provides an important mechanism of inhibition of fusion cross sections for transitional systems ("pre-SHE" systems) between medium-heavy and SHE regions. For SHE systems, we have argued that the deep-inelastic collision needs to be taken into account explicitly in theoretical models. However, it is still an open problem of how to incorporate the competition between fusion and DIC quantum mechanically, and further developments will be required. We have also discussed the large surface diffuseness problem observed in the recent high precision measurements of fusion cross sections for medium-heavy systems in connection with the fusion hindrance in massive systems. We have argued that the competition between fusion and deep-inelastic collision may provide a promising avenue to explain this anomaly. In this connection, it would be interesting to study, both experimentally and theoretically, light systems near to the fusion barrier to see whether fusion was inhibited by the presence of DIC. Especially, theoretical calculations involving friction for light systems would be of great interest, since experimental measurements are likely to be difficult when the fusion suppression factor appoaches unity.

REFERENCES

1. M. Dasgupta, D.J. Hinde, N. Rowley, and A.M. Stefanini, Annu. Rev. Nucl. Part. Sci. **48**, 401 (1998).
2. K. Hagino, N. Rowley, and A.T. Kruppa, Comp. Phys. Comm. **123**, 143 (1999).
3. W. Reisdorf, J. Phys. **G20**, 1297 (1994).
4. C. Shen, G. Kosenko, and Y. Abe, Phys. Rev. C**66**, 061602(R) (2002); B. Bourquet, Y. Abe, and G. Kosenko, e-print: nucl-th/0308019.
5. V.I. Zagrebaev, Y. Aritomo, M.G. Itkis, Yu. Ts. Oganessian, and M. Ohta, Phys. Rev. C**65**, 014607 (2002).
6. Y. Aritomo, T. Wada, M. Ohta, and Y. Abe, Phys. Rev. C**59**, 796 (1999).
7. V.Yu. Denisov and S. Hofmann, Phys. Rev. C**61**, 034606 (2000).
8. C.H. Dasso, S. Landowne, and A. Winther, Nucl. Phys. **A405**, 381 (1983); **A407**, 221 (1983).
9. M. Dasgupta *et al.*, Nucl. Phys. **A539**, 351 (1992).
10. Y. Nagame *et al.*, Japan Atomic Energy Res. Inst. Tandem VDG Ann. Rept., 1993, p. 45 (1994); T. Ohtsuki *et al.*, *ibid* p. 46.
11. J.O. Newton, C.R. Morton, M. Dasgupta, J.R. Leigh, J.C. Mein, D.J. Hinde, H. Timmers, and K. Hagino, Phys. Rev. C**64**, 064608 (2001).
12. K. Hagino, M. Dasgupta, I.I. Gontchar, D.J. Hinde, C.R. Morton, and J.O. Newton, *Proc. Fourth Italy-Japan Symposium on Heavy-Ion Physics, Tokyo, Japan* (World Scientific, Singapore, 2002), p. 87; e-print:nucl-th/0110065.
13. K. Hagino, N. Rowley, and M. Dasgupta, Phys. Rev. C**67**, 054603 (2003).
14. J.O. Newton, R.D. Butt, M. Dasgupta, I.I. Gontchar, D.J. Hinde, C.R. Morton, A. Mukherjee, and K. Hagino (to be published).
15. A.C. Berriman, D.J. Hinde, M. Dasgupta, C.R. Morton, R.D. Butt, and J.O. Newton, Nature **413**, 144 (2001).
16. D.J. Hinde, M. Dasgupta, and A. Mukherjee, Phys. Rev. Lett. **89**, 282701 (2002).
17. R.N. Sagaidak *et al.*, Phys. Rev. C**68**, 014603 (2003); A. Yu. Chizhov *et al.*, Phys. Rev. C**67**, 011603 (2003).
18. K. Hagino, N. Takigawa, M. Dasgupta, D.J. Hinde, and J.R. Leigh, Phys. Rev. C**55**, 276 (1997).
19. K. Hagino, N. Takigawa, M. Dasgupta, D.J. Hinde, and J.R. Leigh, Phys. Rev. Lett. **79**, 2014 (1997).
20. A.M. Stefanini *et al.*, Phys. Rev. Lett. **74**, 864 (1995).
21. N. Rowley, G.R. Satchler, and P.H. Stelson, Phys. Lett. **B254**, 25 (1991).
22. K. Hagino, N. Takigawa, and A.B. Balantekin, Phys. Rev. C**56**, 2104 (1997).
23. T. Ohtsuki, H. Ikezoe, Y. Nagame, J. Kasagi, and K. Hagino (to be published).
24. F.L.H. Wolfs, W. Henning, K.E. Rehm, and J.P. Schiffer, Phys. Lett. **B196**, 113 (1987); F.L.H. Wolfs, Phys. Rev. C**36**, 1379 (1987).
25. H. Esbensen, C.L. Jiang, and K.E. Rehm, Phys. Rev. C**57**, 2401 (1998).
26. C.H. Dasso and G. Pollarolo, Phys. Rev. C**39**, 2073 (1989).
27. R.A. Broglia and A. Winther, *Heavy Ion Reactions* (Addison-Wesley, Redwood City, CA, 1991).
28. C.R. Morton *et al.*, Phys. Rev. C**60**, 044608 (1999).
29. G. Russo *et al.*, Phys. Rev. C**39**, 2462 (1989).

Superheavy element production, nucleus-nucleus potential and μ-catalysis

V.Yu. Denisov

Gesellschaft für Schwerionenforschung (GSI), Planckstrasse 1, 64291 Germany, Germany
Institute for Nuclear Research, Prospect Nauki 47, 03680 Ukraine

Abstract. The semi-microscopic potential between heavy nuclei is evaluated for various colliding ions in the approach of frozen densities in the framework of the extended Thomas-Fermi approximation with $\hbar^2$ correction terms in the kinetic energy density functional. The proton and neutron densities of each nucleus are obtained in the Hartree-Fock-BCS approximation with SkM* parameter set of the Skyrme force. A simple expression for the nuclear interaction potential between spherical nuclei is presented. It is shown that muon bound with light projectile induces the superheavy elements production in nucleus-nucleus collisions.

INTRODUCTION

Knowledge of the nucleus-nucleus interaction potential is a key ingredient in the analysis of nuclear reactions. By using the potential between nuclei we can estimate the cross sections of different nuclear reactions [1].

The nucleus-nucleus interaction potential related to the Coulomb repulsion force and the nuclear attraction force has, as a rule, the barrier and the capture potential well near a touching point. The Coulomb part of the ion-ion potential is well-known. In contrast, the nuclear part of the nucleus-nucleus potential is less defined. There are many different approaches to the nuclear part of the interaction potential [1-6]. Unfortunately, barriers evaluated within different approaches for the same colliding system differ considerably, especially when both nuclei are very heavy or one nucleus is very heavy and another is light [7, 8]. The uncertainty of the interaction potential between heavy ions near the touching point gives rise to a variety of proposed nuclear reaction mechanisms. So, there is a need to reduce the uncertainty of the interaction potential around the touching point, especially between heavy nuclei used for the synthesis of superheavy elements (SHEs).

The production cross section of SHEs with $Z \geq 112$ is very low and close to the limit of current experimental possibility [9-13]. Due to this it is of interest to find new types of reactions, which can induce fusion of two heavy nucleus.

In second section of the paper we briefly discuss our semi-microscopic approach for the nucleus-nucleus interaction potential and present some new numerical results. The simple analytical expression for the nuclear potential between two heavy spherical nuclei is presented in section 3. We show in section 4 that muon bound with light nucleus induce SHE formation during nucleus-nucleus fusion reaction.

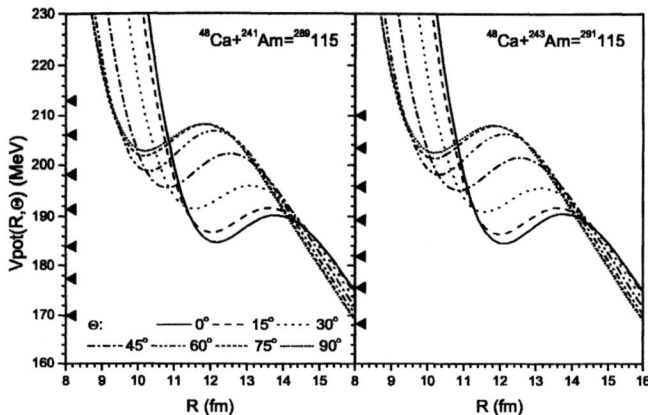

FIGURE 1. The SMPs for the collisions ^{48}Ca on ^{241,243}Am evaluated with the SkM* Skyrme force. The SMPs are evaluated for different angular orientations of the heavy deformed nuclei. The ground-state Q-values are indicated by the lowest triangle at the left vertical axis. The other 6 triangles mark are, respectively, the thresholds for the emission of 1, 2, 3, 4, 5 and 6 neutrons.

SEMI-MICROSCOPIC POTENTIAL AND SHE PRODUCTION

We evaluate the nuclear part of interaction potential between heavy nuclei in the semi-microscopic frozen density approximation due to a short reaction time [7]. The frozen (or sudden) approximation is good for evaluation of the nucleus-nucleus potential near the touching point at collision energies above the barrier height. The shape of each ion cannot appreciably change and the energy of relative motion cannot be strongly transferred to another degrees of freedom during the short reaction time.

The interaction energy between ions is obtained with the help of a local energy density functional. The extended Thomas-Fermi (ETF) approximation with $\hbar^2$ correction terms is used for the evaluation of the kinetic energy density functional [14]. The Skyrme and Coulomb energy density functionals are employed for the calculation of the potential energy. These energy density functionals depend on the proton and neutron densities. These densities in each nuclei are obtained in the microscopic Hartree-Fock-BCS approximation with the Skyrme force. Our approximation is semi-microscopic because we use the microscopic density distributions and the ETF approximation for the calculation of the interaction energy of nuclei. Note that the binding energies of nuclei evaluated in the ETF approximation with the help of microscopic density distributions well agree with those obtained in the fully microscopic Hartree-Fock-BCS model [7, 8]. Therefore, our semi-microscopic method for evaluation of the interaction potential between various nuclei is quite accurate. The details of our method and results for various even-even projectiles and targets are discussed in [7-8,15-16]. Due to this we present here only some new results obtained in our approximation and discussions.

At the beginning we consider the semi-microscopic potentials (SMPs) for hot fusion

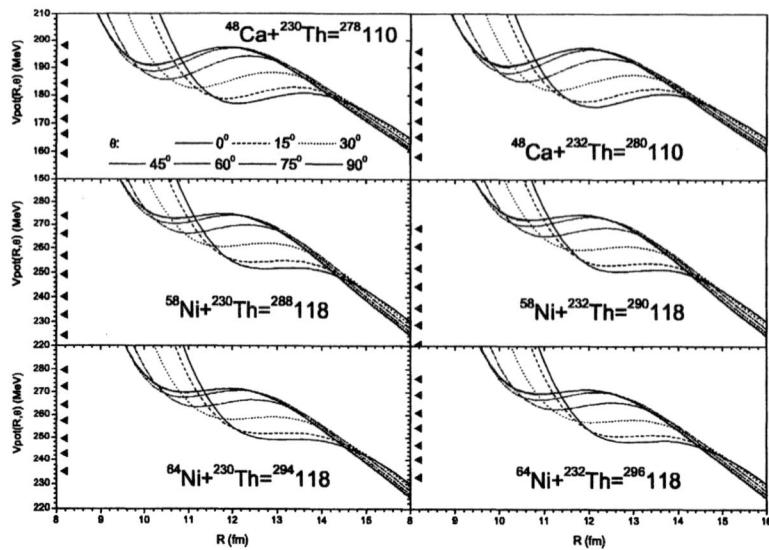

FIGURE 2. The SMPs for the collisions ^{48}Ca on ^{230,232}Th and ^{58,64}Ni on ^{230,232}Th evaluated with the SkM* Skyrme force. The notations are the same as in Fig. 1.

reactions which was used for the formation of element with 115 protons in Dubna very recently [11]. In Fig. 1 we presented SMPs for reactions ^{48}Ca+^{241,243}Am=^{289,291}115 evaluated for SkM* Skyrme force [17]. We see that the potential shape (radial positions of both the barriers and the capture wells for various orientations of deformed ^{241,243}Am and the depths of capture wells) for these reactions are very similar to the ones for reactions between ^{48}Ca and other closest even-even targets ^{238}U, ^{236,238,240,242,244}Pu, ^{248}Cm and ^{252}Cf [7, 15, 16]. However the positions of both the barriers and the capture wells relatively the ground state of compound nucleus for various orientations ^{241,243}Am are slightly lower then the ones evaluated for nearest even-even projectiles ^{240,242,244}Pu, see Fig. 1 in [16]. This is favorable for the SHE formation in reaction with ^{241,243}Am.

A choice of light isotope is very important for the SHE production [7, 15, 16]. A SHE formation reactions induced by ^{48}Ca projectile are limited by availability of heavy transuranic elements in the nature. Due to this it is practically impossible to synthesize element with $Z \geq 120$ by using ^{48}Ca beam. By using Fig. 2 and Figs. 4-7 in [7] and analyzing the chapter of nuclides we may conclude that ^{64}Ni is the nearest nuclide heavier then ^{48}Ca, which we may recommend to use for the SHE production, because of *(1)* ^{64}Ni is easily available as a projectile, *(2)* ^{64}Ni is sufficiently neutron reach, *(3)* ^{64}Ni produces relatively low-excited compound nuclei during fusion with easily available set of target nuclei and *(4)* the number of protons in ^{64}Ni is magic as that in ^{48}Ca. Therefore, reactions with ^{64}Ni projectile (^{64}Ni+X→SHE) should play similar role as reactions with ^{48}Ca (^{48}Ca+X→SHE), which are successfully used now in Dubna [11]. However the capture well for reactions with ^{64}Ni is more shallow then the one for reactions with ^{48}Ca, see Fig. 2. This is may reduce the capture probability for reactions with ^{64}Ni projectile.

By analyzing SMPs for reactions presented in Figs. 1,2 and for other reactions considered in [7, 15, 16] we observe the common features of SMPs for various reactions and make conclusions for reactions leading to the SHEs:

(1) The depth of the pockets is important for the fusion probability. We observe correlation between pocket depth and experimentally observed reduction of SHE formation with increasing size of the projectile.

(2) The pocket depth should be as large as possible for the SHE production, because the deeper pocket has larger the capture window and, therefore, better chance for fusion. Due to this hot fusion reactions has better chance for capture then cold fusion reactions leading to the same SHE.

(3) For the subsequent formation of a compound nucleus it is best to have a most compact capture configuration. Due to this both the cold fusion systems are more preferable then more symmetric systems leading to the same compound nucleus and the side orientation of deformed nucleus is more preferable for the SHE formation then the tip orientation of deformed nucleus in the case of the hot fusion reactions.

(4) The observed fusion windows lie systematically about 5 to 10 MeV below our barriers.

(5) The isotopic composition of heavy transuranic nuclei only weakly affects both the shape of the capture well and barrier heights for different orientations relatively the ground-state fusion reaction Q-value.

(6) The capture properties of SMP and compound-nucleus excitation energy in the hot fusion reactions strongly depend on the isotopic composition of light nuclide.

(7) The difference between the capture barrier position and the ground-state fusion reaction Q-value decreases with increasing as charge as number of neutrons of the projectile for reactions with the same target.

(8) Pocket depth is vanished with increasing size of the projectile. For example in the case of ^{208}Pb target, there are no pocket for ^{96}Zr+^{208}Pb and more heavy projectiles.

(9) The potential pockets for system leading to SHE are much shallower than the ones for more lighter colliding systems.

(10) The interaction potentials obtained by using different standard expressions [1-6] are spread over large interval for heavier systems.

ANALYTICAL EXPRESSION FOR POTENTIAL

The numerical evaluation of SMP is very accurate for determination of interaction potential between nuclei around the touching point. Unfortunately, it is not so convenient for any practical application because one needs to evaluate numerically the microscopic Hartree-Fock-BCS nucleonic densities of interacting nuclei, derivatives of these densities and integrals [7, 8]. It is better to find analytical expression for the potential. To solve this task we choose 119 spherical or near spherical nuclei along the β-stability line from ^{16}O to ^{212}Po and perform numerical calculations of the interaction potentials between all possible nucleus-nucleus combinations in the semi-microscopic approximation. We evaluate potential for any nucleus-nucleus combinations at 15 distances between ions around the touching point. By using database for 7140 nucleus-nucleus potentials at 15 points each, we find a simple analytical expression for the potential between spherical nuclei in the form [8]

$$V(R) = -1.989843 \, C \, f(R - R_{12} - 2.65) \qquad (1)$$
$$\times \left[1 + 0.003525139(A_1/A_2 + A_2/A_1)^{3/2} - 0.4113263(I_1 + I_2)\right],$$

where R is the distance between mass centers of colliding nuclei, $C = R_1 R_2 / R_{12}$, R_i is the effective nuclear radius, $R_{12} = R_1 + R_2$,

$$f(s) = \left\{1 - s^2 \left[0.05410106 \, C \exp\left(-\frac{s}{1.760580}\right) - 0.5395420 \, (I_1 + I_2)\right.\right. \qquad (2)$$
$$\left.\left.\times \exp\left(-\frac{s}{2.424408}\right)\right]\right\} \times \exp\left(\frac{-s}{0.7881663}\right), \quad \text{for } s \geq 0,$$

$$f(s) = 1 - \frac{s}{0.7881663} + 1.229218 s^2 - 0.2234277 s^3 - 0.1038769 s^4 \qquad (3)$$
$$- C(0.1844935 s^2 + 0.07570101 s^3)$$
$$+ (I_1 + I_2)(0.04470645 s^2 + 0.03346870 s^3), \quad \text{for } -5.65 \leq s \leq 0,$$

A_i is the number of nucleon in nucleus i ($i = 1, 2$), $I_i = (N_i - Z_i)/A_i$, Z_i and N_i are numbers of protons and neutrons in nucleus i. The effective nuclear radius is given by

$$R_i = R_{ip}(1 - 3.413817/R_{ip}^2) + 1.284589(I_i - 0.4 A_i/(A_i + 200)), \qquad (4)$$

where the proton radius is determined as in [18] $R_{ip} = 1.24 A_i^{1/3}(1 + 1.646/A_i - 0.191 I_i)$. The last term in (4) takes into account deviation of the nuclear radius from the proton radius when the neutron number in nucleus deviates from the β-stability value for given A. The line of β-stability is described by Green's approximation $I = (N - Z)/A = 0.4 A/(A + 200)$ [19]. Note that the potentials obtained by means of the analytical expression well agree with semi-microscopic one [8].

The "empirical" fusion barrier B_{empir} and "empirical" barrier radius R_{empir} between heavy nuclei are extracted by means of a special analysis of the experimental data for subbarrier fusion reactions in Ref. [20]. The absolute and relative differences between "empirical" fusion barrier and barrier evaluated by using various analytical expressions [1-6,8] B_{theor} are presented in Fig. 3. In Fig. 4 we present the absolute and relative differences between "empirical" barrier radius and radii of barrier evaluated by using various analytical expressions [1-6,8] R_{theor}. The barriers B_{theor} and radii R_{theor} evaluated by using Eqs. (1)-(4) well agree with "empirical" the ones respectively, see Figs. 3-4. The distributions of deviations $B_{empir} - B_{theor}$, $(B_{empir} - B_{theor})/B_{empir}$, $R_{empir} - R_{theor}$ and $(R_{empir} - R_{theor})/R_{empir}$ are almost symmetric with respect to the lines $B_{empir} - B_{theor} = 0$ or $R_{empir} - R_{theor} = 0$ correspondingly for the case of using analytical expression for SMP, see Figs. 3-4. This also suggests the reliability of A- and Z-dependencies of our expression for the nucleus-nucleus potential. In contrast to this similar distributions

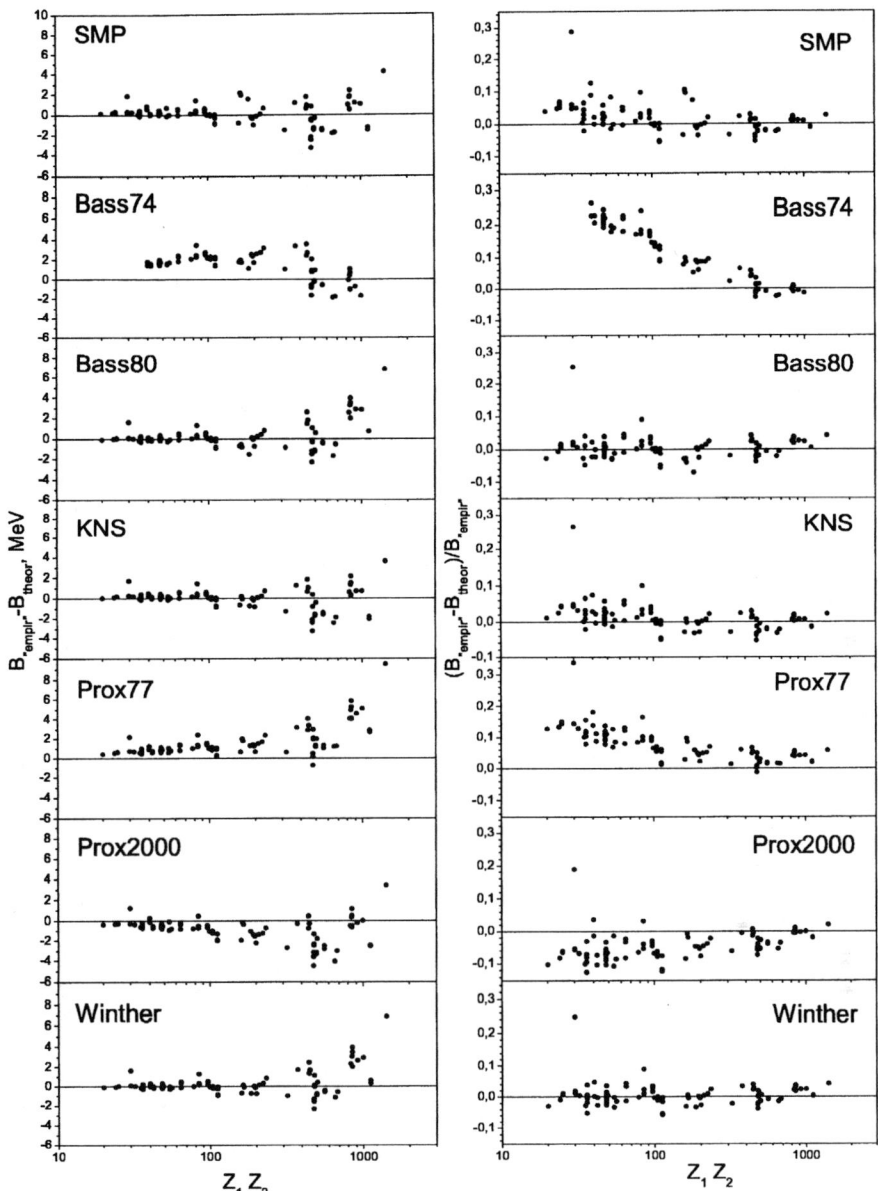

FIGURE 3. The absolute (left panels) and relative (right panels) differences between the "empirical" fusion barrier and the barrier evaluated by using various analytical expressions (SMP - [8], Bass74 - [2], Bass80 - [1], Prox77 - [3], Prox2000 - [5], KNS - [4] and Winther - [6]).

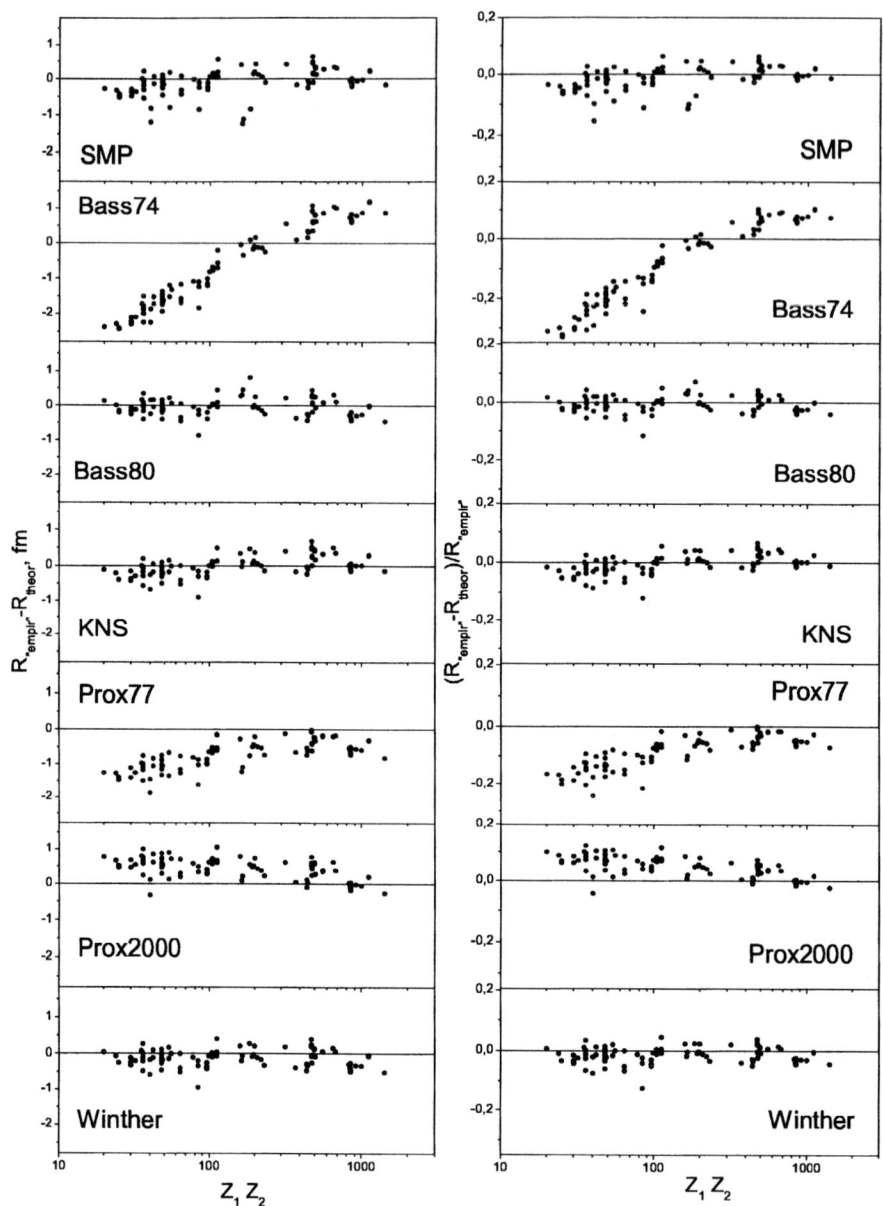

FIGURE 4. The absolute (left panels) and relative (right panels) differences between the "empirical" radius of barrier and the barrier radius evaluated by using various analytical expressions (SMP - [8], Bass74 - [2], Bass80 - [1], Prox77 - [3], Prox2000 - [5], KNS - [4] and Winther - [6]).

for the proximity-1977 [3], proximity-2000 [5] and Bass-1974 [2] potentials have no symmetry with respect to the lines $B_{\text{empir}} - B_{\text{theor}} = 0$ and $R_{\text{empir}} - R_{\text{theor}} = 0$, see corresponding panels in Figs. 3-4. Note that it is impossible to evaluate both the depth and the width of capture well by using Bass-1974 [2], Bass-1980 [1], Krappe-Nix-Sierk [4] and Winther [6] potentials because the shape of these potentials are unrealistic or unknown at distances smaller then the touching point distance of two nuclei.

CATALYSIS OF THE SHE SYNTHESIS BY MUON

It is easy to understand qualitatively a influence of muon μ^- on the SHE fusion process, if we recollect that the wave function of $1s$ state of μ^- in a very heavy nucleus is located inside the nucleus [21]. Therefore, negatively-charged muon inside heavy nucleus should effectively reduce the Coulomb repulsion between protons. Due to this the forces, inducing fission of compound nucleus and preventing fusion of two nuclei should decrease. Consequently, the SHE formation probability should rise due to μ^-.

Let's we consider in detail the catalysis of the SHE production by μ^-. The process of SHE formation is subdivided into three steps [22-27]: *(1)* The capture of two nuclei in an entrance-channel potential well and formation of a common nuclear system of two touching nuclei. *(2)* The formation of a spherical or nearly spherical compound nucleus during shape evolution from the common nuclear system of two touching nuclei to a compound nucleus. *(3)* The surviving of the excited compound nucleus due to evaporation of neutrons and γ-ray emission in competition with fission.

The capture process depends on both the barrier thickness and pocket shape of entrance-channel potential between nuclei [7]. The shape evolution step is determined by a potential-energy landscape between the touching configuration of two colliding nuclei and the compound-nucleus shape [22-27]. Decay properties of the compound nucleus drastically depend on the fission barrier height [22-27]. Therefore, enhancement of the SHE production in fusion reaction may be achieved by processes which *(1)* make capture pocket dipper and barrier of entrance-channel potential thinner, *(2)* increase the slope or reduce both barrier height and thickness of the potential-energy landscape between touching configuration of two colliding nuclei and compound nucleus, *(3)* increase the fission barrier height. Below we show that these three conditions can be met in a reaction between a light nucleus with captured μ^--meson L_μ and a heavy nucleus T.

The potential energy of $L_\mu + T$ system before touching can be approximated as

$$E_{L_\mu T}(R) = B_L + B_T + B_{L\mu} + V_{LT}(R) + V_{T\mu}(R), \qquad (5)$$

where B_L and B_T are the binding energies of light L and heavy T nuclei, respectively, $B_{L\mu}$ is the binding energy of muon in the light nucleus L, $V_{LT}(R)$ is the interaction potential between the light and heavy nuclei related to Coulomb and nuclear forces at distance R between their mass centers [7, 8], and $V_{T\mu}(R) = -e^2 Z_T/R$ is the Coulomb interaction between Z_T protons in the heavy nucleus and the muon. The potential energy of the compound nucleus with bound μ^- is connected with the binding energy of the compound nucleus B_{CN} and with that of muon in the compound nucleus $B_{CN\mu}$, i.e., $E_{CN} = B_{CN} + B_{CN\mu}$.

The potential-energy landscape evaluated relatively to the ground state of compound nucleus with bound μ^-, which formed during $L_\mu+T$ fusion reaction, is related to

$$\delta(R) = E_{L_\mu T}(R) - E_{CN} = \delta_N(R) + \delta_{N\mu}(R). \tag{6}$$

Here we split $\delta(R)$ into contributions of pure nuclear $\delta_N(R)$ and muon-nuclear $\delta_{N\mu}(R)$ subsystems, where

$$\delta_N(R) = B_L + B_T - B_{CN} + V_{LT}(R), \qquad \delta_{N\mu}(R) = B_{L\mu} - B_{CN\mu} + V_{T\mu}(R). \tag{7}$$

We see that the Coulomb interaction between muon and protons modifies the potential-energy landscape of fusing system. (Note that realistic landscape of potential-energy surface of fusing system depends on a great number of various collective coordinates. However, we take into account only the most important collective coordinate, which describes the distance between mass centers of separated nuclei or elongation of fusing system upon the capture step.)

Note that $\delta_{N\mu}(R) = B_{L\mu} - B_{CN\mu} - Z_T e^2/R$ is decreased with reduction of R. At distance R_{CN}, which corresponds to the distance between left and right mass centers of compound nucleus, $\delta_N(R_{CN}) = \delta_{N\mu}(R_{CN}) = 0$ because we evaluate $\delta(R)$ relatively to the ground state of compound nucleus with bound μ^-. Note that $B_{L\mu} - B_{CN\mu} > 0$, therefore $\delta_{N\mu}(R)$ is mainly reduced with R. Consequently if $\delta_{N\mu}(R)$ continuously decreases with reducing of R, then muon induces the SHE formation due to three effects:

(1) A more dipper capture pocket is formed as a result of such R dependence of $\delta_{N\mu}(R)$. Therefore, the capture state formation probability increases.

(2) The potential-energy landscape of the muon-nuclear system becomes more favorable for shape evolution from captured states of two touching nuclei to the compound nucleus.

(3) The muon-nuclear system exhibits a larger fission barrier height as compared to pure nuclear system, see also [28] and papers cited therein. Note that variation of the fission barrier height on 0.3 MeV lead to change of SHE production cross section on approximately 30% [22]. Due to muon the fission barrier of heavy nucleus is increased near to 1 MeV [28]. Consequently, the fission or quasi-fission probability of muon-nuclear system get reducing approximately on one order as compared to the pure nuclear system. Therefore only due to this effect the SHE production cross section may rise to one order.

μ^- is a convenient particle for inducing compound-nucleus formation in reactions $L_\mu + T \to SHE + xn + e^- + \overline{\nu}_e + \nu_\mu$, because its lifetime ($\approx 2.2 \times 10^{-6}$s [21]) is sufficient for making 1s bound state with a light projectile nucleus just before the collision with a target and induce fusion reaction. The process of SHE formation during nucleus-nucleus collision is fast relatively typical μ^- dynamic time. Therefore there is high probability of population of 1s bound state of μ^- in SHE during nuclear reaction time. The compound nucleus relatively rarely excited during the decay μ^- ($\mu^- \to e^- + \overline{\nu}_e + \nu_\mu$ [21]). It is possible to make beam of muonic projectile L_μ by merging beams of strongly ionized projectile nucleus L and of μ^- at the same velocities before the target.

Note that muon catalysis of thermonuclear reactions is also related to effective reduction of the Coulomb repulsion between protons and is well studied both theoretically and experimentally (see [21] and papers cited therein). Muon catalysis of thermonuclear reactions between two hydrogen isotopes is mainly related to reduction of both fusion barrier heights and thickness. In contrast to this muon catalysis of SHE production is connected with more complex processes as reduction of fusion barrier thickness, modification of capture pocket, variation of potential-energy landscape between capture and compound-nucleus shapes and rising of the fission barrier height.

The author would like to thank W. Nörenberg and Yu. Ts. Oganessian for useful discussions and communications. Author gratefully acknowledges support from GSI.

REFERENCES

1. R. Bass, *Nuclear reactions with heavy ions* (Springer-Verlag, Berlin 1980).
2. R. Bass, Nucl. Phys. **A231**, 45 (1974).
3. J. Blocki, J. Randrup, W.J. Swiatecki, C.F. Tang, Ann. Phys. (N.Y.) **105**, 427 (1977).
4. H.J. Krappe, J.R. Nix, A.J. Sierk, Phys. Rev. **C20**, 992 (1979).
5. W.D. Myers, W.J. Swiatecki, Phys. Rev. **C62**, 044610 (2000).
6. A. Winther, Nucl. Phys. **A 594**, 203 (1995).
7. V.Yu. Denisov, W. Nörenberg, Eur. Phys. J. **A15**, 375 (2002).
8. V.Yu. Denisov, Phys. Lett. **B526**, 315 (2002).
9. S. Hofmann, G. Münzenberg, Rev. Mod. Phys. **72**, 733 (2000).
10. P. Armbruster, Ann. Rev. Nucl. Part. Sci. **35**, 135 (1985); **50**, 411 (2000).
11. Yu.Ts. Oganessian, et al., Phys. Rev. **C62**, 041604 (2000); Phys. Rev. **C63**, 011301 (2001); submitted to Phys. Rev. C.
12. P.A. Wilk, et al., Phys. Rev. Lett. **85**, 2697 (2000); W. Loveland, et al., Phys. Rev. C66, 044617 (2002); K.E. Gregorich, et al., Phys. Rev. C 67, 064609 (2003).
13. K. Morita, et al., *Proc. Symposium on Nuclear Clusters: from Light Exotic to Superheavy Nuclei*, Rauischholzhausen Castle, Germany, Eds. R. Jolos, W. Scheid, (EP Systema, Debrecen, Hungary, 2003), 359; RIKEN Accelerator Progress Report 2002, p. 90, 91 (2003)
14. M. Brack, C. Guet, H.-B. Hakanson, Physics Reports **123**, (1985) 275.
15. V.Yu. Denisov, *Proc. 5 Int. Conf. on Dynamical Aspects of Nuclear Fission*, Casta-Papiernicka, Slovak Republic, Eds. J. Kliman, M.G. Itkis, S. Gmuca (World Scientific, Singapore, 2002) p. 99.
16. V.Yu. Denisov, *Proc. Symposium on Nuclear Clusters: from Light Exotic to Superheavy Nuclei*, Rauischholzhausen Castle, Germany, Eds. R. Jolos, W. Scheid, (EP Systema, Debrecen, Hungary, 2003), 427.
17. J. Bartel, Ph. Quentin, M. Brack, C. Guet, H.-B. Hakansson, Nucl. Phys. **A386**, 79 (1982).
18. B. Nerlo-Pomorska, K. Pomorski, Z. Phys. **A348**, 169 (1994).
19. A.E.S. Green, *Nuclear Physics* (McGraw-Hill, New York, 1955).
20. L.C. Vaz, J.M. Alexander, G.R. Sachler, Phys. Rep. **69**, 373 (1981).
21. Y.N. Kim, *Mesic atoms and nuclear structure*, North-Holland Publ.Co., Amsterdam, 1971.
22. V.Yu. Denisov, S. Hofmann, Phys. Rev. **C61**, 034606 (2000); Acta Phys. Pol. **B31**, 479 (2000).
23. V.Yu. Denisov, *Proc. Int. Workshop on Fusion Dynamics at the Extremes*, Dubna, 25-27 May 2000, ed. by Yu.Ts. Oganessian, V.I. Zagrebaev (World Scientific, Singapore, 2001), 203.
24. V.Yu. Denisov, *Proc. Tours Symposium on Nuclear Physics. IV*, Tours, France, 4-7 September, 2000, ed. by M. Arnould et al., (AIP Conference Proceedings, New York, 2001) vol. **561**, 433.
25. V.Yu. Denisov, Prog. Part. Nucl. Phys. **46**, 303 (2001).
26. V.Yu. Denisov, J. Nucl. Radiochem. Sci., **3**, 23 (2002).
27. V.Yu. Denisov, *Proc. The Nuclear Many-Body Problem 2001*, NATO Advanced Research Workshop, Brijuni, Croatia, Eds. W. Nazarewicz, D. Vretenar, (Kluwer Academic Publishers, Dordrecht, 2002) NATO Science Series, 2. Matematics, Physics and Chemistry - Vol. 53, 305.
28. G. Leander, P. Möller, Phys. Lett. **57B**, 245 (1975).

Shell Stabilization in Compound Nucleus Survival

A.R. Junghans[1], K.-H. Schmidt[2], A.M. Heinz[3], A.V. Ignatyuk[4],

[1] Forschungszentrum Rossendorf, Postfach 510119, 01324 Dresden, Germany
[2] GSI, Planckstr. 1, 64291 Darmstadt, Germany
3) Wright Nuclear Structure Laboratory, Yale Univ., 272 Whitney Ave., New Haven, CT 06520, USA
[4] IPPE Obninsk IPPE, Bondarenko sq.1, Obninsk, Kaluga region, 249033, Russia

Abstract. The formation of evaporation residues from compound nuclei is ultimately limited by the competition of fission and particle evaporation. In the super-heavy elements (SHE) nuclear structure effects play a paramount role in the stabilization against spontaneous fission, e.g. the fission barrier consists entirely of the ground-state shell effect. This work investigates the influence of this stabilization on the formation of SHE. This shell stabilization against fission for the predicted SHE island around N≈184 cannot be investigated directly. The best probing ground reachable is the transition region from spherical to deformed actinides around N=126. There, the liquid-drop part of the fission barrier and shell-effects are of comparable magnitude. The production of these nuclei has been thoroughly investigated at GSI in fusion-evaporation reactions of different projectile-target combinations leading to thorium isotopes, in projectile fragmentation of uranium, and in electromagnetic-induced fission of proton-deficient actinides between astatine and uranium. In all these experiments the expected stabilization against fission through the spherical 126-neutron shell is not found. Collective effects in the level density tend to cancel the shell stabilization. We conclude that spherical SHE loose at least great part of their stability against fission at rather low excitation energies and angular momenta.

INTRODUCTION

Nuclear structure plays a significant role in the existence of SHE as their mere stability against spontaneous fission is a product of the shell stabilization. There is no liquid-drop barrier. In the fusion reaction, the minimum excitation energy E^*_{min} of the compound nucleus (CN) is determined by shell effects in the projectile, target and CN. Stronger binding in projectile and target reduces E^*_{min} while a strong shell effect in the compound nucleus will increase the minimum excitation energy of the compound system.

But nuclear structure is also paramount after the amalgamation of projectile and target. The competition of fission and particle evaporation in the deexcitation of the CN is determined by details in the nuclear level densities that are governed by the nuclear structure of the nuclei in question. The fission probability P_f is given by the

$$P_f = \frac{\Gamma_f}{\Gamma_f + \Gamma_n}; \frac{\Gamma_n}{\Gamma_f} \propto \frac{\rho_n(E-S_n)}{\rho_f(E-B_f)}$$

quotient of the fission decay width Γ_f and the sum of fission and particle (neutron) emission decay widths.

The ratio of the decay widths is proportional to the level densities of the two neighboring nuclei in very different configurations: the CN at its (strongly) deformed saddle point, ρ_f, and the daughter nucleus after neutron evaporation, ρ_n in its spherical or deformed ground state.

The level-density at the saddle-point depends on the height of the fission barrier B_f. If the fission-barrier height increases, the level density at saddle will decrease, the fission width will be reduced, and a smaller fission probability would result.

Self-consistent mean-field models predict a region of spherical SHE between $N = 172$ and $N = 184$ as well as $Z \approx 120$, for an overview see [1]. These SHE have been elusive to experiment so far. But spherical fissile nuclei with a large ground-state shell effect exist also in the region of neutron-deficient actinides around the 126-neutron shell. In this range, the nuclei are not entirely shell stabilized, but the liquid-drop fission barrier and the ground-state shell effects are of comparable magnitude. A ground-state shell effect will raise the fission barrier by about 6 MeV for nuclei around 126-neutron shell, and therefore the fission probability is expected to be significantly reduced. So these nuclei can be used as a test case for the effect of shell stabilization against fission.

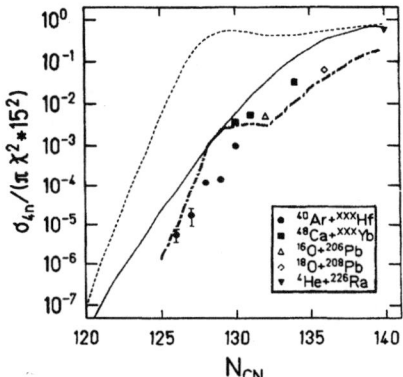

FIGURE 1. Maximum 4n evaporation-residue cross sections for Th-isotopes as a function of CN neutron number from fusion reactions with varying asymmetry [2]. The geometrical scaling of the fusion cross section is divided out. The ordinate corresponds approximately to the survival probability for central collisions. The dotted line corresponds to an evaporation calculation with shell effects including damping with excitation energy. The full line shows a calculation using liquid-drop properties only. The dash-dotted line shows an evaporation calculation with shell effects and rotational levels included.

In several heavy-ion fusion studies, for the complete references see[2], thorium compound nuclei have been produced with different projectile and target combinations, and the evaporation residue cross sections were measured.

Figure 1 shows the 4n neutron evaporation cross sections from different reactions with Ar, Ca, O and He projectiles. The geometrical scaling of the fusion cross section has been divided out, so that cross sections from different projectile-target combination should be comparable. The cross sections tend to decrease with decreasing neutron number of the CN nucleus. The standard evaporation calculation clearly overestimates the evaporation residue cross sections around the 126-neutron shell. An unrealistic calculation - based on the liquid drop properties only – disregarding shell effects in masses and level density does describe the trend of the data much better. A calculation including rotational enhancement in the level density does describe the data quite well, which is indicating that collective enhancement of the level density is significant for describing fission of nuclei around the 126-neutron shell. Collective enhancement is the increase in the level density due to collective vibrations and rotations that is not included in the intrinsic level density.

PRODUCTION CROSS SECTIONS OF ^{238}U PROJECTILES

Neutron-deficient actinides around the 126-neutron shell can be produced by projectile fragmentation of relativistic ^{238}U. In this reaction long chains of isotopes are produced as projectile fragments from a collision with the target nuclei.

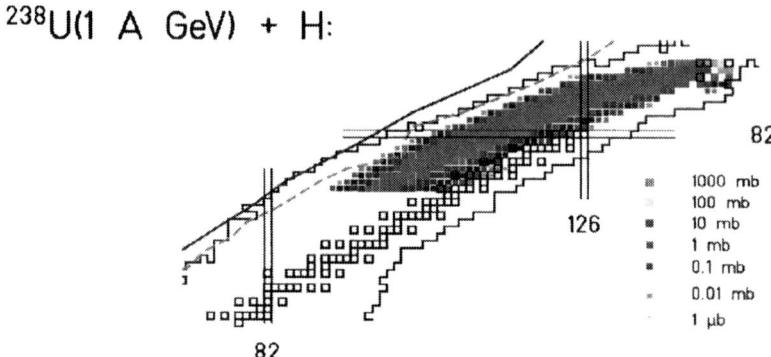

FIGURE 2. Measured production cross section of projectile fragments from ^{238}U on a hydrogen target[3] shown as clusters on a chart of nuclides.

Figure 2 shows the measured production cross sections from ^{238}U on a hydrogen target on a chart of nuclides[3]. The data have been measured at the fragment separator FRS at GSI, Darmstadt. A ^{238}U beam of about 10^7/s particles impinges on a target in front of the magnetic spectrometer FRS. Mass and charge of the projectile fragments are determined by measuring in flight the magnetic rigidity of the fragments, the energy loss in an intermediate energy degrader and the time of flight in the second stage of the FRS ($B\rho - \Delta E - tof$ method). Fragmentation does induce high excitation energy into the fragments of 30 MeV per abraded nucleon on the average, while the angular momentum induced is on the order of 10-15 $\hbar$. These values of the average excitation energy and angular momentum were deduced with the abrasion-ablation model, intranuclear cascade models tend to give higher values by approximately a factor of 2. Nuclear-structure effects are observable nevertheless, as the fragment looses energy through particle evaporation or fission in the deexcitation process until the final, observed nucleus is formed. The survival probability is the product over all decay steps, and nuclear-structure effects influence the final steps of each evaporation chain. Fragmentation is independent of entrance-channel effects that can play a role in fusion studies, like incomplete fusion. The data in Figure 2 do not show an enhancement of the production cross sections for nuclei around the 126-neutron shell which would be expected from a shell stabilization against fission.

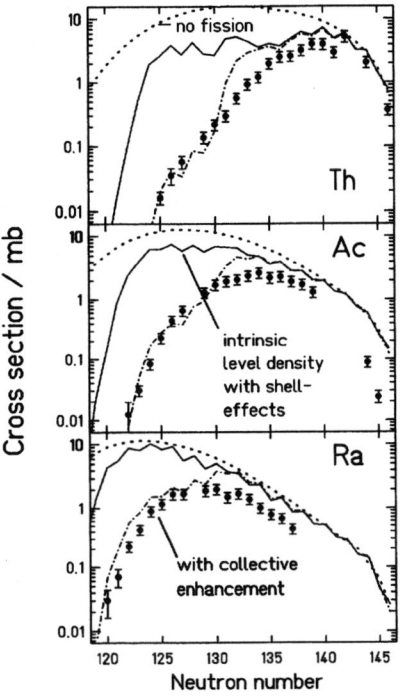

FIGURE 3. Production cross sections of thorium, actinium and radium isotopes from the reaction ^{238}U(950 A MeV) + Cu. compared to abrasion-ablation model calculations. The full lines are calculated with shell effects included in masses and level densities. The dash-dotted curve is with shell effects and collective enhancement in the level density. The dotted lines show a calculation where fission is artificially switched off in the evaporation part.

Production cross sections from the reaction ^{238}U(950 A MeV) + Cu have been compared with model calculations of the abrasion-ablation model including fission [4]. A calculation with fission switched off, which describes the fragmentation of e.g. Pb projectiles quite well, cannot describe the fragmentation of U. Fission clearly reduces the production cross sections of the uranium fragments. The standard calculation that includes shell effects in masses and level densities predicts too high production cross sections for the neutron-deficient nuclei in the neighborhood of the 126-neutron shell, due to the increase of the fission barriers. When collective enhancement is introduced in the level densities, then the model can describe the data sufficiently well [4]. In deformed shapes the collective enhancement is due to rotational levels and enhances the level density by a factor of 60 - 80. In spherical shapes the enhancement is due to vibrations and only amounts to a factor of 1 – 10. In the model there is an artificial step. Nuclei with a quadrupole deformation $\beta_2 \leq 0.15$ are treated as spherical nuclei, at this deformation the collective enhancement is not continuous. Figure 4 shows the collective enhancement factors deduced for all the nuclei investigated experimentally. One can distinguish the regions of ground-state deformed neutron-rich actinides above $N = 135$ and the region of spherical nuclei around $N = 126$.

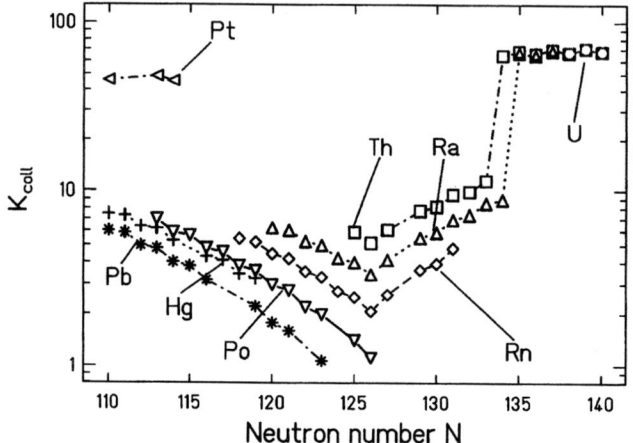

FIGURE 4. Collective enhancement for the ground states of the nuclei investigated experimentally from the reaction ^{238}U(950 A MeV) + Cu [4]. The large collective enhancement for deformed shapes is due to rotations, while the smaller enhancement for spherical shapes is due to vibrational excitations.

The rotational enhancement is calculated assuming axially symmetric deformations, the magnitude of the vibrational enhancement is adjusted to describe the production cross sections. It therefore can be regarded as an empirical information on the level density in spherical nuclei. The damping of the collective enhancement with excitation energy is found to be independent of deformation with a critical energy of the damping described by a Fermi function $E_{cr} = 40$ MeV, for details of the calculations, see ref. 4).

ELECTROMAGNETICALLY INDUCED FISSION CROSS SECTIONS OF SECONDARY PROJECTILES

In a secondary-beam experiment fission cross sections of 58 neutron-deficient actinides have been measured at GSI, Darmstadt [5]. The identified secondary beams, as described in the previous section, were investigated behind the fragment separator FRS with a setup shown in Figure 5.

Fission was induced in-flight in an active lead target, which has a high cross section for electromagnetic excitation of the incoming secondary beam. It worked as a subdivided ionization chamber allowing to determine the location where fission took place (lead foil or plastic scintillator) by the different energy-loss signals of the unfissioned secondary projectile and the two fission fragments. Both fission fragments were counted in a double plastic scintillator giving a trigger on fission events, while the charge number $(Z/\Delta Z \approx 120)$ of the fission fragments was measured through their energy loss in a twin multi-sampling ionization chamber. A time-of-flight measurement of the fission fragments with a plastic-scintillator ToF wall was used to measure the average total kinetic energy and also required to deduce the charge number of the fission fragments from their energy loss. As the fission fragments are kinematically focused in the forward direction, the set up had a total detection efficiency of 90%.

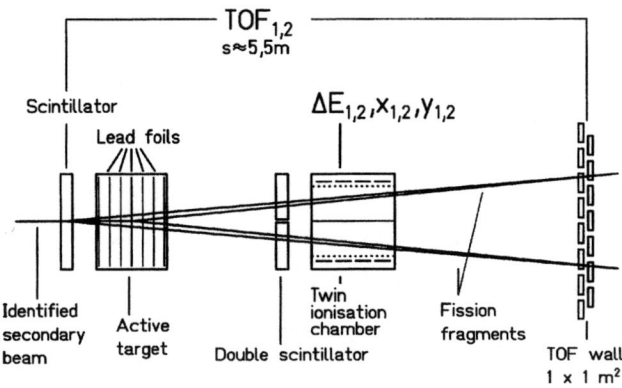

FIGURE 5. Experimental setup behind the Fragment Separator FRS for the secondary beam fission experiment. Fission of the identified secondary projectiles is induced in flight at approx. 420 A MeV in the lead foils and the charges of both fission fragments are determined by the energy loss in the twin ionization chamber and time of flight.

Figure 6 shows sum spectra of the fission-fragment charges as measured with the twin ionization chamber for different secondary beams. The spectrum of the ^{233}U beam that is fissioning in lead is dominated by a large peak at $Z_{sum}=92$ which corresponds to the charge number of the beam ($Z_{pro}=92$). This is typical for fission taking place after electromagnetic excitation of the Giant Dipole Resonance (GDR). The excitation energy where fission takes place from the decay of the GDR is around 11 MeV, which is not sufficient for the nucleus to evaporate a proton. The peaks at $Z_{sum}< 92$ are very small. They are due to nuclear collisions with the lead target nuclei, where higher excitation energies were induced and protons were lost, before fission took place. If the ^{233}U beam fissions in the plastic scintillator, the Z_{sum} spectrum does not have a peak that is much larger than the peak at $Z_{sum}<Z_{pro}$. The electromagnetic excitation cross section in collisions with hydrogen or carbon nuclei in the plastic is vanishing due to the small charge of the target nuclei, so this spectrum is dominated

by nuclear-induced fission events. The ^{214}Ra Z_{sum} spectrum shows that electromagnetic-induced fission is much weaker for this nucleus on the 126-neutron shell as the peak in the lead spectrum is not so dominant anymore. The Z_{sum} spectra of nuclear-induced fission in the plastic scintillator were used to scale and subtract the nuclear-induced fission in the lead to get pure electromagnetically induced fission cross sections. The total fission cross sections were determined with the help of the double scintillator requiring that fission occurred in the lead foils. For details of the separation of nuclear and electromagnetically induced fission, see ref. 5).

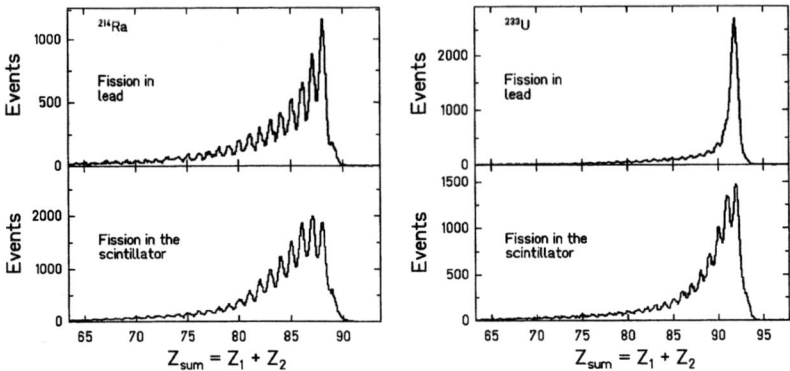

FIGURE 6. Sum spectra of fission-fragment charges for the secondary beams ^{214}Ra and ^{233}U. The active target allows to gate the spectra on fission events occurring in the lead foils or in the plastic scintillator in front of the target.

In this way electromagnetic fission cross sections and total fission cross sections could be determined.

Figure 7 shows the results for all nuclei measured. The cross sections increase with increasing fissility of the nuclei. They show a smooth trend over the 126-neutron shell. There is no decrease in the electromagnetic fission cross section due to the increasing fission barrier at the shell. The total fission cross sections agree within errors with the results from a previous experiment. The electromagnetic fission cross sections test the level densities at a few MeV above the barrier. Only first- and second-chance fission contribute here. The model description of these data is quite complicated as they depend strongly on the magnitude and the washing out of the shell effects and the collective enhancement.

With the help of the electromagnetic excitation cross section given by the method of virtual photons an average fission probability can be deduced from the measured

$$P_{fis} = \frac{\sigma_{fis}^{em}}{\int_{B_f}^{\infty} \frac{d\sigma}{dE^*} dE^*}$$

electromagnetic fission cross sections σ_{fis}^{em}. The denominator is the integral over the theoretical electromagnetic excitation cross section starting at the fission barrier and includes excitation of the GDR, the two phonon excitation of the GDR and other giant resonances. It is assumed that sub-barrier fission can be neglected. Details are given in ref. 7 and references therein. The present results indicate that shell effects in spherical nuclei do not decrease the fission probability from excited states, even if situated only slightly higher than the fission barrier. These nuclei practically behave like fictive nuclei with liquid-drop binding energies and fission barriers and Fermi-gas level densities. The stabilizing influence resulting from the higher fission barrier seems to be compensated or even over-compensated by the destabilizing effect of the lower collective enhancement in the spherical ground-state shape. Low-lying states in spherical nuclei, belonging to deformed configurations that are not shell stabilized, may also enhance the fission probabilities [8].

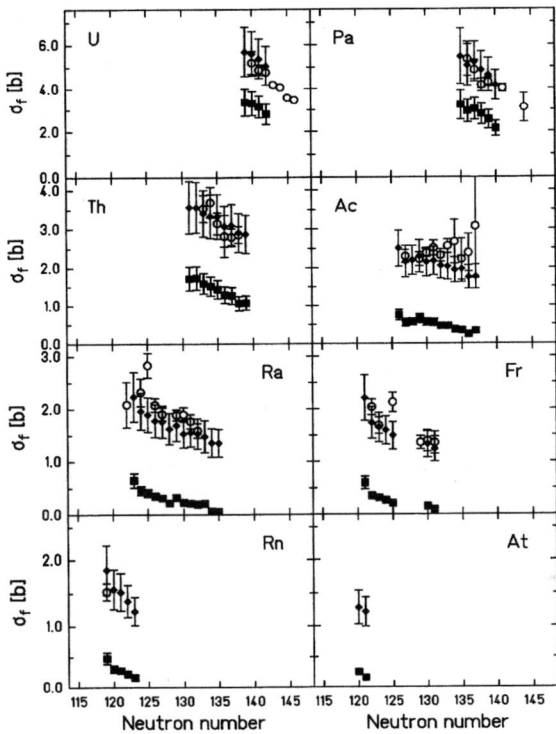

FIGURE 7. Measured fission cross sections of secondary beam from uranium to astatine, induced at 420 A MeV in a lead target. The circles correspond to the total fission cross sections, while the squares depict electromagnetically induced fission cross sections at approx. 11 MeV excitation energy. The open symbols are data from a previous experiment [6].

CONCLUSION

We conclude that the electromagnetic excitation of secondary projectiles at relativistic energies is the first promising tool to investigate the fission cross sections of exotic nuclei near the 126-neutron shell at low excitation energies. The data obtained up to now with this new method lead to the same conclusions as those deduced from previous experiments performed with conventional methods in which these nuclei could only be produced at appreciably higher excitation energies in heavy-ion fusion and fragmentation reactions.

Within the experimental resolution, the data do not show any stabilizing influence of the 126-neutron shell, which would reduce the fission competition in the deexcitation process. This finding might have significant consequences for the production of spherical super-heavy nuclei.

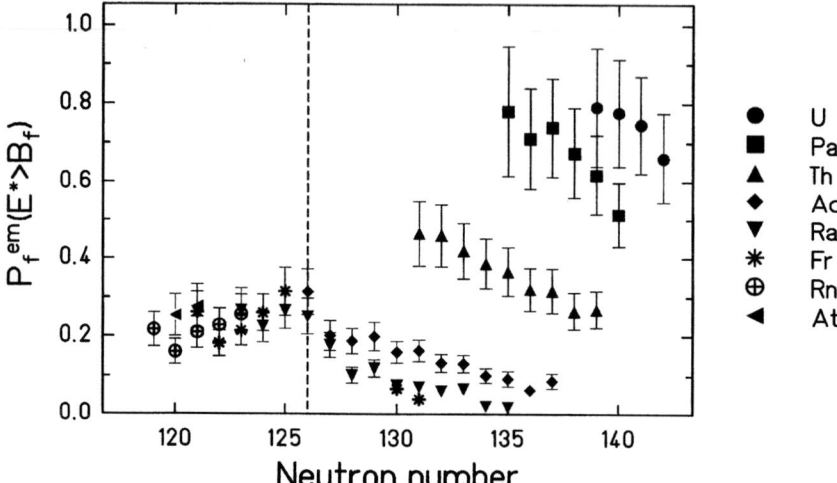

FIGURE 8. Deduced fission probabilities of secondary projectiles at 420 A MeV in a Pb target due to electromagnetic excitation. The error bars represent the uncertainties of the measured fission cross sections. The assumptions used for the analysis may increase the uncertainties, especially for the lighter systems.

REFERENCES

1. Armbruster, P., *C. R. Physique* **4**, 571-594 (2003).
2. Schmidt, K.-H. and Morawek, W., *Rep. Prog. Phys.* **54**, 949-1003 (1991).
3. J. Taieb, K.-H. Schmidt, L. Tassan-Got, P. Armbruster, J. Benlliure, M. Bernas, A. Boudard, E. Casarejos, S. Czajkowski, T. Enqvist, R. Legrain, S. Leray, B. Mustapha, M. Pravikoff, F. Rejmund, C. Stephan, C. Volant, W. Wlazlo, *Nucl. Phys.* **A 724** 413-430 (2003).
4. A. R. Junghans, M. de Jong, H.-G. Clerc, A. V. Ignatyuk, G. A Kudyaev, K.-H. Schmidt, *Nucl. Phys.* **A 629** 635-655 (1998).
5. A. Heinz, K.-H. Schmidt, A.R. Junghans, P. Armbruster, J. Benlliure, C. Böckstiegel, H.-G. Clerc, A. Grewe, M. de Jong, J. Müller, M. Pfützner, S. Steinhäuser, B. Voss, *Nucl. Phys.* **A 713**, 3-23 (2003).
6. K.-H. Schmidt, A. Heinz, H.-G. Clerc, B. Blank, T. Brohm, S. Czajkowski, C. Donzaud, H. Geissel, E. Hanelt, H. Irnich, M.C. Itkis, M. de Jong, A. Junghans, A.Magel, G.Münzenberg, F. Nickel, M. Pfützner, A. Piechaczek, C. Röhl, C. Scheidenberger, W. Schwab, S. Steinhäuser, K. Sümmerer, W. Trinder, B. Voss, S.V. Zhdanov, *Phys. Lett.* **B 325** 313-316 (1994).
7. K.-H. Schmidt, S. Steinhäuser, C. Böckstiegel, A. Grewe, A. Heinz, A.R. Junghans, J. Benlliure, H.-G. Clerc, M. de Jong, J. Müller, M. Pfützner, B. Voss, *Nucl. Phys.* **A 665** 221 (2000).
8. K.-H. Schmidt, J.G. Keller, D. Vermeulen, *Z. Phys.* **315** 159 (1984).

Multi-modal nuclear fission in the actinide nuclei

T. Asano[1], T. Wada[1], M. Ohta[1], T. Ichikawa[2], S. Yamaji[3] and H. Nakahara[4]

1) Department of Physics, Konan University, 8-9-1 Okamoto, Kobe 658-8501, Japan
2) Japan Atomic Energy Research Institute, Tokai-Mura, Ibaraki 319-1195, Japan
3) Cyclotron Center, RIKEN, Wako, Saitama 351-0198, Japan
4) Graduate School of Science, Tokyo Metropolitan University, Tokyo 192-0397, Japan

Abstract. Multi-modal fission of the uranium nucleus at the low excitation energy has been dynamically investigated. The multi-dimensional Langevin equation is used for the dynamical calculation. We used the potential energy taking account of the microscopic energy which depends on the excitation energy. We obtained the peaks of the mass distribution and the total kinetic energy (TKE) distribution. The former is consistent with experimental results and the latter is consistent with experimental systematics. We study the evolution of the mass-asymmetry of the fragments in the dynamical process from the second saddle to the scission.

1. INTRODUCTION

The mass-asymmetric peak has been observed for the fission in the light actinide region including uranium [1]. It is clear that we can explain only mass-symmetric fission from the liquid drop model (LDM), so the mass-asymmetric fission is thought to be associated to the shell effect. In fact, we expect the following mass-separations for the fission of ^{236}U using the concept of magic numbers, i.e., the combination of the heavy fission fragment mass $A_H \sim 130$ ($Z_H=50$, $N_H=82$) and $A_H \sim 155$ ($N_L=50$). But such a simple consideration of the magic number does not give the heavy fission fragment mass $A_H \sim 140$ which is obtained in experiments. Therefore, we need more elaborated consideration.

On the other hand, the total kinetic energy (TKE) is an important observed quantity for the nuclear fission, since we obtain the information about the scission configuration. Zhao et al. investigated the systematics of the TKE in the actinide region [2,3]. It is reported that in the actinide region we can identify the mass-asymmetric fission concerning with a certain type of the scission configuration which does not depend on nuclide. That is, the quantity defined by $\dfrac{1}{TKE} \dfrac{Z_1 Z_2 e^2}{R_1 + R_2}$ systematically takes the value of 1.53 where $Z_{1,2}$ are the atomic number of the fission fragments and $R_{1,2}$ are the radii of the fission fragments.

In order to understand the multi-modal nuclear fission, many theoretical efforts have been done so far. One is the method conjecturing the mass distribution from the potential energy surface (PES) and another is dynamical calculation. Recently, P. Möller et al. [4] give a description for this mass-asymmetric fission by searching the probable fission paths in the static feature of the PES. This approach has been applied for many nuclei, but dynamical effects, such as the dissipation, are not included. Dynamical calculation using the multi-dimensional Langevin equation has been applied at the high excitation energy and given the explanation for experimental results like pre-scission neutron multiplicity, but no shell effect is included [5].

Since the shell energy is not smeared out completely at the low excitation energy, e.g. lower than 30 MeV, the inclusion of the shell energy is very important in the dynamical calculation in the study of the multi-modal nuclear fission. This is pointed out for the fission of trans-actinide nuclei [6,7] and for the fusion-fission of the superheavy nuclei [8]. In this paper the dynamical calculation with the potential energy including the shell effects is performed to understand the multi-modal fission of uranium. It will be clarified in the following sections that the fragment deformation degree of the freedom plays an important role in the multi-modal nuclear fission.

Now, we have the following two questions to answer. Can we reproduce the experimental mass peak from the dynamical calculation? Is the TKE value obtained from the dynamical calculation consistent with the experimental systematics? Section 2 is a reminder of the multi-dimensional Langevin equation and the potential energy. Calculated results are presented in Sect. 3. Summary is given in Sect. 4.

2. FRAMEWORK

Dynamical approach based on the multi-dimensional Langevin equation has been successfully applied to nuclear reactions, i.e., fission and fusion-fission [9,10]. In this study, we apply this approach to investigate the multi-modal fission in the uranium nucleus. The fission motion in the multi-dimensional deformation space is traced starting from the ground state of the fissioning nucleus via saddle points to various scission configurations.

The multi-dimensional Langevin equation has the following form;

$$\frac{dq_i}{dt} = (m^{-1})_{ij} p_j,$$
$$\frac{dp_i}{dt} = -\frac{\partial V}{\partial q_i} - \frac{1}{2}\frac{\partial}{\partial q_i}(m^{-1})_{jk} p_j p_k - \gamma_{ij}(m^{-1})_{jk} p_k + g_{ij} R_j(t),$$
(1)

where q_i denote a coordinate in the deformation space and p_i is its conjugate momentum. The summation from 1 to n over the repeated indices is assumed with n being the number of collective degrees of freedom. $V(q)$ is the potential energy and $m_{ij}(q)$ and $\gamma_{ij}(q)$ are the shape-dependent collective inertia and dissipation tensors, respectively. The hydrodynamical inertia tensor is adopted with the Werner-Wheeler approximation for the velocity field [11]. The dissipation tensor is calculated with the one-body type wall-and-window formula [12]. The normalized random force $R_i(t)$ is assumed to be a white noise, i.e.,

$$\langle R_i(t) \rangle = 0, \quad \langle R_i(t_1) R_j(t_2) \rangle = 2\delta_{ij}\delta(t_1 - t_2) \tag{2}$$

The strength of the random force g_{ij} is given by $\Sigma_k g_{ik} g_{jk} = T\gamma_{ij}$, where T is the temperature of the compound nucleus. It is calculated from the excitation energy E_X as $E_X = aT^2$, where a is the level density parameter for a spherical nucleus. The potential energy is calculated using the macroscopic-microscopic method. The macroscopic part of the energy is calculated with the Yukawa plus exponential model [13] and the microscopic part is calculated with the Strutinsky shell correction method [14,15] using the two-center harmonic oscillator single-particle potential [16-19]. The potential energy is expressed as a sum of the two terms,

$$V(q, E_X) = E_{Macro}(q) + E_{Micro}(q, E_X). \tag{3}$$

The macroscopic energy consists of the nuclear potential energy $E_{Nuclear}$ and the Coulomb energy $E_{Coulomb}$[13]. The microscopic energy is calculated as a sum of the shell correction energy ΔE_s and the pairing correlation correction energy ΔE_{pc} [20]. The microscopic energy depends on the excitation energy (temperature) since the occupation probability of the single particle level depends on the excitation energy. The excitation energy dependence of the microscopic energy is introduced in the following form based on Ignatyuk's suggestion [21];

$$E_{Micro}(q, E_X) = E_{Micro}(q, E_X = 0)\exp(-E_X/E_d), \tag{4}$$

where E_X is the excitation energy and E_d is the shell damping energy, respectively. We take E_d as 20.0 MeV in this study.

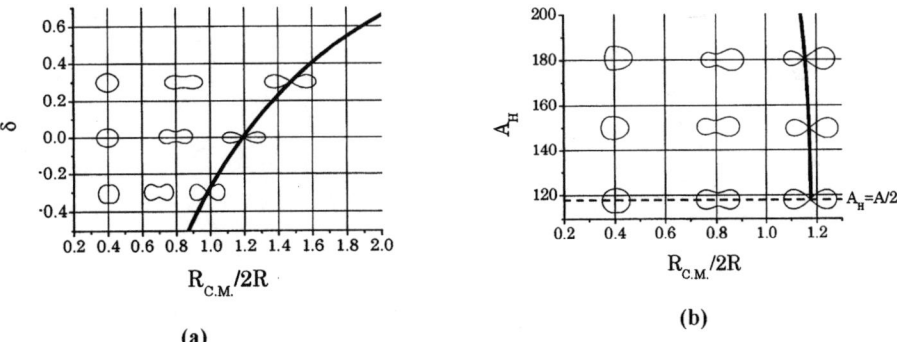

(a)

FIGURE 1 (a) Various nuclear shapes in the deformation space for mass-symmetric fission of ^{236}U. The transverse axis denotes the distance between the centers of mass of future fission fragments and the longitudinal axis denotes the deformation parameter δ. R is the radius of the fissioning nucleus. Solid line shows the scission line. **(b)** Nuclear shapes in the deformation space with $\delta=0.0$ for ^{236}U. The transverse axis denotes the distance between the centers of mass of future fission fragments and the longitudinal axis denotes the mass number of the heavy fission fragment. R is the radius of the fissioning nucleus. Solid line shows the scission line.

The nuclear shape is expressed by three parameters. We treat $R_{C.M.}$ (the distance between the centers of mass of future fission fragments), δ (the deformation of the fission fragments) and A_1 (the mass number of a fission fragment with A_2 being the mass number of the other fission fragment) as the three collective parameters. In order to obtain the deformation of the two fission fragments by a single parameter δ. Figures 1(a) and (b) show various nuclear shapes that can be described with the three collective parameters. Figure 1(a) shows the nuclear shapes in the mass-symmetric fission and Figure 1(b) shows the nuclear shapes for the mass-asymmetric division with $\delta=0.0$.

In the present study, we used TWOCTR of two-center shell model code to calculate the potential energy surface [18,19,22]. The origin of the potential energy is set so that the macroscopic energy for the spherical shape vanishes. We do not take account of the effect of the angular momentum or the particle evaporation in this study.

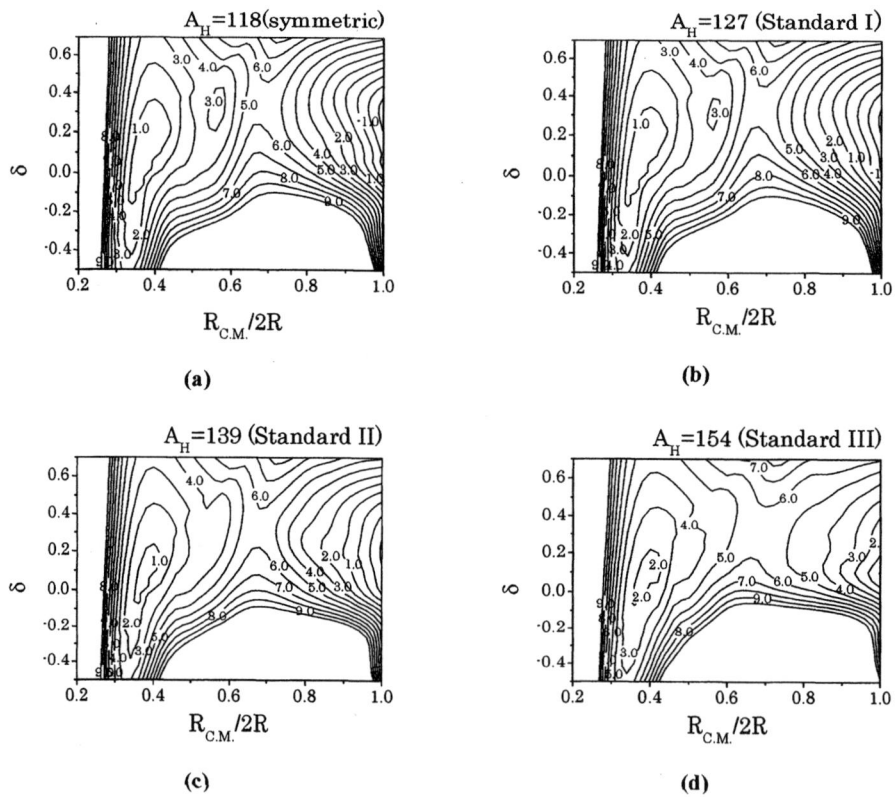

FIGURE 2 The PES including the shell energy for the fission of ^{236}U. Each figure shows the contour plot for a fixed value of A_H: (a) $A_H=118$, (b) $A_H=127$, (c) $A_H=139$ and (d) $A_H=154$. The transverse axis denotes the distance between the centers of mass of future fission fragments and the longitudinal axis denotes the deformation of fission fragment. R means radius of the fissioning nucleus.

3. RESULTS

Static calculation

We start the discussion with the static features of the PES for the fission of ^{236}U. It is known that only the mass-symmetric fission path exists in the LDM. Then, the question is whether we can obtain fission valleys that lead to mass-asymmetric fission when we take account of the shell energy. Figures 2 (a), (b), (c) and (d) show the PES including the shell energy for A_H=118 (mass-symmetric fission), 127 (Standard I), 139 (Standard II) and 154 (Standard III), respectively. One finds the second saddle point around $\delta \sim 0.30$ and a fission valley shifting to a smaller value of δ as $R_{C.M.}$ increase in all cases in Figure2. The saddle height does not depend very much on the mass separation. Figure 3 shows the PES for δ=0.30. This time the longitudinal axis denotes the mass number of a fission fragment. We find the saddle point at $A_H \sim 140$. We can not, however, predict where the resulting mass peak will locate only from the information of the PES, since the fission valley is rather flat along the mass-asymmetric degree of freedom. Thus we need the dynamical calculation.

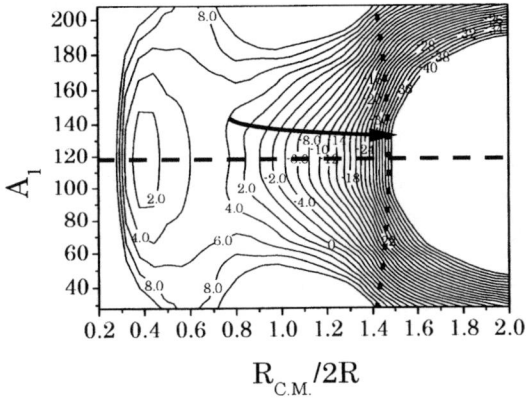

FIGURE 3. The contour plot of the PES at δ=0.30 for ^{236}U. The transverse axis denotes the distance between the centers of mass of future fission fragments and the longitudinal axis denotes the mass number of a fission fragment. R is the radius of the fissioning nucleus. Dashed line shows the mass-symmetric path. Dotted line shows the scission line. Arrow shows an expected fission path.

Dynamical calculation

In this section, we discuss the results of the dynamical calculation. First, Figure 4 shows the TKE distribution for the fission of ^{236}U at the excitation energy E_x=20.0 MeV. A single peak can be seen and we obtain the location of the peak at 169 MeV. Two experimental systematics for the average TKE are known, i.e., Viola's systematics [23] and Zhao's systematics [2,3]. We compare our results with the two experimental systematics: Viola's gives 170.1 MeV and Zhao's gives 170.2 MeV (for mass-asymmetric fission). The deviations of our peak TKE from the systematics are 1.1 and 1.2 MeV, respectively.

Next Figure 5 shows the mass distribution of the heavier fragment for the fission of ^{236}U. Solid squares denote our results for the excitation energy E_x=20.0 MeV and open circles denote the experimental data for the fission of ^{235}U induced by 14 MeV neutron [1]. We obtain a peak at A_H=136 in our calculation and the experimental data are rescaled at this point. The position of the mass-asymmetric peak is well reproduced, but the yield of the symmetric fission is bigger than the experimental one.

In order to study the fissioning motion from the second saddle to the scission, we study the deformation of the fission fragments at the scission. In Figure 6, we show the distribution of the deformation parameter δ at scission configuration for the fission of ^{236}U at the excitation energy E_x=20.0 MeV. We obtain a peak at δ=0.22 that is different from δ~0.30 which we obtained at the second saddle. We plot the potential energy along the scission line at δ=0.22 in Figure 7. There is a potential energy minimum at A_H=130 at the scission. It is to be noted that the second saddle is located around A_H~140 as shown in Figure 3. Therefore, it is clarified that the dynamical variations in the mass-asymmetry and the deformation of the fragments occur from the second saddle to the scission, and consequently the resultant peak of the mass distribution has been shifted to A_H=136.

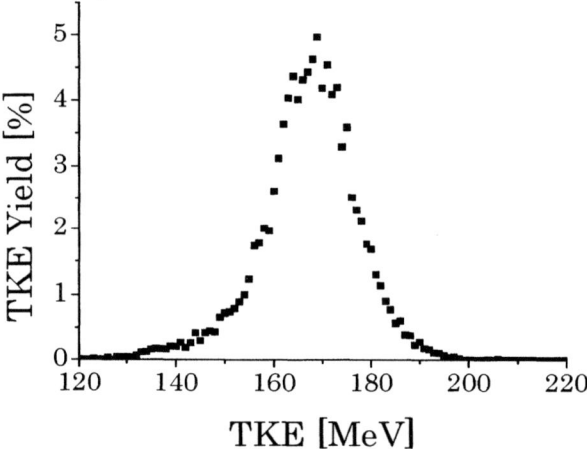

FIGURE 4. Distribution of the TKE of the fragments for the fission of ^{236}U at the excitation energy E_x=20.0 MeV. The transverse axis denotes the TKE and the longitudinal axis denotes the yield.

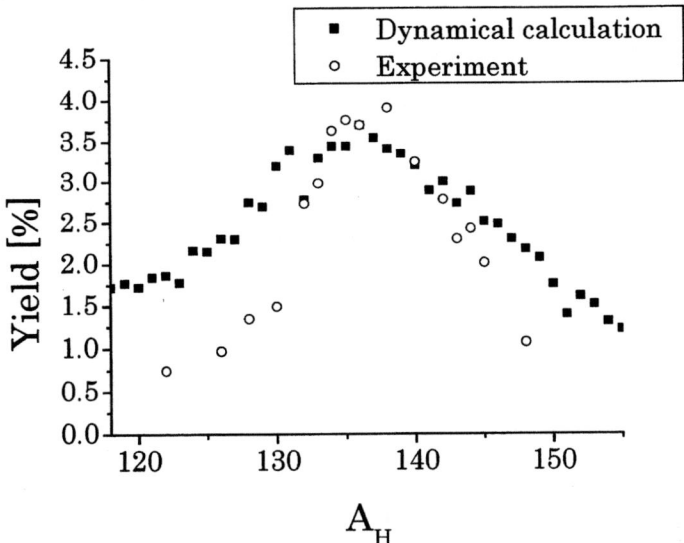

FIGURE 5. Distribution of the heavier fragment mass number for the fission of ^{236}U at the excitation energy E_x=20.0 MeV. Open circles denote experimental result [1] and solid squares denote our result. The transverse axis denotes the mass number of the heavy fragment and the longitudinal axis denotes the yield. The experimental data is shown so as to fit calculation data around A_H=136.

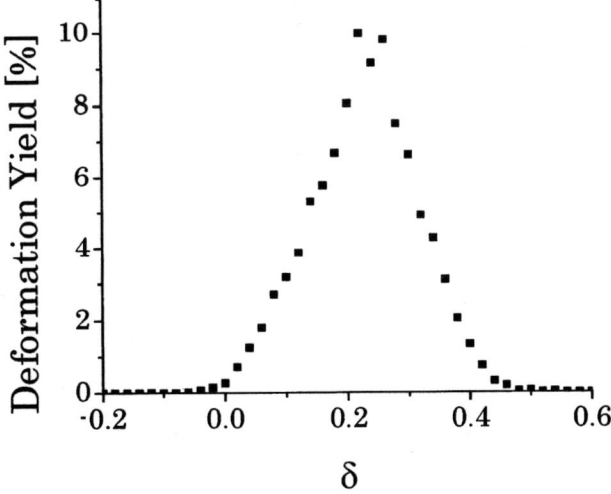

FIGURE 6. Distribution of the deformation parameter δ at scission for the fission of ^{236}U at the excitation energy E_x=20.0. The transverse axis denotes the deformation of the fission fragments at scission and the longitudinal axis denotes the yield.

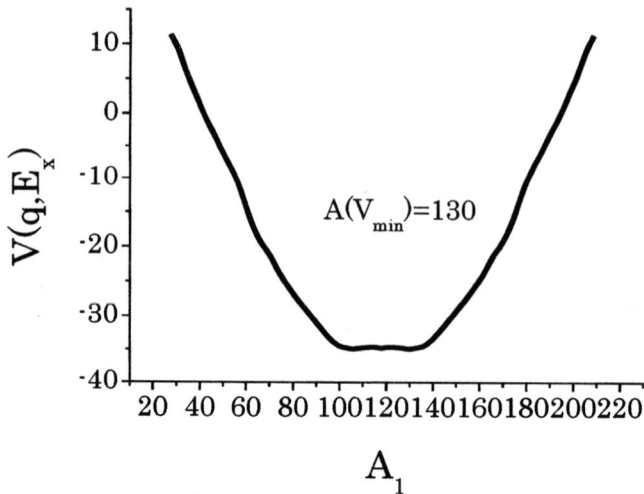

FIGURE 7. The potential energy of $\delta=0.22$ along the scission line. The transverse axis denotes the mass number of fission fragment and the longitudinal axis denotes the potential energy.

4. SUMMARY

We performed a dynamical calculation for the multi-modal nuclear fission of ^{236}U. We obtained the following three results:
1. We obtain a mass-asymmetric peak which is consistent with experimental data.
2. We obtain a TKE peak which is consistent with the two experimental systematics, i.e., Viola's systematics and Zhao's systematics.
3. We found that that mass-asymmetric fission comes not only from the structure of the potential energy but also from the dynamical effect, especially in the process from the second saddle to the scission.

In the present work, the remaining problem is that the yield of the mass-symmetric fission is bigger than that in the experiment. In order to overcome this problem, we are planning to extend this calculation to the four-dimensional space, i.e., $R_{C.M.}$ (the distance between fission fragments), A_1 (the mass number of a fission fragment), δ_H (the deformation of the heavy fission fragments) and δ_L (the deformation of the light fission fragments). We think that this procedures contribute to understanding the multi-modal nuclear fission more realistically.

ACKNOWLEDGMENTS

The authors wish to express their gratitude to Dr. Y. Nagame and Dr. Y. L. Zhao for their useful discussions.

REFERENCES

1. K. H. Flynn and L. E. Glendenin, Rep. ANL-7749 (1970). Argonne Nat. Lab., Argonne, Illinois.
2. Y. L. Zhao et al., *Phys. Rev. Lett.* **82**, 3408-3411 (1999).
3. Y. L. Zhao, Y. Nagame, I. Nishinaka, K. Sueki and H. Nakahara, *Phys. Rev.* **C62**, 014612-1-9(2000).
4. P. Möller et al., Nature 409 (2001) 785-790.
5. T. Wada et al., Phys. Rev. Lett. 70 (1993) 3538-3541.
6. T. Ichikawa, Doctoral thesis. Konan University (2003).
7. T. Ichikawa, et al., J. of Nucl. And Radiochemical Sci. 3(2002) 542-545.
8. Y. Aritomo, et al., Phys. Atom. Nucl. 66 (2003) 1105-1113.
9. Y. Abe et al., J. Physique 47 (1986) C4-329.
10. P. Fröbrich and S.Y. Xu, Nucl. Phys. A477 (1988)143-161.
11. K. T. R. Davis, A. J. Sierk and J. R. Nix, *Phys. Rev.* **C13**, 2385-2403 (1976).
12. J. Randrup and W. J. Swiatecki, *Nucl. Phys.* **A429**, 105-115 (1984).
13. H. J. Krappe, J. R. Nix and A. J. Sierk, *Phys. Rev.* **C20**, 992-1013 (1979).
14. V. M. Strutinsky, *Nucl. Phys.* **A95**, 420-442 (1967).
15. V. M. Strutinsky, *Nucl. Phys.* **A122**, 1-33 (1968).
16. D. Scharnweber, W. Greiner and U. Mosel, *Nucl. Phys.* **A164**, 257-278 (1971).
17. J. Maruhn and W. Greiner, *Z. Phys.* **251**, 431-457 (1972).
18. S. Suekane, A. Iwamoto, S. Yamaji and K. Harada, *JAERI-memo*. 5918 (1974), unpublished.
19. A. Iwamoto, S. Yamaji, S. Suekane and K. Harada, *Prog. Theor. Phys.* **55**, 115-130 (1976).
20. M. Bolsterli, E. O. Fiset, J. R. Nix and J. L. Norton, *Phys. Rev.* **C5**, 1050-1077 (1972).
21. A. V. Ignatyuk et al., *Sov. J. Nucl. Phys.* **21**, 255-257 (1975).
22. K. Sato, S. Yamaji, K. Harada and S. Yoshida *Z.Phys.* **A290**, 149-156 (1979).
23. V. E. Viola, K. Kwiatkowski and M. Walker, *Phys. Rev.* **C31**, 1550(1985).

Effects of nuclear structure in the transport coefficients of large-scale collective motion

F. Ivanyuk[†]

GSI, Planckstrasse 1, D-64291 Darmstadt, Germany,
Institute for Nuclear Research, Prospect Nauki 47, 03028 Kiev, Ukraine

Abstract. We study the collective motion of iso-scalar type at finite excitations and concentrate on slow motion, due to the presence of a strong friction force. In the present talk the extension of approach to the case of low excitation energies, where shell effects and pairing correlation are important, is reviewed. The the case of rotating nuclei is also included. As an application of the theory, the numerical results are presented for the transport coefficients for few composite systems formed in the so called warm fusion reactions used for the synthesis of the super heavy systems.

INTRODUCTION

The synthesis of the superheavy elements is one of the most challenge problems of nuclear physics. Intensive experiments in this direction are carried out at GSI, Darmstadt, JINR, Dubna and RIKEN, Tokyo. The theoretical description of fusion-fission reaction is often based on the Langevin equation [1, 2, 3] for the collective variables Q_μ which parameterize the shape of the composite system,

$$\begin{aligned}\frac{dP_\mu}{dt} &= -\frac{\partial V}{\partial Q_\mu} - \frac{1}{2}\frac{\partial}{\partial Q_\mu}(M^{-1})_{\rho\nu}P_\rho P_\nu - \gamma_{\mu\nu}(M^{-1})_{\rho\nu}P_\nu + g_{\mu\nu}R_\nu(t), \\ \frac{dQ_\mu}{dt} &= (M^{-1})_{\mu\nu}P_\nu \qquad \rho,\mu,\nu = 1,2,...N, \end{aligned} \qquad (1)$$

where $R_\nu(t)$ is the random force obeying conditions

$$\langle R_\nu(t)\rangle = 0, \quad \langle R_\nu(t)R_\nu(t')\rangle = 2\delta(t-t') \quad \text{and} \quad \sum_\rho g_{\mu\rho}g_{\nu\rho} = T\gamma_{\mu\nu}. \qquad (2)$$

To solve the Langevin equation (1) for $Q_\mu(t)$ one would need first of all its coefficients - the potential energy $V(Q_\mu)$, friction $\gamma_{\mu\nu}$- and mass $M_{\mu\nu}$-tensors.

The potential energy is commonly calculated within the microscopic-macroscopic shell correction method [4, 5] which describes the collective energy in the quasi-static picture rather accurately. The tensors of friction $\gamma_{\mu\nu}$ and mass $M_{\mu\nu}$ are usually computed within the macroscopic approaches (the wall and window formula for friction [6] and the Werner-Wheeler (WW) method for the inertia [7]). These methods provide rather simple expressions for the collective friction and mass coefficients. However the deformation and temperature dependence of macroscopic transport coefficients $\gamma_{\mu\nu}$ and $M_{\mu\nu}$ is not very reliable, at least at small excitation energies. For example, the observed in [8] increase of the damping parameter $\eta = \gamma/2\sqrt{M|C|}$ with the excitation energy is impossible to explain neither with the wall formula value γ^{wall} for friction (where the friction parameter practically does not depend on the temperature T or excitation energy) nor within the hydrodynamical model (friction parameter decreases as $1/T^2$). Also the shell and pairing effects which are very important at low excitation energies are completely ignored in macroscopic models. Thus the necessity of having a microscopic theory for the collective transport coefficients becomes quite evident.

LINEAR RESPONSE THEORY FOR THE COLLECTIVE MOTION

It is supposed [9] that the nuclear many-body Hamiltonian can be approximated by

$$\hat{H}(\hat{x}_i, \hat{p}_i, Q_\mu) = \hat{H}_{mf}(\hat{x}_i, \hat{p}_i, Q_\mu) + \hat{V}_{res}^{(2)}(\hat{x}_i, \hat{p}_i), \tag{3}$$

where the mean field Hamiltonian $\hat{H}_{mf}$ depends explicitly on one or few collective variables Q_μ which specify the shape of nuclear surface and the residual two body interaction $\hat{V}_{res}^{(2)}$ is assumed to be *independent* of the collective coordinates Q_μ. As the consequence the generators for the collective motion, namely,

$$\hat{F}_\mu \equiv \frac{\partial \hat{H}(\hat{x}_i, \hat{p}_i, Q_\mu)}{\partial Q_\mu}\bigg|_{Q_\mu = Q_\mu^0} \equiv \frac{\partial \hat{H}_{mf}(\hat{x}_i, \hat{p}_i, Q_\mu)}{\partial Q_\mu}\bigg|_{Q_\mu = Q_\mu^0} \tag{4}$$

are the one-body operators, what allows to apply the independent particle model. It is shown in [9] that the slow collective motion can be described locally in terms of so-called collective response function $\chi_{coll}(t)$ whose Fourier transform has the form

$$\chi_{coll}(\omega) = \kappa(\kappa + \chi(\omega))^{-1}\chi(\omega). \tag{5}$$

In (5) the $\chi(\omega)$ is the Fourier transform of the causal response function,

$$\chi_{\mu\nu}(t-s) = \Theta(t-s)\frac{i}{\hbar}\text{tr}\left(\hat{\rho}_{qs}(Q,T)[\hat{F}_\mu^I(t), \hat{F}_\nu^I(s)]\right) \equiv 2i\Theta(t-s)\chi''_{\mu\nu}(t-s). \tag{6}$$

Here $\hat{F}_\mu^I(t)$ is the interaction representation of operator (4) and $\hat{\rho}_{qs}$ represents the thermal equilibrium $\rho_{qs}(Q_\mu, T) \propto \exp(-\hat{H}(Q_\mu)/T)$. The inverse of coupling tensor $\kappa_{\mu\nu}$ in (5) is defined by the quasi-static properties of the system,

$$\kappa_{\mu\nu} = -\chi_{\mu\nu}(0) - C_{\mu\nu}(0), \tag{7}$$

where $\chi_{\mu\nu}(0)$ is the static response and $C_{\mu\nu}(0)$ is the stiffness of quasi-static free energy, $C_{\mu\nu}(0) \equiv \partial^2 F/\partial Q_\mu \partial Q_\nu$.

It turns out that the transport coefficients for the average collective motion can be expressed in terms of $\chi_{coll}(\omega)$. As shown in [10] one has to approximate the quantity $\kappa\chi_{coll}^{-1}(\omega)\kappa$ by the second order polynomial in frequency

$$(\kappa\chi_{coll}^{-1}(\omega)\kappa)_{\mu\nu} \implies -M_{\mu\nu}\omega^2 - i\gamma_{\mu\nu}\omega + C_{\mu\nu}. \tag{8}$$

Evidently, the coefficients $M_{\mu\nu}$, $\gamma_{\mu\nu}$ and $C_{\mu\nu}$ stand for the elements of the tensors for the mass, friction and stiffness. For slow collective motion the transport coefficients can be deduced by expanding the left hand part of (8) around $\omega = 0$. In this way one gets

$$\gamma \approx i\frac{\partial}{\partial \omega}(\kappa \chi_{\text{coll}}^{-1}(\omega)\kappa)\Big|_{\omega=0} = \kappa \chi^{-1}(0)\gamma(0)\chi^{-1}(0)\kappa,$$

$$M \approx -\frac{\partial^2(\kappa \chi_{\text{coll}}^{-1}(\omega)\kappa)}{2\partial\omega^2}\Big|_{\omega=0} = \kappa \chi^{-1}(0)[M(0) + \gamma(0)\chi^{-1}(0)\gamma(0)]\chi^{-1}(0)\kappa. \quad (9)$$

The friction $\gamma(0)$ and mass $M(0)$ tensors are expressed in terms of first and second derivatives of the intrinsic response function $\chi_{\mu\nu}(\omega)$ at $\omega = 0$,

$$\gamma_{\mu\nu}(0) = -i\frac{\partial \chi_{\mu\nu}(\omega)}{\partial \omega}\Big|_{\omega=0}, \quad M_{\mu\nu}(0) = \frac{1}{2}\frac{\partial^2 \chi_{\mu\nu}(\omega)}{\partial \omega^2}\Big|_{\omega=0}. \quad (10)$$

For obvious reasons, expression (10) are referred to as the "zero frequency limit".

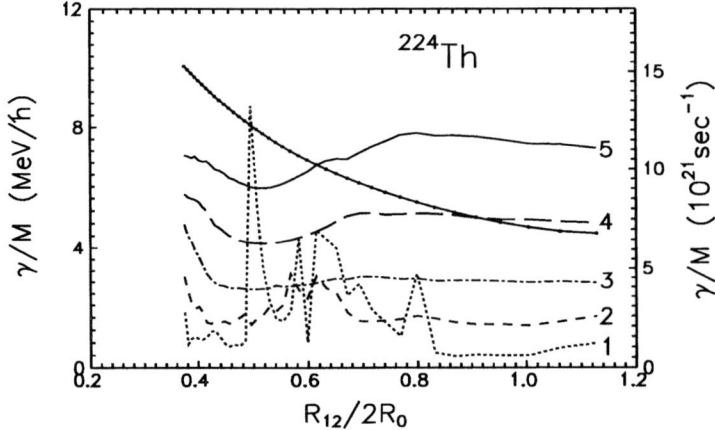

FIGURE 1. The reduced friction coefficient γ/M as function of deformation and temperature $T = (1-5)$ MeV (indicated in the Figure). The curve with dots marks γ^{wall}/M_{irr}.

An example of numerical results for the ratio of the friction coefficient to the mass parameter is shown in Fig.1. These results were obtained with the deformed Woods-Saxon shell model potential. The shape of nuclear surface was parameterized in terms of Cassini ovaloids [11]. As the compound nucleus the system $^{208}Pb + ^{16}O \Longrightarrow ^{224}Th$ was chosen here for which the rapid increase of dissipation with the excitation energy was found [12].

It is seen from Fig.1 that γ/M is essentially constant over the whole deformation region, for all computations but $T = 1\ MeV$, a case for which the fluctuations are seen. However, there is a marked dependence on excitation: γ/M increases strongly with T. This is in clear distinction to the result one gets from applying the wall formula γ_{wall} for friction and that of irrotational flow M_{irr} for the inertia. Note also that the numerical values of γ/M are in agreement with those deduced from the analysis of experimental data [13].

THE EFFECT OF PAIRING

The pair correlations are vital for understanding of many elementary features of nuclear physics at small thermal excitations. It is of great interest to account for pair correlations also in the description of typical transport problems of dissipative systems. On general grounds one could expect that pairing will greatly diminish nuclear dissipation.

To account for the pairing interaction we add to the mean field Hamiltonian the pairing part

$$\hat{H}_{mf} \Longrightarrow \hat{H}_{mf} - G\hat{P}^\dagger \hat{P}, \qquad \hat{P}^\dagger = \sum_k a_k^\dagger a_{\bar{k}}^\dagger, \qquad (11)$$

where the coupling constant G is assumed to be state independent and $a_k^\dagger$ and a_k being the creation and annihilation operators. The Hamiltonian (11) is solved then within the independent quasiparticle approximation.

The explicit expression for the response function $\chi_{\mu\nu}(\omega)$ is found directly from (6) after straightforward though somewhat lengthy derivation

$$\chi_{\mu\nu}(\omega) = \sum_{kj}{'} (n_k^T - n_j^T)\xi_{kj}^2 \frac{2(E_k - E_j)^2}{(\hbar\omega + i\Gamma_{kj})^2 - (E_k - E_j)^2} F_\mu^{kj} F_\nu^{jk}$$

$$+ \sum_{kj} (n_k^T + n_j^T - 1)\eta_{kj}^2 \frac{2(E_k + E_j)^2}{(\hbar\omega + i\Gamma_{kj})^2 - (E_k + E_j)^2} F_\mu^{kj} F_\nu^{jk}, \qquad (12)$$

where $n_k^T = 1/(1 + \exp(E_k/T))$, $\eta_{kj} \equiv u_k v_j + v_k u_j$, $\xi_{kj} \equiv u_k u_j - v_k v_j$ and u_k, v_k are coefficients of Bogolyubov-Valatin transformation.

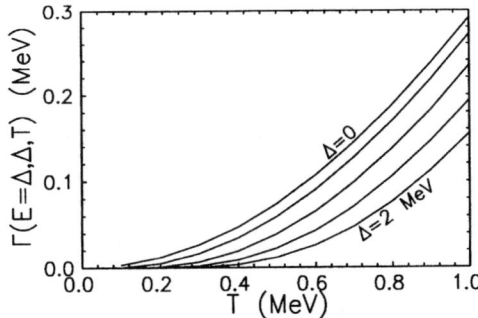

FIGURE 2. The collisional width (13) taken at the Fermi energy ($E = \Delta$) as function of the temperature. Different curves correspond to different values of pairing gap Δ indicated in the figure.

In (12) the width Γ_{kj} is the average width of the two-quasiparticle state, $\Gamma_{kj} = (\Gamma(E_k, \Delta, T) + \Gamma(E_j, \Delta, T))/2$. The calculation of Γ with pairing included is discussed in detail in [10]. We would mention only that in the presence of pairing no analytical expression is available for the width $\Gamma(E_k, \Delta, T)$ and this quantity is computed numerically, $\Gamma(E_k) = \Gamma_d(E_k)/(1 + \Gamma_d(E_k)\Gamma_0/c^2)$, with

$$\Gamma_d(E) = \frac{2}{\Gamma_0} \int d\varepsilon_2 d\varepsilon_3 d\varepsilon_4 \delta(E + E_2 + E_3 + E_4)[n_2^T n_3^T n_4^T + (1 - n_2^T)(1 - n_3^T)(1 - n_4^T]. \qquad (13)$$

For the Γ_0 and c the values are taken $\Gamma_0 = 33$ MeV, $c = 20$ MeV, (see [9]).

The dependence of collisional width $\Gamma(E = \Delta, \Delta, T)$ on the temperature for few fixed values of pairing gap Δ is shown in Fig.2. One can see that with increasing Δ the value of Γ gets smaller. Eventually, this leads to the suppression of the collective friction by pairing.

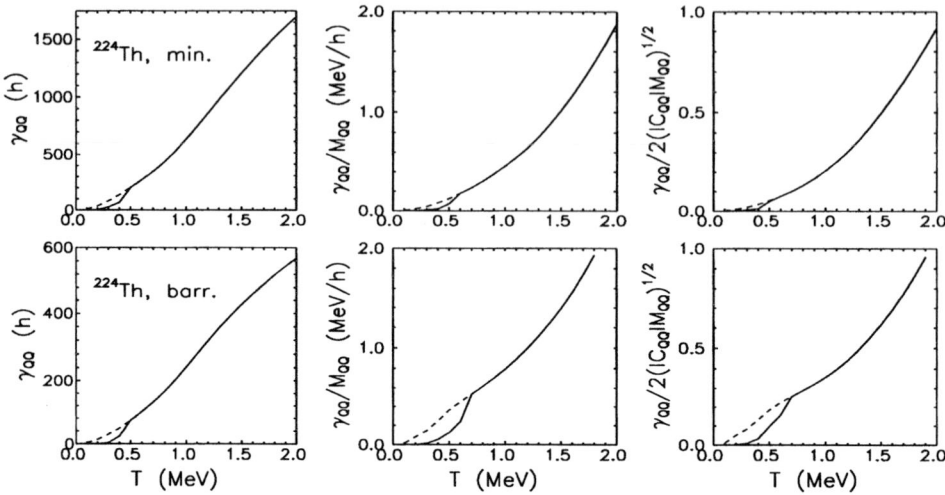

FIGURE 3. The QQ-friction coefficient γ_{QQ}, reduced friction coefficient $\beta_{QQ} \equiv \gamma_{QQ}/M_{QQ}$ and the damping factor $\eta_{QQ} = \gamma_{QQ}/2\sqrt{C_{QQ}M_{QQ}}$ at the ground state (top) and fission barrier (bottom) of ^{224}Th as function of temperature. The dash curves show the results obtained neglecting pairing.

This effect is clearly demonstrated in Fig.3 where the friction coefficient, reduced friction coefficient and the damping factor for QQ-mode are shown as function of the temperature. As it is seen from Fig.3 the friction coefficients demonstrate kind of super fluidity. It is negligibly small until the pairing gap would disappear at critical temperature $T_c \approx (0.5 - 0.6)$ MeV. These results are in qualitative agreement with the "onset of dissipation" found experimentally in [12].

The results of detailed numerical calculation of the friction tensor for the system $^{48}Ca + ^{244}Pu \Longrightarrow ^{292}114$ given by three deformation parameters α, α_3 and α_4 are shown in Fig.4. Because of the lack of space only most important diagonal components of friction tensor are shown. For convenience the friction coefficients were divided by the wall formula value. Otherwise the details of the structure of $\gamma_{\alpha\alpha}, \gamma_{\alpha_3\alpha_3}$ and $\gamma_{\alpha_4\alpha_4}$ would not be seen on the scale of Figure. The most important conclusion from Fig.4 is: the values of $\gamma_{\alpha\alpha}, \gamma_{\alpha_3\alpha_3}$ and $\gamma_{\alpha_4\alpha_4}$ differ substantially from their wall formula counterparts. Unlike γ^{wall}, $\gamma_{\alpha\alpha}, \gamma_{\alpha_3\alpha_3}$ and $\gamma_{\alpha_4\alpha_4}$ demonstrate not regular dependence on the deformation. The period of fluctuations is approximately the same as in case of the deformation energy, thus one can attribute these fluctuations to the effect of the shell structure. The value and deformation dependence of the corresponding components of mass tensor also differ very much [14] from their macroscopic counterparts (obtained within Werner-Wheeler method [7]). Thus the results of dynamical computations using macroscopic friction and mass parameters [6, 7] are not reliable at least at low excitation energies.

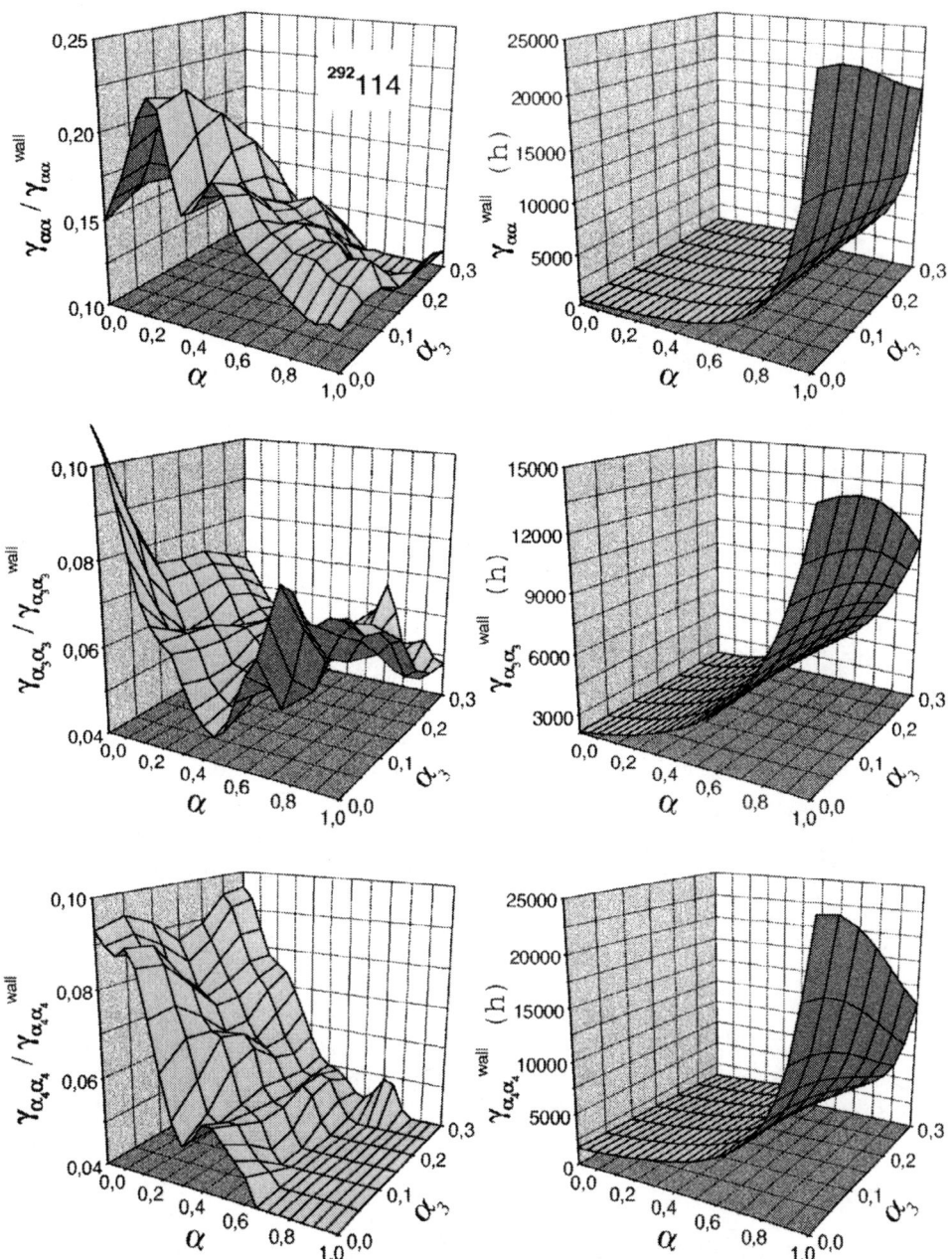

FIGURE 4. Left: The diagonal components $\gamma_{\alpha\alpha}$, $\gamma_{\alpha_3\alpha_3}$ and $\gamma_{\alpha_4\alpha_4}$ of friction tensor as function of deformation parameters α and α_3 in units of the wall formula value, see right part of the Figure. The calculations are done for the compound system $^{292}114$ at the temperature $T = 0.5$ MeV.

The microscopic transport coefficients were used recently in [15] where the two stage approach to the description of fusion-fission reactions is suggested. On each stage (fusion or fission) the three-dimensional Langevin equation for the variables describing the shape of nuclear system was solved. The results obtained on the first stage are used as the input data for description of fission dynamics. In this way it turned out possible to describe for the reaction $^{18}O+^{208}Pb$ both fusion and fission cross sections, the energy and mass distribution of fission fragments, the probability of the evaporation residue formation, the dependence of pre-fission neutron multiplicities on the fragment mass number.

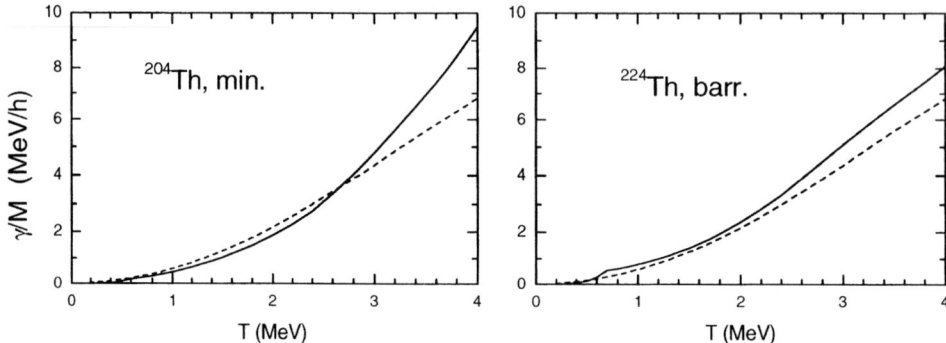

FIGURE 5. The reduced friction coefficient γ/M as function of temperature: the microscopic results (solid curves) are compared to the approximation (14) (dotted curves).

The computations of the transport coefficients shown above are rather time consuming. For practical use in codes based on the Langevin equation the simple analytical approximation is highly desirable. The one of the most important quantities in such calculations is the reduced friction coefficient γ/M. In Fig.5 we show it on the left hand panel as function of T together with the following approximation, see [16],

$$\frac{\gamma}{M}\hbar = \approx 2\Gamma_{sp}(\mu,T) = \frac{2}{\Gamma_0}\frac{\pi^2 T^2}{1+\pi^2 T^2/c^2} \approx \frac{0.6 T^2}{1+T^2/40} \text{ MeV} \qquad (T \text{ in MeV}). \qquad (14)$$

The approximation (14) represents the microscopic result quite well.

The analytical approximations of [16] were used recently by [17] to describe the formation probability of super heavy system. By examining the long time behavior of the Fokker-Planck equation for the distribution function it was shown that the formation probability increase by few orders of magnitude if microscopic transport coefficients are used rather than those of the common picture.

THE TRANSPORT COEFFICIENTS FOR ROTATING NUCLEI

The nuclear compound system formed in the result of fusion of heavy ions are commonly formed with non zero angular momentum. The effect of rotation on the fusion or fission probability is included at most in the calculation of the macroscopic part of the deformation energy. The possible dependence on rotation of the shell correction as well as friction and inertia is completely ignored. However one might expect some dependence of the transport coefficients on rotation since the rotation changes considerably the single-particle spectrum. To clarify this problem we have carried out the calculation of transport coefficients for rotating nuclei [18]. The computations are carried out with two-center shell model which allows for rather flexible parametrization of the shape around the touching point and which was used earlier in dynamical computations [19]. Due to technical reasons we had to limit ourselves to the excitations above $T = 1$ MeV where the pairing can be neglected.

By describing the rotating nuclei one usually transforms the Hamiltonian from the laboratory co-ordinate system to the body fixed (or intrinsic) co-ordinate system. In the result, instead of the Hamiltonian $\hat{H}(Q_\mu)$ one has to consider the Routhian operator

$$\hat{R}(Q_\mu, \omega_{rot}) = \hat{H}(Q_\mu) - \omega_{rot}\hat{J}_x, \quad (15)$$

with ω_{rot} being the rotational frequency and $\hat{J}_x$ - the projection of angular momentum on the rotation axes (x-axes).

Like in the case without rotation we will use for calculation of the potential energy the Strutinsky shell correction method [4, 5]. Following [20, 21] one can express the intrinsic energy $E(Q_\mu, I)$ as

$$E(Q_\mu, I) = E_{LDM}(Q_\mu, I) + \delta R(Q_\mu, I), \quad (16)$$

where $E_{LDM}(Q_\mu, I)$ is the liquid drop energy of rotating nucleus and $\delta R(Q_\mu, I)$ is the shell correction. In the case of finite temperature instead of the shell correction to the intrinsic energy one has to consider the shell correction to the free energy $\delta R \Longrightarrow \delta F = \delta R - T\delta S$, where δS is the shell correction to the entropy.

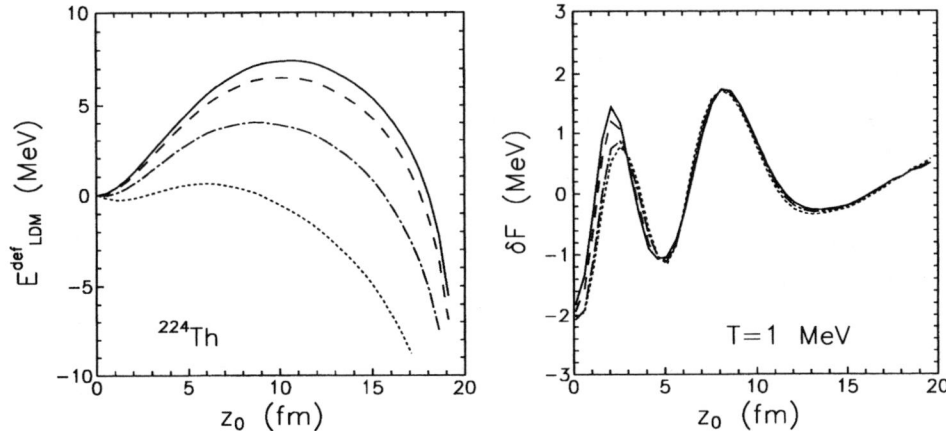

FIGURE 6. The liquid drop deformation energy (left) and the shell correction $\delta F = \delta R - T\delta S$ to the free energy (right) for temperature $T = 1$ MeV as function of the deformation parameter z_0. The solid, dash, dotted-dash and dotted lines correspond to the values of angular momentum equal to 0, 20, 40 and $60\hbar$.

The left-hand-side part of Fig.6 shows the rotational dependence of the liquid drop part of deformation energy. As it is seen the rotational dependence of the deformation energy is rather strong. The fission barrier disappears completely at $I \approx 60\hbar$ for the nucleus ^{224}Th shown in the figure. The effect of rotation on the fission barriers is known for decades. The rotational dependence of the shell correction is less clear. It is assumed usually that this dependence is weak and the shell correction is computed at $\omega_{rot} = 0$ only. To clarify this point we have computed the shell correction for several values of I as a function of deformation along the liquid drop fission valley of ^{224}Th. Indeed, see right-hand-side of Fig.6, the fluctuation of δF is less then 1 MeV for variation of I from zero to $I = 60\hbar$. Very likely such weak dependence of δF on I can be neglected.

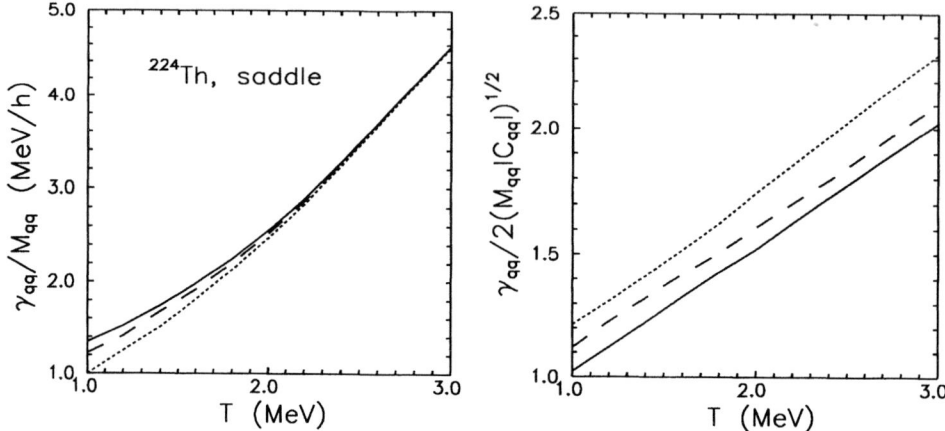

FIGURE 7. The reduced friction coefficient $\beta_{qq} = \gamma_{qq}/M_{qq}$ (left) and the damping factor $\eta_{qq} = \gamma_{qq}/2\sqrt{|C_{qq}|M_{qq}}$ (right) versus temperature. The dotted, dash and solid curves correspond to the values of angular momentum equal to 0, 40 and 60 $\hbar$.

Like in the case without rotation the transport coefficient of collective motion, can be derived within the linear response theory [9] replacing the mean field Hamiltonian by the Routhian (15). It turns out however [18] that the friction γ and mass M parameters for rotating nuclei are rather sensitive to such fine effects as the violation of rotational symmetry by Coriolis term $-\omega_{rot}\hat{J}_x$. For the ground state deformation the spurious contributions to collective friction and mass are (at least) as large as those of physical importance. In order to remove the spurious contributions we had to modify the model of "stationary rotation" and to introduce the time-dependent rotational frequency. In the result we obtained the friction and the mass parameters which demonstrate rather weak dependence on the rotational frequency ω_{rot}.

Fig.7 shows the reduced friction coefficient $\beta_{qq} = \gamma_{qq}/M_{qq}$ and the damping factor $\eta_{qq} = \gamma_{qq}/2\sqrt{|C_{qq}|M_{qq}}$ at the saddle of ^{224}Th as the function of temperature. Both β_{qq} and η_{qq} shown in Fig.7 increase with the temperature. This behaviour is in a qualitative agreement with the one found in [12]. The rather weak dependence of η_{qq} on I seen from Fig.7 is mainly due to some dependence of liquid drop stiffness on rotation.

SUMMARY

• The microscopic approach for the transport coefficients (tensors of friction and inertia) for slow collective motion is reviewed, which accounts in a natural way for the pairing interaction, shell effects and collective rotation.

• As an application of the theory, the numerical results for the transport coefficients are presented for few composite systems formed in the so called warm fusion reactions. It is demonstrated that both friction and inertia show a sensible dependence on the configurations of the mean field caused both by the shell effects as well as by avoided crossings of single-particle levels. The dissipation decreases with decreasing temperature and growing pairing gap and falls well below the values of common "macroscopic models".

• The semianalytical expressions are suggested for the temperature dependence of those combinations of the transport coefficients which govern the fission process.

• At the excitations corresponding to the temperatures $T \geq 1$ MeV the shell correction to the energy practically does not depend on nuclear rotation. The friction and mass parameters obtained within the linear response theory for the same excitations are rather stable with respect to rotations *provided* that the contributions from the spurious states arising due to the violation of rotational symmetry are removed.

Acknowledgments. The author wishes to acknowledge the very fruitful collaboration with Profs. H.Hofmann, V.V.Pashkevich and S.Yamaji throughout which the main results presented here were obtained.

REFERENCES

1. Y. Abe, S. Ayik, P. Reinhard and E. Suraud, *Phys. Rep.*, **275**, 49–196 (1996).
2. P. Fröbrich and I. I. Gontchar, *Phys. Rep.*, **292**, 131–237 (1998).
3. K. Pomorski, J. Bartel, J. Richert and K. Dietrich, *Nucl. Phys. A*, **605**, 87–121 (1996).
4. V. M. Strutinsky, *Nucl. Phys. A*, **95**, 420–442 (1967); **122**, 1–33 (1968).
5. M. Brack, J. Damgaard, A. S. Jensen et al, *Rev. Mod. Phys.*, **44**, 320–405 (1972).
6. J. Blocki, Y. Boneh, J. Nix et al, *Ann.Phys.*, **113**, 330–386 (1978).
7. K. Davies, A. Sierk and J. Nix, *Phys. Rev. C*, **13**, 2385–2403 (1976).
8. D. Hofman, B. Back, I. Diószegi et al, *Phys. Rev.Let.*, **72**, 470–473 (1994).
9. H. Hofmann, *Phys. Rep.*, **264**, 137–380 (1997).
10. F. A. Ivanyuk and H. Hofmann, *Nucl. Phys. A*, **657**, 19–58 (1999).
11. V. V. Pashkevich, *Nucl. Phys. A*, **169**, 275–293 (1971).
12. D. J. Hofman, B. B. Back and P. Paul, *Phys.Rev. C*, **51**, 2597–2605 (1995).
13. D. Hilscher, I. Gontchar and H. Rossner, *Physics of Atomic Nuclei*, **57**, 1187–1199 (1994).
14. F. A. Ivanyuk, in *Nuclear Shells - 50 Years*, edited by Yu. Ts. Oganessian and R. Kalpakchieva, World Scientific, Singapore-New Jersey-London-Hong Kong, 2000, pp. 456–465.
15. G. I. Kosenko, F.A. Ivanyuk and V. V. Pashkevich, *Journal of Nuclear and Radiochemical Science*, **3**, 71–76 (2002).
16. H. Hofmann, F. A. Ivanyuk, C. Rummel and S. Yamaji, *Phys. Rev. C*, **64**, 054316-1–16 (2001).
17. C. Rummel and H. Hofmann, *Nucl. Phys. A*, **727**, 24–40 (2003).
18. F. A. Ivanyuk and S. Yamaji, *Nucl. Phys. A*, **694**, 295–311 (2001).
19. Y. Aritomo, T. Wada, M. Ohta and Y. Abe, *Phys.Rev. C*, **59**, 796–809 (1999).
20. V. V. Pashkevich and S. Frauendorf, *Sov.J.Nucl.Phys.*, **20**, 1122–1130 (1975).
21. K. Neergaard, V. V. Pashkevich and S. Frauendorf, *Nucl.Phys. A*, **262**, 61–90 (1976).

Stochastic Model of the Tilting Mode in Nuclear Fission

Vadim A. Drozdov, Dmitri O. Eremenko, Olga V. Fotina,
Sergey Yu. Platonov, Oleg A. Yuminov

Institute of Nuclear Physics, Moscow State University, 119992 Moscow, Russia,

Abstract. A model of induced nuclear fission is developed with consideration of thermodynamically fluctuating orientational degree of freedom of deformed nuclei. This model was applied in analysis of the experimental angular anisotropy of fission fragments in the $^{16}O + ^{232}Th$, ^{238}U and ^{248}Cm reactions at the oxygen energies from 90 to 160 MeV. Information on the equlibrating time of the orientational mode was obtained.

INTRODUCTION

The time scales of equilibrating various degrees of freedom in deformed nuclei are of fundamental importance in the theory of nuclear fission. This work focuses on the rotational degrees of freedom of fissioning nuclei. It is common knowledge that energy partition for various rotational degrees of freedom manifest itself in angular distributions [1] and spins [2] of fission fragments. Usually these observables are described within the statistical approaches, namely the model of the transition states in the saddle point [1] or the scission point model [3]. In the saddle point model (SPM) it is assumed that the fission fragment angular distribution reflects the orientation of the nuclear symmetry axis relative to the total angular momentum (J) in the saddle point of the fission barrier. It is practice to express the orientation in terms of the component K of J onto the symmetry axis (see fig. 1a). The model postulates that the equilibrium K distribution is achieved at the saddle point and frozen beyond the saddle. In other words, the equilibrating time for the K degree of freedom is shorter than the time spent by the nucleus near the saddle point and larger than the time between the saddle and scission points. In SPM the fragment angular distribution for fission of the nucleus produced in fusion of spin zero nuclei ($M=0$ is the projection of J onto the beam axis) is given by

$$W(\theta) \propto \sum_{J=0}^{\infty}(2J+1)\, T_J \sum_{K=-J}^{K=J} \frac{1}{2}(2J+1)\left|D^{J}_{M=0,K}(\theta)\right|^2 \rho(K) \tag{1}$$

where T_J is a transmission coefficient for a partial wave J, $D^{J}_{0,K}(\theta)$ is a symmetric top wave function, $\rho(K)$ is the K distribution in the saddle point of the fission barrier. Using arguments based on the statistical theory of nuclear reactions, it can be shown [1] that

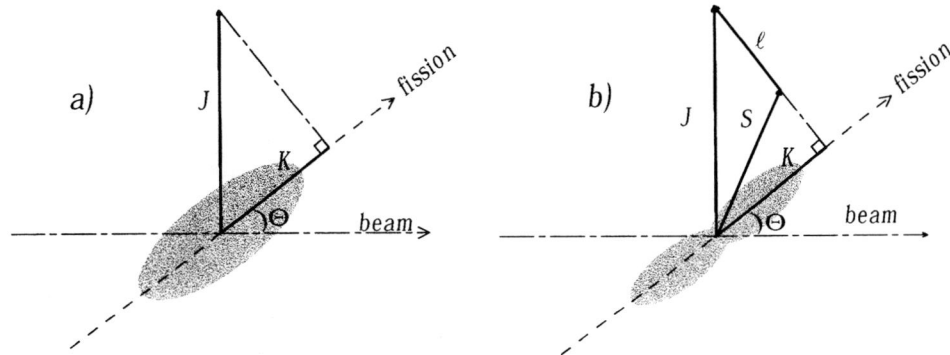

FIGURE 1. Schematic illustration of the relationship between J and K ($M=0$) for the saddle point model (a) and the relationship between J, ℓ, S and K for the scission point model (b).

$$\rho(K) = \exp\left(\frac{-K^2}{2K_0^2}\right), \qquad (2)$$

where

$$K_0^2 = \frac{\Im_\perp \Im_\parallel}{\hbar^2 (\Im_\perp - \Im_\parallel)} T, \qquad (3)$$

T is a nuclear temperature, $\Im_\parallel$ and $\Im_\perp$ are the parallel and perpendicular moments of inertia at the saddle point. SPM had a great success in description of the experimental data for the light particle induced reactions [1]. However, SPM loses its validity for heavy fissioning nuclei with high J, especially when the angular momentum reduces the fission barrier to a value similar or smaller than the nuclear temperature. For the last case, the scission point model (SCPM) was developed [3]. The principal assumption of the SCPM is the statistical partition of the total angular momentum J into orbital angular momentum ℓ and channel spin S (vector sum of the two fragment spins) at the scission point. Therefore, in SCPM the angular distribution is determined by the phase space available at the scission point, and the model is based on the assumption of very quick equilibrating the rotational degrees of freedom (the equilibrating time is substantially shorter than the saddle-scission time). In practice, within the SPCM the fragment angular distributions are calculated by the same relation as in SPM (1). However, K is treated as the projection of S onto the symmetry axis of the fissioning system (see fig.1b). For the simple case of a symmetric split of nucleus consisting of two spheroids joined on the long axis,

$$\rho(K) = \exp\left(\frac{-K^2}{2S_0^2}\right), \qquad (4)$$

here

$$S_0^2 = \frac{2\Im_{\parallel} T(\Im_{\perp} + \mu R_c^2)}{\hbar^2 (\mu R_c^2 + 2\Im_{\perp} - 2\Im_{\parallel})} , \qquad (5)$$

and $\Im_{\parallel}$ and $\Im_{\perp}$ are the parallel and perpendicular moments of inertia of the spheroids, R_c is the distance between the centers of mass of the fission fragments, and μ is the reduced mass of the fission channel. Here it should be noted that in heavy ion induced reactions experimental anisotropies of the fission fragment angular distribution are spread in between the predictions of both SPM and SCPM [4,5]. Perhaps this experimental data may be explained assuming that the equilibrating time for the rotational degrees of freedom is smaller than the saddle-scission time. In this situation, it is desirable to develop a new model of the fission fragment angular distribution, which takes into account both the time evolution of the rotational degrees of freedom and the dynamical aspects of the fission processes. In this work, we suggest a dynamical model of the fission fragment angular distribution, which takes into account the stochastic aspects of nuclear fission.

MODEL

In the frames of our model the dynamics of induced nuclear fission is considered in the stochastic approach [6,7] by the Langevin equations for the collective coordinate r (distance between the centers of mass of the forming fission fragments) and the corresponding momentum p:

$$\frac{dr}{dt} = \frac{p}{m(r)}$$

$$\frac{dp}{dt} = -\frac{1}{2}\left[\frac{p}{m(r)}\right]^2 \frac{dm(r)}{dr} - \frac{\partial F}{\partial r} - \beta(r)p + f(t) , \qquad (6)$$

In eq. (6) $f(t)$ is the random force with the following properties: $<f(t)>=0$, $<f(t_1)f(t_2)>=2D\delta(t_1-t_2)$; D is expressed, through the Einstein relation, in terms of the nuclear temperature T and the nuclear viscosity coefficient γ as $D = T\gamma$; $\beta = \gamma/m$ is the damping coefficient in the fission mode and m is the inertial parameter, which is calculated in the frames of the Werner–Wheeler approach. The conservative forces are calculated by the free energy of the excited nuclear system, $F(r, T, J, K) = V(r, J, K) - a(r)T^2$. The nuclear temperature is defined as $T = (E_{int}/a(r))^{1/2}$ with $E_{int} = E^* - p^2/(2m) - V(r) - E_{rot}(J,K)$, where E^* is the total excitation energy, $E_{rot}(J,K)$ is the rotational energy. In this work, the level density parameter is chosen in the form $a(r) = a_1 A + a_2 A^{2/3} B_s(r)$, where $B_s(r)$ is the surface energy of the deformed nucleus, a_1 and a_2 are taken from [8]. The potential energy $V(r)$ is calculated within the liquid drop model with the Myers-Swiatecki parameters by using the procedure proposed in [9] and taking into account its dependence on the K value:

$$V(r,J,K) = B_S(r)E_S^0 + B_C(r)E_C^0 + \frac{[J(J+1)-K^2]\hbar^2}{2\Im_\perp} + \frac{K^2\hbar^2}{2\Im_\parallel}. \quad (7)$$

Here, $B_c(r)$ and $B_s(r)$ are dimensionless functionals of Coulumb and surface energies, E_s^0 and E_c^0 are the surface and Coulomb energies of the spherical system.

Initial r, p, J and K are generated for each Langeven sample on the base of the next distribution:

$$W(r,p,J,K) = \frac{1}{\sqrt{2\pi \, mT}} \exp\left(-\frac{p^2}{2mT}\right) \delta(r-r_{eq}) \, P_J(K) \frac{d\sigma}{dJ}, \quad (8)$$

where r_{eq} is a collective coordinate at the equilibrium deformation, $P_J(K)$ is an initial K distribution at given J (it is described bellow). Light particle (n,p and α) emission is simulated by means of the Monte-Carlo method [7,10]. More detailed description of the method is presented in [10].

In this work, K is treated as a thermodynamically fluctuating value. Thus, K experiences few jumps during the evolution of the fissioning system and at the scission point every fission event is characterized by J (changes because of light particle emission) and K. In order to simulate such evolution of K we generate a random number ξ uniformly distributed in the interval $[0,1]$ for every time step h of integration of the Langeven equations. Then if the condition $\xi < h/\tau$ is satisfied we chouse a new K value from the next distribution

$$P(K) \propto \exp\{-\Delta F(r,T,J,K)/T\} \quad (9)$$

taking into account that $-J \le K \le J$. Finally, the fission fragment angular distribution is calculated by the relation:

$$W(\theta) = \frac{1}{N_f} \sum_{i=1}^{N_f} \frac{1}{2}(2J_i+1) \left| d_{0,K_i}^{J_i}(\theta) \right|^2, \quad (10)$$

where N_f is a number of Langeven samples which have fissioned, J_i is the angular momentum for the i-th Langeven sample. It should be noted that the described calculations reproduce the well-known random process of Kubo-Andersen [11,12]. It is simple to show that the process is characterized by the next conditional probability

$$P(K,\Delta t | K_0) = e^{-\Delta t/\tau} \delta(K-K_0) + (1-e^{-\Delta t/\tau}) P(K), \quad (11)$$

As one can see from (11) this process has a typical relaxation nature. To illustrate it we simulated the evolution of K by the method described above for the ^{248}Cf nucleus ($J = 30\hbar$, $T = 1.4$ MeV) and $\beta = 5 \times 10^{21}$ s^{-1}. Thereupon the probability of a jump to a new K value during the time interval Δt was reconstructed from the all data file of the points in time for the K jumps (see Fig. 2).

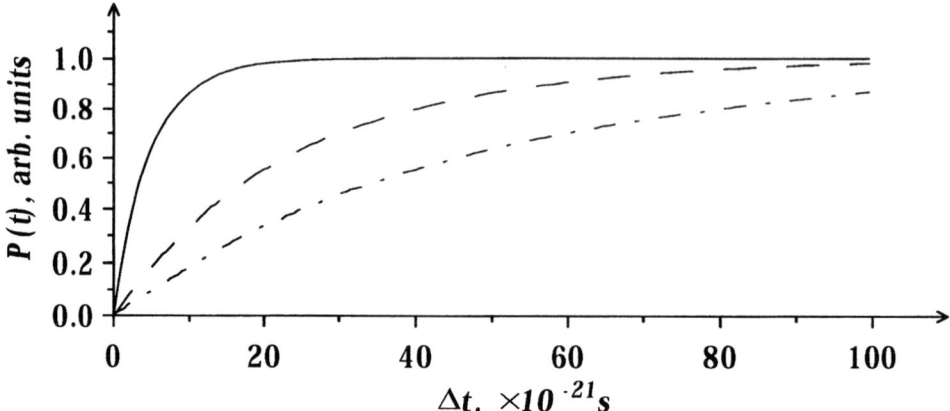

FIGURE 2. Probability of a jump to a new K value versus Δt with $\tau = 5. \times 10^{-21}s$ (solid line), $\tau = 25.\times 10^{-21}s$ (dashed line) and $\tau = 50.\times 10^{-21}s$ (dashed-dotted line).

As another illustration of our method, Fig. 3 presents the potential energy versus of r and K with typical Langeven samples for different equilibrating times. These calculation were also performed for the ^{248}Cf nucleus ($J=30\hbar$, $T = 1.4$ MeV and the fission barrier $B_f = 1.5$ MeV) and for $\beta = 5 \times 10^{21}$ s^{-1}. The fission times for every Langeven sample are listed in the figure caption. Here for the smallest relaxation time one can observe a lot of jumps through the whole evolution up to the scission. It means that the K distribution is formed near the scission point. Evidently, this is situation of SCPM. As the equilibrating time increases, the jumps have been grouped in the range before the saddle more and more. In these cases, the system keeps a memory about the K distribution formed before the scission. For the biggest equilibrating time, presented here, all jumps proceeds before the saddle. So, for comparable equilibrating and fission times the K distribution will be formed near the equilibrium deformation. Here, it should be noted, the important requirement imposed on the new model implies that it has to reproduce the experimental data not worse than SPM throughout the wide ranges of nuclear temperatures, angular momenta and atomic numbers. In other words for fission of nuclei with the temperatures under the fission barriers it is desirable that the system keeps a memory about the K distribution, which was near the saddle point. This requirement may be fulfilled assuming the temperature dependent τ, namely decrease of τ with T. In the last case if τ was smaller than the saddle-scission time for $T<B_f$, the fissioning system should keep a memory on the K distribution of the saddle point of the fission barrier. Another attractive possibility is the deformation dependent τ (small before and large beyond the saddle) [13]. For lack of such information, we employed the assumption on independence of the equilibrating time for the tilting mode on deformation and nuclear temperature.

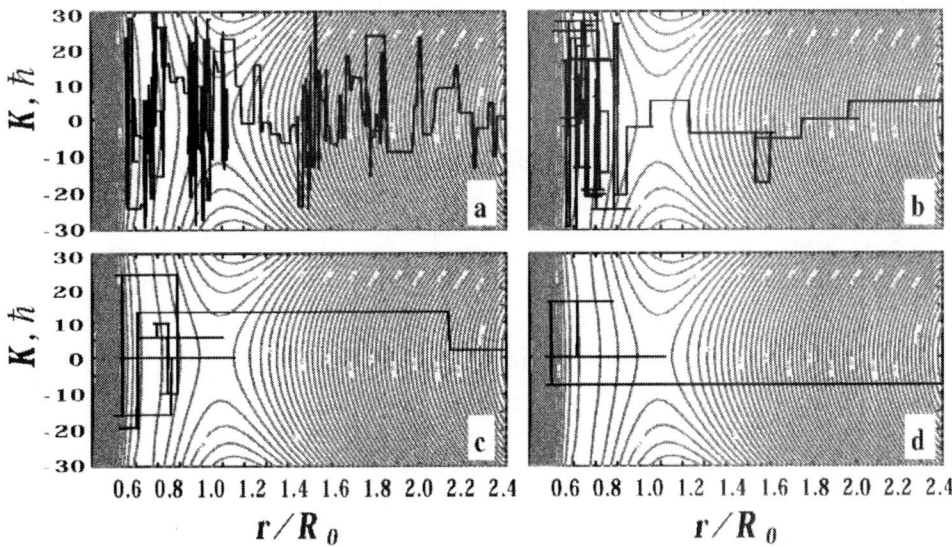

FIGURE 3 The potential energy versus of r (R_0 is a radius of the spherical nucleus) and K with typical Langeven samples for different equilibrating and fission times: (a) – $\tau = 0.1 \times 10^{-21}$ s and $\tau_f = 18. \times 10^{-21}$ s; (b) – $\tau = 1.0 \times 10^{-21}$ s and $\tau_f = 48.0 \times 10^{-21}$ s; (c) – $\tau = 10.0 \times 10^{-21}$ s and $\tau_f = 66.0 \times 10^{-21}$ s; (d) – $\tau = 100.0 \times 10^{-21}$ s and $\tau_f = 214.0 \times 10^{-21}$ s.

ANALYSIS OF THE EXPERIMENTAL DATA

In the present work, we analyze the experimental data on the anisotropy of the fragment angular distribution for the $^{16}O + ^{232}Th$, ^{238}U and ^{248}Cm reactions with the oxygen energies from 90 to 160 MeV. The initial J distributions ($d\sigma / dJ$) were calculated using the parameterization based on the surface friction model [7]. The examples of the J distributions are presented in Fig. 4. Such choosing of the reactions was dictated by the requirements of the negligibly small contribution of quasifission to the total fragment yields. Here, it is necessary to stress, that the quasifission is characteristic process for the reactions that are induced on heavy target nuclei by the $A \geq 20$ projectiles [5]. An addition point to emphasize is that the reactions lead to formation of the heavy compound nuclei with the fission barriers reduced by the angular momentum to values similar or smaller than the nuclear temperatures for the projectiles energy considered here. Fig. 5 shows the ratio of the fission barrier $B_f(J,K=0)$ to the nuclear temperature $T_{eq}(J)$ at the equilibrium deformation for the partial wave distributions of Fig. 4. As is seen from comparisons of Fig. 4 and Fig 5 essential parts of the partial wave distributions correspond to $T(J)>B_f(J,K=0)$. It is the situation when the saddle point no longer controls the fission. So, use of SPM in analysis of the experimental data on the angular distribution of the fission fragments for the reactions under study is not proper.

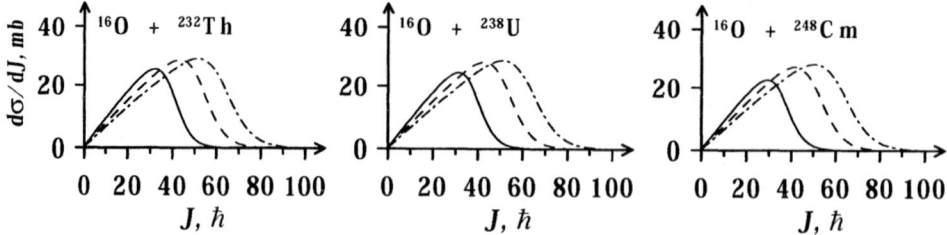

FIGURE 4. The example of the initial J distributions calculated by the parameterization of [7] for E_{cm} = 100 MeV (solid lines), E_{cm} = 120 MeV (dashed lines) and E_{cm} = 140 MeV (dashed-dotted lines).

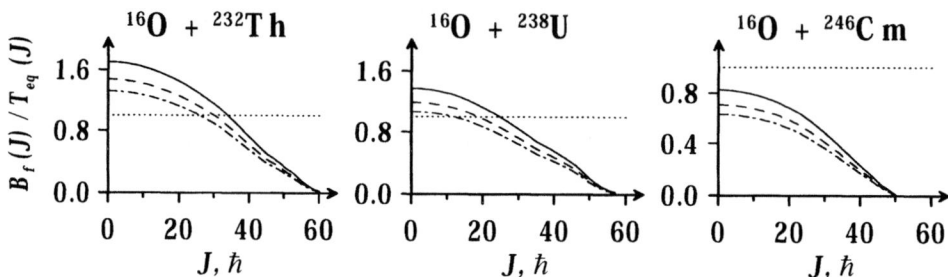

FIGURE 5. The ratio of fission barrier to the nuclear temperature at the equilibrium deformation versus J. The solid lines are calculated for E_{cm} = 100 MeV, the dashed lines for E_{cm} = 120 MeV and the dashed-dotted lines for E_{cm} = 140 MeV. The dotted lines mark the $B_f(J,K=0)/T_{eq}(J) = 1$.

For the damping coefficient β, we used the one body dissipation model with reduction factor for the wall formulae $k_s = 0.26$ [14]. Such value of k_s was obtained in the analysis of the kinetic energy of fission fragments [14] and in the analysis of the prescission neutron multiplicities [15]. Fig. 6 presents comparison of our calculations with the experimental prescission neutron multiplicity for the $^{16}O+^{232}Th$ reaction (for other reactions such data is absent). The good coincidence of the experimental data with the calculations allows us to hope that our description of the fission time is correct.

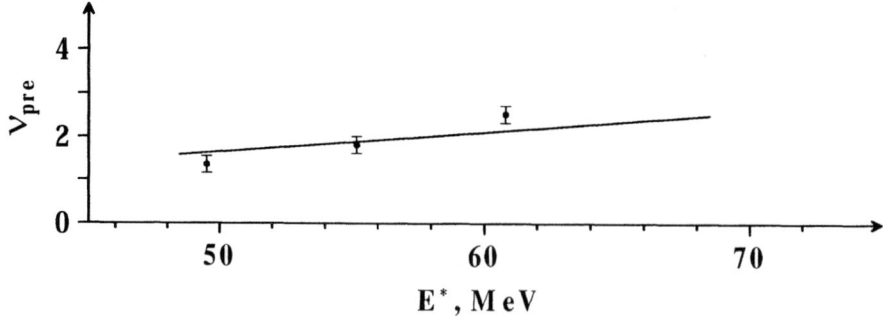

FIGURE 6 Prescission neutron multiplicity for the $^{16}O + ^{232}Th$ reaction. The points are the experimental data from [16], the solid line presents our calculation.

In order to describe the experimental data on the angular anisotropies of the fission fragment yields as well as possible we varied the τ value. Fig. 7 presents comparisons of the calculation results with the experimental data. We achieved the best description of the experimental angular anisotropy with $\tau = 21.0 \times 10^{-21}$ s.

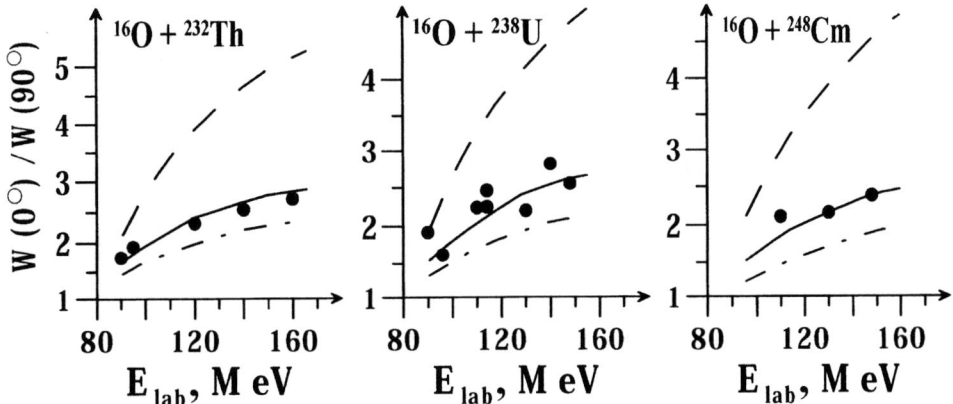

FIGURE 7. Fission fragment angular anisotropy. The points are the experimental data from [17], the solid lines present the calculations performed by the present model; the dashed and dotted–dashed lines are the predictions of the saddle and scission point models, correspondingly.

In addition, the predictions of both SPM and SCPM are also shown in Fig 7. It is seen that the experimental data are between two calculations performed within the statistical models. As was mentioned above, the reason of the underestimation of the angular anisotropy by SPM is that the model losses its validity for the cases under study $(B_f(J,K) \leq T)$. Possibly, the disagreement of the experimental data with SCPM points to the fact that the assumption about quick equilibrating of the orientation mode in the saddle–scission range is not proper. Here, it should be recalled, that in the frames of our approach there is no assumptions about the special points on the potential energy surface, which are responsible for the K distribution and consequently for the angular distribution of the fission fragments. Instead of it, the K values at the scission are determined by frequency and magnitude of their fluctuations during previous history of the fissioning system. In such way, the K distribution is directly controlled by the equilibrating time of the tilting mode. The last allows us to hope that appropriate choose of the deformation (and / or energy) dependence of τ will results in the description of the experimental data in more wide nuclear temperature range including $T < B_f(J,K)$ and also for lighter fissioning nuclei (such work is under progress).

CONCLUSION

The dynamical model of the fission fragment angular distributions is suggested. The model takes into account thermal fluctuations of the orientation degree of freedom and the stochastic aspects of nuclear fission. The model may be a useful tool in analysis of the experimental data for fission of hot and rotating heavy compound nuclei when a validity of the traditional statistical models is questionable. Analisys of the experimental data for a set of complete fusion-fission reactions allowed us to obtain the equlibrating time for the tilting mode. We hope that an appropriate deformation and temperature dependences of the equilibrating time will allow the consideration of SPM as a special case of our model.

ACKNOWLEDGMENTS

This work was supported by the Russian Foundation for Basic Research (02-02-17077) and the State Program "Russian Universities" (UR.02.03.014).

REFERENCES

1. Vandenbosh, V., Huizenga J.R., in Nuclear fission, N.Y., Acad. Press, 1973, pp. 179-215.
2. Schmitt, R.P., Cooke, I., Dejbakhsh, H., Haenni, D.R., Shut, T., Srivastava, B.K., Utsunomiya, H., Nucl. Phys., **A592**, pp.130-150 (1995).
3. Rossner, H., Huizenga, J.R., Schroder, W.U., Phys. Rev. Lett., **53**, pp. 38-41 (1986).
4. Freifelder, R., Prakash, M., Alexander, J.M., Phys. Rep. **133**, pp. 315-355 (1986).
5. Newton, J., Sov. J. Particles and Nuclei **21,** pp. 349-383 (1990).
6. Abe, Y., Auik, S., Reinhardand, P.-G., Suraud, E., Phys. Rep. **275,** pp. 49-196 (1996).
7. Fröbrich, P., Gontchar, I.I., Phys. Rep. **292**, pp. 131-237 (1998).
8. Ignatyuk, A.V., Itkis, M.G., Smirenkin, G.I., Tishin., S.A., Soviet Journal of Nucl. Phys. **21**, pp. 1185-1205 (1975).
9. Lestone, J.P., Phys. Rev. C **51**, pp. 580-585 (1995).
10. Drozdov, V.A., Eremenko, D.O., Platonov, S.Yu., Fotina, O.V., Yuminov, O.A., Physics of Atomic Nuclei **64**, pp.179-185 (2001).
11. Bharucha-Reid, A.T., in Elements of the Theory of Markov Processes and their Application, New York, McGraw-hill, 1960.
12. Ormand, W.E., Camera, F., Bracco, A., Phys. Rev. Lett. **69**, pp. 2905-2907 (1992).
13. Leston, J.P., Phys. Rev.C **59**, 1540-1544 (1999).
14. J. R. Nix, A. J. Sierk, in Proceedings of the Intern. School-Seminar on Heavy Ion Physics, Dubna 1986, p. 453-464 (Dubna, 1987).
15. Drozdov, V.A., Eremenko, D.O., Platonov, S.Yu, Fotina, O.V., Yuminov, O.A., Bulletin. of Russian Academi of Science (physics) **64**, pp.84-87 (2000).
16. Saxena, A., Chatterjee, A., Choundrhury, R.K., Phys. Rev. **C49**, pp. 932-940 (1994).
17. Back, B.B., Bets, R.P., Gindler, J.E., Saini, S., Tsang, M.B., Gelbke, C.K., Lynch, W.G., McMahan, M.A., Baisden., P. A. , Phys. Rev.C **32**, pp. 195-213 (1985).

Tracking dissipation in capture reactions

T. Materna*, V. Bouchat *, V. Kinnard*, F. Hanappe*, O. Dorvaux†,
C. Schmitt†**, L. Stuttgé†, K. Siwek-Wilczynska‡, Y. Aritomo§¶,
A. Bogatchev§, E. Prokhorova§ and M. Ohta∥

*PNTPM, Université Libre de Bruxelles, Brussels, Belgium
†Institut de Recherches Subatomiques, Strasbourg, France
**GSI, Darmstadt, Germany
‡Warsaw University, Warsaw, Poland
§Flerov Laboratory of Nuclear Reactions, Joint Institute for Nuclear Research, Dubna, Moscow region, Russian Federation
¶Tokyo University, Tokyo, Japan
∥Kobe University, Kobe, Japan

Abstract. Nuclear dissipation in capture reactions is investigated using backtracing. Combining the analysis procedure with dynamical models, the difficult and long-standing problem of competition and mixing of quasi-fission and fusion-fission is solved for the first time. At low excitation energy a new protocol able to handle low statistics data gives access to the prescission neutron multiplicity in two different systems ^{48}Ca + ^{208}Pb, Pu. The results are in agreement with a domination of fusion-fission in the case of ^{256}No and an equal mixing of quasi-fission and fusion-fission in the case of Z = 114. The nature of the relevant dissipation is determined as one-body dissipation.

INTRODUCTION

Introduced by Kramers[1] very early after the discovery of nuclear fission, the role of the dissipation was only recognized in the 1970s with the emergence of deep inelastic collisions. Its nature, one-body dissipation (OBD) or two-body dissipation (TBD) and its magnitude and evolution with different parameters as the shape and the temperature, is still a matter of controversy.

A lot of experimental data obtained in most of the cases by the observation of the pre- and postscission emission of particles or γ-rays have been devoted to the determination of the dissipation. But until recently, depending on the experiments but also deeply on the models used to extract the dissipation coefficient (Kramer's γ coefficient), a dispersion of the results covering at least two orders of magnitude is observed. Different behaviours are extracted for the evolution with the temperature and no definitive conclusions can be drawn.

Even if coherent experimental data[2] are selected and analysed by the same dynamical model[3], a large spreading of the values for the dissipation coefficient is observed. For example, in figure 1 adapted from reference[4], the evolution with the temperature of the dissipation coefficient is presented for different reactions leading to very different fissioning systems. As pointed out by the authors, on one hand, for true fusion-fission systems ($Z_1 Z_2 \ll 1600$), the deduced γ coefficient ranges from 2 to 10 and is clearly

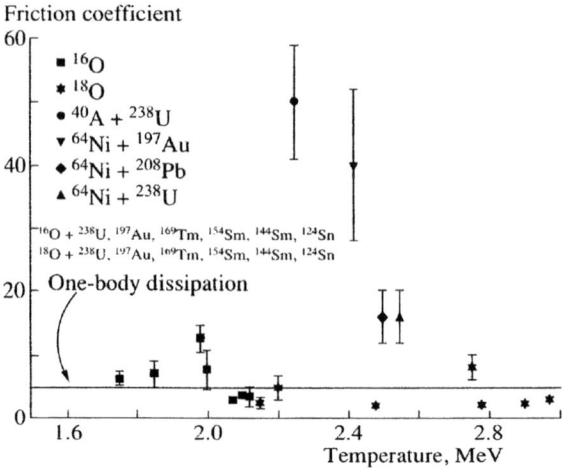

FIGURE 1. Evolution of the dissipation coefficient as a function of the temperature for different reactions. Data from Hinde et al[2].

compatible with OBD. No particular evolution with the temperature is observed. On the other hand, for systems with $Z_1 Z_2 > 1600$, the friction coefficient values are considerably larger and clearly not compatible with OBD. This puzzling behaviour for these systems is assumed to be due to TBD or to the expected mixing of mass-symmetric fragments coming from the different reaction mechanisms of capture reactions: quasi-fission and fusion-fission. Indeed, no separation between the two mechanisms was available and only the mean value of the multiplicity for neutron pre- and postscission emission could be obtained using the classical χ^2 minimization.

In this report we will show how a powerful analysis protocol, the backtracing[5], which is able to produce not only mean values but also correlations and distributions, can help us to solve this long-standing problem.

Recently applied to the Ni + Pb[6] and Ca + Th[7] reactions leading to isotopes of Z = 110, the backtracing procedure allowed us for the first time to clearly disentangle, at least intuitively, the contributions or quasi-fission and fusion-fission in the neutron prescission multiplicity distribution. A complete description of the backtracing procedure and its application to our case is described in references[6, 7, 8].

A SIMPLE CASE: FUSION-FISSION LEADING TO ^{126}BA

In order to validate the backtracing application to the determination of the distribution of pre- and postscission neutron multiplicity, we will first present results obtained for the ^{28}Si + ^{98}Mo reaction at 204 MeV.

This system has been investigated[8] at the VIVITRON, IReS, Strasbourg, using, as for all the data presented here, the DEMON neutron detector associated to parallel plates or CORSET setups for the detection of the reaction fragments.

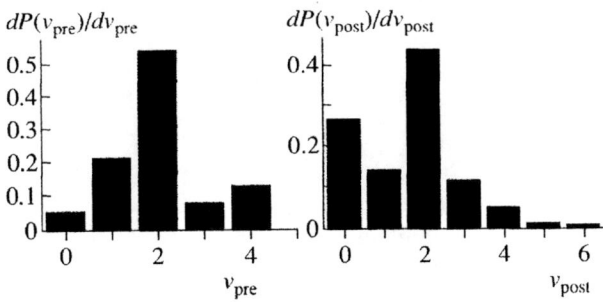

FIGURE 2. Pre- and postscission neutron multiplicity distribution for ^{126}Ba obtained by backtracing.

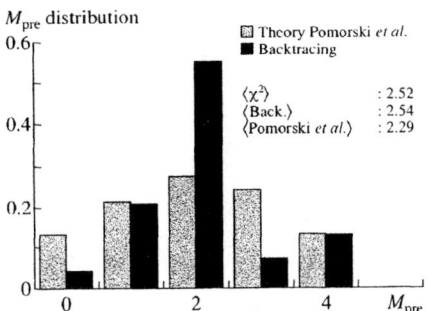

FIGURE 3. Comparison between the experimental prescission neutron multiplicity obtained by backtracing for the $^{28}Si + ^{98}Mo$ reaction at 204 MeV and the prediction of the model of Pomorski[8].

In such a low-mass system, obtained in a reaction with a low Z_1Z_2 product, only the fusion-fission mechanism is expected to contribute to the mass-symmetric fragment distribution.

Figure 2 shows the distributions of the pre- and postscission neutron multiplicities as obtained by the backtracing protocol. It must be noted that the mean values obtained here are in complete agreement with those deduced from the conventional χ^2 minimization. For instance the mean values of the neutron prescission multiplicity are 2.52 and 2.54 for the χ^2 and the backtracing, respectively (see figure 3). Figure 3 presents, in addition to the backtraced distribution, the theoretical results obtained using the dynamical model of Pomorski et al[9]. This model, based on the resolution of the one-dimensional Langevin equation, is able to describe only the fusion-fission process and is thus well adapted to the ^{126}Ba system. The excellent agreement observed between experimental data and model calculations is obvious. In particular one can note that the zero-neutron multiplicity channel is well reproduced. The model uses OBD (wall and window formula) and the agreement confirms that, at least when only the fusion-fission process is concerned and even if the model is not perfect – for instance, no dependence on temperature is included – there is no need to introduce TBD to reproduce the experimental data.

To our knowledge, this is the first time that the neutron prescission multiplicity distribution has been experimentally observed and compared with such a good agreement with dynamical calculations.

A MORE COMPLEX CASE: Z = 110

In this case, the fused Z = 110 nucleus is obtained by two different entrance channels (^{40}Ca or ^{58}Ni projectiles on ^{232}Th or ^{208}Pb targets) leading to the same high excitation energies ranging from around 60 MeV to more than 160 MeV. For such (super)heavy systems, the competition between quasi-fission and fusion-fission is expected to populate the symmetric part of the fragment mass distribution and until now there was no real way to disentangle these two contributions at these excitation energies. The experiments were carried out at SARA, Grenoble[6, 7].

Neutron pre- and postscission multiplicities (mean values) were first obtained by a χ^2 minimization and, as usually, led to the same difficulties as before: large values for γ are deduced using dynamical models. It must be noted that, when it was possible, our experimental results were compared to and found to be in complete agreement with those of Hinde et al[2].

In a second step, backtracing was applied and provided us with neutron pre- and postscission multiplicity distributions at least for the highest excitation energy for both systems. Indeed backtracing required very high statistics and has been applied only to these two cases.

Figure 4 presents the backtracing results for Ni + Pb (left) and Ca + Th (right). In both cases, the neutron pre- and postscission correlations exhibit two well-defined regions corresponding essentially to two different distributions for the prescission neutrons. Intuitively, we can think that each separated distribution can be attributed to each of the two capture processes. As expected, a low mean value (of the order of 4) of the neutron prescission multiplicity distribution can be associated to quasi-fission (faster mechanism), a larger mean value (around 7) to fusion-fission (slower mechanism).

Then, using the same HICOL + DYNSEQ code as the one used by K. Siwek-Wilczynska to deduce the dissipation coefficient for the Hinde et al data, we can reproduce our experimental backtraced correlations by two different scenarii: quasi-fission and fusion-fission (see rectangles and squares on fig. 4). They correspond to different angular momentum ranges obtained in HICOL by the comparison with the experimental mass distribution. In both systems, the same dissipation coefficients are needed to reproduce the experimental distributions (γ= 5 for quasi-fission and ranges from 5 to 11 for fusion-fission). This spectacular and first-time agreement with OBD (or with a value between OBD and two times OBD) reconciles completely these data with those corresponding to fusion-fission only and clearly supports the conclusion that, as soon as one is able to distinguish between quasi-fission and fusion-fission, no discrepancy remains and OBD is large enough to reproduce the experimental data.

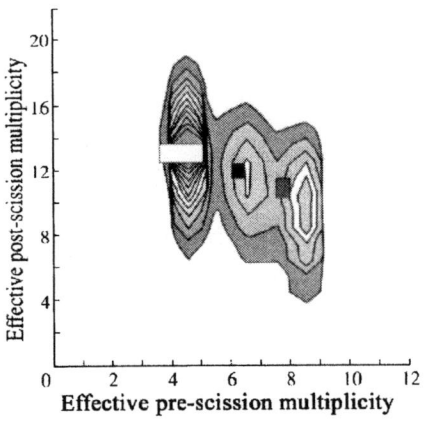 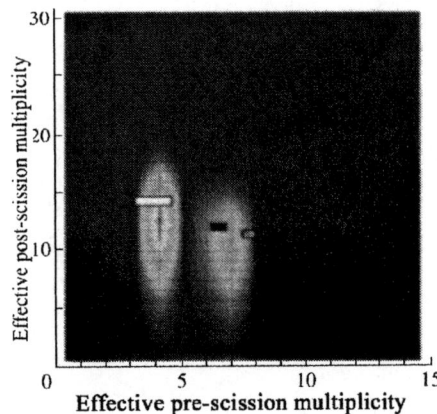

FIGURE 4. Pre- and postscission neutron multiplicity correlations for the systems $^{58}Ni + {}^{208}Pb$ (left) and $^{40}Ca + {}^{232}Th$ (right) at 186 MeV and 166 MeV excitation energy, respectively. Results of calculations using HICOL + DYNSEQ from Siwek-Wilczynska et al[3] are also shown. The rectangle to the left represents quasi-fission (30<l<120, where l is the angular momentum) with only one-body dissipation. The two squares stand for fusion-fission (0<l<30) with OBD (left) and two times OBD (right).

COMPARISON WITH LANGEVIN EQUATION MODELS

Figure 5 shows the distribution of the prescission multiplicity for the Ni + Pb system. Two free Gaussian curves have been fitted on the experimental backtraced distribution (full line) and are now considered to represent the distribution to be attributed to the quasi-fission and fusion-fission mechanisms. Calculations for the fusion-fission part, performed by Schmitt[10] using the Pomorski model, are also given in this figure (dotted curve). One can note that the mean value and the width obtained in this model are slightly too high, but the agreement can be considered as satisfactory with this OBD (only) symmetric fission model if one takes into account that it does not contain the dissipation in the entrance channel and that no dependence of friction on shape nor temperature is considered. These points are under consideration.

Recently, a three-dimensional Langevin equation model has been developed and applied to the dynamics of capture reactions, in particular, in the superheavy region[11]. A first comparison of these calculations with our experimental result for the Ni + Pb reaction is presented in figure 6[12].

If one takes into account that these preliminary calculations include all the events associated not only with quasi-fission and fusion-fission but also with deep-inelastic processes, the overall agreement is satisfactory. It seems obvious that, as soon as the experimental mass and total kinetic energy cuts are included in the calculations, the agreement will be better. Indeed, deep-inelastic processes are known to be low prescission multiplicity events and the symmetric fusion-fission process corresponds to the largest muliplicity.

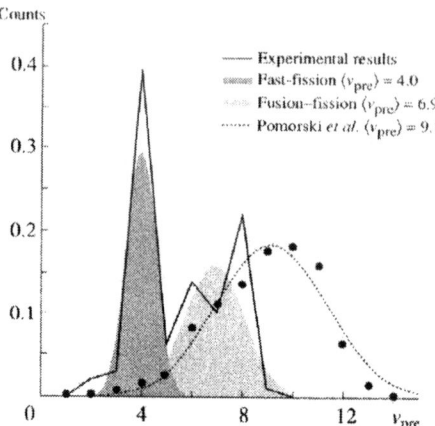

FIGURE 5. Comparison between the experimental prescission neutron multiplicity distribution obtained by backtracing for the $^{58}Ni + {}^{208}Pb$ system at 186 MeV excitation energy (black curve) and the prediction of the Pomorski model[8] (points). The full Gaussian curves are a fit to the experimental distribution, whereas the dashed Gaussian one is a fit to the model.

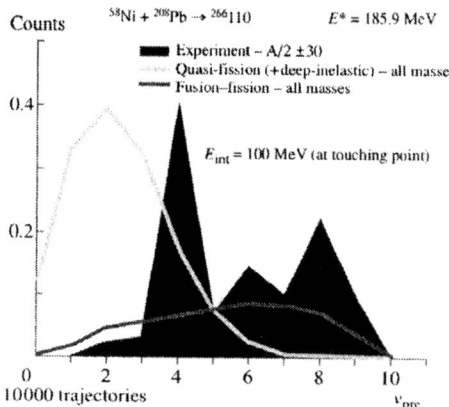

FIGURE 6. Comparison between the experimental prescission neutron multiplicity distribution obtained by backtracing for the ^{58}Ni + ^{208}Pb system at 186 MeV excitation energy for fragment masses in the range of $\frac{A}{2} \pm 30$ (full black curve) and the prediction of the Aritomo model[11] for quasi-fission and fusion-fission for the whole mass range of deep-inelastic, quasi-fission and fusion-fission processes.

AT LOW EXCITATION ENERGY

Low excitation energies, which means the ones required for the superheavy synthesis, represent a very serious problem in the determination of the dynamical aspects of capture reactions. Indeed, regardless of the decreasing of the cross-section, which is already a severe limitation, the prescission multiplicity will also drop and can represent

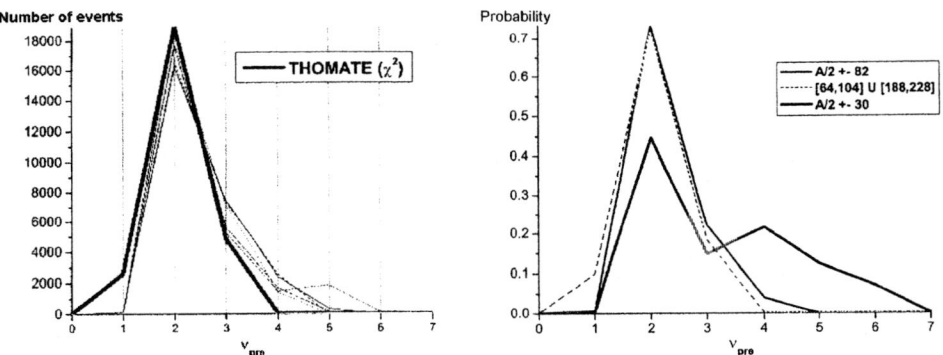

FIGURE 7. Prescission neutron multiplicity distributions for Ca + Pb (left) and Ca + Pu (right). In the Pu case, two distributions appear, a one-bump distribution corresponding to quasi-fission events with mass around 208 and a two-bump distribution corresponding to symmetric-mass fragments.

a real difficulty for the backtracing since it requires very high statistics for the neutron experimental observables.

A new protocol (THOMATE), able to handle low statistics data[13], has been developed and applied to ^{48}Ca + ^{208}Pb, Pu at 40 MeV of excitation energy. The neutron prescission multiplicity distributions obtained in both cases are presented on figure 7 for the symmetric part of the mass distribution. In the ^{256}No case, as expected, the prescission multiplicity distribution exhibits only one bump, confirming the domination of the fusion-fission process. In the Z = 114 case, the distribution shows two bumps in agreement with 50-50 % of quasi-fission and fusion-fission. A first preliminary comparison with Langevin calculations using OBD is in agreement with the experimental results[12].

ACKNOWLEDGMENTS

This work has been partly supported by INTAS 97-11929 and 00-655. It's also a pleasure to thank all our colleagues from the DEMON-CORSET collaboration for all the experimental work performed at the U400 of FLNR, Dubna and at the VIVITRON, Strasbourg.

REFERENCES

1. H. Kramers, *Physica* **7**, 284 (1940).
2. D. Hinde et al., *Phys. Rev.* **C45**, 1229 (1992).
3. K. Siwek-Wilczinska et al., *Phys. Rev.* **C51**, 2054 (1995).
4. J. Wilczinski et al., *Phys. Rev.* **C54**, 325 (1996).
5. P. Désesquelles et al., *Nucl. Phys.* **A604**, 183 (1996).
6. L. Donadille et al., *Nucl. Phys.* **A656**, 259 (1999).
7. B. Benoit, Ph.D. Thesis, Univ. of Brussels, Brussels, 2000.

8. E. de Góes Brennand, Ph.D. Thesis, Univ. of Brussels, Brussels, 2000.
9. K. Pomorski *et al.*, *Nucl. Phys.* **A679**, 25 (2000).
10. C. Schmitt, Ph.D. Thesis, Univ. of Strasbourg, Strasbourg, 2002.
11. Y. Aritomo, in *Exotic Nuclei, EXON-2001 — Proc. Lake Baikal Int. Symp. on Exotic Nuclei*, ed. Yu. E. Penionshkevich and E. A. Cherepanov (World Scientific, Singapore, 2002), p. 106.
12. Y. Aritomo and T. Materna, work in progress.
13. T. Materna, Ph.D. Thesis, Univ. of Brussels, Brussels, 2003 and submitted to NIM.

Dynamics of fusion-fission process in superheavy mass region

Y. Aritomo*, M. Ohta[†], T. Materna**, F. Hanappe** and L. Stuttge[‡]

*Department of Physics, University of Tokyo, Hongo, Bunkyo-ku, Tokyo, Japan; Flerov Laboratory of Nuclear Reactions, JINR, Dubna, Russia
[†]Department of Physics, Konan University, 8-9-1 Okamoto, Kobe, Japan
**Universite Libre de Bruxelles, 1050 Bruxelles, Belgium
[‡]Institut de Recherches Subatomiques, F-67037 Strasbourg Cedex, France

Abstract. The fusion-fission process for the synthesis of superheavy elements is discussed on the basis of the fluctuation-dissipation model. Recently the experiments at Dubna on fission of superheavy nuclei were carried out, and the mass and total kinetic energy distributions of fission fragments were measured. By analyzing the mass distribution of fission fragments, we can distinguish between fusion-fission process and quasi-fission process. We employ three-dimensional Langevin equation. We find almost all of the mass symmetric fission events come from the quasi-fission process in the superheavy mass region. In order to classify the fusion-fission paths and compare with the experimental data directly, we analyze the pre-scission neutron emission in the correlation with fission fragments. The neutron multiplicity depends on the travelling time of the trajectory. It is useful to investigate the fusion-fission process.

INTRODUCTION

In the superheavy mass region, the nuclear shell model predicted the existence of the Island Stability [1]. The most recent experiments on the synthesis of superheavy elements are approaching to the region and supports this hypothesis. The production of new superheavy elements are reported at Dubna and GSI [2, 3, 4]. Riken group succeeded to reproduce $Z=110$ and 111 in cold fusion reactions and measured the excitation function with high accuracy [5].

Also many theoretical papers have been published and they reproduced the excitation function of evaporation residue cross section. However, up to now we are still at the stage of refining the model and investigating the property of undecided parameters. In order to estimate the possibility of synthesis of new elements, we have to clear up the fusion-fission (FF) mechanism. We do not directly discuss the evaporation residue cross section here, because there are many uncertainty parameters, especially in statistical model in superheavy mass region. In FF process, a lot of relevant experimental data become available recently. For example, the mass and total kinetic energy distributions of fission fragments were extensively measured at Dubna [6]. The emission of neutrons in coincidence with fission fragments has been studied by DeMon group [7]. In this paper, using the concomitant data with fusion-fission process we try to investigate this process dynamically for aiming the more precise determining of the fusion-fission cross section in superheavy mass region.

In order to treat the process, we use the fluctuation-dissipation model. We employ the Langevin equation [8, 9, 10, 11, 12]. By analyzing the mass distribution of fission fragments, we can distinguish between fusion-fission process and quasi-fission process (QF). We find almost all of the mass symmetric fission events come from the quasi-fission process in the superheavy mass region. However, by the experimental data of mass distribution of fission fragments, we can not know where mass symmetric fission event comes from, fusion-fission process or quasi-fission process.

In order to compare the calculation with the experimental data directly, we analyze the pre-scission neutron emission in the correlation with fission fragments. The neutron multiplicity depends on the travelling time of the trajectory. It is useful to investigate the fusion-fission process. We introduce the effect of the neutron emission with the Langevin calculation. We combine the Langevin code with the statistical code, and analyze the experimental data.

The main purpose of the present paper is to elucidate the mechanism of fusion-fission process.

In section 2, we explain our framework and model. In section 3, we discuss the mass distribution of fission fragment. By analyzing the Langevin trajectories in three dimensional coordinate space, we discuss the fusion-fission mechanism. We present the results for the excitation function of fusion-fission cross section in the reactions ^{48}Ca+^{244}Pu. In section 4, we present the neutron multiplicity in correlation with fission fragments.

MODEL

Usually, the fusion-fission cross section σ_{ER} is expressed as;

$$\sigma_{ER} = \frac{\pi \hbar^2}{2\mu_0 E_{cm}} \sum_{l=0}^{\infty} (2l+1) T_l(E_{cm}, l) P_{CN}(E^*, l) W(E^*, l), \qquad (1)$$

where μ_0 denotes the reduced mass in the entrance channel. E_{cm} and E^* denote the incident energy in the center-of-mass frame and the excitation energy of the composite system, respectively. $P_{CN}(E^*, l)$ is the probability of forming a compound nucleus in competition with quasi-fission. $T_l(E_{cm}, l)$ is the barrier penetration coefficient of the lth partial wave through the potential barrier. $T_l(E_{cm}, l)$ is calculated with the empirical coupled channel model which is suggested by Zagrebaev [13]. $W(E^*, l)$ denotes the survival probability of compound nuclei during deexitation, which is calculated by statistical model.

Our final goal is to estimate σ_{ER}. However, it is very difficult now because there are many unknown parameters at each stages of the reaction. Here, we focus our attention to the FF process which is rather complicated. We discuss on the FF probability $T_l P_{CN} (1 - W)$.

In order to calculate P_{CN}, we introduce the fluctuation-dissipation model and employ the Langevin equation. We adopt a three-dimensional nuclear deformation space with two-center parametrization [14, 15]. The neck parameter ε is fixed to 1.0 in the present calculation. The three collective parameters to be described by the Langevin equation are treated as follows: z_0 (distance between two potential centers), δ (deformation of fragments) and α (mass asymmetry of the colliding partner); $\alpha = (A_1 - A_2)/(A_1 + A_2)$, where A_1 and A_2 denote the mass number of target and projectile, respectively.

The multi-dimensional Langevin equation is given in the following form,

$$\frac{dq_i}{dt} = (m^{-1})_{ij} p_j,$$
$$\frac{dp_i}{dt} = -\frac{\partial V}{\partial q_i} - \frac{1}{2}\frac{\partial}{\partial q_i}(m^{-1})_{jk} p_j p_k - \gamma_{ij}(m^{-1})_{ik} p_k + g_{ij} R_j(t), \quad (2)$$

where summation over repeated indicates is tacitly assumed. V is the potential energy, m_{ij} and γ_{ij} are the shape-dependent collective inertia and dissipation tensors, respectively. The normalized random force $R_i(t)$ is assumed to be a white noise, i.e., $\langle R_i(t) \rangle = 0$ and $\langle R_i(t_1) R_j(t_2) \rangle = 2\delta_{ij}\delta(t_1 - t_2)$. The strength of random force g_{ij} is given by $\gamma_{ij} T = \sum_k g_{ij} g_{jk}$, where T is the temperature of the compound nucleus calculated from the intrinsic energy of the composite system as $E_{int} = aT^2$ with a denoting the level density parameter. Intrinsic energy of the composite system E_{int} is calculated for each trajectory as,

$$E_{int} = E^* - \frac{1}{2}(m^{-1})_{ij} p_i p_j - V(q, l, T), \quad (3)$$

where E^* is the excitation energy of the composite system which is given as $E^* = E_{cm} - Q$ with Q denoting the Q-value of the reaction. At $t = 0$, each trajectory starts from the contact configuration with the initial velocity in the z_0 direction.

A hydrodynamical inertia tensor is adopted with the Werner-Wheeler approximation for the velocity field, and the wall-and-window one-body dissipation is adopted for the dissipation tensor [16, 17, 18].

RESULTS

mass distribution of fission fragments and fusion-fission cross section

First, we try to reproduce the experimental data of the mass and the total kinetic energy (TKE) distribution of fission fragment [6]. By analyzing the mass distribution, we can distinguish between FF process and QF process. The mechanism of FF and QF are clarified by analyzing the trajectory on the three-dimensional potential energy surface.

Figure 1 show the potential energy surface of liquid drop model with shell correction energy in nuclear deformation space for $^{256}102$, which is calculated using the code [19, 20]. The coordinate z is defined as $z = z_0/(R_{CN} B)$, where z_0 and R_{CN} denote the distance between two potential centers and the radius of the spherical compound nucleus, respectively. The parameter B is defined as $B = (3 + \delta)/(3 - 2\delta)$. By this scaling, we can save a great deal of computation time. The position at $z = \alpha = 0$ corresponds to a spherical compound nucleus. Due to the shell structure of Pb and Sn, we can see the valley on the potential surface. Black arrow denotes the injection point.

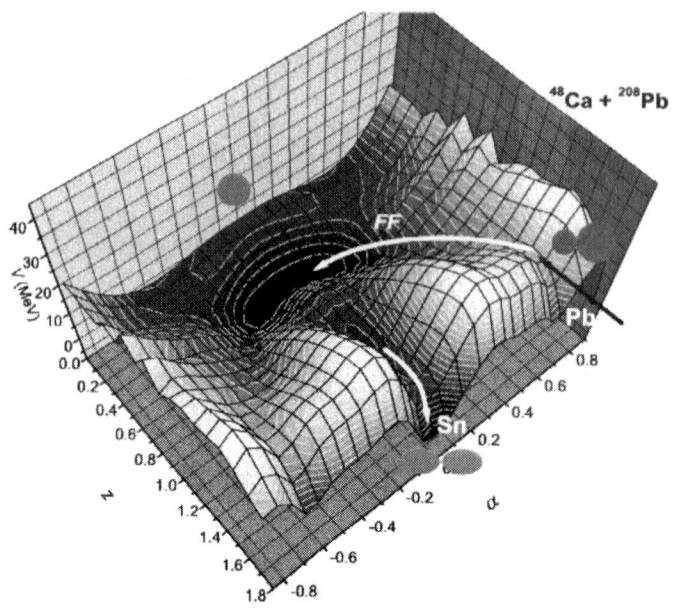

FIGURE 1. The potential energy surface of liquid drop model with shell correction energy in nuclear deformation space for $^{256}102$. z and α denote the separation between two potential center and the mass asymmetry, respectively. The white arrows indicate the fusion-fission paths.

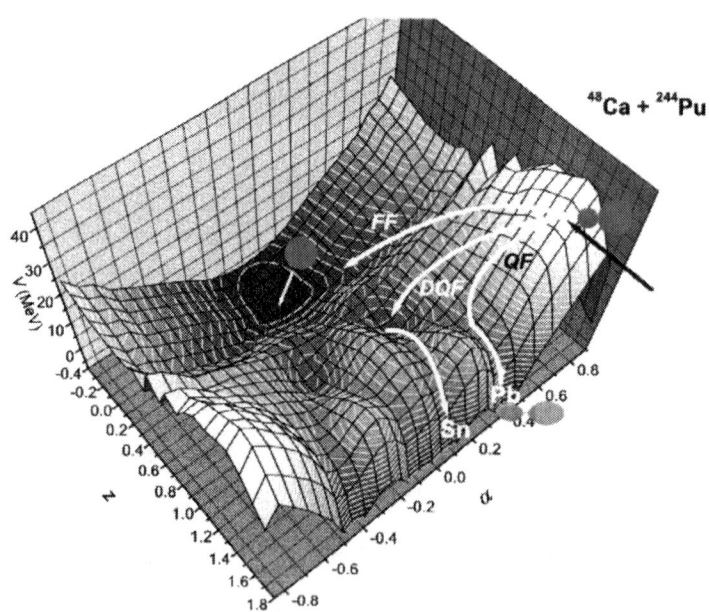

FIGURE 2. The potential energy surface of $^{292}114$. The QF, DQF and FF are denoted by white arrows.

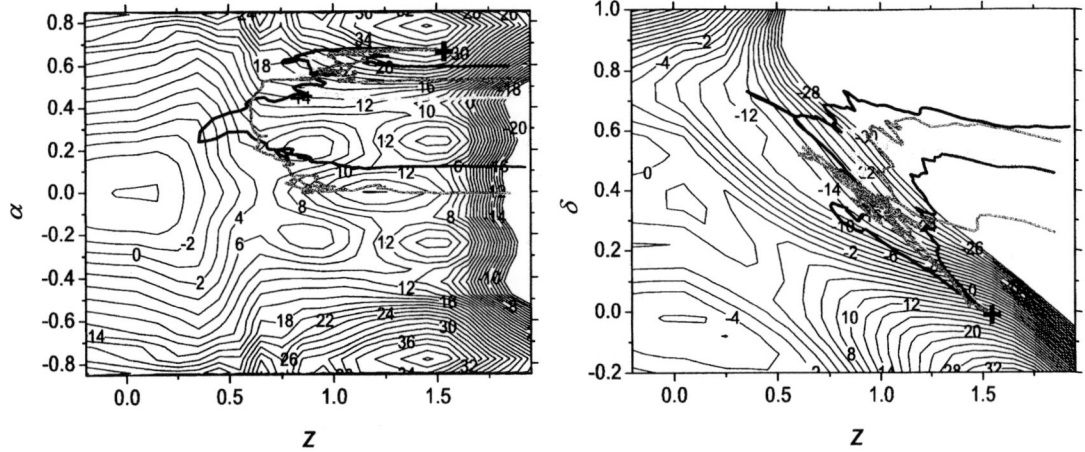

FIGURE 3. Samples of the trajectory which are projected on $z - \alpha$ plane (a) and $z - \delta$ plane, at $E^* = 33$ MeV in the reaction ^{48}Ca+^{244}Pu. The trajectories go to $+\delta$ direction, and can not enter the region of spherical nucleus.

We can see the FF process is dominant and mass symmetric fission events are detected. The white arrows indicate the fusion-fission paths.

However, in the case of ^{48}Ca+^{244}Pu the situation changes. The potential energy surface of ^{292}114 are shown in Fig. 2. In this case, QF process is dominant. Only few probability enters the spherical region.

At Dubna the experiments on the fission of superheavy nuclei in the reaction ^{48}Ca+^{244}Pu carried out and they present the FF cross section of compound nuclei which is derived from the mass symmetric fission fragments ($A/2 \pm 20$) [6]. The subsequent important question is whether all of the mass symmetric fission fragments come from the compound nuclei or not. As the final results of the experiments the mass symmetric fission fragments are detected, but there exists two possibilities where it comes from. One is that the mass symmetric fission fragments come from the compound nuclei and the other is that they come from QF.

We try to check them by using three-dimensional trajectory calculation with Langevin equation. In the results, about 90~99 % of mass symmetric fission fragments are found to come from QF process, which we call *"deep quasi-fission process" (DQF)*. We can see it more precisely in the reference [21, 22].

In heavy mass region, due to the large Coulomb repulsion force, potential energy surface has very steep slope in the direction of $+\delta$. Therefore, even if trajectories overcome the fusion barrier, almost all of them flow down to the $+\delta$ direction and

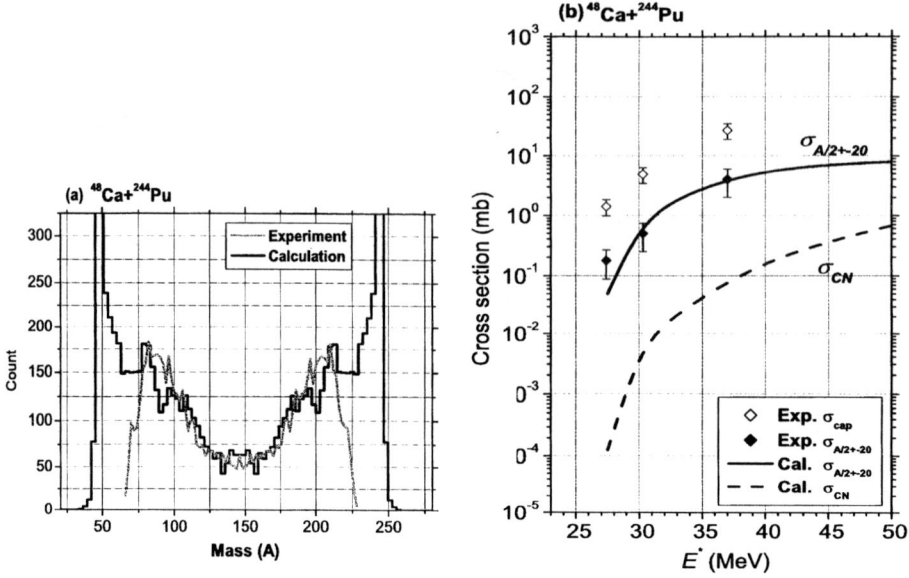

FIGURE 4. (a) Mass distribution of fission fragments in the reaction ^{48}Ca+^{244}Pu at E^*=37 MeV. The black line and gray line denote the calculation and experiment, respectively. (b) The cross section by the calculation and experiment. Lines and diamonds are given in the text.

can not reach the spherical region. Figure. 3 shows the samples of trajectory which are projected on the $z-\alpha$ and $z-\delta$ plane, in the reaction ^{48}Ca+^{244}Pu at $E^* = 33$ MeV. The cross point (+) denotes the touching point of the system. We can see that trajectories go to $+\delta$ direction, and can not enter the region of spherical nucleus. That is reason why the formation of compound nucleus is extremely reduced in heavy mass region. We can say that fusion is hindered by the flow of the trajectory to the δ degree of freedom which leads to quasi-fission. We would like to emphasis that one- or two-dimensional calculation is not enough to describe the fusion-fission process in superheavy mass region. Three- or more-dimensional model should be used to calculate the fusion-fission probability.

Figure 4 (a) shows mass distribution of fission fragments in the reactions ^{48}Ca+^{244}Pu at $E^*= 37$ MeV. The mass asymmetric fission events is dominant. The calculation agrees with experimental data. The excitation function of the cross section are shown in Fig. 4 (b). The open and closed diamonds denote the capture cross section σ_{cap} and the cross section $\sigma_{A/2\pm20}$ which derived by the yield of the mass symmetric fission frag-

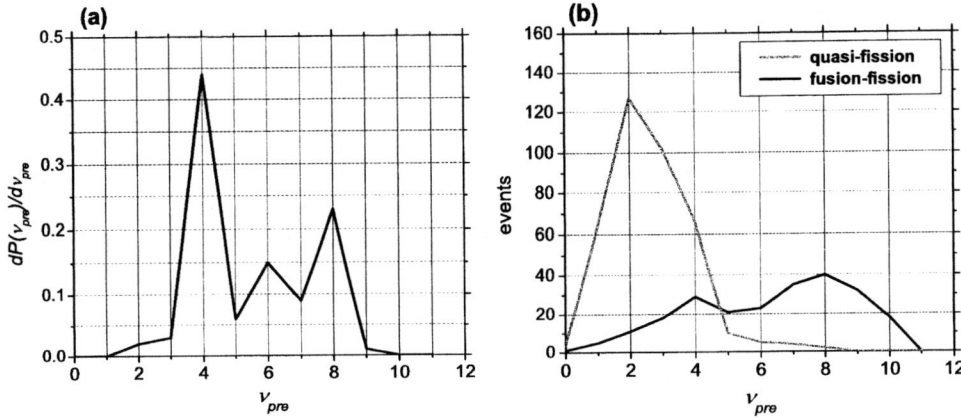

FIGURE 5. Pre-scission neutron multiplicity in the reaction ^{58}Ni+^{208}Pb at $*E^*$=189.5 MeV. (a) The experimental data. (b) Calculation.

ments with $\frac{A}{2} \pm 20$ in the experiments, respectively [6]. The theoretical value of $\sigma_{A/2\pm20}$ is denoted by the solid line. We can see very good agreement with the experimental data and our calculations. The calculated fusion-fission cross section σ_{CN} is denoted by the dashed line. This cross section σ_{CN} is derived from the trajectory crossing the three-dimensional fusion box. The fusion-fission cross section σ_{CN} by the calculation is one or two order magnitude smaller than the cross section $\sigma_{A/2\pm20}$ beyond the Bass barrier region. We see that the cross section $\sigma_{A/2\pm20}$ includes the deep quasi-fission events. Such an information is very important to estimate the evaporation residue cross section.

pre-scission neutron multiplicity

In order to classify the fusion-fission process more precisely, we compare calculation with the experimental data of pre-scission neutron multiplicity [6, 7]. The neutron multiplicity directly depends on the time scale of trajectory, so we can investigate the different class of the dynamical process more precisely.

In the reaction ^{58}Ni+^{208}Pb at $E^* = 185.9$ MeV which was measured by DeMon group [7], the neutron multiplicity has two peaks (near 4 and 8 neutron emission), which is shown in Fig. 5 (a). We suppose that the first peak is connected with the trajectory of QF process and second one is connected with FF process. The multiplicity depends on the travelling time of trajectories from the contact point to the scission one. Figure 6 shows the sample trajectories of QF process and FF process on $z - \alpha$-plane, which are denoted by gray and black lines, respectively. On the FF process, the trajectory is trapped in the

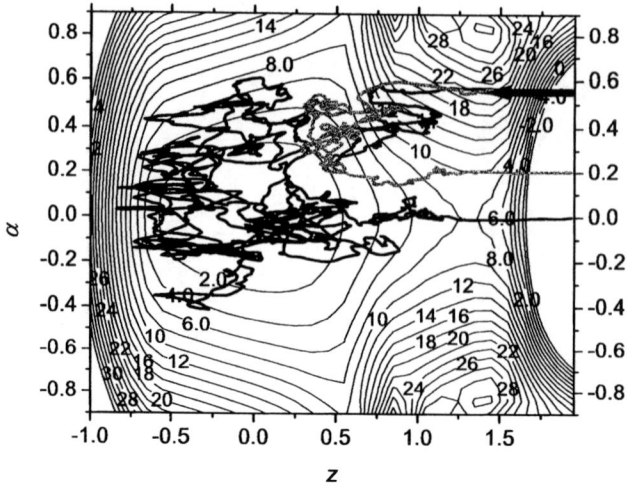

FIGURE 6. Samples of the trajectory projected on $z-\alpha$ plane at $E^* = 189.5$ MeV in the reaction ^{58}Ni+^{208}Pb. The trajectory of QF and FF are denoted by gray and black lines, respectively. The arrow denotes the injection point.

pocket around spherical region. It takes a long time to stay in the pocket and it has a large chance to emit neutrons. The calculation is shown in Fig. 5 (b). We can see the two peaks which come from QF process and FF process. Actually the travelling time scale of each process from contact point to scission point are clearly distinguished like as Fig. 7 (a). The time scale of FF process is longer than that of QF process.

We investigate the neutron multiplicity in the reactions ^{48}Ca+^{208}Pb and ^{48}Ca+^{244}Pu [23]. The calculation and experimental data are shown in Fig. 8, which excitation energy is about 40 MeV. In the reaction ^{48}Ca+^{208}Pb, the neutron multiplicity shows the single peak near 2 neutron emission, which comes from FF process. In the three dimensional Langevin calculation, we can see that the trajectory of FF process is dominant and single peak is reproduced.

In the ^{48}Ca+^{244}Pu case, in the Langevin calculation, we found mass symmetric fission events come from not only FF process but also DQF process. The experiment of neutron multiplicity shows the two peak near 2 and 4 neutron emission. Such two peaks are originated by two different fusion-fission mechanism. It looks that the first peak comes from QF process and the second one comes from mainly DQF or FF. The calculation can not reproduce the two peaks. We check the travelling time of both case, and they overlap and can not be distinguished, which is shown in Fig. 7 (b). In the calculation, the lifetime of neutron emission is rather short, because the excitation energy is rather lower. The time scale of trajectory fluctuation is same order of the life time of neutron emission, so we can not distinguish QF and DQF process clearly. Also the event of FF process is quit low. We see the only few FF events. To reproduce the experimental data,

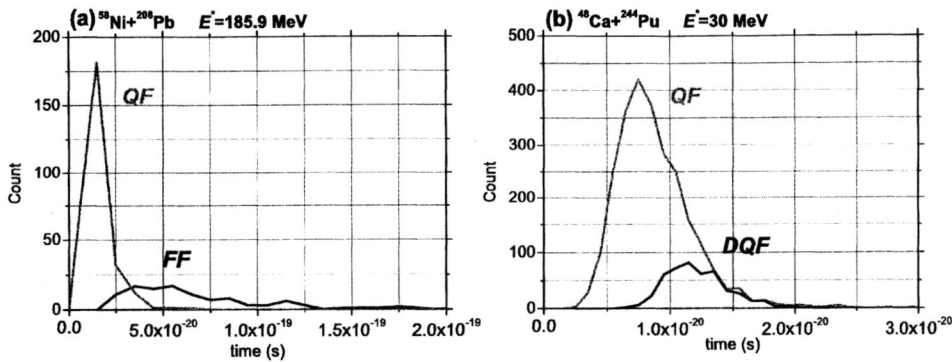

FIGURE 7. The travelling time from the contact point to the scission point in the reaction (a) ^{58}Ni+^{208}Pb and (b) ^{48}Ca+^{244}Pu (b).

we have to investigate the some parameters, for example friction tensor, level density parameter and neutron binding energy near di-nucleus configuration etc. We try to use the friction tensor which is derived from liner response theory [24, 25, 26]

Our analysis is useful for identifying the detail of the FF process in connection with the neutron multiplicity. The time scale of FF process depends on the strength of friction tensor. Also the time scale of neutron emission depends on the value of level density parameter and neutron binding energy at each nuclear shapes. The measured neutron multiplicity can be used for the determination of such values.

REFERENCES

1. W.D. Myers and W.J. Swiatecki, Nucl. Phys. 81 (1966) 1; A. Sobiczewski *et. al.*, Phys. Lett. 22 (1966) 500.
2. Yu.Ts. Oganessian et al., Nature **400** (1999) 242 ; Yu.Ts. Oganessian et al., Phys. Rev. Lett. **83**(1999) 3154.
3. Yu.Ts. Oganessian et al., Phys. Rev. C **63** (2001) 011301(R).
4. S. Hofmann and G. Munzenberg, Rev. Mod. Phys. 72 (2000) 733; S. Hofmann et al., Eur. Phys. J. A 14 (2002) 147.
5. K. Morita, Proceedings of the VIII International Conference on Nucleus-Nucleus Collisions (NN2003), Moscow, Russia, 2003, to be published.
6. M.G. Itkis et al., Proceeding of *Fusion Dynamics at the Extremes* (World Scientific, Singapore, 2001) p93.
7. L. Donadiile et al, Nucl Phys, A656 (1999) 259.
8. C.E. Aguiar *et al.*, Nucl. Phys. **A491**, 2301 (1989); C.E. Aguiar et al. Nucl. Phys. **A514**, 205 (1990).

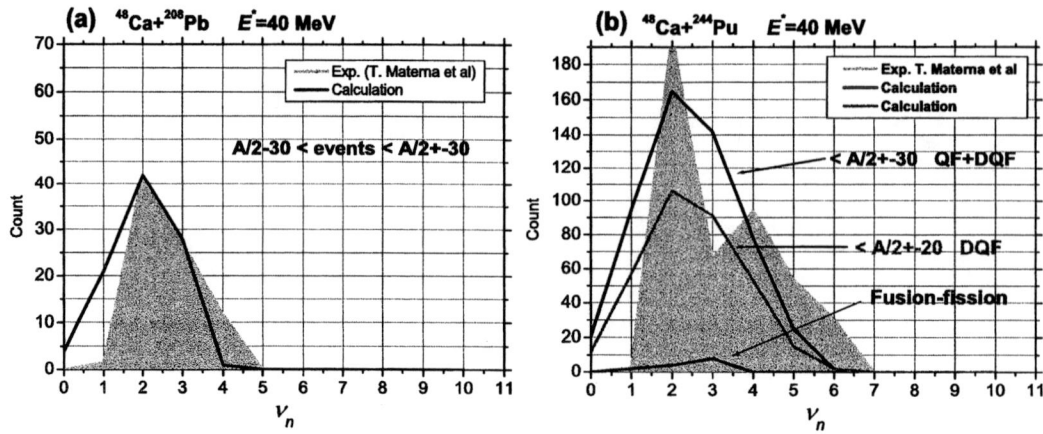

FIGURE 8. Pre-scission neutron multiplicity in the reaction (a) ^{48}Ca+^{208}Pb and (b) ^{48}Ca+^{244}Pu at E^*=40 MeV. Experiments are carried out by T. Materna. The absolute value of calculations are normalized.

9. T. Wada, Y. Abe and N. Carjan, Phys. Rev. Lett. **70**, 3538 (1993).
10. Y. Aritomo, T. Wada, M. Ohta and Y. Abe, Phys. Rev. **C55**, R1011 (1997); T. Wada, T. Tokuda, K. Okazaki, M. Ohta, Y. Aritomo and Y. Abe, Proceeding of DANF98, Slovakia 1998 (World Seientific)pp. 77; Y. Aritomo, T. Wada, M. Ohta and Y. Abe, Phys. Rev. **C59**, 796 (1999);
11. T. Tokuda, T. Wada and M. Ohta, Prog. Theor. Phys. **101**, 607 (1999).
12. C. Schmitt, J. Bartel, A. Surowiec and K. Pomorski, Acta Phy. Polon. B32 (2001) 841
13. V.I. Zagrebaev, Phys. Rev. C **64** (2001) 034606.
14. J. Maruhn and W. Greiner, Z. Phys. **251**, (1972) 431.
15. K. Sato, A. Iwamoto, K. Harada, S. Yamaji, and S. Yoshida, Z. Phys. **A288** (1978) 383.
16. J. Blocki, Y. Boneh, J.R. Nix, J. Randrup, M. Robel, A.J. Sierk and W.J. Swiatecki, Ann. Phys. **113**, 330 (1978).
17. J.R. Nix and A.J. Sierk, Nucl. Phys. **A428**, 161c (1984).
18. H. Feldmeier, Rep. Prog. Phys. **50**, 915 (1987).
19. S. Suekane, A. Iwamoto, S. Yamaji and K. Harada, JAERI-memo, 5918 (1974).
20. A. Iwamoto, S. Yamaji, S. Suekane and K. Harada, Prog. Theor. Phys. **55**, 115 (1976).
21. Y. Aritomo, Proceedings of *Int. Conf. on Nuclear Physics at Border Lines*, Lipari, Italy, 2001 (World Scientific, Singapore, 2002) p38.
22. Y. Aritomo, Proceedings of *International Conference on Exotic Nuclei*, Bikal, Russia, 2001 (World Scientific, Singapore, 2002)
23. T. Materna et. al, Proceedings of the VIII International Conference on Nucleus-Nucleus Collisions (NN2003), Moscow, Russia, 2003, to be published.
24. H. Hofmann, Phys. Rev. **284** (1997) 137.
25. S. Yamaji, F.A. Ivanyuk, H. Hofmann, Nucl. Phys. **A612** (1997) 1.
26. F.A. Ivanyuk, H. Hofmann, V.V. Pashkevich, S. Yamaji, Phys. Rev. **C55** (1997) 1730.

PHYSICS WITH EXOTIC NUCLEI (PEN)

- Physics of Nuclei Far from Stability
- New Results with RI Beam
- Present Status and Future Plan of RIB Facilities
- New Aspects of Clustering Structure of Light Nuclei
- Physics of Cluster Structure of Nuclei

Two-proton radioactivity - a curiosity of Nature?

B. Blank

CEN Bordeaux-Gradignan, Le Haut-Vigneau, F-33175 Gradignan Cedex, France

Abstract. In recent experiments at GANIL and GSI, we studied the decay of ^{45}Fe, according to theoretical predictions one of the most promising cases for 2p radioactivity. Our results show for the first time clear evidence of this new radioactivity. In another experiment at GANIL, 2p emission was observed from excited states in ^{17}Ne. In these complete kinematics measurements performed at the SPEG facility of GANIL, the angle between the two protons has been measured evidencing a ^{2}He emission pattern. Both results will be presented and future studies are discussed.

INTRODUCTION

An ensemble of protons and neutrons can form a nucleus stable against any radioactive decay only if a subtle equilibrium between the number of protons and neutrons is respected. For light nuclei up to an atomic mass number A=40, this equilibrium corresponds to a roughly equal number of protons (Z) and neutrons (N). For heavier nuclei, more neutrons than protons have to be added to stabilize a nucleus to overcome the Coulomb repulsion of the positively charged protons. These stable nuclei form the valley of stability. If this equilibrium condition is not respected in a nucleus, it becomes radioactive. For small deviations from equilibrium, the nuclei decay by β^+ (neutron-deficient nuclei) or β^- (neutron-rich nuclei) decay. For increasing asymmetry between proton and neutron numbers, the nuclei become more and more unstable and will decay by more exotic decay modes like β-delayed proton or β-delayed neutron emission.

The limits of stability, the drip lines, are reached if the nuclear forces are no longer able to bind an ensemble of nucleons with a too large neutron or proton excess. On the proton-rich side of the valley of stability, these unstable nuclei decay by emission of one proton for odd-Z nuclei or two protons for even-Z nuclei from their ground states. The one-proton radioactivity has been observed for the first time at GSI Darmstadt in 1981 [1, 2]. Meanwhile more than 20 cases of proton radioactivity have been identified for odd-Z nuclei beyond the proton drip line between Z=53 and Z=83 [3] allowing to establish the sequence of shell-model single-particle levels beyond the proton drip line as well as to determine masses and to test mass models and mass predictions for very exotic nuclei.

Two-proton (2p) radioactivity is predicted since 1960 [4, 5, 6] to occur for even-Z proton-rich nuclei beyond the proton drip line. Due to the pairing energy, the 2p candidates cannot decay by a sequential emission of two protons as the one-proton daughter is energetically not accessible. Therefore, only a simultaneous two-proton emission is possible which can take place in two different ways: i) by an isotropic emission of the two protons which fill the whole phase space available, but in order to easily penetrate through the Coulomb and centrifugal barrier of the daughter nucleus share most probably equally the decay energy available. Such a picture might be altered by final-state interaction between the two protons; ii) by a correlated emission where in the decay the ^{2}He resonance is formed which decays either already under the Coulomb and centrifugal barrier or outside the nucleus. In both cases, an energy difference between the two protons of zero is most likely. However, for a ^{2}He emission, a small relative angle between the two protons might be observable. In addition, in a simple picture, two independent protons with half the total decay energy each would penetrate through the barrier in one case, whereas in the other case one ^{2}He particle with the total decay energy has to tunnel through the barrier.

Recent theoretical predictions [7, 8, 9, 10, 11] pointed out that ^{45}Fe, ^{48}Ni, and ^{54}Zn are the best candidates for two-proton ground-state decay as their 2p Q values is about 1.1-1.8 MeV, whereas the one-proton emission is either energetically forbidden or extremely disfavoured due to small one-proton decay energies and very narrow intermediate states as a consequence of the rather high combined Coulomb and centrifugal barrier. Therefore, half-lives in the 100μs to a few millisecond range were predicted for these medium-mass nuclei.

For lighter nuclei, the Coulomb and centrifugal barriers for proton emission are smaller and the ground states of these nuclei become larger. Thus, one-proton emission is energetically possible through the tails of broad intermediate states even for 2p candidates. Such cases have been observed experimentally for ^{6}Be [12] and ^{12}O [13]. In both cases, the emission pattern is compatible with an uncorrelated decay into the available phase space. Two-proton emission from excited states has been observed in eight cases after β decay (see [14] for a recent reference) as well as after inelastic excitation from ^{16}O [15] and from ^{18}Ne [16]. In all cases, the experimental data are in nice agreement with predictions assuming a sequential decay pattern. In these light nuclei a correlated emission might be expected only, if nuclear structure strongly favors 2p emission over one proton emission which is favored by barrier penetration arguments.

With the advent of projectile-fragmentation facilities equipped with powerful separators for in-flight isotope separation, proton drip-line nuclei came into experimental reach and two of the above mentioned most promising candidates, ^{45}Fe [17] and ^{48}Ni [18], could be observed for the first time as quasi-stable isotopes with half-lives longer than a few hundred nanoseconds. Part of the decay strength of ^{45}Fe was observed [19] in the experiment designed to observed for the first time ^{48}Ni [18]. In addition, complete kinematics experiments allow now to study efficiently nuclei beyond the limit of nuclear binding.

THE TWO-PROTON RADIOACTIVITY OF ^{45}FE

The GANIL experiment

In an experiment [20] at the SISSI-LISE3 facility of GANIL, we used the projectile fragmentation of a ^{58}Ni primary beam at 75 MeV/nucleon to produce proton-rich nuclei in the range Z=22-28. After production in a natural nickel target (240 μm in thickness) located in the SISSI device, the fragments of interest were selected by the Alpha spectrometer and the LISE3 separator equipped with an intermediate beryllium degrader. At the focus of the LISE3 separator, a set-up was mounted to identify and stop the fragments as well as to study their radioactive decays. This set-up consisted in two channel-plate detection systems for timing purposes mounted at a first LISE focal point 20m upstream from the final focus and a sequence of four silicon detectors. The silicon detectors were equipped with two parallel electronic chains with different gains, one for heavy-fragment identification and the other for decay spectroscopy.

The fragments of interest were stopped in the third silicon detector of the telescope and identified on an event-by-event basis by means of their flight times, by their energy loss in all detectors of the telescope, and by their position in a position-sensitive detector. All in all, 22 ^{45}Fe implantations were identified. Radioactive decay events were triggered either by the implantation detector or by the adjacent silicon detectors. For any event triggering the data acquisition, all channels were read and written on tape. The geometrical efficiency to observe a β particle in the fourth detector for a β decay occurring in the implantation detector is about 30%.

Figure 1a shows the decay-energy spectrum correlated with implants of ^{45}Fe where only decay events occurring less that 15 ms after a ^{45}Fe implantation were analysed. The spectrum exhibits a pronounced peak at (1.14±0.06) MeV with only a very few other counts. In contrary, in the spectrum in figure 1b conditioned by a decay time in the interval between 15 ms and 100 ms, the 1.14 MeV peak has almost completely disappeared and other events higher in decay energy show up. These counts are consistent with the decay-energy spectrum of ^{43}Cr [19], the 2p daughter of ^{45}Fe. The 1.14 MeV peak, however, originates only from the fast decay of ^{45}Fe. In addition, the events in

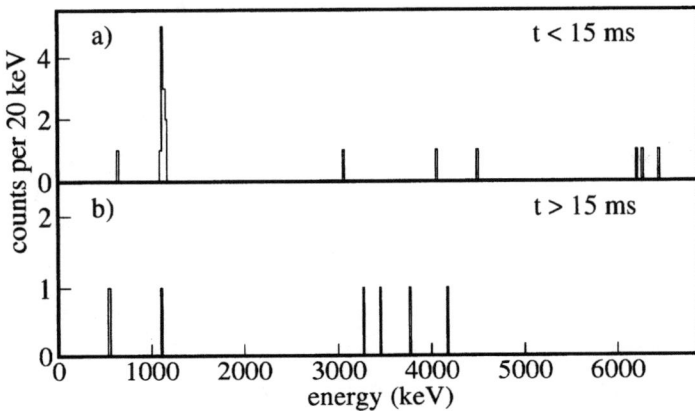

FIGURE 1. Decay-energy spectrum correlated with ^{45}Fe implantation. The upper spectrum is obtained under the condition that the radioactive decay occurs faster than 15 ms after implantation, whereas the lower spectrum contains decay events with decay times between 15 ms and 100 ms. The peak at 1.14 MeV is clearly identified as being due to a fast decay.

this peak have no coincident β-particle signals in the adjacent detector (see figure 2a) beyond the noise level, whereas these coincident β particles can be observed nicely for ^{46}Fe implants analysed with a similar condition in energy for the implantation detector.

The non-observation of β particles in coincidence with the 1.14 MeV peak alone is a strong indication that this peak originates from a direct two-proton ground-state decay of ^{45}Fe. The probability to miss all β particles for the 12 events in the peak is as low as 1.4%.

The observation of only twelve events in the peak at 1.14 MeV indicates that the branching ratio for 2p decay of ^{45}Fe is not 100%. However, dead-time losses (the data-acquisition dead time is somewhat less than 1 ms) as well as losses due to decays in the dead zone between two strips of the implantation detector are other possible explications.

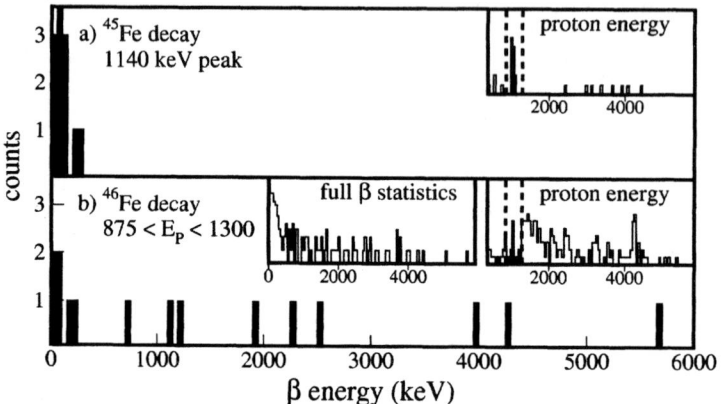

FIGURE 2. Upper part: β-particle spectrum from the 6mm-thick silicon detector in coincidence with events in the peak at 1.14 MeV in figure 1. Lower part: Similar spectrum obtained after ^{46}Fe implantation conditioned by a decay-energy range in the implantation detector of $E = 1.0 - 1.4$ MeV. To allow for a direct comparison, the same statistics is taken as in the ^{45}Fe case. The inset shows the decay energy spectrum for the two nuclei with the shaded area corresponding to the energy condition applied to generate the β spectra as well as the full-statistics β spectrum from ^{46}Fe.

The distinct difference between the decay spectrum of ^{45}Fe (figure 1a) and of its neighbor ^{46}Fe [19] is another hint that we are not dealing with the same decay type. If ^{45}Fe would decay by a β-delayed decay, it seems to be reasonable to expect a similar decay-energy spectrum as for ^{46}Fe and for ^{45}Fe. The observation of a single peak from the decay of ^{45}Fe is a strong hint that ^{45}Fe is not decaying by a β-delayed mode as does ^{46}Fe.

The decay-time spectrum of ^{45}Fe gated by the 1.14 MeV peak fitted with a one-component exponential yields a half-life for ^{45}Fe of $T_{1/2} = (4.7^{+3.4}_{-1.4})$ ms. The decay-time spectrum of events up to 100 ms after a ^{45}Fe implantation can be fitted by taking into account the decay of ^{45}Fe and its 2p daughter ^{43}Cr. The half-life is then $(5.7^{+2.7}_{-1.4})$ ms.

The GSI data

In a GSI experiment [21], 6 ^{45}Fe implantations were observed in a six-day experiment. Decay events were observed after ^{45}Fe implantation in a stack of seven large-size silicon detectors surrounded by a cylindrical NaI barrel to search for the 511 keV annihilation

γ rays from positrons. Figure 3 shows the decay energy spectrum from this experiment. For the four events around 1 MeV, we deduce a total decay energy of (1.1±0.1) MeV, in nice agreement with the GANIL data which yielded an energy of (1.14±0.06) MeV. No γ rays have been observed in coincidence with the four events which we attribute to 2p radioactivity of ^{45}Fe. The half-life was determined to be $(3.2^{+2.6}_{-1.0})$ms, again in agreement with the GANIL data.

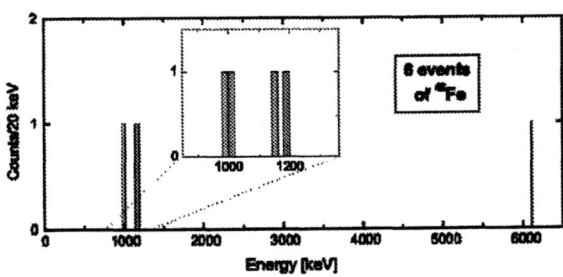

FIGURE 3. Decay-energy spectrum correlated with ^{45}Fe implantation from the GSI experiment. For six implantation events, four decays are observed with an energy release of 1.1 MeV, whereas one event at about 6 MeV is visible.

Comparison with theoretical predictions

The energy of the peak observed at 1.14 MeV agrees nicely with 2p Q value predictions from Brown [7] of 1.15(9) MeV, from Ormand [8] of 1.28(18) MeV, and from Cole [9] of 1.22(5) MeV. These models use the isobaric multiplet mass equation and shell-model calculations to determine masses of proton-rich nuclei from their neutron-rich mirror partners. It is worth noting that these model calculations have been tailor-made for the region of medium-mass nuclei. Q-value predictions from models aiming at predicting masses for the entire chart of nuclei are in less good agreement with our present result. All the 2p Q-value predictions are summarized in figure 4, where they are used in barrier-penetration calculations using the simple di-proton model for two-proton emission.

These calculations predict a di-proton barrier penetration half-life of 0.02 ms, if one assumes a spectroscopic factor of unity. It is now well known that such a model yields only a lower limit of the half-life.

Recent theoretical studies of the decay of ^{45}Fe were performed by Grigorenko and co-worker [28] and by Brown and Barker [29]. Brown and Barker used an R-matrix model which explicitly includes the decay of ^{45}Fe via the ^{2}He resonance. Their calculations yield half-lives for 2p decay of 4 - 41 ms, depending on the decay energy they use (the experimental uncertainty enters exponentially) and on correction factors for the shell-model calculations they determined from mirror nuclei. These results therefore indicate that the decay could indeed proceed via the ^{2}He resonance.

Grigorenko et al. [11] use their three-body model to determine the 2p decay half-life as a function of the Q value. As shown in figure 5 their theoretical results are in

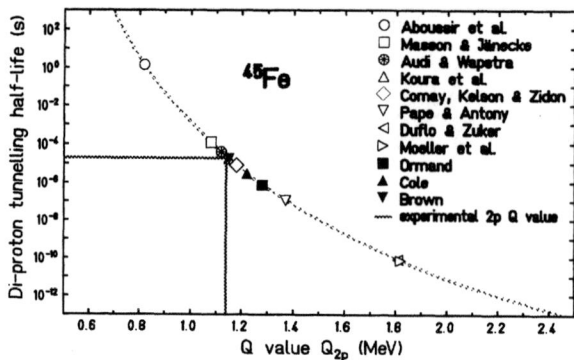

FIGURE 4. Barrier-penetration half-life as a function of the two-proton Q value, Q_{2p}, for ^{45}Fe. The barrier penetration was calculated by assuming a spectroscopic factor of unity. Different model predictions [7, 8, 9, 22, 23, 24, 25, 26, 27] were used for Q_{2p}. The experimentally observed Q value of ^{45}Fe implies a di-proton barrier-penetration half-life of 0.02 ms.

agreement with the experimental data, if they assume that the protons are emitted from a $p_{1/2}$ state rather than from a $f_{7/2}$ state as expected from the mirror nucleus. However, a small low-ℓ contribution in the wavefunction might be sufficient to dominate the decay.

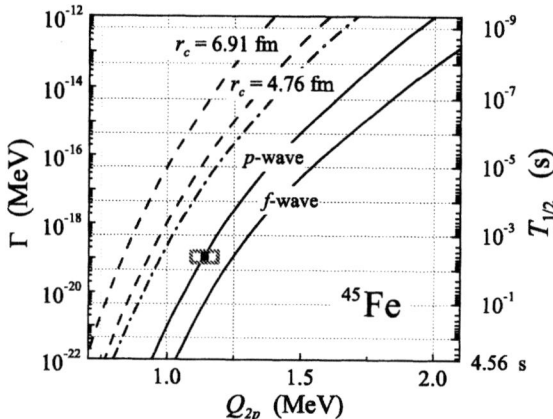

FIGURE 5. Widths and decay half-lives as calculated in different models. The dashed lines correspond to the di-proton model with channel radii of 6.91 fm and 4.76 fm. The dashed-dotted line originates from an R-matrix calculation assuming simultanuous, but independent emission of two s-wave protons. The solid lines correspond to results from the 3-body model using p-wave or f-wave protons [28]. The experimental point and its error bar is also shown.

Clearly further theoretical work is still needed to get a consistent description of 2p emission. On the experimental side, higher-statistics data on ^{45}Fe and search for 2p emission from the other top candidates ^{48}Ni and ^{54}Zn are the next steps. A study of the decay mechanism should be possible by measuring the individual decay energies of the protons emitted and their relative angle as already done in the complete-kinematics measurements described below. However, for long-lived 2p emitters like ^{45}Fe new techniques, such a time-projection chambers, have to be developped to get a three-dimensional view of the decay.

TWO-PROTON EMISSION FROM EXCITED STATES IN ^{17}NE

In an experiment on SPEG at GANIL [30], the decay of excited states in ^{17}Ne has been studied by reconstructing the invariant mass from a complete-kinematics measurement. A ^{24}Mg primary beam at 95 MeV/nucleon was fragmented in a carbon target in the SISSI device. By means of an achromatic degrader in the Alpha spectrometer, a secondary cocktail beam containing ^{17}F, ^{18}Ne and ^{20}Mg was selected. These isotopes interacted again in a beryllium target at the entrance of the SPEG spectrometer to produce short-lived proton-rich products by e.g. one-neutron stripping reactions.

The decay of these unstable nuclei was observed by means of the MUST array [31], which detected protons from decays in the target, and the SPEG spectrometer, which was used to select the reaction channel and to determine the momentum vector of the heavy recoil. In this way, the momentum vectors of all decay products were measured which allowed us to determine the invariant mass of the decaying nucleus as well as, in the case of 2p emission, the proton-proton emission angle.

To study the decay of excited states in ^{17}Ne, one-neutron stripping reactions of ^{18}Ne nuclei incident on the secondary beryllium target were used. Figure 6 shows the invariant mass spectrum as determined from ^{15}O + 2p events. The peak at a mass excess of 18.5 MeV is due to the decay of the second and third excited states in ^{17}Ne which are not resolved in our experiment (we estimate our experimental resolution to be about 250 keV). Above 20 MeV events from higher-lying states are visible.

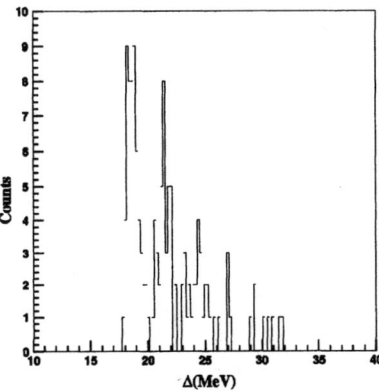

FIGURE 6. Invariant mass spectrum for ^{15}O + 2p events. The peak at 18.5 MeV is interpreted as the decay of the second and third excited states ($E^* = 1.76$ MeV and 1.91 MeV) in ^{17}Ne by two-proton emission, whereas the activity above 20 MeV arises from the decay of higher lying states.

The angle between the two protons in the center of mass of ^{17}Ne is shown in figure 7. A more or less continuous spectrum is observed, however, with an excess of counts at about 50°. Monte Carlo simulations, which include experimental acceptances, straggling effects, energy-loss etc., show that neither a sequential picture nor a pure ^{2}He emission pattern can reproduce the experimental data. The best fit is obtained by assuming a 50% ^{2}He branch and 50% of sequential decay. Details concerning the physics used in the simulations can be found in [32].

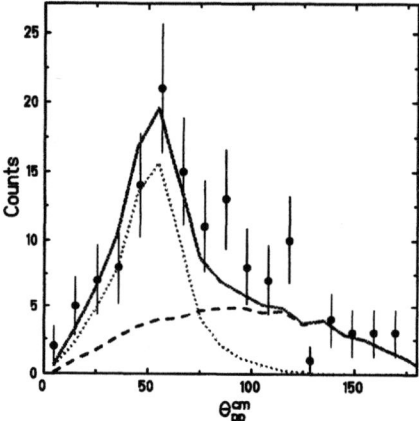

FIGURE 7. Angular distribution in the center of mass for all 2p events correlated with the observation of a ^{15}O recoil. The solid curve is a simulation of a correlated two-proton emission via a ^{2}He resonance, whereas the dashed line corresponds to a sequential emission pattern via the ground state of ^{16}F.

A more detail analysis is possible by performing energy cuts. Figure 8a shows the same angular distribution for decays from the second and third excited states. This angular distribution is in agreement with the expected sequential emission pattern. A similar results was recently obtained by Chromik et al. [33] at MSU where only states up to 2 MeV excitation energy were produced due to the use of Coulomb excitation.

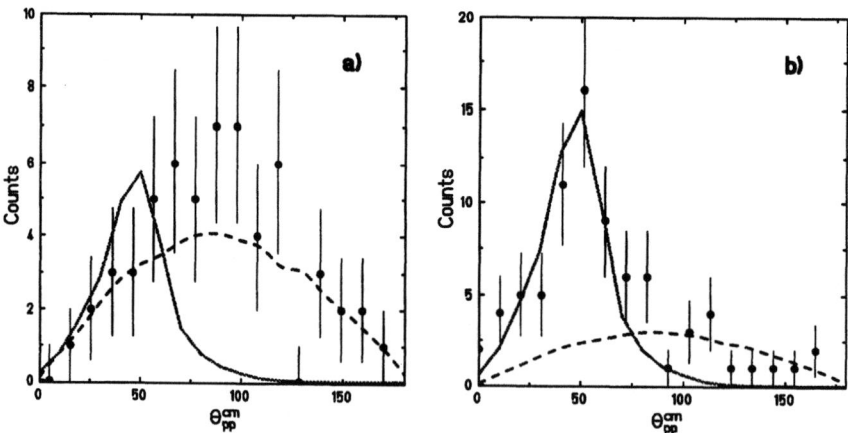

FIGURE 8. Angular distribution for 2p events with a ^{15}O recoil yielding a mass excess of less than 20 MeV (a) and above 20 MeV (b). The solid and the dashed curves are again simulation of a correlated two-proton emission via a ^{2}He resonance and a sequential emission pattern, respectively.

However, for the events from states with a mass excess above 20 MeV, a significantly different angular distribution is obtained. Although the statistics is low, an enhancement of events around $50°$ is observed. Again the distribution is in agreement neither with a simulation for a sequential decay nor with a ^{2}He picture alone. However, in these events the ^{2}He portion has to be raised to 50% to get agreement with the experimental data. This seems to indicate that one or several high-lying states could decay by a correlated 2p emission, despite the fact that the sequential branch is widely open.

Much higher statistics data are clearly needed to confirm or infirm the present finding. New experimental studies can take profit from the high-intensity radioactive neon beams now available at the SPIRAL facility of GANIL. In addition, theoretical calculations which try to locate states in ^{17}Ne, where the wavefunction has very little overlap with states in the 1p daughter ^{16}F, but large overlap with the ground state in ^{15}O, the 2p daughter, are most wellcome.

CONCLUSIONS

In experiments at GANIL and GSI, we have identified for the first time a ground-state two-proton emitter, ^{45}Fe. Clear evidence for this decay mode has been gained by measuring its decay energy, its decay time, and, in particular, due to the absense of any β particle in the decay of the 1.14 MeV peak. This observation is in nice agreement with theoretical predictions which pointed to ^{45}Fe as one of the most promising 2p emission candidates.

In an experiment at the SPEG facility of GANIL, complete kinematics experiments allowed for the first time to observe a proton-proton angular correlation in the 2p emission from excited state in ^{17}Ne. This angular correlation is in agreement with the formation of a ^{2}He resonance in the decay.

ACKNOWLEDGMENTS

The work presented here was performed in collaboration with collegues from CEN Bordeaux, University of Warsaw, IPN Orsay, GANIL, GSI, CEA Saclay, NSCL-MSU, IAP Bucharest, University of Tennessee, University of Liverpool, ORNL, and the University of Edinburgh.

REFERENCES

1. Hofmann, S., Reisdorf, W., Münzenberg, G., Hessberger, F., Schneider, J., and Armbruster, P., *Z. Phys.*, **A305**, 111 (1982).
2. Klepper, O., Batsch, T., Hofmann, S., Kirchner, R., Kurewicz, W., Reisdorf, W., Roeckl, E., Schardt, D., and Nyman, G., *Z. Phys.*, **A305**, 125 (1982).
3. Woods, P. J., and Davids, C. N., *Annu. Rev. Nucl. Part. Sci.*, **47**, 541 (1997).
4. Goldansky, V. I., *Nucl. Phys.*, **19**, 482 (1960).
5. Goldansky, V. I., *Nucl. Phys.*, **27**, 648 (1961).
6. Jänecke, J., *Nucl. Phys.*, **27**, 648 (1965).

7. Brown, B. A., *Phys. Rev. C*, **43**, R1513 (1991).
8. Ormand, W. E., *Phys. Rev. C*, **53**, 214 (1996).
9. Cole, B. J., *Phys. Rev. C*, **54**, 1240 (1996).
10. Nazarewicz, W., Dobaczewski, J., Werner, T., Maruhn, J., Reinhard, P.-G., Rutz, K., Chinn, C., Umar, A., and Strayer, M., *Phys. Rev. C*, **53**, 740 (1996).
11. Grigorenko, L., Johnson, R., Mukha, I., Thompson, I., and Zhukov, M., *Phys. Rev. Lett.*, **85**, 22 (2000).
12. Bochkarev, O. V., Korsheninnikov, A. A., Kuz'min, E. A., Mukha, I. G., Chulkov, L. V., and Yan'kov, G. B., *Sov. J. Nucl. Phys.*, **55**, 955 (1992).
13. Azhari, A., Kryger, R., and Thoennessen, M., *Phys. Rev. C*, **58**, 2568 (1998).
14. Fynbo, H., Borge, M., Axelsson, L., Äystö, J., Bergmann, U., Fraile, L., Honkanen, A., Hornshoj, P., Jading, Y., Jokinen, A., Jonson, B., Martel, I., I, M., Nilsson, T., Nyman, G., Oinonen, M., Piqueras, I., Rissager, K., Siiskonen, T., Smedberg, M., Tengblad, O., Thaysen, J., and Wenander, F., *Nucl. Phys. A*, **677**, 38 (2000).
15. Bain, C., Woods, P., Coszach, R., Davinson, T., Decrock, P., Gaelens, M., Galster, W., Huyse, M., Irvine, R., Leleux, P., Lienard, E., Loiselet, M., Michotte, C., Neal, R., Ninane, A., Ryckewaert, G., Shotter, A., Vancraeynest, G., Vervier, J., and Wauters, J., *Phys. Lett.*, **373B**, 35 (1996).
16. Gomez del Campo, J., Galindo-Uribarri, A., Beene, J., Gross, C., Liang, J., Halbert, M., Stracener, D., Shapira, D., Varner, R., Chavez-Lomeli, E., and Ortiz, M., *Phys. Rev. Lett.*, **86**, 43 (2001).
17. Blank, B., Czajkowski, S., Davi, F., Moral, R. D., Dufour, J. P., Fleury, A., Marchand, C., Pravikoff, M. S., Benlliure, J., Boué, F., Collatz, R., Heinz, A., Hellström, M., Hu, Z., Roeckl, E., Shibata, M., Sümmerer, K., Janas, Z., Karny, M., Pfützner, M., and Lewitowicz, M., *Phys. Rev. Lett.*, **77**, 2893 (1996).
18. Blank, B., Chartier, M., Czajkowski, S., Giovinazzo, J., Pravikoff, M. S., Thomas, J.-C., de France, G., de Oliveira Santos, F., Lewitowicz, M., Borcea, C., Grzywacz, R., Janas, Z., and Pfützner, M., *Phys. Rev. Lett.*, **84**, 1116 (2000).
19. Giovinazzo, J., Blank, B., Borcea, C., Chartier, M., Czajkowski, S., de France, G., Grzywacz, R., Janas, Z., Lewitowicz, M., de Oliveira Santos, F., Pfützner, M., Pravikoff, M., and Thomas, J.-C., *Eur. Phys. J. A*, **10**, 73 (2001).
20. Giovinazzo, J., Blank, B., Chartier, M., Czajkowski, S., Fleury, A., Jimenez, M. L., Pravikoff, M., Thomas, J.-C., de Oliveira Santos, F., Lewitowicz, M., Maslov, V., Stanioiu, M., Grzywacz, R., Pfützner, M., Borcea, C., and Brown, B., *Phys. Rev. Lett.*, **89**, 102501 (2002).
21. Pfützner, M., Badura, E., Bingham, C., Blank, B., Chartier, M., Geissel, H., Giovinazzo, J., Grigorenko, L., Grzywacz, R., Hellström, M., Janas, Z., Kurcewicz, J., Lalleman, A., Mazzocchi, C., Mukha, I., Münzenberg, G., Plettner, C., Roeckl, E., Rykaczewski, K., Schmidt, K., Simon, R., Stanoiu, M., and Thomas, J.-C., *Eur. Phys. J.*, **A14**, 279 (2002).
22. Haustein, P., *At. Data Nucl. Data Tab.*, **39**, 185 (1988).
23. Möller, P., Nix, J. R., Myers, W. D., and Swiatecki, W. J., *At. Data Nucl. Data Tab.*, **59**, 185 (1995).
24. Duflo, J., and Zuker, A., *Phys. Rev. C*, **52**, R23 (1995).
25. Aboussir, Y., Pearson, J., Dutta, A., and Tondeur, F., *At. Data Nucl. Data Tab.*, **61**, 127 (1995).
26. Audi, G., and Wapstra, A. H., *Nucl. Phys.*, **A625**, 1 (1997).
27. Koura, H., Uno, M., Tachibana, T., and Yamada, M., *Nucl. Phys.*, **A674**, 47 (2000).
28. Grigorenko, L., Johnson, R., Mukha, I., Thompson, I., and Zhukov, M., *Proc. ENAM* (2001).
29. Brown, B. A., and Barker, F., *Phys. Rev. C*, p. submitted for publication (2003).
30. Zerguerras, T., Blank, B., Blumenfeld, Y., Suomijärvi, T., Thoennessen, M., Beaumel, D., Brown, B., Chartier, M., Fallot, M., Giovinnazzo, J., Jouanne, C., Lapoux, V., Lhenry-Yvon, I., Mittig, W., Roussel-Chomaz, P., Savajols, H., Scarpaci, J. A., and Shrivastava, A., *submitted to EPJA* (2003).
31. Blumenfeld, Y., Auger, F., Sauvestre, J., Maréchal, F., Ottini, S., Alamanos, N., Barbier, A., Beaumel, D., Bonnereau, B., Charlet, D., Clavelin, J., Courtat, P., Delbourgo-Salvador, P., Douet, R., Engrand, M., Ethvignot, T., Gillibert, A., Khan, E., Lapoux, V., Lagoyannis, A., Lavergne, L., Lebon, S., Lelong, P., Lesage, A., Ven, V. L., Lhenry, I., Martin, J., Musumarra, A., Pita, S., Petizon, L., Pollacco, E., Pouthas, J., Richard, A., Rougier, D., Santonocito, D., Scarpaci, J., Sida, I., Soulet, C., Stutzmann, J., Suomijärvi, T., Szmigiel, M., Volkov, P., and Voltolini, G., *Nucl. Instr. Meth.*, **A421**, 471 (1999).
32. Kryger, R. A., Azhari, A., Hellström, M., Kelley, J. H., Kubo, T., Pfaff, R., Ramakrishnan, E., Sherrill, B. M., Thoennessen, M., Yokoyama, S., Charity, R. J., Dempsey, J., Kirov, A., Robertson, N., Sarantites, D. G., Sobotka, L. G., and Winger, J. A., *Phys. Rev. Lett.*, **74**, 860 (1995).
33. Chromik, M., Thirolf, P., Thoennessen, M., Brown, B., Davinson, T., Gassmann, D., Heckman, P., Prisciandro, J., Reiter, P., Tryggestad, E., and Woods, P., *Phys. Rev. C*, **66**, 024313 (2002).

The Search for Neutral Nuclei

F. Miguel MARQUÉS MORENO

Laboratoire de Physique Corpusculaire,
IN2P3-CNRS, ENSICAEN et Université de Caen, F-14050 Caen cedex, France

Abstract. The long debate on the possible existence of neutral nuclei has been recently revived by the application of a new experimental approach, which lead to the observation of few events consistent with the formation of a bound ^{4}n in the breakup of ^{14}Be. The description of these events by a resonant (low-energy) ^{4}n state is discussed. Finally, the preliminary results of an experiment measuring the breakup reaction (^{12}Be, $2\alpha 4n$) are presented.

INTRODUCTION

The very lightest nuclei have long played a fundamental rôle in testing nuclear models and the underlying nucleon-nucleon interaction. In this context the study of systems exhibiting very asymmetric N/Z ratios may provide new perspectives on the nucleon-nucleon interaction and few-body forces. In the case of the light, two-neutron halo nuclei such as ^{6}He, insight is already being gained into the effects of the three-body force [1]. Very recently evidence has been presented that the ground state of ^{5}H exists as a relatively narrow, low-lying resonance [3]. In the case of the lightest $N = 4$ isotone, ^{4}n, nothing is known despite experimental searches over the past 40 years [4,2]. Theoretically it is difficult to produce a bound 4 neutron cluster (or "tetraneutron") [4–8]. The discovery of such neutral systems as bound states would therefore, as discussed by Timofeyuk [5] and Pieper [8], have fundamental implications for our understanding of nuclear forces.

It is, therefore, interesting to speculate that multineutron halo nuclei and other very neutron-rich systems may contain components of the wavefunction in which the neutrons present a relatively compact cluster-like configuration. If this were to be the case, then the dissociation of beams of such nuclei may offer a means to produce bound neutron clusters (if they exist) and, more generally study multineutron correlations.

To date the majority of searches for multineutron systems have relied on very low (typically $\sim$1 nb) cross section double-pion charge exchange (DπCX) and heavy-ion transfer reactions (see, for example, refs [14,13]). In the case of dissociation of an energetic beam of a very neutron-rich nucleus, relatively high cross sections

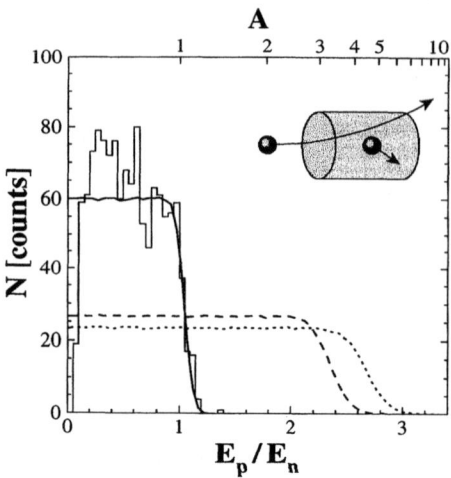

FIGURE 1. Distribution of the ratio of proton energy, E_p (MeV), to the energy derived from the flight time, E_n (MeV/N), for data from the reaction (^{14}Be, ^{12}Be+n) and for simulations of elastic scattering of ^{1,3,4}n (solid, dashed and dotted lines, respectively) on protons [9].

(typically ~100 mb) are encountered. Thus, even only a small component of the wavefunction corresponding to a multineutron cluster could result in a measurable yield with a moderate secondary beam intensity. Furthermore the backgrounds arising in DπCX and heavy-ion transfer reactions from target impurities and complex many-body phase space reactions are obviated in breakup.

EVENTS CONSISTENT WITH A TETRANEUTRON

The difficulty in this approach lies in the direct detection of a ^{A}n cluster. The avenue that has been explored here is to detect the recoiling proton in a liquid scintillator [9]. One of the principle advantages of a liquid scintallator is that neutrons may be discriminated with good efficiency from the γ and cosmic-ray backgrounds using pulse-shape analysis. Careful calibrations, employing sources, cosmic rays and the maximum proton recoil energy for a given E_n, permit the charge deposited and hence the energy (E_p) of the recoiling proton to be determined. This may be compared to the energy derived from the measured time-of-flight (E_n): for a single neutron and an ideal detector, $E_p/E_n \leq 1$; for a realistic detector with finite resolution the limit is ~1.4. In the case of a multineutron cluster (^{A}n) E_p can exceed the incident energy per nucleon and E_p/E_n will take on a range of values extending beyond 1.4 — up to ~3 in the case of $A = 4$ (Fig. 1).

The data already at hand from the study of the disociation of ^{14}Be and ^{11}Li [10,12,11] was examined with a view to testing the method outlined above. The

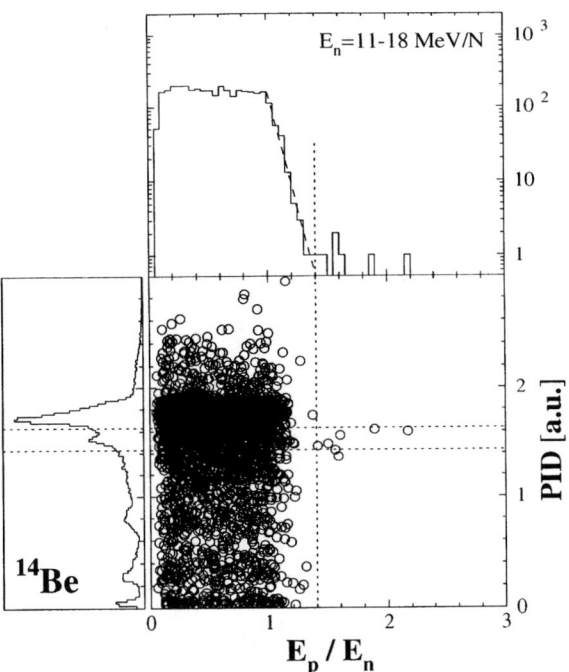

FIGURE 2. Scatter plot, and the projections onto both axes, of the particle identification parameter PID vs E_p/E_n for the data from the reaction (^{14}Be, $X+n$). The dotted lines correspond to $E_p/E_n = 1.4$ and to the region centred on the ^{10}Be peak [9].

details of the analyses carried out may be found in ref. [9]. The essential results are provided by Fig. 2 which displays the charged fragment particle identification (PID) derived from the Si-CsI detector telescope versus E_p/E_n. The E_p/E_n distributions (upper panel) exhibit a general trend below 1.4: a plateau up to 1 followed by a sharp decline, which may be fitted to an exponential distribution (dashed line). In the region where ^An events may be expected to appear some 7 events with E_p/E_n ranging from 1.4 to 2.2 are observed for ^{14}Be. In the case of ^{11}Li, despite the greater number of neutrons detected (factor of 2.4), only 4 events appear which lie just above threshold. Turning to the coincidences with the charged fragments, the 7 events produced by the ^{14}Be beam fall within a region centred on ^{10}Be. In the case of the 4 events produced in the reactions with ^{11}Li no correlation appears to exist with any particular fragment.

The most obvious source of events that may mimic a multineutron cluster is the detection, in the same event, of more than one neutron in the same module. The rates at which such pileup is expected to occur have been examined in detail employing both simulations which reproduce the observed neutron angular, energy and multiplicity distributions, together with an analysis based on the measured neutron-neutron relative angle distributions [9]. The two methods provide consistent results which are in line with the numbers of events observed for the channels (^{11}Li,X+n) and (^{14}Be,^{12}Be+n). In the case of (^{14}Be,^{10}Be), less than one event arising from pileup is estimated to occur with $E_p/E_n > 1.4$, compared to some 6 observed events. It may be concluded, therefore, that a signal consistent with the production of a multineutron cluster in the breakup of ^{14}Be – most probably in the channel ^{10}Be+^{4}n – has been observed at a level some 2-sigma above that attributable to background processes.

The average flight time of the 6 events from the target to the neutron array is ~ 100 ns. Unless the decay of the 4 neutron system takes place via the emission of highly correlated neutrons, the lifetime of the putative tetraneutron is of this order or longer; suggesting a particle bound system or a very narrow resonance. The conditions applied in the analysis make an estimate of the production cross-section rather difficult. Nonetheless, if it is assumed that these conditions affect the number of neutrons and ^{4}n events in a similar manner, the cross-section measured for the production of ^{10}Be [12] may be scaled by the relative yields, resulting in a cross section $\sigma(^4\text{n}) \sim 1$ mb.

THE UNBOUND STATE SCENARIO

In light of these results, several calculations have been developed in order to investigate the possibility of existence, and its implications, of a bound tetraneutron [5–8]. The overall consensus is that it should not exist according to our present knowledge of the nuclear interaction. However, one ot these works [8] suggests that the tetraneutron could exist as a low-lying resonance, only about 2 MeV above the 4 neutron threshold. Results from a transfer experiment point towards the same direction [15].

Therefore, we have considered the hypothesis of the decay of a tetraneutron resonance and check whether it would be able to explain the observed events. In [9], the probability of pileup — two neutrons scattering on two different protons in the same module — was estimated through the emission of 4 independent neutrons following the breakup of ^{14}Be. Obviously, if a low-energy ^{4}n resonance is formed the 4 neutrons will flight separetely to the neutron array, but the low energy shared between them should increase the probability that 2 of them go through the same module.

We have simulated the breakup of ^{14}Be into ^{10}Be and a ^{4}n resonance of given energy E and width Γ. The decay of the resonance can lead to very complex final states, as one can consider, for example, a two-step process going through two "dineutrons", or a n-n final state interaction (FSI) between some of the possible n-n pairs. In all these cases the result will depend on many additional parameters or hypothesis.

We have thus chosen the decay given by pure 4-body phase space, which depends

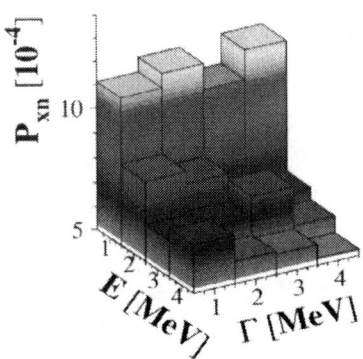

FIGURE 3. Pileup probability of more than 1 neutron in a DEMON module as a function of the energy E and width Γ of a ^{4}n resonance formed in the breakup of ^{14}Be [16].

only on E and Γ. This will provide a lower limit of the pileup probability, as any FSI between neutrons will get them closer in momentum space [16]. The evolution of the pileup probability P_{xn} for E and Γ between 1 and 4 MeV exhibits the expected trend (Fig. 3): an increase towards low energies of the resonance.

The pileup probability beyond $E \sim \Gamma \sim 4$ MeV corresponds to the one estimated with 4 independent neutrons in [9], $P_{xn} \sim 5 \times 10^{-4}$. This probability increases towards $E = 1$ MeV, but only by a factor 2. The increase factor needed to explain the 6 events observed is ~ 20 (the probability associated to the 6 events was 10^{-2} [9]).

If we explore resonance energies below 1 MeV, such an increase factor can be reached. For example, for $E = \Gamma = 500$ keV the increase factor is 15, and for $E = \Gamma = 100$ keV it is 50. Of course this should have to be decreased by the probability of the resonance being formed, but increased by any interaction between the neutrons in the final state. Therefore, we conclude that a low-energy ^{4}n resonance can also explain the data presented in [9].

AN IMPROVED EXPERIMENT

A more intense beam of ^{12}Be, at 41 MeV/N, was used in order to investigate the $-4n$ channel, i.e. breakup into two α particles [17]. With respect to the experimental setup used in [9], the zero degree Si-CsI telescope was replaced by 16 CsI crystals and two Si strip detectors, allowing for the detection of the two α particles in coincidence with neutrons in DEMON.

The energy distribution of the $\alpha\alpha$ system exhibits clear peaks (thin histogram in Fig. 4) that can be assigned to states in ^{8}Be: the ground state (left), the first excited state (right), and the decay of an excited state in ^{9}Be into an excited state

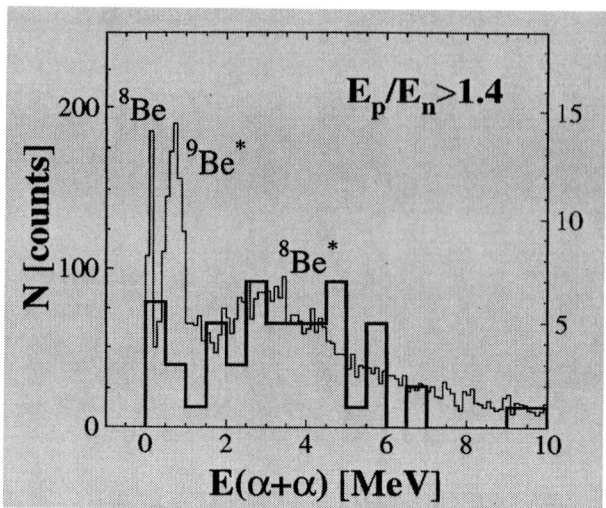

FIGURE 4. Energy of the $\alpha\alpha$ system detected in the breakup of ^{12}Be in coincidence with one neutron. The thin histogram corresponds to all the data, the thick one to the events with $E_p/E_n > 1.4$ [17].

in ^{8}Be (middle). For all three the final state is the same, $2\alpha 4n$, but only for the peak in the middle the emission of the four neutrons cannot be direct: the passage through ^{9}Be makes that at most three neutrons can be emitted at once.

Having a look at the same distribution gated on $E_p/E_n > 1.4$ in DEMON, if these abnormal recoils in DEMON are associated with a tetraneutron state, bound or not, the peak at ~ 700 keV should thus disapear. And this seems to be the case (thick histogram in Fig. 4).

CONCLUSIONS

We have described a new approach to the long search for neutron clusters: the breakup of very neutron-rich beams and the detection of the neutron cluster liberated by the recoil induced on a proton in a liquid scintillator. The first application to data has revealed 6 events consistent with the production of a tetraneutron in the breakup of ^{14}Be. We have seen that the formation of a ^{4}n state, bound or resonant, is needed for the description of these events.

The application of the technique to the breakup of ^{12}Be into $2\alpha 4n$ has revealed more events, which seem to be correlated with the direct emission of the four neutrons from ^{12}Be to ^{8}Be. The analysis of these data is still in progress [17]. Other experiments with a more intense ^{8}He beam are being planned, both on the breakup into $\alpha 4n$ or using a proton target and measuring the $(p, p\alpha)$ channel.

ACKNOWLEDGMENTS

I am grateful to my colleagues at LPC, specially to Nigel Orr who is the main responsible of the experimental programme briefly described here, and Marc Labiche and Guillaume Normand for the data analysis of the two DEMON experiments.

REFERENCES

1. M.V. Zhukov et al., Phys. Rep. **231**, 151 (1993).
2. A.A. Ogloblin, Y.E. Penionzhkevich, "Treatise on Heavy-Ion Science (vol. 8): Nuclei Far From Stability", Ed. D.A. Bromley, Plenum Press, New York (1989), p. 261 and references therein.
3. A.A. Korsheninnikov et al., Phys. Rev. Lett. **87**, 092501 (2001).
4. D.R. Tilley, H.R. Weller, G.M. Hale, Nucl. Phys. **A541**, 1 (1992) and references therein.
5. N. Timofeyuk, J. Phys. G **29**, L9 (2003), nucl-th/0301020.
6. C. Bertulani, V. Zelevinsky, nucl-th/0212060.
7. J. Carbonell, private communication.
8. S.C. Pieper, nucl-th/0302048.
9. F.M. Marqués et al., Phys. Rev. C **65**, 044006 (2002).
10. F.M. Marqués et al., Phys. Lett. B **476**, 219 (2000).
11. F.M. Marqués et al., Phys. Rev. C **64**, 061301 (2001).
12. M. Labiche et al., Phys. Rev. Lett. **86**, 600 (2001).
13. H.G. Bohlen et al., Nucl. Phys. **A583**, 775 (1995).
14. J. Gräter et al., Eur. Phys. J. 4, 5 (1999).
15. D. Beaumel, private communication.
16. F.M. Marqués et al., in preparation.
17. G. Normand et al., in preparation.

Recent results from β-decay studies in the ^{100}Sn region

Z. Janas*, L. Batist†, A. Blazhev**‡, W. Brüchle**, J. Döring**, M. Gierlik*,
M. Górska**, H. Grawe**, T. Faestermann§, S. Harissopulos¶,
A. Jungclaus‖, M. Karny*, M. Kavatsyuk**, O. Kavatsyuk**, R. Kirchner**,
M. La Commara††, C. Mazzocchi**, I. Mukha**, C. Plettner**,
A. Płochocki*, E. Roeckl**, M. Romoli‡‡, M. Schädel**, R. Schwengner§§
and J. Żylicz*

Institute of Experimental Physics, University of Warsaw, Warsaw, Poland
†*St. Petersburg Nuclear Physics Institute, Gatchina, Russia*
**Gesellschaft für Schwerionenforschung mbH, Darmstadt, Germany*
‡*University of Sofia, Sofia, Bulgaria*
§*Technical University, Munich, Germany*
¶*Inst. of Nuclear Physics, NCSR "Democritos", Athens, Greece*
‖*Universidad Autónoma de Madrid, Madrid, Spain*
††*Universita "Federico II", Napoli, Italy*
‡‡*INFN, Napoli, Italy*
§§*Inst. für Kern- und Hadronenphysik, Forschungszentrum Rossendorf, Dresden, Germany*

Abstract. The decays of the neutron deficient nuclei ^{94,95}Ag, ^{100,102}In and ^{102}Sn were investigated at the GSI on-line mass separator. The Gamow-Teller decay of ^{100,102}In and ^{102}Sn was studied by using an array of germanium detectors and a total absorption spectrometer. In ^{94}Ag and ^{95}Ag, spin-gap isomers were identified and their decays were investigated. Results obtained in these studies are discussed and compared to predictions of shell model calculations.

INTRODUCTION

Studies of nuclei in the ^{100}Sn region offer the possibility to test nuclear models describing properties of nuclei in which protons and neutrons occupy identical orbitals near a double shell closure. A variety of phenomena are predicted to occur in such systems. Nuclei with N≈Z are expected to show enhanced neutron-proton correlations giving rise e.g. to a new pairing mode and high-spin isomers. An insight into the role of the proton-neutron interaction and/or core excitation in the shell model structure of N=Z nuclei close to ^{100}Sn can be gained, e.g. by studying decay properties of their ground and isomeric states.

Beta decay of nuclei in the ^{100}Sn region proceeds mainly via the Gamow-Teller (GT) transformation of a $g_{9/2}$ proton into a $g_{7/2}$ neutron. Since the N=Z=50 shell closure occurs far from the β stability line, isotopes in this region have relatively large Q_{EC} values and the GT strength can be investigated and confronted with theoretical predictions over a broad range of excitation energies.

Several high-spin isomers are predicted to occur in nuclei close to ^{100}Sn as a result of the attractive interaction of $\pi g_{9/2} - \nu g_{9/2}$ holes in the upper part of the $g_{9/2}$ subshell [1]. This interaction lowers the energy of stretched configurations creating long-lived spin-gap isomers which may disintegrate via high-multipole electromagnetic transitions or/and via β decay. Studies of spin-gap isomers characterized by very specific configurations provide a valuable test of residual interactions and truncation schemes in the shell model calculations.

From the experimental point of view, decay studies of exotic nuclei at ISOL systems require the application of efficient and chemically selective ion sources and an adequate spectroscopic approach. Such methods have been developed and applied at the GSI-ISOL facility to investigate decays of very neutron-deficient isotopes of silver, indium and tin.

IDENTIFICATION OF HIGH-SPIN ISOMERS IN SILVER ISOTOPES

The ^{94,95}Ag nuclei were produced in the ^{58}Ni(^{40}Ca,pxn) reaction. The reaction products were stopped in a graphite catcher of a FEBIAD B2C ion source [2, 3]. This ion source is characterized by a very fast release of silver isotopes while palladium isotopes are suppressed by a factor of 30 by trapping in two cold pockets that are attached to the ion-source. The intensity of the mass-separated ^{94}Ag and ^{95}Ag beams amounted to about 1.5 and 28 atoms/s, respectively.

The secondary beam was implanted into a transport tape surrounded by a β-particle detector. The latter consisted of a prism-shaped array of silicon detectors in the ^{94}Ag studies and a plastic scintillator tube in the ^{95}Ag measurement. The efficiencies of these detectors amounted to 65 and 85%, respectively. An array of germanium detectors, comprising Cluster, Clover and single germanium detectors, was used to register γ-rays emitted in the decays of ^{94,95}Ag.

In earlier decay studies of ^{94}Ag performed at the GSI-ISOL facility an evidence for β-decaying low-spin $I^{\pi}=(7^+)$ and high-spin ($I \geq 17$) isomers in ^{94}Ag was reported [4]. In the recent studies of ^{94}Ag the existence of the high-spin isomer has been firmly established (for details, see C. Plettner et al. [5]). Figure 1 shows the decay scheme of ^{94}Ag resulting from the analysis of β-γ-γ coincidences. A cascade of four transitions was placed above the known 0.53(1) μs 14^+ isomer in ^{94}Pd. The spins and parities of the new high-spin states in ^{94}Pd were inferred from the level systematics in the N=48 isotones guided by empirical shell model calculations. According to this systematics, the highest state observed in ^{94}Pd at excitation energy of 7.7 MeV is most probably a $I^{\pi}=20^+$ state. This leaves possibilities of $(19^+$-$21^+)$ for the spin and parity of the β-decaying high-spin isomer in ^{94}Ag. By comparing the observed β-feeding and the measured half-life to shell model predictions the spin and parity of (21^+), resulting from the stretched coupling of the $(g_{9/2}^{-3})_{21/2}$ proton and neutron configurations, has been tentatively assigned to the high-spin isomer in ^{94}Ag. The excitation energy of the ^{94m}Ag(21^+) isomer was estimated to be about 6.3 MeV, which yields a Q_{EC} value of about 18.9 MeV for the β-decay of this state. Such a large Q_{EC} value opens a wide energy window for β-delayed particle emission. Indeed, β-delayed proton decay of

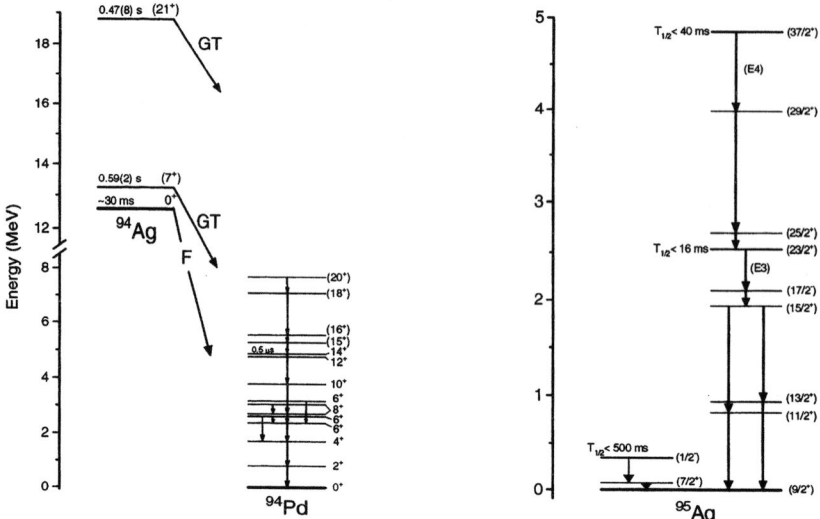

FIGURE 1. Decay schemes of ^{94}Ag [5] and ^{95}Ag [10] isomers. The excitation energies of ^{94}Ag isomers result from the shell model calculations.

^{94m}Ag(21^+) was identified and γ-ray cascades in ^{93}Rh were observed that are known from in-beam measurements to connect high-spin states (for details, see I. Mukha et al. [7]).

The shell model calculations for ^{94}Pd and ^{94,95}Ag were performed by considering ^{88}Sr as an inert core and by restricting the valence proton particles and neutron holes to the $(p_{1/2}, g_{9/2})$ orbitals. The two body matrix elements and the single-particle energies fitted to the experimental data for N=48-50 isotones by Gross and Frenkel [8] were employed. The calculations in the $(p_{1/2}, g_{9/2})$ space remarkably well reproduce energies and systematic trends of states in ^{90}Mo, ^{92}Ru and ^{94}Pd but fail to predict the existence of the 21^+ isomer in ^{94}Ag. This state is calculated to lie 224 keV above the 19^+ level and thus no β-decaying isomer is predicted.

The inversion of the 21^+ and 19^+ states leading to the isomerism of the 21^+ level can only be reproduced by the large scale shell model calculations performed within the (gds) model space comprising the $g_{9/2}, g_{7/2}, s_{1/2}, d_{5/2}, d_{3/2}$ orbitals for protons and neutrons [5, 6]. The effective interaction was derived by employing a perturbative many-body scheme starting from the free nucleon-nucleon interaction. For ^{94}Ag states calculations were performed allowing up to $4p - 4h$ excitations across the ^{100}Sn shell closure. In these calculations the isomerism of the 21^+ in ^{94}Ag was reproduced with a 21^+ - 19^+ inversion of 34 keV.

In spite of several experimental attempts, the $23/2^+$ spin-gap isomer in ^{95}Ag predicted by the shell model calculations of Ogawa [1] has been observed only recently [9, 10]. In the experiment performed at the GSI-ISOL facility the decays of ^{95}Ag isomers were searched for in the γ-ray spectra registered in anticoincidence with positrons (for details, see J. Döring et al. [10]). Such condition enhances γ-rays following the internal deexcitation of an isomer with respect to the β-delayed radiation. Figure 1 shows the level scheme of ^{95}Ag resulting from the analysis of the γ-γ coincidences. The proposed spin and parity assignments are based on a comparison with shell model predictions. A $(37/2^+)$ high-spin isomer in ^{95}Ag has been identified which decays to the $(23/2^+)$ isomer. The latter disintegrates most likely via an $E3$ transition to the $(17/2^-)$ state in ^{95}Ag and further down to the $(9/2^+)$ ground state. Another sequence of γ transitions has been assigned to the decay of the $(1/2^-)$ isomer in ^{95}Ag.

The existence of the $(1/2^-)$ and $(23/2^+)$ isomers in ^{95}Ag can be well described by the shell model calculations within the $(p_{1/2}, g_{9/2})$ space using the empirical interaction. However, the $(37/2^+)$ isomer is not predicted by these calculations. One may expect that, as in the case of ^{94}Ag, large scale shell model calculations allowing $p-h$ excitations across the N=Z=50 shell closures will reproduce the isomerism of the $(37/2^+)$ state which shows close relation to the (21^+) isomer in ^{94}Ag, namely $|^{94}\text{Ag}, 21^+\rangle = |^{95}\text{Ag}, 37/2^+\rangle \times v g_{9/2}^{-1}$.

DECAY STUDIES OF INDIUM ISOTOPES

The neutron-deficient isotopes ^{100,102}In were produced in the ^{50}Cr(^{58}Ni, αpxn) reaction. The reaction recoils were stopped inside a graphite catcher of the thermal ion source [11]. The suppression of the isobaric silver and cadmium isotopes was achieved due to the differences in their ionization potentials with respect to that of indium. The intensity of the mass-separated ^{100,102}In beams amounted to about 1.6 and 210 atoms/s, respectively [12, 13].

The decay of ^{100,102}In was investigated by using the high-resolution germanium detector array, described above, and the total absorption spectrometer (TAS) [14]. The analysis of the high-resolution data enabled us to establish coincidence relations between the observed γ transitions and to construct partial decay schemes of the isotopes studied. However, as it was demonstrated e.g. in decay studies of silver isotopes [15, 16], high-resolution, low efficiency measurements are generally unable to record all of the many weak γ transitions depopulating states at high excitation energy in the decay product. As a consequence, only apparent β-feeding of daughter states can be determined.

The TAS measurements provide a way to overcome this limitation of the low efficiency γ-ray measurements. In the TAS all members of each γ-ray cascade depopulating an excited state are added up to yield an output signal corresponding to the excitation energy of this state. Such a signal provides in principle an unambiguous signature for each individual β transition of the decay under consideration. However, due to the limited efficiency of the TAS, the determination of the β-feeding distribution from the experimental spectra requires thorough knowledge of the detector response and application of deconvolution procedures [13, 17, 18]. Partial decay schemes resulting from the high-resolution studies are used as a constraint in the unfolding procedures. Figure 2 shows the comparison of the β-intensity distribution for the decay of ^{102}In obtained from the high-resolution and the TAS measurement [13]. It is evident that in the decay scheme built only on the basis of the high-resolution data, β-feeding is assigned to levels which are in fact fed by the deexcitation of the higher-lying states. Due to the very strong

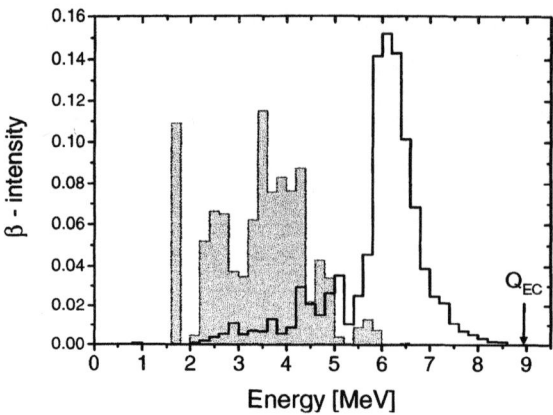

FIGURE 2. Beta intensity distribution for the decay of ^{102}In obtained from the high-resolution (shaded area) and the TAS (solid line histogram) measurement [13]. The intensities are normalized to 1.

dependence of the β-decay rate function on the transition energy, the distortion of the β-intensity distribution has severe impact on the resulting GT strength (B_{GT}) values. These effects point to the prime importance of the TAS measurements in obtaining reliable B_{GT} distributions for the decays of isotopes with large Q_{EC} values, where highly excited daughter states might be populated. Figure 3 shows the B_{GT} distributions for ^{100}In and ^{102}In decay resulting from the TAS measurements. The shape of the distributions is dominated by the prominent concentration of the GT strength at an excitation energy of about 6.5 MeV. Within a simple single-quasiparticle model such decay characteristics can be interpreted as the $\pi g_{9/2} \to \nu g_{7/2}$ GT transformation of one of eight paired $g_{9/2}$ protons to four-quasiparticle daughter configurations.

A more realistic description of the GT strength distribution can be obtained in the shell model approach. In the model space used proton holes were allowed to occupy $p_{1/2}$ and $g_{9/2}$ states and the neutron particles occupy $d_{5/2}$, $g_{7/2}$, $s_{1/2}$, $d_{3/2}$ and $h_{11/2}$ orbitals. The effective interaction was derived from the free nucleon-nucleon potential. The GT-strength distributions was calculated using the free nucleon GT operator and assuming $6^+(\pi g_{9/2}^{-1} \nu d_{5/2}^1)$ configuration for the ground state of ^{100}In and ^{102}In, as predicted by the calculations [12].

In Figure 3 the GT strength for ^{100}In and ^{102}In resulting from shell model calculations is confronted with the experimental B_{GT} distributions. For ^{100}In the position and the width of the measured distribution are reasonably well reproduced by the theory. However, in the case of ^{102}In the calculations fail to reproduce the high energy part of the measured B_{GT} distribution. Most probably the spreading of the experimental strength in the high energy region is due to the mixing with configurations outside the model space. The total experimental GT strength for ^{100}In and ^{102}In is by a (hindrance) factor of 4.1(9) and 2.8(6), respectively, smaller than the shell model predictions. These new values extend the systematics of hindrance factors for indium isotopes which shows a decrease from the value of 7 for ^{106}In [19] to the value of about 4 for ^{100}In [12]. The discrepancy between the total experimental and calculated GT strength can be reduced

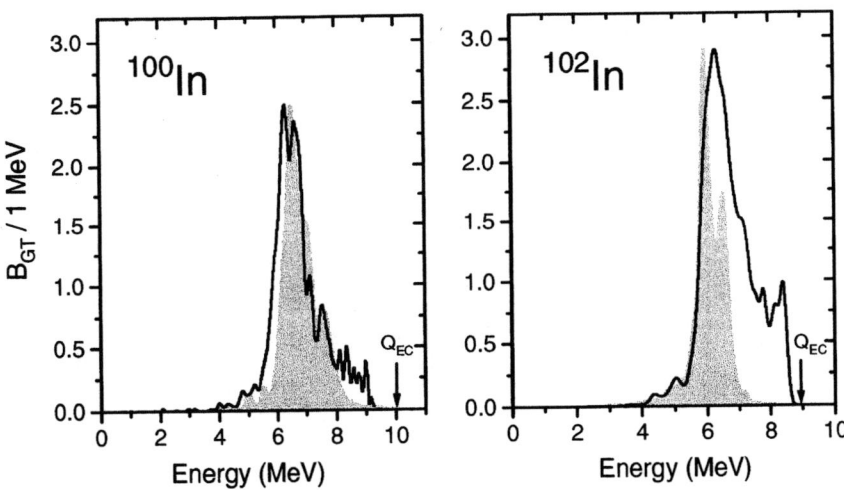

FIGURE 3. B_{GT} distribution for the decay of ^{100}In and ^{102}In derived from the TAS measurements (blank histograms) and resulting from the shell model calculations (shadowed histograms). The theoretical distributions were normalized to the maxima of measured B_{GT} functions.

if one assumes a universal character of the hindrance mechanisms observed in the $p-$, $sd-$, and $pf-$ shells and adopts for the ^{100}Sn region the global hindrance factor of 1.8, which was introduced to account for the higher-order effects in the pf-shell. The remaining disagreement can be related to the truncation of the configuration space in the shell model calculations.

DECAY STUDIES OF TIN ISOTOPES

The decay spectroscopy of the most neutron-deficient ($T_{1/2}<$ 20 s) tin isotopes at ISOL systems was so far hindered by the dominance of the simultaneously produced isobaric indium, cadmium, and silver isotopes in the mass-separated samples. The radiation from these contaminants overwhelmed e.g. the measured γ-ray spectra and disabled identification and studies of the decay of the lightest tin isotopes. The dominant contamination stems from indium isotopes which are produced with much higher cross-section than the isobaric tin isotopes.

Further studies of the most neutron-deficient tin isotopes required the development of an efficient ISOL ion source with very high chemical selectivity for tin. This requirement was achieved by the separation of tin isotopes as SnS$^+$ molecules formed in the FEBIAD-B2 ion source with sulphur vapours addition [20]. This on-line chemistry permitted a powerful separation of tin from isobaric indium, cadmium, silver and palladium isotopes. Thus, only the strongly produced silver activities were traced in the γ-ray spectra. The on-line measurements showed that about 50% of the tin ions is converted into SnS$^+$ molecules. For short lived tin isotopes (e.g. ^{100}Sn) the $\sim$50% loss due to the SnS$^+$ formation is compensated by the faster release speed of SnS$^+$ molecules compared to that reached by using atomic tin ions [20].

The ^{102}Sn nuclei were produced in the fusion-evaporation reactions, induced by a ^{58}Ni beam impinging on the ^{50}Cr target. The reaction products were stopped in a ZrO$_2$ fiber catcher inside the FEBIAD source. The intensity of the mass-separated ^{102}SnS$^+$ beam amounted to about 1900 atoms/hour. This value is about 45 times higher than that observed for ^{102}Sn in the ^{112}Sn beam fragmentation experiment at FRS [21].

The mass-separated beam was implanted into a tape surrounded by the β detector. To avoid buildup of daughter activities, the tape periodically removed the source from the measurement position. The β-delayed γ-rays were measured by the setup of germanium detectors described above. This detection set-up allowed us to establish coincidence relations between transitions in the decay of ^{102}Sn. Figure 4 shows, as an example, a γ-ray spectrum gated by the 322 keV line assigned to the decay of ^{102}Sn by Stolz et al. [21] and the partial decay scheme of ^{102}Sn, constructed according to the observed coincidence relations. It includes two 1$^+$ states fed in the decay of ^{102}Sn and depopulated by

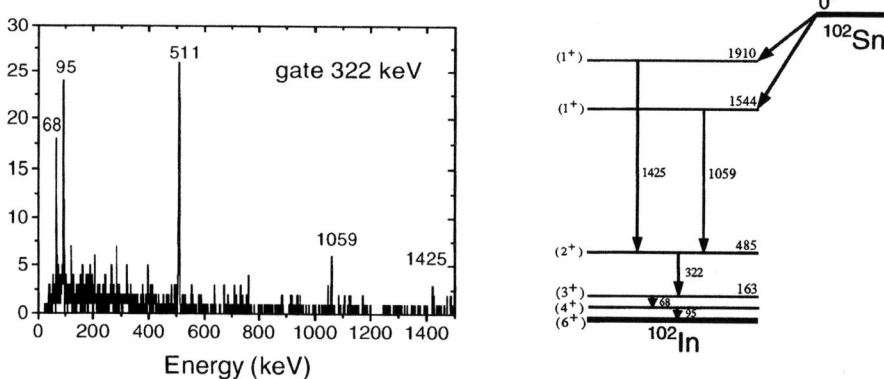

FIGURE 4. Left part: γ-ray spectrum gated by the 322 keV line from ^{102}Sn decay. Right part: Partial decay scheme of ^{102}Sn. Proposed spins and parities of ^{102}In states were inferred from the comparison with shell model predictions.

the 1059 and 1425 keV transitions to the (2$^+$) level. The latter deexcites to the ground state by the cascade of 322, 68 and 95 keV transitions.

In a complementary measurement the ^{102}SnS$^+$ beam was collected on a tape and periodically moved between two silicon detectors placed in the center of the TAS. The high efficiency of the TAS allowed us to considerably improve the sensitivity limit reached in the high-resolution measurement. Figure 5 shows the TAS energy spectrum of ^{102}Sn measured in coincidence with positrons. The solid line shows the TAS spectrum simulated according to the partial decay scheme presented in Figure 4. Good agreement between the measured and simulated TAS spectra corroborates the main features of the decay scheme proposed in Figure 4. However, a peak visible in the TAS spectrum at energy of about 3.2 MeV reveals feeding of another 1$^+$ state in ^{102}In which lies at an excitation energy of about 2.2 MeV. A more sophisticated analysis of the high resolution data, including add-back procedure, is in progress to identify γ-rays deexciting this state.

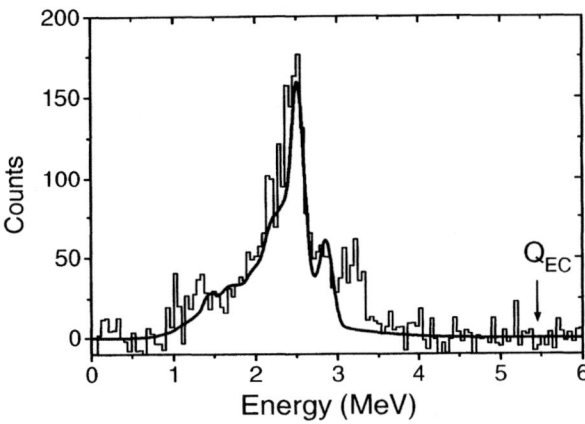

FIGURE 5. Measured (histogram) and simulated (line) TAS spectrum for the positron decay of ^{102}Sn. The simulation was based on the partial decay scheme shown in Figure 4.

SUMMARY

Fusion-evaporation reactions were used at the GSI-ISOL facility to synthesize very neutron-deficient isotopes of silver, indium and tin. Successful decay studies of these isotopes were possible thanks to the development of fast, efficient and chemically selective ion-sources. The decay properties of studied nuclei were investigated by using an array of germanium detectors and/or a total absorption spectrometer.

The GT-strength distribution for ^{100}In decay, the closest neighbour of ^{100}Sn, was reliably determined by combining results from the high resolution and a total absorption measurements. Significant progress has been made in studies of very neutron deficient tin isotopes. In particular, the decay scheme of ^{102}Sn has been firmly established on the basis of $\beta\gamma\gamma$ coincidence measurements. Decay studies of ^{101}Sn and ^{100}Sn are planned for the near future. However, with the expected rate of about 8 atoms/hour, an observation of the ^{100}Sn decay will be a very challenging task. Investigations of silver isotopes lead to the identification of high-spin isomers in ^{94}Ag and ^{95}Ag. The Q_{EC} and spin values of the β-decaying ^{94}Ag(21^+) isomer have no precedence in the whole nuclide chart.

New data obtained for nuclei close to ^{100}Sn provide a stringent test of the theoretical models. The shell model calculations employing empirical interaction in the restricted model space reasonably well describe systematical trends of level energies but fail e.g. to predict the ^{94m}Ag(21^+) and ^{95m}Ag($37/2^+$) isomers and to describe the B_{GT} strength measured for the ^{102}In. The isomerism of the (21^+) in ^{94}Ag is reproduced by the large-scale shell model calculations performed in the gds space using a realistic interaction. The shell model calculations within the complete gds oscillator shell are expected to reproduce the shape and the total value of the B_{GT} distributions measured for ^{100,102}In.

ACKNOWLEDGMENTS

This work was supported in part by the European Community under Contracts ERBFMGECT950083 and HPRI-CT-1999-00001. The authors from Poland acknowledge support by the Polish Committee of Scientific Research under project No. 2P03B03523. A. J. acknowledges financial support from the Deutsche Forschungsgemeinschaft (DFG) within the Heisenberg program.

REFERENCES

1. K. Ogawa, Phys. Rev. **C28** (1983) 958.
2. R. Kirchner, Nucl. Instr. Meth. **B70** (1992) 186.
3. R. Kirchner, Nucl. Instr. Meth. Phys. Res. **B20** (1997) 204.
4. M. La Commara *et al.*, Nucl. Phys. **A708** (2002) 167.
5. C. Plettner *et al.*, to be published.
6. F. Nowacki, Nucl. Phys. **A704** (2002) 223c.
7. I. Mukha *et al.*, *Proceedings of the International Symposium on Proton-Emitting Nuclei, Legnaro, Italy, 2003* edt. E. Maglione and F. Soramel, AIP Conf. Proc. No. 681, 2003, p. 209.
8. R. Gross and A. Frenkel, Nucl. Phys. **A267** (1976) 85.
9. N. Mărginean *et al.*, Phys. Rev. **C67** (2003) 061301.
10. J. Döring *et al.*, Phys. Rev. **C68** (2003) 034306.
11. R. Kirchner, Nucl. Instr. and Meth. **186** (1981) 295.
12. C. Plettner *et al.*, Phys. Rev. **C66** (2002) 044319.
13. M. Gierlik *et al.*, Nucl. Phys. **A724** (2003) 313.
14. M. Karny *et al.*, Nucl. Instr. Meth. **B126** (1997) 411.
15. Z. Hu *et al.*, Phys. Rev. **C60** (1999) 024315.
16. Z. Hu *et al.*, Phys. Rev. **C62** (2000) 064315.
17. M. Karny *et al.*, Nucl. Phys. **A640** (1998) 3.
18. D. Cano-Ott *et al.*, Nucl. Instr. and Meth. **A430** (1999) 333.
19. M. Karny *et al.*, Nucl. Phys. **A690** (2001) 367.
20. R. Kirchner, Nucl. Instr. and Meth. **B204** (2003) 179.
21. A. Stolz *et al.*, GSI Report 2002-1, p. 7.
    T. Faestermann *et al.*, Eur. Phys. J. **A15** (2002) 185.

Transfer Reaction Studies with Exotic Nuclei

W.N. Catford[1], R.C. Lemmon[2], C.N. Timis[1], M. Labiche[3], L. Caballero[4] and R. Chapman[3]

[1] *University of Surrey, Guildford, Surrey GU2 7XH, UK*
[2] *CCLRC Daresbury Laboratory, Warrington, Chesire WA4 4AD, UK*
[3] *University of Paisley, Paisley, Scotland PA1 2BE, UK*
[4] *IFIC, CSIC/University of Valencia, Valencia E-46071, Spain*

Abstract. Transfer reactions offer the possibility to study single-particle structure in exotic nuclei, including the structure of ground states, the structure of excited states, and the location and distribution of single particle strength. The last of these means that the evolution of single particle orbitals away from stability, and the consequent changes in shell structure and collectivity, can be studied in detail. The kinematics of transfer reactions initiated by protons and deuterons, in inverse kinematics, have characteristic features that mean a general-purpose design of array can be applied to a wide range of experiments. Due to resolution considerations, coincident gamma-ray detection is highly desirable or even essential. A new silicon detector array called TIARA has been designed so that it can be used together with the segmented germanium detectors of the EXOGAM array. The setup for TIARA is now complete, having been built by several UK groups, and it was recently commissioned by the extended UK-France-Spain collaboration at GANIL. The EXOGAM gamma-ray detectors can be placed as close as 50mm to the target, covering 2/3 of 4π, so as to give the maximum detection efficiency. Simultaneously, the VAMOS magnetic spectrometer can be coupled to the system at zero degrees, to separate the beam from reaction products and give precise energy and angle measurements. The features of TIARA, as presently installed, are described.

THE PROMISE OF TRANSFER REACTIONS

Transfer reactions initiated by light ions are well established, through experiments over many years using stable beams, as a clear probe of single particle structure in nuclei [1-4]. In particular, transfer reactions in which a single nucleon is transferred between the beam and target particles can allow the orbital angular momentum in single-particle states to be deduced. This, together with the excitation energies of the levels involved, and the extent to which they exhaust the full single particle strength (i.e. the spectroscopic factor), are the main properties to be measured. In the future, in experiments using radioactive beams, transfer reactions again promise to be a rich source of information concerning single-particle energy levels and the precise shell model structure of nuclei far from stability. Typical experiments for transfer are performed at collision energies of order 10 MeV/A. Transfer reactions initiated by heavy ions (^{6}Li and heavier) are also established as useful spectroscopic tools. In the peripheral collisions of heavy ions at these energies, an additional strong selectivity is observed [5] that allows the spin of the final orbital to be deduced, in terms of whether it is $j = \ell + \frac{1}{2}$ or $j = \ell - \frac{1}{2}$. This follows from conservation of linear and angular momentum for the system including the transferred nucleon, plus angular momentum selection and combination rules [6]. In studies of exotic nuclei (see for example [7]) this selectivity has allowed the spins of levels to be identified with some confidence, even when the yield is too low to measure an angular distribution.

For studies of ground states of very exotic nuclei, a complementary technique has been developed that allows the orbitals populated by valence nucleons to be deduced, along with their occupancy, or spectroscopic factor. The process of single-nucleon knockout is found to have a cross section of one to two orders of magnitude higher than the typical 1-10 mb/sr cross section of a transfer reaction populating a reasonably strong single-particle state. The simple picture of the sudden removal of a nucleon from its valence orbital, which is employed in the analysis of knockout, works rather well at energies of order 100 MeV/A [8,9]. Because the core survives in these experiments, and is detected at very forward angles, very peripheral collisions are automatically selected. During these collisions, if the target nucleus is imagined as a black disk, a "hole" is bored out of the wavefunction of the valence nucleon. In other words, for the valence nucleon to be removed, only a certain part of its wavefunction will be involved – namely, the part that extends beyond the core and overlaps with the black disk. The angular momentum content of this part of the wavefunction can be computed. The removal of this angular momentum affects in turn the magnetic substate population of the surviving core and hence its longitudinal momentum. The result is that high-resolution measurements of the core momentum (using a magnetic spectrometer typically, but not necessarily) can reveal the angular momentum of the removed nucleon. An elegant example of this type of analysis is an experiment [10] using a "cocktail beam" of many different species, which shows the change in angular momentum of the valence nucleons for nuclei across a range of the nuclear chart. From this brief description, it is evident that knockout will be particularly effective when the valence nucleons are weakly bound, because then the wavefunction will typically extend significantly beyond the core. Thus, knockout as a spectroscopic technique is especially valuable to study nuclei near the drip-line, where simultaneously the reaction mechanism is at its most pure and the cross sections highest. For the removal of nucleons that are more tightly bound, the cross sections become closer to those for transfer.

An interesting feature of transfer reactions such as (d,p) or (α,^{3}He) is that a neutron can be added to a neutron-rich projectile to make states in a nucleus even further from stability. Ideally, spins and spectroscopic factors of various states will be measured, along with their excitation energies. It is sometimes the case, however, that the relative spectroscopic factors of different final states in a given reaction can be extracted with more confidence than the absolute values. Even in the cases where the absolute values are not so well defined, the transfer data will clearly indicate the nature and energies of the states with the strongest single-particle character, and this is itself very valuable information to have. The exploitation of (d,p) reactions looks to be a particularly important area for future work with radioactive beams.

Single particle levels are of interest close to closed shells because they can be used to fix the parameters for shell model calculations and also to test the predictions. However, it should not be forgotten that transfer can also be an important tool to study deformed nuclei. If a deformed nucleus has a single valence particle in a particular

Nilsson orbital, the wavefunction of that nucleon can be expanded in terms of a basis comprising the spherical shell model orbitals in the parent shell: $|\psi\rangle = \Sigma_j\ c_j\,|\,j\,\rangle$ where the $|\,j\,\rangle$ represent the spherical orbitals with good angular momentum quantum number, j. If a single-nucleon transfer reaction is performed using a target of the even-even core nucleus, then the band in the deformed product is populated by the transferred nucleon entering the Nilsson orbital $|\,\psi\,\rangle$. The state in the band that has spin j can only be formed (at least, in a single-step transfer) via the component $|\,j\,\rangle$ of the expansion of the Nilsson orbital, and the magnitude of the cross section will reflect the magnitude of the coefficient c_j. In the sense that the set of coefficients c_j can be thought of as a sort of DNA (or actual) fingerprint of the Nilsson orbital, the relative populations in transfer, of the various states in a rotational band, represent a fingerprint that identifies the orbital on which the band is built. Knowing this orbital can, in turn, allow the deformation to be inferred. This application of transfer reactions is discussed by Bohr and Mottelson [11] and reviewed extensively in ref. [12].

SPECIFIC ISSUES CONCERNING TRANSFER WITH RADIOACTIVE BEAMS

What has been done with radioactive beams

The literature of transfer experiments which use radioactive beams to perform nuclear spectroscopy, exploiting the otherwise traditional methods of angular distribution measurements, is growing. Amongst the first were a measurement of the astrophysically interesting ^{56}Ni(d,p)^{57}Ni reaction [13] using a beam of radioactive ^{56}Ni ions accelerated in the Argonne tandem, and a measurement of the structure of the halo nucleus ^{11}Be via the ^{11}Be(p,d)^{10}Be reaction [14,15] using a secondary beam produced at GANIL by fragmentation. Because ^{10}Be is rather collective, multistep reactions needed to be considered in the ^{11}Be case, in a full theoretical treatment [15]. Together with the halo orbit of the valence nucleon, this also implied that a careful treatment of the form factor for transfer was required [14,15]. The form factor, or in other words the dependence of the transfer probability on the radial variable, is not an issue in knockout reactions because it is clear that in that case the outer parts of the halo wavefunction give rise to the whole cross section. However, the possibility of multistep reactions is an issue for knockout just as it is for transfer, in principle. In fact, transfer with a light, deformed halo nucleus is exceptionally complicated theoretically (see also ref. [16]) and it is encouraging that the results in the ^{11}Be case [14,15] could be interpreted reliably.

Heavy ion induced transfer has also begun to be used, and at Oak Ridge the reactions (^{13}C,^{12}C) and (^{9}Be,^{8}Be) have been compared using targets of ^{13}C and ^{9}Be with a radioactive ^{134}Te beam and clear $j = \ell \pm \frac{1}{2}$ selectivity was observed [17].

What is different when using radioactive beams

The key difference, experimentally, when using radioactive beams (apart from the much lower beam intensity, of course) is that the kinematics are inverted, so that typically the lighter particle is the target. For heavy ion induced transfer, this means that the target-like particle may be very difficult to observe, since it may not escape from the target. For light ion induced reactions, the light target-like particle will typically escape from the target, and hence the experimenter has a choice of observing the beam-like particle, the target-like particle, or both. In the ^{56}Ni(d,p)^{57}Ni example given above [13], the beam-like particle was detected as a tag of the correct reaction and the energy-angle systematics of the light, target-like particle were exploited to identify different excited states in ^{57}Ni. In the ^{11}Be(d,p)^{10}Be example [14,15] the approach was almost the opposite: the target-like particle was used as a tag and the energy and angle of the beam-like particle were analysed to isolate states in ^{10}Be and measure their angular distributions.

The advantages and disadvantages of the various different approaches to light-ion induced transfer have been analysed in some detail [18] and summarised elsewhere [19,20]. In brief, the main difficulty with detecting the target-like particle is that rather a large angular coverage is required for the detection system. If detecting the beam-like particle, the kinematic focussing is into a small solid angle and a high resolution magnetic spectrometer can be used. The problem here is that exceptional resolution in angle is required, which is attainable for light beams, but which gets progressively more difficult as the mass and energy of the beam are increased. In either case, the resolution in excitation energy for excited states is limited (at least for practical target thicknesses, and the focussing conditions generally expected for radioactive beams) to several hundred keV. Active targets may help to overcome this, but for a conventional passive target it is more-or-less required that gamma-ray coincidence data be collected, if better resolutions are to be achieved.

Considering specifically light-ion induced transfer, the kinematical focussing of reaction products can easily be seen to be characteristic of the transfer type [19,20]. That is, for reactions such as (p,d) or (d,t) or (d,^{3}He) in inverse kinematics, where the radioactive projectile loses a neutron or proton, the light target-like ejectiles are confined to come out within a cone of forward angles in the laboratory frame. Furthermore, elastic scattering will always produce target-like ejectiles close to 90° and reactions such as (d,p) will have the target-like ejectiles distributed over the whole range of angles but with backward angles near 180° in the laboratory being of particular interest. The beam-like particles, for which the mass hardly changes with the transfer, are hardly deflected by the collision with such a light target and are found within a few degrees of the incident beam direction, as would intuitively be expected. It is the significant change in mass of the target-like particle that turns out to be the dominant effect in producing the characteristic kinematics. In the centre-of-mass frame, the light recoil has to compensate the momentum of the beam-like ejectile. Since the mass and velocity of the heavy particle are hardly changed by the transfer, the centre-of-mass velocity of the light recoil has to change dramatically in proportion to its dramatic percentage change in mass. The velocity observed in the laboratory frame will be the resultant sum of this centre-of-mass velocity of the light particle and the velocity *of the centre of mass* in the laboratory frame. Details are given elsewhere [19,20] but basically the ratio of these two vector lengths is dominated by a factor $\sqrt{f} = \sqrt{m_{target}/m_{ejectile}}$ related to the masses of the light particles before and after the transfer. For a reaction such as (p,d) or (d,^{3}He) where $f<1$, the velocity *of the centre of mass* dominates and focusses the light products into the forward hemisphere, typically forward of 40°. Of the two kinematic solutions at each forward angle, it is the one with the lower energy for the light particle that corresponds to small-angle scattering in the centre-of-mass frame. Normally, in a transfer reaction, this will be the solution with the higher yield. For a reaction such as (d,p) where $f>1$, the velocity of the light ejectile will overcome the focussing effect of the moving centre of mass, and reaction products corresponding to small scattering angles in the centre-of-mass frame will emerge at backward angles in the laboratory frame.

In addition to these simple considerations of the important angular ranges, the mapping of the solid angle in three dimensions between the centre-of-mass and laboratory frames is also relevant to experimental design. In the case of (d,p) reactions, the formula describing this solid angle mapping [21] implies that coverage of the extreme backmost angles is not vital. In this region, the centre-of-mass angle changes so rapidly with laboratory angle that the effective solid angle is very small and both the yield and the structure in the angular distribution are only slight in the backmost 10° in the laboratory frame. However, it emerges that the whole range of angles between 170° and 90° should ideally be covered with good efficiency.

Finally, as long as the transferred mass is small compared to the projectile mass, it turns out that the energy-angle systematics of the light, target-like particle in inverse kinematics have little dependence on the precise mass or velocity (or MeV/A) of the incident beam particles. Combining all of the above information, it can be seen that a dedicated charged particle array can be designed, which will have wide applicability in the study of transfer reactions.

DESCRIPTION OF THE TIARA DETECTOR ARRAY

The TIARA array of silicon strip detectors has been designed for transfer studies, to give accurate position information for charged particles over a solid angle of $\approx 4\pi$. In addition, it is extremely compact, so that gamma-ray detectors can be mounted in close geometry and hence have a high absolute efficiency. The compact design also allows for a magnetic spectrometer to be coupled at zero degrees and for beam tracking detectors to precede the target.

The detector arrangement in the first implementation of TIARA at GANIL, which is described here, differs slightly from the original design [22] in the forward angle region. It was decided to use a staged system of two concentrically mounted single-wafer annular detectors. These detectors are each divided lithographically to have 16 annular rings and 16 azimuthal sectors. Furthest from the target, at 150mm, one annular detector spans the angular range from 3.8° to 13.1°. The very outer part of that detector is actually obscured because the second detector is mounted at 90mm from the target and spans angles from 12.8° to 28.1°. The barrel spans the angular range from 35.6±0.4° to143.4±0.4° (where the errors indicate the variations due to the octagonal shape). In accordance with the original design, the backward angle region is spanned by an array of 6 wedge-shaped strip detectors, covering angles from 137° to 169.4°. Here, there are a total of 16 different annular bins of about 2° each, and 48 different azimuthal strips. The gap at forward angles, between 28° and 35°, will be filled in the near future.

Apart from the barrel, all other detectors are of a standard double-sided strip design. By combining information from the annular strips on the target side, and the azimuthal strips on the back, the pixel of the hit can be deduced. So far, no attempt at particle identification or time-of-flight has been implemented; just the angle and the deposited energy are recorded. In the backward-angle array, the full energy of the particle will be recorded, but at more forward angles the particles may often punch through and deposit a ΔE signal in the silicon. All back-angle and barrel detectors are 400 μm thick, being fabricated using 6-inch wafer technology [23], whilst the forward-angle detectors are 500 μm thick. For the future, thicker detectors are actively being studied.

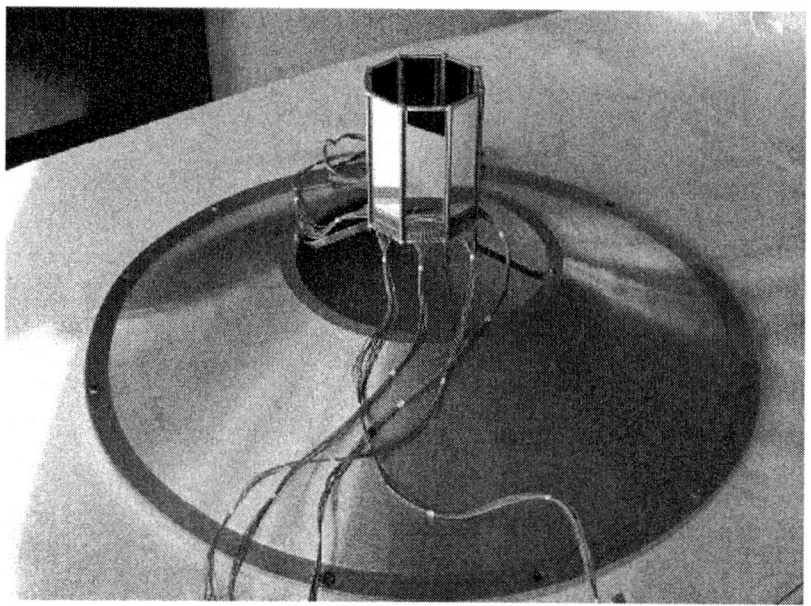

FIGURE 1. The fully assembled barrel of TIARA with its forward angle support structure.

The range of angles from 35.6° to 143.4°, as mentioned, is covered by the barrel part of TIARA. This comprises 8 individual detectors of length 96.8mm and width 24.6mm, each mounted on its own printed circuit board and then assembled into an octagonal barrel (see Fig. 1). Each detector in the barrel has 4 lengthwise resistive strips to measure the scattering angle. By means of tracks laid on both the silicon and the double-sided circuit board, all signals from both ends of the strips and from the back of the detector are brought out to connections at the forward-angle end of the barrel. As shown in Fig. 1, the connections are then made by directly joining micro-coaxial cable to the circuit board, which is due to geometrical constraints. In the azimuthal direction, the position resolution is given by the strip width of 5.6mm or in angular terms (360/32=) 11.25°. For transfer, this is not important. In the longitudinal direction, the angular measurement depends on combining signals from the resistive strips. At present, the deposited energy is determined by adding the signals taken from each end of a given strip, and the position is determined by the difference of the signals. Thus, preamplifiers are connected to each end of each strip, which gives particular electronic properties described in the literature (see [24] and refs. therein). The position resolution has been measured to be < 1mm FWHM using a 5.5 MeV α-particle source. This is shown in Fig. 2, where the observed image of a narrow slit is shown together with various fits; the best fit was derived by convolving a gaussian of 0.5mm FWHM with the projected geometrical image. The position resolution for this type of detector improves as the deposited energy is increased [24]. The energy resolution obtained by adding the signals from the two ends is 120 keV FWHM, and in fact slightly better resolution of 110 keV is obtained by directly taking the signal from the back face of the detector. These figures are typical for this type of detector and are small compared with resolution contributions expected to arise in reaction data due to target thickness effects.

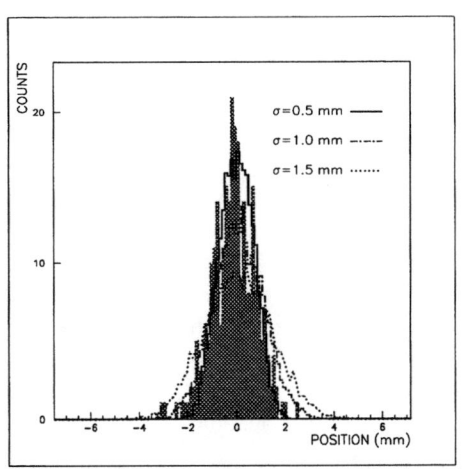

FIGURE 2. A position resolution of < 1mm FWHM measured in the resistive strips with [241]Am.

IMPLEMENTATION WITH OTHER DETECTORS

TIARA can operate as a stand-alone device, or just with gamma-ray detectors, but it is advantageous also to couple a magnetic spectrometer to separate or identify reaction products and un-reacted beam particles. In its first implementation at GANIL, the TIARA array is coupled with four gamma-ray detectors from the EXOGAM array and also with the VAMOS magnetic spectrometer, as described in the following.

The EXOGAM gamma-ray array

The EXOGAM array [25,26] was designed specifically for experiments with radioactive beams. Each "clover" detector comprises 4 separate, tapered Ge crystals with a common cryostat, and then each Ge "leaf" is segmented electrically by means of 4 outer, corner contacts. These outer contacts can be used to infer in which quarter of the crystal the first interaction took place, for a given γ-ray. A separate, central contact in each crystal gives higher resolution energy information. The EXOGAM detectors are designed with symmetric 45° tapers on all edges, so that 18 detectors could be put together to make a complete sphere. With TIARA, either a complete octagonal ring of 8 detectors at 90° can be used or, as presently installed, a compact cube of 4 detectors centered at 90° and touching at their front edges, covering 2/3 of 4π at a distance of 50mm from the target. The absolute full-energy peak efficiency at 1 MeV is around 20%. In this compact configuration, only certain escape-suppression shields at the rear and sides can be mounted, but for low γ-ray multiplicities this is not a significant limitation. The segmentation information allows a correction to be applied for the Doppler shift of γ-rays emitted by beam-like particles travelling along the beam axis. The relatively high velocities of the particles mean that the Doppler broadening, due to the finite geometrical acceptance of the detector, is what limits the γ-ray energy resolution. Depending on the energy, a resolution of order 50 keV is expected, which represents an order of magnitude improvement in excitation energy resolution relative to that achievable using charged particle energies and angles alone.

The VAMOS magnetic spectrometer

The VAMOS spectrometer ("variable mode spectrometer") is installed at GANIL for experiments with either fragmentation beams or ISOL beams from SPIRAL. It consists [27] of a pair of large-bore quadrupoles at the entrance, followed by an optional electrostatic section to give a Wien filter mode and then an optional dipole. At present, as set up with TIARA, it is operated using the quadrupoles and dipole. It has reasonable first-order focussing, a dispersion of 25mm/%$\delta p/p$ and a momentum bite of close to 10%. These properties are suitable for transfer reaction studies in inverse kinematics. The large angular acceptance of ±10° is not strictly needed, and indeed the silicon array is set up to detect smaller angles than this and therefore limits the acceptance into the spectrometer. VAMOS is designed for aberrations to be removed in software using fitted parameters, and this can be done on-line to recover the in-plane and out-of-plane entrance angles and the momentum. A physical beam-stop finger can be moved across the focal plane region so as to intercept the direct beam. This finger is made from plastic scintillator and can monitor and count the intense group that it intercepts.

The merging of the three separate data acquisition systems

The master trigger signal for a transfer experiment with TIARA is produced whenever any strip in any silicon detector records a signal. This signal is produced sufficiently quickly that it can also be included in the fast-trigger logic of EXOGAM and VAMOS. Each of these three data acquisition (DAQ) systems is thus enabled by a charged particle in TIARA. The TIARA system also includes an event counter in a VXI module called Centrum. When the EXOGAM and VAMOS DAQs record an internal trigger in overlap with the master trigger from TIARA, their slave Centrum units request the contents of the event counter kept in TIARA's master Centrum. The events from the three separate DAQs then include this event number and are broadcast in buffers independently onto a local data network, addressed to a processor that reads the buffers. A programme then correlates events, according to the event number, and builds combined events. These combined events are re-broadcast to a tape-server machine for storage. The combined data stream can also be sampled and analysed by on-line sorting programmes. The Centrum hardware units and the software for the on-line combination of the different data streams were developed by the GANIL data acquisition group. The method was developed and first implemented during the beam tests and commissioning of the TIARA array. Figure 3 shows how the three pieces of apparatus and DAQs are physically distributed around the beam area but connected by computer networks. Note that time-stamping, as shown in the figure, can be used instead of "stamping" events with the event counter. The key feature is that the final data as stored on tape includes all parameters from all acquisition systems in a single event structure for each trigger that is processed.

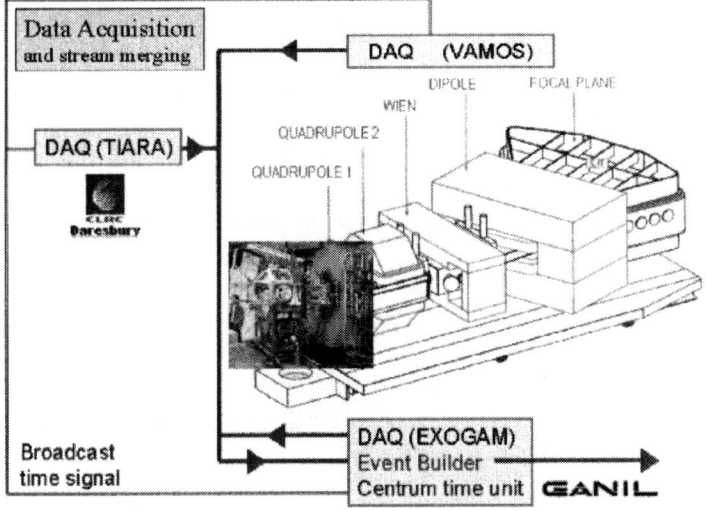

FIGURE 3. Three different data acquisition systems are merged on-line, using event stamping.

COMMISSIONING AND FIRST EXPERIMENTS AT GANIL

TIARA has been successfully commissioned using the full silicon array described here, with 4 EXOGAM detectors and VAMOS. Commissioning runs were performed from June 2003 using beams of ^{20}Ne, ^{36}S and ^{14}N from the CIME cyclotron, to initiate scattering and transfer on normal and deuterated polyethylene $(CH_2)_n$ targets. The first experiment with a radioactive beam from SPIRAL was performed in September 2003 to study the reactions $^{24}Ne(d,^3He)^{23}F$ and $^{24}Ne(d,p)^{25}Ne$ in inverse kinematics. The analysis of these data is in progress.

ACKNOWLEDGMENTS

The TIARA project is the work of many people, including further collaborators at our own institutions. We wish also to acknowledge important input from Nigel Orr at LPC, Caen, France, our collaborators at Birmingham and Liverpool, UK, Christophe Theisen at Saclay, and our collaborators at GANIL, especially Hervé Savajols and Maurycy Rejmund. The development and construction of TIARA was funded by the EPSRC(UK) and the installation and commissioning was supported also by GANIL.

REFERENCES

1. Austern, N., *Direct Nuclear Reaction Theories*, New York: John Wiley, 1970.
2. Satchler, G. R., *Direct Nuclear Reactions*, Oxford: Oxford University Press, 1983.
3. Glendenning, N., "One- and Two-Nucleon Transfer Reactions" in *Nuclear Spectroscopy and Reactions*, Part D, edited by J. Cerny, New York: Academic Press, 1975, pp. 319-344
4. Macfarlane, M. and Schiffer, J., "Transfer Reactions" in *Nuclear Spectroscopy and Reactions*, Part B, edited by J. Cerny, New York: Academic Press, 1975, pp. 169-194
5. Bond, P. D., *Phys. Rev* **C22**, 1539 (1980).
6. Anyas-Weiss, N. et al., *Phys. Reports* **12**, 201-272 (1974).
7. Catford, W. N., Fifield, L. K., Orr, N. A. and Woods, C. L., *Nucl. Phys.* **A503**, 263-284 (1989).
8. Hansen, P. G., *Phys. Rev. Lett.* **77**, 1016 (1996).
9. Hansen, P. G. and Sherrill, B., *Nucl. Phys.* **A693**, 133 (2001).
10. Sauvan, E., et al., *Phys. Lett.* **B491**, 1-7 (2000).
11. Bohr, A., and Mottelson, B. R., *Nuclear Structure*, Vols. 1 and 2, Singapore: World Scientific, 1997 (original editions New York: W. A. Benjamin Inc., 1969 and 1975).
12. Elbeck, B., and Tjorn, O., *Adv. Nucl. Phys.* **3**, 259 (1969).
13. Rehm, K. E., et al.,. *Phys. Rev. Lett.* **80**, 676-679 (1998).
14. Fortier, S., et al.,. *Phys. Lett.* **461B**, 22-27 (1999).
15. Winfield, J. S., et al., *Nucl. Phys.* **A683**, 48-78 (2001).
16. Timofeyuk, N. K., and Johnson, R. C.,. *Phys. Rev.* **C59**, 1545-1554 (1999).
17. Liang, F., Radford, D., et al., K., Oak Ridge National Laboratory, private communication.
18. Winfield, J. S., Catford, W. N. and Orr, N. A., *Nucl. Instr. and Meth.* **A396**, 147-164 (1997).
19. Catford, W. N., *Acta Phys. Pol.* **B32**, 1049-1060 (2001).
20. Catford, W. N., *Nucl. Phys.* **A701**, 1-6 (2002).
21. Schiff, L. I., *Quantum Mechanics*, third edition, New York: McGraw Hill, 1968, p. 113.
22. Catford, W. N., Timis, C.N., Labiche, M., Lemmon, R.C., Moores, G. and Chapman, R., in *Application of Accelerators in Research and Industry 2002*, edited by J.L. Duggan and I.L. Morgan, AIP Conference Proceedings 680, New York: American Institute of Physics, 2003, pp. 329-332.
23. Micron Semiconductor Ltd., Lancing, Sussex, UK.
24. Yanagimachi, T., et al., *Nucl. Instr. and Meth.* **A275**, 307 (1989).
25. Catford, W. N., *J. Phys. (London)* **G24**, 1337 (1998).
26. Simpson, J., et al., *Acta Physica Hungaria: Heavy Ions* **11**, 159 (2000).
27. Savajols, H., et al., *Nucl. Phys.* **A654**, 1027c (1999).

Study of ^{19}Na at SPIRAL

François de Oliveira Santos and the E400S collaboration

GANIL, BP 55027 F-14076 Caen Cedex 5, FRANCE

Abstract. The excitation function for the elastic scattering reaction p(^{18}Ne,p)^{18}Ne was measured with the first radioactive beam from the SPIRAL facility at GANIL. Several resonances have been observed, corresponding to new excited states in the compound nucleus ^{19}Na.

INTRODUCTION

Sodium isotopes have been produced in a wide range of the neutron number, from the most neutron rich isotope N=26: ^{37}Na identified for the first time in a recent experiment done at GANIL [1] with the new facility LISE 2000, to the most proton rich isotope, two steps beyond the proton drip line, N=7: ^{18}Na [4]. The study of ^{19}Na has been the subject of the first SPIRAL experiment at GANIL.

The Motivations

It is undoubtedly an important objective to undertake the study of the light and proton rich nuclei by the measurement of the ground and excited states properties. This allows us to compare them to the properties of the mirror nuclei and with theoretical calculations. In those nuclei several effects are sought: proton distribution effects, Thomas-Ehrman shift, breaking of isospin symmetry, multi particle emission, etc.

The isotope ^{19}Na was for the first time observed in 1969 by Cerny et al. [2] via a transfer reaction. They observed only one peak at the mass excess of 12.974 ± 0.070 MeV, a value close to that of 12.90 MeV predicted with the Isobaric Mass Multiplet Equation. This extracted mass of ^{19}Na implies that this nucleus is not bound for the proton emission, the ground state being only E_R = 366 ± 70 keV above the proton emission threshold. Since that first measurement only three new experiments have been dedicated to the study of ^{19}Na [3] [4] [5]. Today, only the ground state and two excited states have been seen experimentally. The ground state has been observed several times, the adopted value for its energy is E_R = 320 ± 10 keV, but its spin has never been assigned, neither width measured. The first excited state has been observed only once and with a low statistics, 120 keV above the ground state. Nothing more is known about this state. The second excited state has been seen recently in a precise resonant elastic scattering measurement [5], it is 1066 ± 2 keV above the proton emission threshold. Its spin is 1/2+ and its width Γ = 101 ± 3 keV. This state corresponds without any doubt to the known second excited state in the mirror nucleus ^{19}O. It is 727 keV down shifted from its analogue, one of the highest observed values. No more is experimentally known about ^{19}Na.

The Principle Of The Experiment

In the present experiment we have measured the excitation function of the elastic scattering reaction: ^{18}Ne(p,p)^{18}Ne. It is well known that the excitation function for the elastic scattering at low energy is a combination of Rutherford contribution and anomalies (resonances) that correspond to states in the compound nucleus. There is a straightforward correspondence between those anomalies and the properties of the excited states of the compound nucleus. The energy of the state can be extracted from the position of the resonance, its width from the width and intensity of the resonance, its spin and parity from the shape and angular distribution of the resonance. The interest of using this kind of measurement for spectroscopy is manifest: this measurement is simple, give a lot of information about the compound nucleus states and the corresponding cross sections are high. Because ^{18}Ne is radioactive, we have measured this elastic scattering in inverse kinematics: p(^{18}Ne,p)^{18}Ne. The experiment has been done at GANIL with a radioactive beam produced by the new SPIRAL facility. The mean intensity of the beam was 5 10^5 pps. It is very time consuming to use a thin target and to change the energy of the incoming beam by small steps in order to measure the excitation function. To solve this difficulty we have used the thick target method. The idea of using thick target has been developed successfully in several experiments with stable and radioactive beams [6] [7]. If the implanted particle (^{18}Ne), at some point along its slowing down trajectory inside the target, has a center of mass energy that corresponds to a state in the compound nucleus (^{19}Na=^{18}Ne+p), the probability for the elastic scattering change (resonance). The scattered proton can escape the target and can be detected at forward angles in the laboratory frame. We have used a simple silicon detectors telescope (E versus Δ E) to identify and to measure the energy of the light charged particles (see figure 1).

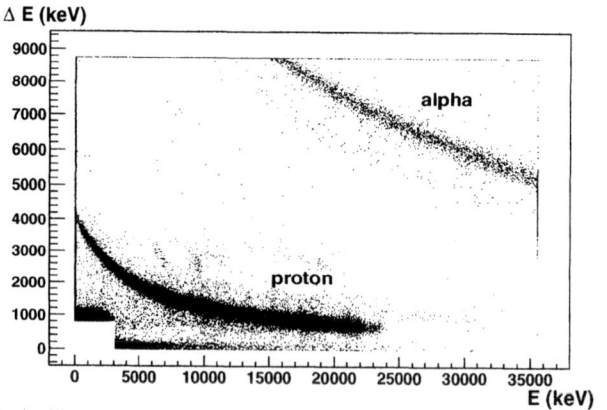

FIGURE 1. Standard E versus Δ E plot, shown here for the p(^{18}Ne,p)^{18}Ne reaction.

The thick target method makes it possible to obtain a complete and continuous excitation function over a wide energy range, by simply detecting the scattered protons and measuring their energies, without changing the energy of the incoming beam.

The Cryogenic Target

We have used a solid hydrogen cryogenic target for two important reasons. The first, the use of compound targets (e.g. $(CH_2)_n$) introduces atoms (e.g. Carbon) in which nuclear reactions can occur and can pollute the measurement. The second, the use of a pure hydrogen target maximizes the counting rate because the higher stoechiometric ratio leads to the highest effective target thickness, this effect is developed hereafter. The main requirements imposed for solid hydrogen targets usable under vacuum with a particle beam are: thickness about 1 mm, very thin windows, and uniform thickness and density. A special cryogenic system has been designed to make the target used in the present experiment. It is composed of a H_2 target cell and a He cell on either side of the target. During the hydrogen target production phase, equivalent He pressure is maintained on either side of the target windows up to the complete formation of the solid H_2 target. Once the target is formed, the helium gas is evacuated.

TESTS AND CALIBRATIONS WITH $^{18}O(P,P)^{18}O$

The raw measured spectrum (number of counts as a function of the detected particles energies) has to be corrected with two functions in order to produce the final spectrum (differential cross section as a function of the center of mass energy). The first correction (Figure 2a) is applied for the energy loss of the protons in the hydrogen target, between the positions of the scattering events up to the silicon detector.

The second correction (Figure 2b) is due to the fact that the effective target thickness depends on the energy loss of the incident ions inside the target. In fact the higher is the energy loss of the incoming ions, the lower is the effective target thickness, and so the lower is the counting rate. This effect explains why it is very important to use a pure hydrogen target. In that case the energy loss is only due to the interaction with the hydrogen atoms of the target. With the pure hydrogen target we get the lowest energy loss, and the highest proton density, which both increase the counting rate.

We have performed a measurement with a stable beam to evaluate the quality of the setting (mainly the cryogenic target) and to calibrate the data. This test was accomplished with an ^{18}O stable beam, produced at the same energy as ^{18}Ne. The compound nucleus $^{18}O + p = ^{19}F$ is quite well known at the corresponding excitation energies. In the Figure 3 we have performed two comparisons:

- We have compared our results with experimental data from a very precise measurement ($\sigma \sim 2$ keV) published by Orihara et al. [9]. They have measured the cross section at a different angle ($\Theta_{CM} = 168.7°$) however the agreement is very good, in the normalization and also in the energy position.

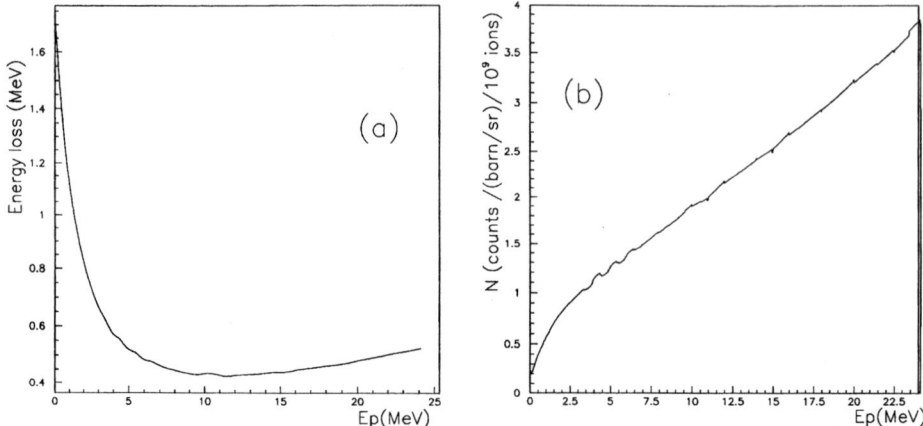

FIGURE 2. Those two figures are results of a simulation of the experiment. (a) The energy loss of the proton inside the hydrogen target is plotted as a function of the detected proton energy. This energy loss is a function of the energy of the proton when it is scattered, and of the distance the proton has to pass in the target. (b) The number of counts (per barn, per steradian and per 10^9 incident ions) is plotted as a function of the detected proton energy. This demonstrates the influence of the effective target thickness as a function of the proton energy, using constant cross section the counting rate increases as the proton energy increases.

- We have compared our results with the results of a R-matrix calculation (using 40 keV resolution) we have done with the code Anarki [8] and using the known states properties in ^{19}F (about 40 states). Again, the agreement is very good.

From those comparisons it was possible to determine the hydrogen target thickness of 1050 ± 50 μm (density of 88.5 mg/cm3). It was also possible to determine the solid angle of the silicon detector, it is $d\Omega = 10 \pm 2$ msr.

RESULTS FOR ^{19}NA

The same analysis of the data has been applied to produce the ^{19}Na spectrum in figure 4.

The energy range of the spectrum is limited by experimental cut-offs. The high energy limit at about 6 MeV is close to the expected maximum excitation energy calculated using the full energy of the incident ^{18}Ne ions E $\approx$ 7 MeV A. The low energy limit E $\approx$ 1.5 MeV is mainly due to the selection (proton banana in a E versus ΔE silicon telescope) we have done to identify protons among different particles detected after the target (proton, alpha and electrons). It is also due to the fact that protons have to pass some distance inside the hydrogen target to exit, resulting in some energy loss as already shown. In order to analyze the experimental data we have used the known properties of ^{19}Na, as well as the known properties of the mirror nucleus ^{19}O. Finally, we have done shell model calculations (in the spsdpf space and with the WBT interaction) with the program Oxbash. The first complete set of shell model calculations for the proton

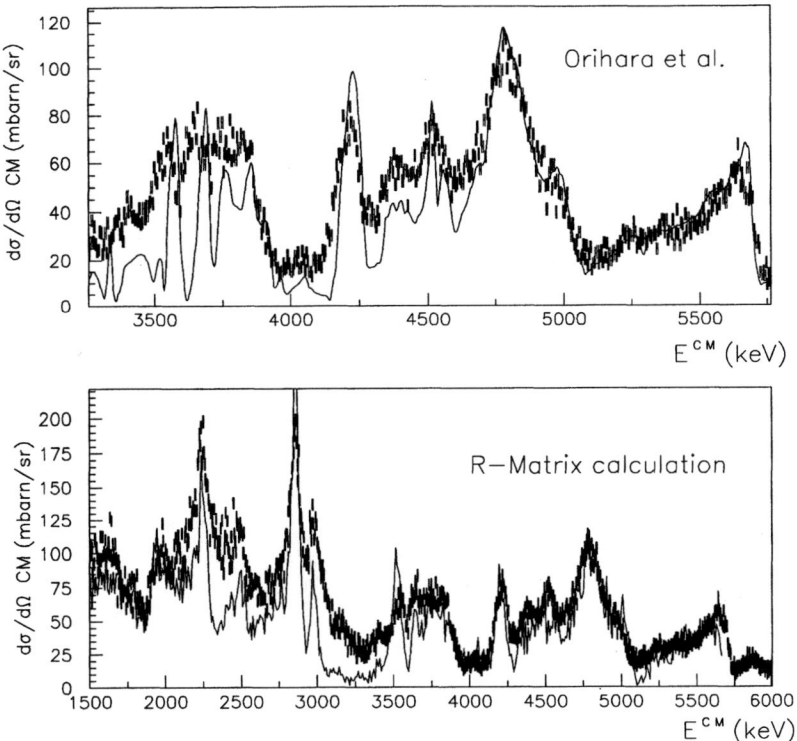

FIGURE 3. Top figure : A part of our reconstructed differential cross section (CM, direct kinematics, Θlab = 180°) for the elastic scattering reaction $^{18}O(p,p)^{18}O$ is plotted as a function of the proton energy in the center of mass system in direct kinematics. The error bars are the statistical error. The continuous line represents the data from the Orihara et al. experiment [9]. Bottom figure: The full reconstructed differential cross section (same conditions). The continuous line represents the R-Matrix calculation using the known properties of ^{19}F. The overall agreement is excellent. Our experimental resolution is $\sigma \simeq 40$ keV.

elastic and inelastic scattering has been performed for the first time in this work. The dotted line in figure 4 represents our R-matrix calculation. The first peak, labelled 1 in figure 4, corresponds to the known second excited state in ^{19}Na. Unfortunately, we were not able to observe this state in that spectrum because of the low energy cut-off. However a new analysis using time of flight selection has provided clear evidence for that resonance. Those results will be shown in the reference [10]. Even if this resonance is not visible in the figure 4, this state also influences the general shape and normalization of the spectrum at higher excitation energy. If one takes this effect into account, our measurement is in a good agreement with the calculation. In a second step of the calculation we have introduced three new states in the R-matrix analysis: corresponding

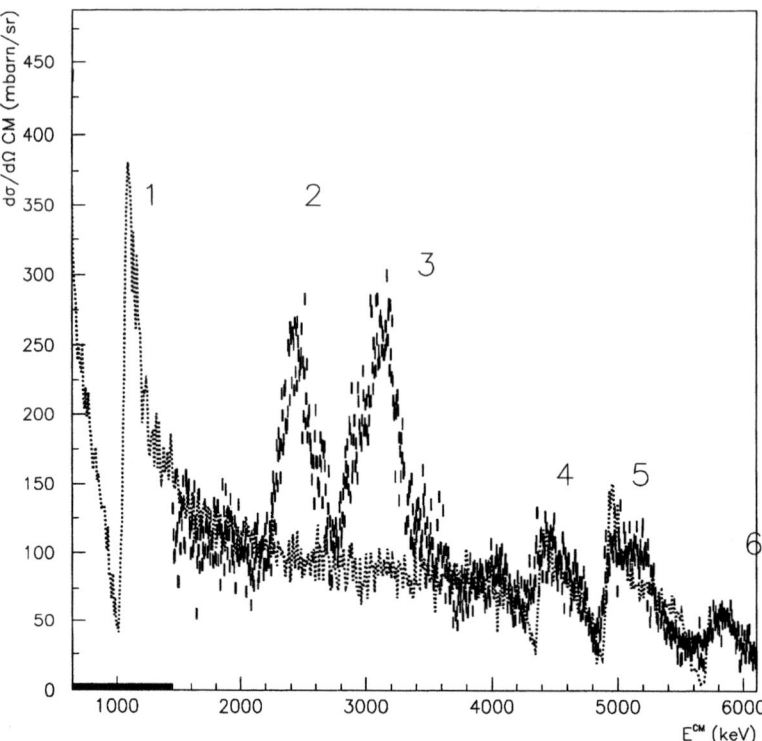

FIGURE 4. The reconstructed differential cross section (CM, direct kinematics, $\Theta_{lab} = 180°$) for the elastic scattering reaction ^{18}Ne(p,p)^{18}Ne as a function of the proton energy in CM is shown. The points are experimental data; errors are statistical, the dotted line represents a calculation from R-Matrix model using the known and expected properties of ^{19}Na. The numbers correspond to the peaks described in the text.

to the states 4, 5 and 6. In fact our shell model calculation has shown only three states with width larger than 5 keV for elastic scattering, they all have spin 3/2-. Starting from the predicted properties for those states we have fitted the experimental spectrum with the R-matrix analysis, the result is shown in the figure 4.

The measured properties of the peak 4 E_x=4371 keV and Γ=30 keV and the peak 5 E_x=4903 keV and Γ=50 keV are very close to calculated values: peak 4 E_x=4258 keV and Γ=61 keV; peak 5 E_x=4667 keV and Γ=47 keV. The used values for the state 6: E_x=5466 keV and Γ=160 keV, are indicative. In fact, states at energy higher than 5.5 MeV are difficult to analyze because at those energies there is a high density of broad states not easy to disentangle. Even if we have found a good agreement in those fits, the two main observed peaks 2 and 3 in the figure 4 couldn't be explained by the elastic scattering. In the same shell model calculations we have performed, several broad states have been predicted for the inelastic scattering. In that case the raw spectrum has to be re-analyzed with new correction functions, in order to take into account different kinematics and energy loss. This results in the following properties for the two peaks labelled 2 and 3, if interpreted as inelastic scattering: peak 2 E_x=3560 keV and $\Gamma \approx 130$ keV; peak 3 E_x=4540 keV and $\Gamma \approx 150$ keV, in good agreement with our predictions: E_x = 3167 keV, Γ=187 keV and J^π=5/2+; E_x=3746 keV, Γ=305 keV and J^π=3/2+.

A full description of the experimental details, calculations and final results will be published in the following article [10].

ACKNOWLEDGMENTS

A large part of the success of this experiment is due to the quality of the SPIRAL beam, and we would like to thanks again all the GANIL staff for this result. This work has partially been supported by the EU Access to Large Scale Facilities program.

REFERENCES

1. S.M. Lukyanov, Yu. E. Penionzhkevich, R. Astabatyan, S. Lobastov, Yu. Sobolev, D. Guillemaud-Mueller, G. Faivre, F. Ibrahim, A.C. Mueller, F. Pougheon, O. Perru, O. Sorlin, I. Matea, R. Anne, C. Cauvin, R. Hue, G. Georgiev, M. Lewitowicz, F. de Oliveira Santos, D. Verney, Z. Dlouhy, J. Mrazek, D. Baiborodin, F. Negoita, C. Borcea, A. Buta, I. Stefan and S. Grevy, J.Phys.G. **28**, (2002) L41-L45.
2. J. Cerny, R.A. Mendelson, G.J. Wozniak, J.E. Esterl and J.C. Hardy, Phys. Rev. Let. **V22**, (1969) 612
3. W. Benenson, A. Guichard, E. Kashy, D. Mueller, H. Nann and L.W. Robinson, Phys. Let. **V58B N1**, (1975) 46
4. Thomas Zerguerras, Thesis IPNO T01-05 (2001) IPN Orsay France
5. C. Angulo, G. Tabacaru, M. Couder, M. Gaelens, P. Leleux, A. Ninane, F. Vanderbist, T. Davinson, P.J. Woods, J.S. Schweitzer, N.L. Achouri, J.C. Angélique, E. Berthoumieux, F. de Oliveira Santos, P. Himpe, and P. Descouvemont, Phys. Rev. **C67**, (2003) 014308
6. K. Markenroth, L. Axelsson, S. Baxter, M.J.G. Borge, C. Donzaud, S. Fayans, H.O.U. Fynbo, V.Z. Golberg, S. Grévy, D. Guillemaud-Mueller, B. Jonson, K.-M. Källman, S. Leenhardt, M. Lewitowicz, T. Lönnroth, P. Manngard, I. Martel, A.C. Mueller, I. Mukha, T. Nilsson, G. Nyman, N.A. Orr, K. Riisager, G.V. Rogatev, M.-G. Saint-Laurent, I.N. Serkov, N.B. Shul'gina, O. Sorlin, M. Steiner, O. Tengblad, M. Thoennessen, E. Tryggestad, W. H. Trzaska, F. Wenander, J.S. Winfield, and R. Wolski., Phys. Rev. **C62**, (2000) 034308
7. V.Z. Goldberg, V.I. Dukhanov, A.E. Pakhomov, G.V. Rogachev, I.N. Serikov, M. Brenner, K-M. Källman, T. Lönnroth, P. Manngård, L. Axelsson, K. Markenroth, W. Trzaska and R. Wolski, Physics of Atomic Nuclei (Yadernaja Fizika) **60**, (1997) 1186-1193
8. E. Berthoumieux, B. Berthier, C. Moreau, J.P. Gallien, A.C. Raoux, Nucl. Instr. and Meth. **B136-138**, (1998) 55-59
9. H. Orihara, G. Rudolf and Ph. Gorodetzky, Nucl. Phys. **A203**, (1973) 78
10. F. de Oliveira Santos et al. in preparation

The EXODET Apparatus And Its First Experimental Results: ^{17}F Scattering By ^{208}Pb Below The Coulomb Barrier

M. Romoli[1], M. Mazzocco[2], E. Vardaci[1,3], R. Bonetti[4],
A. De Francesco[1], A. De Rosa[1,3], M. Di Pietro[1], T. Glodariu[2,5],
A. Guglielmetti[4], G. Inglima[1,3], M. La Commara[1,3], B. Martin[1,3],
V. Masone[1], P. Parascandolo[1], D. Pierroutsakou[1], M. Sandoli[1,3],
P. Scopel[2], C. Signorini[2], F. Soramel[6], L. Stroe[7], J. Greene[8],
A. Heinz[8], D. Henderson[8], C. L. Jiang[8], E. F. Moore[8], R. C. Pardo[8],
K. E. Rehm[8], A. Wuosmaa[8] and J. F. Liang[9]

[1] INFN, Sezione di Napoli, Via Cintia, I-80125, Napoli, Italy.
[2] Università di Padova and INFN, Sezione di Padova, Via Marzolo 8, I-35131, Padova, Italy.
[3] Università degli studi di Napoli "Federico II", Dip. Sc. Fis., Via Cintia, I-80125, Napoli, Italy.
[4] Università di Milano and INFN, Sezione di Milano, Via Celoria 16, I-20133, Milano, Italy.
[5] National Institute for Physics and Nuclear Engineering, 76900 Magurele-Ilfov, Romania.
[6] Università di Udine and INFN, Sezione di Udine, Via Delle Scienze 208, I-33100, Udine, Italy.
[7] INFN, Laboratori Nazionali di Legnaro, Viale Dell'Università 2, I-35020, Legnaro (Padova), Italy.
[8] ANL, Physics Division, Argonne National Laboratory, Argonne, Illinois 60439, USA.
[9] ORNL, Physics Division, Oak Ridge National Laboratory, Oak Ridge, Tennessee 37831, USA.

Abstract. A new detector apparatus has been designed and developed to be used in experiments performed with radioactive ion beams. It consists of 16 highly segmented silicon strip detectors arranged in two-layer telescopes and subtending a large solid angle (about 70% of 4π sr). An innovative readout system for the position information that uses highly integrated electronics (ASIC chips) has been implemented. A first successful experiment has been performed at the Argonne National Laboratory (USA) to study the ^{17}F scattering by a ^{208}Pb target at 90.4 MeV of incident energy. The ^{17}F angular distribution has been analyzed and the optical model potential best-fit parameters determined. The same analysis performed on ^{17}F data taken at higher incident energy, in completely different experimental conditions, gives consistent results. The comparison with experiments performed with stable beams (^{19}F, ^{16}O, ^{17}O) indicates a behavior for the ^{17}F more similar to that of the Oxygen isotopes than to the ^{19}F one. Despite of the short data collection time, also the cross section for the ^{17}F $\rightarrow$ ^{16}O + p break-up process has been estimated.

INTRODUCTION

The study of radioactive and exotic nuclei is a useful tool to get information about the behavior of the nuclear potential when the interaction radius, due to the external loosely bound nucleons, can be very different from that of the stable nuclei in the same mass region. The effect of such a spatially extended nuclear density (nuclear halo or skin) on the reaction mechanism cross sections is not clearly understood. The possibility to shed light on the presence of new typical features for such nuclei seems to increase around the Coulomb barrier energy region, where several mechanisms are in competition (elastic and inelastic scattering, break-up and stripping break-up, complete and incomplete fusion).

The interest for the experiments involving radioactive ion beams (RIBs) is growing up at the moment, but practical limitations rise up from the very low beam intensities (10^5-10^6 pps) presently available at the first generation RIB accelerator facilities. To partially overcome such limitations, the project of a new RIB measurement requires the utilization of a charged particle detector system subtending the largest possible solid angle. Moreover high energy resolution and position sensitivity, that means high granularity of the system, are also required in order to reconstruct the ejectile momentum distribution.

In this framework, we have developed a new experimental apparatus for charged particle detection and identification, named **EXODET** (an acronym for **EXO**TIC **DET**ECTOR) and designed to be used in RIB experiments. In the next section, the main features of such apparatus will be described. Due to the high segmentation of the detector array (counting 1600 elements), it has been necessary to use a non-standard highly integrated readout system for the position information (ASIC chips), while for the energy information treatment we used standard nuclear electronic chains. In the following, a section is dedicated to the readout electronics description.

The EXODET apparatus has been successfully used, for the first time, at the Argonne National Laboratory (USA), to study the scattering of the exotic ^{17}F beam from ^{208}Pb at an incident energy just below the Coulomb barrier. The experimental layout of the performed measurement, the analysis of the data collected, the comparison of the obtained results with theory and data analyses on different nuclei will be the arguments of the last two sections, together with final comments and considerations.

THE EXODET APPARATUS

The basic EXODET apparatus module consists of a large active area silicon detector, produced and delivered by MICRON Semiconductors Ltd, on the basis of a Naples custom design. The detector active area is 50 x 50 mm^2 large and the total occupancy of the die is 53 x 53 mm^2, due to the guard rings and the passivated region surrounding the active area. A G-10 support is placed under such 1.5 mm wide unusable region to ensure a safer handling of the die and the transparent transmission through the detector stack with minimum dead zone.

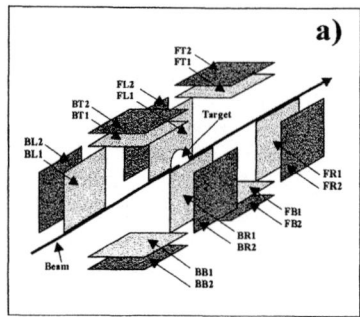 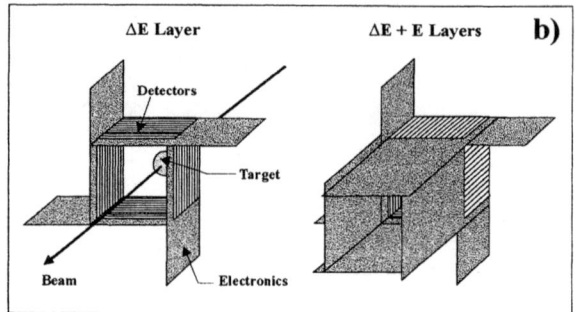

FIGURE 1. a) Layout of the EXODET detectors. b) Assembling of the EXODET detectors. Note the direction of the strips of the ΔE and E layers with respect to the beam one.

The detector front side active area is segmented in 100 strips, having a pitch size of 0.5 mm and a separation distance of 50 μm. Each strip is wire bonded to an external gold-plated cupper pin. The ohmic rear side of the detector is not segmented and is used to get the particle energy loss signals. Detectors with different thickness are available (40, 60, 100 μm for the ΔE layer and 150, 500, 1000 μm for the E layer). In the standard configuration 60 μm and 500 μm thick detectors are mounted as ΔE and E detectors, respectively.

16 of such detectors are arranged in 8 two-layer telescopes, allowing the identification in atomic number of the particle passing through the first layer by means of the usual ΔE-E technique. In fig.1a) is shown the displacement of the telescopes around the target in the forward and backward hemispheres with respect to the beam direction and the target position. We have introduced a code for the identification of the detectors which consists of three characters XYN, with the following meaning: X=F (Forward) or B (Backward), Y=T (Top) or B (Bottom) or R (Right) or L (Left), N=1 (ΔE detector) or 2 (E detector). The strips of the ΔE layer are placed perpendicularly to the beam direction, while the E detector strips are in the same direction of the beam, as shown in fig.1b). In such a way, it is possible to determine the position of a particle passing through the first layer and reaching the second one with an indetermination of 0.5 mm x 0.5 mm. The detection pixels so defined subtend different solid angles and scattering angle ranges depending on their position and distance from the target. We used Monte-Carlo calculations to evaluate the geometrical efficiency for each part of the EXODET apparatus. Also for the particles stopping in the first layer, the ΔE strip displacement and the utilization of such Monte-Carlo codes allow to extract information about the particle angular distribution, although with a higher indetermination, due to the non-spherical shape of the EXODET detectors. The total solid angle subtended by the whole EXODET apparatus is about 70% of 4π sr.

In fig.2a) is shown a photo of two EXODET detectors before their assembling to form a single telescope. The detectors are already glued and connected to the PCBs containing the detector-chip interface electronics and the preamplifier used for the energy signal treatment. A photo of the whole EXODET apparatus mechanics is shown in fig.2b). The reduced size of the apparatus, 350 mm in the beam direction and 230 mm in the other two, without cabling, and its compactness allow an easy transportation to different laboratories and the possibility to be used in more complex experimental set-up.

A cooling system, that uses Peltier cells and water as cooling fluid, ensures the appropriate detector temperature control and stabilization.

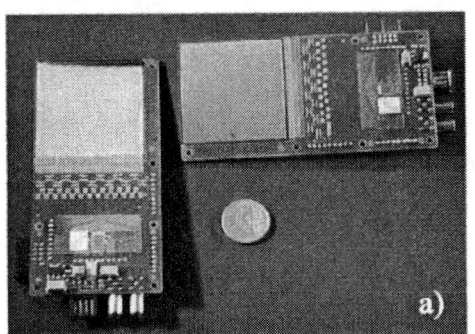

FIGURE 2. a) The two detectors of one EXODET telescope before their assembling. b) The whole EXODET apparatus mounted on its own mechanical support.

THE ELECTRONIC READOUT

The information about the energy released by the impinging particles is obtained from the unsegmented rear side of each detector. The analog energy signal treatment is carried out using a standard nuclear electronic chain, consisting of a low noise preamplifier, a spectroscopy amplifier and an ADC. The timing output of the amplifier is sent to a CFD in order to obtain a logic signal, used for the trigger analysis and generation (see fig.4a)). The most interesting part of this typical electronic chain is the new low-cost hybrid charge preamplifier, named CHAPLIN (CHArge Preamplifier Low-noise INFN Napoli), which has been completely designed and developed by the Electronic Service of the INFN – Sezione di Napoli. The main features of such device are: 1) a sensitivity of 28 mV/MeV (or 5 mV/MeV in a different version); 2) a noise at C_{source}=0 pF of 300 e rms; 3) a Risetime at C_{source}=2 pF of 22 ns; 4) a Falltime at C_{source}=2 pF of 3.3 µs; 5) a power dissipation of 180 mW. This preamplifier has been already successfully used for several different experimental setups.

For the readout of the position information we did not use standard electronic chains because of the prohibitive costs due to the large number of strips to be analyzed (1600 for the whole apparatus). Therefore, we developed a new readout system based on a highly integrated electronic circuitry, namely an ASIC chip. This chip, originally developed for high-energy particle physics experiments [1], has been found suitable also for our purposes, with an opportune design of the detector-chip interface. In particular, we have implemented a resistive attenuator, with an attenuation factor of 70, to match the strip signals into the dynamic range of the chip and designed a pitch adapter to reduce the pitch size from 0.5 mm at the detector end to about 80 µm at the chip end. On the detector-chip interface board the connectors for the communication lines of the chip I/O, for the energy signal output, and for the power supplies are also located. Each chip has a size of 5.7 x 8.3 mm^2 and drives up to 128 input channels. Hence we have used one chip for each detector module, connecting the strips of the detector to the first 100 lines. Inside the chip, both the analogical and the digital treatments of the signals, coming from each strip, are performed contemporarily and separately. In fig. 3 is shown a scheme of the single channel electronics contained into the chip. The input signals are pre-amplified and shaped. Afterwards a sampling of the input line voltage is performed at a frequency of 15 MHz and compared to the threshold voltage V_{THRESH}, that is externally adjustable via a 6-bit DAC (with a 7.5 mV step). When the input signal is higher than V_{THRESH}, a bit of the circular memory buffer is set to 1. If an external trigger command arrives to the chip, this sampling procedure is stopped and the digital circuitry of the chip analyzes the memory buffer to search for a bit set to 1. If found, the chip gives in output a data stream containing the event identification header, the identification number of the strips hit and, for each of them, the ToT (Time over Threshold) and the JT (Jitter Time). ToT is the time spent by the input signal over the threshold, measured in clock hit number, with a sensitivity of 67 ns; JT is the time interval, expressed in clock hit number, between the input signal and the trigger arrivals. The ToT is roughly proportional to the energy lost into the detector and the JT is a sort of correlation time useful to disentangle among spurious events and the trigger correlated ones, as shown in the following section. In the used set-up, the chip conversion gain was 150 mV/fC, the signal memory buffer length was 12 µs (193 cells) and the signal analysis window was 2 µs (32 cells).

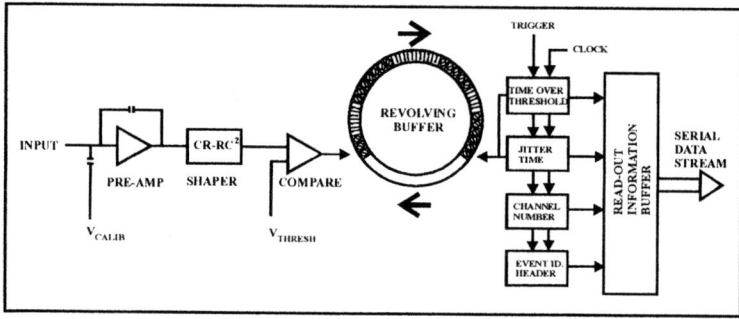

FIGURE 3. Simplified electronic scheme of the ASIC chip used for the positional information readout of the EXODET apparatus.

In fig.4a), the scheme of the electronic chain used for one of the EXODET detectors is reported. We have developed several new front-end electronic boards, interfaced to the VME bus, and a new data acquisition system, named **VIPER** (**V**ME **I**nterfaced to **P**CI **E**XODET **R**eadout), has been designed from scratch to manage the readout from homemade and commercial electronic modules. We decided to base the new developments on the VME bus because of its widespread use in nuclear physics laboratories and for the large variety of front-end modules (ADC, TDC, QDC, scalers...) commercially available. The dark colored boxes in fig.4a) represent the new hardware and software components developed. The interface between the ASIC chips and the VME bus is ensured by the AVI boards (ASIC to VME Interface), one card for each chip. The TS board (Trigger Supervisor) is a highly versatile board and is the trigger filter and deliverer for the whole front-end. The AVI board manages the communication channels between the chip and the VME bus for operations like sending setup or calibration commands to the chip. When the acquisition is running and the TS board asserts a valid trigger signal, the AVI sends the trigger command to the chip and makes available to the VME bus the data stream received from the chip. Signals regarding the internal status of the AVI board and/or the chip connected to it are sent, via the control bus, back to the TS board which, in turn, gives the occurrence of errors, FIFO half-full signals, and other important control signals.

Another important task performed by the TS board is the logical combination (OR/AND) of all the input channels devoted to provide a trigger signal. When a channel proposes a trigger signal, before promoting it to a valid trigger, the TS board controls that the whole system is ready (i.e., no AVI is busy, no error occurred, acquisition is running,...). It is possible to mask or sample one or more of the input channels and to force a valid trigger assertion for testing purposes. An external inhibit signal can also be used to integrate commercial boards in the readout system. This is the case for the ADC used to encode the energy signal from the silicon detector: the inhibit signal comes from the ADC, and is kept asserted by the ADC itself until the data digitization cycle has been completed. To evaluate the dead time, the TS board has been equipped with rate meters and counters for the proposed and accepted triggers. Furthermore, dividers by 10/100/1000 can also be used for the input channels. Finally, the TS board is also used to synchronize all the front-end boards in order to guarantee a common time stamp.

To ensure a complete reconfiguration of the developed electronics and flexibility with respect to wider experimental requirements, FPGAs have been extensively used in the design of the boards along with advanced techniques to optimize the connection arrangement between them. The AVI and TS boards have been developed by the Electronic Service of the INFN – Sezione di Napoli.

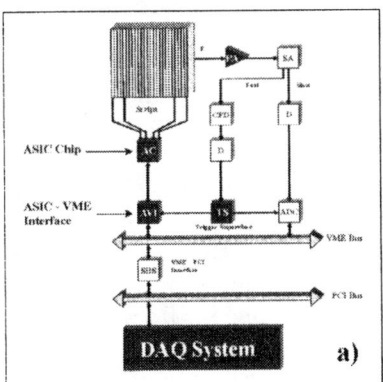 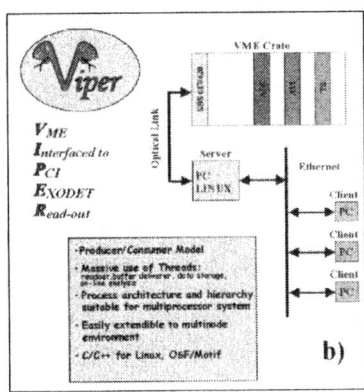

FIGURE 4. a) Scheme of the electronic chain for one of the EXODET detectors. The dark colored boxes represent homemade electronics or software, whereas the white ones represent commercial equipments. b) Schematic drawing of VIPER. The front-end VME crate is linked to a personal computer via a VME-PCI bridge board. The DAQ software runs on a Linux environment and the whole system is accessible over the network.

The VME bus is connected to a PC via a commercial VME-PCI bridge. This solution reduces the problems due to the CPU obsolescence, allows a progressive low-cost enhancement of the DAQ system performances with the upgrade of the CPUs available on the market, and does not link the DAQ software to a specific platform. The choice of transferring the CPU from the VME bus to the PC also allows taking the most advantage from the continuous improvement of the I/O devices (disks, RAM, DVD writer...) of the PC industry. The whole system, schematically shown in fig.4b), is accessible over the network because of a *Server/Client* architecture. The *Server* is the process that manages the hardware and runs on the PC directly linked to the front-end. The *Clients* operate on the front-end and monitor the data acquisition through the *Server*. The DAQ system can manage more than one VME crate and includes an innovative and general method to setup the front-end modules, and the relative on-line analysis, which eliminates the burden due to the implementation of software drivers for new modules. The user can quickly operate a new module without the need of intervention on the DAQ core software. The whole software has been developed in the C/C++ languages for the Linux operating system platform, and using the OSF/Motif environment for the graphical user interface.

THE ^{17}F SCATTERING EXPERIMENT

The first experiment performed using the EXODET apparatus has been the study of the ^{17}F scattering by a ^{208}Pb target at 90.4 MeV of incident energy. The ^{17}F beam was produced as secondary beam at the RIB facility of the Argonne National Laboratory (USA), using the inverse kinematics reaction p(^{17}O, ^{17}F)n, with an average intensity of about 10^6 pps. The target was a self-supporting 1 mg/cm^2 thick ^{208}Pb foil. The main beam contamination came from the ^{17}O, having the same magnetic rigidity but with an energy about 64/81 times the ^{17}F one. At these energies both the ^{17}O and ^{17}F scattered ions are completely stopped into the ΔE layer of the telescopes. For this experiment only a section of the whole EXODET apparatus, consisting of two telescopes, was used. From the telescope placed in the backward hemisphere, subtending the scattering angular range θ_{lab}=98°-154°, both the position and energy information were taken. The detector placed in the forward hemisphere (θ_{lab}=26°-82°) was used for data normalization purpose, being the scattering purely Rutherford at the angle and energy ranges here considered. The trigger for the data acquisition system was obtained by the OR of all the energy signals. In the upper panel of fig.5a) is reported the energy spectrum of the ΔE backward detector. The two bumps in the high energy side correspond to the ^{17}F (at the higher energy) and ^{17}O scattered ions. Their broadening is mainly due to the kinematic spread of the particle hitting the detector, which subtends a solid angle of about 110 msr, and to the energy lost in the target. The overall energy resolution was evaluated to be about 6-8%. A better resolution can be achieved selecting a limited number of strips at the cost of a lower statistics. On the left side of the spectrum a peak is visible at around 8 MeV and is mainly originated from light particles. By gating the ΔE spectrum with conditions on the information obtained from the strips, it is possible to select the contributions coming from ^{17}F ions and from the light particles, as shown in the middle and lower panel of fig.5a), respectively. In the upper panel of fig.5b) is reported the JT spectrum of the ΔE backward detector, showing a sharp peak for JT=10, which evidences the time correlation of the collected events and is very useful to avoid spurious and/or uncorrelated contributions. In the middle and lower panels of fig.5b) are also shown two ToT spectra corresponding to the scattered ^{17}F ions and to the light particles respectively. In the present case, we clearly see that the ^{17}F scattering events have a ToT sharply peaked around 400 ns (6 x 67 ns), while for light particles the ToT is smaller than 266 ns (4 x 67 ns). The ToT is only roughly proportional to the energy released by the particles into the strip, but it is very useful to disentangle the hitting positions when two particles, with very different energy ranges, impinge on the same detector in the same event.

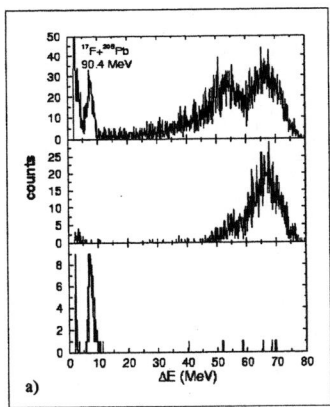 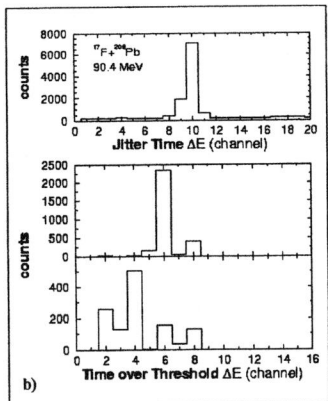

FIGURE 5. a) ΔE spectrum collected from the backward detector of the EXODET apparatus. Total spectrum (upper panel), gated by JT=10 and ToT=6 (middle panel), gated by JT=10 and ToT ranging from 2 to 4 (lower panel). b) Jitter Time (JT) spectrum of the backward ΔE detector (upper panel). Time over Threshold (ToT) spectra of the backward ΔE detector collected with the following gates: JT=10 and ΔE>59 MeV, corresponding to the ^{17}F peak, (middle panel); JT=10 and ΔE<10 MeV, corresponding to the light particles range, (lower panel).

The strip distribution of the scattered ^{17}F ions was corrected for the strip geometrical efficiency. We performed a Monte Carlo simulation to evaluate the solid angles and the scattering angle ranges subtended by each strip, with an uncertainty estimated to be less than 1%. The normalization of the cross section was obtained considering the ^{17}F scattering in the forward detector as purely Rutherford.

RESULTS AND COMMENTS

In fig.6a) the angular distribution of the scattered ^{17}F ions is presented. Because of the target thickness and of the target frame screening, only the experimental points above 115° could be evaluated. For comparison, the data for the same system at 98 MeV of beam energy have been also reported. This experiment [2] was performed at the Holifield Radioactive Ion Beam Facility, ORNL, Oak Ridge (USA), with a ^{17}F beam produced through the direct reaction ^{16}O(d, n)^{17}F. The continuous lines are optical model best fits obtained with the code SFRESCO [3], using the same parameter set ($a_v=a_w=0.53$ fm, $V_0=52.7\pm6.2$ MeV, $r_{0v}=r_{0w}=1.24$ fm) and leaving free only the depth of the imaginary potential ($W_0=7.1\pm1.8$ MeV and $W_0=15.9\pm2.9$ MeV for the 90.4 MeV and 98.0 MeV data, respectively). We could not distinguish experimentally the contribution of the inelastic scattering to the total scattering cross section, since an energy resolution better than 0.7% is required. So we did not include in our calculations the excitation of the first excited state of ^{17}F (at 0.4953 MeV). We checked *a posteriori* the validity of our assumption, running the code FRESCO [3] within the DWBA, using the potential parameters obtained by the best fit and including the first excited state with the experimental transition probability B(E2)↑=21.64 e^2fm^4 [4]. We found that the contribution of this channel to the quasi-elastic cross section was less than 2%, in the energy range here considered, and therefore it could be neglected in a first approximation. Additional details of the performed analysis will be published into an outcoming paper.

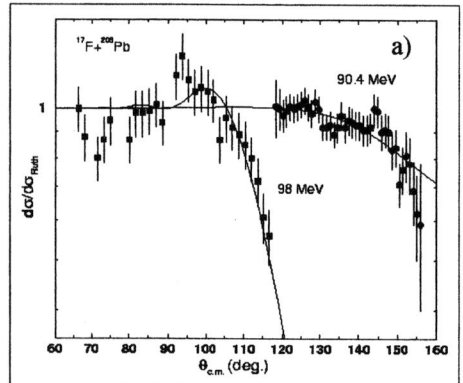 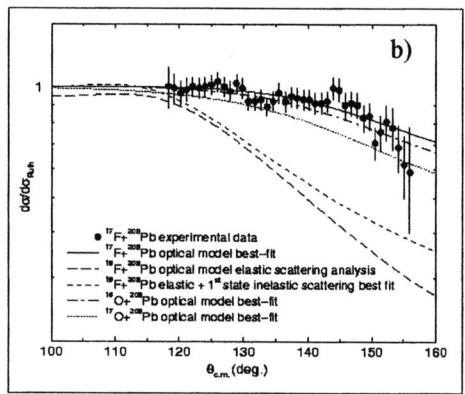

FIGURE 6. a) Angular distribution of the ^{17}F at 90.4 MeV incident energy (present experiment) and at 98 MeV (ORNL experiment [2]). The solid lines represent optical model best fits. b) The data and the optical model best-fit for ^{17}F are compared with those of other nuclei. The ^{17}F behavior results more similar to those of the ^{16}O and ^{17}O than to the ^{19}F one.

A comparison of the scattering angular distributions for four different systems is shown in fig.6b) [2,5,6]. In table 1 the optical model potential parameters obtained from the best-fit analyses are reported. The behavior of the ^{17}F angular distribution results more similar to those of the Oxygen isotopes than to the ^{19}F one, also if we consider the quasi-elastic contribution coming from the first excited state of the ^{19}F (at 0.197 MeV). The difference between the ^{17}F and ^{19}F projectiles is also remarked by the reaction cross section that is three times as large as for ^{19}F. The explanation of such a different behavior might be found in the completely different structure of the two nuclei: ^{17}F is a weakly bound radioactive nucleus, while ^{19}F is well bound with a rotational band. The ^{16}O and ^{17}O nuclear structures are quite similar to that of ^{17}F even if their binding energies are much higher.

TABLE 1. Optical model potential best-fit parameter.

reaction	E_{lab} (MeV)	E_{cm}/V_c	V_0 (MeV)	r_{0v} (fm)	a_v (fm)	W_0 (MeV)	r_{0w} (fm)	a_w (fm)	χ^2/pt	σ_R (mb)
$^{17}F+^{208}Pb$	90.4	0.96	52.7 ± 6.2	1.24	0.53	7.1 ± 1.8	1.24	0.53	0.525	77
$^{19}F+^{208}Pb$	91	0.96	107.6 ± 6.1	1.24	0.53	20.1 ±3.0	1.24	0.53	1.111	269
$^{16}O+^{208}Pb$	78	0.93	78.28	1.215	0.65	17.11	1.162	0.623	0.99	47
$^{17}O+^{208}Pb$	78	0.93	82.21	1.226	0.65	9.93	1.226	0.60	1.25	91

Despite the small beam time used to collect data (about 17 hours), we performed also a search for $^{17}F \rightarrow p+^{16}O$ break-up events. We considered only the events in which: a) two strips of the ΔE detector and only one of the E detector are hit; b) the JT of all the strips are in the correlation peak; c) the ToT of one strip of the ΔE detector is in the Fluorine-Oxygen range; d) the ToT of the other strip of the ΔE detector is lower; e) the total energy released in the ΔE detector is in the spectrum region of the ^{17}F-^{17}O peaks (corresponding to the total energy of the ^{16}O plus the energy lost by the proton passing through the ΔE). A value σ_{BU}=(3.1±1.4) mb was obtained for the break-up cross section in good agreement with that extracted from literature [7] (σ_{BU}=1.6 – 3.2 mb).

In summary, we reported about the features and the performances of the new experimental apparatus EXODET, designed to be used in RIB experiments. The use of an ASIC chip for the readout electronics allows high performances at a relatively low cost. The integration of energy and chip information gives the possibility of strong data selection. In the first experiment, successfully performed using the EXODET apparatus, the $^{17}F+^{208}Pb$ scattering at 90.4 MeV has been studied. The ^{17}F angular distribution has been measured and an optical model analysis performed to extract the best-fit potential parameter set. We have found that the ^{17}F scattering behavior is consistent with the ^{17}F data taken at higher energy. By the comparison with other systems, we have seen that such behavior is very different from that of the ^{19}F one, in the same energy range, but it is quite similar to those of ^{16}O and ^{17}O. This might be interpreted as a consequence of the different nuclear structure of the ^{17}F with respect to ^{19}F.

REFERENCES

1. Perazzo, A., et al., *BABAR Note #501* (1999) and references therein.
2. Lin, C. J., et al., *Phys. Rev. C* **63**, 064606 (2001).
3. Thompson, I. J., *Comput. Phys. Rep.* **2**, 167 (1998).
4. Tilley, D. R., et al., *Nucl. Phys. A* **565**, 1 (1993).
5. Thompson, I. J., et al., *Nucl. Phys. A* **505**, 84 (1989).
6. Lilley, J. S., et al., *Nucl. Phys. A* **463**, 710 (1987).
7. Rehm, K. E., et al., *Phys. Rev. Lett.* **81**, 3341 (1998).

Subbarrier fusion in the systems ^{11,10}Be+^{209}Bi: the experimental data

C.Signorini[1], A.Yoshida[2], Y.Watanabe[2], D.Pierroutsakou[3], L.Stroe[4], T.Fukuda[5], M.Mazzocco[1], N.Fukuda[2], Y.Mizoi[5], M.Ishihara[2], H.Sakurai[6], F.Soramel[7]

[1] *Physics Department and INFN, via Marzolo 8, I-35131 Padova*
e-mail: *signorini@pd.infn.it*
[2] *RIKEN, Hirosawa 2-1, Wako, Saitama 351-0198, Japan*
[3] *Physics Department and INFN, via Cinthia, I-80125 Napoli*
[4] *INFN, Laboratori Nazionali di Legnaro, I-35020 Legnaro (Padova)*
[5] *KEK/IPNS, Oho, Tsukuba, Ibaraki, Japan*
[6] *Physics Department, The University of Tokyo, Hongo, Bunkyo-ku, Tokyo, Japan*
[7] *Physics Department and INFN, Udine, Italy*

Abstract The subbarrier fusion cross section has been measured in the system ^{11,10}Be+^{209}Bi. The data were collected at the RIKEN Ring Cyclotron with a total beam intensity of ~10^{+5} Hz typical for such radioactive nuclear beams of first generation. The cross section accuracy is around 10 to 15 %. The two excitation functions look, within the statistical uncertainty very similar as well as the ^{9}Be+^{209}Bi one at difference with the expectations. This is triggering further theoretical analysis.

INTRODUCTION

The present contribution is focused primarily onto the experimental part of measurements and the relative evaluations of fusion cross sections at barrier energies in the systems ^{11,10}Be+^{209}Bi at the RIPS beam course of the RIKEN Ring Cyclotron.

The main motivation of this experiment was the study of the behavior of the subbarrier fusion cross sections of ^{11}Be loosely bound and with a well established halo structure, in comparison with ^{10}Be. It is expected that these ^{11}Be feature produce a behavior of the fusion cross section quite different compared to a well bound projectile like ^{10}Be particularly in the subbarrier region.

In the present article we will discuss mainly the experimental data and analysis details relevant to the cross sections evaluation; the theoretical interpretation of the results will be published successively.

EXPERIMENTAL

The background to this experiment are the papers on fusion cross sections measured by our group in the system $^{11,9}Be+^{209}Bi$ [1] and $^{9}Be+^{209}Bi$ [2].

The goal of the experiment was to measure the ^{11}Be system with better energy resolution and possibly higher statistics compared to [2] and to get data with the reference nucleus ^{10}Be.

The $^{11,10}Be$ beams were delivered by the RIPS beam course [3] via fragmentation of a primary ^{13}C beam with 100*A MeV energy onto a beryllium production target and subsequent heavy energy degradation of a factor ~10 according to an already well established scheme [1,2]. The continuous beam energy so produced, ranging from 35 to 55 MeV was tagged in energy event by event in a ~6m flight path; the final energy resolution was around 1 MeV with a total intensity in the whole energy range of ~10^{+5} Hz. This intensity clearly affects the statistics that one can collect in some days of experiment.

The fusion cross sections were deduced from the yield of the evaporation residues populated by xn reactions and by the fission fragments; these two processes are expected to be the strongest channels according to statistical model realistic estimates. The cross sections for the evaporation channels with one charged particle are at least one order of magnitude smaller.

Large area (50*50 mm^2) Silicon detectors were closely packed around a multiple target set up in a cube-like structure which could achieve a large solid angle coverage of ~50%. The fission fragments were identified by coincidences between two opposite detectors. The evaporation residues were detected tagging on the alpha particles emitted from their ground state as follows:

^{215}Fr, $T_{1/2}$=90ns, E_α=9360 keV, 5n (4n) from ^{11}Be (^{10}Be) fusion,
^{216}Fr, $T_{1/2}$=700ns, E_α= 9005 keV, 4n (3n) from ^{11}Be (^{10}Be) fusion.

The electronic set up allowed to explore events delayed up to Δt=400 ns after the prompt ones. This time interval was a compromise between an acceptable pile up rate into the energy tagging event by event (at 0.5 MHz rate) and a gate long enough for the 700 ns lifetime.

In the case of ^{10}Be the main evaporation channels are 4n, 3n, and 2n (this last one expected to be weak) for ^{11}Be 5n, 4n and 3n (this last one expected to be weak); these channels have one neutron more than ^{10}Be at the same beam energy since in ^{11}Be the last neutron is bound only by 0.5MeV.

For ^{10}Be we can realistically observe only complete fusion; the incomplete fusion of ^9Be is to be excluded since ^{10}Be is well bound. This is not the case for ^{11}Be. In this case both complete fusion of ^{11}Be and incomplete fusion of ^{10}Be fragments produced by a possible ^{11}Be breakup into ^{10}Be+1n are in principle included since in both cases the same evaporation residues are produced. However from our previous analysis [1] it resulted that the incomplete fusion contributes no more than 25% to the total fusion process. We concluded that within the accuracy of our data, estimated around 10 to 15%, we deal essentially with complete fusion

In figs. 1 and 2 we report the cross sections measured as resulted from two different runs, labeled {1} and {2}, done 3 years apart from each other.

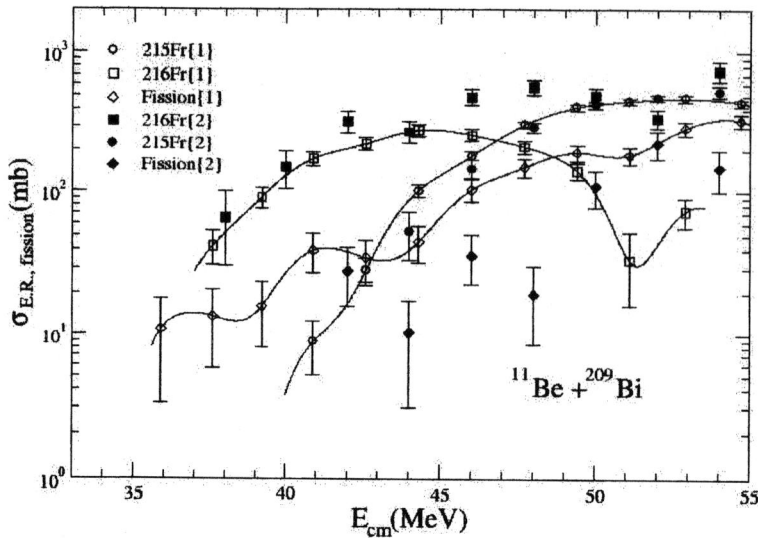

FIGURE 1. Excitation function for the production of evaporation residues and fission in the reaction $^{11}Be+^{209}Bi$. The lines are drawn only to guide the eye.

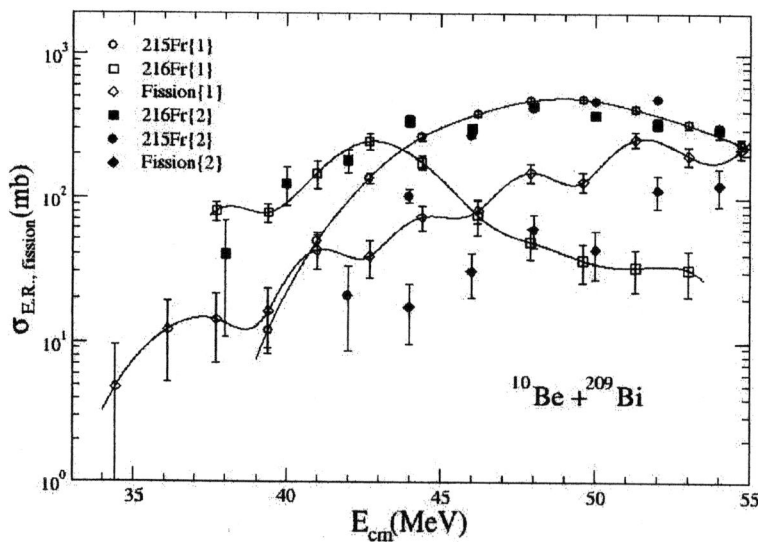

FIGURE 2. Excitation function for the production of evaporation residues and fission in the reaction $^{10}Be+^{209}Bi$. The lines are drawn only to guide the eye.

The scattering of the data, unfortunately disappointing in some (few) cases, is due to the very limited statistics of some data originating by the radioactive beam intensity which is 4 to 5 order of magnitude weaker than a typical beam with stable isotopes.

However if we sum up the various channels the total cross sections shown in fig. 3 have a relatively smooth behavior similar to a "normal" fusion cross section.

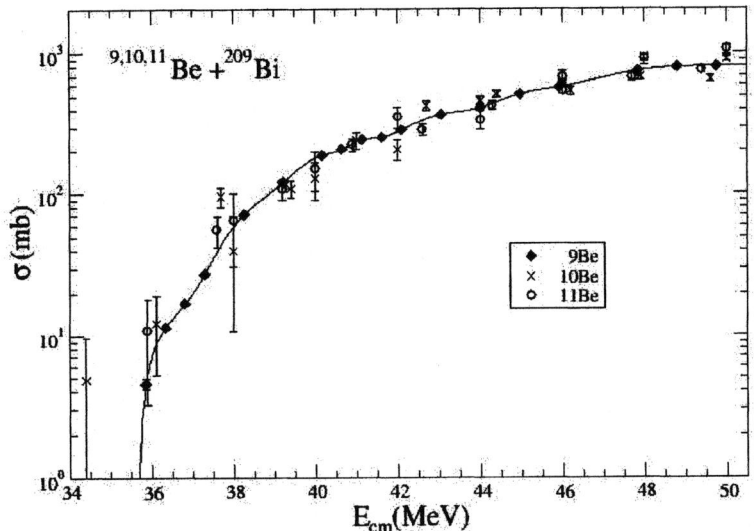

FIGURE 3. Fusion cross section for the three systems ^{9,10,11}Be+^{209}Bi.

We checked also an other very important point namely the contribution to the total cross section of the 2n (3n) channel from $^{10(11)}$Be beam. This channel produces ^{217}Fr $T_{1/2}$=16μsec, E_α=8315 keV. Such channel could not be observed in our experiment even if considerable experimental effort was placed into the realization of a 20 μsec delayed gate. The adopted solution was to make a realistic estimate of these two cross sections with the evaporation code PACE2 which is supposed to handle quite well also the fission process. The PACE2 code parameters adopted were standard.

The low energy data were normalized in order to reproduce the 3n (4n) channel with ^{10}Be(^{11}Be) up to ~44 MeV; an energy range were the results of the two different runs are well coinciding.. The results are shown in fig. 4,5. The continuous lines labeled "norm." are the normalized results. From these results we notice that the cross section for the "missing" ^{217}Fr channel: for ^{11}Be (^{10}Be) beam are 15 mb, 22 mb, 50 mb (15 mb, 10 mb, 6 mb) at E_{cm} 34, 35, 36 MeV respectively.

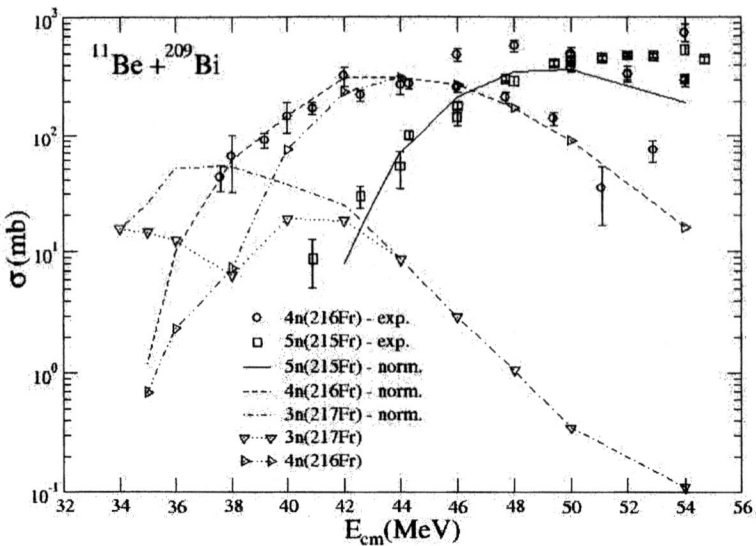

FIGURE 4. PACE2 predictions for the reaction ^{11}Be+^{209}Bi (curves with no label) compared with the experimental data. For the curves "norm." see the explanations in the text.

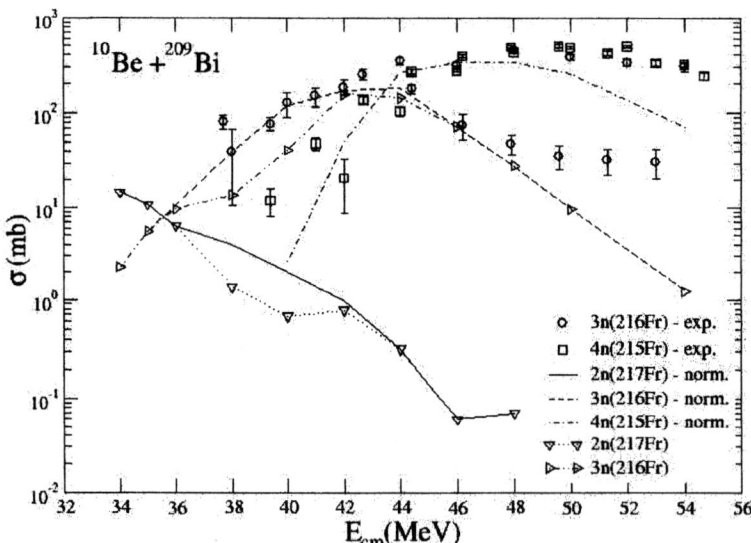

FIGURE 5. Same data as in FIGURE 4 for the reaction ^{10}Be+^{209}Bi..

COMMENTS

The results shown in fig. 3 to our opinion give an indication for the moment that the three systems behave in a very similar way. Significant is in particular the comparison between ^{10}Be and ^{11}Be beams cross sections since the two sets of data have been collected with the same set up during the same runs so that any unforeseen systematic error cancel in the relative comparison.

A priori one would have expected a different behavior of the two cross sections since ^{11}Be is very weakly bound and moreover it has a well established halo structure. This has triggered extensive theoretical work in the frame of Coupled Channel approach with coupling to a Discretized Continuum (due to the breakup process of the weakly bound ^{11}Be) so-called CDCC.; the results will published in a forthcoming paper..

Of course the statistical model evaluation of the "missing" channel indicate (but only) at E_{cm}=36 MeV for ^{11}Be an additional cross section predicted of 46 mb with respect to ^{10}Be; only an other experiment can eliminate or confirm this theoretical estimate.

REFERENCES

1. C.Signorini et al., Eur.Phys.J. **A2**(1998) 227.
2. A.Yoshida et al., Phys.Lett. **B389**(1996)457
2. T.Kubo et al., Nucl.Instr.Meth.Phys.Res. **B70**(1992)309

A non perturbative approach to neutron and proton halo breakup

Angela Bonaccorso[1]

Istituto Nazionale di Fisica Nucleare, Sez. di Pisa, 56127 Pisa, Italy.
email : bonaccorso@pi.infn.it

Abstract. In this paper we show how effective parameters such as effective binding energies can be defined for a proton in the combined nuclear-Coulomb potential, including also the target potential, in the case in which the proton is bound in a nucleus which is partner of a nuclear reaction. Using such effective parameters the proton behaves similarly to a neutron. In this way some unexpected results obtained from dynamical calculations for reactions initiated by very weakly bound proton halo nuclei can be interpreted. Namely the fact that stripping dominates the nuclear breakup cross section which in turn dominates over the Coulomb breakup even when the target is heavy at medium to high incident energies. Our interpretation helps also clarifying why the existence and characteristics of a proton halo extracted from different types of data have sometimes appeared contradictory.

INTRODUCTION

This paper is concerned with the differences which might arise in reactions initiated by a neutron halo nucleus like ^{11}Be and a proton halo nucleus like ^{17}F or ^{8}B. Halo nuclei are a special case of radioactive beams for which the last nucleon is very weakly bound, with separation energies of the order of 0.5 MeV or less, and in a state of low angular momentum (l=0,1). They exhibit extreme properties like very large total and breakup cross sections. Nuclear and Coulomb breakup of neutron halo nuclei have been studied in great detail both experimentally as well as theoretically and are now quite well understood processes [1]. On the other hand proton halo nuclei such as ^{8}B and ^{17}F are still under investigation. Their behavior as projectiles of nuclear reactions needs to be understood better in particular as ^{8}B is partner in (p,γ) radiative capture reactions of great astrophysical interest for the understanding of the neutrino flux from the sun (see for example the discussion and references of [2]). Also the existence of a proton halo has sometimes been questioned [3] and results from different experiments might seem to be contradictory [4]. For those nuclei Coulomb breakup reactions in the laboratory have been used to get indirect information on the radiative capture, since it has been shown that the Coulomb breakup cross section is proportional to the radiative capture cross section [5].

In the case of neutrons the Coulomb breakup cross section is largest for heavy targets and the interplay with nuclear breakup is well understood both experimentally as well as

---

[1] In collaboration with D. M. Brink, J. Margueron, C. Papineau, C. A. Bertulani.

theoretically, in particular thanks to the measurements of angular distributions for both processes [6, 7, 8, 9]. Then ^{208}Pb and ^{58}Ni have been used as targets with beams of ^{8}B or of ^{17}F at various energies [2][10]-[15]. Data on lighter targets such as ^{9}Be and ^{28}Si [3, 17, 18] also exist. At the same time a number of theoretical papers have appeared dealing with the problem of the accuracy necessary to interpret the data [19]-[23]. In particular the problems of higher order effects in Coulomb breakup, of the inclusion of E0, E1, and E2 multipolarities in the Coulomb field and of the relative magnitude of nuclear and Coulomb contributions and of their interference have been discussed at length.

A number of experimental papers [18, 15] have shown that for a ^{8}B projectile it is the nuclear breakup and in particular the stripping (or absorption) [17, 18] component that dominates the experimental cross section. In [17, 18] a ^{28}Si target was used and different beam energies around 40A.MeV were explored. The data of Table 1 of [18] show that stripping (110± 9 mb) is very close to the total diffraction (112± 12mb) which contains both nuclear and Coulomb components. On the other hand at the same beam energy and on the same target the one neutron breakup of ^{11}Be, measured by [24] and calculated by [25] gave a stripping cross section of 220 mb and a total diffraction of 300 mb of which 120mb from Coulomb breakup. These results could be considered rather astonishing in view of the fact that the proton in ^{8}B has a separation energy of 0.14MeV while the neutron separation energy in ^{11}Be is larger and equal to 0.5MeV. On the other hand the data of [15] for the breakup of ^{8}B on ^{208}Pb at 142A.MeV provided a one-proton removal cross section of 744±9 mb of which about 300-450 mb were estimated to be due to nuclear breakup and 311 mb to Coulomb breakup. This is again a surprising result because for the system ^{11}Be+^{208}Pb at 120A.MeV it was calculated in [9] that the cross sections would be 321 mb for nuclear breakup and 1050 mb for Coulomb breakup, the model of [9] being very reliable as it agrees with exclusive data [6]. Similarly, the recent data from GSI [16] at the relativistic beam energy of 936 A.MeV give for the one proton removal cross section of ^{8}B on ^{208}Pb and ^{12}C, 662±60 mb and 94±8mb respectively while at a similar energy (790A.MeV) the one neutron removal from ^{11}Be on the same targets was 960±60mb and 169±4 respectively [26].

On the other hand, very recently a new theoretical work has appeared where the authors treat the nuclear breakup of ^{17}F to first order [27], contrary to what has been established in the literature, namely that halo breakup should be treated to all orders of the neutron-target interaction. The approach of [27] can be justified with the results of another theoretical work by Esbensen and Bertsch [22] on the proton halo nucleus ^{8}B, where it was shown that starting from about 40 A.MeV in the reaction ^{8}B +^{208}Pb, dynamical calculations and first order perturbation theory with or without far field approximation, yields nearly the same Coulomb breakup cross sections for distances of closest approach for the core target trajectory of 20 fm or larger. Also in [21] the same authors found that for ^{17}F nuclear diffraction and Coulomb breakup have very similar probabilities to occur and the values are also close to those for nuclear stripping. An earlier calculation by Esbensen and Hencken [20] showed that nuclear one proton removal cross sections for a ^{8}B projectile would be larger than Coulomb cross sections up to target mass A_T=100. Similar conclusions were reached by Dasso, Lenzi and Vitturi [28].

In order to get some insight into the peculiarities of the proton halo reactions, in particular in comparison to neutron halos, we introduce here an effective treatment of proton single particle states which simplifies their treatment and makes them behaving as neutrons. Related approaches have recently been introduced by other authors [29]. We do not propose our method as opposed to dynamical calculations, but we are simply concerned with the understanding of the underlying physics and the interpretation of numerical results from more sophisticated methods such as direct solutions of the Schrödinger equation or coupled channels.

PROTON VS. NEUTRON: EFFECTIVE POTENTIAL

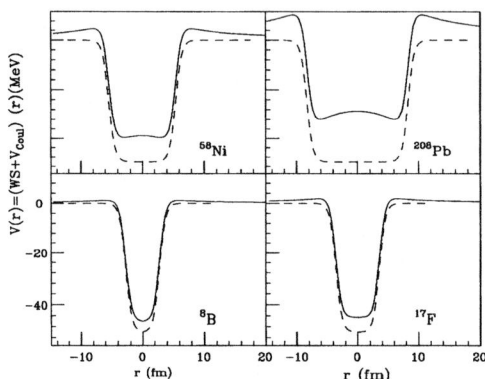

FIGURE 1. Nuclear (dashed) and nuclear-Coulomb (solid) potentials for ^8B, ^{17}F, ^{58}Ni and ^{208}Pb.

We consider the breakup of a proton halo nucleus like ^{17}F consisting of a proton bound to a ^{16}O core in a collision with a target nucleus. The system of the halo nucleus and the target is described by Jacobi coordinates $(\mathbf{R}, \mathbf{r})$ where $\mathbf{R}$ is the position of the center of mass of the halo nucleus relative to the target nucleus and $\mathbf{r}$ is the position of the neutron relative to the halo core, and the coordinate $\mathbf{R}$ is assumed to move on a classical path. This allows target recoil to be included in a consistent way. The Hamiltonian of the system is

$$H = T_R + T_r + V_{pc}(\mathbf{r}) + V_2(\mathbf{R}, \mathbf{r}) \qquad (1)$$

where T_R and T_r are the kinetic energy operators associated with the coordinates $\mathbf{R}$ and $\mathbf{r}$ and V_{pc} is the potential describing the interaction of the proton with the core, and it contains nuclear and Coulomb parts. The potential V_2 describes the interaction between the projectile and the target. It is a sum of two parts depending on the relative coordinates of the proton and the target and of the core and the target

$$V_2(\mathbf{R}, \mathbf{r}) = V_{pt}(\beta_2 \mathbf{r} + \mathbf{R}) + V_{ct}(\mathbf{R} - \beta_1 \mathbf{r}) \qquad (2)$$

Here $\beta_1 = m_p/m_P$, $\beta_2 = m_c/m_P = 1 - \beta_1$, where m_p is the proton mass, m_c is the mass of the projectile core and $m_P = m_p + m_c$ is the projectile mass. Both V_{pt} and V_{ct} are represented by complex optical potentials. The potentials V_{pt} and V_{ct} also include the Coulomb interaction between the proton and the target and the halo core and the target. This part of V_{ct} is responsible for Coulomb breakup. The mass ratio β_1 is small for a halo nucleus with a heavy core. For example $\beta_1 \approx 0.06$ and $\beta_2 \approx 0.94$ in the case of ^{17}F. This property is used here to approximate the proton-target and proton-core potentials by $V_{pt}(\beta_2 \mathbf{r} + \mathbf{R}) \approx V_{pt}(\mathbf{r}+\mathbf{R})$ and $V_{ct}(\mathbf{R} - \beta_1 \mathbf{r}) \approx V_{ct}(\mathbf{R}) + V_{eff}(\mathbf{r},\mathbf{R})$

The halo breakup is caused by the direct proton target interaction V_{pt} or by a recoil effect due to the core-target interaction. Coulomb breakup of a one-proton halo nucleus is mainly a recoil effect due the Coulomb component V_{ct} of the core-target interaction and is contained in $\mathbf{V}_{eff}(\mathbf{r},\mathbf{R})$. It is proportional to the mass ratio β_1. As discussed in Ref.[9] the main effect of $V_{ct}(\mathbf{R})$ is to give an absorption for small core-target impact parameters and thus it reduces the core survival probability.

Then the wave function $\phi(\mathbf{r},\mathbf{d},t)$ describing the dynamics of the halo proton satisfies the time-dependent equation

$$i\hbar \frac{\partial \phi(\mathbf{r},\mathbf{d},t)}{\partial t} = (H_r + V_{pt}(\mathbf{r}+\mathbf{R}(t)) + V_{eff}(\mathbf{r},\mathbf{R}(t)))\phi(\mathbf{r},\mathbf{d},t) \qquad (3)$$

where $H_r = T_r + V_{pc}(\mathbf{r})$ is the Hamiltonian for the halo nucleus. In the present paper we neglect the nuclear part of final state interactions between the proton and the halo core, but include the Coulomb proton-core interaction $V_{pc}(\mathbf{r}) = \frac{Z_h Z_C e^2}{|\mathbf{r}|}$ and the final state interactions between the proton and the target. This approximation should be satisfactory unless there are resonances in the proton-core final state interaction which are strongly excited during the reaction. The proton-core potential does not act dynamically and it cannot cause breakup. It gives the maximum contribution at the top of the proton-core barrier where $|\mathbf{r}| = R_i$. Therefore we take it constant as $V_{pc} = \frac{Z_h Z_C e^2}{R_i}$.

When the nuclear proton-core final state interactions are neglected we can define a potential

$$\bar{V}_2(\mathbf{r},t) = V_{pt}^N(\mathbf{r}+\mathbf{R}(t)) + V_{pt}^C(\mathbf{r}+\mathbf{R}(t)) + \mathbf{V}_{eff}(\mathbf{r},\mathbf{R}(t)) = V_{pt}^N(\mathbf{r}+\mathbf{R}) + V_{\text{Coul}}, \qquad (4)$$

and V_{pt}^N and V_{pt}^C are the nuclear and Coulomb parts of the proton-target interaction.

Thus in the case of a proton breakup and for a heavy target, besides the proton target nuclear potential it is necessary to include in the total Hamiltonian the proton-core, proton-target and core-target Coulomb potentials. The proton-target potential and the core-target potential are included dynamically but the effect on the center of mass has to be subtracted. We have then

$$V_{\text{Coul}} = Z_T e^2 \left(\frac{Z_h}{|\mathbf{R}+\beta_2 \mathbf{r}|} + \frac{Z_C}{|\mathbf{R}-\beta_1 \mathbf{r}|} - \frac{Z_C}{|\mathbf{R}|} - \frac{Z_h}{|\mathbf{R}|} \right) \qquad (5)$$

where charges and masses are : core (A_C,Z_C), halo (A_h,Z_h), target (A_T,Z_T). Because $Z_h Z_C$ and $Z_h Z_T$ are much smaller than $Z_T Z_C$ appearing in the other terms, the interactions $Z_T e^2 \left(\frac{Z_h}{|\mathbf{R}+\beta_2 \mathbf{r}|} - \frac{Z_h}{|\mathbf{R}|} \right)$ give the maximum contribution around the distance of closest approach d where $|\mathbf{R}| \approx d = R_i + R_f$, $|\mathbf{r}| \approx R_i$ and $|d + \beta_2 \mathbf{r}| \approx R_f$. We neglect then the time and velocity dependence of $\mathbf{R}(t) = \mathbf{d} + vt\hat{\mathbf{z}}$ only in the term $Z_T e^2 \left(\frac{Z_h}{|\mathbf{R}+\beta_2 \mathbf{r}|} - \frac{Z_h}{|\mathbf{R}|} \right)$.

This should be a reasonable approximation at low incident energies when $v \approx 0$ (adiabatic approximation) and at high incident energies (sudden approximation) because then $Z_T e^2 \left(\frac{Z_h}{|\mathbf{R}+\beta_2 \mathbf{r}|} - \frac{Z_h}{|\mathbf{R}|} \right)$ will be negligibly small anyway. Thus we obtain that the proton can be treated as a neutron described by the Scrödinger equation (3) and initial condition that as $t \to -\infty$ the wave function tends to the initial halo nucleus wave-function provided the separation energy is given by $\tilde{\varepsilon}_i = \varepsilon_i - \frac{Z_P e^2}{R_i} - Z_T e^2 \left(\frac{1}{R_f} - \frac{1}{d} \right)$.

The effect of the proton-core and proton-target Coulomb potentials can also be discussed qualitatively. Fig. (1) shows the potentials felt by a neutron (dashed line) and a proton (full line) in ^{8}B, ^{17}F, ^{58}Ni and ^{208}Pb. Supposing the two particles have the same binding energy $\varepsilon_i < 0$ the proton wave function inside the potential of Fig.(1) is like a neutron wave function with binding energy $\varepsilon_i - Z_P e^2/R_i$ up to the radius R_i. The proton potential is like the neutron potential pushed up by $Z_P e^2/R_i$ where R_i is the barrier radius. For any given nucleus this radius is rather larger than the nuclear or Coulomb radius values usually quoted in the literature. But from Fig. (1) one can see that it is the value corresponding to the barrier peak. We give these values in Table 1, together with the experimental binding energies of the halo state in ^{8}B and of two states in ^{17}F.

TABLE 1. Barrier radii from Fig. (1) and initial binding energies.

	^{8}B	J^π	^{17}F	J^π	^{58}Ni	^{208}Pb
$R_{i,f}(fm)$	6.0		6.5		7.5	10.5
$\varepsilon_i(MeV)$	-0.14	$1p_{3/2}$	-0.6	$1d_{5/2}$	–	–
$\varepsilon_i^*(MeV)$	–		-0.1	$2s_{1/2}$	–	–

But as it is shown in Ref.[30], in a scattering process there is also an effect due to the Coulomb potential of the projectile. It can be understood by looking at Fig. (2) which shows the nuclear-Coulomb potentials for ^{8}B$+^{58}$Ni (top) and ^{17}F$+^{208}$Pb (bottom) at several distances. Short and long dashed lines are the separate projectile and target potentials respectively. Full line is the projectile-target combined potential.

TABLE 2. Effective parameters.

	^{8}B$+^{58}$Ni	^{8}B$+^{208}$Pb	^{17}F$+^{58}$Ni	^{17}F$+^{208}$Pb
$\Delta_i(MeV)$	-3.6	-5.3	-4.5	-6.3
$|\Delta_f|(MeV)$	6.03	12.01	6.44	12.47
$\tilde{\varepsilon}_i(MeV)$	-3.74	-5.44	-5.1	-6.9
$\tilde{\gamma}_i(fm^{-1})$	0.4	0.48	0.48	0.56
$\tilde{C}_i(fm^{-1/2})$	1.33	1.79	2.11	3.43
$\tilde{\varepsilon}_i^*(MeV)$	–	–	-4.6	-6.4
$\tilde{\gamma}_i^*(fm^{-1})$	–	–	0.47	0.54
$\tilde{C}_i^*(fm^{-1/2})$	–	–	5.8	6.4

The effect of the target potential on the projectile potential is actually twofold:

- the center of the projectile potential shifts up by an amount $Z_T e^2/d$ where d is the distance of closest approach between the two nuclei.
- the height of the barrier on the side near the target goes up by an amount $Z_T e^2/R_f + Z_P e^2/R_i$ relative to the center. The same happens to the target. For each trajectory we shall take $R_f \approx d - R_i$.

This suggests that the true binding energy ε_i could be replaced by:

$$\varepsilon_i \to \tilde{\varepsilon}_i = \varepsilon_i - \frac{Z_P e^2}{R_i} - Z_T e^2 \left(\frac{1}{R_f} - \frac{1}{d}\right). \tag{6}$$

In the case of a bound final state one would also have

$$\varepsilon_f \to \tilde{\varepsilon}_f = \varepsilon_f - \frac{Z_T e^2}{R_f} - Z_P e^2 \left(\frac{1}{R_i} - \frac{1}{d}\right). \tag{7}$$

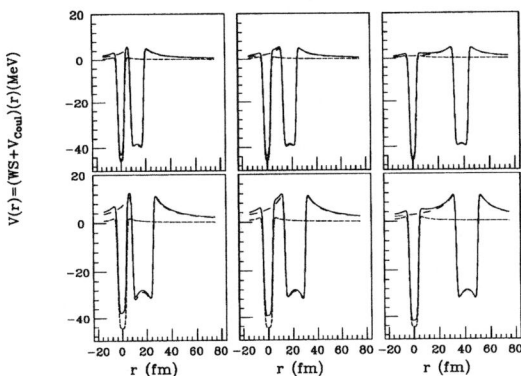

FIGURE 2. Nuclear-Coulomb potentials for ^{8}B $+^{58}$Ni (top) and ^{17}F $+^{208}$Pb (bottom) at distances between the centers equal to $d = 1.4(A_P^{1/3} + A_T^{1/3})$ fm $+ s$, with $s = 5$, 15 and 30 fm. Short and long dashed lines are the projectile and target potentials respectively. Full line is the projectile-target combined potential.

In the reactions we are discussing the initial states are always bound. According to Eq.(6) they will be shifted down by a Δ_i equal to $\Delta_i = \frac{Z_P e^2}{R_i} + Z_T e^2 \left(\frac{1}{R_f} - \frac{1}{d}\right)$. On the other hand for continuum final states of positive energy Eq.(7) means that states with $\varepsilon_f < \Delta_f$ where Δ_f is an effective shift defined by $\Delta_f = \frac{Z_T e^2}{R_f} + Z_P e^2 \left(\frac{1}{R_i} - \frac{1}{d}\right)$ will behave as bound states in an effective potential which is deeper than the real proton potential by a Δ_f. Therefore the phase space for breakup states will be reduced and thus breakup probabilities for protons will be smaller than for neutrons having the same binding energy. Furthermore there will be an important target dependence.

Then we conclude that some features of proton transfer and breakup could be understood by analogy with neutron transfer by using effective parameters in the following way:

- use effective $\tilde{\gamma}_i$ and $\tilde{\gamma}_f$ calculated from $\frac{\hbar^2 \tilde{\gamma}_{i,f}^2}{2m} = |\tilde{\varepsilon}_{i,f}|$
- calculate the normalization constants $\tilde{C}_{i,f}$ of asymptotic wave functions [32] as for neutron wave functions with binding energies $\tilde{\varepsilon}_{i,f}$.

The approximation of using effective parameters for protons so that they could be treated similarly to neutrons has a long history in direct reaction theories, see for example

[36]. The definitions Eqs.(6) and (7) used here represent however a generalization and improvement with respect to those used in [33, 34, 36]. This is because we not only take into account the effect of the Coulomb barrier in the projectile and target potentials, but also consider the "polarization" effect that the target Coulomb potential has on the projectile and vice versa. As Fig. (2) (bottom part) shows, in the case of a light projectile and a heavy target, the long range effect of the Coulomb potential gives a considerable shift upwards of the projectile potential.

At the medium to high energies we are dealing with in this work, the most favorite final energies will be larger than the top of the barrier and the usual treatment of breakup should be valid. From Fig. (2) one sees clearly that the effect of the barrier is very important even at distances as large as 30 fm. In order to quantize the effects discussed above we give in Table 1 the barrier radii, called R_i and R_f, for two nuclei usually used as projectiles ^{8}B, ^{17}F, and for ^{58}Ni, ^{208}Pb, which have been used as targets. In this case R_f has been taken as the barrier radius which can be deduced from Fig.(1). We give also the effective energy shift Δ_i and Δ_f, the effective binding energies for the possible projectile-target combinations studied in this paper. For completeness we add the effective length parameters $\tilde{\gamma}_i$ and asymptotic normalization constants $\tilde{C}_i$ of the initial asymptotic wave functions. It is indeed the tail of the wave function which determines the main characteristics of the breakup mechanism ([38] and references therein).

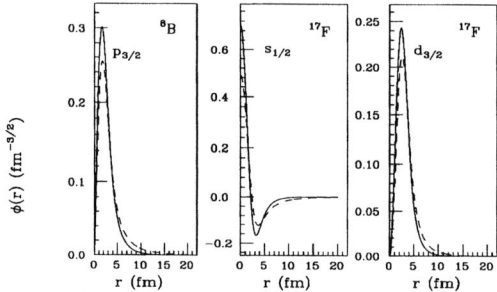

FIGURE 3. Proton (dashed) and neutron (solid) wave functions for ^{8}B, ^{17}F as indicated. Neutron wave functions obtained for effective energies as in Table 2.

To illustrate the last point we give in Fig.(3), by the dashed lines, the proton single particle wave functions corresponding to the three initial states of Table 1, calculated in a Woods-Saxon plus spin-orbit [38] plus Coulomb potential with parameters: $r_0 = 1.27 fm$, $a = 0.65 fm$, $V_{so} = 7 MeV$, $r_c = 1.3 fm$. The Woods-Saxon depth is fitted to give the correct binding energy. The solid line shows the neutron wave functions calculated with the effective binding energies. One sees clearly that in each case the true proton wave function is very close to the "effective energy" neutron wave function from the distances $R \approx R_i + R_f$ at which both nuclear and Coulomb breakup are most sensitive. We remind the reader that at small distances the breakup is strongly reduced due to the core-target absorption into more complicated reaction channels.

From the values shown it is clear that the proton halo behaves in a breakup reaction with a heavy target as a neutron state bound with a "normal" energy of several MeV, for which it is very well known that the nuclear breakup is comparable to the Coulomb breakup and on the other hand that the stripping is dominant on diffraction [37].

To give an idea of the orders of magnitude involved, we have calculated total breakup cross sections for two reactions: $^{11}\text{Be} \rightarrow {^{10}\text{Be}} + n$ and $^{17}\text{F}(1/2^+) \rightarrow {^{16}\text{O}} + p$, both at 40 A.MeV on a ^{208}Pb target. Nuclear and Coulomb breakup of ^{11}Be have been studied in many experiments on heavy targets and absolute breakup cross sections are very well known [6, 7]. For Coulomb breakup we used first order perturbation theory and for nuclear breakup we used the transfer to the continuum model [31, 35]. Our aim here is only to give some order of magnitude estimates. For the breakup of ^{17}F we used a neutron wave function and the "effective parameters" of Table 2. The values obtained are given in Table 3. One sees that for the proton "halo" state in ^{17}F there is a reduction of about one order of magnitude. The reduction is stronger for Coulomb breakup and diffraction because both require the neutron to be in a final free particle state, which is obviously less probable the stronger the "effective binding" of the nucleon in the initial state. It is interesting to note that the reduction in the proton removal cross sections from ^{17}F as compared to ^{11}Be calculated here and in [27] would be stronger than the reduction already seen in the data for ^{8}B discussed in the Introduction. This is because ^{17}F has a larger Z_P than ^{8}B and therefore as shown in Fig. (2) and Table 2 its effective separation energies are larger.

TABLE 3. Cross sections in mb.

	σ_C	σ_S	σ_D
^{11}Be + ^{208}Pb	2190	266	215
^{17}F + ^{208}Pb	83	26	7

It appears also clear that under such conditions Coulomb and nuclear breakup could not need to be calculated to all orders. However both should be included in a coherent way [9, 21].

Dipole vs. Quadrupole

Another problem which is important to solve in order to understand the differences between the characteristics of a proton halo with respect to a neutron halo has to do with the relative importance of the dipole term vs. the quadrupole term in the Coulomb potential. In order to discuss this point we go back to Eq.(5) and make the usual expansion

$$V_{\text{eff}} = C^{(D)} \frac{\mathbf{r} \cdot \mathbf{R}}{|\mathbf{R}|^3} + C^{(Q)} \left(3 \frac{(\mathbf{r} \cdot \mathbf{R})^2}{|\mathbf{R}|^5} - \frac{|\mathbf{r}|^2}{|\mathbf{R}|^3} \right) \quad (8)$$

where $C^{(D)} = Z_T e^2 (\beta_1 Z_C - \beta_2 Z_h)$ and $C^{(Q)} = \frac{1}{2} Z_T e^2 (\beta_1^2 Z_C + \beta_2^2 Z_h)$. V_{pc} is kept fixed as before such that the effective binding energy is now simply given by $\tilde{\varepsilon}_i = \varepsilon_i - \frac{Z_P e^2}{R_i}$.

By using the formalism of Ref.[9] we have shown that in the sudden approximation there is no interference between dipole and quadrupole contributions and that the quadrupole is negligible for impact parameters larger than about 30fm, in agreement with the recent results from Ref.[40]. An example for the case ^{8}B+^{208}Pb is shown in Fig.(4). We used an effective initial binding energy of -3MeV and an incident energy of 46.5 A.MeV. Both dipole and quadrupole contributions were calculated to second order according to [39].

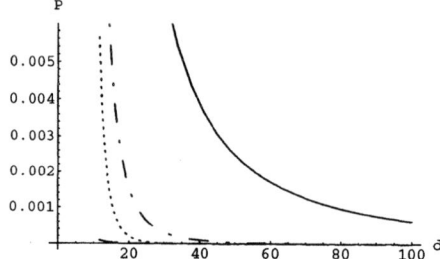

FIGURE 4. Breakup probabilities as a function of the impact parameter in fermi, for various components of the Coulomb potential: blue curves are for the dipole, full curve first order and dashed curve second order (hardly visible); red curves are for the quadrupole, dotdashed first order, short dashed second order.

CONCLUSIONS

In this paper we have tried to discuss the physical origins of the differences in the behavior in a reaction of a proton halo nucleus as compared to a neutron halo. We have shown that if the target is heavy, but also if the projectile is heavier, as in the case of ^{17}F vs. ^{8}B there is an effective barrier which makes the proton "effectively" bound by several MeV, so that some typical halo features might change in breakup reactions. In particular nuclear breakup and its stripping component could be larger than Coulomb breakup. This could explain the apparent discrepancy in the interpretation in terms of halo structure between data from different types of experiments. Also first order calculations are not completely unjustified. However the effect of the "effective parameters" introduced here has to be studied in more detail and results should be compared to full dynamical calculations. The proton-target Coulomb potential has not been considered dynamically in this discussion but only via the adiabatic approximation and it would be interesting to understand if and how a more accurate treatment would change the pattern discussed above. Therefore approaches of the type used in [27, 28] although not very accurate would give reasonable order of magnitude predictions for weakly bound protons interacting with a heavy target but not for interactions with light targets or in the case of neutron breakup.

It is known that Coulomb breakup on a heavy target can be useful to simulate the (p,γ) reactions of astrophysical interest. However, exclusive measurements need to be done to separate Coulomb from nuclear breakup. Measuring proton angular distributions as done in [6] for neutron would help disentangling the dominant reaction mechanism, but also separating the large core-target impact parameter contributions as done in [2, 7] is very useful.

REFERENCES

1. P. G. Hansen and B. M Sherrill, Nucl. Phys. **A 693** (2001) 133.
2. B. Davids et al., Phys. Rev. **C 63** (2001) 65806 and references therein.
3. J. H. Kelley et al., Phys. Rev. Lett. **77** (1996) 5020 and references therein.
4. M. Gai, Proc. 19th Winter Workshop on Nuclear Dynamics, nucl-ex/0303009.
5. G. Baur, C. A. Bertulani and H. Rebel, Nucl. Phys. **A 458** (1986) 188.
6. R. Anne et al., Nucl. Phys. **A 575** (1994) 125.
7. T. Nakamura, Phys. Lett. **331** (1994) 296, and Proc. of the 4th Italy-Japan Symposium on Perspectives in Heavy Ion Physics, edited by K. Yoshida, S. Kubono, I. Tanihata, C. Signorini, (World Scientific, 2003), p. 25, and to be published.
8. A. Bonaccorso and D. M. Brink, Phys.Rev. **C 57** (1998) R22.
9. J. Margueron, A. Bonaccorso and D. M. Brink, Nucl. Phys. **A 703** (2002) 105, Nucl. Phys. **A720** (2003) 337.
10. T. Motobayashi et al. Phys. Rev. Lett. **73** (1994) 2680.
11. T. Kikuci et al., Eur. Phys. J. **A3** (1998) 213.
12. N. Iwasa et al. Phys. Rev. Lett. **83** (1999) 2910.
13. K. E. Rehm et al., Phys. Rev. Lett. **81** (1998) 3341.
14. J. F. Liang et al., Phys. Lett. **491** (2000) 23.
15. B. Blank et al., Nucl. Phys. **A 624** (1997) 242.
16. D. Cortina-Gil et al., Nucl. Phys. **A** (2003) in press.
17. C. Borcea et al., Nucl. Phys. **A 616** (1997) 231c.
18. F. Negoita et al., Phys. Rev. **C 54** (1996) 1787.
19. F. M. Nunes and I. J. Thompson, Phys Rev. **C 59** (1999) 2652.
20. H. Esbensen and K. Hencken, Phys. Rev. **C 61** (2000) 054606.
21. H. Esbensen and G. F. Bertsch, Nucl. Phys. **A 706** (2002) 383.
22. H. Esbensen and G. F. Bertsch, Phys. Rev. **C 66** (2002) 044609.
23. J. Mortimer, I. J. Thompson, and J. A. Tostevin, Phys. Rev. **C 65** (2002) 064619.
24. F. Negoita et al., Phys.Rev. **C 59** (1999) 2082.
25. A. Bonaccorso and F. Carstoiu, Phys. Rev. **C 61** (2000) 034605.
26. T. Kobayashi, Proc. First Int. Conf. on Radioactive Beams (Berkley,1989), pag. 524.
27. C. A. Bertulani and P. Danielewicz, Nucl. Phys. **A 717** (2003) 199.
28. C. H. Dasso, S. M. Lenzi and A. Vitturi, Nucl. Phys. **A 639** (1998) 635.
29. C. Angulo et al., Nucl. Phys. **A 716** (2003) 211.
30. A. Bonaccorso, D. M. Brink, C. A. Bertulani, submitted to Phys. Rev. C.
31. A. Bonaccorso, Phys. Rev. **60** (1999) 054604 and references therein.
32. A. Bonaccorso, G. Piccolo and D. M. Brink, Nucl. Phys. **A 441** (1985) 555.
33. H. Hashim, D. Phil. Thesis, University of Oxford, 1986, unpublished.
34. L. Lo Monaco, D.Phil. Thesis, University of Oxford, 1985, unpublished.
35. A. Bonaccorso and D. M. Brink, Phys. Rev. **C 38** (1988) 1776.
36. R. A. Broglia, and A. Winther, *Heavy Ion Reactions*, Benjamin, Reading, Mass, 1981. pag. 338
37. J. Enders et al., Phys. Rev. **C 65** (2002) 034318.
38. A. Bonaccorso and D. M. Brink, Phys. Rev. **C 44** (1991) 1559.
39. C. Papineau, A. Bonaccorso, D. M. Brink, in preparation.
40. F. Schümann et al., Phys. Rev. Lett. **90** (2003) 232501.

The new Radioactive Ion Beam Facility at GSI

H. Weick

Gesellschaft für Schwerionenforschung, D-64291 Darmstadt, Germany

Abstract. The new radioactive ion beam facility is planned as part of the project of a new international accelerator facility. These new accelerators shall serve several user communities from different fields of physics, namely nuclear structure and nuclear astrophysics, hadron studies with antiprotons, investigations of compressed nuclear matter, compressed macroscopic matter in plasmas, atomic and applied physics. The existing GSI accelerators will be extended by two new synchrotrons SIS 100 and SIS 300 for high intensity and higher energy primary beams of protons and heavy ions. Radioactive ion beams will be produced and separated in-flight with a new fragment separator called Super-FRS providing fragments to three experimental areas: fast in flight beams, slowed down or stopped rare isotopes and radioactive ions stored in one of the new storage rings CR and NESR.

INTRODUCTION

The science that can be explored with the planned facility spans a wide range of research areas. The following list can only represent key features of the research programme. A more detailed discussion can be found in the conceptual design report [1].

- The investigation of rare isotopes will answer important questions about the nuclei far from stability. Astrophysical aspects like nucleosynthesis in the r and rp-process and testing fundamental symmetries using radioactive nuclides.
- The study of hadronic matter at the sub-nuclear level with beams of antiprotons. Key aspects are the confinement of quarks and the generation of hadron masses intimately linked to to existence (and breaking of chiral symmetry). Search for the existence of other forms of hadronic matter than baryons and mesons, like "glueballs" consisting only out of gluons or bound systems of quark-antiquark pairs and gluons (hybrids). The heavier quarks can be created in proton-antiproton annihilation. For example the spectroscopy of mesons built of charmed quark-antiquark pairs provides favorable conditions as interpretation based on perturbative Quantum Chromo Dynamics is possible. Hypernuclei including charm quarks offer a new degree of freedom on the chart of nuclides.
- The study of compressed, dense hadronic matter in nucleus-nucleus collisions at high energies. Here the emphasis is on an area of the temperature-density phase diagram of nuclear matter reachable with lower energies than at the RHIC (BNL) or LHC (CERN) program but of higher baryon density.
- The investigation of high energy density in bulk matter leading to the plasma state. The combination of intense heavy ion beams and the strong PHELIX laser [2] will facilitate novel beam-plasma interaction studies.

- The study of Quantum Electrodynamics in extremely strong electro-magnetic fields like in highly charged ions and intense laser beams. Applications in ion-matter interaction.

THE ACCELERATOR SIS100/300

The goals for the accelerator are to provide beams of heavy ion up to uranium at higher intensity and higher energy and intense proton beams for production of antiprotons. Storage rings will be used for a large part of the experimental program. Up to now the only efficient driver accelerator to be coupled to storage rings are synchrotrons. They also allow efficient acceleration to higher energies.

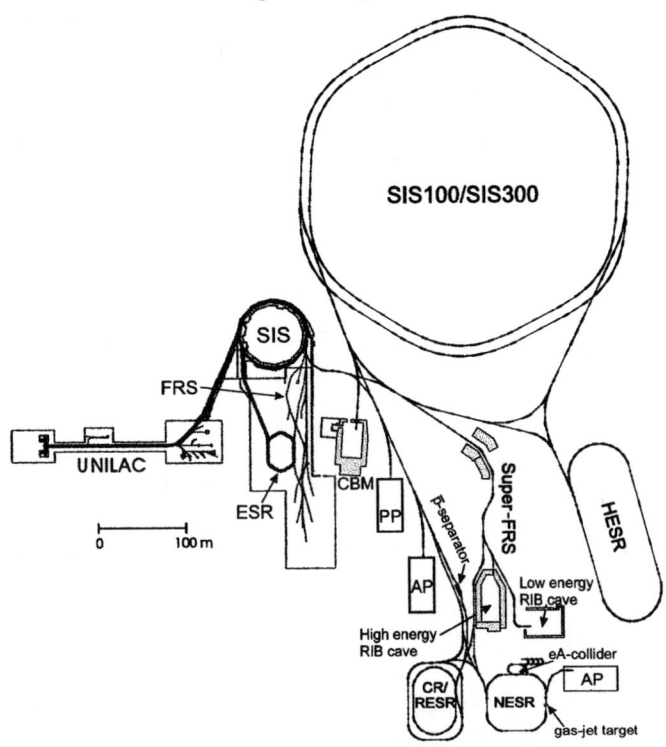

FIGURE 1. Layout of the existing (left) and the planned accelerator facility (right). Marked are the new experimental places for atomic physics (AP), compressed baryonic matter research (CBM), plasma physics (PP), the HESR for stored antiprotons, and the three branches of the Super-FRS leading to caves for high and low energy radioactive beams and to the storage rings CR/RESR, NESR.

The present SIS18 has basic limitations for primary beam intensities due slow ramping, space-charge effects and the fixed maximum magnetic rigidity ($B\rho_{max}$) of 18 Tm. These shortcomings can be reduced if the SIS18 is operated in a fast cycling mode of 4 Hz up to only 12 Tm [1]. Then a fast ramping larger synchrotron (SIS100) will take over and accelerate heavy ions in a lower charge state, for example $^{238}U^{28+}$, to avoid the space charge limit. Fast here means one complete cycle for $^{238}U^{28+}$ to 1000 MeV/u takes $1s$. The intensity thus will reach $10^{12}/s$ of uranium ions. The SIS100 will also be used to provide intense proton beams of $2x10^{13}$ per cycle at 29 MeV for production of antiprotons.

A second synchrotron (SIS300) can accelerate heavy ions even further up to $B\rho_{max} = 300$ Tm. This corresponds to 34 GeV/u for $^{238}U^{92+}$. An average intensity of about $1x10^9/s$ will be available for nuclear collision experiments. The SIS300 can also be used as a stretcher ring for experiments requiring a DC beam, like all nuclear physics experiments with targets not inside storage rings (external targets). Actually, the SIS100/300 accelerator consists of two synchrotron rings positioned in the same tunnel.

The total accelerator chain then works as follows: The linear accelerator UNILAC feeds the SIS12, from there the beam will be injected into SIS100 and may be further to SIS300. From there the beam goes to the experimental caves or storage rings. Fig.1 gives an overview of the planned accelerator complex.

The facility intended for users from many fields can be operated to a large extend truly in parallel [1]. For example heavy ion beams can be accelerated in SIS100 for production of rare isotopes while at the same time SIS300 can accelerate other heavy ions to higher energies in SIS300. As the cycle for higher energy is much longer only from time to time a pulse from SIS12 has to be taken for SIS300. Another scenario would be production of antiprotons with proton beams from SIS100. The antiprotons will be cooled and collected in the CR/RESR. Again this is a longer cycle and in between other beams can be accelerated in SIS100 and slowly extracted as a quasi DC beam via SIS300. Finally, plasma physics experiments typically need only one short intense pulse from time to time.

THE NEW FRAGMENT SEPARATOR SUPER-FRS

The technique of producing radioactive beams at rather high energies up to 1 GeV/u was already successful at the existing GSI fragment separator FRS [3]. Here it was also discovered that in-flight fission can be an advantageous source of neutron rich nuclei [4]. However, the much larger emittance of the fission fragment beam after the target compared to projectile fragmentation requires a separator with much increased momentum and angular acceptance to use them efficiently. Therefore, a new separator was designed with an aperture size twice the one of the FRS. The angular acceptance should be 40*mrad* times 20*mrad* with a momentum acceptance of $\pm 2.5\%$ [5]. This requires also increased quadrupole field strength which can only be achieved with superconducting coils. For this reason all magnets should be of super-ferric type [6] (superconducting coil with iron yoke) which stimulated the name "Super-FRS". Fig.2 compares the aperture and length of the existing and planned fragment separator.

The enlarged acceptance together with the increased primary beam intensities at the new facility will provide secondary beam intensities orders of magnitude higher than at the existing GSI facility. But besides the fragment intensities, the selectivity and sensitivity are crucial parameters that strongly influence the success of an experiment with

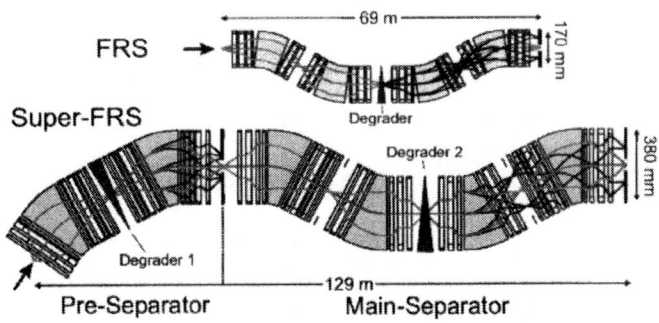

FIGURE 2. Comparison of the layout of the FRS and the Super-FRS. The ion-optical plot represents a projection onto the dispersive plane. The transverse scale is enlarged compared to the longitudinal scale for better visibility.

very rare nuclei. One prerequisite for a clean isotopic separation is that the fragments are fully ionized to avoid cross contamination from different ionic charge states. The maximum magnetic rigidity of the Super-FRS is 20 Tm which corresponds to a reference energy of 1500 MeV/u for ions of mass to charge ratio $m/q = 2.67$. This is higher than at any other in-flight facility in order to provide even for uranium beams mostly fully stripped ions [5].

The Super-FRS consists out of a pre- and a main-separator. One can imagine that with the increased intensity the rate of fragments hitting the first degrader is similar to the primary beam intensities at the FRS target today. Therefore, a large number of secondary fragments will be produced in the degrader. Two separation stages are necessary to reduce the background from these contaminants efficiently.

The proposed Super-FRS is a powerful in-flight facility which will provide spatially separated isotopic beams up to elements of the heaviest projectiles. The high secondary rates will allow experiments with nuclei so far not accessible like those important in the astrophysical r-process. After separation the radioactive ion beam can be delivered to three experimental areas.

THE EXPERIMENTAL AREAS

The "High-Energy Branch"

The ALADIN/LAND setup at the present GSI facility allows to study nuclear reactions in complete kinematics. For the new facility a much more versatile set up is foreseen, which was studied and proposed within the framework of a European collaboration (Reactions with Relativistic Radioactive Beams [R^3B]) [7]. Within this R^3B project, a new large-acceptance superconducting dipole magnet of high bending power was designed, along with advanced detectors for fragment identification and tracking with high-rate capability. In this dipole magnet the fragments coming from a reaction target in front are momentum analyzed. Neutrons can pass straight through and be detected in an upgraded LAND detector. Light charged particles are tracked by wire chambers. Protons are bend even more and will be analyzed in separate wire chambers. Each of these three arms will also use time-of-flight for identification. A new Cherenkov counter will be used to determine the fragment velocity. The target can be surrounded by γ-detectors.

The experiments foreseen in the high-energy branch cover a large variety of different reactions, such as elastic scattering, knockout reactions, electromagnetic and nuclear excitations, charge-exchange reactions, fission studies, in-beam γ-ray spectroscopy, or multifragmentation.

The "Low-Energy Branch"

The low-energy branch, designed for a maximum magnetic rigidity of 10 Tm, makes use of a high-resolution dispersive separator stage. In combination with a wedge-shaped energy degrader the energy spread of the fragment beam can be reduced considerably [8]. This will lead to a reduced range distribution and allow to couple the Super-FRS efficiently to a gas cell in which the ions will be stopped and extracted from as ions [9]. From there they will be transferred at keV energies to various experiments like ion or atom traps for precision spectroscopy. In principle, these are experiments like at a typical ISOL facility but with the advantage of a production and separation independent of the chemical properties.

Another possibility is not to stop the ions completely but to slow them down to Coulomb-barrier energies at which gamma- and particle-spectroscopy research can be done. These investigation methods are otherwise presently not available for many rare isotopes.

The "Ring Branch"

The combination of fragment separator and storage rings has proven to be an excellent tool [10] but new kind of experiments require a better coupling in terms of transmission as well as shorter cooling times after which the beam can be used for experiments. Both can be achieved with the concept of a high acceptance collector ring (CR), a separate ring for deceleration (RESR), if needed, and a new experimental storage ring (NESR). In the current design the CR is characterized by an acceptance of $\pm 1.75\%$ in momentum spread and 200 mm mrad in horizontal and vertical direction. Stochastic cooling at a fixed beam energy of 740 MeV/u will reduce this emittance in less than a second to about 0.05 % and 0.5 mm mrad [1].

For short-lived isotopes ($T_{1/2} \leq 1$s down to the μs range) experiments can be performed directly in the CR without cooling. For example, mass spectrometry using the CR in an isochronous mode will greatly expand our knowledge of ground state properties of short-lived nuclei close to the driplines [11].

The NESR is equipped with an electron cooler to achieve the highest phase-space density for fragment beams and with an internal gas jet target for nuclear reaction studies. Direct-reaction studies with stored fragment beams are difficult with the present ESR, but will be possible with the many orders of magnitude increased luminosity due to higher primary beam intensity and much increased acceptance. The advantage of internal targets is that the target itself is very thin and thus does not disturb the detection of reaction products. For example at an external target the region of very small momentum transfer to the target like ejectile in a scattering experiment is usually not accessible even though the cross section at the corresponding scattering angle is the highest. Due to the many revolutions in the ring at a typical frequency of MHz, the resulting luminosity is similar to an external target or can be even higher [12].

The most challenging part of the experimental setup at the NESR is the incorporation of a small electron ring, which allows scattering of electrons on exotic nuclei [1, 13] (eA-Collider). Such a device is also being investigated for the RIKEN radioactive beam factory [14].

One of the main physics questions to be addressed in light-ion scattering experiments on exotic nuclei concern matter density distributions, e.g. the question of neutron skins of neutron-rich nuclei. The nuclear density distributions can be studied in detail by elastic proton scattering. In addition, elastic electron scattering in the eA-collider yields the charge distribution. Inelastic excitations, induced either by light ions or by electrons, give access to a wide field of magnetic and electric giant resonance studies.

ACKNOWLEDGMENTS

The work presented here is based on the common effort of many people within GSI and its international collaborators as listed in ref.[1]. I am indebted very much to all of them in particular to the Super-FRS collaboration as in ref.[5].

REFERENCES

1. An International Accelerator Facility for Beams of Ions and Antiprotons, Conceptual Design Report, GSI (2001). http://www-new.gsi.de/zukunftsprojekt
2. PHELIX, Petawatt High-Energy laser for Heavy-Ion eXperiments, GSI-Report 98-10 (1998). http://www.gsi.de/~phelix
3. H. Geissel et al., Nucl. Instr. and Meth. B 70 (1992) 247.
4. C. Engelmann et al., Z. Phys. A 352 (1995) 351.
5. H. Geissel et al., Nucl. Instr. and Meth. B 204 (2003) 71.
6. A.F. Zeller et al., Adv. in Cryogenic Engineering, Vol.43 (1998) 245.
7. T. Aumann et al., GSI Scientific Report 2001 (2002) 212. (http://www.gsi.de/annrep2001/); http://www-land.gsi.de/r3b/
8. H. Weick et al., Nucl. Instr. and Meth. B164/165 (2000) 168.
9. G. Savard et al., Nucl. Instr. and Meth. B204 (2003) 582.
10. H. Geissel, et al., Nucl. Phys. A 685 (2001) 115c.
11. M. Hausmann, et al., Nucl. Instr. and Meth. A 446 (2000) 569.
12. P. Egelhof et al., Physica Scripta T104 (2003) 104.
13. G. Münzenberg, et al., Nucl. Phys. A626 (1997) 249c.
14. RIKEN Radioactive Isotope Factory, T. Katayama et al., Design Report November 2000. http://www.rarf.riken.go.jp/ribf/index.html

The SPIRAL2 project at GANIL

A.C.C. Villari for the SPIRAL2 group

GANIL
B.P. 55027 Caen Cedex 5 France

Abstract. Based on the "LINAG Phase 1" conceptual design, a two years detailed study on a ISOL-type facility for the production of high intensity exotic beams, named SPIRAL2, has been launched. The radioactive isotope beams are produced via the fission process, with the aim of 10^{13} fissions/s at least, induced in a UC_x target either by fast neutrons from a C converter or by direct bombardment of fissile material. Fusion-evaporation residues, using heavy ions beams in different targets can also be produced in this facility. The driver, with an acceleration potential of 40 MV will accelerate deuterons (5 mA) and q/A =3 ions (1 mA) Even heavier ions will be possible in a later stage. The driver consists in high-performance ECR sources, an RFQ cavity and independent phase superconducting resonators. As it is a linear accelerator, further upgrade will be possible in future.

I. INTRODUCTION

The systematic and very successful use of high-energy fragmentation at GANIL, (the first operational high intensity heavy ion accelerator in the 50-100 MeV/nucleon domain) for exploring the structure of nuclei far from stability, triggered the question how to proceed even further in this domain. The study of nuclei far from stability has become one of the major activities at GANIL, and is one of its domains of excellence. It turns out from the principle of production and separation using a spectrograph – the so-called in-flight method, which corresponds to the use of SISSI or LISE at GANIL – that the optimum efficiency of the process is reached when the radioactive beam has a velocity similar to that of the primary beam. This production process, however, does imply losses in intensity and/or quality of the secondary beam, which become increasingly important as the beam is slowed down. The ISOL (Isotopic Separation On-Line) method, used at SPIRAL [1] since November 2001, provides the production and separation of radioactive ion beams, with subsequent acceleration by a K=265 cyclotron CIME [2] between 1.7A and 25A MeV, thus allowing the study of nuclear reactions around the Coulomb barrier with radioactive ion beams. While SPIRAL is well suited for the production of light masses (A<80), the new project SPIRAL2 [3] will be devoted to the production of fission-like and fusion-evaporation radioactive ion beams, with subsequent acceleration by the CIME cyclotron. An important part of this project is the production of radioactive beams via fission induced by neutrons. The neutrons will be produced by the break-up of deuterons in a carbon wheel converter. The intensity of the deuteron beam (40 MeV and 5 mA) allows to reach 10^{13} fissions per second at least using a standard UC_x target of density 2.3 g/cm^3. A possible encrease of this yield can be obtained by changing the size or the density of the UC$_x$ of the target. If one considers the high density UC$_2$ with 11 g/cm^3, the yield will be enhanced by the same factor, i.e. a yield of 5 10^{13} fissions per second can be reached. In this contribution, other uses of the driver of SPIRAL2, e.g. for production of neutron deficient nuclei via fusion-evaporation reaction or the use of in-flight production for the study of superheavies are not discussed.

The SPIRAL2 detailed design study will lasts up to November 2004 with IN2P3/CNRS, DSM/CEA and the Lower Normandy Region support.

In a second phase, the full accelerator LINAG (Linear Accelerator at GANIL) could allow to accelerate light and heavy ions to energies of around 100A MeV at intensities as high as 1 mA, corresponding to an enhancement of the radioactive beam intensities of more than two orders of magnitude in comparison with the present GANIL accelerators.

II. PRODUCTION OF FISSION FRAGMENT RADIOACTIVE BEAMS

One of the techniques proposed for SPIRAL2 has been already discussed in the EU-RTD report [4], consisting in the use of energetic neutrons to induce fission of ^{238}U. The neutrons are generated by the break-up of deuterons in a thick target, the so-called converter, of sufficient thickness to prevent charged particles from escaping. The energetic forward-going neutrons impinge on a thick production target of fissionable material, i.e. Uranium carbide UC$_x$. The resulting fission products accumulated in the target diffuse to the surface from which they evaporate, are ionised, mass-selected, eventually charge breeded and finally post-accelerated. This method has several advantages and was suggested by Gerald Nolen [5] in the framework of RIA. The material of the highly activated converter can be chosen to withstand the power of the beam without constraint concerning the diffusion of radioactive atoms. Moreover, the temperature of the converter does not affect the neutron flux. As projectiles, neutrons do not contribute to the heating of the target material directly neither the entrance window, which can be very thick and do not present any special security issue. Neutrons then bombard the target loosing energy only in useful nuclear interactions and having a high penetrating power, which allows very thick targets to be used.

The choice of the Deuteron bombarding energy – 40MeV –has been made taken into consideration basically four main factors:

I) the production rate of neutrons at forward angles as a function of energy,
II) the angular distribution of the neutrons,
III) the excitation energy of uranium, which defines the fission fragment distribution,
IV) the cost of the project.

The strong forward peaking of the yield of high-energy neutrons from the deuteron break-up at 40 MeV favours a compact geometry, consisting of a converter to produce the neutrons followed by a second target containing the fissionable material. The energy distribution of the neutrons produced in the deuteron break-up (that determines the excitation energy of Uranium) is centred at about 40% of the energy of the incident deuterons and has at 0° a width between one-third and half of the energy of the incident deuterons. This energy is optimal to produce neutron rich fission fragments.

II.A. The rotating target/converter

During the studies of SPIRAL2 EU RTD, simulations of neutron spectra for several beam-converter configurations have been performed with the LAHET high-energy transport code combined with Monte Carlo N-Particle MCNP code for low-energy transport were performed. Deuteron Coulomb dissociation has been added to the more standard processes, the forward peaked break-up and direct reactions and the rather isotropic evaporation of low-energy neutrons [6]. A comparison with data, in the energy regions where it was available, has shown an agreement of around 30%. Three different converters were studied: lithium, beryllium and carbon.

The neutron yield is only one of the factors to be taken in consideration for the choice of the converter material. Other aspects in the evaluation are:

i) Thermal properties that allow a compact geometry of the converter and the production target
ii) Toxicity and material properties of the converter;
iii) Production of long-lived radioactive nuclides;
iv) Cost of operation.

The conclusions we can infer, regarding these aspects, are summarised below:
- A Beryllium converter produces the largest amount of neutrons, however, its low melting point (1278° C), does not allow high-intensity deuteron beams, nor its location very close to the hot target.
- Liquid lithium is a more robust converter with respect to deposited beam power than Beryllium or Carbon. However, the flow of hot liquid Lithium containing some amount of radioactive products requires special care of design especially due to safety considerations. A converter designed along the lines originally described by Grand and Goland [7] is probably not to be considered in the context of SPIRAL2, but could be of interest for a next generation facility, e.g. EURISOL since it can stand extremely high beam power.
- The above mentioned properties clearly favour Carbon as converter material. It is non-toxic, easy to handle and has a high melting point of 3632 °C. These excellent properties allow high beam intensities with a rotating wheel cooled mainly by thermal radiation.

For SPIRAL2 the main parameters of the rotating Carbon wheel have been obtained by simulations using the code SYSTUS [8]. For the simulations, an infinite rotation velocity was taken, in a first approach. Once the main characteristics chosen, the temperature variation with respect of the beam impact was calculated for a real angular velocity of 1000 RPM. The thermal shock for which graphite achieves the ultimate strength (brake-down condition) is for about 50°C in our conditions, therefore a maximum temperature variation of 20-30°C has been considered.

The evaporation ratio of the carbon is dependent of the graphite saturated vapour pressure. Experiments performed at GANIL and IPN-Orsay [9] for a specific carbon from POCO and Carbon Lorraine industries show that the evaporation ratio of carbon is in agreement with the values found in literature. The evaporation rate for evaporating 1mm thickness of the carbon wheel in a period of 2000 hours is 2.6×10^{-7} Kg/m^2s. It corresponds to a temperature of 2085°C. This consideration fixes the sizes of the wheel and the beam spot on the carbon converter; i.e. 350 mm of radius for a beam spot of 10 x 35 mm.

A similar study with equal results [LOG00] has been made in the framework of the SPES project, Legnaro, Italy. The difference between both projects is that, in the latter case, the beam considered were protons of 10 MeV, with a total beam power of 100kW.

II.B. The target and ion source production system

The target and ion source production system is placed just behind the rotating carbon converter. With 1×10^{13} fission per second, the total power produced in the UC_x target is 310W. As mentioned above, the production target perceives no influence on the primary beam. The target temperature is completely controlled by an independent heating system, with a power of about 5 kW.

The production target:

Two possibilities have been considered in this study for the fissile targets. The first one of UC_x with 2.3 g/cm^3, using the technology developed and used for many years at CERN-ISOLDE [10]. Oxide of uranium is mixed with carbon powder in a small container, pressed and heated at 2000°C during approximately one day. The chips produced have generally around 1 mm thickness and a diameter of 20 mm. Larger diameter do not seem to be a problem, but smaller thickness could be very difficult to produce. We assumed in our calculations of production yields a thickness of 1 mm. Any development in order to reduce this value would be welcome.

The second possibility is the use of high-density UC_2 target (11 g/cm^2) already developed at Gatchina – Russia. The high density UC_2 allows either to reduce considerably the size of the target or to increase by a factor of around 5 the yields for a constant geometry. Preliminary results show that the diffusion properties of high-density UC_2 are similar to the low-density. The following yield estimations were done for both possibilities.

The geometry adopted in the simulations is not the optimum. The best would be to have a UC_x targets of conical shape, with an angle of approximately of 30-40°, as proposed in ref. [11]. In order to simplify the simulations, a simple cylindrical geometry has been adopted in all cases. Moreover, a reasonable size of the UC_x target has been adopted. Therefore, the in-target yields can be considered as **lower limits** for all cases.

In the simulations the beam profile was 2 cm of diameer. The UCx fission target of 6 cm diameter is placed at 2 cm behind the converter. It is constituted by slices of 1 mm thickness, separated by 0.5 mm, distributed over a length of 8.5 cm, corresponding to 360 g with low-density UC_x or 1.8 kg of uranium for the high-density material. This target could be made of self-supported disks (if mechanically possible) or with a combination of several smaller targets in a suitable geometry. A prototype of both low-density [12] (1.6 cm diameter) and high-density (1 cm diameter) already exists and has been tested at the PARRNe2 and at the Gatchina on-line mass separators.

Ion sources:

The ionisation source will be installed in a module as close as possible to the target. The chemical features of the selected radioactive element will define the type of the ion source regarding its efficiency. The main methods considered are surface ionisation for alkali elements, an electron cyclotron resonance ion source for noble gases or for volatile mono-atomic or molecular elements, and a laser ion source for refractory elements. A new kind of electron beam ion source, developed at Gatchina is another attractive possibility. Correspondingly existing sources for radioactive ion production are respectively described in the references [13, 14, 15, 16].

It will be possible to install any one of the proposed sources in the production module, the ECRIS being the largest one. In all cases, the lifetime to be considered for the production system is 3 months. The replacement of the whole production system will be performed in a hot cell, specially designed for such an operation.

Charge breeding.

The charge breeding (1+/n+ transformation) will increase the charge-state of the singly charged incoming ion to a charge state compatible with the acceleration by the CIME cyclotron.

An ECR charge booster has been developed by the SSI group at ISN Grenoble (France) [17] based on the use of a Phoenix ion source at 14 GHz. The present results with this source are summarised in the table 1:

	Charge Breeding Efficiency on the most abundant charge state	Overall Efficiency
Noble gas	10%	50 to 70%
Condensable elements	6%	45%

TABLE 1: Order of magnitude of charge breeding efficiency with stable elements.

Other solutions could also be considered, like the use of super-conducting ion sources (SERSE or GYROSERSE) in the future.

II.C. Production rates

Yield calculations of fission fragments with the LAHET+MCNP+CINDER code for ~5 mA deuteron beam of 40 MeV energy in a carbon converter, followed by a UC_x target has been performed.

Two UC_x densities were considered:

- Case 1: An UC_x target with a high density $\rho(^{238}U) = 11 g/cm^3$ and 1 atom of U for 2 atoms of C as developed by V. Panteleev, Gatchina.
- Case 2: An UC_x target with a low density $\rho(^{238}U) = 2.3 g/cm^3$ and 1 atom of U for 9 atoms of C as used in the first PARRNe2 experiment.

The following geometry was used:
The deuteron beam of 2 cm diameter hits a carbon converter of 1.8 g/cm^3 density with an effective length of 0.7 cm, in order to stop deuterons and protons from stripping reactions. The UC_x fissile target of 6 cm diameter is placed at 2 cm behind the converter. The target is composed of 56 slices of 1 mm thickness spaced by 0.5 mm, distributed over a length of 8.5 cm. It contains 360g of ^{238}U in the case of the low density $r(^{238}U) = 2.3 g/cm^3$ and 1800g for higher density $r(^{238}U) = 11 g/cm^3$. Effectively and for simplicity, the target was assimilated to one cylinder of 8.5cm length with density reduced by a factor 2/3 in the LAHET code. This gives the correct solid angle for neutron impact and the correct ^{238}U quantity.

For such a target, 10^{13} fission/s could be obtained with a deuteron beam of 3.8mA for the low-density UC_x target and of 1mA for the higher density, as resumed in table 2. The gain in the number of fission is not exactly proportional to the density ratio. This effect is probably due to high-energy neutrons. These neutrons are produced in the converter and are probably mostly absorbed in the higher U density target. A more important proportion of lower energy neutrons are directly produced inside the UC_x target from fission of U, corresponding to a neutron source inside the target, and are then less sensitive to the effective length of the target. For short living isotopes, a shorter high-density/reduced-volume UC_x target equivalent to 25 mm length will produce 10^{13} fission/s and will ensure faster diffusion out of the target.

Target density	2.3 g/cm³	11 g/cm³
Intensity for 10^{13} fission/s	3780 µA	947 µA
Fission per 5 mA	1.32×10^{13}	5.28×10^{13}

TABLE 2: Required primary beam intensity in order to obtain 10^{13} fission/s and total number of fission for 5 mA deuterons for 2 different densities.

II.D. Release of the products:

Theoretically, the diffusion, effusion and ionisation processes are well known phenomena, provided the particular properties of the selected element are known. However, despite of a large experimental and theoretical effort, the behaviour of diffusion and effusion of ions implanted into some materials, including the details of their thermal transport or their trapping in the temperature range relevant to ISOL system, remain unknown. In particular, the Arrhenius diffusion coefficients are measured for numerous elements mainly in W, Ta, Re or C-matrix. To our knowledge, these coefficients are not known for different tracers in uranium carbide matrices of different densities. A European RTD project "TARGISOL" N° HPRI-2001-50063 has been proposed and accepted recently in order to progress in this critical field.

The expected radioactive beam intensities (after diffusion, effusion, ionisation and acceleration) are shown in the plots of Fig. 1 for some elements: Zn, Kr, Sr, Sn Sb and Xe. The in-target production yields are those calculated using the 11 g/cm^3 UC$_2$ target in the geometry described in section 2.3. The Arrhenius coefficients used in this calculation were supposed to be the same as for C and Ta, both tabulated in the literature. The assumed 1^+ and $1^+/N^+$ ionisation efficiencies are adopted as 90% (1^+) and 12% ($1^+/N^+$) for Kr and Xe, 30% (1^+) and 4% ($1^+/N^+$) for Zn, Sr, Sn, I and Cd. The assumed acceleration efficiency in the CIME cyclotron is 50%.

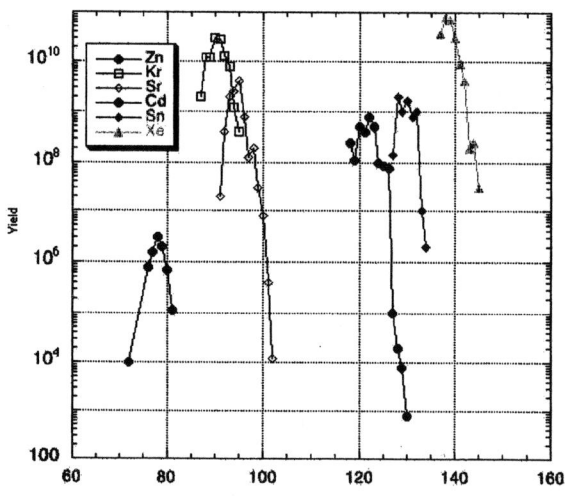

FIGURE 1: Yield after acceleration. See text.

II.E. Radio-protection and target handling

The dose rate just after the stop of the primary beam is estimated to be 32 Sv/h at 1 m from the target, after an irradiation time of 3 months. One month later, the radiation rate is still 34mSv/h. This high level of radiation does not permit manual intervention on the target itself. Therefore, automatic handling is necessary. The high level of radiation also imposes strong protection of critical parts of the production system, like insulators, o-rings, etc.

The plug solution:

The target and the ion source will be placed in a parallelepiped module (called plug) which is surrounded by at least 2 m of concrete and iron shielding. The same principle has been applied on the ISAC facility at TRIUMF (Vancouver, Canada), for 100 µA of 500 MeV protons. The 2 m thickness of concrete reduces the dose rate on the top of the plug down to 7.5 µSv/h when the beam is stopped, allowing people to come and work on the equipment located at this place.
After three month of irradiation, the different connections of the plug (electrical connections, primary pumping, water connections, etc.) can be manually removed. The plug is then remotely removed from the production cave and is evacuated to a storage cell. After a delay and depending on the state of the target ion source system, the plug can be re-used or moved to a hot cell where the elements at the bottom of the module can be replaced with master/slave manipulators.

III. THE DRIVER ACCELERATOR

The proposed LINAG1 driver for the SPIRAL2 project has the capability to accelerate a 5 mA d^+ beam up to 40 MeV; nevertheless, the different parameters are optimised for q/A=1/3 ions up to 14.5A MeV in order to preserve a long term evolution towards an heavy ion driver. It is a continuous wave (CW) mode machine, getting a maximum efficiency in the intensity transmission for heavy ions. It consists in an injector (ECR source + radio-frequency quadrupole), which accelerates the beam up to 0.75 A.MeV, followed by a superconducting linear accelerator based on independently phased resonators.

III.A. THE RFQ injector

The RFQ must operate in a CW mode. Its frequency has been chosen equal to 87.5 MHz, sub-harmonic frequency of 350 MHz. This quite low value has been determined for the following reasons:
- the RF power density is quite low at this frequency, and allows a solution based on a formed metal technology, leading to a cheap mechanical solution.
- at lower frequency, the inter-vane distance is larger, and allows a higher margin for the mechanical tolerances.

The RFQ output energy, 0.75 A.MeV, has been preliminarily determined by the fact that the first cavities of the SC linac must allow a possible evolution of the machine for q/A= 1/5 or 1/6 ions, which means that their beta value has to remain quite low ($\approx$ 0.04). This value should already be optimised during the detailed study phase.

Table 3 presents a summary of the main design parameters. In particular, the maximum peak field value is kept to a conservative level, lower than LEDA and Chalk River RFQs, which also work in a CW mode.

Error simulations have been performed, considering mechanical tolerances of +/- 0.1 mm on the vanes machining and +/- 0.2 mm for misalignments. The results confirm that the deuteron beam transmission remains very close to 100%. This gives a quite comfortable situation from the safety point of view: losses of 3% have been considered for the estimation of the protections.

Parameters	Values
Length	5 m
Aperture	8 - 10 mm
Modulation (m)	1 - 2
Frequency	87,5 MHz
Voltage	100 -113 kV
Transmission (q/A = 1/2 and 1/3)	99.9 %

TABLE 3: Main design parameters of RFQ-LINAG

III.B. The superconducting linear accelerator

The linac must have the capability to accelerate D^+ and q/A=1/3 ions with the maximum energy gain, as well as to extend its performances to heavier ions in the future. A linac based on independent phase superconducting resonators is thus proposed. The Linac design requires accelerating voltages of the order of one MV/cavity and two beta values, around 0.07 and 0.14, at sub-harmonic frequencies of 350 MHz (availability of power sources). The starting frequency is 87.5 MHz, not too high for the lowest beta cavity and not too low for the RFQ. Low beta super-conducting (SC) cavities in the beta range 0.04 to 0.2 are typically quarter wave resonators (QWR), operated at 4.2 °K as the frequency is less than 500 MHz.

The beams dynamics calculations have been performed with the codes TraceWin, PARTRAN and LIONS. Emittances behaviour in the linac is shown in Figure 2. All these codes are able to use 3D field maps. Meanwhile, it is important to notice that the steering effect, induced by QWRs, is not included in the simulation shown by the Figure 2. This problem is studied at present time using maps computed by the code SOPRANO. The first studies shows that vertical displacements are sufficient to reduce the emittance growth induced by the dependent phase deflection for the first family. For the second family, HWR may be used as a fall back solution if steering of 176 MHz QWR is too important.

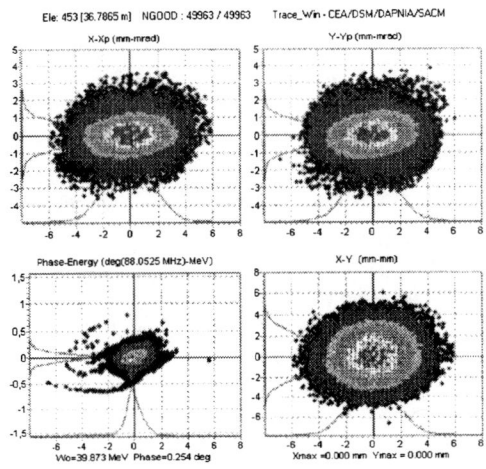

FIGURE 2: Deuterons beam in phase space at linac exit.

IV. POST ACCELERATION

The recently commissioned cyclotron CIME will perform the post acceleration in the SPIRAL2 project. It allows acceleration of heavy ions in the energy range of 1.7A MeV and 25A MeV, depending on the q/A. For fission fragments and considering presently performances of the charge booster, optimal energies would be of the order of 8A MeV.

V. SUMMARY

In the present report we have described the technical aspects of the SPIRAL2 project and commented more precisely the aim of adding medium-mass nuclei to those available in the present GANIL. Fission induced by light particles (e, p, d, etc.) was proposed to produce the radioactive ions, **with an aim of 10^{13} fissions/s at least**.

We have shown here that the SPIRAL2 project can reach even higher fission rates using proven technologies. Using a C-converter and a 5 mA deuteron beam, neutron-induced fission will be $1\ 10^{13}$ fissions/s using standard-density UC_x, and $5\ 10^{13}$ fission/s for high-density UC_2. For both cases, a very small volume (240-cm^3) ion source was selected, in order to have relatively fast diffusion-effusion times for short-lived nuclei. In principle, larger volumes could result in even higher fission rates, up to $2\ 10^{14}$ fissions/s, for which the heat produced by fission in the ion source reaches 6 kW, the present limit for the SPIRAL targets.

The linear accelerator as a driver in this project belongs to the technology of high intensity accelerators, which are of strong current interest for various domains and are a domain of rapid technological development.

As a consequence of the high production rates, the radioprotection constraints become a major factor in the project. This implies a change of technology compared with SPIRAL, with higher costs for the target/ion-source and associated infrastructure. The technology of target "plugs", as used at TRIUMF-ISAC, was chosen. It offers the guarantee of safe handling of the high levels of activity produced.

The SPIRAL2 project is part of a multi-beam policy of GANIL. Another aspect is its possible synergies with EURISOL. We note that the most promising possibility examined for EURISOL for both driver and post-accelerator is a superconducting linac, precisely the same technology for the linac as proposed for the SPIRAL2 driver. These two machines could in fact be one and the same, by adding an appropriate RFQ fo higher M/Q ratios. The proposed (final) LINAG driver should be able to accelerate ions up to mass 100 to energy of 100A MeV.

Even though the EURISOL proposal considers MW beam power, the experience with the 200 kW beams proposed for SPIRAL2 would be very relevant. A demonstrated competence in these areas would be a big advantage to any laboratory proposing to host such a facility.

REFERENCES:

[1] A.C.C. Villari, the SPIRAL group, Nucl. Instr. Meth. B204 (2003) 31.
[2] E. Baron, 14th Int. Conf. Cycl. Applic., Cape Town, South Africa (1995), Word Scientific.
[3] LINAG Phase I, a technical report, version 1.3, GANIL, June 27, 2002, GANIL R 02 08.
[4] M.G. Saint Laurent et al., SPIRAL phase II European RTD report, GANIL R 01-03 2001.
[5] J. Nolen, "A target concept for intense radioactive beams in the 132Sn region", Proc. Third Intern. Conf. On Radioactive Nuclear beams, ed. J. Morrissey, East lansing, Mi, May 24-27, 1993.
[6] D. Ridikas thesis: Optimisation of beam and target combination for hybrid reactor systems and radioactive ion beam production by fission, GANIL T99-04, 1999.
[7] P. Grand and A.N. Goland, Nucl. Instr. Meth. 145 (1977) 49.
[8] SYSTUS code, www.esi-group.com/Products/Systus/index_html
[9] J.C. Putaux et al., Nucl.Instr. Meth., B126 (1997) 113.
[10] P.V. Logatchev, L.B. Tecchio et al.; Graphite neutron target for exotic beams, SPES internal report, Legnaro 2000
[11] D. Ridikas, W. Mittig and A.C.C. Villari, Nucl. Phys. A701 (2002) 343c.
[12] Roussière, B. et al., Release properties of UCx and molten U targets preprint IPNO-DR-2002-002. - 2002.
[13] J-L.Biarotte, IPNO, Private Communication
[14] J.Lesrel, IPNO, Internal Note: Etude préliminaire des systèmes RF de LINAG 1 (2003).
[15] L.Dalesio, EPICS: Recent applications and future directions, Proceedings of the 2001 PAC, Chicago, 2001
[17] N. Chauvin, thesis, Université Joseph Fourier Grenoble (2000)

Perspective of the RIKEN Radioisotope Beam Factory Project

T. Motobayashi

RIKEN, Hirosawa, Wako, Saitama 351-0198, Japan

Abstract. By coupling the Ring Cyclotron and RIPS fragment separator, RIKEN has performed various studies with fast radio-isotope (RI) beams produced by heavy-ion projectile fragmentation. Various studies on properties of nuclei far from the stability line have been performed with the RI beams. To provide more extended research opportunities, a project to build a new accelerator complex called "RI Beam Factory (RIBF)" has been started. RIBF in its first phase is already under construction, and the first RI beams should be supplied during the year of 2006.

INTRODUCTION

Since 1990, RIKEN (the Institute of Physical and Chemical Research) has operated an accelerator complex which provided intermediate-energy heavy-ion beams. Light RI (radioisotope) beams have been produced by the projectile fragmentation reactions at energies of around 100 MeV/nucleon accelerated by a ring cyclotron called RRC, the major accelerator of the facility, with $K=540$ MeV. The fragment separator RIPS [1] has a large angular- and momentum-acceptance together with a high bending power, which enables to provide intense RI beams. By nuclear reaction studies in reversed kinematics at several tens MeV/nucleon energies, interesting features have been found for structure in neutron-rich nuclei: neutron halo, magicity loss in the $N=8$ and $N=20$ regions, new magic numbers, decoupling of neutron and proton motion, and so on. Nuclear astrophysics studies have also been made with direct and indirect methods for nuclear burning processes involving unstable nuclei.

Based on these progress, RIKEN has planned to extend the research with RI beams by building a new experimental facility called "RI Beam Factory (RIBF)". The RIBF project in its first phase includes three ring cyclotrons in cascade to accelerate intense heavy-ion beams up to uranium to 350 MeV/nucleon, and RI beams with high intensities and wider range of nuclide will be produced via the projectile fragmentation or in-flight fission of uranium ions. Its commission is expected to be in the year 2006.

STUDIES WITH FAST RI BEAMS

New isotope search

One of the most basic and straight-forward research is search for new isotopes. Recently, an 18-GHz ECR ion source (ECRIS-18) [2] and a frequency variable RFQ (radio frequency quadrupole) linac [3] have been newly installed. For better matching between the RILAC and RRC, a booster linac system called CSM (Charge State Multiplier) has also been built. These developments were made to provide intense beam for RIBF accelerators, However they can open new opportunities of studies in the present facility. For example, an increase of energy and intensity of ^{48}Ca beams enabled finding of the very neutron-rich isotopes, ^{34}Ne, ^{37}Na and ^{43}Si [4]. These nuclei were identified in RI beams separated by RIPS.

Reactions with fast RI beams

Experiments with fragmentation-based RI beams have several characteristics. Most of the time, measurements are made by so called reversed kinematics, where the target and beam respectively serves the probe and the nucleus to be studied. Difficulty is in its poor quality of the RI beam based on projectile fragmentation. The energy resolution of the separated beam is poor (a few %) and the intensity is low. The latter condition requires a thick target for secondary reactions. Therefore the missing-mass measurement is limited in use for resolving nuclear levels in the final state. In an alternative method, deexcitation γ rays from excited levels are measured in coincidence with the outgoing particles to select the reaction channel. In this scheme, the final state resolution is independent of the beam energy spread, though the Doppler shift of measured γ-rays from fast ejectiles should be corrected for. To overcome the low intensity, reactions with large cross sections should be chosen. There have been various studies of inelastic scattering including the Coulomb excitation, charge exchange reaction, and projectile fragmentation, which have no mass-transfer in the reaction process and hence have good kinematical matching even at high incident energies. Recently transfer reactions are also studied at relatively low energies. The reaction channel is identified by particle identification of the ejectiles. To accommodate these condition, detection systems with high efficiency have been developed. For example, Fig. 1 shows the newly developed DALI2 (Detector Array for Low Intensity radiation) detector with about 160 NaI(Tl) scintillators, which has been used for various γ-ray measurements. Its high granularity enables one to measure the γ-ray emission angle and hence to perform Doppler correction.

FIGURE 1. Schematic view of the γ-ray detector DALI2. The size of each NaI(Tl) crystal is $4 \times 8 \times 16$ cm^3.

Coulomb dissociation of astrophysical interest

The Coulomb dissociation method is useful in studying radiative capture process of astrophysical interest, especially for the cases involving unstable nuclei. The residual nucleus A of the reaction of interest, B(x,γ)A, bombards a high-Z target and is Coulomb excited to its unbound state that decays to the B+x channel. The cross sections of this process can be related since the dissociation process is regarded as absorption of a virtual photon, $i.e.$ A(γ,x)B, the inverse of the radiative capture of interest [5, 6, 7]. In addition to the advantage of possibility to use thick targets, the Coulomb dissociation method enhances the original capture cross section by a large factor due to the large virtual-photon number and the large phase space for the final channel.

In Coulomb dissociation experiments, the relative energy of the two fragments B and x, which corresponds to the center-of-mass energy of the B(x,γ)A capture reaction, is obtained from their invariant mass extracted from the velocities of B and x and their relative angle. The relative energy resolution depends on the accuracy of energy and angle determination in detecting outgoing particles, whereas it is independent of the energy and angular spreads of the incident beam. Figure 2 shows the nuclear chart relevant to astrophysical hydrogen burning. We have performed several Coulomb dissociation experiments to determine the rates of (p,γ) reactions involving unstable nuclei: ^{7}Be(p,γ)^{8}B for solar neutrino production [8, 9, 10]; ^{12}C(p,γ)^{13}N, ^{13}N(p,γ)^{14}O [11] and ^{14}N(p,γ)^{15}O [12] for regular and hot CNO cycles; ^{8}B(p,γ)^{9}C, ^{11}C(p,γ)^{12}N and ^{12}N(p,γ)^{13}O for hot pp mode.

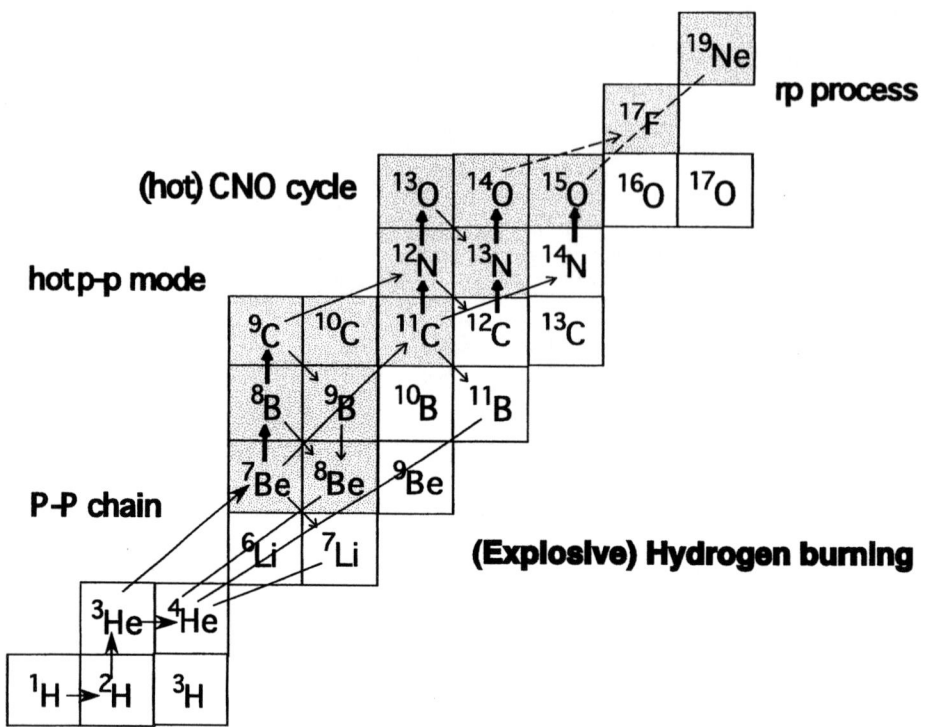

FIGURE 2. Reaction networks for astrophysical hydrogen burning. The thick arrows indicates the (p,γ) reactions studied by the Coulomb dissociation method at RIKEN.

Spectroscopy of neutron rich nuclei

Disappearance of the $N=20$ shell closure in ^{32}Mg was confirmed by a Coulomb excitation experiment at RIKEN [13], which measured a large a reduced transition probability $B(E2;0^+ \to 2^+)$ extracted from the yield of γ-ray measured in coincidence with ^{32}Mg produced in the ^{32}Mg+^{208}Pb inelastic scattering. This new method of Coulomb excitation now becomes one of the standard techniques of spectroscopy for low-lying states in unstable nuclei, and many experiments have been made [14].

Behavior of nuclear structures around $N=20$ has been studied with various reactions including the Coulomb excitation. The 4^+ state in ^{34}Mg could be populated by projectile fragmentation of the unstable ^{36}Si beam [15]. Its location was determined together with that of 2^+ by this newly developed method. The result clearly indicates that a well developed rotational band is formed in ^{34}Mg. This picture was supported by the Coulomb excitation measurement on the same nucleus [16]. The nucleus ^{30}Ne is an $N=20$ nuclei more proton-deficient than ^{32}Mg. The location of the first 2^+ state has been determined by an inelastic proton scattering experiment with a liquid hydrogen target [17]. Figure 3 shows the γ ray spectra obtained with

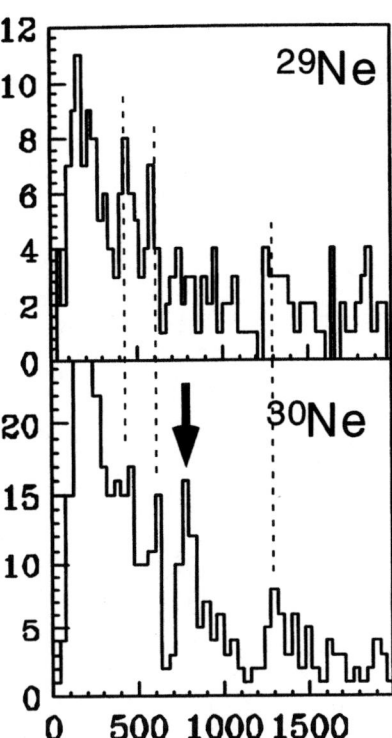

FIGURE 3. Energy spectra of γ rays measured in coincidence with incident particles of ^{29}Ne (upper panel) and ^{30}Ne (lower panel). The location of the $2^+ \to 0^+$ transition in ^{30}ne is indicated by an arrow.

^{29}Ne and ^{30}Ne beams. A distinct peak is seen only for ^{30}Ne case (indicated by an arrow), indicating that the peak corresponds to the $2^+ \to 0^+$ transition in ^{30}Ne. The experiment was performed with the ^{30}Ne beam intensity of 0.2 cps, indicating haw powerful the method is. The result clearly shows that the shell closure is obliterated also in ^{30}Ne. Another $N=20$ nucleus ^{34}Si was studied by γ-γ coincidence spectroscopy with the help of high detection efficiency of the DALI setup [18].

Recently an interesting feature was found in the neutron rich nucleus ^{16}C. The measured lifetime of its first 2^+ state is very long, corresponding to about 0.3 W.U. of $B(E2)$ [19]. This is smaller than any of the transition probabilities known for even-even nuclei. On the other hand, neutron excitation strength evaluated from the ^{16}C+^{208}Pb inelastic scattering data [20] is as large as the one expected from the excitation energy. These results suggests that the neutron and proton motions in ^{16}C are decoupled. This phenomena can be characteristics for certain neutron-rich nuclei, and further studies are desirable.

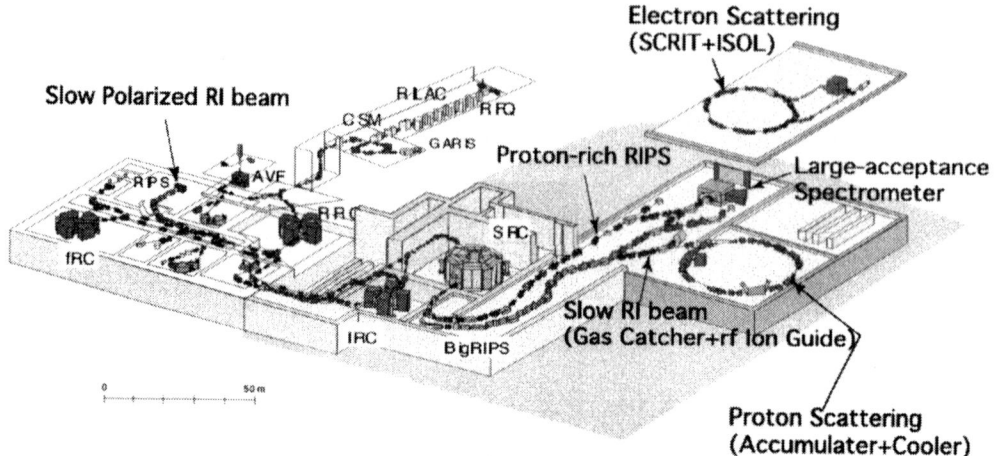

FIGURE 4. Schematic view of the RIKEN accelerator complex. The shaded areas are for RIBF.

RIBF PROJECT

A schematic view of the RIKEN accelerator complex is shown in Fig. 4. The existing facility consists of two injectors, an AVF (azimuthally varying field) cyclotron for ions up to Kr and a linac (RILAC) for heavier ions, and a main ring cyclotron (RRC) with $K=540$ MeV. The RI-beam production is performed with RIPS seen in the left of the figure. Ions with $Z<30$ are mainly used, because the beam energy for heavier ions are not strong enough for efficient RI production.

In RIBF, heavy ion beams from the RILAC-RRC combination are injected to the new ring cyclotrons, fRC (fixed-energy Ring Cyclotron), IRC (Intermediate-stage Ring Cyclotron) and SRC (Superconducting Ring Cyclotron). With the use of the fRC (fixed-frequency Ring Cyclotron), beams with 350 MeV/nucleon energy are available for entire range of atomic mass, whereas higher-energy ions up to 400 MeV/nucleon can be provided by bypassing fRC for light mass up to Ar. The beam intensity is expected to be 1 pμA for light ions, and a little lower for very heavy ions. RI beams are produced by projectile fragmentation or in-flight fission of the accelerated ions from SRC using a new fragment separator Big RIPS [21]. The energy of produced RI beams is 200-300 MeV/nucleon.

The fabrication and basic tests of the IRC and SRC have been made in a factory, and the construction of them on the RIKEN site was started. The building for the accelerator part is already completed and the one for the experimental hall is under construction. The first RI beam through the SRC and BigRIPS will be in 2006 and first experiments are expected to start in 2007. The first phase of RIBF construction includes all the cyclotrons, BigRIPS and RI-beam transport including (at least) a part of the zero-degree spectrometer which identifies products of secondary reactions. Various experimental devices are being developed. They include a γ-ray detector array for in-beam spectroscopy of fast exotic nuclei, a gas

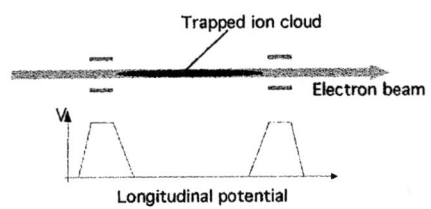

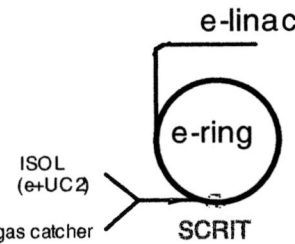

FIGURE 5. Schematic diagrams for electron scattering using SCRIT and an electron storage ring

catcher system [22], a compact TOF system for mass measurements, and so on.

Further extension beyond the first phase is under consideration. A large acceptance spectrometer will be designed for correlation measurements including neutron detection. Slow RI beams provided by the gas catcher system based on the RF ion-guide technique will be coupled with several devices. The TOF mass spectrometer and ion trap systems will be used to measure masses and radii of unstable nuclei. A plan to use beams from IRC for RI beam production with the present RIPS facility is being considered. Compared with the present accelerators, heavier RI beams will become available. Material science with implanted RI, especially spin-oriented ones, is a major subject of research. A new idea of Self-Confining RI Target (SCRIT) with intense electron beams is being studied. Unstable nuclei may be trapped by the electron beam caused by the electric attraction. By longitudinal confinement by potentials, a reasonable luminosity for electron-RI collisions can be achieved. The slow RI beams mentioned above can be the source of the unstable nuclei for SCRIT as well as the beams extracted from an ISOL system (see Fig. 5). Possibility of proton-RI collision experiments is also considered. A detection system for measurements with a hydrogen target with precise angle measurements of recoils are being developed. Another possibility of circulated RI and a internal target in a storage ring is also studied. By triggering the injection to the ring, rare isotope beams (below 100 particle per second) might be able to be circulated in a high injection efficiency close to 100%.

SUMMARY

The RIKEN RI beam factory (RIBF) will provide various possibilities for research using radioactive beams. Construction of the three ring cyclotrons (fRC, IRC and SRC) will be completed in 2006. Then first experiments will start in 2007 with 200-300 MeV/nucleon beams of various unstable nuclei. Plans of constructing experimental devices are under consideration in order to fully exploit research opportunity provided by the intense RI beams with variety of nuclide.

REFERENCES

1. T. Kubo et al., Nucl. Instr. Meth. B 70 (1992) 309.
2. T. Nakagawa et al., Jpn. J. Appl. Phys. **33** (1996) 378.
3. O. Kamigaito et al., Jpn. J. Appl. Phys. **33** (1996) L537.
4. M. Notani et al., Phys. Lett. **B542** (2002) 49.
5. G. Baur, C.A. Bertulani, and H. Rebel, Nucl. Phys. A458 (1986) 188.
6. T. Motobayashi, Nucl. Phys. A693 (2001) 258.
7. G. Baur and H. Rebel, J. Phys. G20 (1994) 1; Ann. Rev. Nucl. and Part. Sci. 46 (1996) 321.
8. T. Motobayashi et al., Phys. Rev. Lett. 73 (1994) 2680; N. Iwasa et al., J. Phys. Soc. Jpn. 65 (1996) 1256.
9. T. Kikuchi et al., Phys. Lett. B391 (1997) 261.
10. T. Kikuchi et al., Eur. Phys. J. A3 (1998) 209.
11. T. Motobayashi et al., Phys. Lett. B264 (1991) 259.
12. K. Yamada et al., Phys. Lett. B, in print.
13. T. Motobayashi et al., Phys. Lett. B346 (1995) 9.
14. T. Glasmacher, Ann. Rev. Nucl. Part. Sci. 48 (1998) 1.
15. K. Yoneda et al., Phys. Lett. B499 (2001) 233.
16. H. Iwasaki et al., Phys. Lett. B522 (2001) 227.
17. Y. Yanagisawa et al., Phys. Lett. B566 (2003) 84.
18. N. Iwasa et al., Phys. Lett. B566 (2003) 84.
19. N. Imai et al., Phys. Rev. Lett. in print.
20. Z. Elekes et al., to be published.
21. T. Kubo, Nucl. Instr. Meth. B204 (2003) 97c.
22. M. Wada et al., Nucl. Instr. Meth. B204 (2003) 570c.

Tri-nucleon cluster structure in ^{6}He and ^{6}Be

H. Akimune*, T. Yamagata*, S. Nakayama†, M. Fujiwara**, K. Fushimi†,
K. Hara**, K.Y. Hara*, K. Ichihara†, K. Kawase**, K. Matsui†,
K. Nakanishi**, A. Shiokawa*, M. Tanaka‡, H. Utsunomiya* and M. Yosoi§

*Department of Physics, Konan University, Kobe, Hyogo 658-8501, Japan
†Department of Physics, University of Tokushima, Tokushima 770-8502, Japan
**Research Center for Nuclear Physics, Osaka University, Ibaraki, Osaka 567-0047, Japan
‡Kobe Tokiwa College, Kobe, Hyogo 654-0838, Japan
§Department of Physics, Kyoto University, Kyoto 606-8502, Japan

Abstract. Tri-nucleon molecular structures in ^{6}He and ^{6}Be were investigated by using the ^{6}Li(^{7}Li,^{7}Be)^{6}He reaction at 455 MeV and ^{6}Li(^{3}He,t)^{6}Be reaction at 450 MeV, respectively. Binary decay into t+t from a broad state at $E_x = 18.0 \pm 1.0$ MeV in ^{6}He and into ^{3}He+^{3}He from one at $E_x = 18.0 \pm 1.2$ MeV in ^{6}Be, were observed by measuring tri-nucleon cluster-decays in coincidence with reaction-particles. The branching ratios for binary decay were estimated to be about 0.7 for both ^{6}He and ^{6}Be. From the angular correlations of decay particles, these states are assigned to be P states.

INTRODUCTION

Over many years, an enormous amount of effort has been devoted to understand excitation energy spectra of light nuclei. Both the molecular-like picture and independent particle picture are essentially important to understand the structure of light nuclei. For example, the microscopic α+d cluster model successfully describes the low-lying states of ^{6}Li. For high-lying resonance states, one naive question naturally arises; are there any resonances including ^{3}He, or ^{3}H particles acting as a cluster ? Such resonant states are known to exist at low excitation energies [1, 2]. Tri-nucleon cluster states were predicted by Thompson and Tang [3], who claimed that the "molecular" resonance with two tri-nucleon clusters should exist in the A=6 triad, ^{6}He, ^{6}Li, and ^{6}Be.

In the past, tri-nucleon resonances were experimentally reported in ^{6}Li and ^{6}Be on the basis of radiative capture reactions [4, 5, 6], and of the phase shift analysis on the ^{3}He+^{3}H and ^{3}He+^{3}He elastic scattering data [7, 8]. In the case of ^{6}Li, Ventura et. al. [5] found evidence for the $^{33}P_2$ resonance at E_x = 18.3 MeV. On the other hand, Vlastou et al., [7] reported that the $^{33}P_2$ and $^{33}P_0$ resonances exist at 21.0 and 21.5 MeV, respectively. Concerning the $^{33}P_2$ resonance in ^{6}Li, there was a serious discrepancy by about 3 MeV in excitation energy. Mondragòn an Hernàndez[10] performed a combined reanalysis for the ^{3}He+t elastic scattering data and the radiative capture data, and gave a consistent understanding about the $^{33}P_2$ resonance at 17.984 MeV. Recently, Ohkura et al. [9] calculated the excitation energy of the ^{33}P resonance at 16.9 MeV using complex-scaled ^{3}He+t resonating group method (CSRGM). Their theoretical results are in good agreement with the combined analysis.

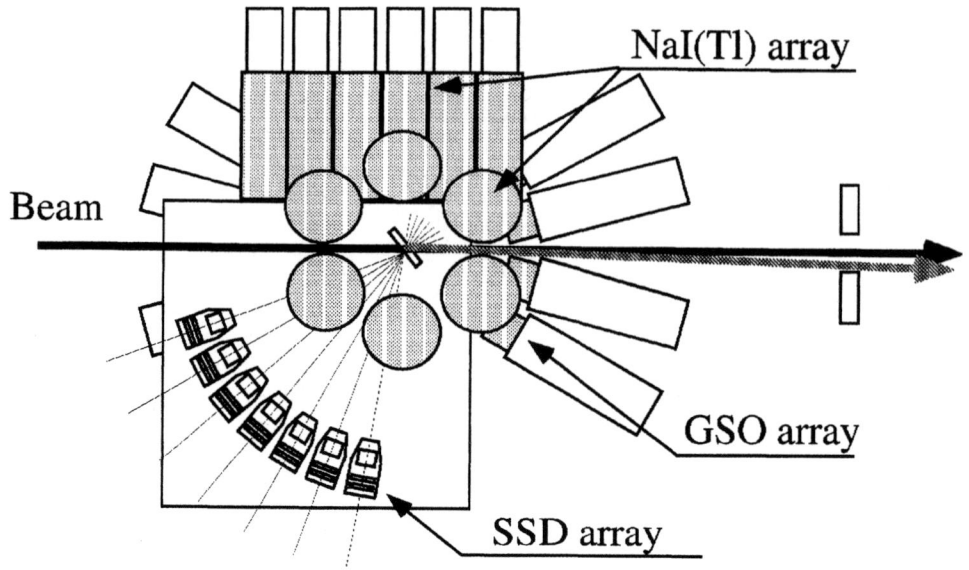

FIGURE 1. Schematic diagram of the experimental set-up used for detecting charged particles and γ rays.

In the case of ^{6}Be, contradictory results were reported about the tri-nucleon cluster resonance. Ventura *et al.* assigned a broad resonance at E_x=23 MeV in ^{6}Be to be the $^{33}F_3$ resonance from the radiative capture reaction of ^{3}He on ^{3}He [6]. However, Vlastou *et al.* did not observe this state in the phase shift analysis of elastic scattering of polarized ^{3}He on ^{3}He, but they observed the $^{33}F_4$, $^{33}F_2$ and $^{33}F_3$ resonances located at E_x=23.4, 26.2 and 26.7 MeV, respectively [8]. Thus, experimental information on the tri-nucleon clustering in A=6 nuclei appears highly contentious.

In order to confirm theoretical prediction, we investigated the t+t molecular state in ^{6}He via the ^{6}Li(^{7}Li,^{7}Be)^{6}He reaction at 455 MeV by measuring binary triton decay in coincidence with ^{7}Be particles [12]. We observed the t+t cluster state at $E_x = 18$ MeV in ^{6}He. From the basis of the isospin symmetry, the analog of the resonance in ^{6}He is expected to exist around $E_x = 18$ MeV in ^{6}Be. In this work, the tri-nucleon cluster-state in ^{6}He and ^{6}Be were searched for via the ^{6}Li(^{7}Li,^{7}Be)^{6}He reaction at 455 MeV and ^{6}Li(^{3}He,t)^{6}Be reaction at 450 MeV, respectively. Unbound state with t+t and ^{3}He+^{3}He cluster configurations are expected to binary decay into t and ^{3}He. The tri-nucleon cluster-states in ^{6}He and ^{6}Be are uniquely deduced by measuring angular correlation of decay-particles in coincidence with reaction-particles. The tri-nucleon cluster-state observed in ^{6}Be will be compared with that in ^{6}He and the structure with tri-nucleon cluster configuration in the A=6 isobars will be discussed.

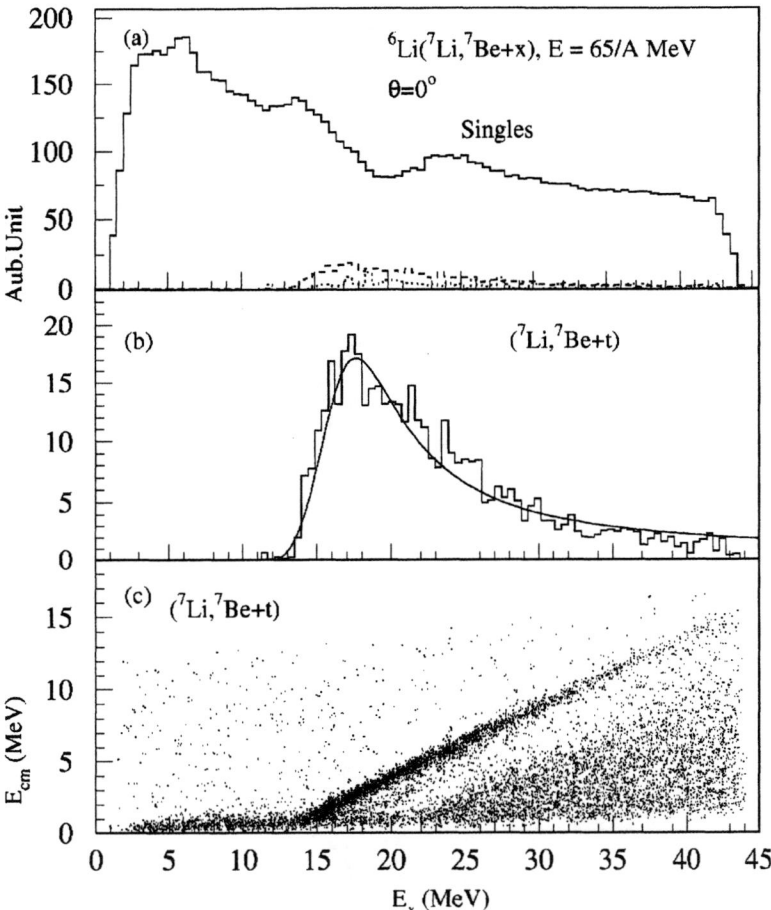

FIGURE 2. (a) Singles spectrum for the ^{6}Li(^{7}Li,^{7}Be) reaction at 65 MeV/A. A gated spectrum for the binary triton decay is also shown in the figure. (b) Coincidence spectrum for the ^{6}Li(^{7}Li,^{7}Be-t)^{3}H reaction. The solid histogram is the experimental data. The solid curve shows the peak shape calculated with the Breit-Wigner one level formula. (c) Two-dimensional scatter plot for coincidence events. The horizontal and vertical axes are the excitation energy in ^{6}He and the energy of tritons, respectively.

EXPERIMENT AND RESULTS

The experiment was performed at the RCNP cyclotron facility of Osaka University with a ^{7}Li^{3+} beam of 65 MeV/nucleon and a ^{3}He^{2+} beam of 150 MeV/nucleon. The target used was a self-supporting foil of an enriched ^{6}Li isotope (95.2%) with a thickness of 0.7 mg/cm^2. Spectra for the ^{6}Li(^{7}Li,^{7}Be) and ^{6}Li(^{3}He,t) reactions were measured with the magnetic spectrometer "Grand Raiden"[14]. The angular acceptance of the spectrometer was ±20 mr horizontally and vertically. The reaction particle, ^{7}Be particles and tritons

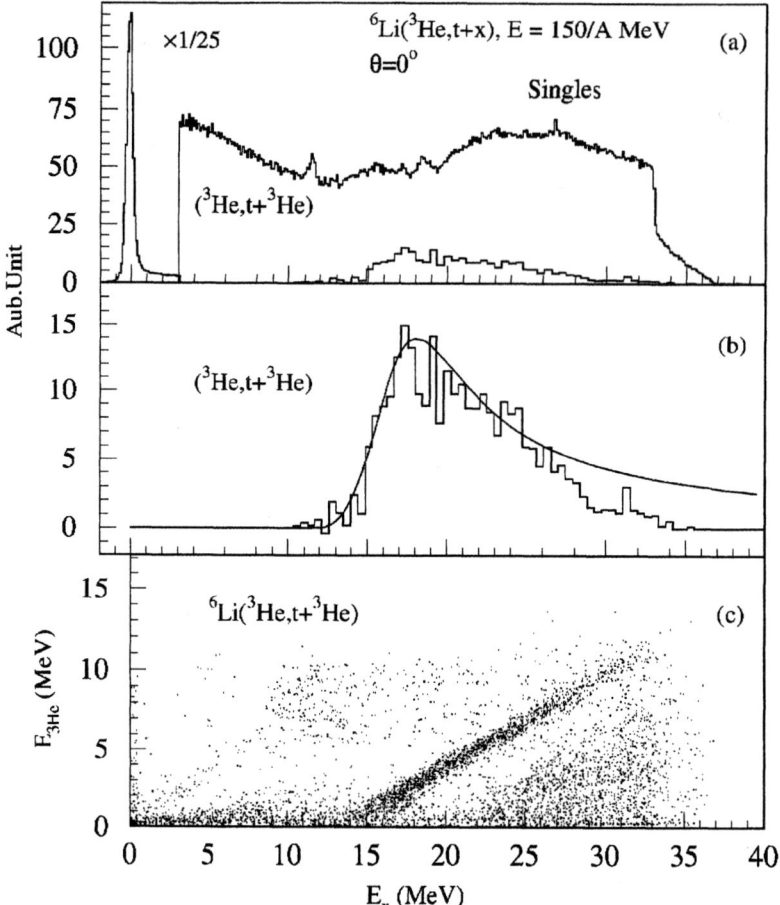

FIGURE 3. (a) Singles spectrum for the ^{6}Li(^{3}He,t) reaction at 150 MeV/A. (b) Coincidence spectrum for the ^{6}Li(^{3}He,t-^{3}He)^{3}He reaction. (c) Two-dimensional scatter plot for coincidence events. The horizontal and vertical axes are the excitation energy in ^{6}He and the energy of ^{3}He, respectively.

were detected by using a focal-plane detector system, which consisted of two multi-wire drift chambers backed by a $\Delta E - E$ plastic-scintillator telescope. Charged particles form the excited stats in ^{6}He and ^{6}Be formed in the ^{6}Li(^{7}Li,^{7}Be) and ^{6}Li(^{3}He,t) reactions were detected using an array of 8 surface barrier type Si solid-state detector (SSD) telescopes. The thickness of the ΔE detectors were 500 μm. Each ΔE detector was backed by a 300 μm thick E detector. These telescope were located at intervals of 10° form 90° to 160° ($\phi = 0°$), and at the distance of 25 cm from the target. The time of flight technique was utilized for particle identification for the decay particles.

Figure 2(a) shows the singles spectrum for the ^{6}Li(^{7}Li,^{7}Be)^{6}He reaction at 65 MeV/nucleon. Three broad resonances were observed at E_x = 5, 15, and 25 MeV.

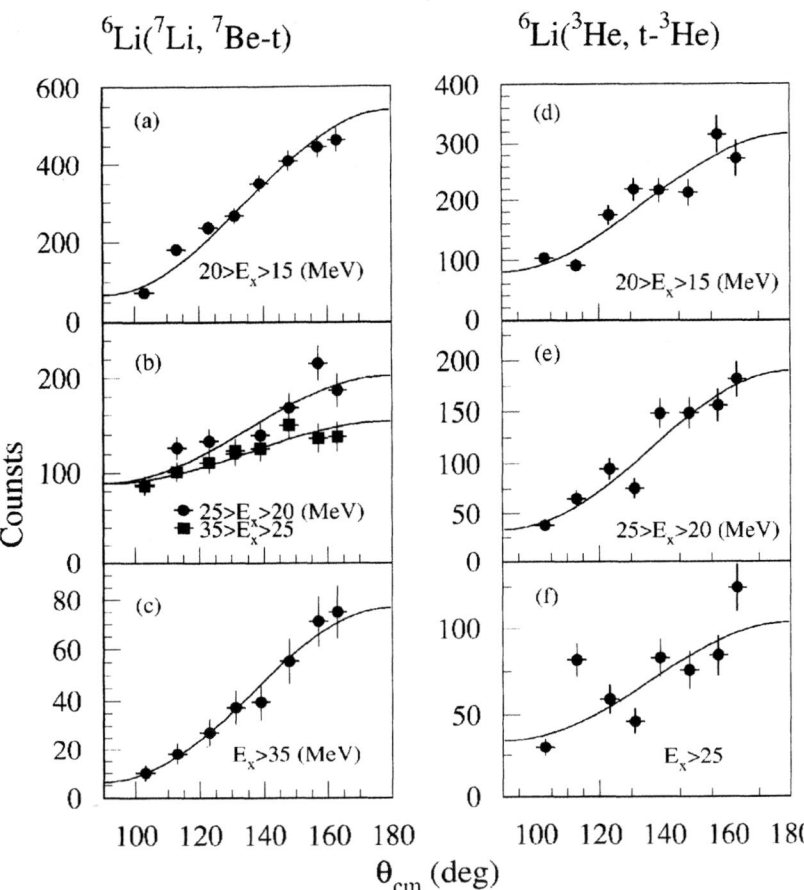

FIGURE 4. Angular correlation patterns for binary triton decay from the excitation energy regions at $15 < E_x < 20$ MeV (a), $20 < E_x < 25$ MeV (b), $25 < E_x < 35$ MeV (b), and 35 MeV $< E_x$ (c) in ^{6}He. Angular correlation patterns for binary ^{3}He decay from the excitation energy regions at $15 < E_x < 20$ MeV (d), $20 < E_x < 25$ MeV (e), and 25 MeV $< E_x$ (f) in ^{6}Be. The $0°$ axis is taken as the direction of incident particles. The abscissa is the angle of particle-emission in the rest frames of ^{6}He and ^{6}Be. The solid curves are the result of χ^2 fit with the Legendre polynomials.

These resonances have been observed in the ^{6}Li(n,p) reaction [15], and the ^{6}Li(^{7}Li,^{7}Be) reaction [13, 16, 17] The resonance at $E_x \simeq 5$ and 25 MeV were assigned by Nakayama *et al.* to be the soft dipole resonance [13] and the analog of dipole resonance caused by an excitation of the α cluster in the ^{6}Li nucleus [16, 18], respectively. The coincidence spectrum in the ^{6}Li(^{7}Li,^{7}Be t)^{3}H reaction obtained by gating on the binary triton-decay events is shown in Fig. 2(b). Only one broad peak was observed at $E_x = 18 \pm 1.0$ MeV. The width (full width at half maximum, FWHM) was 9.5 ± 1.0 MeV. Figure 2(c) shows a two-dimensional scatter plot of ^{7}Be-t coincidence events. Here, energies of tritons were expressed in the energies of the rest system for ^{6}He, taking into account the recoil effect. The coincidence events corresponding to triton decay are clearly separated from other charged particles decay, such as proton, deuteron, and alpha decay. The solid curves in Fig. 2(b) shows the result of peak fitting procedure with the Breit-Wigner one level formula [19]. Detail of this procedure will be discussed in the next section.

Figure 3(a) shows the singles spectrum for the ^{6}Li(^{3}He,t)^{6}Be reaction at 150 MeV/nucleon. One broad resonance was also observed at $E_x = 25$ MeV. The coincidence spectrum in the ^{6}Li(^{3}He,t ^{3}He)^{3}He reaction obtained by gating on the binary triton-decay events is shown in Fig. 3(b). Only a broad peak was observed at $E_x = 18 \pm 1.2$ MeV. The width (full width at half maximum, FWHM) was 9.2 ± 1.3 MeV. Figure 3(c) shows a two-dimensional scatter plot of t-^{3}He coincidence events.

Figure 4(a), (b), and (c) show the angular correlation patterns for binary triton decaying from excitation energies of $15 < E_x < 20$ MeV (a), $20 < E_x < 25$ MeV (b), $25 < E_x < 35$ MeV (b), and 35 MeV $< E_x$ (c) in ^{6}He. Figure 4(d), (e), and (f) show the angular correlation patterns for binary ^{3}He decaying from excitation energies of $15 < E_x < 20$ MeV (d), $20 < E_x < 25$ MeV (e), and 25 MeV $< E_x$ (f) in ^{6}Be. The emission angels of decaying particles were represented in the rest system for ^{6}He and ^{6}Be, respectively.

DISCUSSION

In order to deduce the resonant parameters, the spectral shape of the 18-MeV resonances shown in Fig. 2(b) and 3(b) were fitted by using the Breit-Wigner one level formula. We assumed the triple differential cross section $d^3\sigma(E_x)/d\Omega_{^7Be}d\Omega_t dE_x$ as

$$\frac{d^3\sigma(E_x)}{d\Omega_{^7Be}d\Omega_t dE_x} \propto \frac{\Gamma_3}{(E_x - E_R)^2 + (\Gamma/2)^2},$$

were E_R, Γ, and Γ_3 are the resonance energy, the total width, and the partial width for triton or ^{3}He decay, respectively. The width Γ_3 is expressed as [20]

$$\Gamma_3 = \frac{2\hbar}{R}\left(\frac{2E}{\mu}\right)^{1/2} \theta^2 P_l,$$

where R, E, μ, θ^2, and P_l are the interaction radius of two tri-nucleon clusters, the decay particle energy in the center of mass system, a reduced mass, a dimensionless reduced width, and the penetrability with an angular momentum l, respectively. The interaction radius is given by $R = r_0(A^{1/3} + A^{1/3})$ with $r_0 = 1.4$ fm. The total Γ is expressed as $\Gamma = \Gamma_3 + \Gamma'$, where Γ' is a partial width for neutron, proton or other particle decay. Since the neutron-decay threshold energy for ^{6}He (0.97 MeV) and the proton-decay threshold energy for ^{6}Be (-1.37 MeV) is much lower than the peak position of $E_x = 18$ MeV, the width Γ' was assumed to be constant within the resonance region.

As shown in Fig. 2(b) and Fig. 3(b), the shape of the coincidence spectrum is well reproduced with the Breit-Wigner one level formula assuming $l=1$. In Table 1, parameters for the resonances in this work are summarized. When we use the resonance parameters

TABLE 1. Experimental excitation energy E_x, width (FWHM), reduced width, and Γ'

channel	E_x (MeV)	FWHM (MeV)	θ	Γ' (MeV)
t+t	18.0±1.0	9.5±1.0	1.0±0.4 [1]	3.5 [1]
^{3}He+^{3}He	18.0±1.2	9.2±1.3	1.0±0.4 [1]	3.5 [1]

* 1 Fixed in the fits.

TABLE 2. Experimental angular correlation coefficients α_2

E_x (MeV)	^{6}He α_2	^{6}Be α_2
15 - 20	0.82±0.12	0.66±0.21
20 - 25	0.46±0.07	0.75±0.24
25 - 35	0.32±0.04	0.57±0.18
25 -	0.88±0.15	-

in Table I, the branching ratio R for tri-nucleon decay,

$$R = \int \frac{d^2\sigma}{d\Omega dE} \frac{\Gamma_3}{\Gamma} dE \bigg/ \int \frac{d^2\sigma}{d\Omega dE} dE$$

amounts to 70% both for ^{6}He and ^{6}Be.

In order to determine the angular momentum of the 18-MeV resonance, the angular correlation patterns $W(\theta)$ of decaying triton and ^{3}He was investigated. A satisfactory fitting was obtained with $W(\theta) = 1 + \alpha_2 P_2(\cos\theta)$ for all the angular correlations. This fact indicates that the t+t and ^{3}He+^{3}He cluster state at $E_x = 18$ MeV are assigned to be ^3P state. The obtained α_2's for several excitation energies in ^{6}He and ^{6}Be are listed in Table 2. The result of the fits are shown in the solid curves in Fig. 4.

SUMMARY

Binary decays into t+t from a broad state at $E_x = 18.0 \pm 0.4$ in ^{6}He and into ^{3}He+^{3}He from one at $E_x = 18.0 \pm 0.6$ in ^{6}Be, respectively, were observed by measuring tri-nucleon cluster-decays in coincidence with reaction-particles. The branching ratios for the binary decay were estimated to be 0.7 for ^{6}He and ^{6}Be. These large branching ratio shows that a tri-nucleon cluster state exists as an isobaric partner around $E_x = 18$ MeV in ^{6}He and ^{6}Be.

ACKNOWLEDGMENTS

This experiment was performed at RCNP, Osaka University under program E172, E184, and E190. We are grateful to the RCNP cyclotron crew for preparing a stable and clean beam.

REFERENCES

1. Y.M. Shin et al., Phsy. Lett. **B55**, 297 (1975).
2. F. Werby et al., Phys. Rev. C **8**, 106 (1973).
3. D.R. Thompson and Y.C. Tang, Nucl. Phys. **A106**, 591 (1968).
4. S.L. Blatt et al., Phys. Rev. **176**, 1147 (1968).
5. E. Ventura et al., Phys. Lett. **B46**, 364 (1973); E. Ventura et al., Nucl. Phys. **A173**, 1 (1971) and the related references therein.
6. E. Ventura et al., Nucl. Phys. A **219**, 157 (1974).
7. R. Vlastou et al., Nucl. Phys. A **292**, 29 (1977).
8. R. Vlastou et al., Nucl. Phys. A **303**, 368 (1978).
9. H. Ohkura, T. Yamada, and K. Ikeda, Prog. Theor. Phys. 94, 47 (1995).
10. A. Mondragón, and E. Hernández, Phys. Rev. C **41**, 1975 (1990).
11. D.R. Tilley et al., Nucl. Phys. **A708**, 3 (2002).
12. H. Akimune et al. Phys. Rev. C **67**, 051302(R) (2003).
13. S. Nakayama et al., Phys. Rev. Lett. **85**, 262 (2000).
14. M. Fujiwara et al., Nucl. Instrum. Methods Phys. Res., Sect. A **422**, 484 (1999).
15. F.P. Brady et al., J. Phys. G **10**, 363 (1984).
16. S. Nakayama et al., Phys. Rev. Lett. **87**, 122502 (2001).
17. J. Jänecke et al., Phys. Rev. C **54**, 1070 (1996).
18. T. Yamagata et al., Proc. Tours Symposium on Nuclear Physics V, in print.
19. J.M. Blatt and V.F. Weisskopf, "Theoretical Nuclear Physics" (Jhon Wiley & Sons, N.Y., 1952), p. 391.
20. C.E. Rolfs and W.S. Rodney, "Cauldrons in the Cosmos" (Univ. Chicago Press, Chicago and London, 1988), p. 178.

Search for excited α-cluster resonances and their analogs in A=6 and 7 nuclei

T. Yamagata[*], S. Nakayama[†], H. Akimune[*], M. Fujiwara[**], K. Fushimi[†],
M.B. Greenfield[‡], K. Hara[**], K.Y. Hara[*], K. Hashimoto[**], K. Ichihara[†],
K. Kawase[**], M. Kinoshita[*], Y. Matsui[†], K. Nakanishi[**], M. Tanaka[§],
H. Utsunomiya[*] and M. Yosoi[¶]

[*]*Department of Physics, Konan University, Kobe 658-8501, Japan*
[†]*Department of Physics, University of Tokushima, Tokushima 770-8502, Japan*
[**]*Research Center for Nuclear Physics, Osaka University, Osaka 567-0047, Japan*
[‡]*Department of Physics, International Christian University, Tokyo 181-8585, Japan*
[§]*Kobe Tokiwa College. Kobe 654-0838, Japan*
[¶]*Department of Physics, Kyoto University, Kyoto 606-8502, Japan*

Abstract. The isovector giant dipole resonance (GDR) of an α-cluster in ^{6}Li and ^{7}Li, and their analogs in ^{6,7}He and ^{6,7}Be were searched for by using the ^{6,7}Li(p,p'), ^{6,7}Li(^{3}He,t) and ^{6,7}Li(^{7}Li,^{7}Be) reactions at the incident energies of 300, 450 and 455 MeV, respectively. New resonances were found for the first time at E_x=27.0±1.5 MeV in ^{6}Li and at E_x=29.0±1.5 MeV in ^{7}Li, both with widths (FWHM) of 12±2 MeV. In ^{6}He, ^{6}Be, ^{7}He and ^{7}Be resonances with widths of 12±2 MeV were observed via the charge exchange reactions at E_x=24.0±1.5, 23.5±1.5, 18.0±1.5, and 28.0±1.5 MeV, respectively. All of these resonances were assigned to be dipole resonances based on the measured angular distributions of differential cross sections. The resonance shapes were reproduced well with the resonance shape of the GDR of ^{4}He reported in the (γ,n) reaction. The averaged ratios of the cross sections for these resonances in A=6 to those in A=7 for each isotope were estimated to be nealy unity, 1.2±0.3. The averaged value of the respective cross section ratios of these resonances to the GDR in their respective target nuclei was 0.44±0.08, which was consistent with the ratios of the cross sections of the GDR in ^{4}He to those of the GDR in ^{6,7}Li reported in the photo-nuclear reactions. The excitation energies of these dipole resonances in A=6 and 7 nuclei relative to the separation energies of an α-particle in ^{6}Li and ^{7}Li, respectively, agreed well with the excitation energy of the GDR in ^{4}He ($\approx$26 MeV). We conclude that the resonances observed in ^{6,7}Li are the GDR of the α-cluster and the resonances in ^{6,7}He and ^{6,7}Be are their analogs.

INTRODUCTION

Clusters in nuclei play important roles in nuclear structure and nuclear reactions. α-particle clustering is one of the most general nuclear phenomena exhibited over a wide mass range from light to heavy nuclei. It is considered that clusters in nuclear systems are spatially localized subsystems composed of strongly correlated nucleons [1]. Therefore, we can expect two types of nuclear excitation known to exhibit clustering structure in their ground states.

One is excitation due to an inter-cluster relative motion. A typical example of this excitation is the rotational excitation of a clustering nucleus [2]. The other is due to the intrinsic excitation of the cluster itself, which has not been well known. In 1963, Costa *et al.* experimentally suggested possible excitation of an α-cluster. They observed two resonances at excitation energies $E_x \approx 11.5$ and 26 MeV in spectra obtained from the ^{6}Li(γ,n) reaction [3]. They concluded that the 11.5 MeV resonance was the isovector giant dipole resonance (GDR) in ^{6}Li and the one at 26 MeV was the resonance due to the excited α-cluster, the GDR of ^{4}He in the ^{6}Li nucleus. After their work, many experiments on the ^{6}Li(γ,n) reaction have been done. However, no such evidence for the excited α-cluster has been obtained [4, 5, 6]. We note that in the ^{nat}Li(γ,n) spectrum a peak at about E_x=30 MeV was seen [7] in addition to the peak due to the GDR at $E_x \approx 17$ MeV, though this 30 MeV peak was not observed in other ^{7}Li(γ,n) work [4, 5].

Such α-cluster excitation is very interesting, since it may provide a new concept of nuclear excitation [8]. Furthermore, the possible excited α-cluster may have characteristics different from those exhibited by an excited free α-particle due to nuclear medium effects. Resonances due to the excited α-cluster seem to have been observed by Brady *et al.*, dipole resonances at $Q \approx -30$ MeV in the ^{6,7}Li(n,p)^{6,7}He reaction [9]. Jänecke *et al.* reported a study of the ^{6}Li(^{7}Li,^{7}Be) reaction [10] in which a structure similar to that observed in the (n,p) reaction [9] was observed. In the ^{6,7}Li(p,n) reaction, Yang *et al.* reported high-lying dipole resonances at $E_x \approx 25$ MeV in ^{6}Be and at $E_x \approx 30$ MeV in ^{7}Be [11]. The observed excitation energies and widths of these resonances are very similar to those for the GDR of ^{4}He reported in the ^{4}He(γ,n) reaction [4]. However, neither of the above authors had considered the possibility that the observed resonances might be due to the excited α-cluster in nuclei.

We speculated that these resonances are due to the excited α-cluster and investigated these resonances in ^{6}He and ^{7}He via the (^{7}Li,^{7}Be) reaction [12]. As a result, we demonstrated, for the first time, that the resonances at $E_x \approx 24$ in ^{6}He and at 20 MeV in ^{7}He stem from the α-cluster excitation, i.e., the analog of the GDR of ^{4}He in nuclei [12]. From this result we supposed that the resonances observed in ^{6}Be at $E_x \approx 25$ MeV and in ^{7}Be at $E_x \approx 30$ MeV in the (p,n) reactions [11] might also be the analogs of the excited α-cluster and that the resonances of these analogs should be present at $E_x \approx 26$-30 MeV in ^{6}Li and ^{7}Li, though there had been no report about such highly excited resonances in ^{6,7}Li [13, 14].

In this paper, the details of our experimental study on the excited α-clusters in ^{6,7}Li target nuclei are presented. The ^{6}Li and ^{7}Li nuclei have large d+α and t+α components in their ground states, respectively [13]. The excited α-cluster was searched for in the GDR of ^{4}He in ^{6,7}Li by using the ^{6,7}Li(p,p') reaction at 300 MeV, and their analogs in the ^{6,7}He and ^{6,7}Be by the charge exchange reactions of the ^{6,7}Li(^{7}Li,^{7}Be) reaction at 455 MeV [12] and ^{6,7}Li(^{3}He,t) reaction at 450 MeV, respectively. These reactions were considered suitable for excitation of isovector resonances [15, 16].

The GDR of ^{4}He is known to be a strongly excited resonance. Its excitation energy of about 26 MeV is much higher than the excitation energies of GDR in the target nuclei, $E_x \approx 12$ and 17 MeV in ^{6}Li and ^{7}Li, respectively [4]. Therefore, if there is a resonance due to the excited α-cluster, it should be clearly distinguished as a compact resonance in the continuum region, much higher in excitation energy than the GDR of the target nuclei. The spin dipole resonance (SDR) is also excited at the excitation region of the GDR, because the SDR in the light nuclei has been observed to be similar to the GDR in both excitation energy and the resonance shape [17]. Thus, a single resonance peak composed of the GDR and SDR is observed.

The cross sections in each reaction depend strongly on both the reaction mechanism and nuclear structure. However, the ratio of the cross section deduced from the same reaction dose not depend on the reaction mechanism. Therefore, the excited α-cluster in nuclei may be identified by using three very different reactions: The cross section ratios for the possible excited α-cluster resonances in each isotope should approch unity, since the same constituent, an α-cluster, is excited. The cross section ratio of the possible excited α-cluster resonance to the GDR in the target nucleus deduced in these reactions should nearly be equal to the cross section ratio, $\sigma(^4\text{He})/\sigma(^{6,7}\text{Li})$ obtained for the GDR in the photo-nuclear reactions [18] for all reaction channels.

Since the experimental details of the $^{6,7}\text{Li}(^7\text{Li},^7\text{Be})$ reaction are similar to those reported by Nakayama et al. [12], the present paper will focus on the results of this reaction and will compare these results with other reactions.

EXPERIMENT

The experiment was carried out by using the 300 MeV-proton, 450 MeV-^{3}He, and 455 MeV-^{7}Li beams from the ring cyclotron at the Research Center for Nuclear Physics (RCNP), Osaka University. Targets used were self-supporting metallic foils of enriched ^{6}Li (95.4%) and ^{7}Li (99.9%). Reaction particles were analyzed by using the magnetic spectrometer "Grand Raiden" and were detected with the focal plane detector system, which consisted of two multi-wire drift chambers backed by a ΔE-E plastic scintillator telescope [19]. The typical values of the thickness of these targets used and energy resolution were 15, 20, and 1.0 mg/cm^2 and 120, 340, and 800 keV for the (p,p'), (^{3}He,t), and (^{7}Li,^{7}Be) reactions, respectively.

Energy spectra were measured in the angular ranges of θ_L= 2.5°- 16° for the (p,p') reaction, θ_L= 0°-4.1° for the (^{3}He,t) reaction, and θ_L= 0°-2.3° for the (^{7}Li,^{7}Be) reaction where the cross sections with a ΔL=1 were expected to have the first maximum. Since the momentum range covered with this spectrometer was about 5%, measurements were done with several different settings of the magnetic field to cover a wide excitation-energy range of interest. In the measurement for the (p,p') reaction at forward angles $\theta_L \leq 10°$, the elastically scattered protons were blocked by a thick lead plate in front of the focal plane detector to reduce the counting rate. Additional measurements for the ground state and the low-lying discrete states were done without the plate by using a low intensity beam. The calibration of the excitation energies was carried out by observing known states with ^{12}C targets. The experimental details are presented elsewhere [12, 20].

RESULTS AND ANALYSIS

Energy spectra

Figure 1 shows typical energy spectra for the $^{6,7}\text{Li}(p,p')$, $^{6,7}\text{Li}(^3\text{He},t)$, and $^{6,7}\text{Li}(^7\text{Li},^7\text{Be})$ reactions at θ_L=8°, 2.7°, and 0°, respectively. The states at E_x=3.56 MeV in ^{6}Li and the ground states in ^{6}He, and ^{6}Be are the isobaric analogs of the 0$^+$,

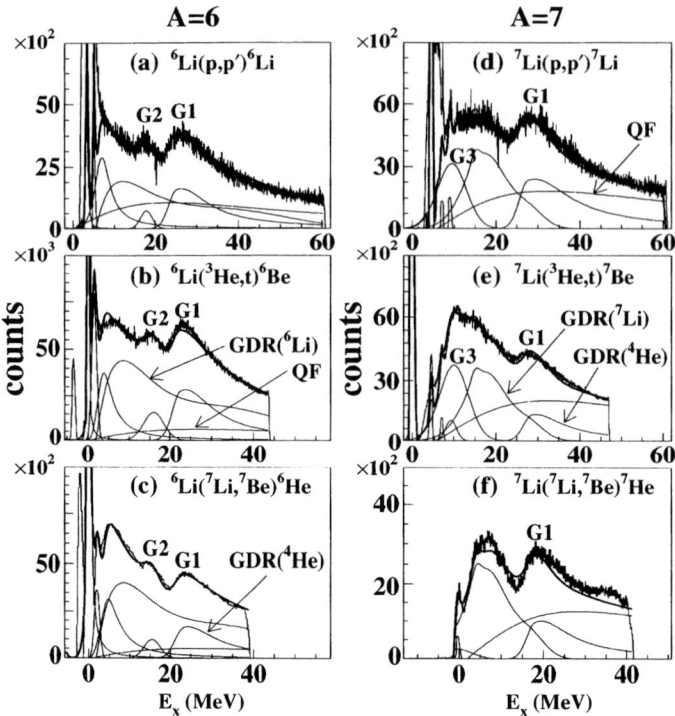

FIGURE 1. The spectra for the reactions of (p,p') at $\theta_L=8°$, (^{3}He,t) at 2.7°, and (^{7}Li,^{7}Be) at 0°. Thin solid lines show the peak fitting results. See text.

$T=1$ state, and similarly the ground state in ^{7}He and the states at E_x=11.25 MeV in ^{7}Li and at E_x=11.01 MeV in ^{7}Be are the isobaric analogs of the $3/2^-$, $T=3/2$ state [14]. In order to clearly show the relationship between the analog states excited in these nuclei, each spectrum in Fig. 1 is shifted such that the location of these analog states coincide horizontally. The respective spectral shapes are very similar in A=6 and A=7.

A common feature of the spectra for A=6 is the appearance of two resonances at $E_x \approx 27$ MeV and ≈ 18 MeV (denoted herein as G1 and G2, respectively). The resonance at $E_x \approx 27$ MeV in ^{6}Li was observed for the first time in the present work. Also in the spectra for A=7, resonances similar to those observed in A=6 nuclei are evident at $E_x \approx 30$ MeV in ^{7}Li and ^{7}Be and at 20 MeV in ^{7}He. The resonance at $E_x \approx 30$ MeV in ^{7}Li was also observed for the first time in the present work. These resonances in A=7 were observed to be G1 type resonances. Based on the fact that the excitation energies of G1's measured with respect to the excitation energies of ^{6}Li and ^{7}Li are in close agreement with the excitation energy of the GDR of ^{4}He reported at $E_x \approx 26$ MeV in the (γ,n) reaction [4], G1's are candidates for an excited α-cluster resonance.

In order to obtain the excitation enrgies, widths and cross sections of these resonances, the measured spectra were decomposed into various peaks and continuum components. We restrict our attention to the excitation energy region above $E_x \geq 10$ MeV in ^{6,7}Li except for some discrete states in the low-excitation energy region where the presence of the soft-dipole resonance had been observed [21].

In the analysis of the spectra obtained in the ^{6}Li(p,p') reaction, the resonances at $E_x\approx 18$ (G2) and 27 MeV (G1), the GDR of ^{6}Li reported in the (γ,n) reaction at ≈ 12 MeV [4] and the quasi-free continuum were taken into account. These spectral shapes were fitted with a Gaussian function, the shape of the GDR referred from the ^{4}He(γ,n) reaction [4], that of the GDR from ^{6}Li(γ,n) reaction [4], and the quasi-free spectra calculated following the prescriptions given by Erell [22] and Jänecke [10], respectively. Since in these (γ,n) reactions the resonance shapes of the GDR in ^{6,7}Li and ^{4}He were reported at $E_x \leq 32$ MeV [4], we should extrapolate their shapes beyond this energy. The resonance shapes employed, thus involved some uncertainties at $E_x \geq 40$ MeV. In order to reproduce the low-excitation energy region, an asymmetric Lorentz function was employed [21].

In the analysis of the ^{7}Li(p,p') spectra, the GDR of ^{7}Li reported in the (γ,n) reaction at $E_x\approx 17$ MeV [4], a resonance at $E_x\approx 30$ MeV (G1), and a quasi-free continuum were also taken into account [10, 22]. The peak shapes for the GDR and the resonance at 30 MeV (G1) were taken from the shapes of the GDR reported in the ^{7}Li(γ,n) and the ^{4}He(γ,n) reactions [4], respectively. To reproduce the spectral shape around $E_x\approx 10$ MeV an additional Gaussian peak at $E_x=9.5$ MeV with a width of ≈ 9 MeV was introduced.

Since the respective A=6 and 7 nuclei exhibited very similar spectra, the spectra for ^{6}He and ^{6}Be, and for ^{7}He and ^{7}Be were analyzed in a manner similar to that used for ^{6}Li and ^{7}Li, respectively, by taking into account the differences in Coulomb displacement energies. In the (^{7}Li,^{7}Be) reaction, only the $T=3/2$ analog-component of the GDR in ^{7}Li can be excited, while the (γ,n) reaction may excite both $T=3/2$ and 1/2 components. Therefore, only the high excitation side of the GDR shape reported in the (γ,n) reaction [4] was used to fit the resulting ^{7}He spectral shape of the GDR, .

The fitted curves are shown in Fig. 1 by the solid lines. Each spectral shape was reproduced well by the present simplified analysis. The resonance energies and widths (FWHM) are summarized in Table 1. The resonances at $E_x=9.5$ MeV in A=7 are assumed to be G3 resonances.

Differential cross sections

In order to assign the ΔL values for the resonances observed in this work, their angular distributions of the differential cross sections observed in the (p,p') and (^{3}He,t) reactions were analyzed, and they are shown in Fig. 2 and Fig. 3, respectively. Error bars are mainly due to the peak fitting procedure. Those for the (^{7}Li,^{7}Be) reactions have been reported in Ref. [12]. In Fig. 2 and Fig. 3, the cross sections are plotted as a function of momentum transfer q. Since the distortion effect and the Coulomb interaction for the present incident energies are less important, the plane-wave Born-approximation (PWBA) was sufficient for a qualitative discussion, as it was well applied for the (^{3}He,t) reaction at 450 MeV [23] and the (t,^{3}He) reaction at 350 MeV [24].

In the PWBA the cross section is simply given as $\sigma(q)\approx |j_L(qR)|^2$, where $j_L(qR)$ is the spherical Bessel function of the angular momentum L, and R is the interaction radius.

TABLE 1. Excitation energies and full width half maximum values of the resoanances deduced in this work.

	^{6}He	^{6}Li	^{6}Be	^{7}He	^{7}Li	^{7}Be	
E_x (MeV)				-	9.5±1.0	9.5±1.0	G3
Width (MeV)				-	9.0±1.0	9.0±1.0	
E_x (MeV)*	8.5±1.0	12.0±1.0	8.5±1.0	5.0±1.0†	17.0±1.0	17.0±1.0	GDR**
Width (MeV)‡	20	20	20	14	14	14	
E_x (MeV)	15.0±1.0	18.0±1.0	15.0±1.0				G2
Width (MeV)	3.0±1.0	6.0±1.0	4.0±1.0				
E_x (MeV)*	24.0±1.5	27.0±1.5	23.5±1.5	18.0±1.5	29.0±1.5	28.0±1.5	G1§
Width (MeV)‡	12	12	12	12	12	12	

* Peak energy.
† See text.
** Taken from the (γ,n) reactions [4].
‡ Fixed.
§ Taken from the ^{4}He(γ,n) reaction [4].

The values of R were chosen to be $1.4A^{1/3}$ and $1.2A^{1/3}$ fm for the (p,p') and (^{3}He,t) reactions, respectively. The typical angular distributions calculated with the PWBA for $\Delta L=0$, 1, and 2 are shown in Figs. 2 and 3.

The differential cross sections for the ^{6,7}Li(p,p') reactions are shown in Fig. 2. The observed angular distribution of the 3.562 MeV, 0^+ state in ^{6}Li peaked sharply at 0°. The PWBA calculation with a $\Delta L=0$ roughly reproduced this trend. The 2.186 MeV, 3^+ state in ^{6}Li (a), and 0.48 MeV, $3/2^-$ and 4.652 MeV, $7/2^-$ states in ^{7}Li (e) have a maximum around $q=1$ fm^{-1}. The PWBA calculation for these states is not inconsistent with a $\Delta L=2$, but the corresponding fit to the observed angular distributions is marginal. The angular distributions of the GDR in ^{6}Li (b) and ^{7}Li (g) were better reproduced with a $\Delta L=1$. Thus we assign the resonances at 18 MeV (G2) and 26 MeV (G1) in ^{6}Li and at 9.5 MeV (G3) and 27.5 MeV (G1) in ^{7}Li to be dipole resonances.

The differential cross sections for the ^{6,7}Li(^{3}He,t) reactions are shown in Fig. 3. The PWBA calculation for $\Delta L=0$ reproduced well the angular distributions for the ground state (g.s.) of ^{6}Be (a), g.s.+ 0.43 MeV-state, and the state at 9.9 MeV ($3/2^-$) in ^{7}Be (c). Those for the 4.5 MeV ($7/2^-$) and 4.71 MeV ($5/2^-$) in ^{7}Be (e) were also qualitatively reproduced by the $\Delta L=2$ calculation. PWBA calculations are appropriate for determing ΔL's in the analysis of the present (^{3}He,t) reactions. Furthermore, the angular distributions for the GDR in ^{6}Be and ^{7}Be are well explained by the $\Delta L=1$ calculations, as shown in Figs. 3(b) and 3(d). Since the resonances at $E_x=15$ MeV (G2) and 23.5 MeV (G1) in ^{6}Be, and those at 9.5 MeV (G3) and 28 MeV (G1) in ^{7}Be were well reproduced by a $\Delta L=1$ calculation, we assigned these resonances to be the dipole resonance.

In the (^{7}Li,^{7}Be) reaction, in which the $\Delta S=0$ and $\Delta S=1$ transitions were distinguished, the resonances at $E_x=24$ MeV (G1) in ^{6}He and at $E_x=18$ MeV (G1) in ^{7}He were both

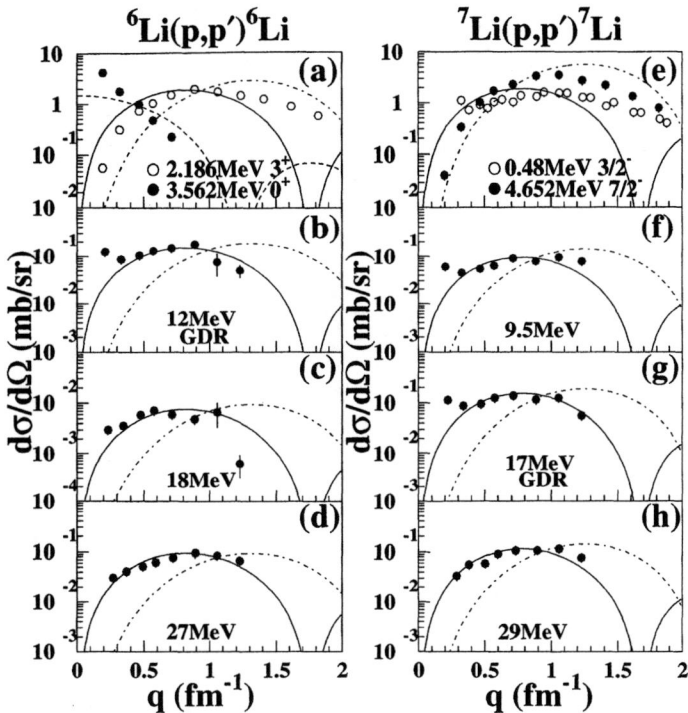

FIGURE 2. Differential cross sections for the (p,p') reactions. The dotted, solid, and dotted-chain lines show the PWBA calculations with $\Delta L=0, 1$, and 2, respectively.

assigned to be consistent with a dipole resonance with components of the $\Delta S=0$ and $\Delta S=1$ [21].

DISCUSSION

Comparison with other results

G1 resonance

In the highly excited energy region of ^{6}Li and ^{7}Li, no resonance states have been reported [14], except for three states in ^{6}Li at E_x=24.779, 24.890, and 26.590 MeV [14, 25]. These three are stated to be the t+^{3}He resonances with the ^3F configuration in the L-S coupling scheme. Therefore, they are all different from the dipole resonance newly observed in the present work at E_x=27MeV in ^{6}Li.

In ^{6}Be and ^{7}Be, Yang *et al.* reported the high-lying dipole resonances at E_x=25.0 and 30 MeV, respectively, from (p,n) reactions at 186 MeV [11]. Their results agree well with the present result.

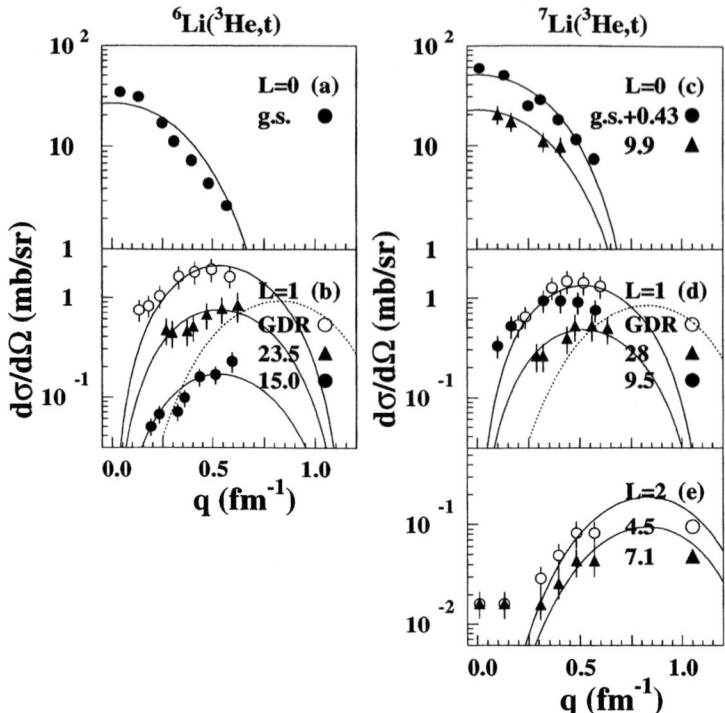

FIGURE 3. Differential cross sections for the (^{3}He,t) reactions. Lines show the PWBA calculations. The dotted curves in (e) and (d) show the PWBA calculation with a $\Delta L=2$.

Brady *et al.* reported the dipole resonance in ^{6}He at $E_x=25\pm1$ MeV ($\Gamma=8\pm2$ MeV) and in ^{7}He at $E_x=20\pm1$ MeV ($\Gamma=9\pm2$ MeV) via the (n,p) reaction at an incident energy of 60 MeV[9]. In the (^{7}Li,^{7}Be) reaction at 350 MeV, Jänecke *et al.* observed the resonance at $E_x=23.3\pm1.0$ MeV ($\Gamma=14.8\pm2.3$ MeV) in ^{6}He [10]. However, from the measurement of the $\Delta S=0$ and $\Delta S=1$ cross sections, they assigned the 23.3 MeV resonance to be a high spin state populated via $\Delta S=1$ transfer. In the present (^{7}Li,^{7}Be) reaction, resonances at $E_x=24.0$ MeV in ^{6}He and at $E_x=18.0$ MeV in ^{7}He are excited via both $\Delta S=0$ and $\Delta S=1$ and are consistent with a $\Delta L=1$ [12].

G2 and G3 resonances

In ^{6}Li, a resonance with a width of 1-2 MeV was observed at $E_x=16$ MeV in the ^{6}Li(γ,n) reactions, as a fine structure of the GDR [4, 6]. The excitation energy reported in their work is lower than 18 MeV observed for the G2 resonance of ^{6}Li in the present work by 2 MeV. Hernández *et al.* claimed a resonance at $E_x=17.985$ MeV [14, 25] with a pure ^3P (2^-, $T=1$) t+^{3}He-clustering structure. However, Akimune *et al.* [20, 26] and Nakayama *et al.* [27] recently observed the isobaric analog triplets of this predicted ^3P,

$T=1$ state in ^{6}He, ^{6}Be, and ^{6}Li, at higher excitation energies of E_x=18.0, 18.0 and 21.5 MeV, respectively. At the excitation energy relevant to $E_x\approx18$ MeV in ^{6}Li, Nakayama et al. observed the ^1P (T=0) resonance with a width of Γ=5 MeV [27]. These experimental results seriously are in contradiction to the result by Hernández et al. [25]. As discussed below, the G2 resonances observed in the present work are supposed to be an isobaric triplet in A=6. Therefore, the isoscalar resonance assigned by Nakayama et al. at 18 MeV [27] is different from the 18 MeV resonance (T=1) observed in the present (p,p') reaction.

In ^{6}Be, Yang et al. reported a dipole resonance at E_x=15 MeV via the (p,n) reaction. In ^{6}He, the 15 MeV resonance has been observed and assigned to be the dipole resonance via the (n,p) reaction, (E_x=15.5±0.5 MeV and Γ=4±1.5 MeV) [9], in the (^{7}Li,^{7}Be) reaction, (E_x=14.6±0.7 MeV and Γ=7.4±1.0 MeV) [10], in the (t,^{3}He) reaction at 336 MeV, (E_x=14.6±0.2 MeV and Γ=5.9±0.7 MeV) [28], and our previous results of the (^{7}Li,^{7}Be) reaction (E_x=15.0±0.5 MeV and Γ=3.0±0.5 MeV) [20].

In both ^{7}Li and ^{7}Be we observed the dipole resonance at E_x=9.5 MeV (G3). In the (p,n) reaction, Yang et al. also observed a dipole resonance at $E_x\approx10$ MeV in ^{7}Be [11]. These resonances have not been observed in the (γ,n) reactions [4, 5, 7], and no positive parity states have been reported in this excitation energy region [13, 14].

Excited α-cluster

Cross section ratios

The ratios of the cross section of the G1 resonance in A=6 to that in A=7 in each isotope are listed in Table 2. The value of the cross section itself depends strongly on both the reaction employed and nuclear structure. However, the ratio of the cross sections using the same reaction may depend only on the nuclear structure. Therefore, if the G1 resonance is due to the excitation of the α-cluster these values should become unity, because the same constituent (^{4}He) in each isotope is excited via the same reaction. Though these values scatter considerably due to ambiguity for the peak fitting, the average value turns out to be 1.2±0.3, which agrees with unity within the error. In the (^{7}Li,^{7}Be) reaction we measured the ΔS=1 and ΔS=0 cross sections for the G1 resonance in ^{6,7}He [12]. The ratio of the ΔS=1 cross section to ΔS=0 cross section has been found to be 1.2±0.4 and 1.4±0.7 for ^{6}He and ^{7}He, respectively. This suggests that the observed peak of G1 consists of the SDR and the GDR in comparable strength to that observed in light nuclei [17].

The cross section ratios for the G1 resonance to the GDR for every nucleus measured herein are shown in Table 3. The averaged value of the present work is 0.44±0.08 The ratios of the GDR cross-sections for ^{4}He to those for ^{6,7}Li reported in the (γ,n) reaction [4] are also shown. For the GDR in ^{4}He, the GDR strength in the (γ,n) reaction was reported to be nearly equal to that in the (γ,p) reaction [18]. Since this GDR strength was missing in the (γ,n) reaction, we estimated the GDR cross section as twice of that of the (γ,n) reaction, as shown in Table 3. If the G1 resonance is due to the excitation of the α-cluster, the present ratios should be the same as those obtained in the photo-nuclear

TABLE 2. The cross-section ratios of the G1 resonance in A=6 to that in A=7 for each isotope.

	He	Li	Be	Averaged value
$\sigma(A=6)/\sigma(A=7)$	1.2±0.3	0.8±0.2	1.5±0.7	1.2±0.3

TABLE 3. The cross section ratios of the G1 resonance to the GDR.

	^{6}He	^{6}Li	^{6}Be	^{7}He	^{7}Li	^{7}Be
$\sigma(G1)/\sigma(GDR)$*	0.23±0.08	0.50±0.15	0.40±0.16	0.43±0.2	0.70±0.20	0.36±0.14
$\sigma(^4He)/\sigma(Li)^\dagger$		0.23			0.26	
$\sigma(^4He)/\sigma(Li)$**		0.46			0.52	

* Present work. The averaged value is 0.44±0.08.
† (γ,n) reaction Ref. [4].
** $\sigma(\gamma,n) \approx \sigma(\gamma,p)$ is assumed for ^{4}He.

reactions ((γ,n)+(γ,p)). The present averaged value is consistent with the ratios obatined in the photo-nuclear reactions.

Excitation energy systematics

Figure 4 shows the energy diagrams of A=6 and 7 nuclei observed in the present work. The energy diagram for ^{4}He is also shown as reference. In order to clarify the isobaric analog relations in A=6 nuclei and in 7 nuclei, the analog states of 0^+, $T=1$ in A=6 and of $3/2^-$, $T=3/2$ in A=7 are shown. Furthermore, the vertical positions of ^{6}Li and ^{7}Li are shifted such that the excitation energies for the separation energy of an α-particle in these nuclei coincide with that of the ground state of ^{4}He.

As seen in Fig. 4, it is remarkable that the resonance positions of G1 in every nucleus agree well with that of the GDR in ^{4}He, i.e. $E_x \approx 26$ MeV, irrespective of mass, although their excitation energies themselves have very different values.

CONCLUSION

In order to search for the excited α-cluster in nuclei, we investigated the ^{6,7}Li(p,p'), ^{6,7}Li(^{3}He,t) and ^{6,7}Li(^{7}Li,^{7}Be) reactions at the incident energies of 300, 450 and 455 MeV, respectively.

Resonances were observed in ^{6}Li, ^{6}He, and ^{6}Be at E_x=27.0, 24.0, and 23.5 MeV, respectively, and in ^{7}Li, ^{7}He, and ^{7}Be at E_x=29.0, 18.0, and 28 Mev, respectively. All these resonances were assigned to be dipole resonances and the resonance shapes were reproduced well with the resonance shape of the GDR of ^{4}He measured in the (γ,n) reaction. The ratios of the cross sections of the resonance in A=6 to that in A=7 for each isotope turned out to be nearly unity, 1.2±0.3. The averaged value of the respective

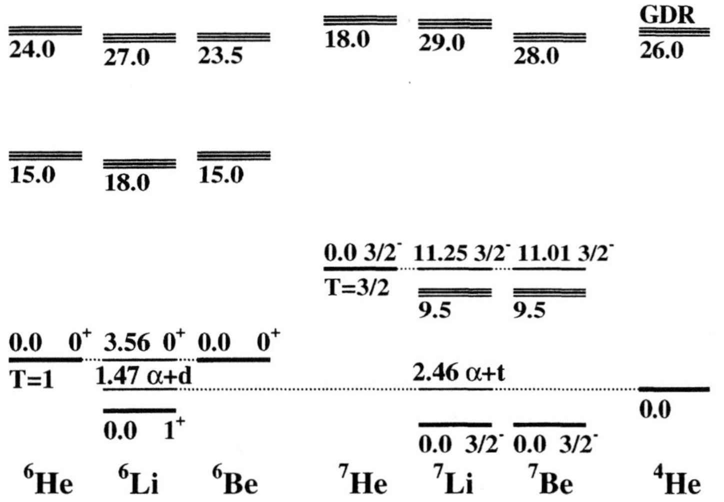

FIGURE 4. Energy diagrams for A=6 and 7 nuclei observed in the present work. The diagram for ^{4}He is also shown. The solid-thick lines show the ground state of each nucleus. The vertical position in this figure is set such that the positions corresponding to the separation energies of α-particle in ^{6}Li and ^{7}Li coincide to the that of the ground state of ^{4}He.

cross section ratios of these resonances to the GDR in target nuclei were 0.44±0.08, which was found to be consistent with the ratios of the cross sections of the GDR in ^{4}He to those of the GDR in ^{6,7}Li estimated from the photo-nuclear reactions [4, 18]. The excitation energies of these dipole resonances in A=6 and 7 nuclei measured from the separation energies of α-particle in ^{6}Li and ^{7}Li, respectively, agreed well with the excitation energy of the GDR in ^{4}He, i.e. about 26 MeV, by taking into account the Coulomb displacement energies. We conclude that the resonances observed in ^{6,7}Li are the GDR of α-cluster and the resonances in ^{6,7}He and ^{6,7}Be are their analogs.

Not only the resonance due to α-cluster, but also those due to the GDR of d and/or t clusters should be present in the spectra. However, since the excitation energies of these resonance are low [5, 29], they could not be distinguished from the GDR in the target nuclei.

As byproducts of the present work, in both ^{7}Li and ^{7}Be a new dipole resonance was found at E_x=9.5 MeV, which might be isobaric analogs of one another. The three dipole resonances were observed in ^{6}He, ^{6}Li, and ^{6}Be at E_x= 15, 18, and 15 MeV, respectively. Based on the similarity of the excitation energies and widths we suppose that these dipole resonances are members of the A=6 isobaric triplet.

ACKNOWLEDGMENTS

These experiments were performed at the Research Center for Nuclear Physics, Osaka University under the Program Nos. E172, E184, E190, and E202. The authors are grateful to the staff of the RCNP cyclotron for their support. This work was supported partly by the Grant-in-Aid (No. 13640304) by the Japan Ministry of Education, Culture, Sports, Science, and Technology, and by the Japan Society for Promotion of Science (JSPS).

REFERENCES

1. K. Ikeda, J. Phys. Soc. Jpn. **58**, 277 (1989) Suppl.
2. H. Horiuchi, J. Phys. Soc. Jpn. **58**, 7 (1989) Suppl.
3. S. Costa, S. Ferroni, V. Wataghin, and R. Malvano, Phys. Lett. **4**, 308 (1963).
4. B.L. Berman and S.C. Fultz, Rev. Mod. Phys. **47**, 713 (1975).
5. S.S. Dietrich and B.L. Berman, At. Data Nucl. Data Tables **38**, 199 (1988).
6. N. Dytlewski, S.A. Siddiqui, and H. H. Thies, Nucl. Phys. **A430**, 214 (1984).
7. J. Ahrens et al, Nucl. Phys. **A251**, 479 (1975).
8. S. Nakayama et al., Prog. Theor. Phys. Suppl. **146**, 603 (2002).
9. F.P. Brady et al., J. Phys. **G10**, 363 (1984).
10. J. Jänecke et al., Phys. Rev. **C54**, 1070 (1996): T. Annakkage et al., Nucl. Phys. **A648**, 3 (1999).
11. X. Yang et al., Phys. Rev. **C52**, 2535 (1995).
12. S. Nakayama et al., Phys. Rev. Lett. **87**, 122502-1 (2001).
13. F. Ajzenberg-Selove, Nucl. Phys. **A490**, 1 (1988).
14. D.R. Tilley et al., Nucl. Phys. **A708**, 3 (2002).
15. W.G. Love and M.A. Franey, Phys. Rev. **C24**, 1073 (1981).
16. F. Petrovich and W.G. Love, Nucl. Phys. **A354**, 499c (1981).
17. S. Nakayama et al., Nucl. Instrum. Methods Phys. Res., **A402**, 367 (1998); S. Nakayama et al., Phys. Rev. **C60**, 047303-1 (1999).
18. A.N. Gorbunov, JETP (Sov. Phys.) **8**, 88 (1968).
19. M. Fujiwara et al., Nucl. Instrum. Methods Phys. Res., **A422**, 484 (1999).
20. H. Akimune et al., Phys. Rev. **C67**, 051302-1 (2003).
21. S. Nakayama et al., Phys. Rev. Lett. **85**, 262 (2000).
22. A. Erell et al., Phys. Rev. **C34**, 3 (1986).
23. H. Akimune et al., Phys. Rev. **C64**, 041305-1 (2001).
24. B.M. Sherrill et al., Nucl. Instrument Methods in Phys. Res. **A432**, 299 (1999).
25. E. Hernández and A. Mondragón, J. Phys. **G16**, 1339 (1990); A. Mondragón and E. Hernández, Phys. Rev. **C41**, 1975 (1990).
26. H. Akimune et al., in Proc. Tours Symp. on Nuclear Physics V, 2003.
27. S. Nakayama et al., To be published in Nucl. Phys. **A**.
28. T. Nakamura et al., Phys. Lett. **B493**, 209 (2000).
29. K.Y. Hara et al., Phys. Rev. **D68**, 072001 (2003).

^{6}Li excitation above the breakup threshold in the ^{6}Li+^{208}Pb system at Coulomb barrier energies

M. Mazzocco*, P. Scopel*, C. Signorini*, L. Fortunato*, F. Soramel[†],
I.J. Thompson**, A. Vitturi*, M. Barbui[‡], A. Brondi[§], M. Cinausero[‡],
D. Fabris*, E. Fioretto[‡], G. La Rana[§], M. Lunardon*, R. Moro[§], A. Ordine[§],
G. Prete[‡], V. Rizzi*, L. Stroe[‡], M. Trotta[‡], E. Vardaci[§] and G. Viesti*

*Physics Department and INFN, via Marzolo 8, I-35131 Padova, Italy
[†]Physics Department and INFN, via delle Scienze 208, I-33100 Udine, Italy
**Physics Department, University of Surrey, Guilford GU2 7XH, United Kingdom
[‡]INFN, Laboratori Nazionali di Legnaro, viale dell'Università 2, I-35020 Legnaro (Padova), Italy
[§]Physical Science Department and INFN, Complesso Universitario di Monte Sant'Angelo, via Cinthia, I-80126 Napoli, Italy

Abstract. The excitation energy of the ^{6}Li+^{208}Pb system has been studied at several energies around the Coulomb barrier with a large solid angle detector array. The $\alpha + d/^6$Li* excitation above the breakup threshold (S_d = 1.475 MeV) has been deduced from the invariant mass of the $\alpha + d$ system. All the collected data show a similar behaviour: the cross sections are peaked at around 1 MeV above the threshold and they have an exponential decay on the high energy side, suggesting a direct breakup process. A Monte-Carlo simulation of the whole reaction mechanism has been performed in order to check the validity of data analysis procedure and to study the influence of the set-up geometry onto the experimental results.

Two different theoretical approaches, the former based on fully quantum mechanical Coupled Channel calculations and the latter based on simpler semi-classical calculations, treating the relative motion of the $\alpha + d$ cluster structure along a classical trajectory, reproduce both in shape and absolute value the experimental excitation energy distributions.

INTRODUCTION

Over the last years, there has been a growing interest for the nuclear reactions induced by Radioactive Ion Beams (RIBs). Many experiments have been performed up to now with light loosely bound RIBs in the energy range around the Coulomb barrier. In particular, the interaction of these nuclei with heavy targets has been largely studied. Some experiments were focused onto the study of the elastic scattering (see, for example, [1] for the ^{6}He+^{209}Bi system and [2] for ^{8}Li+^{208}Pb reaction), others onto the fusion process (^{11}Be+^{209}Bi [3] and ^{17}F+^{208}Pb [4]) or onto the breakup reaction (^{17}F+^{208}Pb [5, 6]).

Moreover for this kind of projectiles there are still some open questions in this energy range. In fact light RIBs typically have r.m.s larger than systematics with a consequent lowering of the Coulomb barrier and an expected enhancement of the fusion cross section. However, on the other side, these projectiles are very loosely bound and they could easily break before the fusion process could take place. In this case the fusion cross section should be hindered.

Another very important information about these nuclei is the projectile excitation function, *i. e.* how the continuum spectrum above the breakup threshold is populated during a breakup process. Similar works have been already performed for three different reactions involving light RIBs: ^{11}Be [7], ^{8}B [8] and ^{11}Li [9]. All these studies were performed at energies much higher than the Coulomb barrier: 72, 46 and 43 AMeV, respectively. However, in spite of the low intensity of these beams and their poor energy resolution, the experimental results show almost the same behaviour: the excitation energy distributions are peaked at around 0.5 - 1 MeV above the breakup threshold and they present, on the high energy side, an exponential decay, suggesting a direct breakup process.

THE EXPERIMENT

The ^{6}Li + ^{208}Pb system

On this research line, we have undertaken the study of the excitation energy distribution in the system ^{6}Li + ^{208}Pb. This reaction was chosen since the ^{6}Li is the most weakly bound nucleus among all the stable ones, its binding energy being only 1.475 MeV. This value is larger in comparison with the few hundreds keV binding energies of the previously mentioned RIBs, but it is much lower than all the nuclei close to the stability valley. Moreover, ^{6}Li is a stable nucleus and it is possible to obtain very high beam intensities with very good energy and position resolution. Then, ^{6}Li usually breaks into at least two charged particles ($\alpha + p$ or, in case, $\alpha + p + n$), that could be easily detected with high efficiency. Finally the elastic scattering of this reaction has been studied at many different beam energies across the Coulomb barrier with very high accuracy [10].

The experiment was performed at the Legnaro National Laboratories (LNL) of the INFN, with a ^{6}Li beam delivered from the Tandem Van de Graaff accelerator. The target was an enriched, self supporting, 200 μg/cm^2 ^{208}Pb foil. The light charged particles produced during the reaction were detected with the experimental apparatus 8πLP [11], only the so-called "BALL" portion of the whole array was used. This portion is made up by 126 two-stage telescopes and covers the polar angles θ ranging from 34° to 163°. The inner detector is a Si 300 μm thick, whereas the outer one is a CsI 5 mm thick. Each telescope spans a solid angle coverage of 17° in θ direction and 20° in ϕ direction.

Inclusive and exclusive alpha particles cross sections

We first investigated the inclusive alpha particle production cross section at several energies around the Coulomb barrier [12]. These cross sections resulted to be as large as in other reactions involving light weakly bound projectiles: ^{9}Be + ^{209}Bi [13], ^{6}He + ^{209}Bi [14] (see Fig. 1). This fact suggests the presence of a very strong reaction channel even at energies below the Coulomb barrier and it should influence all other open channels.

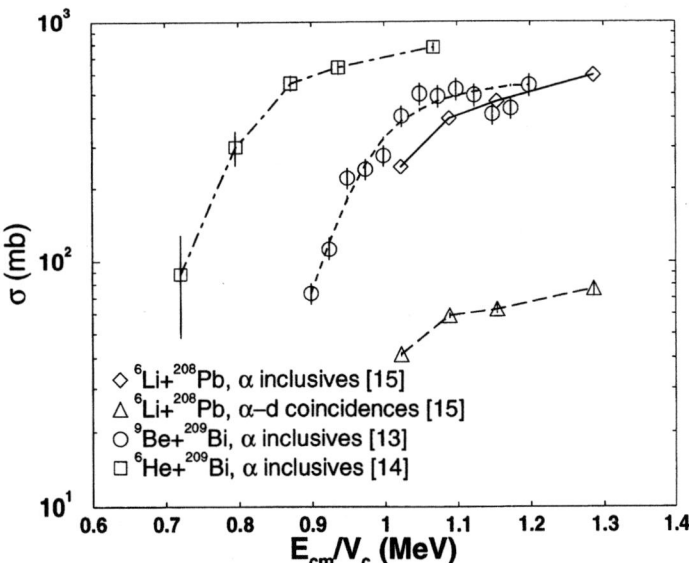

FIGURE 1. Experimental cross sections of the α inclusive production and the $\alpha + d$ coincidence events in the ^{6}Li + ^{208}Pb system and α inclusive cross sections in the reactions: ^{9}Be + ^{209}Bi and ^{6}He + ^{209}Bi. For a better comparison among the different systems, the data are plotted *versus* $E_{c.m.}/V_c$, with V_c Coulomb barrier.

In order to get a deeper understanding on this topic, we measured the exclusive breakup cross sections, analyzing the $\alpha + d$ coincidence events [15]. In this energy range, because of the breakup kinematics, the two fragments are emitted into narrow cones around the ^{6}Li scattering direction. The openings of these cones are strongly connected with the excitation energy of the breakup process. For instance, for breakup processes passing through the first ^{6}Li excited state above the threshold ($E_{rel} = 0.711$ keV), the openings of the two cones are around $\sim 7°$ for the alpha particles and $14°$ for the deuterons, respectively. Due to the granularity of the "BALL" portion of the 8πLP apparatus ($17°$ (θ) $\times$ $20°$ (ϕ)), in this first phase we limited our analysis to coincidence events of an alpha particle and a deuteron into two adjacent detectors.

Fig. 1 shows both the inclusive alpha particle cross sections and the $\alpha + d$ coincidence event cross sections. The inclusive data are much larger than the exclusive ones, even if the trends with the energy are quite similar. This difference evidences the presence of a reaction mechanism where one of the fragments, in this case the deuteron, is captured by the target. The comparison between the exclusive breakup cross sections and the total deuteron production cross sections [16] has also shown the presence of the reciprocal channel, where the alpha particle is captured by the target. This process, already theoretically predicted [17] and sometimes observed in reactions involving weakly bound projectiles [5, 18, 19], has been called "stripping breakup".

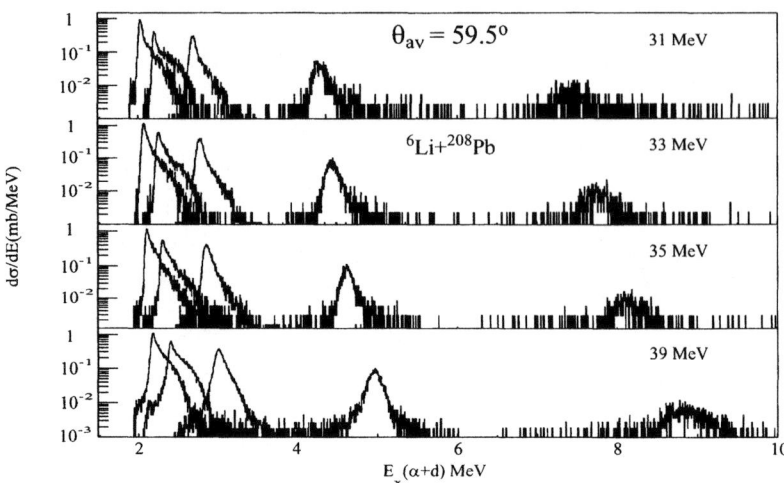

FIGURE 2. Invariant mass spectra of the $\alpha + d/^6$Li* system at four beam energies. The five peaks inside each panel correspond to the five different geometrical combinations of detectors hit by the two fragments, as explained in the text.

The excitation energy distribution

The experiment was also aimed to determine the excitation energy distribution of the $\alpha + d/^6$Li* system [20]. For this purpose from the invariant mass M of the system:

$$M = \sqrt{\left(\sum_i \frac{E_i}{c^2}\right)^2 - \left(\sum_i \frac{\mathbf{P}_i}{c}\right)^2} = \frac{E_{rel}}{c^2} + \sum_i m_i \qquad (1)$$

where E_i, $\mathbf{P}_i$, m_i are, respectively, the total energy, the momentum and the mass of the two fragments (α and deuteron), we can get the ^{6}Li excitation energy E_x through the equation $E_x = E_{rel} + S_d$ (S_d being the deuteron binding energy).

In order to provide a correct reconstruction of ^{6}Li excitation energy, we need the detailed knowledge of the angle τ between the two hitting particles. Because of the limited resolution of the experimental apparatus, we assume for each coincidence event an average angle $<\tau>$ equal to the angle between the centers of the two hit detectors. Obviously, only a given number of this angle $<\tau>$ is accessible with this setup. The average angles we considered were: 17°, 20°, 26°, 40° and 60°. No significant coincidences were found at relative angles larger than 60°.

The results of the performed analysis at four different beam energies are shown in Fig. 2. The five different peaks correspond to the five different average angles $<\tau>$ between the two hitting fragments. The analysis was performed for three central rings of

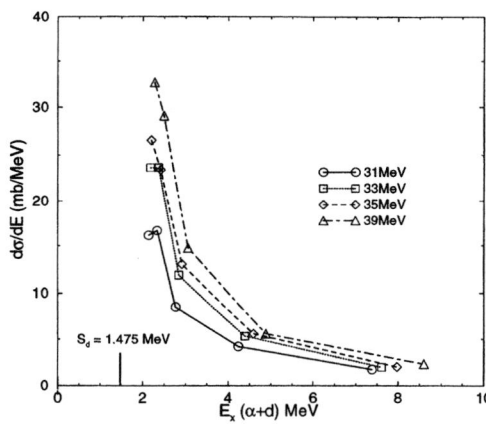

FIGURE 3. Excitation energy distribution of the $\alpha + d / ^6\text{Li}^*$ system at four beam energies.

the "BALL" of the 8πLP apparatus and the behaviour is practically the same for the three considered polar angles θ. The energy resolution of the experimental points is about 0.4 MeV at low energy and around 1.7 MeV for the highest energy point, mainly because of the limited granularity of the present set-up. Fig. 3 shows that the cross section trend is very similar to the ones obtained at much higher beam energy for the light RIBs previously mentioned [7, 8, 9]. The exponential decay at around 1 MeV above breakup threshold suggests a direct breakup process.

MONTE-CARLO SIMULATION

We performed a Monte-Carlo simulation of the whole experimental analysis procedure in order to check the validity of the assumptions made in the evaluation of the invariant mass of the $\alpha + d / ^6\text{Li}^*$, namely that the angle τ between the two fragments were equal to the average angle $<\tau>$ between the centers of the two hit detectors. The breakup process was schematized as follows:

1. we assumed that the breakup occurs along the Rutherford trajectory, during the ^{6}Li elastic scattering. So we always deal with a two-body process: we have first the ^{6}Li scattering and then the excitation with the subsequent breakup;
2. the scattered ^{6}Li nuclei are randomly emitted in both θ and ϕ direction in the laboratory frame. This assumption was done considering the quite flat behaviour of the $\alpha + d$ angular distributions [15];
3. the angle between the ^{6}Li and α and d momenta is randomly chosen in the center of mass frame, since no privileged emission angle realistically exists in this system.

Fig. 4 shows the distributions of the relative angle τ between the alpha particle and the deuteron for four different excitation energies of the breaking ^{6}Li. The maxima of the curves are below 20° up to $\sim$ 1 MeV ^{6}Li excitation energy. This explains why the

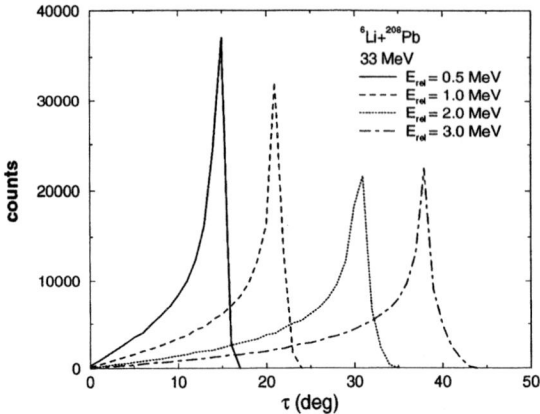

FIGURE 4. Distribution of the relative angle τ between an alpha particle and a deuteron coming from the breakup of ^{6}Li for four different excitation energies. The calculations were performed at 33 MeV beam energy for an average detection angle $\theta = 76.5°$.

TABLE 1. Percentage of α particles and deuterons simultaneously hitting the same detector with respect to all the coincidence events detectable by each ring of the 8πLP apparatus for $E_{rel} = 1$ MeV at two different ^{6}Li beam energies.

Ring	θ_{av} (°)	E = 33 MeV	E = 39 MeV
G	42.5	15%	18%
F	59.5	14%	17%
E	76.5	20%	26%
D	103.5	19%	25%
C	120.5	13%	15%
B	137.5	10%	12%
A	154.5	11%	12%

distributions shown in Figs. 2 and 3 do not start at the breakup threshold. In fact, because of the granularity of the apparatus it is not possible to detect the low E_{rel} coincidence events, where the two particles most likely hit the same detector. However, the simulation helps us to evaluate the fraction of these lost coincidence events with respect to total amount of available coincidences and Table 1 summarizes the results of the simulation for two beam energies.

In the Monte-Carlo simulation we reproduced all the experimental conditions in order to evaluate the average angle between an alpha particle and a deuteron hitting two different detectors. Fig. 5 shows the distributions of the τ angle for the five previously discussed geometrical combinations of detectors hit by the two fragments. The simulation confirmed the validity of our assumptions in the analysis procedure. In fact the average values reported in Fig. 5 agree within a few degrees ($\sim 2-4°$) with the five $<\tau>$ angles used in the calculation of the invariant mass of the $\alpha + d$ system. We can

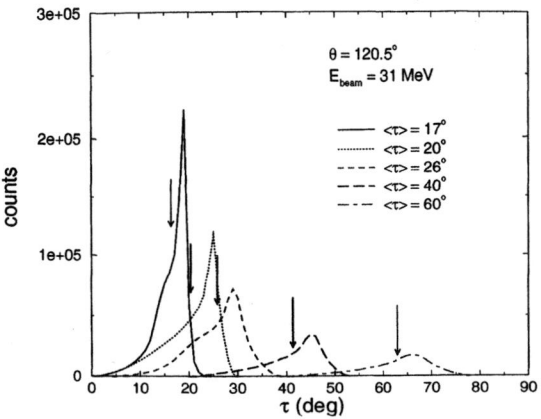

FIGURE 5. Distribution of the relative angle τ between the alpha particle and the deuteron from ^{6}Li breakup for the five different geometrical combinations of hit detectors. The calculations were performed at 31 MeV beam energy for an average detection angle $\theta = 120.5°$. The arrows indicate the average value of each distribution.

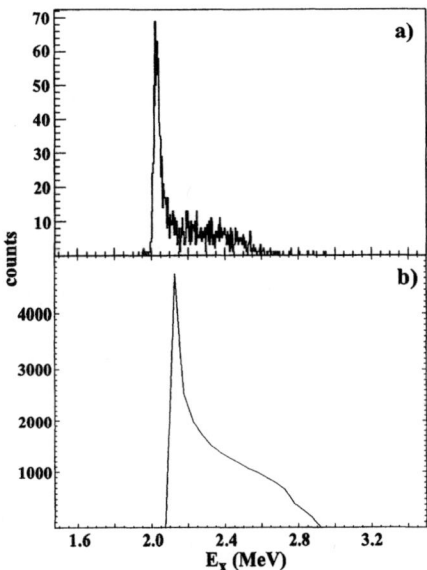

FIGURE 6. Experimental (a) and simulated (b) invariant mass spectrum of the $\alpha + d/^6\text{Li}^*$ system at 33 MeV beam energy for a telescope located at an average polar angle $\theta = 137.5°$.

also appreciate how the calculated τ angle distributions are quite sharply peaked around their average values.

We have also reproduced the experimental invariant mass spectra and Fig. 6 shows the good agreement we achieved between the simulated data (a) and the experimental results (b). However, using a finer description of the ^{6}Li excitation energy distribution a better agreement could easily be reached.

THEORETICAL DESCRIPTIONS

We have also described the breakup process from a theoretical point of view using two different approaches.

CDCC calculations

The first approach was based onto the full quantum mechanical CDCC (Continuum Discretized Coupled Channels) schematization using the code Fresco [21]. As described in a previous work [15], we assumed for the incoming ^{6}Li nucleus an $\alpha + d$ cluster structure. Couplings to discretized continuum states up to 11.5 MeV excitation energy were considered, including both ground state to continuum and continuum-continuum couplings and considering both Coulomb and nuclear excitations. No free parameters were used, they were all deduced from basic theory or from experimental data. The results are shown in Fig. 7 and it is possible to see two low-lying resonances (the former is the $J^\pi = 3^+$ state at 2.19 MeV and the latter is the $J^\pi = 2^+$ level at 4.31 MeV). There is no experimental evidence of these states, mainly because of our limited energy resolution. However the first resonance was well evidenced in some previous experimental results at 60 MeV [22] and 156 MeV ^{6}Li beam energy [23].

Semi-classical coupled channel approach

The second set of calculation was based on a simpler approach. The ^{6}Li nucleus was still assumed to be a dicluster ($\alpha + d$) structure and the projectile excitations were still calculated within the usual coupled channel framework, but in this case the relative motion is treated classically, along the classical trajectory. Therefore, instead of a fully quantum mechanical calculation, the coupled equations are semi-classically solved [24]. Moreover continuum-continuum coupling were not included. Both these features make the calculation simpler than CDCC calculations, allowing for a much finer subdivision of the continuum and for a lower CPU time. Fig. 7 shows that the excitation energy is strongly peaked around the 3^+ resonance state, whereas the non resonant continuum has already smeared the 2^+ states.

The main differences arise form the different bin subdivision of the continuum and the inclusion of continuum-continuum couplings. However, the results for the total $\alpha + d$ breakup cross sections are in good agreement and they both slightly overestimate the experimental cross sections (see Table 2).

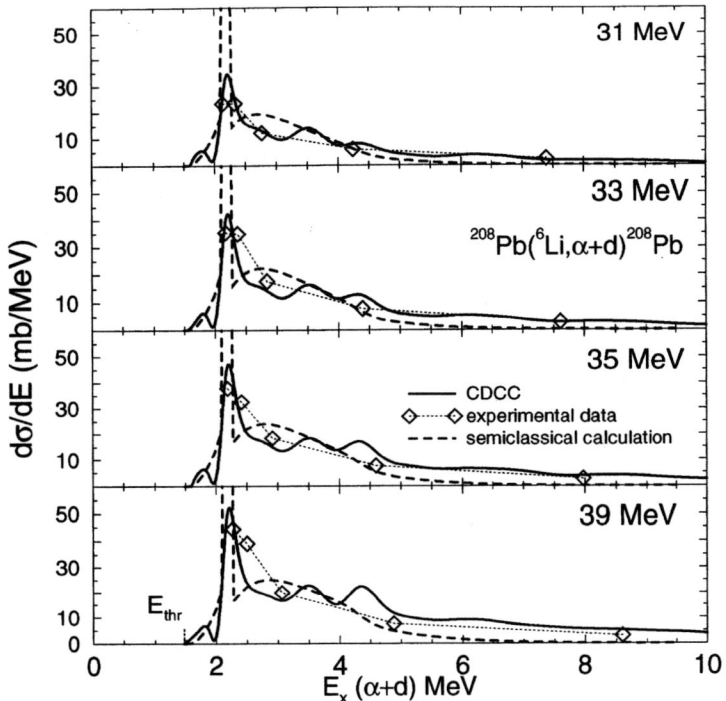

FIGURE 7. Experimental and theoretical excitation energy distribution of the $\alpha+d/^6\text{Li}^*$ system at four beam energies. The maxima of the curves related to the semi-classical calculations, located respectively at 177 mb/MeV (31 MeV), 242 mb/MeV (33 MeV), 276 mb/MeV (35 MeV) and 285 mb/MeV (39 MeV), have been cut for graphical purposes

TABLE 2. Comparison between the experimental ^{6}Li exclusive breakup cross sections and the ones evaluated by the semi-classical coupled channel calculations and by the CDCC approach.

^{6}Li Beam Energy	Experimental Data	Semi-classical calculations	CDCC approach
31 MeV	41(1) mb	55 mb	49 mb
33 MeV	59(1) mb	69 mb	65 mb
35 MeV	63(3) mb	78 mb	81 mb
39 MeV	77(5) mb	85 mb	102 mb

CONCLUSIONS

The excitation energy distribution of the $\alpha+d/^6\text{Li}^*+^{208}\text{Pb}$ system has been measured for the first time at four beam energies around the Coulomb barrier. We extended a previously performed work at higher beam energy [22] up to ~ 8 MeV above the breakup threshold, with an energy resolution ranging form 0.4 to 1.7 MeV.

The distibutions, obtained from the invariant mass of the $\alpha+d/^6\text{Li}^*$ system, result to be very similar at all beam energies and all measured polar angles θ. Their trend is also similar to the ones observed with three light RIBs at much higher beam energy. This supports the interpretation of the breakup as a process mainly originating from a threshold effect. The low-lying resonances could not be evidenced by our experiment because of the experimental apparatus limited resolution. However, Monte-Carlo simulations confirmed the validity of the assumptions made in the analysis procedure and allowed to evaluate the effects of the set-up geometry onto the experimental results.

Two different theoretical approaches without any free parameters, the former based onto a fully quantum mechanical CDCC framework and the latter onto a semi-classical Coupled Channels description, reproduced quite well both the experimental energy distribution and the total breakup cross sections.

REFERENCES

1. E.F. Aguilera et al., *Phys. Rev. C* **63**, 061603(R) (2001).
2. A.M. Moro et al., *Phys. Rev. C* **68**, 034614 (2003).
3. C. Signorini et al., *Eur. Phys. J. A* **2**, 227 (1998).
4. K.E. Rehm et al., *Phys. Rev. Lett* **81**, 3341 (1998).
5. J.F. Liang et al., *Phys. Lett. B* **491**, 051603(R) (2002).
6. J.F. Liang et al., *Phys. Rev. C* **65**, 23 (2000).
7. T. Nakamura et al., *Phys. Lett. B* **331**, 296 (1994).
8. T. Motobayashi et al., *Phys. Rev. Lett* **73**, 2680 (1994).
9. K. Ieki et al., *Phys. Rev. Lett* **70**, 730 (1993).
10. N. Keeley et al., *Nucl. Phys.* **A571**, 326 (1994).
11. G. Prete et al., *Nucl. Instrum. Methods Phys. Rev. A* **422**, 263 (1999).
12. C. Signorini et al., *Eur. Phys. J. A* **10**, 249 (2001).
13. C. Signorini et al., in *Nucleus-Nucleus Collisions*, Proceedings of the International Conference *BO2000*, Bologna, Italy, May 2000, edited by G.C. Bonsignori et al., (World Scientific, Singapore 2001) p. 413.
14. E.F. Aguilera et al., *Phys. Rev. Lett.* **84** 5058 (2000).
15. C. Signorini et al., *Phys. Rev. C* **67**, 044607 (2003).
16. C. Signorini et al., Proceedings of the 10th International Conference on Nuclear Reaction Mechanisms, Villa Monastero, Varenna, Italy, 9-13 June 2003 (to be published) and http://www.mi.infn.it/ gadioli/.
17. G. Baur et al., *Nucl. Phys.* **A458**, 188 (1986).
18. M. Dasgupta et al., *Phys. Rev. Lett.* **82** 1395 (1999).
19. R.I. Badran et al., *Eur. Phys. J. A* **12**, 317 (2001).
20. M. Mazzocco et al., *Eur. Phys. J. A* in press.
21. I.J. Thompson, *Comput. Phys. Rep.* **7**, 167 (1988).
22. J. Hesselbarth et al., *Phys. Rev. Lett.* **67**, 2773 (1991).
23. J. Kiener et al., *Phys. Rev. C* **44**, 2195 (1991).
24. C.H. Dasso et al., *Nucl. Phys.* **A639**, 635 (1998).

Resonance structure of ^{9}Be and ^{10}Be in a microscopic cluster model

Koji Arai

Department of Physics, University of Surrey, Guildford GU2 7XH, UK

Abstract. Structure of the excited states in ^{9}Be and ^{10}Be have been theoretically explored by means of the $\alpha+\alpha+n$ and $\alpha+\alpha+n+n$ microscopic cluster model respectively. Resonance excited states in ^{9}Be and ^{10}Be are localized by solving the two-body scattering problems of $\{^8\text{Be}(0^+, 2^+, 4^+)+n,$ $^5\text{He}(3/2^-, 1/2^-)+\alpha\}$ and $\{^9\text{Be}(3/2^-, 1/2^+, 5/2^-)+n, ^6\text{He}(0^+, 2^+)+\alpha\}$ respectively. Our models have reproduced satisfactorily the experimental low-energy spectrum in ^{9}Be and given various resonance excited states in ^{10}Be. Our model shows that the two and three competing configurations are quite essential to reproduce the anomalous $1/2^+$ state in ^{9}Be and the second 0^+ state in ^{10}Be respectively. One of these configurations has a strong core(^{8}Be) distortion which is induced by the valence neutron and is responsible for lowering the $s_{1/2}$ orbit in these states. Another configuration produces a spatially extended neutron distribution outside the core like the neutron halo.

INTRODUCTION

Recently certain interests for structures of the Be isotope are rapidly growing due to the recent experimental and theoretical discussions for the possible molecular structures such as ^{6}He+α and ^{6}He+^{6}He(^{8}He+α) among the excited states of ^{10}Be and ^{12}Be[1-10]. Considering the beryllium isotopes, ^{8}Be is an ideal example of a nucleus exhibiting α clustering and ^{9}Be has a good $\alpha+\alpha+n$ three-cluster structure since its ground state has a small binding of $E=-1.6$ MeV with respect to the three-body threshold [11]. In ^{9}Be, the $1/2^+$ state is observed as the first exited state and is lower than the $5/2^-$ and $1/2^-$ excited states, despite the last neutron in this $1/2^+$ state presumably occupying the $s_{1/2}$ orbit in the simple shell model configuration whereas the next two excited states with negative parity would have p-shell configurations. This parity inversion, in addition to the strong $E1$ transition to the ground state as seen also in ^{11}Be and ^{13}C [12], naturally invites an analogy with the ground state of ^{11}Be. With regards to the $1/2^+$ state of the mirror nucleus ^{9}B, Barker predicts an inverted Thomas-Ehrman shift [13] because the $1/2^+$ state of ^{9}Be is not bound like the $1/2^+$ neutron states in ^{13}C and ^{17}O. However neither the resonance position nor the total width have been determined experimentally for ^{9}B.

Concerning ^{10}Be, adding one further additional neutron to ^{9}Be, in contrast with ^{9}Be, causes a large binding of $E=-6.8$ MeV with respect to the ^{9}Be+n threshold [11] and therefore the ground state of ^{10}Be is expected to have a good shell model like structure (p shell configuration). Nevertheless some theoretical studies indicate that similar $\alpha+\alpha$ clustering persists even in ^{10}Be[7, 14, 15]. As for the excited states of ^{10}Be, four states with different spins and parities, $J^\pi=2^+_2, 1^-, 0^+_2,$ and 2^- are concentrated within 0.3

MeV just below the ^{9}Be+n and ^{6}He+α thresholds [11]. Among these four states, the second 0^+ state is a well known intruder state in the shell model[16] and a good candidate for an exotic state holding a developed ^{6}He+α clustering. Recent experiments indicate that the 2^+ state at E_x=7.54 MeV has a developed ^{6}He+α structure and could be a member of the $K^\pi=0_2^+$ band[3, 17].

Since some theoretical calculations indicate that $\alpha+\alpha$ clustering persists as a core throughout heavier Be isotopes[14, 15], it could be a good approximation to employ a picture of the two α particles plus valence neutron(s) in order to discuss the cluster and molecular structures in ^{9}Be and ^{10}Be. In fact we employ the $\alpha+\alpha+n$ and $\alpha+\alpha+n+n$ microscopic cluster models for ^{9}Be and ^{10}Be respectively and in addition we consider not only the bound states but also the resonance excited states in these nuclei. More details can be found in Ref.[18, 8, 19].

MODELS

We have studied ^{9}Be and ^{10}Be with the $\alpha+\alpha+n$ and $\alpha+\alpha+n+n$ microscopic cluster models respectively according to the Resonating Group method(RGM). In this method the nucleons are assumed to be arranged in 0s clusters and all nucleons are treated explicitly. The wave function is constructed to satisfy the Pauli principle for all nucleons exactly and is free from spurious center-of-mass motion whilst also having good total angular momentum and parity. The intrinsic wave functions of the constituent clusters, here simply the α particles, are taken to be simple shell-model wave functions built up from 0s harmonic-oscillator states. The wave functions can be obtained as an approximate solution of the 9- or 10-nucleon Schrödinger equations where the Minnesota potential with the spin-orbit force (Reichstein-Tang force) and the coulomb force is employed as the effective nucleon-nucleon force[20].

In this work, we have calculated not only the bound states but also the resonance excited states in ^{9}Be and ^{10}Be. These resonance excited states are localized by solving the two-body scattering problems of $\{^8$Be$(0^+, 2^+, 4^+)+n\}+\{^5$He$(3/2^-, 1/2^-)+\alpha\}$ in ^{9}Be and of $\{^9$Be$(3/2^-, 1/2^+, 5/2^-)+n\}+\{^6He(0^+, 2^+)+\alpha\}$ in ^{10}Be. The wave functions of ^{8}Be and ^{5}He are given by the $\alpha+\alpha$ and $\alpha+n$ two-cluster models respectively and those of ^{9}Be and ^{6}He are by the $\alpha+\alpha+n$ and $\alpha+n+n$ three-cluster models respectively. These wave functions are approximated by a small number of the basis sets and the real square integrable functions. The two-body multi-channel S-matrices are calculated by means of the Microscopic R-matrix method(MRM)[21] and the resonance parameters are determined by the iterative method[22].

RESULTS AND DISCUSSIONS

We have performed our calculations up to $J^\pi=9/2^\pm$ in ^{9}Be and $J^\pi=4^\pm$ in ^{10}Be. Figures 1 and 2 show the energy spectra and band structures in ^{9}Be and ^{10}Be respectively. In ^{9}Be our results by the MRM are compared with those by the $\alpha+\alpha+n$ three-body Complex Scaling method(CSM) and both results give good agreements for the resonance energies

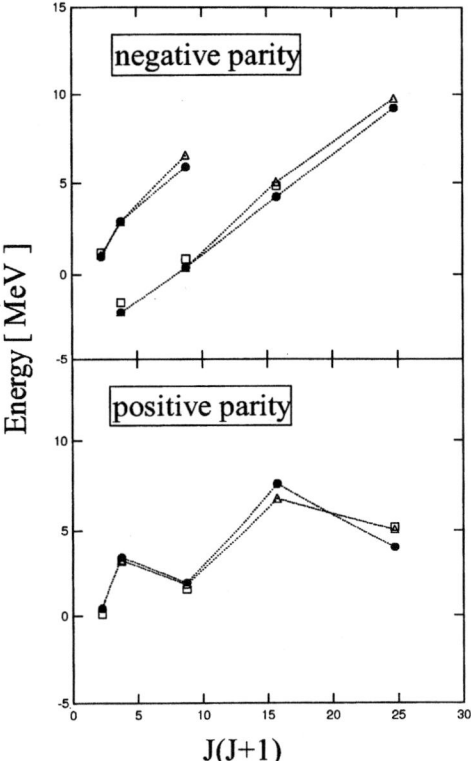

FIGURE 1. Band structure of ^{9}Be. Energies are given relative to the $\alpha+\alpha+n$ threshold. Solid circles and white triangles show our results by the MRM and three-body CSM respectively. White squares show the experimental data[11].

whereas the resonance widths by the MRM incline generally to give smaller widths, naturally because the MRM disregards the three-body direct decay and the sequential decays of ^{8}Be and ^{5}He. However these results prove that the present approximate way employed in the MRM can work satisfactorily at least to reproduce the resonance energy. The anomalous $1/2^+$ first excited state in ^{9}Be cannot be obtained by the CSM whereas the CSM reproduces excellently other resonance states. The Analytic Continuation of the S-matrix to the complex energies(ACS) method[23] with the MRM gives this $1/2^+$ state as the first excited state that agrees with the experimental data and as a ^{8}Be(0^+)+n virtual state where the S-matrix pole has a pure imaginary momentum and lies on the second Riemann sheet only for this channel. A calculation for the mirror nucleus ^{9}B is performed as well and the MRM gives the broad width for the $1/2^+$ state in ^{9}B and the normal Thomas-Ehrman shift though the inverted shift was theoretically predicted by Barker[13]. The present model shows explicitly that ^{9}Be produces three rotational

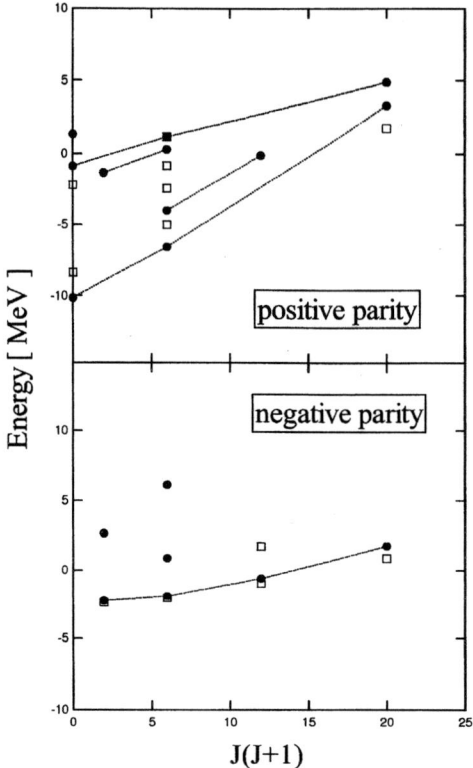

FIGURE 2. Band structure of ^{10}Be. Energies are given relative to the $\alpha+\alpha+n+n$ threshold. Solid circles show our results and white squares show the experimental data[11, 24].

bands; $K^\pi=3/2^-$, $1/2^+$, and $1/2^+$ as shown in Figure 1.

Our model in ^{10}Be reproduces the six states (0_1^+, 2_1^+, 2_2^+, 1^-, 0_2^+, and 2^-) that are observed as bound states while the second 0^+ state is obtained between the ^{9}Be($3/2^-$)+n and ^{6}He(0^+)+α thresholds due to the underbinding of ^{6}He in the present model. Our present model gives also various resonance excited states but the energy spectrum above the ^{9}Be+n threshold is poorly known experimentally so far[11]. Among these states, $J^\pi=0_2^+$, 2_4^+, and 4_2^+ appear to form a $K^\pi=0_2^+$ rotational band where the second 0^+ is a well known intruder state in the shell model[16]. These states have large spectroscopic factors of ^{6}He+α configuration in our model and therefore this rotational band could be a consequence of the large ^{6}He+$\alpha(\alpha+\alpha)$ clustering. As for the ^{9}Be+n configuration, the 0^+ ground state is dominated by the ^{9}Be($3/2^-$)$\otimes v_{p_{3/2}}$ configuration (the spectroscopic factor; $s=2.26$) but the second 0^+ state is by the ^{9}Be($1/2^+$)$\otimes v_{s_{1/2}}$ ($s=0.55$) as shown Figure 3. These show that two neutrons in the second 0^+ state excite into the $s_{1/2}$

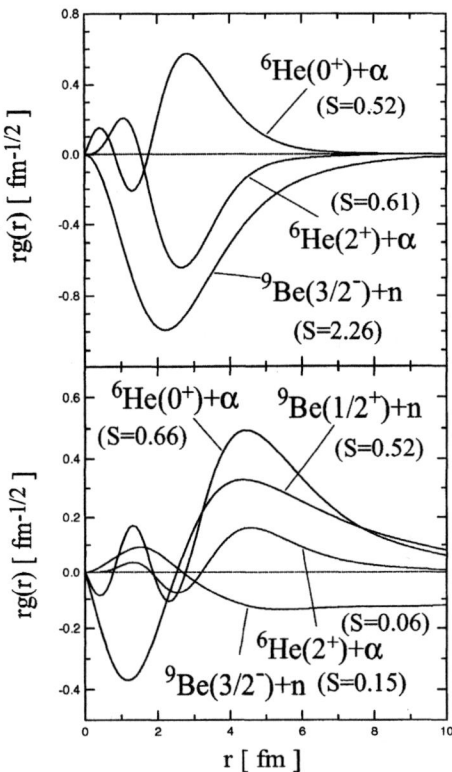

FIGURE 3. Reduced width amplitudes multiplied by r for the two-body decays of ^{9}Be+n and ^{6}He+α for the ground (upper panel) and second (lower panel) 0^+ states. The values of S within the parenthesis indicate the spectroscopic factor.

orbit ($2\hbar\omega$ excitation) while four neutrons fill the $p_{3/2}$ orbit in the 0^+ ground state. The calculated reduced width amplitudes show that the 0_2^+, 2_4^+, and 4_2^+ states are dominated by the $2\hbar\omega$ excitation in the shell model configuration and this supports the idea of these three states being the members of the $K^\pi=0_2^+$ rotational band. It is important to note that above results for the second 0^+ state means an inversion of the $p_{1/2}$ and $1s_{1/2}$ orbits as well as ^{9}Be and ^{11}Be because the second 0^+ state should be dominated by the ^{9}Be($1/2^-)\otimes\nu_{p_{1/2}}$ configuration in a simple shell model picture which could correspond to the third 0^+ state in our model.

Next we have calculated the reduced width amplitudes into the three- and four-body decays in order to clarify the valence neutron motions and $\alpha+\alpha$ clusterings in ^{9}Be and ^{10}Be respectively. Figures 4 and 5 show results of the $1/2^+$ state in ^{9}Be and the second 0^+ state in ^{10}Be respectively. In figure 4, the upper and lower panels are amplitudes for the

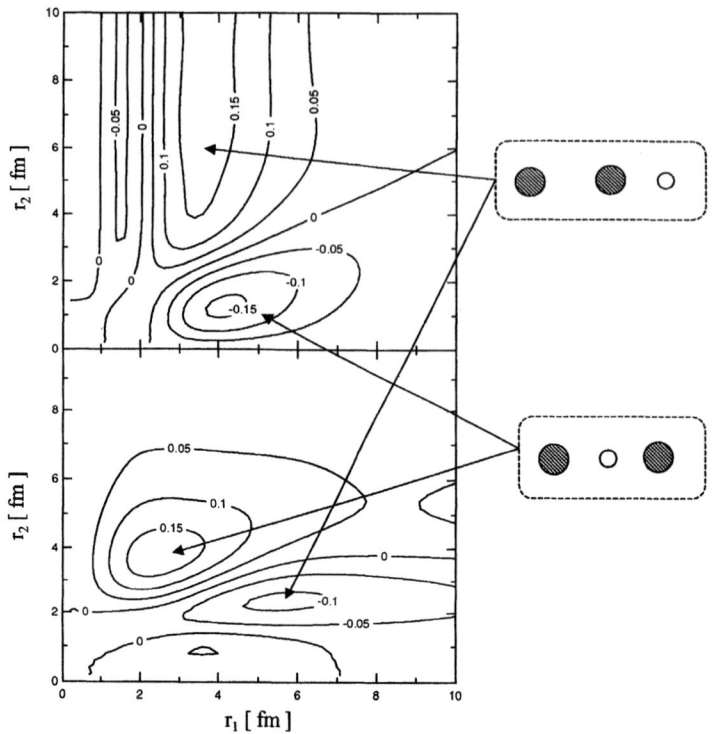

FIGURE 4. Reduced width amplitudes multiplied by $r_1 r_2$ for the three-body decay in the $1/2^+$ state of ^{9}Be. Upper and lower panels are results by the $(\alpha\alpha)n$ and $(\alpha n)\alpha$ Jacobi arrangements respectively. The large and small circles stands for the α particle and valence neutron respectively.

$(\alpha\alpha)n$ and $(\alpha n)\alpha$ Jacobi arrangements respectively. It is clearly recognized that each map has two prominent peaks. The peaks at $(r_1, r_2) \approx (4.1$ fm, 1.2 fm$)$ on the upper panel and at $\approx (2.5$ fm, 3.75 fm$)$ on the lower panel indicate an α-n-α configuration in which the valence neutron is staying between the two α particles. On the other hand, another pair of the peaks indicates an α-α-n configuration where the valence neutron is staying outside of the two α particles but along the axis joining them. The larger r_1 in the former configuration on the upper panel suggests that the $\alpha + \alpha$ clustering is enhanced when the valence neutron stays between the two α particles, mainly due to the Pauli principle and the reduction of the kinetic energy. And the larger r_2 in the latter configuration on the upper panel suggests that the valence neutron is far away from the two α particles. It is quite important to realize that the two competing configurations mentioned above are absolutely essential to reproduce this anomalous $1/2^+$ state in ^{9}Be. One of these has a strong core (^{8}Be) deformation induced by the valence neutron and this distortion effect is

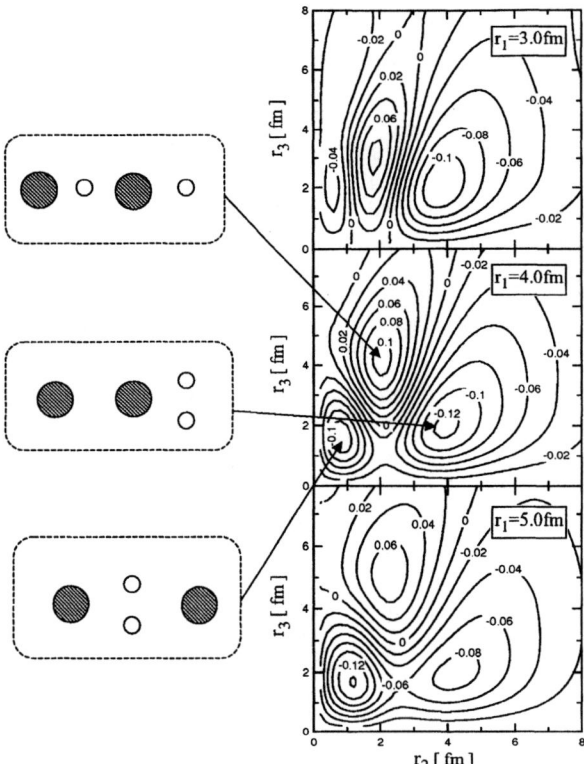

FIGURE 5. Reduced width amplitude multiplied by $r_1 r_2 r_3$ for the four-body decay in the second 0^+ state of ^{10}Be. The large and small circles stands for the α particle and valence neutron respectively.

responsible for lowering the $s_{1/2}$ orbit and causing the parity inversion between the $p_{1/2}$ and $s_{1/2}$ orbits in this nucleus. And another configuration produces the spatially extended neutron distribution outside the ^{8}Be core which could form the neutron halo like ^{11}Be if this state is a bound state (in fact this state is an unbound state). It is interesting to notice that above two configurations appear to be consistent with a picture of the σ-orbit in the molecular orbit[7] where the valence neutron is moving along the $\alpha+\alpha$ axis.

In the second 0^+ state in ^{10}Be the reduced width amplitude is calculated with a single channel where the Jacobi arrangement is selected as a ^{8}Be+n type, that is $(\alpha\alpha)(nn)$. The r_1, r_2, and r_3 indicate the relative distances between the two α particles, between the two α particles and two neutrons, and between the two neutrons respectively. The partial waves on r_1, r_2, and r_3 are set to be zero and this choice of the partial wave on the r_3 allows only zero as the total spin. In Figure 5 three contour maps in the r_2-r_3 plane are plotted for three different $\alpha+\alpha$ distances, $r_1=3.0$, 4.0, and 5.0 fm. These counter maps for the second 0^+ state give three prominent peaks at $(r_2, r_3) \approx$ (0.8 fm, 1.6 fm), (2.1 fm, 4.25 fm), and (4.0 fm, 2.0 fm) with $r_1=4.0$ fm as shown in the middle panel. The first peak indicates an α-(nn)-α configuration where two neutrons have a rather strong correlation and this di-neutron stays roughly around the center of the two α particles. This is superficially similar to the molecular covalent bonding picture. And this peak has a smaller magnitude than the other two peaks at smaller r_1 but becomes the largest at the larger $\alpha+\alpha$ relative distance of $r_1 \approx 5.0$ fm. This indicates that this configuration has a highly developed $\alpha+\alpha$ clustering than other two configurations and than the $1/2^+$ state in ^{9}Be. The second peak emerging at $r_2 \approx 2.0$ fm might have an α-n-α-n configuration in which one neutron stays between two α particles and another neutron stays outside the two α particles but along the $\alpha+\alpha$ axis. This configuration is rather like the ^{5}He+^{5}He configuration, that is, each pair of an α particle plus a valence neutron has a rather strong correlation (the relative distance between the α particle and neutron stays around 2.0 fm at this peak). The third peak might have an α-α-(nn) configuration in which the two neutrons have a rather strong correlation as well as the first configuration while this di-neutron is staying outside the two α particles but along the $\alpha+\alpha$ axis. Among these three different configurations, the first and third configurations could be regarded as ^{6}He(0^+)+α configuration in which ^{6}He has the di-neutron configuration and the second could be regarded as the ^{6}He(0^+)+α configuration in which ^{6}He has the cigar-shape configuration. In addition these three configuration can also be regarded as the ^{9}Be($1/2^+$)$\otimes v_{s_{1/2}}$ configuration according to above discussion for ^{9}Be($1/2^+$). And the valence neutrons are moving along the $\alpha+\alpha$ axis in this state as well as the $1/2^+$ state in ^{9}Be that are similar with the σ-orbit picture in the Molecular Orbital method. The r_1 value at the maximum of the peak, that is, the differences in the $\alpha+\alpha$ clustering among above three configurations indicate that the two valence neutrons staying between the two α particles can more efficiently glue together the two separated α particles than does the single valence neutron. It should be stressed that, just as for the $1/2^+$ state in ^{9}Be, these competing configurations are essential to reproduce this anomalous second 0^+ state in ^{10}Be. On the one hand, the third configuration produces the spatially extended neutron distribution outside the ^{8}Be core. On the other hand, in the first configuration the two valence neutrons staying near or inside the ^{8}Be core cause the highly developed $\alpha+\alpha$ clustering or, in other words, the large deformation of the ^{8}Be core. This is the strong core(^{8}Be) distortion effect induced by the valence neutrons and this strong core deformation could lower the energy of this second 0^+ state deeply. The second configuration has both effects moderately.

CONCLUSION

We have performed theoretical studies of ^{9}Be and ^{10}Be by means of the microscopic $\alpha+\alpha+n$ and $\alpha+\alpha+n+n$ cluster models respectively. The resonance excited states are localized by solving the two-body scattering problems of $\{^8$Be(0^+, 2^+, 4^+)+$n\}$+$\{^5$He($3/2^-$, $1/2^-$)+$\alpha\}$ in ^{9}Be and of $\{^9$Be($3/2^-$, $1/2^+$, $5/2^+$)+$n\}$+$\{^6$He(0^+, 2^+)+$\alpha\}$ in ^{10}Be. Our model in ^{9}Be reproduces satisfactory the experimental energy spectrum and gives good agreements with those by the three-body CSM. The $1/2^+$ first excited state in ^{9}Be, which is not localized by the CSM, is obtained as the ^{8}B(0^+)+n virtual state in the MRM and our results for the mirror nucleus ^{9}B support the normal Thomas-Erhamm shift.

Concerning ^{10}Be, our model reproduces not only the 0^+ ground state but also the second 0^+ state simultaneously where the latter is a well known intruder state in the shell model. Our model produces various resonance excited states above the ^{9}Be($3/2^-$)+n threshold and among these states, $J^\pi=0_2^+$, 2_4^+, and 4_2^+ could be members of the $K^\pi=0_2^+$ rotational band which have rather large spectroscopic factors of the ^{6}He+α configuration and are dominated by the $2\hbar\omega$ excitation in the shell model configuration. Especially in the second 0^+ state, this state are dominated by the ^{6}He(0^+)+α and ^{9}Be($1/2^+$)$\otimes\nu_{s_{1/2}}$ configurations with the spatially extended relative wave functions and the latter indicates that two neutrons are excited into the $s_{1/2}$ orbit as a consequence of the parity inversion between the $p_{1/2}$ and $s_{1/2}$ orbits in ^{10}Be.

Our model shows that two and three competing configurations are necessary in order to reproduce the anomalous $1/2^+$ state in ^{9}Be and the second 0^+ state in ^{10}Be respectively. In one of them the valence neutron stays near or inside the ^{8}Be core and induces the strong core distortion which is responsible for lowering these states and causing the parity inversion between the $p_{1/2}$ and $s_{1/2}$ orbits. And another of them produces the spatially extended neutron distribution outside the core like the neutron halo.

ACKNOWLEDGMENTS

This work was funded by the CHARISSA collaboration grant GR/R38927/01 awarded by the EPSRC.

REFERENCES

1. A. A. Korscheninnikov et al., Phys. Lett. B **343**, 53 (1995).
2. N. Soić et al., Europhys. Lett. **34**, 7 (1996).
3. M. Milin et al., Europhys. Lett. **48**, 616 (1999).
4. M. Freer et al., Phys. Rev. Lett. **82**, 1383 (1999); M. Freer et al., Phys. Rev. C **63**, 034301 (2001).
5. W. von Oertzen, Z. Phys. A **354**, 37 (1996); ibid. **357**, 355 (1997).
6. Y. Kanada-En'yo, H. Horiuchi, and A. Doté, Phys. Rev. C **60**, 064304 (1999); A. Doté, H. Horiuchi, and Y. Kanda-En'yo, ibid. **56**, 1844 (1997).
7. N. Itagaki and S. Okabe, Phys. Rev. C **61**, 044306 (2000); N. Itagaki, S. Okabe, and K. Ikeda, ibid. **62**, 034301 (2000); N. Itagaki, S. Hirose, T. Otsuka, S. Okabe, and K. Ikeda, ibid. **65**, 044302 (2002).
8. Y. Ogawa, K. Arai, Y. Suzuki, and K. Varga, Nucl. Phys. **A673**, 122 (2000).
9. M. Ito and Y. Sakuragi, Phys. Rev. C **62**, 064310 (2000).
10. P. Descouvemont, Nucl. Phys. A **699**, 463 (2002).
11. F. Ajzenberg-Selove, Nucl. Phys. **A490**, 1 (1988).
12. D.J. Millener, J.W. Olness, E.K. Warburton, and S.S. Hanna, Phys. Rev. C **28**, 497 (1983).
13. F.C. Barker, Aust. J. Phys. **40**, 307 (1987).
14. M. Seya, M. Kohno, and S. Nagata, Prog. Theor. Phys. **65**, 204 (1981).
15. Y. Kanada-En'yo, H. Horiuchi, and A. Ono, Phys. Rev. C **52**, 628 (1995).
16. A. G. M. van Hees and P. W. M. Glaudemans, Z. Phys. A **315**, 223 (1984); A. A. Wolters, A. G. M. van Hees, and P. W. M. Glaudemans, Phys. Rev. C **42**, 2062, (1990).

17. J. A. Liendo, N. Curtis, D. D. Caussyn, N. R. Fletcher, and T. Kurtukian-Nieto, Phys. Rev. C **65**, 034317 (2002).
18. K. Arai, Y. Ogawa, Y. Suzuki, and K. Varga, Phys. Rev. C **54**, 132 (1996); K. Arai, P. Descouvemont, D. Baye, and W. N. Catford, Phys. Rev. C **68**, 014310(2003).
19. K. Arai, *submitted to* Phys. Rev. C.
20. D. R. Thompson, M. LeMere, and Y. C. Tang, Nucl. Phys. **A286**, 53 (1977); I. Reichstein and Y. C. Tang, Nucl. Phys. **A158**, 529 (1970).
21. D. Baye, P.-H. Heenen, and M. Libert-Heinemann, Nucl. Phys. **A291**, 230 (1977); H. Kanada, T. Kaneko, S. Saito and Y. C. Tang, Nucl. Phys. **A444**, 209 (1985).
22. P. Descouvemont and M. Vincke, Phys. Rev. A **42**, 3835 (1990).
23. A. Csótó, R.G. Lovas, and A.T. Kruppa, Phys. Rev. Lett. **70**, 1389 (1993); A. Csótó and G.M. Hale, Phys. Rev. C **55**, 536 (1997).
24. N. Curtis, D. D. Caussyn, N. R. Fletcher, F. Maréchal, N. Fay, and D. Robson, Phys. Rev. C **64**, 044604 (2001).

Structure of Continuum States in Unstable Nuclei

K. Katō*, M. Myo† and K. Ikeda**

*Division of Physics, Graduate School of Science, Hokkaido University, Sapporo 060-0810, Japan
†RCNP Osaka, Osaka University, Ibaraki, Osaka 567-0047, Japan
**RI-Beam Science Laboratory, RIKEN, Wako, Saitama 351-0198, Japan

Abstract. Applying the complex scaling method to calculations of strength functions of Coulomb breakup reactions, we study the structure of continuum states of two-neutron halo nuclei ^{6}He and ^{11}Li.

INTRODUCTION

Recently, much interest has been concentrated on the three-body resonances of the so-called Borromean system[1] which has a halo structure[2] of the weakly-bound ground state, but no bound states of any two-body sub-systems. All of excited states in these systems are unbound. Therefore, we must treat resonant and continuum states in addition to the weakly-bound states to study mechanism of binding and excitation of the Borromean systems.

Although those continuum states are obtained as solutions of the Schrödinger equation on each branch cut of the complex Riemann sheets, it is difficult to distinguish them by

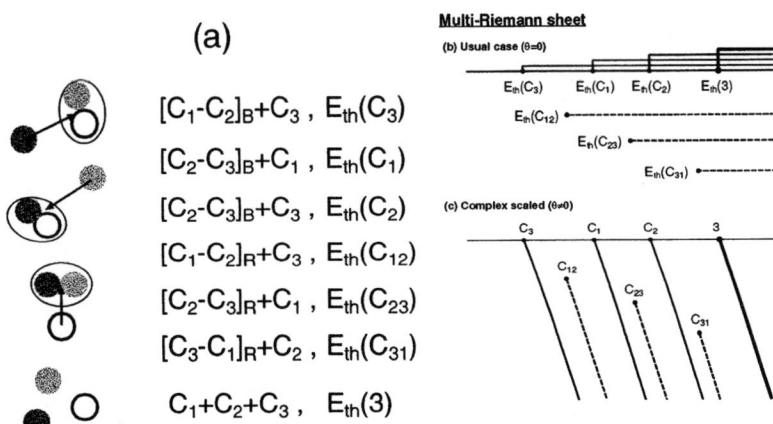

FIGURE 1. Various thresholds (a) of a three-body system and corresponding Riemann sheets in the usual case (b) and the complex scaling method (c).

seeing their energy positions because all branch cuts degenerate on the real energy axis. This difficulty can be solved by applying the complex scaling method[3]. In this method, continuum states on every branch cut are rotated around the corresponding threshold energy by taking a finite scaling angle θ as shown in Fig. 1 (c).

Using these separated continuum states in addition to discrete resonant states, in the complex scaling method, we can investigate the structure and properties of the unbound states. In this talk, we concentrate our attention on the discussion about Coulomb breakup reactions. Coulomb breakup cross sections of ^{6}He and ^{11}Li to three-body systems ^{4}He+n+n and ^{9}Li+n+n are calculated from the electromagnetic strength functions in the complex scaling method, and we study the contribution from three-body resonances, two-body resonances and three-body continuum states.

COULOMB BREAKUP CROSS SECTION IN COMPLEX SCALING METHOD

The strength function for the transition operator $\hat{O}$ is expressed as

$$S(E) = \sum_v \langle \Psi_{gr} | \hat{O}^\dagger | \Psi_v \rangle \langle \tilde{\Psi}_v | \hat{O} | \Psi_{gr} \rangle \delta(E - E_v) \tag{1}$$

$$= -\frac{1}{\pi} \mathrm{Im} R(E). \tag{2}$$

Here $R(E)$ is the so-called response function defined by

$$R(E) = \int dr dr' \tilde{\Psi}^*_{gr}(r) \hat{O}^\dagger G(r,r') \hat{O} \Psi_{gr}(r'). \tag{3}$$

Green's function,

$$G(r,r') = \langle r | \frac{1}{E-H} | r' \rangle, \tag{4}$$

is calculated with the help of the completeness relation. The usual completeness relation of bound states and continuum states,

$$\sum_B |\Psi_B\rangle\langle\Psi_B| + \int_L dE |\Psi_k\rangle\langle\Psi_k| = 1, \tag{5}$$

is proven by taking Cauchy's integral along a closed semi-circle in the above half momentum plane.[5] This proof can easily be applied to solutions of the complex scaling method where the semi-circle is rotated as shown in Fig. 2. Since the rotated semi-circle includes the resonant poles whose number is N_θ, the resonance term is added as follows[4]:

$$\sum_B |\Psi^\theta_B\rangle\langle\tilde{\Psi}^\theta_B| + \sum_R^{N_\theta} |\Psi^\theta_R\rangle\langle\tilde{\Psi}^\theta_R| + \int_{L_\theta} dE_\theta |\Psi_{k_\theta}\rangle\langle\Psi_{k^*_\theta}| = 1. \tag{6}$$

To apply this extended completeness relation, we calculate the response function in the complex scaling method:

$$R(E) = \int dr dr' \tilde{\Psi}^*_{gr}(re^{-\theta}) \hat{O}^\dagger(re^\theta) G^\theta(r,r') \hat{O}(re^\theta) \Psi_{gr}(r'e^\theta), \tag{7}$$

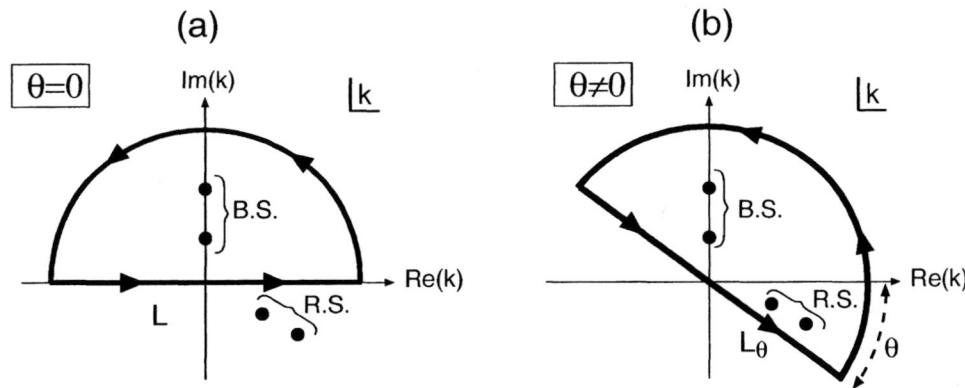

FIGURE 2. Semi-circle contours of Cauchy's integral in the usual case (a) and the complex scaling method (b).

where Green's function is also calculated by using the extended completeness relation as

$$G^\theta(r,r') = \sum_B \frac{\Psi_B(r,k_B)\tilde{\Psi}_B^*(r',k_B)}{E-E_B} + \sum_R^{N_\theta} \frac{\Psi_R(r,k_R)\tilde{\Psi}_R^*(r',k_R)}{E-E_R} \quad (8)$$

$$+ \int dE_\theta \frac{\Psi(r,k_\theta)\Psi^*(r',k_\theta^*)}{E-E_\theta}. \quad (9)$$

We obtain the strength function separated into three-terms of bound-state, resonant-state and continuum-state contributions;

$$S(E) = S_B(E) + S_R^\theta(E) + S_k^\theta(E), \quad (10)$$

where the separation between resonances and continua depends on the scaling angle θ. However, the total strength $S(E)$ is independent on θ.

CONTINUUM STRUCTURES IN COULOMB BREAKUP STRENGTH OF ^{6}HE AND ^{11}LI

Using this method, we study Coulomb breakup reactions of the two-neutron halo systems ^{6}He and ^{11}Li.

Coulomb breakup of ^{6}He to and ^{4}He$+n+n$

We apply the ^{4}He$+n+n$ three-body model to ^{6}He. The ^{4}He$+n+n$ three-body model, where the Pauli principle is taken into account by the orthogonality condition model, is explained in detail in Ref. [6]. In this model, we introduce a ^{4}He-n-n three-body force to include an effect from couplings between motion of valence neutrons and internal degrees of freedom in the ^{4}He-core. This three-body force is expressed as $V_{\alpha nn}(r_1, r_2) = V_3 e^{-\nu(r_1^2 + r_2^2)}$ with $V_3 = -0.218$ MeV and $\nu = (0.1/b)^2$ fm^{-2}, where $b = 1.4$ fm is a harmonic oscillator length parameter of the ^{4}He-core cluster wave function.[7] In Fig. 3 (a), we show the calculated energy spectrum of ^{6}He in comparison with the experimental data. It is seen that the 2_1^+-level observed experimentally as a genuine three-body resonance is well reproduced. Furthermore, the 2_2^+ resonance state is predicted. The detailed discussion of excited resonant states in ^{6}He is given in Ref. [6].

On the other hand, as shown in Fig. 3 (b), this ^{4}He+n+n model does not predict narrow 1^- resonances at the low energy region of ^{6}He. All eigenvalues of calculated 1^- states are distributed along the continuum lines of the ^{4}He+n+n three-body channel and of the two-body channels with ^{5}He(3/2$^-$)+n and ^{5}He(1/2$^-$)+n configurations. The strength distribution of the dipole transition from the 0^+ ground state to the 1^- continuum states provided us with interesting information on a reaction mechanism of the Coulomb breakup process. It is seen from the strength distribution to the ^{4}He+n+n three-body channel or ^{5}He(3/2$^-$)+n and ^{5}He(1/2$^-$)+n two-body channels whether ^{6}He dissociates directly into ^{4}He+n+n or sequentially through ^{5}He resonances.

In Fig. 4 (a), the calculated $E1$ strength distributions are shown. It is found that in the total strength, there is a low energy enhancement at around 1 MeV measured from the three-body threshold energy, which is just above the two body threshold of ^{5}He(3/2$^-$)+n (0.74 MeV),[6] and the $E1$ strength decreases slowly with the excitation energy. The most interesting result is that the dominant transition strength comes from the two-body continuum component of ^{5}He(3/2$^-$)+n with the low energy enhancement, and that the contributions from the other components are relatively very small. This result indicates that the three-body breakup strength of ^{6}He consists of dominantly the sequential breakup of a ^{6}He→^{5}He+n→^{4}He+n+n process.

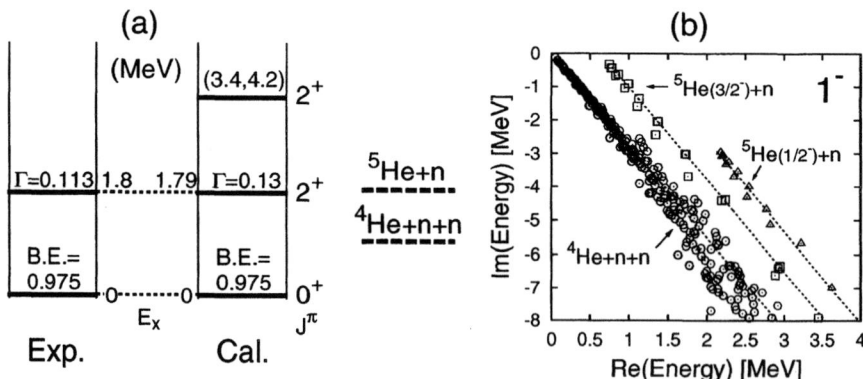

FIGURE 3. (a) Observed and calculated energy levels of ^{6}He. (b) The 1^- energy eigenvalues of the complex scaled ^{4}He+n+n Hamiltonian, where open circles show the 3-body continuum states, open squares and triangles the 2-body continuum states of ^{5}He(3/2$^-$)+n and ^{5}He(1/2$^-$)+n, respectively.

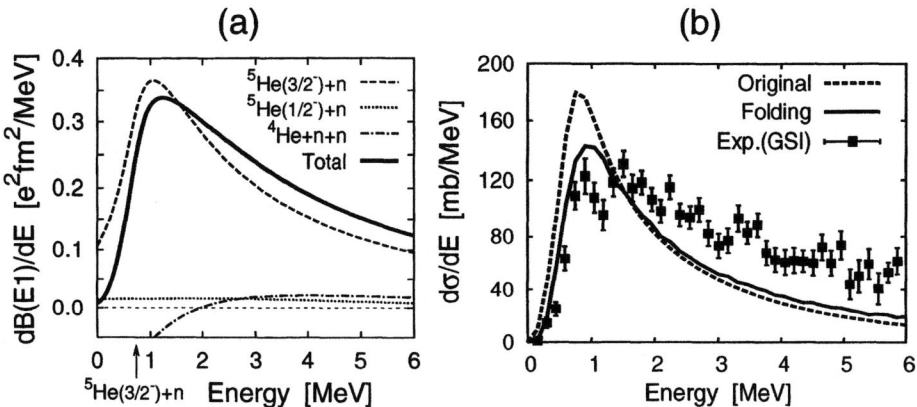

FIGURE 4. (a) E1 transition strength distributions, where dashed, dotted and dash-dotted lines are contributions from the 2-body and 3-body continuum states. The thick solid line shows the total strength distribution. (b) Observed and calculated Coulomb breakup cross sections of ^{6}He. The solid line shows the convoluted cross section in addition to the original calculation (broken line).

This result can be understood with reference to the structure of the initial ground state of ^{6}He. Although ^{5}He($3/2^-$) is a resonance, the wave function ^{5}He($3/2^-$) in the ^{5}He($3/2^-$)+n channel has a large overlap with the ^{4}He+n($p_{3/2}$) configuration which is a dominant component in the ground state of ^{6}He. The result of the large transition strength into ^{5}He($3/2^-$)+n channel indicates that one of the neutrons of $(p_{3/2})^2$ in the ground state of ^{6}He is excited to a continuum state of s- or d-wave orbit by the external $E1$ transition field. Furthermore, we see that three-body continuum states of ^{4}He+n+n do not contribute strongly. This result is also understood by considering that two $p_{3/2}$-neutrons in the ground state of ^{6}He, are hardly excited simultaneously in the breakup process. Another two-body continuum component of ^{5}He($1/2^-$)+n hardly contributes to the strength because the probability of $(p_{1/2})^2$ is very small in the ground state, nearly a few percent.

The total breakup strength is dominated by the $E1$ components, because the $E2$ component is very small in comparison with $E1$. Our result of the $E1$ strength distribution is similar to that of Danilin et al.[8], but their result shows a little bit large enhancement at low energies. The breakup cross section is calculated by multiplying the reduced transition probability $dB(E\lambda,E)/dE$ and the virtual photon number $N_{E\lambda}(E)$ from the equivalent photon method. The obtained results are shown in Fig. 4 (b) comparing with the observed data[9] for ^{6}He. We can see a good agreement.

Coulomb breakup of ^{11}Li to and ^{9}Li+n+n

As a typical two-neutron halo nucleus, ^{11}Li has been studied experimentally[2] and theoretically[1] by many people. In theoretical studies, the ^{9}Li+n+n three-body model has been used widely. However, the ^{9}Li+n+n three-body model for ^{11}Li is less reliable in comparison with the ^{4}He+n+n model for ^{6}He. We discussed this problem in Ref. [10] in detail.

Recently, we proposed a new coupled-channel three-body model beyond the ^{9}Li+n+n model by throwing away the assumption of a single closed sub-shell configuration $(0s)^4(0p_{3/2})^1_\pi(0p_{3/2})^4_\nu$ for the ^{9}Li core. The wave function of ^{11}Li in the present new model is generally given as

$$\Psi(^{11}Li) = \sum_i \Phi_i(^9Li)\chi_i(\vec{r}_1,\vec{r}_2). \tag{11}$$

Here, the suffix i denotes the channel specified by different configurations $\Phi_i(^9Li)$ for ^{9}Li and then $\chi_i(\vec{r}_1,\vec{r}_2)$ is the wave function of the active valence two neutrons in its channel. The old model of ^{9}Li+n+n can be included in the present model as a single channel case by taking the frozen core configuration $(0s)^4(0p_{3/2})^1_\pi(0p_{3/2})^4_\nu$.

Since paying much attention to the importance of the neutron $J^\pi = 0^+$ pairing in the binding of ^{11}Li, we describe the ^{9}Li core in ^{11}Li as

$$\Phi_{gr}(^9Li) = \alpha_0\Phi(C0) + \alpha_1\Phi(C1) + \alpha_2\Phi(C2) + \cdots, \tag{12}$$

where

$$(C0) : (0s)^4(0p_{3/2})_\pi(0p_{3/2})^4_\nu, \tag{13}$$

$$(C1) : (0s)^4(0p_{3/2})_\pi(0p_{3/2})^2_{\nu,J_1=0}(0p_{1/2})^2_{\nu,J_2=0}, \tag{14}$$

$$(C2) : (0s)^4(0p_{3/2})_\pi(0p_{3/2})^2_{\nu,J_1=0}(1s_{1/2})^2_{\nu,J_2=0}. \tag{15}$$

$$\cdots$$

The coefficients $\{\alpha_i;\ i=0,1,2,\cdots\}$ are determined by solving the Schrödinger equation $H_c(^9Li)\Phi_{gr}(^9Li) = \mathcal{E}_{gr}\Phi_{gr}(^9Li)$ when the ^{9}Li core cluster is isolated from the valence neutrons. However, in ^{11}Li, they depend on the motion of valence neutrons in the ^{9}Li+neutrons system. They are described by the wave functions $\{\chi_i\}$ being solutions of the coupled-channel equation. We first applied this model to the ^{9}Li+n system, and successfully investigated the virtual s-state problem in ^{10}Li by solving the coupled-channel equation.[11]

In Ref. [12], this new model was applied to ^{11}Li, and problems of the small binding energy, the large r.m.s. radius and the large amplitude of $(s_{1/2})^2$ in ^{11}Li were discussed consistently with the spectroscopy of ^{10}Li. Thus, it is clarified that the assumption of the frozen core for ^{9}Li does not give the consistent answer to those problems.

Using the obtained wave functions of ^{11}Li in the coupled-channel ^{9}Li+n+n model, we calculate the Coulomb breakup cross sections and compare the results with experiments[14, 15, 16] as shown in Fig. 5. The calculated results depend largely on the s-wave amplitudes of halo neutrons in the ground state of ^{11}Li. In Figs. 5 (a) and (b), comparison between experiments of MSU[15] and GSI[16] and the results calculated with changing the s-wave amplitude. Furthermore, we can see that the data of RIKEN[14] is well reproduced by the calculation with the 40% s-wave amplitude.

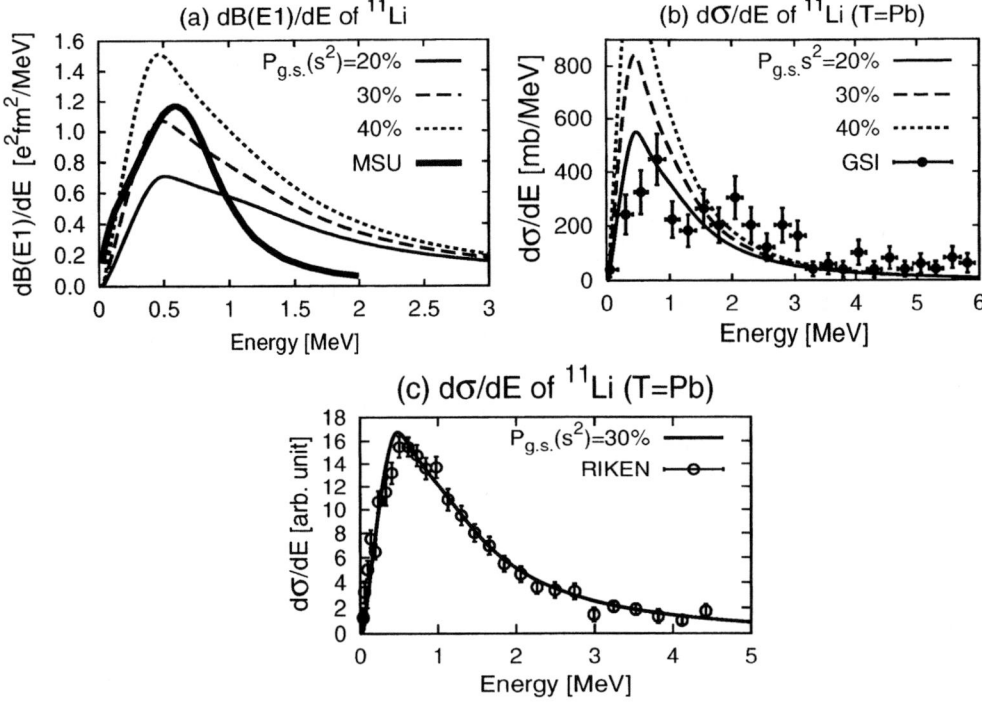

FIGURE 5. Coulomb breakup cross section of ^{11}Li into ^{9}Li+n+n. Experimental data are taken from (a) [15], (b) [16] and (c) [14], and we show the calculations of with different s-wave amptitudes in (a) and (b) and with the 40% fixed amplitude in (c).

CONCLUSION

In this talk, we discussed the continuum structure observed in Coulomb breakup reactions of two-neutron halo nuclei ^{6}He and ^{11}Li. These studies are performed by applying the complex scaling method, and it is shown that the complex scaling method is very useful in not only calculating the resonance energies and widths but also studying the structure of continuum states of many-body systems. Due to the weak binding of drip-line nuclei, those nuclei breakup easily by adding an external field. The breakup spectrum depends on the halo structure of the ground state and also on properties of continuum states.

The breakup cross sections of ^{6}He and ^{11}Li show a very similar behavior of the low energy peak. However, the ground state of ^{11}Li has a large component of 1s-configurations. The size of the low energy peak strongly depends on magnitudes of 1s-configurations. The experimental data of MSU, GSI and RIKEN are not consistent with each other. Our result obtained from the coupled-channel ^{9}Li+n+n model supports the data of RIKEN. However, since absolute values of the cross section have not been obtained in the RIKEN data, further experiments are strongly desired.

REFERENCES

1. M. V. Zhukov et. al.,Phys. Rep. **231** (1993), 151.
2. I. Tanihata, J. Phys. G: Nucl. Part. Phys. **22** (1996), 157.
3. J. Aguilar and J.M. Combes, Commun. Math. Phys. **22** (1971), 269; E. Balslev and J.M. Combes, Commun. Math. Phys. **22** (1971), 280.
4. T. Myo, A. Ohnishi and K. Katō, Prog. Theor. Phys. **99** (1998), 801.
5. R. G. Newton, J. Math. Phys. **1** '1960), 319; B. G. Giraud and K. Katō, Preprint.
6. S. Aoyama, S. Mukai, K. Katō and K. Ikeda, Prog. Theor. Phys. **93** (1995), 99; **94** (1995), 343.
7. T. Myo, K. Katō, S. Aoyama and K. Ikeda, Phys. Rev. **C63** (2001), 054313.
8. B. V. Danilin, I. J. Thompson, J. S. Vaagen, and M. V. Zhukov, Nucl. Phys. **A632** (1998), 383.
9. T. Aumann et al. Phys. Rev. **C59** (1999), 1252.
10. S. Mukai, S. Aoyama, K. Katō and K. Ikeda, Prog. Theor. Phys. **99** (1998), 381.
11. K. Katō, T. Yamada and K. Ikeda, Prog. Theor. Phys. **101** (1999), 119.
12. T. Myo, S. Aoyama, K. Katō and K. Ikeda, Prog. Theor. Phys. **108** (2002), 133.
13. T. Myo, S. Aoyama, K. Katō and K. Ikeda, to be published in Phys. Lett. **B** (2003)
14. S. Shimoura et. al. Phys. Lett. **B348** (1995), 29.
15. D. Sackett et. al. Phys. Rev. **C48** (1992), 118.
16. M. Zinser et. al. Nucl. Phys. **A619** (1997), 151.

Clustering in Exotic Nuclei Studies by Transfer Reactions

[1]R. Wolski*, Yu.M. Tchuvil'sky†, S.D. Kurgalin**, G.M. Ter-Akopian‡, P. Roussel-Chomaz§, L. Giot§ and K. Rusek¶

*Institute of Nuclear Physics, Cracow, Poland
†Moscow State University, Moscow, Russia
**Voronezh State University, Voronezh, Russia
‡Flerov Laboratory of Nuclear Reactions, JINR, 141980 Dubna, Russia
§ GANIL BP5027, 14021 Caen, France
¶A.Soltan Institute for Nuclear Studies, Hoza 69, 00681 Warsaw, Poland

Abstract. Various aspects of the application of shell model approaches to cluster transfer reactions involving exotic nuclei are discussed. A scheme to calculate Spectroscopic Amplitudes for transfer reactions is given. It is calculated for a tetraneutron in ^{8}He as an example. The reduction of the t+t clustering in the ^{6}He nucleus, suggested by an experiment where transfer reactions were studied in the ^{6}He+p collision is confronted with the concept of a supermultiplet for A=6 nuclei.

INTRODUCTION

Direct Nuclear reactions and, especially transfer reactions are an important tool for the nuclear structure study. If dynamics of a process is described properly then the yield of a transfer reaction depends on clustering in nuclei between which the cluster is exchanged. DWBA matrix elements used for the calculation of transfer reaction cross-sections contain the values of spectroscopic amplitude (SA). This quantity is a measure of a cluster pre-formation in the initial and final configurations. Shell model (SM) developed in the past for calculating SA will be reminded in the first section. The model was formulated within the harmonic oscillator basis. Therefore it is convenient for practical use since the calculations can be conducted to the algebraic expressions in that basis. However, one could rise a question about its applicability for exotic nuclei with their large outstretched density distribution.

At energies of few tens of MeV/n, it is difficult to extract the absolute values of SA from experimental data due to the uncertainty of optical model (OM) parameters used. Nevertheless, in some cases, comparative analysis could be done if OM parameters are the same or similar. Among such favorable cases involving reactions with exotic nuclei two examples could be analyzed. As the first example, one could compare transfer reactions in the ^{6}He+p and ^{6}Li+p systems at the same CM energy. The second case

---

[1] e-mail:wolski@lnr.jinr.ru

makes the 2n transfer process occurring in the ^{8}He+p and system leading to ^{6}He either in its ground or excited state. The first case will be discussed leaving the later one is beyond the scope of the paper. The difficulties of the method connected with unusual size parameters of exotic nuclei (EN) and necessary corrections are discussed also.

SPECTROSCOPIC AMPLITUDES IN TRANSLATIONALLY INVARIANT SHELL MODEL

The definition of SA for the shell model (SM) wave functions (WFs) of the mother and daughter nuclei was introduced in Ref. [1]. In Ref. [2] and in the subsequent papers the definition and the formalism were extended to light nuclei, for which the translation invariant shell model (TISM) [3], instead of the ordinary SM, is necessary. A complete solution of the discussed problem for arbitrary clusters within 1p-shell nuclei was given in Ref. [4].

The definition of the discussed quantity is given by the overlap:

$$\Phi(\vec{\rho}) = <\Psi_A^{TISM}|\hat{A}\{\Psi_{(A-X)}^{TISM}\Psi_X\}> \qquad (1)$$

. All WFs in the overlap are many-nucleon internal functions of the initial nucleus A, final nucleus (A-X), and the cluster X, $\hat{A}$ - antisymmetrizer, and $\vec{\rho}$ is the coordinate of the cluster relative motion. This overlap is sometimes called as the cluster form factor. The spectroscopic factor (SF) is defined by the norm of the function (1):

$$W_X = <\Phi(\vec{\rho})|\Phi(\vec{\rho})> \qquad (2)$$

The overlap of the function (1) with a normalized WF of some basis is called spectroscopic amplitude (SA) for a given channel (fragmentation). In the oscillator TISM, which is the most convenient one, the function $\Phi(\vec{\rho})$ has the oscillator form, so the SF is the square of SA. It should be noticed that a new definition of the spectroscopic factor was introduced in Ref. [5, 6] for the treatment of α-decay processes [7]. Nowadays, this definition seems to be preferable for this treatment. At the same time this definition is inapplicable for the cluster transfer reactions [8]. Thus the original definition of SA remains well justified and workable at least for light ($A \leq 20$) nuclei and light clusters ($X \leq 6$).

Question about the validity of TISM for loosely bound light nuclei deserves a special notice. In contrast to habitual 1p-shell nuclei the WFs of such subjects are characterized by two dimensions: the ordinary (or slightly modified) size of the core and the larger dimension, which characterizes the halo extension. However the approximation of a halo nucleus by a TISM WF, where the core and halo dimensions are assumed to be equal, turns out to be generally acceptable for SA calculations. Indeed, let us consider the halo nucleus ^{6}He as an example. One can expand the real internal α-core WF into a series of 4-nucleon WFs with halo oscillator parameter $\hbar\omega_{halo}$. The first term of the expansion is given by:

$$C \equiv <\Psi_\alpha(\hbar\omega_{core})|\Psi_\alpha(\hbar\omega_{halo})> = \left(\frac{2\sqrt{\omega_{core}\omega_{halo}}}{\omega_{core}+\omega_{halo}}\right)^{9/2}. \qquad (3)$$

Taking into account that for ^{6}He, ^{8}He etc. the ratio $\omega_{core}/\omega_{halo} \simeq 2$ one can conclude that the damping factor $C \simeq 0.8$. The overlap of 6-nucleon WFs containing such distinct core WFs is essentially a little larger than C due to the antisymmetrization. Thus, one could expect that the deviation caused by the discussed approximation is not large and remains within the accuracy of any modern approach to matrix element calculations of the multi-nucleon transfer reactions.

An interesting questions concerning the transfer of an unbound system like Xn, ^{4}H, etc. should be discussed. Let us consider one-step mechanism dominating at the energy of few tens of MeV/A. The TISM WF of the motion of the transferred X-nucleon group in the donor (acceptor) nucleus can be expanded into the following series:

$$\Psi^{SM}_{XNL} = \sum_{nLl} K_{NnLL'l}|\Psi_{XN'L'}\varphi_{nl}(R_X):L>, \qquad (4)$$

where Ψ_{XNL} - the WF of X nucleons in the nucleus, $\Psi_{XN'L'}$ - the internal WF, $\varphi_{nl}(R_X)$ - the WF of their center of mass. Indexes N, N', and n (N+N'+n) are the principal quantum numbers characterizing these WFs, L, L', l are the respective angular momenta, $K_{NnLL'l}$ is the expansion coefficient. It can be seen that the lowest ("ground") state of the cluster and the "excited" ones are contained in this expansion. The WF $\varphi_{nl}(R_X)$ which describes the cluster center of the mass distribution becomes more compact with decreasing n. Cluster transfer reactions are of surface type due to the strong absorption, thus they are sensitive to the size of the distribution. For that reason, one can expect that the contribution of an "excited" cluster term to the DWBA amplitude, where $n < n_{ground}$, is small and the "ground" state term dominates even for an unbound system.

We present now a brief summary of the formalism of SA for 1p-shell nuclei in the TISM. The original formalism of Ref. [4, 9] is essentially simplified here by excluding the options for excited cluster transfer. Using definitions presented above and Dirac notation one can write:

$$C_X \equiv <A|\hat{A}\{(A-X)\varphi_{nl}(\vec{\rho})X\}> \qquad (5).$$

By a simple reduction this equation can be rewritten in the form:

$$C_X = \binom{A}{X}^{1/2}(-1)^n\left(\frac{A}{A-X}\right)^{n/2}<A^{SM}|(A-X)^{SM}\varphi_{nl}(\vec{R}_X)X> \qquad (6),$$

where SM denotes that the WFs of initial and residual nuclei are written now in ordinary shell model coordinates. $|X>$ is the internal WF of the cluster as before. The second and the third multipliers appear due to the recoil effect and demonstrate the necessity to use TISM instead of ordinary SM. The last factor in Eq. (6) is in fact the sole subject of calculation procedure. The first step to do it is the presentation of the SM function of the initial nucleus as the superposition of WFs of (A-X) and X nucleons. Taking into account that the former WF in the last overlap of the Eq. (6) is fixed, one can write:

$$< A^{SM}|(A-X)^{SM}\varphi_{nl}(\vec{R}_X)X> = \sum_i < A^{SM}|(A-X)^{SM}X_i^{SM}>< X_i^{SM}|\varphi_{nl}(\vec{R}_X)X> \quad (7).$$

The first multiplier in Eq. (7) is the fractional parentage coefficient (FPC). This is the coefficient of the expansion of the totally antisymmetrized WF $|A^{SM}>$ onto the series of the WF $|(A-X)^{SM}>$ and $|X_i^{SM}>$ products, each of these WFs is antisymmetrized but there is no antisymmetrization between these two groups on the nucleon coordinates [10]. The second multiplier is the so-called cluster coefficient (CC). It is defined as the coefficient of the expansion (4) and was denoted as $K_{NnLL'l}$ there.

Calculation methods of FPCs for 1p-shell nuclei are rather well developed. Particular formalism depends on the version of the shell model. The versions are distinguished in relation to the coupling scheme (jj-, LS, intermediate) and to the choice of either isospin or proton-neutron formalisms. It is not a hard task to transform the WF from one version to another one. We present here two convenient equations for FPC in isospin LS-coupling scheme. They take the forms:

$$< s^4 p^{A-4}[f]LST|s^4 p^{A-X-4}[f']L'S'T', p^X[f_X]L''S''T''> = \binom{A}{X}^{-1/2}\binom{A-4}{X}^{1/2}$$

$$< p^{A-4}[f_p]LST|p^{A-X-4}[f'_p]L'S'T', p^X[f_X]L''S''T''> \quad (8)$$

where the last factor is FPC of 1p-shell. In turn, this coefficient is factorized onto the weight, spin-isospin and spatial multipliers:

$$< p^{A-4}[f_p]LST|p^{A-X-4}[f'_p]L'S'T', p^X[f_X]L''S''T''> = \sqrt{\frac{n_{[f'_p]}n_{[f_X]}}{n_{[f_p]}}}$$

$$< (st)^{A-4}[\tilde{f}_p]ST|(st)^{A-X-4}[\tilde{f}'_p]S'T', (st)^X[\tilde{f}_X]S''T''>$$

$$< p^{A-4}[f_p]L|p^{A-X-4}[f'_p]L', p^X[f_X]L''>, \quad (9)$$

where $[\tilde{f}_p]$ is the Young transposition scheme, $n_{[f_p]}$ is the dimension of the irreducible representation of the permutation group. Spin-isospin FPC and 1p-shell spatial FPC coincide with the Clebsch-Gordan coefficients (CGC) of the reduction chains $SU(4) \supset SU(2) \times SU(2)$ and $SU(3) \supset O(3)$ respectively. There is a number of methods solving completely the problem of calculation of CGC of the $SU(3)$ group. The spin-isospin FPC are usually calculated directly with the use of recursive formulas. Wide range tables are presented for these FPCs in refs. [12, 13, 14, 15]. The values of spatial FPCs can be found in these papers also.

General formulas for CCs are presented in Ref. [4, 9]. We write them here for the case of the cluster in its lowest SM configuration formed by 1p-nucleons only but not by any combination of 1s- and 1p-nucleons:

$$< p^N = X[f]N(\lambda\mu)L|(n0)l, XN - n[f](\lambda_X\mu_X)L_X > = \left(\frac{N!}{(N-n)!N^n}\right)^{1/2}$$

$$\sqrt{\frac{n_{[f']}}{n_{[f]}}} < (\lambda\mu)L|(n0)l,(\lambda_X\mu_X)L_X >, \qquad (10)$$

where $(\lambda\mu)$, $(\lambda_X\mu_X)$, and $(n0)$ are Elliott symbols characterizing SU(3) symmetry of relevant WFs, the last multiplier is CGC of this group in $SU(3) \supset O(3)$ reduction, and $[f']$ is the Young frame of the configuration of 1p-shell nucleons in the cluster. In the discussed case the Young frame is obtained from $[f]$ by skipping the first line. For any configuration of 1p-shell nucleons the number of lines in Young frame is not larger than 3 and Elliott symbols are determined as follows:

$$\lambda = f_1 - f_2$$
$$\mu = f_2 - f_3. \qquad (11)$$

Formulas presented above enable one to calculate SA of interest keeping in mind the simplifications made.

Clustering of tetraneutron in the ^{8}He nucleus

Let us consider the SA of a tetraneutron cluster in the ^{8}He nucleus. The subject is interesting in the context of the recent experimental attempt to measure the excitation energy spectrum of that object in α particle transfer reaction d(^{8}He,^{6}Li)4n [16].

As it was already pointed out the transfer of the four-nucleon system is realized mostly by the transfer of the lowest configuration of tetraneutron. So WF of this system in TISM can be written as:

$$\Psi_{4n} = |X = 4N = 2[f] = [22][f_p] = [2](\lambda\mu) = (20)L = 0S = 0T = 2 >. \qquad (13)$$

It should be noticed that this WF in ordinary SM can be written in the same form with the only necessary addition which is the notation of the nucleon configuration s^2p^2. According to the Bethe-Rose-Elliott-Skirme theorem SM and TISM lower configuration WFs are distinguished by the factor $\Phi_{000}(R)$ determining "zero oscillations" of the center of mass of the system in ordinary SM (TISM is created especially for removing the center of mass oscillations). In the LS-coupling scheme the WF of the ^{8}He nucleus takes the form:

$$\Psi_{^8He} = |A = 8N = 4[f] = [422](\lambda\mu) = (02)L = 0S = 0T = 2 >. \qquad (14)$$

According to (6) the SA expression for tetraneutron contains the trivial binomial (the first in (6)) and recoil (the second in (6)) factors. FPC of ordinary SM contained in (7) and expressed by (8) is reduced to the multiplier $\binom{A}{X}^{-1/2}$ (annihilating with the first one) multiplied by Kroneker's $\delta_{LL'',SS'',TT'',[\tilde{f}_p][\tilde{f}_X]}$. Thus, the only subject to calculate is in fact CC or more precisely CGC of SU(3) group in it. However the above mentioned relation between FPC of 1p-shell and CGC of SU(3) group can be used here reversely – for the calculation of the latter through the former. The tables of various FPCs of 1p-shell are tabulated in [12, 13, 14]. By this way one can find that the value of CGC of our interest is -2/3. Another possibility is to use an available code for this purpose, such as one presented in [11]. Let us present the values of other factors contained in the expression of the discussed SA. Recoil factor is equal to 2, the first multiplier in (10) - $\sqrt{3/4}$, $n_{[2]} = 1$, $n_{[22]} = 2$. Finally CC is equal to $-\sqrt{1/6}$ and $C_{4n} = -\sqrt{2/3}$. Discussion of the size effect is relevant here also. The oscillator parameter of unbound transferred cluster is a "subject of compromise" between a donor and an accepter nuclei. In other words transfer of 4n with $\hbar\omega_{4n}$ resulting by maximal value of the product of SA in a donor and in an acceptor $C_X^{don} C_X^{acc}$ is preferable. However the dependance of this product on $\hbar\omega_{4n}$ is flat thus a weak suppression of the cross section is expected.

Selection rules and ^{7}Li(^{6}Li,^{8}B)^{5}H reaction

The ^{7}Li(^{6}Li,^{8}B)^{5}H reaction has been used twice in order to search for the ground state of the ^{5}H nuclear system [17, 18]. The reaction could be attractive for experimentalists. The ^{8}B nucleus is the lightest, particle stable ejectile, which can be produced in processes of 2-proton transfer. The particles were detected and their energies measured by spectrometers. However, this reaction seems to be less advantageous if the relevant clustering of nuclei involved are concerned. One could expect that the direct di-proton pick-up would be of the largest strength. The value of SA for the overlap of 2p+^{5}H with ^{7}Li is reasonable, see for example Appendix of Ref. [9], whereas the SA of the 2p+^{6}Li overlap with ^{8}B is small. To make this statement it is unnecessary to calculate the SA. According to the selection rules the cross-section of the quoted reaction is small. Indeed, one can see the SM amplitudes for the ^{6}Li and ^{8}B ground states in Table 1. The numbers were taken from a phenomenological Shell Model [19].

A simple inspection of Table 1. is sufficient to conclude that, the direct 2p transition for the main components of the ^{6}Li and ^{8}Be ground states is forbidden either by the parity or by the angular momentum conservation. The reaction can proceed only due to the minor components of the ground state WFs of both nuclei and/or through multi-step transfers.

The above arguments obviously apply for the 2-n +^{6}Li as well as d+^{6}He overlaps with the ^{8}Li ground state Shell Model WF.

TABLE 1. Shell Model wave functions for ground states of ^{6}Li and ^{8}B nuclei [19].

^{6}Li(1^+) configuration $(2T+1)(2S+1)L$	SM Amplitude	^{8}B(2^+) configuration $(2T+1)(2S+1)L$	SM Amplitude
^{13}S	-0.996	^{33}P	-0.951
^{11}P	-0.077	^{31}D	-0.150
^{13}D	0.018	^{33}D	-0.207
		^{33}F	-0.042

Three-nucleon clustering in ^{6}He and ^{6}Li

^{6}He is usually treated as the lightest nucleus having two-neutron halo. The properties of this nucleus have been successfully described by a three-body model [20, 21]. The TISM predictions for SA of t+t and ^{3}He+t clustering in ^{6}He and ^{6}Li nuclei respectively are presented in the next subsection in Table 2. The size effect of the ^{3}He+t clustering in ^{6}Li has been calculated in terms of Antysymmetrized Multi-cluster Dynamical Model (AMDM) [22]. The obtained reduced factor of SA is equal to 0.874. A similar reduction can be expected for the analogous clustering in ^{6}He nucleus since the expansions with oscillator parameter $\hbar\omega_{triton}$ of the exact 3-body ^{6}He and ^{6}Li WF [23] are very close. An extended 3-cluster model for ^{6}He has been developed [24]. In this model the alpha-particle core is described as a 3N+N 2-cluster system. The model reproduces the small ^{6}He binding energy and yields a probability of about 0.5 for the t+t clustering. This value is smaller than the corresponding value of the ^{3}He+t clustering in ^{6}Li and, although non negligible, but substantially smaller than that of TISM.

The pioneering transfer reaction experiment with a ^{6}He beam was done in Dubna [25]. The result is seen in Fig. 1. together with the result obtained for an analogous reaction from the p+^{6}Li collision at the same energy per nucleon [26].

A striking feature is that, in spite of the similarity in the ^{6}He and ^{6}Li structures, the angular distributions for these two reactions differ significantly. In a qualitative approach to the reaction mechanism the forward angle part of the angular distribution is attributed to the 2-nucleon transfer, i.e. the 2-neutron in the first reaction and the deuteron transfer in the second one, whereas the triton transfer in both reactions contributes mostly to the cross-sections at the backward angles.

Thus results shown in Fig. 1. suggest rather smaller t+t clustering in the ^{6}He nucleus than that of ^{3}He+t in the ^{6}Li nucleus. The one-step DBWA calculations, shown in Fig. 1.

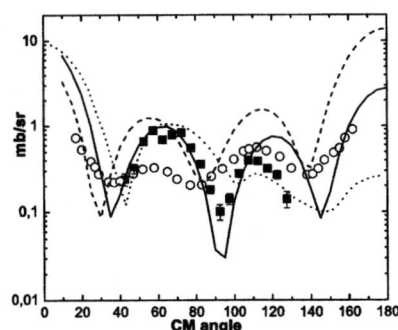

FIGURE 1. Experimental data for the p(^{6}He,^{4}He)t (closed symbols) [25] and the ^{6}Li(p,^{3}He)^{4}He (open symbols) [26] reactions at 25MeV/n. The lines are for 1-step DWBA calculations of the p(^{6}He,^{4}He)t reaction. The dotted curve is for the case when the t+t clustering was ignored, the dashed curve is calculated for the full SA(t+t)=-1.333, the solid one is for SA(t+t) reduced by two times, see text.

by curves, support this suggestion. These calculations and all quoted onward were done using FRESCO code [27]. The dotted line in Fig. 1. corresponds to neglecting the triton transfer process, i.e. SA of t+t in ^{6}He is put equal to zero. The dashed curve is for the full SA for that clustering as predicted by TISM. The solid one is for the SA reduced by two times.

In two more involved attempts to describe the ^{6}He data the transfer of triton was either neglected *a priori* [28] or diminished in respect to the TISM prediction [29].

On the other hand the authors of Ref. [30] claim to succeed in their 2-step DWBA description of the 2-nucleon transfer channel data for ^{6}He+p and ^{6}Li+p reactions. Their t+t spectroscopic factor for the ^{6}He+p reaction appeared to be close to that predicted by TISM, However, their spectroscopic factor for ^{3}He+t clustering in ^{6}Li is significantly smaller than the theoretical prediction.

It is interesting that the calculated differential cross-section of the p(^{6}He,^{4}He)t reaction at the angular region close to 180 deg. reaches a value around 3 mb/sr which is even larger than the corresponding cross-section for the ^{6}Li(p,^{3}He)^{4}He reaction in the same angular region [30]. Therefore, it was quite desirable to extend the measurement of the differential cross section for the p(^{6}He,^{4}He)t reaction up to extremely forward and backward scattering angles. The relevant data were recently obtained at GANIL [31]. The high resolution spectrometer "Speg" was used in this measurement. New numerical data are not shown here prior to the publication of the original experimental work [31]. The obtained data points for the differential cross-section are close to 2 mb/sr at extremely forward scattering angles, whereas in the angular region of 160-180 deg. the cross-section is close to 0.3 mb/sr albeit with large error bars. The later value is in apparent disagreement with the prediction of Ref. [30] for that angular region.

It is important to stress at that the hypothesis of damping the t+t clustering in ^{6}He ought to be reconciled with the earlier experimental works on that subject done prior to the advent of radioactive beams. This subject is discussed in the next subsection.

Supermultiplet symmetry of transfer reactions and DWBA

Under the assumption of spin and isospin independence of the nuclear forces the amplitude of nuclear reaction must possesses Wigner supermultiplet symmetry. Such symmetry has been tested experimentally as early as 1975 year [32]. The authors investigated t + ^{9}Be -> ^{6}He + ^{6}Li reaction in the two exit channels: ^{6}Li in its ground state and ^{6}Li(0^+) excited state. The last is the isospin analog of the ^{6}He ground state. Triton beam energies were 23.5 MeV and 21.5 MeV. ^{6}He and ^{6}Li reaction products were measured simultaneously. Angular distributions for both channels were taken in the angular range of 20-160 deg. They exhibit almost a perfect symmetry in respect to 90 deg. If for the ^{6}Li(0^+) exit channel it implies an ordinary isospin symmetry than for the ^{6}Li(1^+) one this experimentally proves a very good supermultiplet symmetry.

In terms of the direct transfer process the final ^{6}He is produced by the triton pick-up process leaving the target-like ^{6}Li in its ground or excited states. The ^{3}He pick-up produces the ^{6}Li ejectile in the two states and the ^{6}He as a target-like product. The calculated TISM SA values for the leading components of WF of all nuclei in question

TABLE 2. TISM Spectroscopic Amplitudes for the t and ^{3}He pick-up processes in the ^{9}Be(t,^{6}He(^{6}Li))^{6}Li(^{6}He) reactions

^{6}He ejectile nlj transition, SA	^{6}Li residue nlj transition, SA	^{6}Li ejectile nlj transition, SA	^{6}He residue nlj transition, SA
^{6}He = t+t $2S_{1/2}$ = -1.333	^{9}Be = t+^{6}Li(1$^+$) $2p_{1/2}$ = -0.192	^{6}Li(1$^+$) = ^{3}He+t $2S_{1/2}$ = -0.943	^{9}Be = ^{3}He+^{6}He(0$^+$) $2p_{1/2}$ = -0.236
	^{9}Be = t+^{6}Li(1$^+$) $2p_{3/2}$ = -0.215	^{6}Li(0$^+$) = ^{3}He+t $2S_{1/2}$ = -0.943	
	^{9}Be = t+^{6}Li(0$^+$) $2p_{1/2}$ = -0.167		

are presented Table 2. The symmetry for the ^{6}Li(0$^+$) channel is obvious since the products of the relevant SA are equal. The symmetry for the ^{6}Li(1$^+$) channel seems to be not so trivial.

The DWBA calculations for the one step transfer of the ^{9}Be(t,^{6}He)^{6}Li and ^{9}Be(t,^{6}Li)^{6}He reactions with the relevant SA values taken from Table 2. are shown in Fig. 2. For each of the two exit reaction channels 2 transitions were calculated and added coherently. As one could expect the angular distributions obtained are symmetrical in respect to 90 deg.

The obtained picture confirms the fact that DWBA formalism possesses the appropriate supermultiplet properties. Therefore the discussed reaction turns out to be a good test on the relative strength of t + t clustering in ^{6}He respect to that of ^{3}He + t in ^{6}Li(1$^+$).

One should keep in mind that the above test was done without the spin-orbit interactions and with the main components of the shell model WF for ^{9}Be, ^{6}He and ^{6}Li nuclei

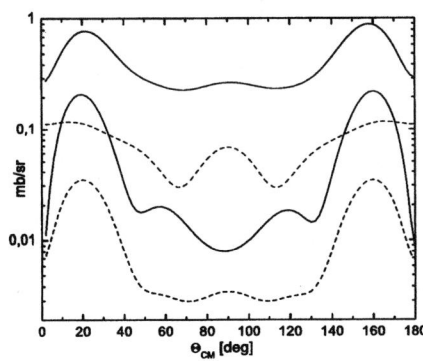

FIGURE 2. 1-step DWBA calculations of the 3-nucleon transfer processes in the ^{9}Be(t,^{6}He(^{6}Li))^{6}Li(^{6}He) reactions at 23.5 MeV. Two different sets of OM parameters and SA from Table 2. were used. The solid lines are for the two transition, added coherently for the ^{6}He+^{6}Li(1$^+$) exit channel, the dashed lines are for the analogous two transitions for the ^{6}He+^{6}Li(0$^+$) exit channel.

only, i.e. the D component of the ^{6}Li(1^+) was ignored.

ACKNOWLEDGMENTS

The partial support of the work by Russian Basic Research Foundation (grant No. 00-02-16683) and collaboration IN2P3-Laboratoires Polonais 02-106 is acknowledged.

REFERENCES

1. H.J.Mang, *Z. Phys.* **148**, (1957) 582.
2. V.V. Balashov, V.G. Neudatchin, Yu.F. Smirnov, N.P. Yudin, *ZhETF* **37**, (1959) 1385.
3. Kurdyumov I.V. Smirnov Yu.F. Shitikova K.V. El-Samarai S.H., *Nucl.Phys.* **A145**, (1970) 593.
4. Yu.F. Smirnov and Yu.M. Thuvil'sky, *Phys. Rev.* **C 15**, (1976) 84.
5. T. Fliessbasch, *Z. Physik* **A 272**, (1975) 39.
6. T. Fliessbasch, H.J. Mang, *Nucl.Phys.* **A 263**, (1976)75
7. R.G. Lovas, R.J. Liotta, A. Insolia, K. Varga, D.S. Delion, *Phys.Rep.* **294**, (1998) 265.
8. R.G. Lovas, *Z. Physik* **A 322**, (1985) 589.
9. O.F. Nemetz, V.G. Neudatchin A.T. Rudchik, Yu.F. Smirnov, Yu.M. Tchuvil'sky, *Nucleonic clusters in nuclei and the many-nucleon transfer reactions,* Naukova Dumka, Kiev, (1988) (in Russian).
10. G. Racah, *Phys. Rev.* **63**, (1943) 367.
11. J.Drieyer *Comp. Phys. Comm.* **5**, (1973) 405.
12. H.A.Jahn, H. van Weiringen *Proc.Roy.Soc.* **A 209**, (1951) 502.
13. J.P. Elliott, J. Hope, H.A. Jahn *Trans.Roy.Soc* **A 246**, (1953) 241.
14. D. Chlebowska, *Acta Phys. Polonica* **25**, (1964) 313.
15. Yu.F. Smirnov and V.G. Neudatchin, *Nucleon clusters in light nuclei,* ed. Nauka, Moscow (1969).
16. D. Beaumel, R. Wolski et al., *Book of Proposals,* Ganil (2002).
17. R.B. Weisenmuller et al., *Nucl. Phys.* **A 280**, (1977) 217.
18. D.V. Aleksandrov et al., *Proc. Int. Conf. on Exotic Nuclei and Atomic Mass,* Arles, France (1995) p.329.
19. A.N. Boyarkina, *Structure of 1p-shell Nuclei,* MSU, Moscow, 1973.
20. V.M. Krasnopolski, V.I. Kukulin, P.B. Sazonov, B.T. Voronchev, *Phys. Lett.* **B121**, (1983) 96.
21. V.I. Kukulin, V.M. Krasnopolski, V.T. Voronchev, P.B. Sazonov, *Nucl. Phys.* **A453**, (1986) 365.
22. G.G. Ryzhikh, Yu.M. Tchuvil'sky, *Izv. Akad. Nauk, ser. phys.* **56**, (1992) 112, (in Russian).
23. V.I. Kukulin, V.N. Pomerantsev, Kh.D. Razikov, V.T. Voronchev, G.G. Ryzhikh, *Nucl. Phys.* **A586**, (1995) 151.
24. K. Arai, Y. Suzuki and R.G. Lovas, *Phys. Rev.* **C 59**, (1999) 1432.
25. R. Wolski et al., *Phys. Lett.* **B 467**, (1999) 8.
26. M.F. Werby et al., *Phys. Rev.* **C 8**, (1973) 106.
27. I.J. Thompson, *Computer Phys. Rep.* **7**, (1988) 167.
28. Yu.Ts. Oganessian, J.S. Vaagen and V.I. Zagrebayev, *Phys. Rev. Lett.* **82**, (1999) 4996.
29. K.Rusek, K.Kemper and R.Wolski, *Phys. Rev.* **C 64**, (2001) 044602.
30. H. Heiberg-Andersen, J.S. Vaagen and I.J. Thompson, *Nucl. Phys.* **A 690**, (2001) 306c
31. L. Giot, *thesis,* GANIL, 2003.
32. W. von Oertzen and E.R. Flynn, *Ann. Phys.* **95**, (1975) 326

PHYSICS FOR
NUCLEAR TRANSMUTATION (PNT)

Nuclear Waste Incineration By ADS And Main Aspects Of The Accelerator Studied Within The European PDS-XADS

Alex C. Mueller

CNRS-IN2P3
Accelerator Division, Institut de Physique Nucléaire, F-91406 Orsay, France

Abstract. A possible issue to the nuclear waste problem is the transmutation of the long-lived radiotoxic components of the spent fuel. The development of subcritical systems with an external neutron source, provided by a high-energy proton beam impinging onto a spallation target are of special interest in this context. This paper describes the accelerator for such an Accelerator Driven System (ADS), studied within the European PDS-XADS programme.

INTRODUCTION

One important aspect of sustained growth is the long-term availability of energy resources and their environmental impact. The European Commission's Green Paper: *towards a European strategy for the security of energy supply* [1] clearly points out the importance to take into consideration all reasonably exploitable resources in Europe. Especially nuclear fission power should take a substantial share, provided that the problems of managing and stocking nuclear waste might be solved. Concerning nuclear energy, it is often advocated as having an (almost) "zero"-contribution to global warming and being economically competitive. However, it is also heavily debated in many European countries because of the long-term environmental burden of nuclear waste from the present-generation light-water reactors. At present, the strategy related to the closure of the fuel cycle depends on national policy. However, in any case, the transmutation of most of the long-lived radioactive waste is a promising solution, which could play a substantial role.

Both critical and sub-critical reactors are potential candidates as dedicated transmutation systems. Critical reactors, however, loaded with fuel containing large amount of minor actinides (Americium and Curium) pose safety problems caused by unfavourable reactivity coefficients and small delayed neutron fraction. With regard to this latter problem, the sub-criticality is particularly favourable and allows a maximum transmutation rate while operating in a safe manner. The sub-critical reactors are Accelerator-Driven Systems (ADS) constituted by the coupling of a proton accelerator, a spallation target and the sub-critical core. A technical working group (TWG) chaired by Carlo Rubbia, issued in 2001 a "European Roadmap" [2] towards ADS technology, advocating, in particular, the construction of an experimental facility "XADS", operational at the 2015 horizon.

PDS-XADS

Triggered by an initiative of the TWG members, the European Commission has meanwhile funded, within its 5th framework programme, the project PDS-XADS [3], where 25 partner organisations, including university groups, large research institutes and industrial firms, work together on a "preliminary design study" (PDS). Its goal is to select the most promising technical concepts, to address the critical points of the whole system, to identify the research & development (R&D) in support, to define the safety and licensing issues, to preliminary assess the cost of the installation, and, then, to consolidate the roadmap of the XADS development. A great amount of R&D and engineering activities have been already performed in Europe to demonstrate separate basic aspects of the ADS concept and to define conceptual reference configurations of the installation. Several different technological and design options have been considered and now studied in more details.

Taking into account that fast neutron spectrum is the a priori solution for transmutation purpose, the R&D efforts are focused on liquid metal-cooled ADS and gas-cooled ADS. The preliminary design studies are concentrated mainly on three concepts of the nuclear reactor part:
- a large gas-cooled concept (about 80 MW$_{thermal}$),
- a large lead-bismuth eutectic (LBE)-cooled concept (about 80 MW$_{thermal}$)
- a smaller LBE-cooled concept (20-50 MW$_{thermal}$); based on the MYRRHA design developed by Belgian organisations.

The spallation target situated in the center of the subcritical assembly is preferably a liquid heavy metal. LBE has been selected for this purpose. Two main concepts are envisaged: liquid heavy metal separated or not from the accelerator by a window in contact with the heavy metal. For the accelerator, the two types of machines to be investigated are the cyclotron and the linac.

The purpose of the PDS-XADS project is to develop all these configurations to a level sufficient to define precisely the supporting R&D needs, perform objective comparisons and recommend the solutions to be engineered in detail and realised. Five Working Packages (WP) cover the relevant issues, where WP3 is dedicated to the design of the high intensity proton accelerator for an XADS. This will be described in some detail in the following sections.

WP 3 OF PDS-XADS

Coordinated by CNRS-IN2P3, the following institutions collaborate on the elaboration of the six deliverables of WP3: ANSALDO (Italy), CEA (France), CNRS-IN2P3 (France), ENEA (Italy), Framatome ANP (France), Framatome GmbH (Germany), FZ Jülich (Germany), IBA (Belgium), INFN (Italy), ITN (Portugal), University of Frankfurt (Germany). WP3 contains the following deliverables, where the organization responsible for issuing the document is given in parenthesis:
- Requirements for the XADS accelerator and the technical answers (CEA)
- Potential for reliability improvement and cost optimization of linac and cyclotron accelerators (INFN)

- Accelerator: feedback systems, safety grade shutdown & power limitation (CEA)
- Accelerator: radiation safety and maintenance (CNRS)
- Definition of the XADS-class reference accelerator concept & needed R&D (CNRS)
- Extrapolation from XADS accelerator to the accelerator of an industrial transmuter (INFN)

In order to accomplish the tasks to be performed within these deliverables, WP3 naturally relies on input from the other working packages and conversely provides output to them. On the technical interfacing aspects, a good interaction is established with the other WP concerned with the design of the main components (and their principal versions) of the XADS. Further, a particular rôle is played by the link with WP1 guaranteeing the overall coherence of PDS-XADS. Today, slightly past mid-term of the contract, the studies of WP3 are already in a quite advanced phase (e.g. certain deliverables are released in final form), and the following sections of this paper aim at giving a synthetic view of the present facts and findings.

XADS ACCELERATOR SPECIFICATIONS

The main technical specifications for the XADS accelerator are summarized in Table 1. These characteristics clearly show that this machine belongs to the category of the so-called HPPA (high-power proton accelerators). HPPA are presently very actively studied (or even under construction) for a rather broad use in fundamental or applied science [4].

TABLE 1. XADS Proton Beam Specifications

Parameter	Required technical specification
Maximum Beam Intensity	up to 6 mA CW on target (accelerator 10 mA rated)
Proton Energy	600 MeV (including 800 MeV upgrade study)x
Beam Entry	Vertically from above preferred
Beam Trips	Less than 5 per year (exceeding 1 second duration)
Beam Stability	Energy: ± 1%, Intensity: ± 2%, size: ± 10%
Beam Footprint on Target	Gas-cooled XADS: circular ⌀ 160
	LBE-cooled XADS: rectangular 10×80
	"small-scale" XADS: circular, "donut" ⌀ 72
Intensity Modulation	0.2ms "holes" in CW beam for neutronis measurements, repetition frequency 0.01-1 Hz

The overall performance of the sub-critical system will be critically determined by a strict adherence of the XADS accelerator to its specifications. Compared to other HPPA, many requirements are similar, but it is to be noted that the reliability specification, i.e. the number of unwanted "beam-trips", is rather specific to the use as driver for an ADS. Therefore, the WP3 studies for the reference design had to integrate this stringent requirement from the very beginning, taking into account that this issue could be a potential "show-stopper" for ADS technology in general.

THE XADS ACCELERATOR REFERENCE LAYOUT

Choice of the basic accelerator concept

With the present state-of-the-art in accelerator technology, only two basic concepts of accelerators have shown to be able to deliver beam intensities in the mA range. These namely are sector-focused cyclotrons and linear accelerators (linacs).

Concerning cyclotrons, a final energy of 600 MeV is well established [5], namely with the experience of the PSI machine, and from this, it is felt in the cyclotron community, that a value of about 5 mA should be considered as safely reachable. However, extrapolating up to 10 mA is more questionable, and might require a complex of at least two cyclotrons with the two beams being funneled together. A (given) cyclotron also cannot be expanded in energy, so that boosting the energy from 600 to 800 MeV, as envisaged by WP1 would require the full replacement of the final and main stage, an absolutely not cost effective operation. For energies of in the GeV range (industrial transmuter), the intrinsic limits of the very working principle of a cyclotron is reached, because the proton is becoming too relativistic. Furthermore, a cyclotron is basically a CW machine and the requirement to provide pulses for neutronics measurements is a major difficulty for a cyclotron of such power. None of all these limitations[1] are present in a linac where intensities can reach above 100 mA without an intrinsic energy limit.

As will be further discussed in the next sections, the strategy to implement reliability relies on over-design, redundancy and fault-tolerance [6]. This approach requires a highly modular system where the individual components are operated substantially below their performance limit. A superconducting linac, with its many repetitive accelerating sections grouped in "cryomodules", conceptually meets this reliability strategy. It further allows keeping the activation of the structures rather low, important for radioprotection and maintenance issues, whereas the extraction channel of high power cyclotrons is in this respect a considerable concern.

For all these reasons, WP3 concluded that the cyclotron solution for an XADS presents a number of difficulties if not impossibilities: funneling, pulsing, beam trips, double-machine scheme, intrinsic current limitation, energy upgrading that precludes this solution despite its advantages such as lower price, proven technology at the MW level as demonstrated by PSI, and compactness. Therefore, the reference solution discussed in the next section is a superconducting linac [7].

Finally, it should be noted this assessment is corroborated by the one of OECD/NEA [8]:

Cyclotrons of the PSI type should be considered as the natural and cost-effective choice for preliminary low power experiments, where availability and reliability requirements are less stringent. CW linear accelerators must be chosen for demonstrators and full-scale plants, because of their potentiality, once properly designed, in term of availability, reliability and power upgrading capability.

---

[1] The released deliverables D9 and D57 elaborate these schematic considerations up to a certain level of detail

A reliable linac

The reliability optimized XADS accelerator, is shown in Figure 1. A "classical" proton injector (ECR source + normal conducting RFQ) feeds additional warm IH-DTL or/and superconducting CH-DTL structures up to a transition energy from where on a fully modular superconducting linac accelerates the beam up to the final energy.

Up to the transition energy, fault-tolerance is guaranteed by means of a "hot stand-by" spare. Above this energy, "spoke" and, from 100 MeV on, "elliptical" cavities are used. Beam dynamic calculations for this part have shown that an individual cavity failure can be handled at all stages without loss of the beam[2]. Besides this fault-tolerance, another remarkable feature of the concept is its validity for a very different output energy range: 350 MeV for the smaller-scale XADS require 9 cryomodules of $\beta=0.65$ elliptical cavities; in order to obtain 600 MeV, simply 10 more cryomodules have to be added (7 with $\beta=0.65$ and 3 with $\beta=0.85$) and 12 additional ($\beta=0.85$) boost the energy to 1 GeV. Therefore, already the small-scale XADS accelerator is fully demonstrative not only of the 600 MeV XADS (and could be converted to it), but even for an industrial machine.

The chosen superconducting cavities are subject of important R&D programmes presently underway, e.g. at the laboratories of the collaborating institutions of WP3. The performance of the prototypes has been measured to exceed the operational characteristics for the XADS by a very comfortable safety margin [8] that ensures the "over-design" criteria imposed by the reliability strategy.

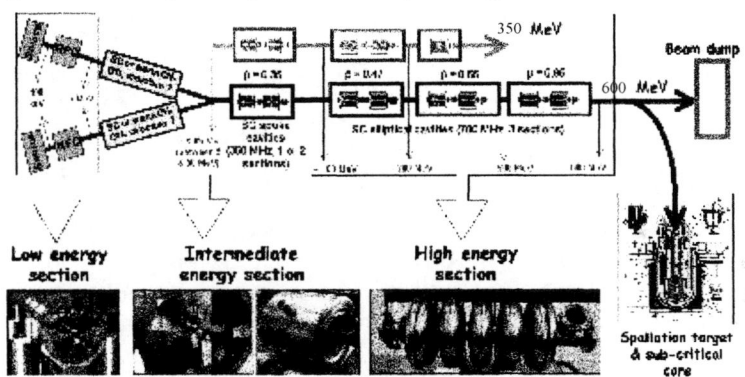

FIGURE 1. XADS reference accelerator layout: a doubled injector accelerator is followed by a fully modular spoke and elliptical cavity superconducting linac. Photos of typical cavity prototypes are shown in the lower part. From left to right: RFQ, CH structure, Spoke, Elliptical 5-cell.

A safe beam transport line

The objective of the final transport line is to safely inject the 600 MeV (or 350 MeV) proton beam onto the spallation target with a footprint of the required size and density distribution. To this end, a doubly achromatic module composed of two 45 degrees bending dipoles and three focusing quadrupoles has been designed[3] and

---
[2] Calculations accomplished within PDS-XADS deliverable D57 (INFN & CNRS-IN2P3)
[3] Details are given in deliverable 9

adopted for the two reference XADS concepts. With scaled magnetic rigidity, the same layout can be used at 350 MeV. Such a system is non-dispersive. Therefore, the beam position at the target does not depend on the energy variations, and, in addition, the beam size is independent of the internal beam energy spread. Thus, residual energy spread and central energy fluctuations from the accelerator will have no effect on the beam stability at the target. The system is however dispersive in between the two dipoles. Thus, the beam position and beam size monitors located in this region will provide information on proton energy variations, and to trigger a feedback system.

To expand the beam onto the target, the so-called raster scanning method has been adopted. It consists in deflecting a pencil-like beam with fast steering magnets operated at frequencies of 50 to a few hundreds Hertz, and acting in the two transverse directions in order to paint the target area. Various shapes (rectangular, circular) and various particle distributions (uniform, parabolic...) are achievable by simply adjusting the rastering frequencies of the steerers.

The scanning system would be very similar (but less demanding) to the one studied for the APT and ATW projects. Four raster magnets will be operated synchronously and independently so that the beam will be always moved at the target if one magnet fails. Redundant fault detection circuits will monitor the magnet current and the magnetic field to ensure proper operation and to shut down the beam in case of necessity. Similar systems are used in cancer therapy with protons or heavy-ion where they meet the stringent requirements for medical use[4].

A shielding for radiation protection

The shielding calculations for the XADS accelerator must be in line with the general radiation protection philosophy, based on the recommendations from the ICRP publication n°60 [10] that have been adopted in the European decree Euratom/96/29; all national legislations of the member states of the European Union must respect this European decree.

The goal of the shielding design of the XADS accelerator is therefore to guarantee that, under normal operational conditions, the added integrated dose to anybody working around the XADS accelerator will be extremely small, i.e. comparable or smaller than the natural background. To obtain this goal, the shielding calculations are made using conservative (= pessimistic) normal beam loss assumptions, and assuming an "occupancy factor" of 1, that means that a person will be present during 2000 hours per year immediately behind the shield wall where maximum dose rates exist. The design dose rate is 0.5 µSv/h, corresponding to the ICRP-60 annual limit of 1 mSv for the public, taking into account an annual working time of 2000 hours. The natural background being of the order of 1mSv/y, the conservatism built into the shielding calculations should therefore guarantee that the effective dose for any person resulting from the operation of XADS accelerator will be smaller than the natural background. This is an important argument because it is not unlikely that future ICRP recommendations will go in the sense of comparing exposure from "man-made" sources to exposure from natural sources.

---

[4] e.g. WP3 participant IBA has such systems developed commercially

The shielding defined for the normal operational conditions, with the conservative criteria explained above, must also guarantee that extra exposure to radiation created from abnormal operational conditions must remain sufficiently low, in order not to jeopardize the main shielding objective, i.e. keep the total integrated dose added by the operation of the XADS accelerator, for any person, below the natural background. The exposure to radiation from activated accelerator components is minimized via a system of administrative measures, by keeping the normal beam losses in the accelerator small and keeping the integrated power of accidental beam losses as low as possible via a powerful accelerator interlock system.

Figure 2 shows, e.g., the required earth profile in the case of an accelerator tunnel with 60 cm (light-concrete) side walls and roof, buried underground (the top of the concrete roof corresponds to zero ground level). The profiles correspond to the minimum earth thickness required to reduce the residual dose levels outside the earth below 0.5 µSv/h for a 1 nA/m linear beam loss at 350 MeV (small-scale XADS case case). For higher energies, it will be sufficient to slightly increase the amount of earth[5] It is remarkable that the shielding profiles derived from the requirements for normal operation will provide sufficient shielding for the planned commissioning period as well as for planned periods of beam tuning and setup, by allowing beam loss rates 100 times higher than during normal operation during significant periods of time, and it will also protect correctly against several types of accident condition.

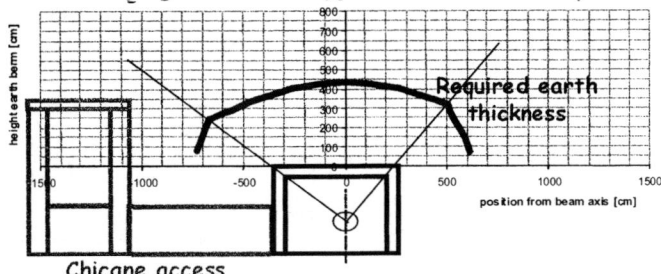

Fig 2. Required earth profile above the 60 cm thick concrete linac tunnel, for at 350 MeV proton energy and 1 nA/m linear beam losses. An access to the linac tunnel (chicane) is also shown.

A maintenance strategy

The foremost goal of the maintenance strategy for the XADS accelerator is to guarantee its reliability and availability for its operation[6] over its anticipated period of life, i.e. an order of magnitude of 30 years. In other words, proper maintenance has to guarantee that the "over-design" margin does not deteriorate and that a proper amount of redundant equipment is regularly restored if partial failures have occurred. Concerning fault-tolerance, there again, lost equipment has to be replaced at a regular basis, and it is of prime importance to ensure that the supervising control system regularly undergoes a complete performance check to ensure that it can replace components-at-fault by readjusting the operational parameters of the overall machine.

---

[5] Calculations performed within deliverable D48 by Paul Berkvens and Serge Palanque
[6] WP3 presently uses as reference an operational scheme of 3 month of uninterrupted beam, followed by a 1 month maintenance period.

These requirements request the development of an expert system able, while the accelerator is running with nominal beam, to precisely identify and locate equipment which is loosing rated performance, and/or is out-of-order and to be replaced or repaired. Thus, the expert system provides the database for the scheduled maintenance periods for planning repair and/or replacement of deteriorated or faulty equipment.

This fastest possible way of preparing the maintenance procedure also is in-line with the ALARA[7] principle for the concerned personnel. Indeed, one may point out that many conditions requested for radiation doses conforming to the ALARA principle, like enough working place and quick disconnection of sub equipments are actually the very much the same that are asked by an optimization of the reliability.

AN R&D PROGRAMME FOCUSED ON RELIABILITY

Generalities

As already mentionned, the broad field of applications covered by HPPA accelerators is at the origin of remarkable R&D effort presently underway world-wide. The study, design and testing of the main components of these new generation linear accelerators have contributed to a good synergy by developing complementary activities between many laboratories (see also [9] and references therein).

The XADS accelerator can profit from these general background and even built on it quite directly. However, a dedicated R&D programme has to be proposed in order to guarantee the requirement for the extremely low number of beam trips. It has to be focused on reliability and fault tolerance design. In this spirit, participant laboratories to WP3 have developped during the last months a collaborative R&D program that could be submitted within the forthcoming 6th FP call (see Table 2).

Table 2. Summary of the R&D topics needed for the XADS accelerator, including the responsibilities of participating laboratories, and an estimate of the necessary EU contribution.

R&D Topic	Low Energy Section	Intermediate Energy Section				High Energy Section	RF System	Global Coherence
	Injector Reliability	Spoke cav. (SC)	CH struct. (SC)	IH struct. (NC)	Other cav. (SC)	Beta 0.5 Cryomodule	Control System	
Responsible	CEA	CNRS				INFN	CEA	CNRS
Participants	CNRS, INFN, U.Fra	CEA, FZJ, IBA, INFN, Framatome - CERCA , U.Fra				CEA, CNRS, IBA	CNRS, INFN, ITN, IBA	CEA, ENEA, IBA, INFN, ITN, U.Fra
Requested EU Contribution	1 M€	2 M€				2.5 M€	0.5 M€	0.5 M€

Specific Issues for the injector

Concerning the injector section, its feasibility has already been demonstrated in the frame of several projects either partly, like in the IPHI (France) and TRASCO (Italy)

---

[7] ALARA: As Low As Reasonably Achievable.

projects, or completely, like in the LEDA project (Los Alamos, USA). But the high reliability required by the XADS accelerator implies a thorough campaign to test the reliability of every component, operated over a long period of time (e.g. a continuous 6 month run) at very high power, which would require additional resources from FP6.

Specific Issues up to 100 MeV

Some basic R&D is required for the subsequent section in order to assess a solution simultaneously reliable and economical. For these reasons, WP3 has developed the vision that superconducting components should in principle be used "early on".

However, "the warm option" has to be carefully studied and prototyped for two reasons. First, it could very well be required for boosting the RFQ-injector energy if the transition energy to the superconducting structures is higher than the RFQ output. Second, although the RF losses are quite high for the warm option, its development risk is low, relying on a well established technology. In fact, the warm option could as well be considered a baseline option to which the new superconducting structures have to be compared with. Indeed, the superconducting resonators considered within WP3 are short and modular. That enables a better and more efficient approach for the implementation of the reliability strategy. First cavity prototypes for the intermediate section are presently built and tested with quite a success [9]. It is therefore important to pursue these developments, in particular for spoke- and CH- structures equipped with their helium tank and power coupler in order to built a complete cryomodule. The final aim of all these developments is to assess the best technical option for the intermediate energy section based on established demonstrated performance.

At present, several participating institutions to WP3 (from Belgium, France, Germany and Italy) are engaged in this R&D using their own funding resources. If the dead-line of 2006 for the decision on an XADS accelerator is to be met the present yearly funding is to be maintained and to be complemented by FP6 resources.

Specific issues at high energy

Concerning the high-energy section, R&D is already going on since a few years in Europe on superconducting elliptical cavities at a frequency of 700 MHz. Nevertheless, the demonstration of the full technology is not yet accomplished, and needs a few more years of additional work. As a matter of fact, it is important, besides the development of the bare superconducting cavity, to build prototypes of each auxiliary system needed for the cavity operation in a real environment (power coupler, RF source, power supply, RF control system, cryogenic system, cryostat...). This full demonstration requires the construction of a real prototype of an accelerating module in which all these elements are included. The construction of such an integrated module at a given beta value (for example $\beta = 0.5$) would thus be considered as a proof of principle of the technology, not only for that particular beta, but also for all others since many similarities exist between them. Moreover, such a module could allow real scale demonstration with RF tests at nominal power (although without beam), and could be used for specific studies dealing with the XADS reliability issue, like the completion of the RF control system procedures in case of a cavity failure.

R&D for an RF system

Finally, the last item concerns the study and development of the RF system where the low-level part needs to be highly specialized for an XADS accelerator. This system must handle beam trips, reacting with enough speed to retune the whole accelerator, in order to recover nominal beam conditions in a short time (less than 1s) and to ensure the fault-tolerance principle. Digital techniques are necessary to meet the speed and software configuration requirements. Unique specialized control software and adapted interfaces to the RF power electronics systems must be developed and tested. Co-funding by FP6 resources would allow to start this essential programme.

CONCLUDING REMARK

Within less than two years, it has been possible to develop a generic and robust technical solution for the XADS accelerator and its associated beam-line. This superconducting linear accelerator can be used for all different versions of XADS studied within the 5th FP, and it is also representative for an industrial machine. The proposed machine is reliable through the rigorous implementation of a highly modular system with de-rated components operated in a fault-tolerant way. The continuation of the present vigorous R&D programme, with a focus on reliability within the 6th FP, places the XADS accelerator on a roadmap in-line with the TWG recommendations.

ACKNOWLEDGMENTS

The author, coordinator of WP3 of PDS-XADS, acknowledges the work of all his colleagues of the WP3 collaboration on which this short summary paper is based.

REFERENCES

1. Commission of the EC Green Paper: "Towards a European strategy for the security of energy supply", *COM* (2000) 769. http://europa.eu.int/comm/energy_transport/library/livre-vert/livre-vert-en.pdf
2. Carlo Rubbia et al., A European Roadmap for Developing Accelerator Driven Systems (ADS) for Nuclear Waste Incineration, published by ENEA, Rome 2001, ISBN 88-8286-008-6.
3. Bernard Carluec in *Proceedings of the International Workshop on P&T and ADS Development*, 6-8 Oct. 2003, Mol, Belgium, edited by W. Haeck et al., http://www.cen.be/sckcen_en/activities/conf/conferences/20031006/topics.shtml
4. See, e.g., *Proceedings of the International Workshops on Utilization and Reliability of High Power Proton Accelerators (HPPA)*, organized by the OECD Nuclear Energy Agency.
   October 13-15, 1998, Mito, Japan, ISBN 92-64-17068-5
   November, 22-24, 1999, Aix-en-Provence, France, ISBN 92-64-18749-9
   May, 12-16, 2002, Santa Fe, NM, USA, ISBN 92-64-10211-6
5. Th. Stammbach et al., *Nucl. Inst. Meth.* **B 113**, 1 (1996)
6. Paolo Pierini, in *InWor on P&T and ADS development*, see reference [3]
7. Henri Safa, in *InWor on P&T and ADS development*, see reference [3]
8. Summary of Working Group Discussion on Accelerators in [4], 3rd Meeting.
9. Tomas Junquera, in *InWor on P&T and ADS development*, see reference [3]
10. ICRP, "Recommendations of the International Commission on Radiological Protection", ICRP Publication 60. Annals of the ICRP 21(1-3) Pergamon Press, Oxford, 1991.

ADS Network in Japan and in Asia

Yasuki Nagai

Research Center for Nuclear Physics, Osaka University,
Mihogaoka 10-1, Ibaraki, Osaka 567-0047, Japan

Abstract. The management of long-lived high-level waste, such as minor actinides (MA) and long-lived fission products (LLFP), is one of the most important issues to be solved for the utilization of the nuclear fission energy. The transmutation project of the radioactive wastes based on an accelerator driven sub-critical (ADS) system is very attractive, since it could reduce the risk of long-lived radioactivity and also increase the capacity of the disposal site. In the development of the ADS project networks of researchers between different fields and between different countries are crucial, since there are many scientific and technical issues, which should be studied. The activities based on these networks are discussed.

INTRODUCTION

Peaceful use of nuclear energy is an important issue in Japan and also in many Asian countries, since there are few energy resources in Japan and there are many developing countries in Asia. Concerning the use of nuclear energy, however, the Japanese people are concerned about the nuclear waste disposal problem in addition to the safety problem of nuclear power plant. While, many researchers working at universities are concerned about the training of young people as well as the disposal problem, since we know that young people are leaving from the field of nuclear energy in the last two decades. Hence, in order to get public understanding of the nuclear waste problem and also to attract young generation to this field it is very important to make medium- and long-term plans with dreams. We know that they show their full talent with enthusiasm when they encounter the project with dreams. An accelerator-driven sub-critical system (ADS) project would be one of the dream plans to meet the requirements mentioned-above. The ADS project would require various basic R&D works on many subjects, such as accelerator, spallation target, beam window, sub-critical reactor physics and so on, since it has not yet been proven that the ADS works well as we expect. Hence, it is essential for the innovative and scientific development of the ADS project to organize collaboration groups, which consist of researchers from different fields. In this paper I report the activities based on the communication between Japanese nuclear physicists and nuclear engineers as an example of such collaboration.

ADS PROJECT IN JAPAN

It is well known that the ADS project has a great potential not only for the transmutation of nuclear waste but also energy production and energy resources production [1]. In

Japan, the research group at Japan Atomic Energy Research Institute (JAERI) [2] and the research group at Kyoto Research Reactor Institute (KURRI) at Kyoto University [3] are carrying out R&D works for the nuclear transmutation project & energy production project, respectively. The transmutation project will be funded under the framework of the J-PARC project in the second phase of the project. Hence, the transmutation project may be regarded as a medium-term project and the energy production project as a long-term project, respectively. These two projects are briefly described in the next subsection.

The nuclear transmutation project

One of the major and growing concerns for the utilization of the nuclear fission energy is the fate of the long-lived radioactive, such as minor actinides (MA) and long-lived fission products (LLFP). The Atomic Energy Commission in Japan encouraged the R&D of Partitioning and Transmutation Technology of the long-lived nuclear waste from the processing processes of spent fuel as a possible technology to reduce the potential radiotoxicity in the radioactive waste disposal. The Accelerator Driven Transmutation System of J-PARC is the R&D project dedicated for transmutation. The ADS plant, which should be constructed after the demonstration of the transmutation of MA etc., is an 800MWth sub-critical reactor with a multiplication factor (k) of 0.95. The reactor will be driven by a superconducting linear accelerator with the beam power of 20-30MW, and Lead-bismuth (Pb-Bi) is considered as the liquid metal spallation target and the core coolant. By using the ADS system, 250kg of MA can be transmuted through fission reactions per year, which corresponds to the amount of MA produced in ten plants of light water reactor (LWRs) with 1GWe. In order to reach the ADS project mentioned above, J-PARC group has made a near-future plan to make the ADS development in a stepwise way, and the plan will be realized within the framework of the J-PARC, Japan Proton Accelerator Research Complex project. Since the principal aim of the plan is the feasibility study on the ADS from physics and engineering aspects, the construction of Transmutation Experimental Facility (TEF) is proposed. The facility consists of two buildings of Transmutation Physics Experimental Facility (TEF-P) and ADS Target Test Facility (TEF-T). The TEF-P is used to research the reactor physics and the controllability of ADS. The TEF-T is a material irradiation facility with use of the Pb-Bi spallation target. The construction of TEF is expected to start from 2006. After this project an experimental ADS with thermal power of about 80MW is necessary in the late 2010s to demonstrate the feasibility of the ADS from engineering aspects. Here, mixed oxide (MOX) fuel will be used at first and then the MA nitride fuel. The demonstration of the MA transmutation will be completed by 2030, since according to the strategy of Japanese government for waste isolation nuclear wastes are buried in geologic repositories and the first geological repository will be constructed in 2030.

The Energy Amplifier project

The Neutron Factory project is proposed at Kyoto University Research Reactor Institute (KURRI). The Factory is a system consisting of a fixed field alternating gradient (FFAG) Synchrotron accelerator and an existing reactor (ADSR). In 2002, the Kyoto group got the fund for the five years project, entitled as "Technology Development for the ADSR", from the Japanese Government to make the feasibility study on the ADSR as a nuclear energy system as well as the future neutron source for scientific research. The FFAG proton synchrotron can provide the intense proton beam from 2.5 MeV to 150 MeV. Using the FFAG, the reactor physics of ADSR will be studied from 2005.

ACTIVITY OF NUCLEAR DATA GROUP

The measurement and evaluation of the neutron capture and fission cross sections of minor actinides are important issues for the R&D of ADS. About 140 people are working in the nuclear data field, and they are responsible for the R&D of Japanese Evaluated Nuclear Data Library (JENDL). In 2002, the Japanese Government approved a newly proposed five years project by the nuclear data group and nuclear physicists. The aim of the project is to develop of advanced measurement technology, measure neutron cross sections of minor actinides, and develop neutron cross-section utilization system. Under the project the Ge-spectrometer system with the geometry of 4π, which consists of 18 HPGe detectors with BGO shields is under construction for the neutron capture cross section measurement of MA. The spectrometer is hopefully used at the spallation neutron source facility at J-PARC in near future.

NUCLEAR PHYSICS COMMUNITY IN JAPAN

The JAERI's original nuclear transmutation project of radioactive waste has been a matter of concern for nuclear physicists in Japan. However, before 1998 most nuclear physicists did not know well essential nuclear physics problems in the project due to little communication between nuclear physicists and nuclear engineering researchers. The situation has changed since 1999, when the joint project between KEK and JAERI (currently called J-PARC) was proposed [4]. Since then two communities of nuclear physics and nuclear engineering have started the communication, and researchers working in these fields organized several workshops on "Accelerator-Driven Sub-critical Reactor System and Nuclear Physics". In these workshops we discussed various problems not only on the nuclear transmutation project but also on a basic R&D project on energy amplifier, and participants exchanged their information. Based on these activities we have built a voluntary association, named "ADS and nuclear physics", in the middle of 2001 to continue the communication and extend the scientific activity on the ADS through the exchange of information. Currently more than 80 researchers from various fields join the association.

What we have learned on the ADS project through these communications was very impressive as described below. First, participants appreciated open discussions about current and future problems on the nuclear energy program and the promotion of scientific communication. Second, nuclear physicists started to show their interests in important issues in the ADS project. Third, we noticed that several experimental and theoretical results, which have been already obtained by nuclear physicists in their works, are quite useful in the design and development of the ADS project. This may suggest that we can expect a cross-cultural synergy effect in the development of the ADS project through the communication between these two research groups, which would be of vital importance in the design of not only the ADS project but also of a new concept of the energy problem.

JOINT PROPOSAL TO THE SPALLATION NEUTRON FACILITY AT J-PARC

Both researchers in the fields of cosmo-nuclear physics and nuclear engineering are very much interested in the high intensity spallation neutron source, since they are measuring cross sections of neutron induced reaction of various nuclei using fast neutrons in the energy region of between a few keV and a few hundred keV. These data are necessary for cosmo-nuclear physicists to construct models of stellar nucleosynthesis and for nuclear engineers to design the system based on the ADS. Especially fission, capture, inelastic, (n,2n) etc. cross sections on MA and capture cross sections on rare abundant elements & unstable nuclei are quite important. Since the sample amounts we could prepare for such measurements are quite small, we need intense neutron beam with wide energy range to obtain accurate data. In Japan the high intensity proton accelerator facility called J-PARC is now under construction at Tokai JAERI campus [4] J-PARC is an accelerator complex consisting of 50 GeV and 3 GeV proton synchrotron and 400 MeV linac. At Material & Life Science Experimental Facility at J-PARC spallation neutrons will be produced by the 3GeV 1MW Proton synchrotron. Since the neutrons are so attractive for the study of ADS and Cosmo-nuclear physics we have made the proposal to construct the beam course at the facility and submitted the proposal to the director of J-PARC.

ASIA ADS NETWORK

The energy problem is one of important issues not only in Japan but also in our neighboring countries, where rapid annual growth of economy development is expected in the coming decades. Actually, several countries have started actively the R&D work of the ADS project, and also the discussion about a possibility to introduce the nuclear power plant. Under such situations of the nuclear power plan in Asian countries the collaborative R&D work of the ADS between Asian countries would be very important. Hence, in order to discuss about the possible collaboration we have organized the International symposium on "Accelerator-Driven Transmutation System and Asia ADS Network Initiative" on March 24 and 25, 2003 in Tokyo. At the symposium current R&D activities of the ADS at United States, Europe, Korea, China and Japan were presented, and then a panel discussion was organized to discuss about the international collaboration on the ADS project. Through the discussions and various comments we came to the conclusion to establish the international network for scientific information exchange among Asian countries including Korea, China, Vietnam and Japan. Here, it should be mentioned that there are many scientific and technical issues to be solved for the development of the ADS. Hence, it is extremely important for the development to study various creative and innovative approaches by different countries. The ADS network would play an important role for the ADS development. It was also agreed to extend the network to the international one near future.

CONCLUSION

The problems of nuclear waste disposal and training of young generation are quite important for the peaceful use of fission energy in Japan. The ADS project could be very attractive to solve these problems, since it has a potential for the transmutation of nuclear waste and also for the energy production. Hence, the ADS project would also play an important role to absorb young generation into the filed of nuclear power plant. In order to develop the ADS project there are many scientific and technical issues, which should be studied through various kinds of approaches. The communication of researchers between different fields and an international ADS network could play an essential role to promote the R&D works of the ADS.

REFERENCES

1. C. Rubbia et al., CERN/AT/95-44(ET), Sept. 29, 1995
2. T. Mukaiyama et al., Prog. in Nucl. Energy, Vol.38,(1995)107
3. S. Shiroya et al., Prog. in Nucl. Energy, Vol.37,(2000)357
4. S. Nagamiya et al., KEK Report 99-4, (JAERI-Tech 99-56)

NUCLEAR ASTROPHYSICS (NAP)

- Nuclear Physics in Space: From Solar Energetic Particles to Ultra-High Energy Cosmic Rays

- Radionuclides in the Galaxy: Some Selected Aspects

- Nuclear Data for Low-Energy Astrophysics and Other Applications

- Low-Energy Nuclear Reactions for Astrophysics

- Neutron Stars and Other Stars

Nucleosynthesis by Spallation Reactions in the Early Solar System
- The Need for Spallation Cross Sections -

Ingo Leya

Institute for Isotope Geology and Mineral Resources, ETH Zürich, NO C60.6, CH8092 Zürich, Switzerland

Abstract. Ever since their discovery in 1960, the origin of the relatively short-lived radionuclides, now extinct but alive in the early solar system, has been under debate. Possible scenarios are either nucleosynthetic production in stellar sources, *e.g.* AGB stars, Wolf-Rayet stars, novae and supernovae, with subsequent injection into the solar nebula, or production by spallation reactions in the early solar system by energetic particles from the young sun. Here we present model calculations for the second scenario, the production of the relatively short-lived radionuclides by solar energetic particle events in the early solar system. For successful modeling the cross sections for proton-, ^{3}He-, and ^{4}He-induced reactions for all relevant nuclear reactions have to be known. In addition, the modeling depends on the relative fluence contributions of protons, ^{3}He, and ^{4}He in the solar energetic particle events as well as on their energy distribution. The ability of the model calculations to simultaneously describe the observed nuclide ratios ^{7}Be/^{9}Be, ^{10}Be/^{9}Be, ^{26}Al/^{27}Al, ^{41}Ca/^{40}Ca, ^{53}Mn/^{55}Mn, and ^{92}Nb/^{93}Nb is presented. Special emphasis is given to the problems arising from ill-known cross sections for ^{3}He- and ^{4}He-induced reactions.

INTRODUCTION

Numerous studies have unambiguously proven the existence of relatively short-lived radionuclides in the early solar system. Since the radionuclides are extinct by now and cannot directly be detected, excesses of their daughter isotopes correlated with mother-daughter elemental ratios indicate their former presence. The most promising samples for such studies are refractory condensates found in some types of meteorites, the so-called calcium-aluminum-rich refractory inclusions (CAIs). CAIs are among the first objects crystallized out of the solar nebula. Due to analytical improvements, the number of relatively short-lived radionuclides that were known to be alive in the early solar system has shown a dramatic increase in the last few decades.

The relatively short-lived radionuclides offer a great potential for chronological studies of early solar system processes with high time resolution, *e.g.*, crystallization of first solids, formation of planets, formation of planetary cores, formation of first atmospheres. However, before they can be used for dating, their origin has to be known, whereas the question of whether they were homogeneously distributed throughout the solar system is of special importance. Only a homogeneous distribution enables useful chronometry.

For many years the most promising hypothesis for the origin of the relatively short-lived radionuclides was production in stellar sources, *e.g.* AGB stars, Wolf-Rayet stars, novae, and supernovae, with subsequent injection into the solar nebula. Connected to this stellar production scenario is the hypothesis that the stellar source may also have locally triggered the collapse of the molecular cloud from which the solar system evolved. However, a self-consistent modeling of the radionuclide abundances in the early solar system is not straightforward, since continuous galactic nucleosynthesis fails to simultaneously describe the short-lived nuclide abundances inferred from meteorite data. To circumvent this problem, Meyer and Clayton [1] proposed nuclide production in three different stellar events. Apart from the crucial timing another difficulty is that the proposed scenario requires that only the outermost layers of a supernova, about 1 Myrs prior to solar system formation, were injected into the collapsing molecular cloud core [1]. However, in recent years the stellar production hypothesis was faced with another difficulty, the finding of life ^{10}Be in early solar system condensates.

In recent years ^{10}B excesses resulting from the *in-situ* decay of ^{10}Be in early solar system condensates have been reported [2,3,4]. The existence of life ^{10}Be is difficult to explain by nucleosynthetic production because in most stellar events Be is destroyed rather than produced. Recently, trapping of some Be from the galactic cosmic-rays (GCR) and ^{10}Be production by interactions of GCR particles with molecular cloud core material is discussed as a possible source for the early solar system ^{10}Be [5]. Another possibility to explain the short-lived nuclide abundances is spallogenic production within the early solar system by energetic particles emitted from the young sun (YSO, Young Stellar Object). Such a scenario is supported by astronomical observations. Systematic studies of X-ray emissions from YSOs indicate that most of them eject significant amounts of their material as solar energetic particle events (SEP) and/or bipolar outflows. A recent systematic study shows that the so-called T-Tauri phase is common for stellar objects with 0.8-1.4 $M_\odot$ and that the particle flux in the YSO phase is up to 10^5 times higher than for the recent sun [6]. The energy of the emitted particles, *i.e.* mostly protons, ^{3}He, and ^{4}He, is as large as some tens of MeV nucleon^{-1} in present-day SEP events. Particles in this energy range induce nuclear reactions in the irradiated material. If SEP production is either near the accretion disk or if the YSO is magnetically connected to the accretion disk, various kinds of nuclear reactions within the disk are induced. In such nuclear interactions, among others, short-lived radionuclides are produced. Depending on the mechanism for SEP production, the projectiles might get focussed on small regions, leading locally to high flux densities and consequently to high nuclear production rates. Therefore, nuclear spallation reactions might well have contributed to the nucleosynthesis of some of the relatively short-lived radionuclides alive in the early solar system.

In recent years various groups tested the local production scenario by comparing modeled nuclear spallation yields with inferred solar system ratios [7,8,9]. However, they all failed, for various reasons, to simultaneously describe the initial solar system radionuclide abundances (for a more comprehensive discussion see [10]). Here we present new model calculations for the production of the relatively short-lived radionuclides by spallation reactions in the early solar system. Assuming a solar chemical composition of the target and using physically reasonable particle spectra and fluences enables for a first time to simultaneously determine the inferred radionuclide abundances of ^{7}Be, ^{10}Be, ^{26}Al, ^{41}Ca, ^{53}Mn, and ^{92}Nb using only one free parameter, the relative fluence contribution of gradual and impulsive SEP events. However, of similar important is the fact that the model fails to describe the solar system initial ^{60}Fe abundances. Here we focus mainly on the need for an improved database for spallation cross sections for proton-, ^{2}H-, ^{3}He-, and ^{4}He-induced reactions needed for accurate modeling. The reader interested in consequences and time constraints for CAI production in the framework of the local production scenario is referred to [10].

MODEL CALCULATIONS

The details of the model calculations as well as the involved uncertainties are discussed in detail in [10]. Briefly, the production rates ($g^{-1} \cdot s^{-1}$) for the relatively short-lived radionuclides in the early solar system, $P_j(\alpha)$, are calculated using:

$$P_j(\alpha) = N_A \sum_i c_i A_i^{-1} \sum_k \int_0^\infty \sigma_{j,i,k}(E) J_k(E,\alpha) dE \qquad (1)$$

where N_A is Avogadro's number, A_i the mass number of the target element i, and c_i is the concentration of i ($g \cdot g^{-1}$). The excitation function is $\sigma_{j,i,k}$, and $J_k(E,\alpha)$ is the differential flux density of the projectile type k.

Flux densities

In the present Sun the emitted particles, which reach energies up to a few hundred MeV nucleon^{-1}, originate in either gradual (G-SEP) or impulsive (I-SEP) solar particle events. For modeling we assume that the particle spectra and chemical compositions of the SEP events from the YSO were similar to those from the recent sun. The parameters of the particle spectra are compiled in Tab. 1. For a detailed discussion of the particle spectra see [10].

TABLE 1. Parameters of the SEP events used for modeling.

Parameter	Impulsive events	Gradual events
Shape	$E^{-2.7}$	$E^{-4.0}$
H / He	10	300
He-3 / He-4	1	4×10^{-4}
Weight	1	1000

Cross Sections

Calculating the production rates with equation (1) requires for each product nuclide detailed excitation functions for all relevant target elements and all relevant projectile types. The present version of model calculations is restricted to proton-, ^{3}He-, and

^{4}He-induced reactions. From detailed studies of nuclear reactions in meteorites it is known that SEPs produce few, if any, secondary neutrons within the target (for references see [10]). Therefore neutron-induced reactions are not included. It is shown below that for some target-product combinations ^{2}H-induced reactions might have contributed to the production rates as well. However, due to the lack of experimental and modeled cross sections ^{2}H-induced reactions are so far not included in the model calculations.

Proton-induced reactions

The cross section database for proton-induced reactions is fairly complete by now due to many years of effort modeling cosmogenic production rates in meteorites and the Moon, e.g. [11,12]. Figure 1 shows as an example the cross sections for the production of ^{53}Mn from iron. The symbols indicate experimental data (for references see [13]) and the solid line is the result from Hauser-Feshbach calculations (GNASH-FKK [14]). However, for some product nuclides, those that are not used in meteorite studies, e.g. ^{92}Nb, no measured cross sections exist and we have to rely on excitation functions determined using theoretical nuclear model codes. Throughout this study I used a Hauser-Feshbach approach in the programmed version GNASH-FKK [14].

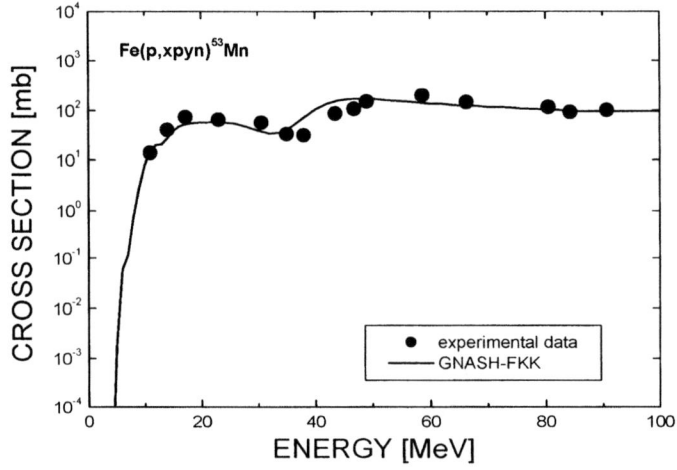

Figure 1. Excitation function for the production of ^{53}Mn from iron. The symbols are experimental data from thin target experiments. The theoretical results (solid line) are obtained from the GNASH-FKK code.

However, the predictive power of nuclear model codes for proton-induced reactions is limited to within a factor of 2 at best [15]. Therefore, if one is interested in the origin of a short-lived nuclide not analyzed in meteorite studies, e.g. ^{92}Nb for the short-lived nuclide system ^{92}Nb-^{92}Zr, it is necessary to measure the cross sections for all relevant proton-induced reactions leading to the production of ^{92}Nb.

^{3}He- and ^{4}He-induced reactions

For ^{4}He-induced reactions only few measured cross sections have been published so far. Furthermore, the existing database shows a large scatter and deducing consistent excitation functions for the relevant nuclear reactions is not straightforward. Even worse, for ^{3}He-induced reactions essentially no measured cross sections exist. We therefore have to exclusively rely on excitation functions determined using nuclear model codes. Again the cross section data are calculated using the code GNASH-FKK [14]. However, the predictive power of nuclear model codes for ^{3}He- and ^{4}He-induced reactions has never been tested (due to the lack of experimental data). We therefore expect the modeled production rates due to ^{3}He- and ^{4}He-induced reactions to be rather uncertain. A detailed discussion of the difficulties resulting from calculated cross sections, the expected uncertainties as well as our approach to circumvent some of the problems is given below.

Uncertainties resulting from the ill-known cross sections

Using a statistical model like the Hauser-Feshbach approach makes the resulting uncertainties in the cross sections strongly dependent on the target mass number. We already demonstrated that the production of ^{7,10}Be from oxygen by proton-induced reactions calculated using either experimental cross sections or GNASH-FKK results differ by about one order of magnitude [10]. For the production of ^{26}Al from Al and Si the differences are up to a factor of 2. For the production of ^{53,54}Mn from Fe modeled and experimental cross sections give production rates in reasonable agreement, *i.e.* to within about 50%.

A major problem for the calculated cross section database is the existence of isomeric states which are not considered in the model codes. An example is the first isomeric state in ^{26}Al ($E^* = 228$ keV, $T_{1/2}$ (β^+) = 6.35 s). Within the Hauser-Feshbach code all excited states are allowed to decay via internal transition into the ground state resulting in a net accumulation of ground state ^{26}Al ($T_{1/2}$ (β^+) = 0.716 Myr). It is this relatively long-living ground state which enables the detection of live ^{26}Al in early solar system condensates. However, in nature the situation is different. Some (most) of the nuclear reactions produce ^{26}Al in excited states, which then decay via γ-cascades. During such γ-cascades the first exited state in ^{26}Al is populated. Due to the large multipole order of the transition from the excited state ($J_p = 0^+$) to the ground state ($J_p = 5^+$) the first excited state has a relatively long half-life of 6.35 s and decays via β^+-decay directly to ^{26}Mg. Therefore, assuming that all nuclear reactions end in the net production of ground state ^{26}Al is not correct. Consequently the modeled cross sections for ^{26}Al are too large because only a (small) fraction of nuclear reactions end in ground state ^{26}Al. Some (most) reactions end in excited states in ^{26}Al and finally decay via isomeric β^+-decay into ^{26}Mg, not resulting in a net accumulation of ^{26}Al.

Systematic studies of proton-induced ^{26}Al production enable estimating that the uncertainty resulting from the neglecting of the isomeric state should not be much larger than about a factor of 2. The production rates for ^{26}Al from various target elements modeled using either experimental or calculated proton-induced cross sections typically differ by less than about a factor of 2. Assuming that the population of the isomeric state relative to the ground state is essentially the same for proton-, ^{3}He-, and ^{4}He-induced reactions, the expected overall uncertainty should therefore not be much larger than the factor of about 2.

However, at present it is not possible to accurately estimate the overall uncertainty of the modeled excitation functions. In order to overcome some of the problems arising from ill-known cross sections, the production rates by ^{3}He- and ^{4}He-induced reactions are not directly modeled. Instead I calculated the excitation functions for proton-, ^{3}He-, and ^{4}He-induced reactions using the GNASH-FKK code. From these data and using equation (1), the production rates $P_{p,GNASH}$, $P_{He3,GNASH}$, and $P_{He4,GNASH}$ are modeled. From this the ratios $P_{He3,GNASH}$ / $P_{p,GNASH}$ and $P_{He4,GNASH}$ / $P_{p,GNASH}$ are calculated. It can be expected that the Hauser-Feshbach approach describes cross section ratios more accurately than absolute cross sections. The final production rates used for further discussion were then determined via:

$$P_{p,final} = P_{p,exp}; \quad P_{He3,final} = \frac{P_{He3,GNASH}}{P_{p,GNASH}} \times P_{p,exp}; \quad P_{He4,final} = \frac{P_{He4,GNASH}}{P_{p,GNASH}} \times P_{p,exp} \quad (2)$$

The applied method reduces the uncertainties of the model calculations due to ill-known cross sections as far as possible. However, for some reactions, e.g. ^{44}Ti from Ca, production is possible by ^{3}He- and ^{4}He-induced reactions but not by protons. In those rare cases one have to assume $P_{He3,final} = P_{He3,GNASH}$ and $P_{He4,final} = P_{He4,GNASH}$.

To summarize, considering the fact that most relevant cross sections are missing, i.e. all data for ^{3}He- and ^{4}He-induced reactions but also some of the proton-induced cross sections, the total uncertainty of the modeled solar system initial ratios can be estimated to be about a factor of 10. *Therefore, to further improve the model calculations, which would allow rigorously proofing or rejecting the local production scenario, the complete cross sections database for ^{3}He- and ^{4}He-induced reactions for all relevant targets elements have to be measured. In addition, there still is some need for more and more precise proton-induced cross sections. The new database should then enable to evaluate and improve nuclear model codes, which in turn would allow to accurately predict level dependent cross sections for proton-, ^{2}H (see below), ^{3}He-, and ^{4}He-induced reactions over a wide range of energies.*

Modeled isotopic ratios

Modeling the production of ^{7}Be, ^{10}Be, ^{26}Al, ^{41}Ca, ^{53}Mn, and ^{60}Fe

For modeling the solar system initial ratios we used the data for the particle spectra compiled in Tab. 1. For references see [10]. Note that using the values given in Tab. 1 implies that we assume that the physics of the SEP events did not change during the evolution of the sun. However, we do allow for some changes, i.e. we do allow that the relative fluence contributions of G-SEP and I-SEP events were different in the YSO compared to the recent sun by treating the ratio G-SEP / I-SEP as a free parameter. For the following discussion we further assume an irradiation time short enough to neglect saturation effects for the production of any radionuclide considered. Some of the modeled results are shown in Fig. 2. Fig. 2a depicts the ratio of modeled to inferred solar system initial ratios for ^{10}Be/^{9}Be ($\equiv 10^{-3}$, because of normalization),

$^{26}Al/^{27}Al$, $^{41}Ca/^{40}Ca$, and $^{53}Mn/^{55}Mn$ as a function of the ratio G-SEP / I-SEP assuming solar composition for the target. Fig. 2b depicts the results for targets having CAI composition throughout the irradiation. For a detailed discussion of the data the reader is referred to [10].

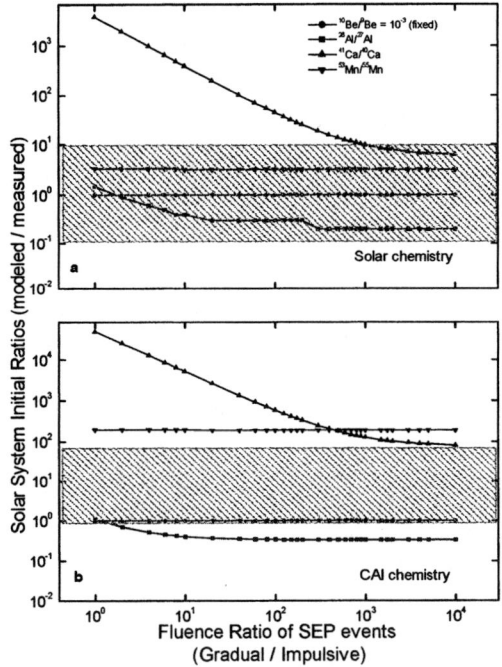

Figure 2. Ratio of modeled and measured solar system initial ratios for $^{10}Be/^9Be$ ($\equiv 10^{-3}$), $^{26}Al/^{27}Al$, $^{41}Ca/^{40}Ca$, and $^{53}Mn/^{55}Mn$ as a function of the ratio G-SEP / I-SEP. The shaded area ranging from 0.1 to 10 indicates the uncertainties expected for the model calculations. (a) Results for targets having solar chemical composition. Assuming a ratio G-SEP / I-SEP > 1000 enables to simultaneously describe the solar system initial ^{10}Be, ^{26}Al, ^{41}Ca, and ^{53}Mn abundances (b) Results for targets with present-day CAI composition. It can be seen that a simultaneous description of the solar system initial abundances is not possible.

Fig 2a demonstrates that the model simultaneously describes the inferred solar system initial abundances for ^{10}Be, ^{26}Al, ^{41}Ca, and ^{53}Mn within a factor of 10, *i.e.* within the uncertainties expected for the model calculations, if the ratio G-SEP / I-SEP is assumed to be greater than 1000 and if the chemical composition of target is assumed to be solar during the irradiation. In contrast, if one assumes present-day CAI composition for the irradiated targets, the model fails to simultaneously describe the solar system initial abundances (Fig. 2b). For an explanation and a detailed discussion see [10].

To shortly summarize, assuming solar chemical composition for the irradiated target, a fluence ratio G-SEP / I-SEP of >1000, and energy spectra as well as the relative proportions of protons, ^3He, and ^4He in the SEPs that are identical to those in present-day SEP events permits for a first time a simultaneous description of the solar system initial abundances of ^{10}Be, ^{26}Al, ^{41}Ca, and ^{53}Mn in the framework of the local production scenario. In addition, the model gives a solar system initial ^{7}Be/^{10}Be of ~70, close to the experimental result of 230±130 given by [16].

There are discrepancies between modeled and inferred values for the solar initial ^{60}Fe/^{56}Fe, a feature already observed in earlier studies [7,8]. The model underestimates the initial ^{60}Fe abundance published recently [17,18] by about 5 or 4 orders of magnitude assuming solar or present-day CAI chemistry, respectively. In general the production of neutron-rich isotopes by spallation reactions is very inefficient. Therefore, the occurrence of live ^{60}Fe but also ^{182}Hf in the early solar system requires another production mechanism for these isotopes. For a discussion about how the new ^{60}Fe data support a nucleosynthetic origin for ^{26}Al in concordance with ^{60}Fe see [10,19].

Modeling the collateral production of ^{14}C, ^{22}Na, ^{36}Cl, ^{44}Ti, ^{54}Mn, ^{63}Ni, and ^{91}Nb

In the previous section it is demonstrated that the new model calculations allow for a consistent description of the abundances of most of the relatively short-lived radionuclides whose presence in the early solar system is established. However, in the framework of a local production scenario, production by spallation is also expected for other radioactive and stable nuclides. I therefore also modeled the expected solar system initial ratios ^{14}C/^{12}C, ^{22}Na/^{23}Na, ^{36}Cl/^{35}Cl, ^{44}Ti/^{48}Ti, ^{54}Mn/^{55}Mn, ^{63}Ni/^{58}Ni, and ^{91}Nb/^{93}Nb as well as the shifts expected in Ti-isotopes (not discussed).

Unfortunately, most of the systems turn out not to be very useful for further proofing or rejecting the local production scenario. A successful candidate is ^{14}C. Assuming a short-term irradiation with a high fluence, as required to explain the measured data for ^{7}Be, the collateral production of ^{14}C would result in measurable isotopic effects in nitrogen. Note that ^{7,10}Be and ^{14}C are produced by spallation reactions on oxygen. Hence, the isotopic effects in lithium, boron, and nitrogen should correlate. Therefore, it might be of interest to search for ^{7}Li and ^{10}B enrichments correlated with high ^{14}N/^{15}N in early solar system condensates.

The decay products of ^{22}Na and ^{36}Cl, ^{22}Ne and ^{36}Ar, are volatile and might have been lost during the last 4.5 Gyrs. This is confirmed by noble gas studies because the predicted amounts of pure ^{22}Ne and ^{36}Ar have never been found in early solar system condensates. The isotopic shift in Ca expected from the radioactive decay of ^{44}Ti (less than 30 ppb for ^{44}Ca/^{40}Ca) is much too small to be detectable with present-day analytical precision. For the modeled ^{91}Nb/^{93}Nb the existing data in ^{91}Zr/^{93}Zr scatter too much to discern any trend. The modeled solar system initial ^{63}Ni/^{58}Ni is 3 orders of magnitude lower than a recent experimental estimate [20]. However, the experimental data is still under some debate.

A second successful candidate, besides ^{14}C, which would allow further proofing or rejecting the local production scenario, is ^{54}Mn. The model calculations predict a solar system initial ^{54}Mn/^{55}Mn ratio about 5 times lower than ^{53}Mn/^{55}Mn. Considering the isotopic abundances of the daughter isotopes, the expected $\varepsilon(^{54}$Cr/^{52}Cr) should be about the same as $\varepsilon(^{53}$Cr/^{52}Cr), where 1ε is the isotopic shift relative to a standard in units of 10^{-4}. In contrast, the experimental data from meteorites suggest $\varepsilon(^{54}$Cr/^{52}Cr) ratios about 4 times higher than $\varepsilon(^{53}Cr/^{52}Cr)$ [21]. Note that the model calculations predict the relative proportions of the ^{53}Mn and ^{54}Mn production rates to within a factor of about 2, because a large part of the uncertainties cancel out when production rate ratios from the same target element are considered. Therefore, the observed difference of a factor of 4 is of significance. A possible explanation for the observed discrepancies is that ^{2}H-induced reactions might have been of importance in the very beginning of the solar system before ^{2}H were fused to ^{3}He. Deuterium-induced reactions on iron preferentially produce ^{54}Mn (via the direct nuclear reaction ^{56}Fe(^{2}H,α)^{54}Mn) and are therefore a significant sources for ^{54}Mn. *Consequently, in order to solve the discrepancy between measured and modeled solar system initial $^{54}Mn/^{55}Mn$ and $^{53}Mn/^{55}Mn$ ratios there is a need for experimental cross sections for ^{2}H-induced reactions and/or for nuclear model codes which accurately predict cross sections for direct nuclear reactions.*

A possible scenario for the short-lived nuclide production

A detailed discussion of the consequences for CAI production in the framework of the local production scenario is given in [10]. Briefly, the major challenge for a local production scenario is the experimental finding that the solar system initial ratios show only modest variations. This in turn requires that the material now forming a large part of the CAIs has experienced surprisingly homogeneous particle fluences. In a recent paper we demonstrated that in a successful scenario a gaseous target with solar composition is irradiated. The irradiation would need to have lasted not considerably longer than 1 Myrs. Such material crystallized with live ^{10}Be, ^{26}Al, ^{41}Ca, ^{53}Mn, and ^{92}Nb close to levels inferred from meteorite studies. However, the modeled ^{7}Be/^{9}Be of 10^{-5} would not allow detecting any ^{7}Li anomalies. Therefore, some of the material must has been irradiated by high-flux SEP events, which lasted no longer than about 1 year. Those CAIs would crystallize with ^{7}Be/^{10}Be of about 100, and ^{10}Be, ^{26}Al, ^{41}Ca, ^{53}Mn, and ^{92}Nb in concentrations close to their inferred solar system initial values.

Note that the fluence needed for the 1 Myrs irradiation scenario of $J(E > 10$ MeV$) \approx 10^6$ cm^{-2} s^{-1} is in accordance with present-day SEP events if the 10^5 times higher flux densities in YSOs compared to the contemporary sun is taken into account [6]. Therefore, the proposed scenario is straightforward and physical reasonable.

CONCLUSIONS AND OUTLOOK

I demonstrated that the model consistently describes the abundances of most of the relatively short-lived radionuclides alive in the early solar system. An important exception is ^{60}Fe, for which the model the inferred solar system abundance underestimates by more than 4 orders of magnitude. Therefore, the occurrence of life ^{60}Fe in the early solar system requires in any case at least some stellar nucleosynthesis. The proposed scenario of a homogeneous irradiation of a gaseous target by energetic protons, ^{3}He, and ^{4}He from the young stellar object is straightforward and the required irradiation times and flux densities are within physical reasonable limits. However, the study suffers from ill-known cross sections for ^{3}He- and ^{4}He-induced reactions. Therefore, in order to improve the model, which would allow further proofing or rejecting the local production scenario, there is a need for experimental cross sections for the relevant ^{3}He- and ^{4}He-induced reactions. Furthermore, if spallation reactions occurred in the very beginning of the solar system, before ^{2}H has been fused to ^{3}He, deuterium-induced reactions might also have contributed and the relevant cross sections should be measured.

A best approach would be to measure level dependent cross sections for all relevant proton-, ^{2}H-, ^{3}He-, and ^{4}He-induced reactions. Such a database would enable, using advanced γ-Astronomy, to investigate young stellar objects and not only qualitatively but also quantitatively detect the amount of produced radionuclides.

ACKNOWLEDGMENTS

I would like to thank the organizers of the Tours symposium for the kind invitation. Numerous discussions with R. Wieler improved this study. Discussions with A.N. Halliday, M. Chaussidon and G. Lugmair are also appreciated. This work was supported by the Swiss National Science Foundation.

REFERENCES

1. B.S. Meyer and D.D. Clayton, *Space Science Reviews* **92**, 133-152 (2000).
2. K.D. McKeegan, M. Chaussidon and F. Robert, *Science* **289**, 1134-1137 (2000).
3. G.J. MacPherson and G.R. Huss, *Lunar Planet. Sci. Conf.* **32**, #1182 (2001).
4. N. Sugiura, Y. Shuzou and A. Ulyanov, *Meteoritcs and Planet. Sci.* **36**, 1397-1408 (2001).
5. S.J. Desch and C. Connolly Jr., *Meteoritics and Planet. Sci. Suppl.* **38**, A133 (2003).
6. E.D Feigelson, G.P. Garmier and S.H. Pravdo, *Astropyhs. J.* **572**, 335-349 (2002).
7. T. Lee, F.H. Shu, H. Shang, A.E. Glassgold and K.E. Rehm, *Astrophys. J.* **506**, 898-919 (1998).
8. M. Gounelle, F.H. Shu, H. Shang, A.E. Glassgold, K.E. Rehm and T. Lee, *Astrophys. J.* **548**, 1051-1070 (2001).
9. A.N. Goswami, K.K. Marhas and S. Sahijpal, *Astrophys. J.* **549**, 1151-1159 (2001).
10. I. Leya, A.N. Halliday and R. Wieler, *Astrophys. J.* **594**, 605-616 (2003).
11. I. Leya, H.-J. Lange, S. Neumann, R. Wieler and R. Michel, *Meteoritics and Planet. Sci.* **35**, 259-286 (2000).
12. I. Leya, S. Neumann, R. Wieler and R. Michel, *Meteoritics and Planet. Sci.* **36**, 1547-1561 (2001).
13. I. Leya et al., *Meteoritics and Planet. Sci.* **35**, 287-318 (2000).
14. P.G. Young, E.D. Arthur and M.B. Chadwick, GNASH-FKK (computer code system) (PSR-0125/07; Paris:NEA) (1998).
15. R. Michel and P. Nagel, International Codes and Model Comparison for Intermediate Energy Activation Yields (NSC/DOC[97]-1; Paris:NEA) (1997).
16. M. Chaussidon, F. Robert and K.D. McKeegan, *Lunar Planet. Sci. Conf.* **33**, #1564 (2002).
17. S. Tachibana and G.R. Huss, *Lunar Planet. Sci. Conf.* **34**, #1737 (2003).
18. S. Mostefaoui, G.W. Lugmair, P. Hoppe and A. ElGoresy, A., *Lunar Planet. Sci. Conf.* **34**, #1585 (2003).
19. G.J. Wasserburg, R. Gallino and M. Busso, *Astrophys. J.* **500**, L189 – L193 (1998).
20. J.M. Luck, D. Ben Othman, J.A. Barrat and F. Albaréde, *Geochim. Cosmochim. Acta* **67(1)**, 143-151 (2003).
21. A. Shukolyukov and G.W. Lugmair, *Meteoritics and Planet. Sci. Suppl.* **36**, A188 (2001).

Updated Big–Bang Nucleosynthesis compared to WMAP results

Alain Coc*, Elisabeth Vangioni-Flam[†], Pierre Descouvemont**, Abderrahim Adahchour[1,**] and Carmen Angulo[‡]

CSNSM, CNRS/IN2P3/UPS, Bât. 104, 91405 Orsay Campus, France
[†]*IAP/CNRS, 98bis Bd. Arago, 75014 Paris France*
**Physique Nucléaire Théorique et Physique Mathématique, CP229, Université Libre de Bruxelles, B-1050 Brussels, Belgium*
[‡]*Centre de Recherche du Cyclotron, UCL, Chemin du Cyclotron 2, B-1348 Louvain–La–Neuve, Belgium*

Abstract. From the observations of the anisotropies of the Cosmic Microwave Background (CMB) radiation, the WMAP satellite has provided a determination of the baryonic density of the Universe, $\Omega_b h^2$, with an unprecedented precision. This imposes a careful reanalysis of the standard Big–Bang Nucleosynthesis (SBBN) calculations. We have updated our previous calculations using thermonuclear reaction rates provided by a new analysis of experimental nuclear data constrained by R-matrix theory. Combining these BBN results with the $\Omega_b h^2$ value from WMAP, we deduce the light element (^4He, D, ^3He and ^7Li) primordial abundances and compare them with spectroscopic observations. There is a very good agreement with deuterium observed in cosmological clouds, which strengthens the confidence on the estimated baryonic density of the Universe. However, there is an important discrepancy between the deduced ^7Li abundance and the one observed in halo stars of our Galaxy, supposed, until now, to represent the primordial abundance of this isotope. The origin of this discrepancy, observational, nuclear or more fundamental remains to be clarified. The possible role of the up to now neglected $^7Be(d,p)2\alpha$ and $^7Be(d,\alpha)^5Li$ reactions is considered.

INTRODUCTION

Big–Bang nucleosynthesis used to be the only method to determine the baryonic content of the Universe. However, recently other methods have emerged. In particular the analysis of the anisotropies of the cosmic microwave background radiation has provided $\Omega_b h^2$ values with ever increasing precision. (As usual, Ω_b is the ratio of the baryonic density over the critical density and h the Hubble constant in units of 100 km·s^{-1}·Mpc^{-1}.) The baryonic density provided by WMAP[1], $\Omega_b h^2 = 0.0224 \pm 0.0009$, has indeed dramatically increased the precision on this crucial cosmological parameter with respect to earlier experiments: BOOMERANG, CBI, DASI, MAXIMA, VSA and ARCHEOPS. It is thus important to improve the precision on SBBN calculations. Within the standard model of BBN, the only remaining free parameter is the baryon over photon ratio η directly related to $\Omega_b h^2$ [$\Omega_b h^2 = 3.6519 \times 10^7 \eta$]. Accordingly, the main source of uncertainties comes from the nuclear reaction rates. In this paper we use the results of a new

---

[1] Permanent address: LPHEA, FSSM, Université Caddi Ayyad, Marrakech, Morocco

analysis[2, 3] of nuclear data providing improved reaction rates which reduces those uncertainties.

NUCLEAR REACTION RATES

In a previous paper[4] we already used a Monte–Carlo technique, to calculate the uncertainties on the light element yields (^4He, D, ^3He and ^7Li) related to nuclear reactions. The results were compared to observations that are thought to be representative of the corresponding primordial abundances. We used reaction rates from the NACRE compilation of charged particles reaction rates[5] completed by other sources[6, 7, 8] as NACRE did not include all of the 12 important reactions of SBBN. One of the main innovative features of NACRE with respect to former compilations[9] is that uncertainties are analyzed in detail and realistic lower and upper bounds for the rates are provided. However, since it is a general compilation for multiple applications, coping with a broad range of nuclear configurations, these bounds had not always been evaluated through a rigorous statistical methodology. Hence, we assumed a simple uniform distribution between these bounds for the Monte–Carlo calculations. Other works [10, 11] have given better defined statistical limits for the reaction rates of interest for SBBN. In these works, the astrophysical S–factors (see definition in Ref. [4]) were either fitted with spline functions[10] or with NACRE S–factor fits and data but using a different normalization[11]. In this work, we use a new compilation[2] specifically dedicated to SBBN reaction rates using for the first time in this context nuclear theory to constrain the S–factor energy dependences and provide statistical limits. The goal of the R-matrix method[12] is to parametrize some experimentally known quantities, such as cross sections or phase shifts, with a small number of parameters, which are then used to interpolate the cross section within astrophysical energies. The R-matrix theory has been used for many decades in the nuclear physics community (see e.g. Ref. [13, 14] for a recent application to a nuclear astrophysics problem) but this is the first time that it is applied to SBBN reactions. This method can be used for both resonant and non-resonant contributions to the cross section. (See Ref. [2] and references therein for details of the method.) The R-matrix framework assumes that the space is divided into two regions: the internal region (with radius a), where nuclear forces are important, and the external region, where the interaction between the nuclei is governed by the Coulomb force only. The physics of the internal region is parameterized by a number N of poles, which are characterized by energy E_λ and reduced width $\tilde{\gamma}_\lambda$. Improvements of current work on Big Bang nucleosynthesis essentially concerns a more precise evaluation of uncertainties on the reaction rates. Here, we address this problem by using standard statistical methods [15]. This represents a significant improvement with respect to NACRE [5], where uncertainties are evaluated with a simple prescription. The R-matrix approach depends on a number of parameters, some of them being fitted, whereas others are constrained by well determined data, such as energies or widths of resonances. As usual, the adopted parameter set is obtained from the minimal χ^2 value. The uncertainties on the parameters are evaluated as explained in Ref.[15]. The range of acceptable p_i values is such that $\chi^2(p_i) \leq \chi^2(p_i^{min}) + \Delta\chi^2$, where p_i^{min} is the optimal parameter set. In this equation, $\Delta\chi^2$

is obtained from $P(v/2, \Delta\chi^2/2) = 1 - p$, where v is the number of free parameters p_i, $P(a,x)$ is the Incomplete Gamma function, and p is the confidence limit ($p = 0.683$ for the 1σ confidence level)[15]. This range is scanned for all parameters, and the limits on the cross sections are then estimated at each energy. As it is well known, several reactions involved in nuclear astrophysics present different data sets which are not compatible with each other. An example is the $^3He(\alpha, \gamma)^7Be$ reaction where data with different normalizations are available. In such a case, a special procedure is used[2].

This new compilation[2] provides 1-σ statistical limits for each of the 10 rates: $^2H(p,\gamma)^3He$, $^2H(d,n)^3He$, $^2H(d,p)^3H$, $^3H(d,n)^4He$, $^3H(\alpha, \gamma)^7Li$, $^3He(n,p)^3H$, $^3He(d,p)^4He$, $^3He(\alpha, \gamma)^7Be$, $^7Li(p,\alpha)^4He$ and $^7Be(n,p)^7Li$. The two remaining reactions of importance, $n \leftrightarrow p$ and $^1H(n,\gamma)^2H$ come from theory and are unchanged with respect to our previous work[4].

SBBN CALCULATIONS

We performed Monte-Carlo calculations using Gaussian distributions with parameters provided by the new compilation and calculated the ^4He, D, ^3He and ^7Li yield range as a function of η, fully consistent with our previous analysis[4]. The differences with Ref. [11] on the ^7Li yield is probably due to their different normalization procedure of the NACRE S-factors. Figure 1 displays the resulting abundance limits (1-σ) [it was 2-σ in Fig.4 of Ref. [4]] from SBBN calculations compared to primordial ones inferred from observations. Using these results and the WMAP $\Omega_b h^2$ range (quoted WMAP+SBBN in the following), it is now possible to infer the primordial ^4He, D, ^3He and ^7Li abundances.

We obtain (WMAP+SBBN) a deuterium primordial abundance of $D/H = (2.60^{+0.19}_{-0.17}) \times 10^{-5}$ [ratio of D and H abundances by number of atoms] which is in perfect agreement with the average value $(2.78^{+0.44}_{-0.38}) \times 10^{-5}$[16] of D/H observations in cosmological clouds. These clouds at high redshift on the line of sight of distant quasars are expected to be representative of primordial D abundances. The exact convergence between these two independent methods is claimed to reinforce the confidence in the deduced $\Omega_b h^2$ value.

The other WMAP+SBBN deduced primordial abundances are $Y_P = 0.2479 \pm 0.0004$ for the ^4He mass fraction, $^3He/H = (1.04 \pm 0.04) \times 10^{-5}$ and $^7Li/H = (4.15^{+0.49}_{-0.45}) \times 10^{-10}$. We leave aside ^3He whose primordial abundance cannot be reliably determined because of its uncertain rate of stellar production and destruction[17].

The ^4He primordial abundance, Y_p (mass fraction), is derived from observations of metal-poor extragalactic, ionized hydrogen (HII) regions. Recent evaluations give a relatively narrow ranges of abundances: $Y_p = 0.2452 \pm 0.0015$[18], 0.2391 ± 0.0020[19] but systematic uncertainties may prevail due to observational difficulties and complex physics[20]. With these ranges, compatibility with the calculated value (WMAP+SBBN) is marginally acceptable.

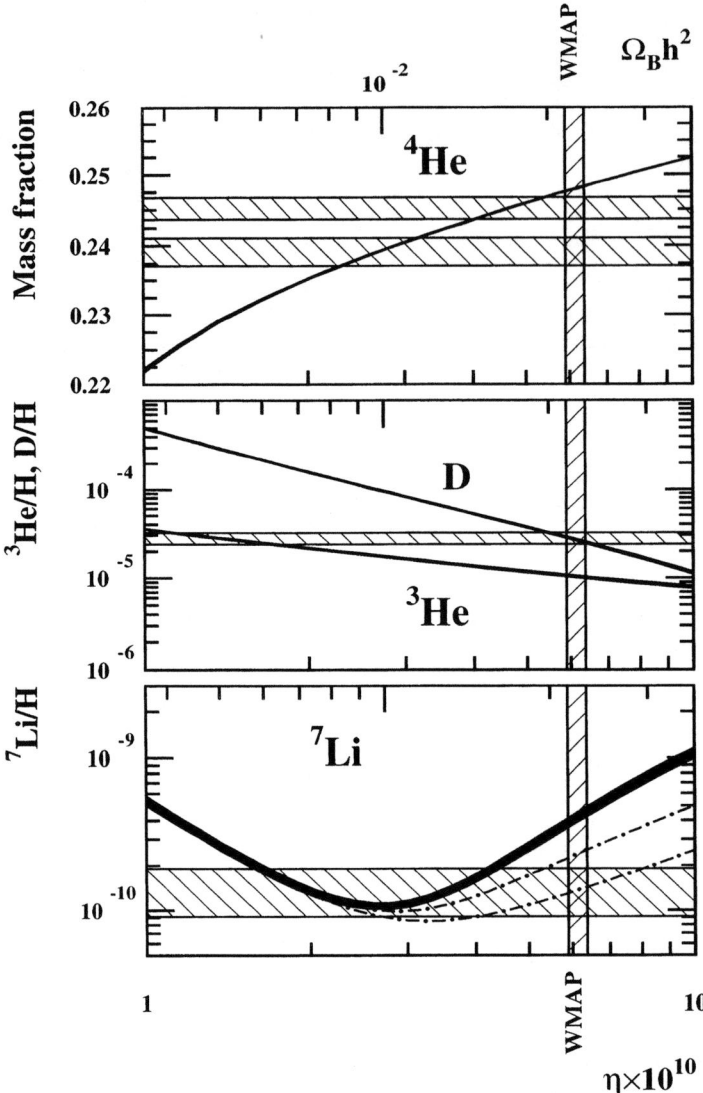

FIGURE 1. Abundances of ^4He (mass fraction), D, ^3He and ^7Li (by number relative to H) as a function of the baryon over photon ratio η or $\Omega_b h^2$. Limits (1-σ) are obtained from Monte Carlo calculations. Horizontal lines represent primordial ^4He, D and ^7Li abundances deduced from observational data (see text). The vertical stripe represents the (1-σ) $\Omega_b h^2$ limits provided by WMAP[1]. For the dash-dotted lines in the bottom panel: see text.

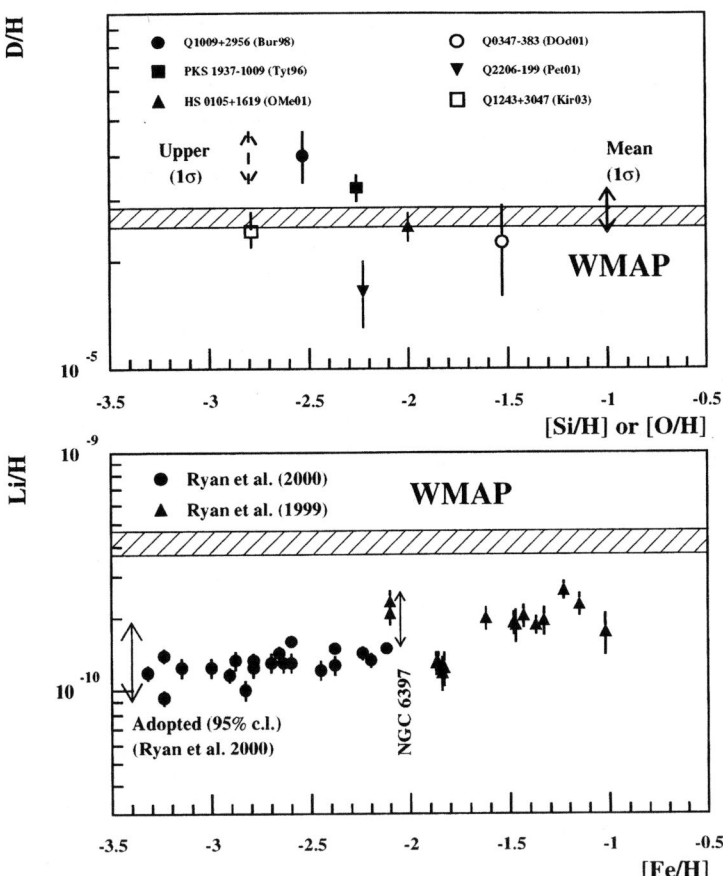

FIGURE 2. Observed abundances as a function of metallicity from objects which are expected to reflect primordial abundances. Upper panel : observed D abundances, from Refs. [22, 23, 24, 25, 26, 27, 16] (parenthesis indicate less established observations.) The mean observational value[16] and the highest observed value used in Ref. [4] are shown by arrows. Lower panel : observed ^7Li abundances, circles[28] and triangles[29] from Ryan et al. and extrapolated primordial abundance[30] (arrow). (The range of values observed[32, 33] in the metal poor globular cluster NGC 6397 is also shown by an arrow.)

The ^7Li abundance measured in halo stars of the Galaxy is considered up to now as representative of the primordial abundance as it display a plateau[31] as a function of metallicity (see definition in Ref. [4]). Recent observations have lead to (95% c.l.) Li/H = $\left(1.23^{+0.68}_{-0.32}\right) \times 10^{-10}$[30]. These authors have extensively studied and quantified the various sources of uncertainty : extrapolation, stellar depletion and stellar atmosphere parameters. This Li/H value, based on a much larger number of observations than the

D/H one was considered as the most reliable constraint on SBBN and hence on $\Omega_b h^2$[4]. However, it is a factor of 3.4 lower than the WMAP+SBBN value. Even when considering the corresponding uncertainties, the two Li/H values differ drastistically. This confirms our[4] and other[11, 21] previous conclusions that the $\Omega_b h^2$ range deduced from SBBN of ^7Li are only marginally compatible with those from the CMB observations available by this time (BOOMERANG, CBI, DASI and MAXIMA experiments). It is strange that the major discrepancy affects ^7Li since it could a priori lead to a more reliable primordial value than deuterium, because of much higher observational statistics and an easier extrapolation to primordial values.

In Fig. 2 we plot the D primordial abundances inferred from SBBN calculations and $\Omega_b h^2$ range from WMAP. The stripe widths represent the uncertainty (1-σ) originating from both the WMAP $\Omega_b h^2$ and nuclear uncertainties. In Fig. 2 (lower panel) are shown the most recent ^7Li observations by Ryan et al. [29, 30] as a function of metallicity for old halo stars together with their extrapolated (2-σ) primordial Li/H. This figure emphasizes the good agreement on D abundances (WMAP+SBBN versus cosmological cloud observations) but a strong incompatibility for ^7Li (WMAP+SBBN versus observations in halo stars).

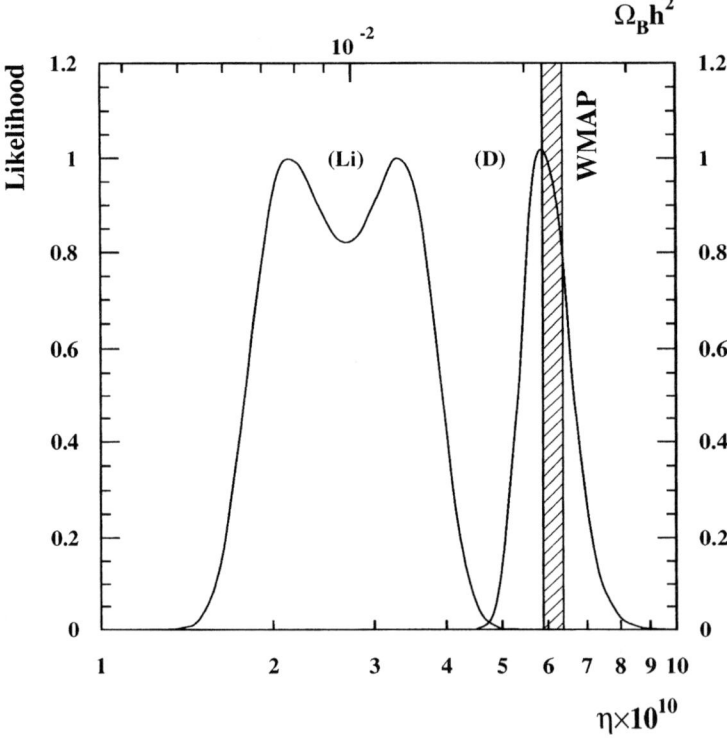

FIGURE 3. Likelihood functions for D and ^7Li obtained from our SBBN calculations and Kirkman et al.[16] and Ryan et al.[30] data for D and ^7Li respectively and WMAP $\Omega_b h^2$ range[1].

Fig. 3 shows a comparison between $\Omega_b h^2$ ranges deduced either from SBBN or WMAP. The curves represent likelihood functions obtained from our SBBN calculations and observed D[16] and ^7Li[30] primordial abundances. These were obtained as in our previous analysis[4] except for the new reaction rates and new D/H primordial abundance[16]. The incompatibility between the D and ^7Li likelihood curves is more obvious than before due to the lower D/H adopted value (Kirkman et al., averaged value). Putting aside, for a moment, the CMB results on the baryonic density, we would deduce the following 95% c.l. intervals: $1.6 < \eta_{10} < 4.3$ [$0.0058 < \Omega_b h^2 < 0.0157$] from ^7Li only or $5.0 < \eta_{10} < 7.3$ [$0.0183 < \Omega_b h^2 < 0.0267$] from D only and favor as in our previous work[4] the lower (^7Li) interval. The WMAP result on the contrary definitively favors the upper (D) one. The origin of this discrepancy on ^7Li remains a challenging issue very well worth further investigations.

POSSIBLE ORIGINS OF DISCREPANCY

Stellar

Both observers and experts in stellar atmospheres agree to consider that the abundance determination in halo stars, and more particularly that of lithium requires a sophisticated analysis . The derivation of the lithium abundance in halo stars with the high precision needed requires a fine knowledge of the physics of stellar atmosphere (effective temperature scale, population of different ionization states, non LTE (Local Thermodynamic Equilibrium) effects and 1D/3D model atmospheres[34]. However, the 3D, NLTE abundances are very similar to the 1D, LTE results, but, nevertheless, 3D models are now compulsory to extract lithium abundance from poor metal halo stars[35].

Modification of the surface abundance of Li by nuclear burning all along the stellar evolution has been discussed for a long time in the literature. There is no lack of phenomena to disturb the Li abundance: rotational induced mixing, mass loss,...[36, 37]. However, the flatness of the plateau over three decades in metallicity and the relatively small dispersion of data represent a real challenge to stellar modeling. In addition, recent observations of ^6Li in halo stars (an even more fragile isotope than ^7Li) constrain more severely the potential destruction of lithium[38].

Nuclear

Large systematic errors on the 12 main nuclear cross sections are excluded[2, 3]. However, besides the 12 reactions classically considered in SBBN, first of all the influence of *all* nuclear reactions needs to be evaluated[3]. It is well known that the valley shaped curve representing Li/H as a function of η is due to two modes of ^7Li production. One, at low η produces ^7Li directly via $^3H(\alpha,\gamma)^7Li$ while ^7Li destruction comes from $^7Li(p,\alpha)^4He$. The other one, at high η, leads to the formation of ^7Be through $^3He(\alpha,\gamma)^7Be$ while ^7Be destruction by $^7Be(n,p)^7Li$ is inefficient because of the lower neutron abundance at high density (^7Be later decays to ^7Li). Since the WMAP results point toward the high η region, a peculiar attention should be paid to ^7Be synthesis. In particular, the ^7Be+d reactions could be an alternative to $^7Be(n,p)^7Li$ for the destruction of ^7Be, by compensating the scarcity of neutrons at high η. Fig. 1 shows (dash–dotted lines) that an increase of the $^7Be(d,p)2^4He$ reaction rate by factors of 100 to 300 would remove the discrepancy. The rate for this reaction[9] can be traced to an estimate by Parker[39] who assumed for the astrophysical S–factor a constant value of 10^5 kev.barn. based on the single experimental data available[40]. To derive this S–factor, Parker used this measured differential cross section at 90° and assumed isotropy of the cross section. Since Kavanagh measured only the p_0 and p_1 protons (i.e. feeding the ^8Be ground and first excited levels), Parker introduced an additional but arbitrary factor of 3 to take into account the possible population of higher lying levels. Indeed, a level at 11.35 MeV is also reported[41]. This factor should also include the contribution of another open channel in ^7Be+d: $^7Be(d,\alpha)^5Li$ for which no data exist. In addition, one should note that *no* experimental data for this reaction is available at energies relevant to ^7Be Big Bang nucleosynthesis, taking place when the temperature has dropped below 10^9 K. A seducing possibility[3] to reconciliate, SBBN, ^7Li and CMB observations would then be that new experimental data below $E_d = 700$ keV ($E_{cm} \approx 0.5$ MeV) for $^7Be(d,p)2^4He$ [and $^7Be(d,\alpha)^5Li$] would lead to a sudden increase in the S–factor as in $^{10}B(p,\alpha)^7Be$[42, 5]. This is not supported by known data, but considering the cosmological or astrophysical consequences, this is definitely an issue to be investigated and an experiment is planned at the Cyclotron Research Centre in Louvain-la-Neuve.

Cosmology

Recent theories that could affect BBN include the variation of the fine structure constant[43], the modification of the expansion rate during BBN induced by quintessence[44], modified gravity[45], or leptons asymmetry[46]. However, their effect is in general more significant on ^4He than on ^7Li.

It may not be excluded that some bias exists in the analysis of CMB anisotropies. For instance, it has been argued[47] that a contamination of CMB map by blazars could affect the second peak of the power spectrum on which the CMB $\Omega_b h^2$ values are based.

Pregalactic evolution

We note that between the BBN epoch and the birth of the now observed halo stars, ≈ 1 Gyr have passed. Primordial abundances could have been altered during this period. For instance, cosmological cosmic rays, assumed to have been born in a burst at some high redshift, could have modified these primordial abundances in the intergalactic medium[48]. This would increase the primordial ^7Li and D abundances trough spallative reactions, increasing in the same time the discrepancy between SBBN calculations and observations instead to reconciliate them.

Another source of alteration of the primordial abundances could be the contribution of the first generation stars (Population III). However, it seems difficult that they could reduce the ^7Li abundance without affecting the D one, consistent with CMB $\Omega_b h^2$.

CONCLUSIONS

The baryonic density of the Universe as determined by the analysis of the Cosmic Microwave Background anisotropies is in very good agreement with Standard Big-Bang Nucleosynthesis compared to D primordial abundance deduced from cosmological cloud observations. However, it strongly disagrees with lithium observations in halo stars (Spite plateau). The origin of this discrepancy, if not nuclear, is a challenging issue.

REFERENCES

1. D.N. Spergel, L. Verde, H.V. Peiris, E. Komatsu, M.R. Nolta, C.L. Bennett, M. Halpern, G. Hinshaw, N. Jarosik, A. Kogut, M. Limon, S.S. Meyer, L. Page, G.S. Tucker, J.L. Weiland, E. Wollack and E.L. Wright, *Astrophys. J. S.* **148**, 175 (2003).
2. P. Descouvemont, A. Adahchour, C. Angulo, A. Coc and E. Vangioni–Flam, preprint, http://pntpm.ulb.ac.be/bigbang.
3. A. Coc, E. Vangioni-Flam, P. Descouvemont, A. Adahchour and C. Angulo, *Astrophys. J.* in press, ArXiv:astro-ph/0309480
4. A. Coc, E. Vangioni–Flam, M. Cassé and M. Rabiet, *Phys. Rev.* **D65**, 043510 (2002).
5. C. Angulo, M. Arnould, M. Rayet et al., *Nucl. Phys.* **A656**, 3 (1999).
6. M.S. Smith, L.H. Kawano, and R.A. Malaney, *Astrophys. J. S.* **85**, 219 (1993).
7. C.R. Brune, K.I. Hahn, R.W. Kavanagh and P.W. Wrean, *Phys. Rev.* **C60**, 015801 (1999).
8. J.-W. Chen, and M.J. Savage, *Phys. Rev.* **C60**, 065205 (1999).
9. G.R. Caughlan and W.A. Fowler, *Atomic Data and Nuclear Data Tables* **40**, 283 (1988).
10. K.M. Nollett and S. Burles (NB), *Phys. Rev.* **D61**, 123505 (2000).
11. R.H. Cyburt, B.D. Fields, and K.A. Olive (CFO), *New Astronomy* **6**, 215 (2001).
12. A.M. Lane and R.G. Thomas, *Rev. Mod. Phys.* **30**, 257 (1958)
13. F.C. Barker and T. Kajino, *Aust. J. Phys.* **44** (1991) 369.
14. C. Angulo and P. Descouvemont, *Nucl. Phys.* **A639** 733 (1998).
15. Particle Data Group, K. Hagiwara et al., *Phys. Rev.* **D66**, 010001 (2002)
16. D. Kirkman, D. Tytler, N Suzuki, J.M. O'Meara, and D. Lubin, submitted to *Astrophys. J. S.*, ArXiv:astro-ph/0302006.
17. E. Vangioni–Flam, K.A. Olive, B.D. Fields and M. Cassé, *Astrophys. J.* **585**, 611 (2003).
18. Y.I. Izotov, F.H. Chaffee, C.B. Foltz, R.F. Green, N.G. Guseva, and T.H. Thuan, *Astrophys. J.* **527**, 757 (1999).
19. V. Luridiana et al., *Astrophys. J.* **592**, 846 (2003).
20. B.D. Fields and K.A. Olive, *Astrophys. J.* **506**, 177 (1998).
21. R.H. Cyburt, B.D. Fields, and K.A. Olive, *Phys. Lett.* **B567**, 227 (2003).
22. S. Burles and D. Tytler, *Astrophys. J.* **499**, 699 (1998).
23. S. Burles and D. Tytler, *Astrophys. J.* **507**, 732 (1998).
24. D. Tytler, X.-M. Fan and S. Burles, S., *Nature* **381** 207 (1996).
25. J.M. O'Meara D. Tytler, D. Kirkman, N. Suzuki, J.X. Prochaska, D. Lubin and A.M. Wolfe, *Astrophys. J.* **552**, 718 (2001).
26. M. Pettini, and D.V. Bowen, 2001, *Astrophys. J.* **560**, 41 (2001).
27. S. D'Odorico, M. Dessauges-Zavadsky, and P. Molaro, *Astron. Astrophys.* **368**, L21 (2001).
28. S.G. Ryan, T. Kajino, T.C. Beers, T.K. Suzuki, D. Romano, F. Matteucci and K. Rosolankova, *Astrophys. J.* **549**, 55 (2001).
29. S.G. Ryan, J. Norris, and T.C. Beers, *Astrophys. J.* **523**, 654 (1999).
30. S.G. Ryan, T.C. Beers, K.A. Olive, B.D. Fields, and J.E. Norris, *Astrophys. J.* **530** L57 (2000).
31. F. Spite, and M. Spite, *Astron. Astrophys.* **115**, 357 (1982).

32. F. Thévenin, C. Charbonnel, J.A. de Freitas Pacheco, T.P. Idiart, G. Jasniewicz, P. de Laverny, B. Plez, *Astron. Astrophys.* **373**, 905 (2001).
33. P. Bonifacio, L. Pasquini, F. Spite, et al., *Astron. Astrophys.* **390**, 91 (2002).
34. M. Asplund, M. Carlsson and A.V. Botnen, *Astron. Astrophys.* **399**, L31 (2003).
35. P.S. Barklem, A.K. Belyaev, and M. Asplund, 2003, astro-ph 0308170.
36. S. Theado and S. Vauclair, *Astron. Astrophys.* **375**, 86 (2001).
37. M.H. Pinsonneault et al., *Astrophys. J.* **574**, 411 (2002).
38. E. Vangioni-Flam, et al., *New Astronomy* **4**, 245 (1999).
39. P.D. Parker, *Astrophys. J.* **175**, 261 (1972).
40. R.W. Kavanagh, *Nucl. Phys.* **18**, 492 (1960).
41. F. Ajzenberg-Selove, *Nucl. Phys.* bf A490, (1988) 1 and TUNL Nuclear Data Evaluation Project, http://www.tunl.duke.edu/nucldata/fas/88AJ01.shtml.
42. C. Angulo, S. Engstler, G. Raimann, C. Rolfs, W.H. Schulte, and E. Somorjai, *Z. Phys.* **A345**, 231 (1993).
43. K.M. Nollett and R.E. Lopez, *Phys. Rev.* **D66**, 063507 (2002).
44. P. Salati, *Phys. Lett.* **B571**, 121 (2003).
45. A. Navarro, A. Serna and J.-M Alimi, Classical and Quantum Gravity, **19**, 4361 (2002).
46. M. Orito, T. Kajino, G.J. Mathews and Y. Wang, *Phys. Rev.*, **D65** 123504 (2002).
47. P. Giommi, S. Colafrancesco, submitted to *Astron. Astrophys.*, ArXiv:astro-ph/0306206.
48. T. Montmerle, *Astrophys. J.* **216**, 620 (1977).

Propagation of Ultra-High-Energy Nuclei in the Extra-galactic Photon Field

Tokonatsu Yamamoto [*] and Masahiro Teshima[†]

[*]*Center for Cosmological Physics, University of Chicago, Chicago IL 60637 USA*
[†]*Max-Planck-Institute für Physik, Föringer Ring 6, 80805 München, Germany*

Abstract.
We present a calculation of nuclei propagation with energies above 10^{18} eV in the intergalactic photon field. The calculation is based on a Monte Carlo approach for the nucleus-photon interaction as well as the intergalactic magnetic field. We then assume that the Ultra-High Energy Cosmic Rays (UHECR) are nuclei which are emitted from extra-galactic point sources. Four bumps are found in the energy spectrum of the UHECR which form clusters in the distribution of their arrival directions. Based on this calculation, the energy distribution of the clustered events is discussed [1].

INTRODUCTION

After a few decades of observation by several projects, the nature of the UHECR is gradually being revealed. Based on the observational results, there is no doubt of the existence of UHECR up to 10^{20} eV [2, 3, 4]. This has led to consideration of a large number of production models. Most of the models assume that UHECRs are protons or photons.

Several studies of anisotropy have been done to reveal the origin of the UHECR [5, 6, 7, 8, 9, 10]. Around 10^{18} eV, a large scale anisotropy which is correlated to the Galactic center was reported by AGASA group [5]. This result, if confirmed, supports an interpretation that cosmic rays below 10^{18} eV are dominated by a Galactic origin. Above this energy, however, no significant large scale anisotropy has been detected by AGASA [7].

On the other hand, AGASA reported clusters in the distribution of arrival directions including 4 doublets and 2 triplets with energy above 4×10^{19} eV[6, 7]. This results indicate that the UHECR come from extra-galactic point sources. The cluster events reported by AGASA are composed of particles with different energies. A possible explanation would be neutral particles (like neutrinos) propagating in straight lines from the source. Another possible explanation would be charged particles with large rigidity , i.e. with small charge like protons.

If there is only the effect of the Galactic magnetic field then protons of different energy can arrive at same direction [11]. However, these two possibilities cannot explained the three bumps in the energy spectrum of the cluster events between $10^{18.8}$ eV and $10^{19.8}$ eV [14] reported by AGASA. This structure is not statistically significant. However, it suggests to us the interesting possibility that UHECR might be nuclei as it will be discussed in this paper. In this scenario, cluster events would be formed by particles with different charge and energy but with the same rigidity. The atomic number of the primary cosmic ray does not affect the energy determination of AGASA which measures the number of charged particles on the ground. However, the capability of AGASA to measure the composition of the primary cosmic rays is not very strong [12]. Nevertheless, the structure of the energy distribution of the cluster events may be used to measure of the composition indirectly as discussed in this paper.

If UHECR come from point sources and their rigidity is large enough or their source close enough so the direction is not lost, a fraction of the particles emitted by the source will arrive within a small angle from the source position. These particles are what we will call in this paper "small deflection angle" particles. Clusters will be formed when the number of particles expected within this angle in our experiment is larger than one. The small deflection angle particles should represent the cluster events observed by AGASA.

UHE protons should interact with Cosmic Microwave Background Radiation (CMBR) and lose energy by photo-pion production. For this reason, 10^{20} eV protons cannot propagate in intergalactic space beyond 50 Mpc, leading to a cut off in the spectrum below 10^{20} eV, the so called GZK cut off. On the other hand heavy nuclei will interact during its propagation with the Intergalactic Infrared Background Radiation(IIBR). Unfortunately the IIBR cannot be measured directly so far because of the strong background of infrared photons of Galactic origin. The UHE nucleus should be disintegrated by the IIBR and lose energy by the photo-disintegration interaction. For this reason, it was believed that the attenuation length of UHE nuclei is shorter than that of protons in intergalactic space [15, 16]. However, recent calculations which are based on empirical data show the IIBR density is much lower than originally thought [17]. The lack of absorption in the energy spectrum of very high energy gamma rays from AGNs strongly supports this calculation. Using this result, the attenuation length of the UHE nuclei in the intergalactic photon field was re-estimated [18] and found to be longer than that of a proton (a few 100 Mpc for 10^{20} eV Fe).

We simulate the propagation of UHE nuclei which are emitted from extra-galactic point sources based on a Monte Carlo approach. Using this simulation, we found structures in the energy distribution of the UHE nuclei which are observed as cluster events on the earth. In the next section we briefly describe the interaction of nuclei with background photons. In following section we describe the scattering of the nuclei by the intergalactic magnetic field. Then we show the result of this simulation. We conclude in the last section and discuss the results briefly.

Interaction of Nuclei with background photons

There are three processes which affect cosmic rays during intergalactic propagation: energy loss by interaction with cosmic background radiation; deflection by magnetic fields; and energy loss by adiabatic expansion of the universe. The effect of the adiabatic expansion is negligibly small if propagation distances are less than 1 Gpc.

There are four interaction processes of cosmic rays with intergalactic background radiation. The first process, Compton interaction, is negligibly small for our purpose. The second process is pair-production of e^+e^-. This interaction can occur if the energy of the background photon is greater than 1 MeV in the rest system of the nucleus. If the Lorentz factor of the nucleus is larger than 10^9, it will efficiently interact with the CMBR. In this case, the energy loss rate (dE/dt) is proportional to the square of the charge of the nucleus. The third process is photo production which mainly affects protons with energy greater than 10^{20} eV. The interaction of heavier nuclei with the CMBR by this process is negligible for the energy range considered in this paper.

The fourth process is photo-disintegration. The nucleus interacts with the background photon and disintegrates to a lighter nucleus. In this paper, we calculate this process by a Monte-Carlo method according to [16, 18]. These authors parametrized the total cross section $\sigma(\varepsilon)$ as a function of photon energy in the rest frame of the nucleus. Then the probability of photo-disintegration per unit length R is calculated by following equation:

$$R = \int_0^\infty n(\varepsilon) \left[\frac{\int_0^{2\gamma\varepsilon} \sigma(\varepsilon') \frac{\varepsilon'}{\varepsilon} d\varepsilon'}{2\gamma\varepsilon} \right] d\varepsilon \tag{1}$$

where ε and ε' are the energies of the photon in the lab frame and in the nucleus rest frame respectively, $n(\varepsilon)$ is the differential number density of photons including the CMBR and the IIBR [18], and $\gamma = \varepsilon'/\varepsilon$ is the Lorentz factor of the nucleus. The term inside the bracket corresponds to the angle-averaged cross-section for a photon of energy ε.

The effect of the CMBR is maximized when the Lorentz factor γ is around 10^{10}. In case of Fe, this value corresponds to an energy of 10^{21} eV, yielding a value of energy loss time $\tau(= E/\frac{dE}{dt}) = 2 \times 10^{14}$ s (2 Mpc). Because the IIBR density is much lower than the CMBR density, τ increases rapidly at lower energy. τ of Fe with 10^{20} eV is about 2.5×10^{17}s (2.5Gpc). In general, one or a few nucleons and a single lighter nucleus are emitted by this interaction. The Lorentz factor is conserved at each photo-disintegration interaction, though it is reduced by the pair-production. Finally, UHE particles will pile up at a Lorentz factor around 10^9. When ^9Be is disintegrated, one proton and two He are emitted. Therefore no nucleus is created with a mass number between 5 and 8.

The interaction probability is calculated for every possible photo-disintegration process involving the emission of one or more nucleons for all nuclei lighter than Fe. Based on these probabilities, the propagation of the nucleus is simulated by the Monte-Carlo method. The effects of pair production, the adiabatic expansion, and photo-pion production for protons are taken into account analytically in this simulation.

SCATTERING BY THE INTERGALACTIC MAGNETIC FIELD

The effect of magnetic fields is described by the rigidity defined as the ratio of energy to charge ($= E/Z$). Particles with small rigidity are deflected by the magnetic field and cannot be observed as a cluster. The deflection of magnetic field increases the propagation time, and therefore particles emitted from distant sources may not reach to the observer. We calculate the effect of the intergalactic magnetic field based on a Monte Carlo method.

To simulate the scattering by the intergalactic magnetic field, we assume a Kolmogorov spectrum for the random magnetic field according to the reference of [19].

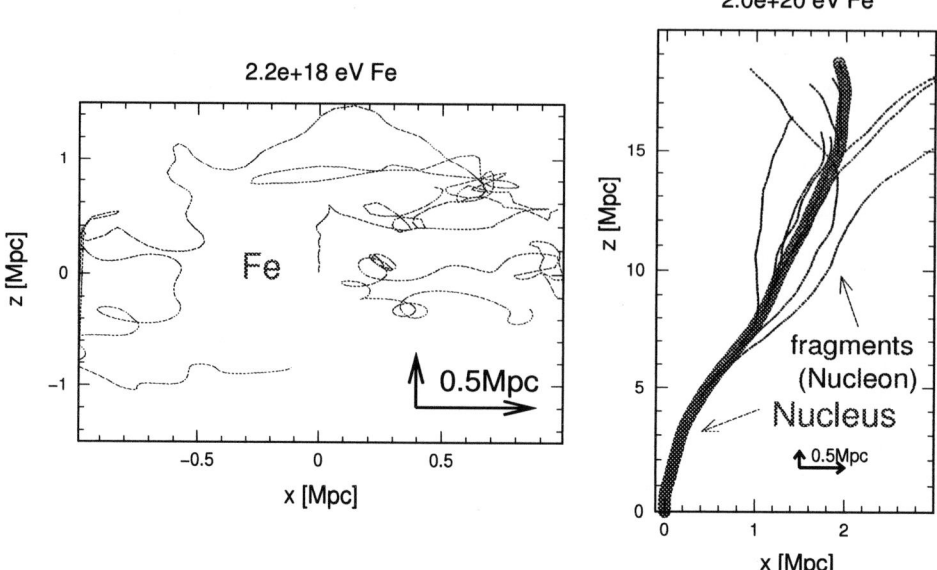

FIGURE 1. Examples of trajectory of UHE *Fe* in the intergalactic magnetic field. A *Fe* emitted in the *z* direction with an energy of 2×10^{18} eV from a source located on the origin in left panel. A projection of the trajectory on x-z plane is shown. At this energy, the attenuation length is sufficiently large. Therefore the energy of the particle is almost conserved and the particle is scattered by the magnetic field significantly. The *Fe* cannot propagate a distance of more than 1 Mpc from the source. In the right panel, a *Fe* with energy of 2×10^{20} eV emitted in the same direction as the left panel. The *Fe* and secondary nuclei are indicated by the wide line and secondary nucleons (proton and neutron) are indicated by the narrow lines. The *Fe* interact with photons rapidly and is disintegrated to lighter nuclei with emission of protons and neutrons. All of the particles are scattered by the magnetic field but deflection angle is comparatively small. Arrows in the panels indicate 0.5 Mpc distance.

In this reference, authors divide space in a lattice of 250 kpc cubes. The lattice is filled with a random magnetic field which follows the Kolmogorov spectrum with three logarithmic scales. Three field vectors of random orientation are sampled at scales $l = 1000$, 500, and 250 kpc with amplitudes proportional to $l^{1/3}$. The final magnetic field in each 250 kpc cube is vectorial sum of these three vectors. The average magnitude of the magnetic field is assumed to be 1 nG. Particles propagate in spiral trajectories until they leave the lattice. Figure 1 shows examples of the trajectories. *Fe* with energy of 10^{18} eV does not rapidly lose energy by the interaction with photons, and is trapped by the magnetic field inside 1 Mpc cubes. In case of 2×10^{20} eV, *Fe* is disintegrated by the photons rapidly and the products propagates close to a straight line in the initial direction of the *Fe*.

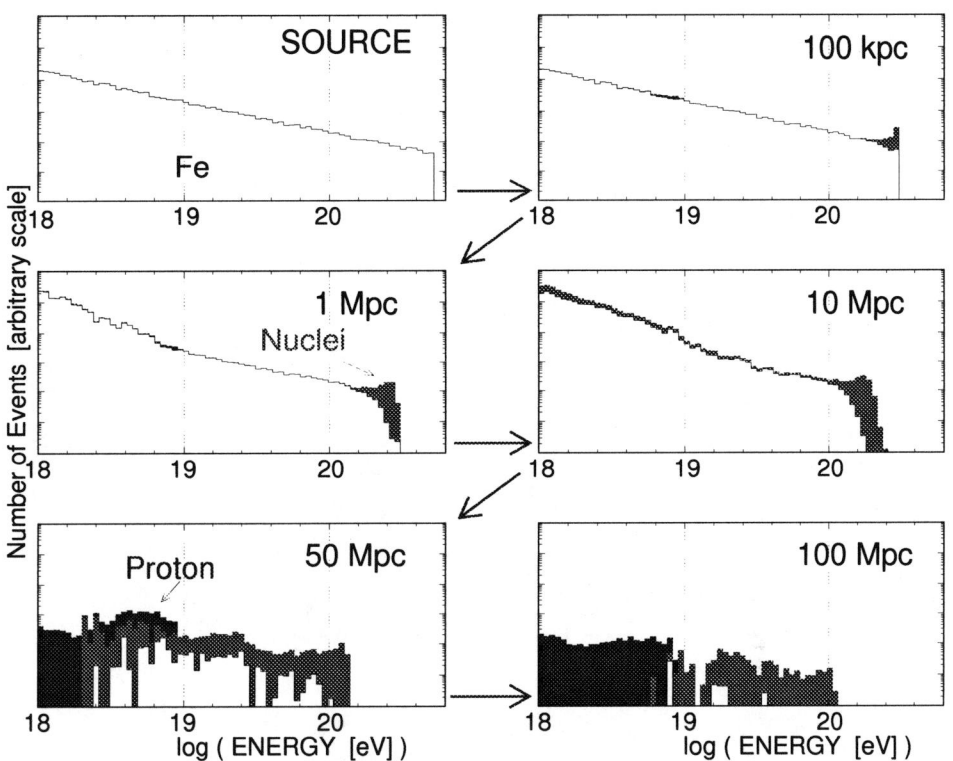

FIGURE 2. Variation of energy distribution of particles that began as Fe after propagation over various distances from a single source. The distances from the source are indicated in each panel. The number of particles (arbitrary scale) are shown as a function of energy. Primary nuclei (Fe), secondary lighter nuclei, and protons are indicated by white, light shade and dark shade in the histograms respectively. The differential source spectrum is assumed to be a power law, $\propto E^{-2}$, with energy cut-off at $Z \times 2 \times 10^{19}$ eV (upper left panel). At 100 Mpc from the source, most of the Fe disappear because of their low rigidity.

ENERGY DISTRIBUTION

In our simulation, the source energy spectrum $dN(E)/dE$ is assumed to have a power law dependence E^{-2} and have a cut-off at the energy of $Z \times 2 \times 10^{19}$ eV, where Z is charge of each primary nucleus. Below 300 MeV/nucleon, the cosmic nuclear composition can be divided to 4 types: He, CNO, $Ne-Si$, and Fe. Nuclear abundance between He and C is relatively low as well as between Si and Fe. We treat the composition as 4 components, $^{2}_{4}He$, $^{7}_{14}N$, $^{12}_{24}Mg$, and $^{26}_{56}Fe$. The fraction of these nuclear abundances in the UHE region are unknown. Therefore we assume equal fractions relative to He at the source. 20000 particles are created for each nucleus.

Figure 2 shows the variation of the energy distribution after propagation from a single source initiated by Fe through the photon and magnetic field. Nuclei with Lorentz factor above 10^{9} interact with the CMBR rapidly. After 1 Mpc propagation a small pile-up

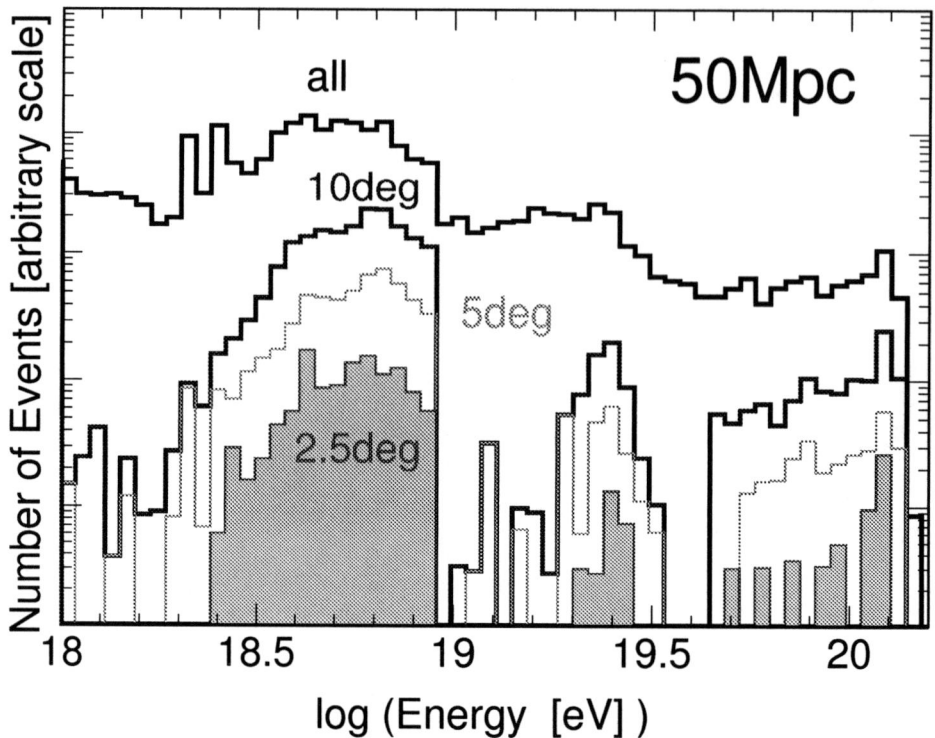

FIGURE 3. Energy distribution of the particles at the 50 Mpc distance from a source as a function of deflection angle by the magnetic field. The assumed source spectrum is same as in the previous figure. The number of particles with the deflection angle smaller than 2.5, 5, 10, 180(same as previous figure) degrees are shown.

is induced below the corresponding energy and the maximum energy of the particle is reduced. Since we assume a cut-off energy in the source spectrum, protons are emitted with energies below $Z/A \times 2 \times 10^{19} (\simeq 10^{19})$ eV where A is mass number of the primary nucleus. The number of low energy nuclei ($< 10^{19}$ eV) increases up to 10 Mpc from the source since these nuclei are trapped by the magnetic field. These low energy nuclei cannot propagate farther than few 10 Mpc. Therefore, the energy distribution becomes steeper for distances below 10 Mpc and flatter for larger distances. No particles remain after propagation of a few 100 Mpc.

Figure 3 is same result as lower left panel of Figure 2. The energy distributions at a distance of 50 Mpc from the source are shown for different deflection angle (angle between the arrival direction and the source position). Particles with the deflection angle smaller than 2.5, 5, 10, 180 degrees are counted (180 degrees corresponds to the entire sample of particles included in Figure 2). Three bumps appear clearly in the distribution of particles with a small deflection angle. The bump below 10^{19} eV is composed of protons which are emitted from higher energy nuclei in the vicinity of the observer who is located at 50 Mpc distance from the source. The bump around $10^{19.4}$ eV is composed of He, and the bump above $10^{19.6}$ eV is composed of nuclei heavier than Be.

These bumps can be explained as follows: The rigidity is proportional to Z^{-1} and E, and Z is approximately proportional to its mass number A while the Lorentz factor is proportional to A^{-1} and E. This means that the Lorentz factor of a particle is approximately proportional to the rigidity. Therefore, if the Lorentz factor of the nucleus is large, the energy is reduced by the photo-disintegration. If the Lorentz factor is small, the particle is deflected by the magnetic field due to its small rigidity. However particles with Lorentz factor of about 10^9 remain after applying the deflection angle cut. As a result, the bumps appear in the energy distribution of the small deflection angle particles at positions dependent on the mass numbers. The combination of the photodisintegration and the magnetic field acts as a "filter" for $\gamma = 10^9$ particles. The energy distribution of the small deflection angle particles shows clearly the spectral features created by interactions with the intergalactic photon background and magnetic fields during their propagation.

If the sources are distributed uniformly in the Universe, the expected energy distribution measured at the earth can be estimated by adding up linearly the energy distributions corresponding to different distances from the source. In Figure 2 the number of particles around 10^{20} eV coming from sources at 100 Mpc is 1.5 orders of magnitude smaller than from sources at 100 kpc. When adding linearly the different distances, the number of particles from sources at 100 Mpc is enhanced 3 orders of magnitude compared to sources at 100 kpc. Therefore, the contribution to the energy spectrum at earth is increasing with the distance to the source till a couple of hundred Mpc. On the other hand, most of the low energy particles emitted by distant sources will be trapped by the magnetic field and they will not reach the earth. However, the low energy secondaries resulting from the photodisintegration of high energy heavy nuclei close to earth will regenerate the low energy spectrum coming from distant sources. Therefore, even at low energies we will still be dominated by the contribution of distant sources. The spectrum of small deflection angle particles will follow the same behavior, i.e. the contribution of sources at large distances to the spectrum exceed largely the one from small distant sources, even if the fraction of small deflected particles for closer sources is slightly larger.

It should be noted that the relation between cluster events and small deflection particles is qualitatively clear but difficult to estimate quantitatively. Current work is in progress for a better comparison with experimental data.

Figure 4 shows the result of the expected energy distribution under the assumption that the sources are distributed uniformly in the Universe. The solid line shows the expected energy distribution of all the particles that reach the earth. The expected energy spectrum is steeper than the source spectrum. Above $10^{20.2}$ eV, the energy spectrum becomes even steeper and the maximum energy is about $10^{20.4}$ eV. The exact details of the upper end of the spectrum will depend on the source spectrum cut off.

The dashed lines in Figure 4 show the expected energy distribution of the small deflection angle particles. This spectrum is much flatter than the source spectrum. Four bumps appear in this figure. The bump just below 10^{19} eV corresponds to protons and strongly depends on the cut off energy at the source. The bump around $10^{19.3}$

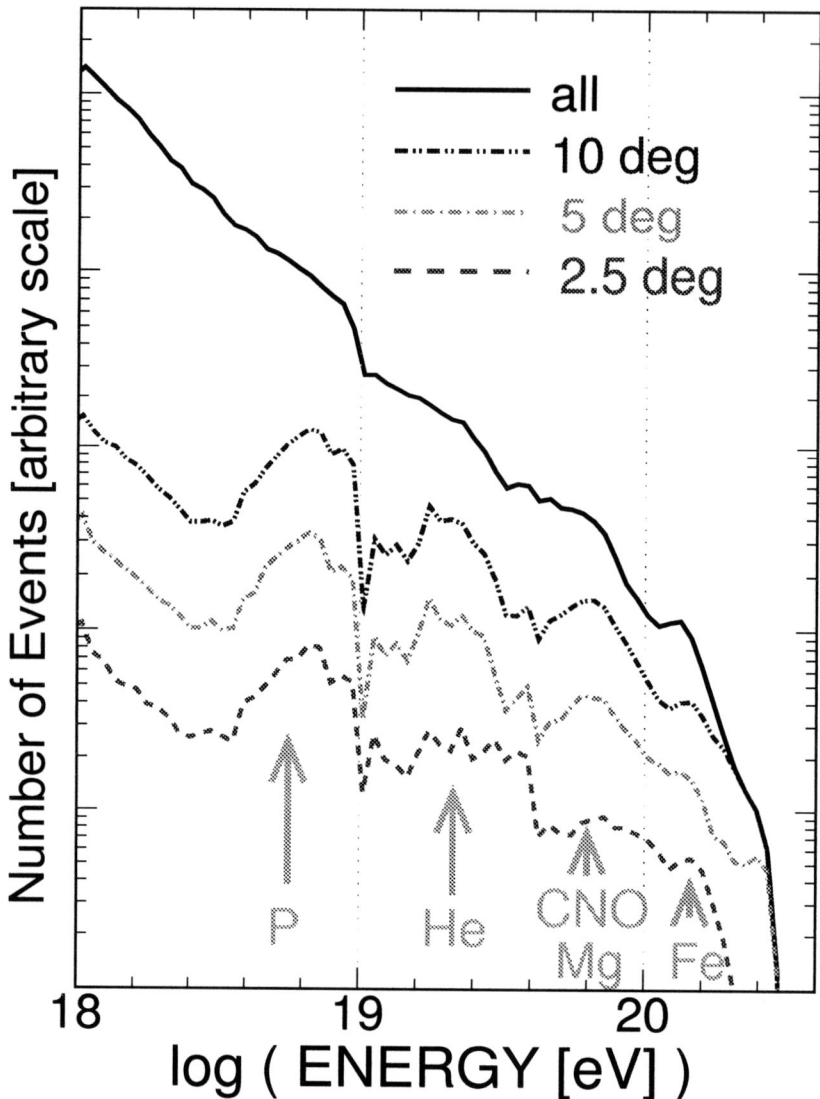

FIGURE 4. The expected energy distribution of particles after propagation under the assumption that point sources are distributed uniformly in the universe. The solid line shows the energy distribution with no deflection angle cut. The other (dashed dots) lines show the energy distributions of the particle with deflection angles smaller than 10, 5, and 2.5 degrees. Particles with lower rigidity are scattered by the magnetic field. Larger rigidity particle are disintegrated by the photon field due to their large Lorentz factor.

eV is composed of *Deuteron* and *He*. Every nucleus with a Lorentz factor around 10^9 contributes to the *He* bump after propagation. The bump between $10^{19.6}$ eV and 10^{20} eV correspond to nuclei with mass between *Be* and *Mg*. The bump above $10^{20.1}$ eV is composed of *Fe* and other nuclei of similar atomic number.

This total spectral structure bears a striking resemblance to the result of AGASA [2, 13, 14].

RESULTS AND DISCUSSION

We have simulated propagation of UHE nuclei which are emitted from extra-galactic point sources using a Monte Carlo approach. Using this simulation, structures in the energy distribution of the small deflection angle particles have been predicted. In this simulation, we assume that UHECR are nuclei which are emitted from extra-galactic point sources. The nuclei interact with cosmic background radiation and lose energy during their propagation. Based on recent estimations of the IIBR, it has been shown that UHE nucleus can propagate longer distances than a proton with same energy. It is also expected that heavier nuclei are accelerated more efficiently to high energy than lighter nuclei or protons.

We have considered the effect of the intergalactic magnetic fields. We use a Kolmogorov spectrum for random magnetic fields that is 1 nG in average intensity with a 1 Mpc correlation length. The energy distribution of the particles with deflection angles smaller than 10, 5, and 2.5 degrees are investigated. Consequently, for this model no cosmic rays can be observed after propagation of a few hundred Mpc. Lower rigidity particles are deflected by the magnetic field relatively rapidly. Particles with larger Lorentz factor lose energy by photodisintegration. Since the rigidity and the Lorentz factor are proportional to the particle energy and inversely proportional to the mass number, particles which have common a Lorentz factor, or similar rigidity, remain after the deflection angle cut. The $Be - Mg$ bump appears at an energy more than 4 times larger than the *He* bump, and the *Fe* bump appear around 6 times larger energy. The rigidity gives a lower bound and energy loss gives an upper bound on the bumps. The energy distribution of the clustered events observed by AGASA can be explained by this model.

In the simulation, we assume a specific intergalactic magnetic field distribution, cut-off energy and source composition of primary nuclei. The Galactic magnetic field is ignored in this simulation. The events with certain deflection angle are selected and shows the effects due to the intergalactic photons and magnetic fields. This effect depends on the strength and correlation of the intergalactic magnetic field. This dependence gives an ambiguity to the analysis since the deflection angle cut is adjustable. Therefore, this angle should be treated carefully.

This result suggests that if we detect few hundred events above 10^{20} eV in future experiments, we should be able to extract the structure of the bumps in the energy distribution of cluster events. The Pierre Auger Observatory is quite sufficient for this purpose and may confirm the signature of nuclei in the spectrum in the near future.

ACKNOWLEDGMENTS

The author thank to Professor Utunomiya and the stuff of the Tours V conference to give this opportunity. This research was supported by grant NSF PHY-0114422 to the Center for Cosmological Physics, University of Chicago, and grant NSF PHY0103717.

REFERENCES

1. Yamamoto, T., et. al, 2004, Astropart.Phys. 20, 405
2. Takeda, M., et. al, 1998, Phys.Rev.Lett. 81, 1163
3. Bird, E. J., et. al, 1994, Astrophys. J. 351, 491
4. T.Abu-Zayyad et.al.:Astro-ph/0208301 (submitted to App)
5. Hayashida, N., et. al, 1999, Astropart. Phys. 10, 303
6. Uchihori, Y., et. al, 2000, Astropart.Phys, 13, 151
7. Takeda, M., et. al, 1999, ApJ 522, 225
8. Farrar, G. R., Biermann, P. L. 1998, Phys.Rev.Lett., 81,3579
9. Farrar, G. R., Biermann, P. L. 1999, Phys.Rev.Lett., 83,1999
10. Tinyakov,P.G., Tkachev,I.I. 2000, Phis.Rev.Lett., 85, 1154
11. Cronin, J. W., 1996, Proceeding of EHECR Astropys. and Future Observations, Tokyo Japan, Sept. 1996,
12. Sinozaki, K., et. al, 2002, Astrophys.J. 571, 117
13. Takeda, M., et. al, 2001, J.Phys.Soc.Jpn.Suppl.B, 70, 15
14. Takeda, M., et. al, 2001, Proceeding of 27th ICRC 2001:341
15. Stecker, F. W. 1968, Phys. Rev., 21, 1016
16. Puget, J. L., Stecker, F. W. 1976, ApJ, 205, 638
17. Malkan, M. A., Stecker, F. W. 1998, ApJ, 496,13
18. Stecker, F. W., Salamon, M. H. 1999, ApJ 512, 521
19. Stanev, T., et. al, 2000, Phys. Rev. D62 093005

Radio Nuclides In The Galaxy Seen In Gamma-Rays

Volker Schönfelder

Max-Plank-Institut für Extraterrestrische Physik, Postfach 1213, D-85741 Garching, Germany
(vos@mpe.mpg.de)

Abstract. From previous gamma-ray missions, especially SIGMA on GRANAT and OSSE and COMPTEL on the Compton Gamma Ray Observatory we have learnt that the sky is rich in gamma-ray line emitting objects and phenomena. Though we have certainly so far seen only the tip the iceberg, the first results are already exciting. COMPTEL produced the first-ever all-sky map in light of a radioactive gamma-ray line, namely that of radioactive ^{26}Al, and OSSE mapped the inner part of the galactic plane in the light of the electron-positron annihilation line. In addition, 1.157 MeV line emission from radioactive ^{44}Ti was detected by COMPTEL for the first time from a supernova remnant, namely Cas-A and later possibly also from the previously unknown remnant RX JO852-4622 as well. OSSE detected ^{57}Co line emission from SN 1987a in the Large Magellanic Cloud after ^{56}Co emission from this supernova had been detected already prior to the Compton mission, and COMPTEL saw hints of ^{56}Co line emission from the extragalactic supernova SN1991 T of type Ia. More recently COMPTEL found also hints for ^{22}Na line emission at 1.275 MeV from the classical nova N Cas 1995 and possible evidence for emission of the same line from the Galactic bulge region which may be due to unresolved novae. Already prior to these Compton observations, SIGMA provided exciting results on positron-electron annihilation features near 511 keV from binary systems with black hole candidates (Nova Muscae and 1 E 1740-2942). Very recently additional results, especially on the widths of the 1.809 MeV and the 511 keV annihilation lines, have been obtained by HESS I and SPI-INTEGRAL.

A review of all previous observations and the prospects of further studies with SPI-INTEGRAL are given.

INTRODUCTION

Radio nuclides provide an important tool to study element formation processes in the Universe. Nowadays it is generally agreed that the creation of the chemical elements is a continuous process, which keeps going even now. Hydrogen and helium were formed very soon after the big-bang, but nearly all of the other elements were formed over billions of years by nuclear processes in the interior of stars or during star explosions. The freshly created elements were then ejected into interstellar space and enriched the interstellar material from which new stars could be formed.

During these nuclear processes in stars not only stable, but also radioactive elements are formed. Some of them are gamma-ray emitters. Each of these isotopes can be characterised by a specific gamma-ray line, and these can be used to study nucleosynthesis processes directly and to test nucleosynthesis theories.

RADIO NUCLIDES FOR GAMMA-RAY ASTRONOMY

The most important gamma-ray emitting isotopes are summarized in Table 1. The decay-times range from months to more than one million years. The short-lived isotopes can only be observed in young or even new events, like a new supernova or new nova. The study of long-lived isotopes - like ^{26}Al and ^{60}Fe - provide information on global aspects of element formation processes over time-intervals of one million years or more.

Table 1.

	Nucleosynthesis Line			
Decay Chain	**Mean Life (Yrs)**	**Emission (MeV)**	**Origin**	**Observed from**
^{7}Be → ^{7}Li	0.145	e-capt, 0.478	Novae	
^{56}Ni → ^{56}Co → ^{56}Fe	0.31	e^+, 0.847, 1.238	Supernova	SN 1987 a, SN 1991 T (?)
^{57}Ni → ^{57}Co → ^{57}Fe	1.1	0.014, 0.122	Supernova	SN 1987 a
^{22}Na → ^{22}Ne	3.8	e^+, 1.275	Novae, WR-stars	NCas 1995 (?), Galactic Bulge (?)
^{44}Ti → ^{44}Sc → ^{44}Ca	90.0	e^+, 0.068, 0.078, 1.156	Supernova	Cas A, RX 0852-4622 (?)
^{60}Fe → ^{60}Co → ^{60}Ni	$2.2 \cdot 10^6$	0.059, 1.173, 1.332	Supernova	Inner Galaxy
^{26}Al → ^{26}Mg	$1.1 \cdot 10^6$	e^+, 1.809	Supernova, WR-stars	Galactic Plane, RX 0852-4622 (?)

Nearly all of these isotopes are also e^+- producers, and therefore – at least to some extent – are sources of e^- - e^+-annihilation radiation. Table 1 is an old one, which has been shown in many review articles before. It is interesting to note that at least hints of line detections have been reported now from all but one of these decay chains. Those observations which are marked with a question mark were either near the sensitivity level of the measuring instrument or they were questionable because of systematic uncertainties. Though we have certainly only seen the tip of the iceberg, it seems that the sky is rich in gamma-ray line emitting objects.

The objects listed in the last column of Table 1 are based on the following nucleosynthetic gamma-ray line observations:

COMPTEL has produced the first ever all-sky map in the light of a radioactive gamma-ray line, namely the 1.809 MeV line from radioactive ^{26}Al [1]. This line had been discovered by HEAO-C [2]. Later is was observed by a number of different instruments from the galactic center region. The profile of the 1.809 MeV line has been measured by GRIS [3], RHESSI [4], and SPI-INTEGRAL [5].

OSSE has produced a map of the inner part of the Galaxy in the light of the $e^+ - e^-$-annihilation radiation [6]. The 511 keV-line was discovered in a balloon flight of the Bell-Sandia group [7]. Later the annihilation radiation was observed by a number of different instruments especially from the Galactic center region. The profile of the 511 keV-line has been measured by Bell-Sandia [7], HEAO-C [8], TGRS [9], and SPI-INTEGRAL [10]. In the early 1990's SIGMA provided exciting results on $e^- - e^+$ annihilation features near 511 keV from binary systems with black-hole candidates. Examples were Nova Muscae [11] and the "Great Annihilator"-source 1 E 1740.7-2942 [12].

COMPTEL for the first time discovered the 1.157 MeV line from radioactive ^{44}Ti from a supernova remnant, namely Cas-A [13] and possibly also from the previously unknown supernova remnant RX 0852-4622, also called Vela-Junior [14], [15], [16]. COMPTEL's discovery of the 1.157 MeV line from Cas A was confirmed by Beppo-Sax observations of the simultaneously emitted lines at 67.9 and 78.4 keV [17].

The COMPTEL 1.809 MeV all-sky map shows a local maximum at the position of Vela-Junior, which therefore could also be an ^{26}Al-source [18].

OSSE for the first time has detected ^{57}Co line emission from the type II supernova SN 1987a in the Large Magellanic Cloud [19], after ^{56}Co line emission from this supernova remnant had been detected already prior to the launch of the Compton Observatory by a number of different experiments, especially by the gamma-ray spectrometer aboard SMM [20].

COMPTEL found hints of ^{56}Co-line emission at 847 keV and 1.238 MeV from the extragalactic type Ia supernova SN 1991 T [21].

COMPTEL found also hints of ^{22}Na line emission at 1.275 MeV from the classical nova Cas-1995 and in addition from the Galactic bulge region, which may be due to unresolved novae [22].

First evidence for line emission from ^{60}Fe-decay at 1.173 and 1.332 MeV from the inner Galaxy has been reported from RHESSI-observations [23].

The scientific relevance of all these observations is discussed in the following sections.

1.809 MEV LINE EMISSION FROM ^{26}AL

The COMPTEL all-sky map in the light of the 1.809 MeV line is bright along the Galactic plane in the inner central radian and in the Cygnus, Carina and Vela-regions [1]. It is therefore concluded that the ^{26}Al is prodominetly produced in massive stars, which are concentrated in those parts of the sky. But the exact origin is not clear. At present we do not know, whether the ^{26}Al is produced by explosive nucleosynthesis processes in supernovae at the end of the life cycle of massive stars or in the hydrostatic nuclear burning phase of massive mass-loosing stars (like Wolt-Rayet stars), or a mixture of both.

Measurements of the 1.809 MeV line profile exist now from three different instruments (GRIS [3], RHESSI [4] and SPI-INTEGRAL [5]). From the HEAO-C measurements [2] only an upper limit to the line width of 3 keV could be determined. The surprisingly broad intrinsic line width of 5.4 keV FWHM found by GRIS could not be confirmed by RHESSI and SPI-INTEGRAL which both measured significantly narrower line profiles. The question, how the ^{26}Al-nuclei could survive over millions of years with velocities of more than 500 km/sec is, therefore, obsolete. Very recently, Kretschmer, Diehl, Hartmann [24] have theoretically investigated, whether the differential rotation of the Galaxy should show a visible Doppler shift of the position of the peak of the line, if the emissions form the eastern and western sides around the Galactic center are compared with each other. The difference should be marginal (of order 0.5 keV), and will be hard to detect, if the intrinsic line widths are folded with the instrumental resolution.

Another very interesting region for studies of the 1.809 MeV line is the Vela region, which was investigated by COMPTEL extensively [25] and which is also a prime target for INTEGRAL observations. This region contains the Vela supernova remnant, γ-Velorum (the closest Wolf-Rayet star), and Vela Junior RX 0852-4622. The peak emission in the 1.809 MeV emission in that region coincides with Vela Junior, and it has to be clarified, whether the emission really originates from the supernova remnant or whether it comes from the molecular ridge in that part of sky.

LINE EMISSION FROM THE ^{60}FE- DECAY CHAIN

The final answer to the question, whether ^{26}Al is produced by explosive nucleosynthesis of core-collapse supernovae or hydrostatic burning of massive mass-loosing stars or a mixture of both may come from the observation of the 1.17 and 1.33 MeV lines from radioactive ^{60}Fe-decay. It is common belief that ^{60}Fe is produced in core-collapse supernovae. As such, the detection of ^{60}Fe line emission and a resulting sky map in the line would provide a tracer for Galactic supernova activity. Therefore, a comparison of the ^{60}Fe and ^{26}Al-maps would allow to identify the supernova contribution to the ^{26}Al-map. The remaining part of the ^{26}Al-map would be due to stellar wind ejection. If the production of ^{60}Fe nuclei by processes other than by core-collapse supernovae should be possible, the separation would not be so simple.

The fluxes of the 1.17 and 1.33 MeV lines are expected to be at the 15% -level of the 1.809 MeV line, each. The preliminary marginal (~3σ) -results from RHESSI on these two lines [23] from the inner Galaxy are indeed consistent with these expectations.

ELECTRON-POSITRON ANNIHILATION RADIATION

The most comprehensive measurements of the e^+ - e^--annihilation radiation has been performed by OSSE, though SPI-INTEGRAL has published first results on the line profile [10] and the morphology of line emission [27] as well. OSSE has performed a large number of pointings, from which maps of the inner Milky Way in the 511 keV annihilation line and in the three-photon continuum emission could be derived. At least, two broadband components can be identified: an extended disk and a central bulge component. The existence of a possible fountain north to the Galactic Center, which appears in the pure line map, needs confirmation [6]. The disk component may be explained by ^{26}Al, ^{44}Sc, ^{56}Co, ^{22}Na positron emitters. Also shorter lived positron emitters may contribute, because the mean annihilation time in interstellar space is of the order of 10^5 years.

The bulge component is more difficult to be explained. Its extent has been measured by SPI-INTEGRAL to be 9° FWHM with a 2σ uncertainty range of 6° to 18° [27]. The bulge line flux was determined to be $9.9 \pm 4.7/-2.1 * 10^{-4}$ cm^{-2}sec^{-1}. To a small fraction the bulge component may be produced by population II-type positron emitters like ^{22}Na and ^{56}Co from type Ia SN. In addition, compact positron emitting sources may play a role as well. We know from SIGMA that sources like Nova Muscae [11] and 1 E 1740.7-2942 [12] are able to produce positrons on a short time scale of 1 day. Such short-time events in principle should be easily detectable by the bi-weekly INTEGRAL Galactic plane scans (so far, however, without success). The contribution to the bulge annihilation line from compact sources must be small (smaller than 4.10^{-4} cm^{-2}sec^{-1}). The profile of the 511 keV line and the positronium fraction depends on the temperature, the density and the ionisation degree of the medium. The most accurate measurements of the line profile are those of TGRS [9]. The line width of 1.8 ± 0.54 keV FWHM and the positronium annihilation fraction of 0.94 are both consistent with the fact that the annihilation either occurs in cold molecular clouds, or in warm neutral or ionised gas. The line profile of the bulge component- as measured by SPI-INTEGRAL [10] - is somewhat broader (2.95+0.45/-0.51keV)

^{44}TI-LINE EMISSION FROM SUPERNOVA REMNANTS

The detection of the 1.157 MeV line from radioactive ^{44}Ti by COMPTEL from Cas-A [13] and possibly RX 0852-4622 [16] needs confirmation and especially uncertainties in the reported fluxes of the Cas-A line flux from different experiments (OSSE for the 1.157 MeV line and RXTE and Beppo Sax for the 67.9 and 78.4 keV lines) have to be removed. Most interesting will be measurements of the line profiles at 67.9 keV, 78,4 keV and 1.157 MeV, from which the expansion velocity of the ^{44}Ti–material can be derived, and from which conclusions on the physical conditions in the inner most layers of the supernova ejecta can be derived. The search for additional young so far unknown supernova remnants (not older than a few hundred years) in the inner Galaxy from their ^{44}Ti-line emission is one of the key objectives of INTEGRAL.

GAMMA-RAY LINES FROM SUPERNOVAE

Supernovae are the most interesting objects for the direct measurement of nucleosynthesis lines. Most promising are gamma-ray lines expected from thermonuclear type Ia supernovae because of their large ^{56}Ni mass-yields and the early visibility of the gamma-ray lines after the explosion (thanks to the thin envelopes). The chances to detect gamma-ray lines from core-collapse supernovae of type II are much smaller (the ^{56}Ni mass-yields are about an order of magnitude smaller and an impressive gas envelope at the beginning of the explosion is practically opaque to gamma-rays). The occurrence of the type II supernova in 1987 in the Magellanic Cloud was a really lucky chance during which for the first time ^{56}Co-gamma-ray lines could be detected [20]. The relatively early detection of the lines by the spectrometer of SMM in spite of a thick envelope required the existence of mixing by convection to remove fresh material out of the burning region.

It is generally accepted that the main energy input to the ejecta of SN 1987 a (after a couple of days) comes from radioactive decay [28]: First from ^{56}Ni- and then from ^{56}Co-decay. After about 1100 days ^{57}Co-decay takes over, and at very late époques (after 2200 days) ^{44}Ti-decay should be the dominant source. The ^{57}Co-line at 122 keV had, indeed, been measured by OSSE [19]. A direct measurement of one of the ^{44}Ti-lines, however, is still lacking; the line fluxes are probably too low for a detection by INTEGRAL.

The only type Ia supernova from which at least hints of the two ^{56}Co-lines at 847 keV and 1.238 MeV have been seen by COMPTEL is SN 1991 T [21]. The ^{56}Ni mass derived from the line-fluxes is in the range 1.15 to 1.95 $M_\odot$ for a Cepheid-based distance of 13 Mpc. This mass is rather high, but it is consistent with estimates from the strength of Fe- and Co-emission lines in the optical and near-infrared [29] and from modelling the bolometric light curve [30].

No detection from ^{56}Co-decay at 847 keV and 1.238 MeV could be achieved from the type Ia SN 1998 bu, though this supernova was relatively close (~ 11 Mpc) and the 2σ-upper limit from COMPTEL to the line fluxes based on an exposure of 14 weeks are $3.3*10^{-5}$ cm^{-2} sec^{-1}, which converts into a ^{56}Ni mass-limit of 0.35 $M_\odot$ [31].

The reason for the non-detection could be either that mixing of the radioactive nuclei is smaller than expected or that the absorption of gamma-rays in the envelope is underestimated in the models of supernova-expansion.

Given the sensitivities of the INTEGRAL-instruments and assuming a ^{56}Ni mass-yield of 1 $M_\odot$ we shall be lucky, if one type Ia supernova with detectable ^{56}Co gamma-ray lines will occur during the INTEGRAL mission life time of 5 years.

CLASSICAL NOVAE

Classical novae originate from binary systems with a white dwarf accreting matter from a companion star. O-Ne-Mg rich novae are most promising for gamma-ray line emission. The most intensive expected gamma-ray lines are at 478 keV (^{7}Be-decay) and at 1.275 MeV (^{22}Na-decay). Prompt 511 keV line emission is expected to occur before the optical nova maximum (usually before the discovery of the nova). The main electron-position sources are ^{13}N and ^{18}F.

So far, no gamma-ray lines have been detected from any nova unambiguously. But hints of the detection of 1.275 MeV line emission from Nova Cas 1995 has been reported, and COMPTEL found also possible evidence for 1.275 MeV line emission from the bulge region of the Galaxy which may be due to unresolved novae [22]. Because of systematic uncertainties in the latter measurements, a confirmation of the result is absolutely necessary.

Given the INTEGRAL sensitivities for the detection of broadened nova lines (typical expansion velocities are around 2000 km/sec), the nova would have to be closer than 0.5 to 1 kpc. Again, we must be lucky to get such a nova in the mission lifetime of INTEGRAL. In case of detection, the evolution of the line width with time will allow detailed studies of the dynamics of the nova.

NUCLEAR INTERACTION LINES FROM INTERSTELLAR SPACE

Nuclear interaction lines from interstellar space (produced by the bombardment of interstellar matter with cosmic ray particles) are still to be discovered. Most promising for detections are the 4.4 MeV and 6.15 MeV lines from excited Carbon and Oxygen nuclei. Newer predictions on the expected line fluxes from the radian of the Galaxy are $2.5*10^{-5}/cm^2 sec$ in the 3 to 7 MeV band [32]. Since these lines are broadened due to the high momentum transfer of the cosmic-ray particle the SPI-INTEGRAL sensitivities for these lines are worse than the SPI narrow line sensitivity. The detection, by INTEGRAL, therefore would be a fascinating surprise.

CONCLUSION

Gamma-ray line astronomy is well suited to study radionuclei in the Galaxy. First promising results have been obtained already in the last decade especially by COMPTEL and OSSE aboard the Compton Observatory. From INTEGRAL we can expect a wealth of new information from interstellar narrow line emission, but – if we are lucky – exciting results can also be expected from line profile measurements of individual objects such as supernovae, supernova remnants and novae.

References:
1. Plüschke, S., et al, ESA-SP 459, 55, 2001.
2. Mahoney, W.A., et al, Ap. J. 286, p. 578, 1984.
3. Naya, J.E., et al, Nature 384,p. 44, 1996.
4. Smith, D., Ap. J. 589. L55, 2003.
5. Diehl, R., et al, A&A accept. for publication, 2003.
6. Milne, P.A., et al, A I P 587. p.11, 2001
7. Leventhal, M., and Callum, C.J.Mac., and Stang, P., Ap. J. 225, L11, 1978.
8. Riegler, G.R., et al, Ap. J. 248, L13, 1981.
9. Harries, M.J., et al, Ap. J. 501, L 55, 1998.
10. Jean, P., et al A&A 407, L55, 2003.
11. Goldwurm, A., et. al, Ap. J. 389, L79, 1992.
12. Bouchet. E., et al, Ap. J. 383, L 45 1991.
13. Iyudin, A., et al, A&A 284, L1, 1994.
14. Iyudin, A., et al, Nature 396, p. 142, 1998.
15. Aschenbach, B., Nature 396, p. 142, 1998.
16. Schönfelder,V., et al, AIP 510, 54, 1999.
17. Vink, J., et al, Ap. J. 560, L79, 2001
18. Aschenbach, B., and Iyudin, A., and Schönfelder, V., A&A 350, 997, 1999.
19. Kurfes, J., et al, Ap. J. 399, L 137, 1992.
20. Leising, M.D, and Share,G.H., Ap. J. 357, p. 638, 1990.
21. Morris, D,. et al, AIP 410, 1084, 1997.

22. Iyudin, A., et al, AIP 587, 508, 2001.
23. Smith, D., New Astr. Reviews, submitted, 2003.
24. Kretschmer, K., and Diehl, R., and Hartmann, D., A&A , submitted 2003.
25. Diehl, R., et al, Astrophys. Lett. Commutation 38, 357, 1999.
26. Timmes, F.X., et al, Ap. J. 449, p. 204,1995
27. Knoedlseder, J., et al, A&A, accepted for publication 2003.
28. Fransson, C., and Kozma,C., New Astron. Rev 46, 487, 2002.
29. Spyromilio, J., et al , MNRAS 258, 53, 1992.
30. Cappelaro, E., et al, A&A 328, 303,1997.
31 Georgii, R., et al, A&A 394, 517, 2002.
32. Ramaty, R., et. al, Ap. J 488, 730, 1997

Radioactivity of the Key Isotope ^{44}Ti in SN 1987A

Yuko MOTIZUKI* and Shiomi KUMAGAI†

*RIKEN, Hirosawa 2-1, Wako 351-0198 Japan[1]
†Department of Physics, Faculty of Science and Technology, Nihon University
Kanda-Surugadai 1-8, Chiyoda-ku, Tokyo 101-0062 Japan

Abstract. We investigate radioactivity from the decay sequence of ^{44}Ti in a young supernova remnant SN 1987A. We perform Monte-Carlo simulations of degradation of the nuclear lines to explain a late-time bolometric luminosity which is estimated from optical and near-infrared observation at 3600 days after the explosion. Assuming the distance to LMC in between 45.5 and 52.1 kpc, we have obtained the initial ^{44}Ti mass of $(0.82 - 2.3) \times 10^{-4} M_\odot$ within the current uncertainty of the physical quantities. The resulting fluxes of γ- and hard X-rays emerged from the ^{44}Ti decay are estimated and compared with the line sensitivity of the INTEGRAL/SPI on board and that of NeXT X-ray satellite planned to be launched in 2010. The effect of ^{44}Ti ionization on the estimated fluxes is briefly remarked.

INTRODUCTION

To detect the nuclear γ-rays from the decay sequence of ^{44}Ti is one of the prime target in the current and future γ-ray and X-ray satellites. In particular, SN 1987A, appeared 16 years ago in the Large Magellanic Cloud (LMC), is providing us with a challenge as the one of such targets.

The important feature of detecting ^{44}Ti nuclear lines from young supernova remnants can be summarized as follows: The initial yield of ^{44}Ti that is synthesized by a single event of a core-collapse supernova explosion is very crucial to constrain dynamics of core-collapse supernova nucleosyntehsis. This is because ^{44}Ti is synthesized at the vicinity of the so-called mass cut, that divides the matter which accretes on a compact object and the ejecta which is scattered into interstellar space. For this, the initial mass of ^{44}Ti depends sensitively on 1) the location of the mass cut, 2) the maximum temperature and the maximum density behind the shock wave, and 3) the internal structure ($\lesssim 2$ M$_\odot$ from the center) of a progenitor.

So far, 1.16 MeV nuclear line that is emitted from the decay-chain of ^{44}Ti (see below) was detected from Cassiopeia A with COMPTEL/CGRO experiment [1, 2] and this was confirmed with BeppoSAX by detecting associated 67.9 and 78.4 keV nuclear lines [3]. It should be noted here that RXJ0852-4622, the "Vela Junior" remnant, was first discovered by the 1.16 MeV γ-ray line as a point source [4], and then discovered in X-rays [5]. In the near future, ^{44}Ti nuclear lines are expected to be detected from the

---

[1] E-mail: motizuki@riken.jp. Spelling of her name (Mochizuki) has changed to Motizuki.

other young ($\lesssim$ a few×1000 yrs) galactic supernova remnants and SN 1987A in LMC. Further, galactic survey of ^{44}Ti nuclear lines may dig out unknown supernova remnants which is difficult to be caught in the other electromagnetic wavelengths, and may give us even a scrap of information on the galactic supernova rate.

The decay sequence of the radioactive ^{44}Ti is the following: ^{44}Ti decays by orbital electron capture mainly to the second excited state of ^{44}Sc (branching ratio of 99.3%). The decay is soon followed by the emissions of 67.9 keV and 78.4 keV nuclear deexcitation lines to the ground state of ^{44}Sc. Although until recently the halflife of ^{44}Ti showed a large uncertainty in those measured in laboratories, compilation of recent 8 experiments which were performed after 1998 (see., e.g., [6] and references therein; [7]) gives weighted mean halflife of $t_{1/2} = 60 \pm 1$ yr (the error is 1 σ, statistical). The daughter nucleus, ^{44}Sc, then decays almost exclusively by positron emission to the first excited state of ^{44}Ca, which emits 1.16 MeV deexcitation line to the ground state. The ground state of ^{44}Ca is stable. The emitted positron in the above sequence ends up with 511 keV annihilation line. It is noted that the halflife of ^{44}Sc is merely 3.93 hrs, so that the timescale and hence the radioactivity of the whole decay chain is regulated by the halflife of ^{44}Ti.

We note that laboratory experiments measure the halflife of *neutral* ^{44}Ti. The crucial point here is that ^{44}Ti decays only by orbital electron capture. This is because the decay Q-value from the ground state of ^{44}Ti to the second excited state of ^{44}Sc is less than twice the electron rest mass, which is at least required for positron emission to be allowed by producing two 511 keV γ-photons when a positron annihilates with an electron (and so does that to the first excited state of ^{44}Sc for the rest of the minor fraction of the branch). Actually the halflife of highly ionized ^{44}Ti becomes longer than that of neutral ^{44}Ti and this affects the radioactivity. Thus, we should be careful to apply the experimental halflife to this problem: The electric environment of ^{44}Ti in a young supernova remnant may be different from that in laboratories.

Previous studies on the ^{44}Ti ionization effect on its radioactivity are found in references [8, 9]; these mainly discuss Cassiopeia A. A recent paper [10] includes a linear analysis in order to simply show why and how the ^{44}Ti ionization affects its radioactivity. As we shall see later, there is a clear possibility of ionization of ^{44}Ti ongoing in SN 1987A. In this article, we first derive the expected nuclear fluxes without consideration of the ^{44}Ti ionization. We then briefly discuss how the possibly-ongoing ^{44}Ti ionization may change this estimate in the near future, using the result of the linear analysis presented in [10].

^{44}TI RADIOACTIVITY IN SN 1987A

Figure 1 depicts observation-based data of bolometric luminosity and theoretically calculated light curves. The observed light curve in early time is first governed by the radioactive decay of ^{56}Ni ($t_{1/2} = 6.1$ d) and then that of its daughter ^{56}Co ($t_{1/2} = 77.3$ d). The synthesized ^{56}Co nuclide decays mainly by positron emission to stable ^{56}Fe. Until ~ 800 days after the explosion, the light curve observed in the wavelength ranging from ultraviolet to infrared has been successfully modeled with the energy supply from the

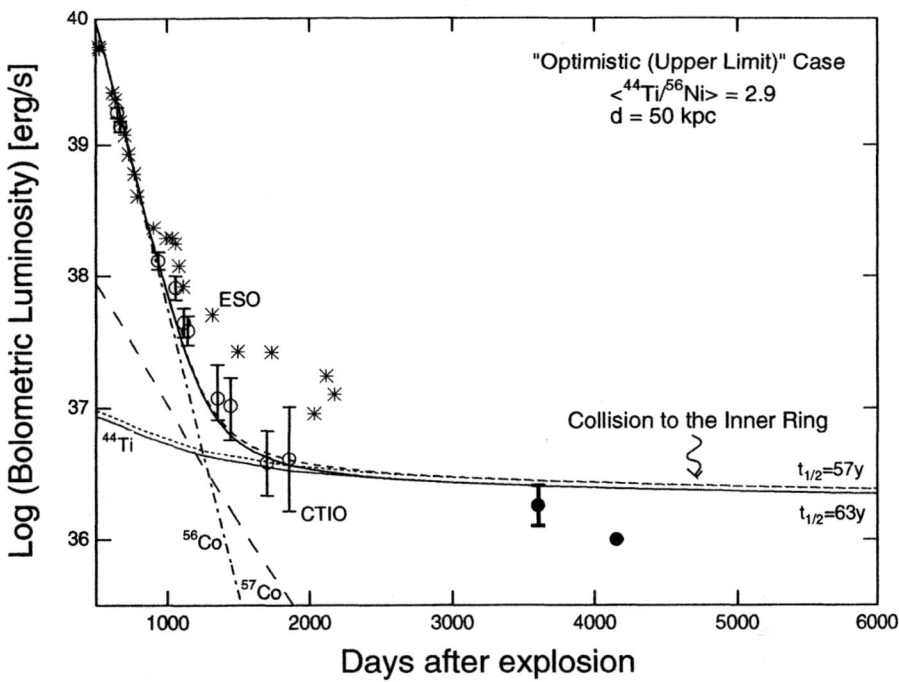

FIGURE 1. Evolution of the bolometric luminosity of SN 1987A. Observed luminosity data are shown (see the text). The thick solid line denotes a theoretical light curve with the ^{44}Ti halflife of $t_{1/2} = 63$ yrs and the short-dashed line denotes that with $t_{1/2} = 57$ yrs. The long-dashed line shows a theoretical contribution from the ^{57}Co decay, and the dash-dotted line that from the ^{56}Co decay. The dotted and thin solid lines denote the contributions from the ^{44}Ti decays that correspond to $t_{1/2} = 57$ yrs and 63 yrs, respectively.

decay of ^{56}Co (see [11] for details).

Afterwards, the decline of the observed light curve apparently slowed down, suggesting the presence of the other heating source. As the next source, ^{57}Co ($t_{1/2} = 272$ d) plays a role in the luminosity. This epoch does not last long, however. The slowness of the decline of the observed light curve becomes distinguished in particular after ~ 1500 days (see Fig. 1). The dominant energy source at this moment is believed to be the ^{44}Ti decay.

Suntzeff [12] reported a bolometric luminosity at 3600 days, i.e., 10 years after the explosion. He used the bolometric corrections to the optical colors VK to estimate a bolometric magnitude under the assumption that the flux distribution was indeed frozen. Under this assumption, he obtained a bolometric luminosity, $L_{S97} = (1.9 \pm 0.6) \times 10^{36}$ erg sec^{-1}, at 3600 days after the explosion. In the following, this bolometric luminosity L_{S97} is referred to as the S97 bolometric luminosity. The author also estimated a bolometric luminosity of $\sim 1.0 \times 10^{36}$ erg sec^{-1} at 4151 days after the explosion under the assumption that the same bolometric corrections from day 1800 [13].

The above-mentioned observational data and observation-based values of bolometric luminosity are shown in Fig. 1.

To explain the upper and the lower bound of the S97 bolometric luminosity at 3600 days after the explosion [12], we performed Monte-Carlo simulations of Compton degradation of the nuclear γ-photons of ^{57}Co (14 keV, 122 keV, 136 keV, etc.) and ^{44}Ti (68 keV, 78 keV, 511 keV, 1.16 MeV). The UV, optical, and IR photons originate from the energy loss of the emitted γ-rays during the radiative transfer in the ejecta.

To determine the velocity distribution of particles, we adopt the explosion model 14E1 [14] whose main-sequence mass, the ejecta mass, and the explosion energy are 20 $M_\odot$, 14.6 $M_\odot$ (4.4 $M_\odot$ core material plus 10.2 $M_\odot$ hydrogen-rich envelope), and 1×10^{51} erg, respectively. This model was derived from a detailed analysis of the plateau shape of the light curve of SN 1987A which was observed until 120 days after the explosion, and well accounts for the earlier optical, X-ray, and γ-ray light curves of SN 1987A [15]. Note that the ^{56}Ni mass in SN 1987A has been constrained as $0.07 M_\odot$ from the intensity during the observed exponential decline.

Note also that at the period of the S97 observation the ionization of ^{44}Ti was not relevant. Later, the supernova blast shock crashed into the dense inner ring, and the shock heating started to ionize the elements (see below). We therefore use the experimental halflife of neutral ^{44}Ti: $t_{1/2} = 60 \pm 3$ yrs (3σ deviation) to explain the S97 bolometric luminosity. The distance to SN 1987A is adopted to be 48.8 ± 3.3 kpc (3σ, see [16]). Other details of our calculation are found in [11]; in the present study, the adopted nuclear decay parameters have been updated.

In Fig. 1, two calculated bolometric light curves which stick to the upper bound of the S97 luminosity are also shown. The difference of the two theoretical curves is in the adopted halflife of ^{44}Ti; the one is calculated with $t_{1/2} = 57$ yrs and the other with $t_{1/2} = 63$ yrs. One sees in Fig. 1 that the current uncertainty in the ^{44}Ti halflife no longer produces a remarkable difference in the light curve.

The upper-bound light curve gives $<^{44}\text{Ti}/^{56}\text{Ni}> = 2.9$, where $<^{44}\text{Ti}/^{56}\text{Ni}>$ is defined as $<^{44}\text{Ti}/^{56}\text{Ni}> \equiv [X(^{44}\text{Ti})/X(^{56}\text{Ni})]_{87A}/[X(^{44}\text{Ca})/X(^{56}\text{Fe})]_\odot$, i.e., the ratio of ^{44}Ti/^{56}Ni in amount in SN 1987A to ^{44}Ca/^{56}Fe in the solar neighborhood. The obtained value 2.9 for the upper bound is considered to be acceptable. One of the reasons for this is that the synthesized amount of ^{44}Ti theoretically tends to be more abundant in aspherical supernovae than in spherical ones; actually recent Hubble Space Telescope images and spectroscopy have revealed that SN 1987A has an aspherical geometry [17]. Translated from the obtained $<^{44}\text{Ti}/^{56}\text{Ni}>$ values, we find the initial ^{44}Ti mass of $(0.82 - 2.3) \times 10^{-4} M_\odot$ within the known uncertainty of the experimental values mentioned above.

Figure 2 shows the expected nuclear fluxes at 6000 days after the explosion (i.e., in 2003) for both the obtained upper and the lower bounds of the initial ^{44}Ti mass. It is worthwhile to mention here that the obtained ^{44}Ti masses depend on the assumed distance to SN 1987A, but the expected fluxes do not. In Fig. 2, the line sensitivity of INTEGRAL/SPI on board and that of the planned NeXT mission are also shown for comparison.

We see in Fig. 2 that, with 10^6 sec observation time, it might be difficult for INTEGRAL/SPI to detect the ^{44}Ti nuclear lines from SN 1987A. We note that our upper

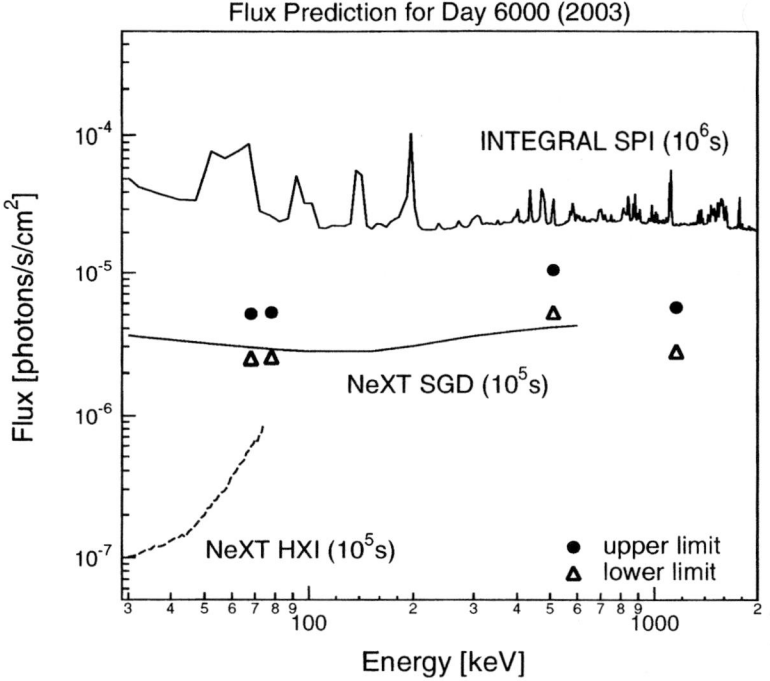

FIGURE 2. Prediction of the nuclear fluxes associated with the ^{44}Ti decay in SN 1987A at 6000 days after the explosion (i.e. in 2003). The line sensitivity of INTEGRAL SPI on board for the typical observation time 10^6 sec and that of the NeXT mission for the typical observation time of 10^5 sec are also shown for comparison. The NeXT satellite is planned to be launched in 2010.

bound estimate relies on the upper bound of the S97 bolometric luminosity, which was estimated under the assumption described previously and corresponds to usual 1 σ deviation: There is a possibility that the starting point, the bolometric luminosity itself, may be larger. Theoretically, on the other hand, the larger bolometric luminosity corresponds to the larger $<^{44}$Ti/^{56}Ni$>$ value. The feasibility of the largest $<^{44}$Ti/^{56}Ni$>$ value remains to be solved, but too large value could not be naturally adopted.

Finally, we point out that the ionization process of ^{44}Ti is considered to be well underway in SN 1987A, due to shock heating caused by the collision of the supernova blast shock with the dense inner ring: H-like and He-like ionization stages of O, Ne, Mg, and Si have been already observed, and also SN 1987A is observed to be a very rapidly evolving remnant (see [18, 19]). If ^{44}Ti reaches the high-ionization stages in the future, the expected fluxes shown in Fig. 2 will become smaller, as discussed in the linear analysis presented in [10]. To get a rough idea, if *all* the ^{44}Ti should be in the He-like (H-like) ionization state, the radioactivity of SN 1987A would suffer a $\sim$ 9 (45) % reduction (see Table 1 of [10]). Under the realistic situation in which the ionization of the elements evolves by the shock heating, various ionization stages will be led. Even in such a case, 68, 78, 511 keV lines appear to be detectable with a planned detector on a future mission such as NeXT (see Fig. 2). Actual conclusion of the effect of the ionization on the radioactivity requires the knowledge of the temperature and the density evolution of the supernova remnant, and this will be a future subject for SN 1987A.

ACKNOWLEDGMENTS

We are grateful to J. Knödlseder for providing us with the SPI sensitivity and T. Takahashi also for providing us with the planned line sensitivity of the NeXT mission. Y.M. would like to thank P. Leleux and A. Gould for helpful comments.

REFERENCES

1. A.F. Iyudin, R. Diehl, H. Bloemen, et al., *Astron. Astrophys.* **284**, L1 (1994).
2. V. Schönfelder, K. Bennett, J.J. Blom, et al., *Astron. Astrophys.* Suppl. **143**, 145 (2000).
3. J. Vink, J.M. Laming, J.S. Kaastra, et al., *Astrophys. J.* **560**, L79 (2001).
4. A.F. Iyudin, V. Schönfelder, et al., *Nature* **396**, 142 (1998).
5. B. Aschenbach, *Nature* **396**, 141 (1998).
6. T. Hashimoto et al., Nucl. Phys. **A686**, 591 (2001).
7. Zs. Fülöp, Y. Wakasaya, et al., AIP Conference Proceedings **529**, 684 (2000).
8. Y. Mochizuki, K. Takahashi, H.-Th. Janka, W. Hillebrandt, and R. Diehl, *Astron. Astrophys.* **346**, 831 (1999).
9. Y. Mochizuki, *Nucl. Phys.* **A688**, 58c (2001).
10. Y. Motizuki and S. Kumagai, New Astronomy Reviews, in press; astro-ph/0311080.
11. S. Kumagai, K. Nomoto, T. Shigeyama, et al., *Astron. Astrophys.* **273**, 153 (1993).
12. N.B. Suntzeff, in SN1987A: Ten Years After, The Fifth CTIO/ESO/LCO Workshop, eds. M.M. Phillips and N.B. Suntzeff; astro-ph/9707324.
13. N.B. Suntzeff, in From Twilight to Hightlight – The Physics of Supernovae, ESO/MPA/MPE Workshop; astro-ph/0212561.
14. T. Shigeyama and K. Nomoto, *Astrophys. J.* **360**, 242 (1990).
15. K. Nomoto, T. Shigeyama, S. Kumagai, and H. Yamaoka, in Supernovae, ed. S.E. Woosley, Springer, Berlin, p. 176 (1991).
16. A. Gould and O. Uza, *Astrophys. J.* **494**, 118 (1998).
17. L. Wang, et al., *Astrophys. J.* **579**, 671 (2002).
18. D.N. Burrows, E. Michael, U. Hwang, et al., *Astrophys. J.* **543**, L149 (2000).
19. E. Michael, S. Zhekov, R. McCray, et al., *Astrophys. J.* **574**, 166 (2002).

BRUSLIB: the Brussels nuclear library for astrophysics applications

S. Goriely[1]

Institut d'Astronomie et d'Astrophysique
Université Libre de Bruxelles
Campus de la Plaine, CP 226
1050 Brussels – Belgium

Abstract. Nuclear reaction rates are obviously quantities of fundamental importance in nuclear astrophysics. Important effort has been devoted in the last decades to measure reaction cross sections and many experimental compilations have now become available. Despite such effort, many nuclear astrophysics applications still require the use of theoretical predictions to estimate experimentally unknown rates. Most of the nuclear ingredients in the calculations of reaction rates need to be extrapolated in an energy or/and mass domain out of reach of laboratory simulations. In addition, important astrophysical applications (in particular the r- and p-processes of nucleosynthesis) often involve a large number (thousands) of unstable nuclei, so that only global approaches can be used. For these reasons, when the nuclear ingredients to the reaction models cannot be determined from experimental data, use is made preferentially of microscopic or semi-microscopic global predictions based on sound and reliable nuclear models which, in turn, can compete with more phenomenological highly-parametrized models in the reproduction of experimental data.

The International Atomic Energy Agency (IAEA) addressed recently through a Coordinated Research Project on the Reference Input Parameter Library (RIPL) the difficult task of collecting, evaluating and recommending the vast amounts of various nuclear parameters of relevance in cross section calculations. The RIPL project is targeted at users of nuclear reaction codes and, in particular, at nuclear data evaluators for applications as large as energy production, accelerator driven waste incineration, production of radioisotopes for therapy and diagnostics, charged particle beam therapy, material analysis and astrophysics. A brief overview of the nuclear input parameters recommended by the RIPL is presented.

Simultaneously, the need to develop more microscopic input parameters has led to new developments that have been compiled in the Brussels Nuclear Astrophysics Library, known as BRUSLIB. The content of the BRUSLIB library is described, with a special emphasis on the latest results concerning the determination of ground state properties within the Hartree-Fock-Bogoliubov approach.

INTRODUCTION

Strong, weak and electromagnetic interaction processes play an essential role in many different applications of nuclear physics, such as accelerator driven waste incineration, production of radioisotopes for therapy and diagnostics, charged-particle beam therapy, material analysis as well as nuclear astrophysics. Although important effort has been devoted in the last decades to measure reaction cross sections, experimental data only covers a minute fraction of the whole set of data required for such nuclear physics

---

[1] S.G. is FNRS senior research assistant

applications. Reactions of interest often concern unstable or even exotic (neutron-rich, neutron-deficient, superheavy) species for which no experimental data exist. Given applications (in particular, nucler astrophysics and accelerator-driven systems) involve a large number (thousands) of unstable nuclei for which many different properties have to be determined. Finally, the energy range for which experimental data is available is restricted to the small range reachable by present experimental setups. To fill the gaps, only theoretical predictions can be used.

For specific applications such as nucler astrophysics or accelerator-driven systems, a large number of data need to be extrapolated far away from the experimentally known region. In this case, two major features of the nuclear theory must be contemplated, namely its *accuracy*, which obviously has always been for most of the application the major (and some time unique) criterion in the model selection, but also its *reliability*. A microscopic description by a physically sound model based on first principles ensures a reliable extrapolation away from experimentally known region. For these reasons, when the nuclear ingredients to the reaction models (e.g Hauser-Feshbach) cannot be determined from experimental data, use is made preferentially of microscopic or semi-microscopic global predictions based on sound and reliable nuclear models which, in turn, is accurate enough to compete with more phenomenological highly-parametrized models in the reproduction of experimental data. The selection criterion of the adopted model is fundamental, since most of the nuclear ingredients in rate calculations need to be extrapolated in an energy and mass domain out of reach of laboratory measurements, where parametrized systematics based on experimental data can fail drastically. Global microscopic approaches have been developed for the last decades and are now more or less well understood. However, they are almost never used for practical applications, because of their lack of accuracy in reproducing experimental data, especially when considered globally on a large data set. Different classes of nuclear models can be contemplated according to their reliability, starting from local parametric systematics up to global microscopic approaches. We find in between these two extremes, approaches like the classical (e.g liquid drop, droplet), semi-classical (e.g Thomas-Fermi), macroscopic-microscopic (e.g classical with microscopic corrections), semi-microscopic (e.g microscopic with phenomenological corrections) and fully microscopic (e.g mean field, shell model, QRPA) approaches. In a very schematic way, historically, the higher the degree of reliability, the less accurate the model used to reproduce the bulk set of experimental data. The classical or phenomenological approaches are highly parametrized and therefore often successfull in reproducing experimental data, or at least much more accurate than microscopic calculations. The low accuracy obtained with microscopic models mainly originates from computational complications making the determination of free parameters by fits to experimental data difficult and computerwise time-consuming. Nowadays, microscopic models can be tuned at the same level of accuracy as the phenomenological models, and therefore could replace the phenomenological inputs little by little in practical applications. In Sect. 2, the major nuclear ingredients of relevance in reaction cross section calculations are recalled, while in Sect. 3, the IAEA recommendations for such nuclear input parameters are described. The new developments towards more microscopic approaches are described in Sect. 4 where the Brussels Nuclear Astrophysics Library, known as BRUSLIB is introduced. A special attention is paid to the recent results obtained in the determination of ground state properties within the

Hartree-Fock-Bogoliubov (HFB) approach.

REACTION MODELS

As far as reaction on heavier nuclei are concerned, most of the low-energy cross section calculations for practical applications can be derived within the statistical model of Hauser-Feshbach[2] [1, 2]. Such a model makes the fundamental assumption that the capture process takes place with the intermediary formation of a compound nucleus in thermodynamic equilibrium. The energy of the incident particle is then shared more or less uniformly by all the nucleons before releasing the energy by particle emission or γ-de-excitation. The formation of a compound nucleus is usually justified by assuming that the level density in the compound nucleus at the projectile incident energy is large enough to ensure an average statistical continuum superposition of available resonances. The statistical model has proven its abitility to predict cross sections accurately. However, this model suffers from uncertainties stemming essentially from the predicted nuclear ingredients describing the nuclear properties of the ground and excited states, and the strong and electromagnetic interaction properties. Clearly, the knowledge of the ground state properties (masses, deformations, matter densities) of the target and residual nuclei is indispensable. When not available experimentally, this information has to be obtained from nuclear mass models. The excited state properties have also to be known. Experimental data may be scarce above some excitation energy, especially for nuclei located far from the valley of nuclear stability. This is why frequent resort to a level density prescription is mandatory. In the Hauser-Feshbach formalism, the probability for particle emission are calculated by solving the Schrödinger equation with the appropriate optical potential for the particle-nucleus interaction. Finally, the electromagnetic de-excitation of the compound nucleus is calculated assuming the dominance of dipole E1 and M1 transitions. Phenomenological, as well as microscopic models are available for each of these ingredients, as described in the next sections.

IAEA RECOMMENDATIONS

The increase use of nuclear reaction theory for predicting cross sections, spectra and angular distributions has led the IAEA to initiate a coordinated research project to develop a library of evaluated and tested nuclear-model input parameters. This first Reference Input Parameter Library (RIPL) for theoretical calculations of reaction cross section published in 1998 [4] summarizes the knowledge on input parameters and recommends the most appropriate models for practical applications. A second coordinated effort was initiated in 1998 and completed in 2002 to further test, update, validate and improve the RIPL-1 compilation. The new library [5] includes a large amount of theoretical re-

---

[2] Direct reactions are also known to play an important role for light, closed shell or heavy exotic neutron-rich systems for which no resonant states are available. This reaction model will not be discussed here; the readers are referred to [3] for further details.

sults suitable for nuclear reaction calculations as well as computer codes for parameter retrieval, determination and use. The Library is targeted at users of nuclear reaction codes interested nuclear applications dealing with incident or outgoing light particles (neutrons, protons, deuterons, tritons, ^{3}He, ^{4}He and photons) up to energies of about 100 MeV. The major content of the RIPL-2 library [5] correponds to

- ground state properties (i.e mainly nuclear masses, but also deformations and nuclear densities) taken either from the experimental compilation of [6] or from two distinct mass formulas, namely the "finite-range droplet model" (FRDM) [7] and the microscopic HFB-2 mass model [8]. The experimental compilation of quadrupole deformation [9] complements the library with the FRDM and HFB deformation parameters. Finally, the HFB-2 nuclear densities are also provided in a table format.
- discrete level schemes (with spins, parities, γ-transition branchings and conversion coefficients) mainly extracted from the Evaluated Nuclear Structure Data File (ENSDF) [10].
- average neutron resonance parameters, i.e the average spacing of resonances required to normalize the level density at the neutron separation energy, the neutron strength function for the optical model potential renormalization at low energies and the average radiative width for the γ-ray strength function.
- an extensive compilation of optical model parameters for incident particles from neutrons to ^{4}He, as well as global phenomenological [11] and microscopic [12] nucleon-nucleus potentials.
- (total as well as partial) level densities within distinct approaches (Constant-Temperature, Back-Shifted Fermi Gas and Microscopic HFBCS models [13]) matching the cumultive number of experimental states at low energies and the average neutron resonance spacing at the neutron separation energy.
- γ-ray strength functions as derived from the experimental giant dipole resonance parameters [14] within the analytical standard or refined temperature-dependent Lorentzian formalism [15]. A compilation of tabulated γ-ray strength obtained within the HFBCS+QRPA model is also included [16].
- fission barriers saddle point deformations from experimental [17] or approximate mean field calculations [18]; the corresponding nuclear level densities at the saddle points determined from the same HFBCS statistical model used for the equilibrium deformation [13] are given in a table format.

A third phase of the RIPL project has now started and should be complete by 2006. This will mainly extend and improve the RIPL-2 library for reaction cross sections involving exotic nuclei with energies beyond 100 MeV and with an emphasis on charged projectiles. A special attention will be paid to the consideration of accurate microscopic models, as described in the next section.

GLOBAL MICROSCOPIC MODELS

BRUSLIB: the Brussels nuclear library for astrophysics applications

As explained in Sect. 1, it is of particular importance for specific applications, and most particularly for nuclear astrophysics applications to develop microscopic models that are accurate enough to compete with parametric approaches in the reproduction of experimental data. The microscopic character of such models ensure a reliable estimate of the various properties for experimentally unknown regions of the nuclear chart and at energies not reachable in the laboratory conditions. For such a purpose, the Brussels nuclear library for astrophysics applications, known as BRUSLIB [19], has been constructed to provide a complete set of reaction cross sections based exclusively on microscopic nuclear input parameters. Such models include

- for the ground state properties, the HFB approach (see below);
- for the nuclear level density, the HFBCS statistical approach corresponding to the exact result that the analytical BSFG-like model tries to reproduce. This approach has the advantage of treating in a natural way shell, pairing and deformation effects on all the thermodynamic quantities. All details can be found in [13];
- for the γ-ray strength, the HFB+QRPA approach which shows clear deviation from the well-known Lorentzian-like formula for exotic neutron rich nuclei. Details can be found in [16, 20].
- for the nucleon-nucleus optical potential, the Brückner–Hartree–Fock approximation based on the Reid's hard core nucleon–nucleon interaction and renormalized on scattering and reaction observables [12, 21];
- for the fission properties, the ETFSI compilations [18] as well as the HFB predictions [22] obtained within the same model as the nuclear masses. In addition to the fission barriers and deformations at the saddle points, the nuclear level densities at these saddle points have also been coherently estimated within the above-described HFBCS statistical model [5, 23].

Such microscopic input data, as well as the corresponding Hauser-Feshbach reaction rates are made available to the scientific community in a table format in BRUSLIB [19]. Some of these models are detailed in the present volume [20, 21, 22, 23] and are consequently not discussed here. We will restrict ourselves to present the recent effort made to derive improved global microscopic mass formulas.

Finally note that BRUSLIB also includes a compilation of experimental reaction rates (among others, [24, 25, 26]) as well as experimental and theoretical β-decay rates [27, 28, 29, 30, 31] of astrophysical interest. All these experimental and theoretical rates are easily accessible through the BRUSLIB *Nuclear Network Generator* (also called Netgen) [32], a tool to generate nuclear-reaction rates for given networks and temperature grids.

Microscopic mass predictions

Until recently the atomic masses were calculated on the basis of one form or another of the liquid-drop model, the most sophisticated version being the FRDM model [7]. Despite the great empirical success of this formula (it fits the 1888 $Z \geq 8$ masses with an rms error of 0.689 MeV), it suffers from major shortcomings, such as the incoherent link between the macroscopic part and the microscopic correction, the instability of the mass prediction to different parameter sets, or the instability of the shell correction. For astrophysics applications, there is an obvious need to develop a mass formula that is more closely connected to the basic nuclear interaction.

It was demonstrated very recently [33, 34] that Hartree-Fock (HF) calculations in which a Skyrme force is fitted to essentially all the mass data are not only feasible, but can also compete with the most accurate droplet-like formulas available nowadays. Such HF calculations are based on the conventional Skyrme force of the form

$$v_{ij} = t_0(1+x_0 P_\sigma)\delta(\mathbf{r}_{ij}) + t_1(1+x_1 P_\sigma)\frac{1}{2\hbar^2}\{p_{ij}^2 \delta(\mathbf{r}_{ij}) + h.c.\}$$
$$+ t_2(1+x_2 P_\sigma)\frac{1}{\hbar^2}\mathbf{p}_{ij}.\delta(\mathbf{r}_{ij})\mathbf{p}_{ij} + \frac{1}{6}t_3(1+x_3 P_\sigma)\rho^\gamma \delta(\mathbf{r}_{ij})$$
$$+ \frac{i}{\hbar^2}W_0(\sigma_i+\sigma_j).\mathbf{p}_{ij} \times \delta(\mathbf{r}_{ij})\mathbf{p}_{ij} \quad , \tag{1}$$

and a δ-function pairing force acting between like nucleons,

$$v_{pair}(r_{ij}) = V_{\pi q}\left[1-\eta\left(\frac{\rho}{\rho_0}\right)^\alpha\right]\delta(r_{ij}) \quad , \tag{2}$$

where ρ is the density and ρ_0 the saturation value value of ρ. A density-independent ($\eta = 0$) zero range pairing force was originally adopted with a strength parameter $V_{\pi q}$ allowed to be different for neutrons and protons, and also to be slightly stronger for an odd number of nucleons ($V_{\pi q}^-$) than for an even number ($V_{\pi q}^+$).

The first competing HFBCS mass table (in which the pairing interaction is treated in the BCS approximation) was obtained with the MSk7 Skyrme and pairing parameters which were determined by fitting to the full data set of 1719 $A \geq 36$ masses [6] with a final rms error of 0.702 MeV. Lately [8, 35], a new Skyrme force has been derived on the basis of HF calculations with pairing correlations taken into account in the Bogoliubov approach, using a density-independent δ-function pairing force. The rms error with respect to the masses of all the 2135 measured nuclei of the 2001 Audi & Wapstra compilation [36] with $Z, N \geq 8$ is 0.674 MeV [8]. The quality of the new predictions is similar to the one obtained with HFBCS. The complete mass table, HFB-2, has been constructed, giving all nuclei lying between the two drip lines over the range $Z, N \geq 8$ and $Z \leq 120$. A comparison between HFB and HFBCS masses shows that the HFBCS model is a very good approximation to the HFB theory provided both models are fitted to experimental masses. The extrapolated masses never differ by more than 2 MeV below $Z \leq 110$. The reliability of the predictions far away from the experimentally known region, and in particular towards the neutron drip line, is however increased thanks to the improved Bogoliubov treatment of the pairing correlations.

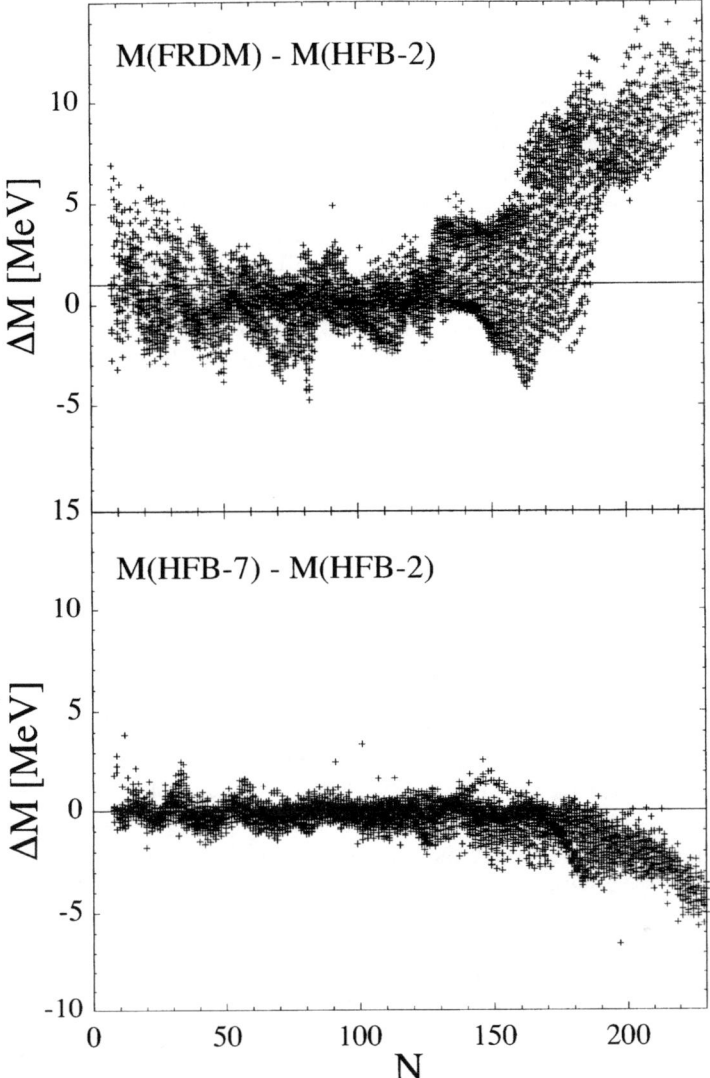

FIGURE 1. Differences between the HFB-2 masses and the FRDM (upper panel) or HFB-7 (lower panel) masses as a function of the neutron number N for all nuclei with $8 \leq Z \leq 110$ lying between the proton and neutron driplines.

Despite the success of the HFB-2 mass formula, it should in no way be regarded as definitive. For this reason, a series of studies of possible modifications to the basic force model and to the method of calculation were initiated all within the HFB framework [37, 38]. The most obvious reason for making such modifications would be to improve the data fit, but there is also a considerable interest in being able to generate different mass formulas even if no significant improvement in the data fit is obtained, since, in the first place, it is by no means guaranteed that mass formulas giving equivalent data fits will extrapolate in the same way out to the neutron-drip line: the closer that such mass formulas do agree in their extrapolations the greater will be our confidence in their reliability. But there is another reason to study different HFB mass models, and that concerns the fact that masses are not the only property of highly unstable nuclei that one might wish to determine by extrapolation from measured nuclei. An understanding of the r-process nucleosynthesis, in particular, requires also a knowledge of the fission barriers, β-decay strength functions, giant dipole resonances, nuclear level densities, etc., of highly unstable nuclei, and it may be that different models that are equivalent from the standpoint of masses may still give different results for other properties. Our intention to develop different HFB mass models is thus motivated also by the quest for a universal framework within which all the different nuclear aspects of the r-process can be treated.

For this reason, a set of additional 6 new mass tables, referred to as HFB-3 to HFB-8 were designed and the sensitivity of the mass fit and extrapolations towards the neutron dripline analysed. These new tables consider modified parametrizations of the effective interaction. In particular HFB-3,5,7 [37, 38] are obtained with a density dependence of the pairing force as inferred from the calculations of the pairing gap in infinite nuclear matter at different densities performed by Garrido et al. [39] using a "bare" or "realistic" nucleon-nucleon interaction (corresponding to $\eta = 0.45$ and $\alpha = 0.47$ in Eq. 2). For the mass tables HFB-4,5 (HFB-6,7) [38], a low isoscalar effective mass $M_s^* = 0.92$ ($M_s^* = 0.8$) is adopted as prescribed by microscopic (Extended Brückner-Hartree-Fock) nuclear matter calculations [40]. The last improvement considered in the HFB-8 model restores the particle number symmetry by applying the projection-after-variation technique to the HFB wave function [22].

The new mass table reproduce the experimental masses with the same degree of accuracy as the HFB-2 mass model. In addition, it is found that globally the extrapolations out to the neutron dripline of all these different HFB mass formulas are essentially equivalent. Figure 1 compares the HFB-2 and HFB-7 masses for all nuclei with $8 \leq Z \leq 110$ lying between the proton and neutron driplines. Although HFB-2 and HFB-7 are obtained from significantly different Skyrme forces (HFB-2 is characterized by an density-independent pairing force and an effective isoscalar mass $M_s^* = 1.04$, while HFB-7 has a density-dependent pairing force and $M_s^* = 0.8$), deviations smaller than about 2 MeV are obtained for all nuclei with $Z \leq 82$. In contrast, higher deviations are seen between HFB-2 and FRDM masses (Fig. 1), especially for superheavy nuclei. For lighter species, the mass differences remain below some 5 MeV, but locally the shell and deformation effects can differ significantly. Most interestingly, the HFB mass formulas show a weaker (though not totally vanishing) neutron-shell closure close to the neutron drip line with respect to droplet-like models as FRDM.

Although complete mass tables have now been derived within the HFB approach, further developments that could have an impact on mass extrapolations towards the neutron drip line need to be studied. Most particularly, all HFB mass fits show a strong pairing effect that need to be re-estimated within the renormalization procedure of [41]. Rotational as well as vibrational correlations need to be studied in more details. More fundamentally, mean field models need to be improved, so that all possible observables (such as giant dipole, Gamow-Teller excitations, nuclear matter properties, fission barriers) can be estimated coherently on the basis of one unique effective force. These various nuclear aspects are extremely complicate to reconcile within one unique framework and this quest towards universality will most certainly be the focus of fundamental nuclear physics research for the coming decade.

Concomitantly, future shell model calculations (e.g in the quantum Monte-Carlo approach) will certainly provide further fundamental insight on the nuclear properties of exotic nuclei. In particular, shell model calculations have shown that the spin-isospin dependent part of the nucleon–nucleon interaction in nuclei could lead to a change of the magic numbers $N = 8, 20$ in the exotic neutron-rich region [42]. These effects are probably underestimated in Skyrme Hartree-Fock calculations because the interaction is truncated to be of the δ-function type. Future global shell-model calculations will certainly reveal interesting properties of heavier exotic neutron-rich nuclei.

CONCLUSIONS

Many nuclear applications involve a large number of unstable nuclei and therefore require the use of global approaches. The extrapolation to exotic nuclei or energy ranges far away from experimentally known regions constrains the use of nuclear models to the most reliable ones. Recently, the RIPL project has achieved an important step of collecting, evaluating and recommending the vast amounts of various nuclear parameters of relevance in cross section calculations. The next step consists in deriving these nuclear parameters on the basis of reliable and accurate microscopic models. This step has been initiated with the Brussels nuclear astrophysics library known as BRUSLIB which includes regularly-updated microscopic models of ground state properties, nuclear level densities, γ-ray strength functions, fission barriers, as well as reaction rates determined on the basis of these microscopic input parameters.

REFERENCES

1. W. Hauser and H. Feshbach, *Phys. Rev.* **87**, 366 (1952).
2. J.A. Holmes, S.E. Woosley, W.A. Fowler and B.A. Zimmerman, *At. Nucl. Data Tables* **18**, 306-412 (2000).
3. S. Goriely, *Phys. Lett.* **B436**, 10-18 (1998).
4. Reference Input Parameter Library, IAEA-Tecdoc-1034 (1998). (also available at *http://iaeand.iaea.or.at/ripl*).
5. Reference Input Parameter Library, IAEA-Tecdoc, in press (2003). (also available at *http://ndsli01.iaea.org/ripl2/*).
6. G. Audi and A.H. Wapstra, *Nucl. Phys.* **A595**, 409-480 (1995).
7. P. Möller, J.R. Nix, W.D. Myers and W.J. Swiatecki, *At. Nucl. Data Tables* **59**, 185-381 (1995).
8. S. Goriely, M. Samyn, P.-H. Heenen, J.M. Pearson and F. Tondeur *Phys. Rev.* **C66**, 024326 (2002).
9. S. Raman, C.W. Nestor, JR. Tikkanen and P. Tikkanen, *At. Nucl. Data Tables* **78**, 1-128 (2001).
10. Evaluated Nuclear Structure Data File, National Nuclear Data Center, BNL, USA
11. A.J. Koning and J.P. Delaroche, *Nucl. Phys.* **A713**, 231-310 (2003).
12. E. Bauge, J.P. Delaroche and M. Girod, *Phys. Rev.* **C63**, 024607 (2001).

13. P. Demetriou and S. Goriely, *Nucl. Phys.* **A695**, 95-108 (2001).
14. S.S. Dietrich and B.L. Berman, *At. Nucl. Data Tables* **38**, 199 (1988).
15. V.A. Plujko, *Nucl. Phys.* **A649**, 209c-213c (1999).
16. S. Goriely and E. Khan, *Nucl. Phys.* **A706**, 217-232 (2002).
17. G.N. Smirenkin, IAEA-Report INDC(CCP)-359 (1993).
18. A. Mamdouh, J.M. Pearson, M. Rayet and F. Tondeur, *Nucl. Phys.* **A644**, 337-358 (1998).
19. BRUSLIB, The Brussels nuclear library for astrophysics applications (2003), available at *http://www-astro.ulb.ac.be*.
20. E. Khan, et al., this volume (2003).
21. E. Bauge, et al., this volume (2003).
22. M. Samyn and S. Goriely, this volume (2003).
23. P. Demetriou, M. Samyn and S. Goriely, this volume (2003).
24. C. Angulo, M. Arnould, M. Rayet, et al., *Nucl. Phys.* **A656**, 3 (1999).
25. G.R.Caughlan and W.A. Fowler, *At. Nucl. Data Tables* **40**, 283 (1988).
26. Z.Y. Bao, H. Beer, F. Käppeler, et al., *At. Nucl. Data Tables* **75**, 1 (2000).
27. T. Horiguchi, T. Tachibana and J. Katakura, Chart of the Nuclides, Japanese Nuclear Data Committee and Nuclear Data Center (1996).
28. K. Takahashi and K. Yokoi, *At. Nucl. Data Tables* **36**, 375 (1987).
29. K. Langanke and G. Martinez-Pinedo, *Nucl. Phys.* **A673**, 481 (2000).
30. T. Tachibana, M. Yamada and N. Yoshida, *Prog. Theor. Phys.* **84**, 641 (1990).
31. I. Borzov and S. Goriely, *Phys. Rev.* **C62**, 035501 (2000).
32. A. Jorissen and S. Goriely, *Nucl. Phys.* **A688**, 508c-510c (2001).
33. F. Tondeur, S. Goriely, J.M. Pearson and M. Onsi, *Phys. Rev.* **C62**, 024308 (2000).
34. S. Goriely, F. Tondeur and J.M. Pearson, *At. Nucl. Data Tables* **77**, 311-381 (2000).
35. M. Samyn, S. Goriely, P.-H. Heenen, J.M. Pearson and F. Tondeur, *Nucl. Phys.* **A700**, 142-156 (2002).
36. G. Audi and A.H. Wapstra, private communication (2001).
37. M. Samyn, S. Goriely and J.M. Pearson, *Nucl. Phys.* **A725**, 69-81 (2003).
38. S. Goriely, M. Samyn, M. Bender and J.M. Pearson, *Phys. Rev.* **C**, in press (2003).
39. E. Garrido, P. Sarriguren, E. Moya de Guerra and P. Schuck, *Phys. Rev.* **C60**, 064312 (1999).
40. W. Zuo, I. Bombaci and U. Lombardo, *Phys. Rev.* **C60**, 024605 (1999).
41. A. Bulgac and Y. Yu, *Phys. Rev. Lett.* **88**, 042504 (2002)
42. T. Otsuka, R. Fujimoto, Y. Utsuno, B.A. Brown, M. Honma and T. Mizusaki, *Phys. Rev. Lett.* **87**, 082502 (2001).

Building better optical model potentials for nuclear astrophysics applications

Eric Bauge, Marc Dupuis

CEA DIF, Département de Physique Théorique et Appliquée, Service de Physique Nucléaire, BP 12, 91680 Bruyères-le-Châtel, France

Abstract.
In nuclear astrophysics, optical model potentials play an important role, both in the nucleosynthesis models, and in the interpretation of astrophysics related nuclear physics measurements. The challenge of nuclear astrophysics resides in the fact that it involves many nuclei far from the stability line, implying than very few (if any) experimental results are available for these nuclei. The answer to this challenge is a heavy reliance on microscopic optical models with solid microscopic physics foundations that can predict the relevant physical quantities with good accuracy. This use of microscopic information limits the likelihood of the model failing spectacularly (except if some essential physics was omitted in the modeling) when extrapolating away from the stability line, in opposition to phenomenological models which are only suited for interpolation between measured data points and not for extrapolating towards unexplored areas of the chart of the nuclides.

We will show how these microscopic optical models are built, how they link to our present knowledge of nuclear structure, and how they affect predictions of nuclear astrophysics models and the interpretation of some key nuclear physics measurements for astrophysics.

INTRODUCTION

Astrophysics is a great consumer of nuclear data at large, and more specifically on optical model potentials (OMPs). It is so because OMPs provide a representation of the direct interaction between a projectile and a target, and such a representation is needed in many nuclear astrophysics modelings. Moreover, the needs of nuclear astrophysics are specific in the sense that they mostly focus on lower energies ($E/A < 10$ MeV), and generally involve a large number of nuclides which can be very far off the stability line. These peculiarities lead to special constraints on the OMP like good low energy accuracy and suitability for extrapolating from stable nuclei towards the drip-lines.

First, let us detail the different uses for OMP derived quantities in nuclear astrophysics. There are two broad families of uses for OMP derived quantities : the first one is related to extracting nuclear structure information from scattering experiments off nuclei of astrophysical interest, the second family is more focused on getting cross sections and related quantities (transmission coefficients) to feed to nucleosynthesis models. Of course, as we will see later on, the cross sections for the second broad family of uses are sensitive to nuclear structure input, so that the nuclear structure of the nuclei is to be taken into account very carefully in any case.

While the first family of OMP uses is closely related to the traditional OMP uses in the nuclear physics community, the second family of uses presents some peculiarities.

For example, in many nuclear astrophysics models, nuclei close to the drip lines play an essential role (p-, and r-processes) whereas next to nothing is experimentally known about those nuclei. Moreover, nucleosynthesis models usually need cross sections for a large number of targets (all the nuclei in the r-, s- or p-process paths), challenging the OMPs to produce a large number of predictions for a large range of target nuclei, which can be very exotic. The obvious response to this challenge resides in using microscopic or semi-microscopic OMPs which can predict observables for systems that have never been experimentally measured, and need inputs only from nuclear structure models.

Finally, since transfer reactions can be extremely useful in nuclear astrophysics, for example to indirectly measure quantities difficult to access in a direct way (like measuring (p,γ) cross sections using (d,n) reactions), OMPs used in interpreting those transfer reactions are needed not only for nucleon projectiles but also for light complex projectiles (deuteron, tritons, ...).

In this paper we will first focus on the construction of our semi-microscopic nucleon-nucleus OMP. We will then show a few examples of calculations performed using this OMP. Then uncertainties related to the model parameters will be discussed followed by a discussion of the limitations of our model. The final section will describe our attempts to use our nucleon-nucleus OMP as a basis to build a deuteron-nucleus OMP usable for reactions off unstable nuclei.

NUCLEON-NUCLEUS OPTICAL MODEL POTENTIAL

Construction of a semi-microscopic OMP

In this section we are going to explain the construction of our semi-microscopic OMP [1, 2] based on the work of Jeukenne, Lejeune and Mahaux in the 70's [3, 4, 5, 6]. This OMP is built from a polynomial parameterization of the on the energy shell mass operator in nuclear matter calculated from the JLM g matrix. In [1, 2] the energy validity of this OMP in nuclear matter is extended to the 1 keV-200 MeV range, and phenomenological normalization factors are applied in order to maximize the agreement between calculations and a large body of experimental nucleon scattering and reaction data. In [2] those normalization factors are reworked into a Lane-consistent form [8] and more constraints ((p,n) reaction to the isobaric analog state) are used to optimize the normalization factor producing the following OMP in nuclear matter

$$U_{NM}(\rho, E) = \lambda_V(E)\left[V_0(\rho, \tilde{E}) \pm \lambda_{V1}(E)\alpha V_1(\rho, \tilde{E})\right] \quad (1)$$
$$+ i\lambda_W(E)\left[W_0(\rho, \tilde{E}) \pm \lambda_{W1}(E)\alpha W_1(\rho, \tilde{E})\right], \quad (2)$$

In (2), E is the incident projectile energy and $\tilde{E} = E - V_c$ is that energy shifted by the Coulomb potential V_c (for incoming protons only). λ_V, λ_W, λ_{V1}, and λ_{W1} are the real, imaginary, real isovector, and imaginary isovector potential depth normalization factors, respectively. The $V_0(\rho, E)$, $V_1(\rho, E)$, $W_0(\rho, E)$, and $W_1(\rho, E)$ quantities are

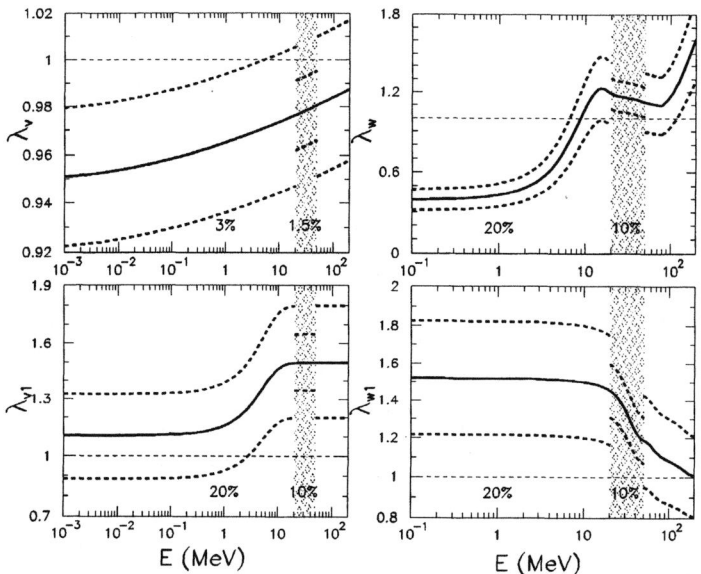

FIGURE 1. Optimized energy dependent potential depth normalization factors λ_V, λ_W, λ_{V1}, and λ_{W1} (solid lines) along with their associated uncertainties bands (dotted lines).

calculated as shown in section II of Ref. [1].[1] Fig. 1 displays the final optimized energy dependent potential depth normalization factors λ_V, λ_W, λ_{V1}, and λ_{W1} (solid lines), along with their associated uncertainties bands (dotted lines). All of the above defines our semimicroscopic OMP in nuclear matter.

In order to perform calculations for finite nuclei (as opposed to nuclear matter) a local density approximation (LDA) must be used. We use the improved LDA prescription given in [6] for spherical OMP. This LDA reads:

$$U_{ILDA}(r, E) = (t\sqrt{\pi})^{-3} \int \frac{U_{NM}(\rho(r'), E)}{\rho(r')} e^{(-|\vec{r}-\vec{r}'|^2/t_r^2)} \rho(r') d^3 r'. \tag{3}$$

Equ. (3) describes the folding of the OMP in nuclear matter $U_{NM}(\rho), E)$ with the radial density $\rho(r')$ using a Gaussian widening form factor. This Gaussian widening form factor can be understood as an effective way of restoring for finite nuclei some of the non-locality that was present in the nuclear matter OMP before being forced on the energy shell. This equation also shows how nuclear structure derived information $(\rho(r))$ enters OMP calculations. This influence of nuclear structure on the OMP is very important since it shows how, by doing nucleon scattering experiments, one actually probes the nuclear radial density in a way that is very similar to probing charge density

---

[1] In table I of [1] the value of D was mistakenly given as 625 MeV^2, the value D=126.25 MeV^2 should be taken instead. This correct value is used in the published MOM code [7].

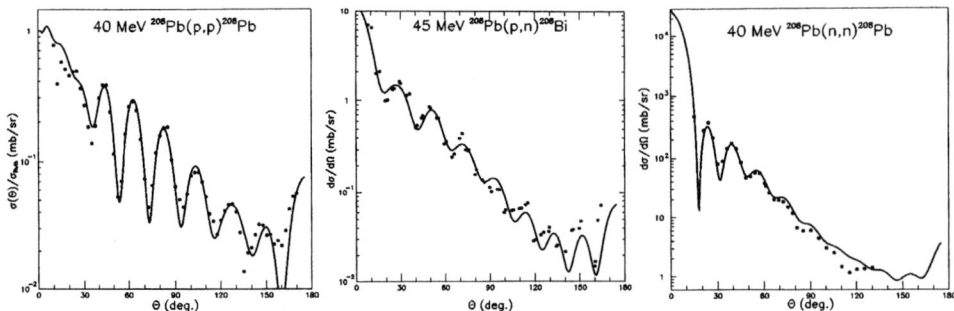

FIGURE 2. Comparisons between calculated and measured cross sections for the reactions 40 MeV ^{208}Pb(p,p)^{208}Pb, 45 MeV ^{208}Pb(p,n)^{209}Bi^{IAS}, and 40 MeV ^{208}Pb(n,n)^{208}Pb.

by electron scattering, the main difference being that the nuclear interaction is much less known and much more difficult to handle than the Coulomb interaction.

For our studies we use structure calculations based on the Gogny D1S [10] interaction along with mean-field (Hartree-Fock and Hartree-Fock-Bogoliubov) and beyond the mean-field (Random Phase Approximation and Generator Coordinate Method) frameworks [11, 12, 13]. These methods have been shown to produce accurate descriptions of a large range of nuclei and observables.

The central nuclear part of the OMP described above is supplemented with a nuclear spin-orbit term as described in [1] and a Coulomb term (in case of proton projectile) and given as input to the ECIS code [9] to calculate cross-sections.

In the case of scattering off target nuclei that cannot be modeled using the spherical OMP, scattering is treated using the coupled-channel approach, with deformed OMPs calculated with appropriate variations of the improved LDA shown in (3) for rigidly rotating [14, 15] or vibrating target nuclei [16, 17]. The successes encountered in modeling such a wide range of reactions is a testimonial of the quality of our semi-microscopic OMP calculation procedure and of its ingredients, namely the mean-field (and beyond) nuclear structure calculations using the Gogny D1S force.

Scattering calculations

Let us show a few comparisons between our OMP calculations and experimental data. First, to illustrate the Lane-consistency of our OMP Fig. 2 shows how (p,p), (p,n)^{IAS} and (n,n) scattering are accounted for simultaneously for a ^{208}Pb target. The good agreement exhibited by all three observable shows that the isoscalar and isovector terms are well under control in the 20-50 MeV energy range where all three types of data ((p,p), (p,n)^{IAS} and (n,n)) are available. Actually, with the advent of the new high quality measurement of the n+^{208}Pb differential cross-section at 96 MeV which was successfully accounted for by our OMP [18], the interval where our OMP is tested against all three types of data could now be extended to 20-100 MeV. In [1, 2, 14, 15, 16, 17, 18, 19] our OMP is tested against a large range of observables, for many

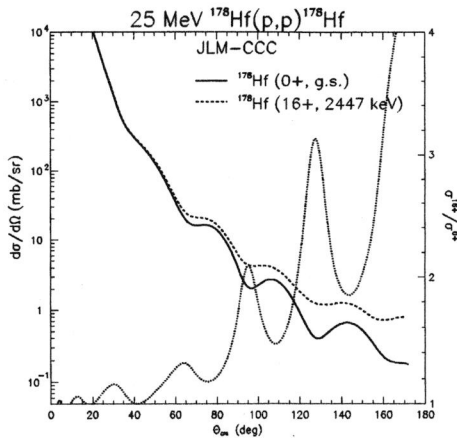

FIGURE 3. Elastic scattering cross section for the ^{178}Hf(p,p) reaction off the ground state (solid line), or 16$^+$ excited state (dashed line) at 25 MeV incident energy. The dotted line represents the ratios of the cross sections.

target nuclei, showing the predictive character as well as the limitations (see section "Limitations" for details) of our approach.

The microscopic character of our OMP allows for calculations to be performed not only for nuclei far from stability, but also for nuclei in a state different from the ground state. Fig. 3 shows the influence of the target state on proton scattering off ^{178}Hf. Such difference between predictions can be of importance, for example when nucleosynthesis occurs at very high temperature (low-lying, long-lived nuclear state excited by being in thermodynamic equilibrium with a high temperature medium) and/or very high flux (several successive nuclear reactions taking place before decay occurs), like in super novae explosion modeling.

Uncertainties on calculated quantities

In [2], not only optimized values of the energy-dependent normalization factors (λ_V, λ_W, λ_{V1}, and λ_{W1}) were given, but uncertainties ($\Delta\lambda$) on these normalization factors were also provided and shown as dotted lines in Fig. 1. The next logical step is to evaluate the influence of these uncertainties on the calculated cross sections. To do so, we did Monte-Carlo sampling of the (λ_V, λ_W, λ_{V1}, λ_{W1}) space assuming a uniform probability distribution between $\lambda - \Delta\lambda$ and $\lambda + \Delta\lambda$, and null probability outside this range. This sampling of the parameter space was then processed to produce cross-section distributions such as the ones shown in Fig. 4 for neutron and proton incident on ^{208}Pb. These distributions were analyzed, and their mean and standard deviation were then calculated. This process was repeated on a mesh of incident energies in order to extract the energy dependence of the standard deviation of the cross-sections as shown in Figs. 4.

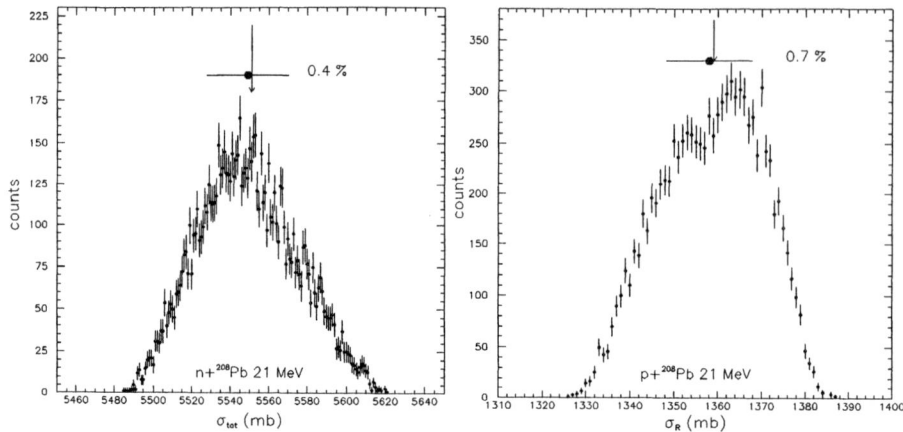

FIGURE 4. Probability distributions of n+^{208}Pb and p+^{208}Pb cross sections at 21 MeV, obtained by sampling the (λ_V, λ_W, λ_{V1}, and λ_{W1}) parameter space using the Monte-Carlo method. The horizontal error bars represent the means, and standard deviations of the distributions, and the arrows represent the cross-sections calculated using the optimized parameter values (see Fig. 1). The standard deviation of the distribution is also given as a percentage of the mean value.

Analyzing these figures show that the relative uncertainties tend to grow with decreasing incident energies. This observed trend is the manifestation of the increased sensitivity of the OMP predictions to its parameters at low energy where small changes of the parameters produce larger changes in the observables. It should be noted that, for the ^{208}Pb case shown here, calculations are performed using the spherical optical model, that exhibits stronger sensitivity to parameter variations than do deformed coupled-channel calculations. For coupled-channel calculations the coupling of the equations stabilizes the system lowering the sensitivity to the potential depth parameters λ_V, λ_W, λ_{V1}, and λ_{W1}. The price to pay for such a lower sensitivity is the introduction of another parameter: the deformation of the target nucleus (assuming rotational model). Yet, even including the extra uncertainties introduced by the deformation parameter, the coupled-channel calculations are shown to be less sensitive to parameter variations than spherical optical model calculations.

Moreover, low energy incident proton reactions seem to be especially sensitive to variations of the parameters, making it very difficult to predict S-Factors with the desired accuracy.

The above uncertainty information and the associated parameter confidence ranges can be very precious when using our optical model for applications. For example, if a disagreement with experimental data stays within the calculation uncertainty range, the calculation and data are actually compatible. It also means that the calculation can be adjusted to match the data while staying within the parameters confidence range : this is very useful when producing evaluated data files for applications since the OMP parameters can be fine-tuned to improve the match with either microscopic or macroscopic (integral) experimental data while retaining the predictive qualities of a microscopic model.

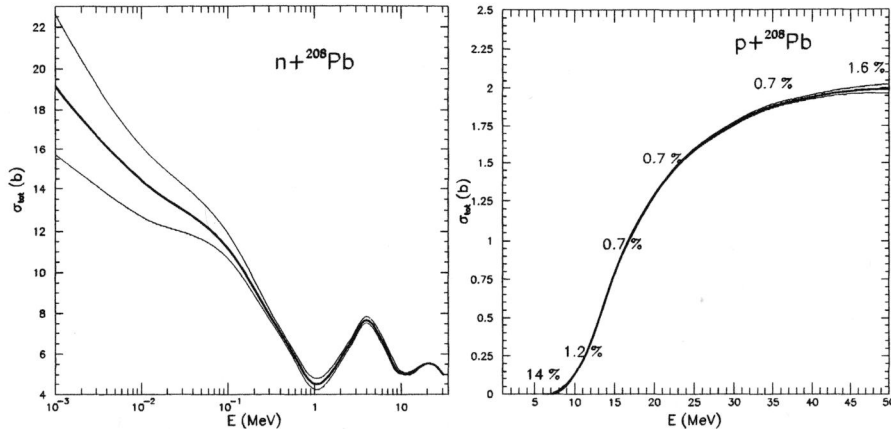

FIGURE 5. Calculated total and reaction cross-sections for n+^{208}Pb and p+^{208}Pb, respectively. Associated "error ranges" (standard deviation of the cross section distributions shown in Fig. 4) are also plotted as envelopes. For incident protons, standard deviations of the cross section distributions are also given as a percentage of their mean value.

Limitations

The OMP described above produces calculations that account for the available experimental data with good accuracy *within its field of validity*. We have found this OMP to be for the 1 keV $\leq E \leq$ 200MeV projectile energy range, and for the $30 \leq A$ mass range. The paper [19] shows an illustration of the limitations of our OMP when extrapolating towards lighter nuclei for incident protons.

Moreover, since the mass operator of JLM is calculated in nuclear matter, it assumes a continuous and unlimited spectrum of excited states, which is an acceptable approximation for nuclei that are not too exotic. However, when the excited levels spectrum become very sparse (like for halo nuclei or for nuclei very close to the drip lines) the approximation of using a nuclear matter OMP in the LDA becomes questionable.

Finally, the uncertainties described in the section on uncertainties do not allow for discriminating between arbitrarily close data : the width of the envelope shown in Fig. 5 can be large enough to encompass data with very different physical meaning (especially at low energy where this width is larger). This might be a problem when very high precision predictions are required in applications.

DEUTERON-NUCLEUS OPTICAL MODEL POTENTIAL

Besides elastic scattering, transfer reaction can provide invaluable information on the single particle structure of nuclei. By putting the transfered particle in a well defined state reconstructed from the kinematics of the scattered target and projectile remnants, positions and occupations of single particle states can be infered. In order to extract structure information from transfer reaction measurements, a DWBA analysis is usually

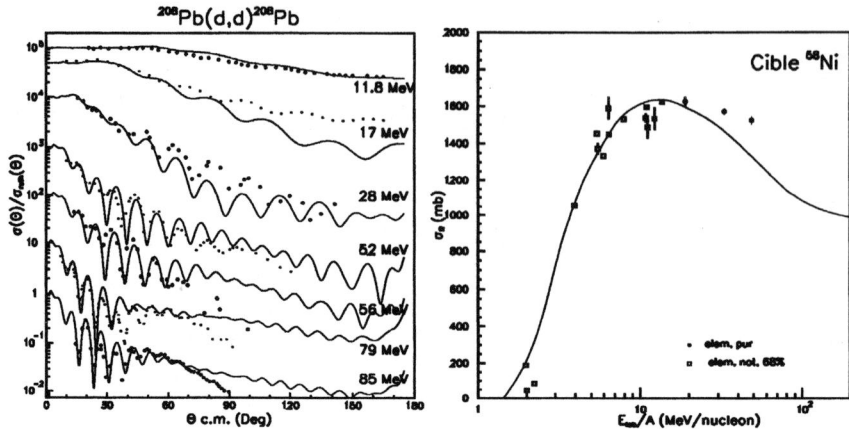

FIGURE 6. Calculated elastic differential cross-sections for the d+^{208}Pb system, and reaction cross sections for the d+^{58}Ni system compared with experimental data

performed. This analysis uses among other ingredients the distorted waves (wave functions in the entrance and exit channels). In the case of deuteron induced transfer reactions ((d,p), and (d,n)), while the exit channel is well described by nucleon-nucleus optical potential, the entrance channel is much less well known, especially if the target nucleus lies far from the stability line, where global deuteron-nucleus OMP are likely to be inappropriate , and very few experimental data is available to calibrate phenomenological OMP. This prompted us to try to build a deuteron-nucleus OMP using our nucleon-nucleus OMP as a base.

Using the double folding procedure of [20]

$$U_d(r, E) = \int [U_n(\vec{r} - \vec{s}) + U_p(\vec{r} + \vec{s})] \Psi^2(\vec{s}) d\vec{s} \qquad (4)$$

with Ψ the S component of the deuteron wave function from [21], and U_n and U_p, respectively the neutron-nucleus and proton-nucleus OMPs described above.

In order to better match the experimental observables, phenomenological normalizations factors are applied to the OMP shown in (4). The real part of the central OMP was multiplied by 0.9 and the imaginary part by 0.92. Fig. 6 shows examples of deuteron elastic scattering and reaction cross sections calculated using the procedure shown above. For the differential cross sections (left panel), the agreement not excellent : while scattering to front angles is reasonablely well accounted for, scattering to large angles is not well predicted at low and high energies. However, the phasing of the angular distribution is well described above the Coulomb barrier. The reaction cross section (right panel) is better accounted for, up to at least 20 MeV/A where our model starts to underestimate the reaction cross section.

To improve the agreement between calculations and data it is possible to introduce a polarization potential that accounts for the breakup of the deuteron. Indeed, such a potential does improves the agreement with data, but it introduces new phenomenological parameters to be adjusted. A better way to proceed, would be to calculate such a potential from first principles in a similar way to what was suggested in [22].

In contrast to the nucleon-nucleus OMP, the study of the deuteron-nucleus potential is still a work in progress both from the theoretical and experimental point of view, since neither the theoretical models nor the relevant experimental data are available in sufficient quality and quantity to adjust the deuteron-nucleus OMP at the precision warranted for deuteron induced transfer reaction studies.

CONCLUDING REMARKS

Although this paper focuses on OMPs based on the JLM approach [4, 5, 6], there are numerous other approaches to the microscopic OMP, the most interesting being that using the Melbourne g-matrix [23]. This Melbourne interaction uses more of the nuclear structure information (one body density matrix vs radial density for JLMs), is non-local, completely microscopic (no adjusted parameters, and thus no uncertainties due to parameters), and contrarily to JLM approaches can be used for very light targets. On the other hand, the limited projectile energy range of this interaction (40-300 MeV) limits its usefulness for some astrophysical applications.

In this paper, we have tried to show how a (semi-)microscopic OMP can and should be used for astrophysical applications, and how the peculiarities of these applications are addressed by microscopic OMPs that can be calculated for any target. Moreover the *strong physical content* of microscopic OMPs and their reliance on predictive theoretical nuclear structure information, guarantees a minimal level of accuracy when used in the uncharted parts of the chart of nuclides. We have also attempted to show the limitations of our semi-microscopic OMP, one such limitation being the uncertainties stemming from the adjusted normalization factors. However, these uncertainties can also be used to give our model some flexibility in the interpretation of existing experimental data while not deteriorating its overall predictive quality.

Finally, we want to highlight the fact the OMPs for composite particles (deuterons, tritons, ^{3}He, alpha) are far from being as well known as nucleon-nucleus OMPs, and that, while there is a renewed interest for transfer reaction studies, at the moment, neither experimental data nor the theoretical models are adequate to properly interpret these new results. We have only started to work on bringing the predictivity of deuteron-nucleus OMPs to a level similar to that of nucleon-nucleus OMPs, but lots of work remains to be done in order to achieve such levels of reliability.

REFERENCES

1. E. Bauge, J.P. Delaroche, M. Girod, Phys. Rev. C **58**, 1118 (1998).
2. E. Bauge, J.P. Delaroche, M. Girod, Phys. Rev. C **63**, 024607 (2001).
3. J.P. Jeukenne, A. Lejeune, and C. Mahaux, Phys. Rep. **25C**, 83 (1976).
4. J.P. Jeukenne, A. Lejeune, and C. Mahaux, Phys. Rev. C **14**, 1391 (1974).
5. J.P. Jeukenne, A. Lejeune, and C. Mahaux, Phys. Rev. C **15**, 10 (1977).
6. J.P. Jeukenne, A. Lejeune, and C. Mahaux, Phys. Rev. C **16**, 80 (1977).
7. E. Bauge, MOM code, RIPL-2, http://ndsli01.iaea.org/RIPL-2/, IAEA, 2003.

8. A.M. Lane, Nucl. Phys. **35**, 676 (1962).
9. J. Raynal, CEA report No CEA-N-2772, 1994.
10. J.F. Berger, M. Girod, and D. Gogny, Comp. Phys. Comm. **63**, 365 (1990) and references therein.
11. J. Dechargé and D. Gogny, Phys. Rev. C **21**, 1568 (1980); J. Dechargé, L. Sips, and D. Gogny, Phys. Lett. **98B**, 229 (1981); J. Dechargé and L. Sips, Nucl. Phys. **A407**, 1 (1983).
12. J. Dechargé, M. Girod, D. Gogny, and B. Grammaticos, Proceedings of the Symposium on Perspectives in Electro- and Photo-Nuclear Physics, Nucl. Phys. **A358**, edited by A. Gérard and C. Samour (North-Holland, Amsterdam, 1981), p. 203c.
13. J. Libert, M. Girod, and J.P. Delaroche, Phys. Rev. C **60**, 054301 (1999).
14. E. Bauge, J.P. Delaroche, M. Girod, G. Haouat, J. Lachkar, Y. Patin, J. Sigaud, and J. Chardine, Phys. Rev. C **61**, 034306 (2000).
15. F.S. Dietrich *et al.*, Phys. Rev C **67**, 044606 (2003).
16. F. Maréchal *et al.*, Phys. Rev. C **60**, 034615 (1999).
17. H. Sheit *et al.*, Phys. Rev. C **63**, 014604 (2000).
18. J. Klug *et al.*, Phys. Rev. C **67**, 031601(R) (2003).
19. A. de Vismes *et al.*, Nucl. Phys **A 706**, 295 (2002).
20. S. Watanabe, Nucl. Phys. **8**, 484 (1958); J. Raynal, CEAReport No CEA-R-2511, 1964.
21. NN-online, http://nn-online.sci.kun.nl/
22. G. Bencze, I. Szentpetery, Phys. Lett. **30B**, 446 (1969).
23. K. Amos, P. J. Dortmans, H. V. von Geramb, S. Karataglidis, and J. Raynal, Adv. in Nucl. Phys. Vol.25 (Plenum, New york, 2000) p. 275, and references cited therein.

Microscopic dipole strength predictions for neutron capture rates

E. Khan*, M. Samyn† and S. Goriely†

*Institut de Physique Nucléaire, IN_2P_3-CNRS, 91406 Orsay, France
†Institut d'Astronomie et d'Astrophysique, ULB - CP226, 1050 Brussels, Belgium

Abstract. Large-scale QRPA calculations of the E1-strength are performed on top of HFB calculations in order to derive the radiative neutron capture cross sections for the whole nuclear chart. The spreading width of the GDR is taken into account by analogy with the second-RPA (SRPA) method. It is shown that the present model allows to constrain the effective nucleon-nucleon interaction with the GDR data and to provide quantitative predictions of dipole strengths.

INTRODUCTION

About half of the nuclei with $A > 60$ observed in nature are formed by the so-called rapid neutron-capture process (or r-process) of nucleosynthesis, occurring in explosive stellar events. The r-process is believed to take place in environments characterized by high neutron densities ($N_n > 10^{20}$ cm^{-3}), so that successive neutron captures proceed into neutron-rich regions well off the β-stability valley forming exotic nuclei that cannot be produced and therefore studied in the laboratory. If the temperatures or the neutron densities characterizing the r-process are low enough to break the $(n,\gamma) - (\gamma,n)$ equilibrium, the r-abundance distribution depends directly on the neutron capture rates by the so-produced exotic neutron-rich nuclei [1]. The neutron capture rates are commonly evaluated within the framework of the statistical model of Hauser-Feshbach (although the direct capture contribution can play an important role for such exotic nuclei). This model makes the fundamental assumption that the capture process takes place with the intermediary formation of a compound nucleus in thermodynamic equilibrium. In this approach, the Maxwellian-averaged (n,γ) rate at temperatures of relevance in r-process environments strongly depends on the electromagnetic interaction, i.e the photon de-excitation probability. The well known challenge of understanding the r-process abundances thus requires that one be able to make reliable extrapolations of the E1-strength function out towards the neutron-drip line. To put the description of the r-process on safer grounds, a great effort must therefore be made to improve the reliability of the nuclear model. Generally speaking, the more microscopic the underlying theory, the greater will be one's confidence in the extrapolations out towards the neutron-drip line, provided, of course, the available experimental data are also well fitted.

Large scale prediction of E1-strength functions are usually performed using phenomenological Lorentzian models [1]. Several refinements can be made, such as the energy dependence of the width and its temperature dependence [1, 2, 3, 4] to describe all available experimental data. The Lorentzian GDR approach suffers, however, from shortcomings of various sorts. On the one hand, it is unable to predict the enhancement of the E1 strength at energies around the neutron separation energy demonstrated by various experiments, such as the nuclear resonance fluorescence. On the other hand, even if a Lorentzian-like function provides a suitable representation of the E1 strength for stable nuclei, the location of its maximum and its width remain to be predicted from some systematics or underlying model for each nucleus.

Recently an attempt was made to derive microscopically the E1-strength for the whole nuclear chart [5]. The dipole response was calculated with the Quasiparticle Random Phase Approximation (QRPA) on top of Hartree-Fock+BCS (HFBCS) description [6]. The only input of this approach was the Skyrme effective interaction injected in the HFBCS model. These microscopic calculations predicted the presence of a systematic low-lying component in the giant dipole resonance (GDR) for very neutron-rich nuclei. This low-lying component influences the neutron capture rate, especially if located in the vicinity of the neutron separation energy S_n. However, the pairing correlation in the BCS model was determined assuming a simple constant-gap pairing interaction. In addition, in the case of the highly neutron-rich nuclei that are of particular interest in the context of the r-process, the validity of the BCS approach to pairing is questionable, essentially because of the role played by the continuum of single-particle neutron states (see [7], and references therein). Therefore the impact of the newly-derived E1 strength functions on the cross section prediction could only be evaluated qualitatively. It was found that the radiative neutron capture cross sections by neutron-rich nuclei were systematically increased by the HFBCS+QRPA calculations [5] with respect to the one obtained using Lorentzian-like strength functions. Predictions with different forces have been compared, but no conclusions could be drawn regarding their intrinsic quality to predict the E1 strength. The final large-scale HFBCS+QRPA calculations performed in [5] were obtained on the basis of the Skyrme force denoted SLy4 [8].

In the present work we calculate the dipole strength with one of the most accurate and reliable microscopic model available to date, namely the Hartree-Fock-Bogoliubov (HFB) and QRPA models [9, 10].

HFB+QRPA CALCULATIONS

The long-term goal of microscopic models is to describe on the same ground a wide variety of nuclear structure properties (in particular, magicity and pairing correlations in open-shell nuclei) for both stable and exotic nuclei. The HFB and QRPA models allows to treat, in a self-consistent way, pairing effects on the ground state as well as collective excitations for nuclei ranging from the valley of stability to the drip-line. The QRPA considers nuclear excitation as a collective superposition of two quasiparticle states built on top of the HFB ground state [12]. This collective aspect of the excitation makes the QRPA an accurate tool to investigate the E1-strength function, in both closed and open shell nuclei.

The HFB calculations considered in the present work are fully detailed in [13, 14, 15, 16]. The derivation of the QRPA response and its numerical application are detailed in

[10]. The Skyrme force is used together with a pairing force of the form :

$$v_\pi(r_{ij}) = V_{\pi q}\left[1 - \eta\left(\frac{\rho}{\rho_0}\right)^\alpha\right]\delta(r_{ij}) \quad , \qquad (1)$$

In the present approach the strength parameter $V_{\pi q}$ is allowed to be different for neutrons and protons. Based on this Skyrme-HFB approach, a number of effective forces have been determined recently [14, 15, 16], the parameters of the underlying forces being fitted *exclusively* to all the 2135 available experimental masses [17], with some additional constraints regarding the stability of neutron matter and the incompressibility of nuclear matter. More details about these BSk2-7 forces can be found in [14, 15, 16].

In the previous HFBCS and QRPA calculation [5], the QRPA strength was folded by an arbitrary Lorentzian to generate the experimentally observed GDR width. We propose to describe the width on more microscopic grounds, by calculating it with a method inspired with the second-RPA (SRPA) framework [11]. In ref. [11], the SRPA allows to take into account the spreading width due to the 2p-2h excitations using $\Delta(E)$ and $\Gamma(E)$, the real and complex part of the self-energy, respectively. The energy dependent width $\Gamma(E)$ is calculated from the measured decay width of particle and hole states. The real part $\Delta(E)$ of the self-energy is obtained from $\Gamma(E)$ by a dispersion relation. This empirical way of determining $\Gamma(E)$ has the advantage of including, in principle, contributions from the excitation beyond 2p-2h [18]. The resulting resonance width can therefore be compared with experimental data, such as photoabsorption cross sections.

COMPARISON WITH EXPERIMENTAL DATA

To illustrate the comparison on with the experimental data, we will emphasize on the low-energy component of the E1 strength. It should be noted that the following results obtained with BSk7 are preliminary and checks are still under progress. For practical astrophysics applications, it is of first importance to describe the tail of the GDR at low energies, i.e around the neutron separation energy, as reliably as possible [1]. When dealing with γ-decay data, a temperature-dependent correction factor is traditionally introduced in the expression of the GDR width to take the collision of quasiparticles into account [3, 20, 21]. In order to guarantee the compatibility with photoabsorption data, we introduce in the damping procedure such a collision term by adding to the width $\Gamma(E)$ a temperature-dependent correction term as

$$\Gamma'(E) = \Gamma(E)[1 + \alpha\frac{4\pi T^2}{EE_{GDR}}] \qquad (2)$$

where T refers to the temperature of the absorbing state, E_{GDR} is the peak energy of the GDR and α a normalization constant. In all calculations performed in the present work, the temperature is derived from the microscopic statistical model of nuclear level densities [22]. As shown below, adopting $\alpha = 3$ gives excellent agreement with most of the available data.

We compare in Fig. 1 the QRPA predictions with the compilation of experimental E1-strength functions at low energies ranging from 4 to 8 MeV [23] for nuclei from

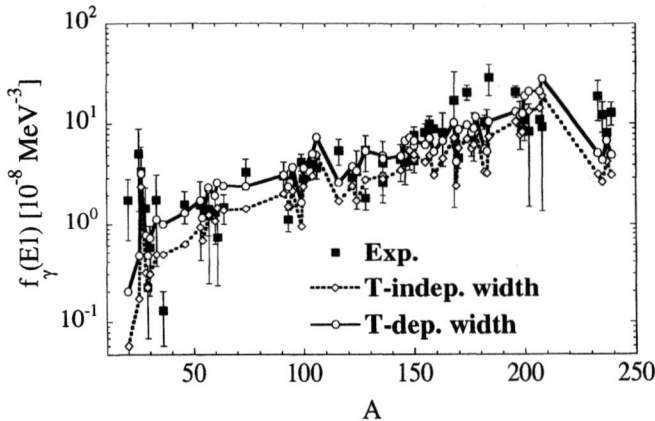

FIGURE 1. Comparison of the QRPA T-dependent and T-independent low-energy E1-strength functions with the experimental compilation [23] including resolved-resonance and thermal-captures measurements, as well as photonuclear data for nuclei from ^{25}Mg up to ^{239}U at energies ranging from 4 to 8 MeV.

^{25}Mg up to ^{239}U. The data set includes resolved-resonance measurements, thermal-captures measurements and photonuclear data. In a certain number of cases the original experimental values need to be corrected, typically for non-statistical effects, so that only values recommended by [23] are considered in Fig. 1. QRPA predictions are globally in good agreement with experimental data at low energies in the whole nuclear chart. The average and rms deviations, on the 62 experimental data have been estimated. The T-independent predictions underpredict the E1 strength by an average factor of 1.6, while on average the T-dependent formula (assuming $\alpha = 3$ in Eq. 2) is in better agreement with the data. The respective rms deviation factors are $f_{rms} = 2.6$ and 2.1 for the T-independent and T-dependent results.

APPLICATION TO THE RADIATIVE NEUTRON CAPTURE

The radiative neutron capture cross section is estimated within the statistical model of Hauser-Feshbach making use of the MOST code [24]. It should be noted that this version makes use of the nuclear ground state properties derived coherently from the same microscopic HFB method with the BSk7 Skyrme force [16]. It also benefits from the improved nuclear level density prescription based on the microscopic statistical model, also used to estimate the nuclear temperature in Eq. (2) [22]. The Maxwellian-averaged radiative neutron capture rate at a temperature $T = 1.5 \ 10^9$ K, typical of the r-process nucleosynthesis, obtained with the QRPA E1-strength are compared in Fig. 2 with those based on the Hybrid Lorentzian-type formula [1]. These rates are sensitive to the neutron capture cross section at incident energies around 130 keV, and therefore depend on the E1 strength in an narrow range of a few hundred keV around S_n. The temperature-dependent Hybrid formula corresponds to a generalization of the energy-

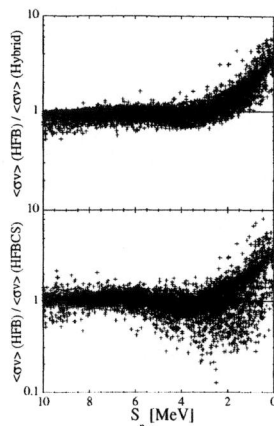

FIGURE 2. Upper panel: Ratio of the Maxwellian-averaged (n,γ) rate (at a temperature of $1.5\ 10^9$ K) obtained with the HFB+QRPA E1 strength to the one using the Hybrid formula [1] as a function of the neutron separation energy S_n for all nuclei with $8 \leq Z \leq 110$. Lower panel: Same as upper panel where the HFB+QRPA neutron capture rates are compared with the HFBCS+QRPA rates of [5].

and temperature-dependent Lorentzian formula including an improved description of the E1-strength function at energies below S_n as derived from [3]. The Hybrid E1 strength differs from the QRPA estimate not only in the location of the centroid energy, but also in the low-energy tail. No extra low-lying strength is included in the phenomenological Hybrid formula, but its temperature dependence increases the E1 strength at low energies and is responsible for its non-zero $E \to 0$ limit. The newly-derived E1 strength gives an increase of the rate by a factor up to 6 close to the neutron drip line. R-process nuclei characterized by $S_n = 3$ MeV are seen to have a neutron capture rate about at least twice faster than the one predicted with the phenomenological Hybrid formula. This is due to the shift of the GDR to lower energies compared with the usually adopted liquid-drop $A^{-1/3}$ rule, as well as to the appearance of some extra strength at low energies as explained above. Both effects tend to increase the E1 strength at energies below the GDR, i.e in the energy window of relevance in the neutron capture process. For less exotic nuclei, the QRPA impact is relatively small, differences being mainly due to the exact position of the GDR energy and the resulting low-energy tail. When compared to our previous HFBCS+QRPA predictions [5], the HFB+QRPA model gives larger neutron capture rates close to the neutron drip line, but lower rates for many of the $4 < S_n\ [\text{MeV}] < 2$ nuclei, as seen in Fig. 2 (lower panel). These differences justify the use of the HFB approach for exotic neutron-rich nuclei.

CONCLUSIONS

The E1-strength function is estimated with one of the most accurate and reliable microscopic model available to date, namely the Hartree-Fock-Bogoliubov (HFB) and QRPA models. The spreading width of the GDR is taken into account by analogy with the SRPA method. In addition to its reliability, it is shown that the HFB+QRPA model also gives accurate predictions and that globally it agrees fairly well with experimental data. The present HFB+QRPA model brings important improvement with respect to our previous HFBCS+QRPA model and can provide quantitative predictions of the dipole strength. Large-scale calculations of the E1-strength function are performed and preliminary results are used to estimate the radiative neutron capture rates of relevance for the r-process nucleosynthesis. A systematic increase of the reaction rates for exotic neutron-rich nuclei is found.

ACKNOWLEDGMENTS

M.S. and S.G. are FNRS Research Fellow and Associate, respectively. This work has been performed within the scientific collaboration (Tournesol) between the Wallonie–Bruxelles Community and France.

REFERENCES

1. S.Goriely, Phys. Lett. B436 (1998) 10.
2. C.M. McCullagh, M.L. Stelts, R.E. Chrien, Phys. Rev. C23 (1981) 1394.
3. S.G.Kadmenskii, V.P. Markushev, V.I. Furman, Sov. J. Nucl. Phys. 37 (1983) 165.
4. J. Kopecky, R. E. Chrien, Nucl. Phys. A468 (1987) 285.
5. S.Goriely, E. Khan, Nucl. Phys. A706 (2002) 217.
6. E. Khan and Nguyen Van Giai, Phys. Lett. B472 (2000) 253.
7. J.Dobaczewski, W.Nazarewicz, T.R.Werner, J.F.Berger, C.R.Chinn, J.Decharge, Phys. Rev. C53 (1996) 2809.
8. E. Chabanat, P. Bonche, P. Haensel, J. Meyer, R. Schaeffer, Nucl. Phys. A635 (1998) 231.
9. M. Grasso, N. Sandulescu, Nguyen Van Giai, R. J. Liotta, Phys. Rev. C64 (2001) 064321.
10. E. Khan, N. Sandulescu, Nguyen Van Giai and M. Grasso, Phys. Rev. C66 (2002) 024309.
11. S. Drożdż, S. Nishizaki, J. Speth and J. Wambach, Phys. Rep. 197 (1990) 1.
12. P. Ring, P. Schuck, *The nuclear many-body problem, Springer-Verlag* (1980).
13. M. Samyn, S. Goriely, P.-H. Heenen, J.M. Pearson, and F. Tondeur, Nucl. Phys. A700 (2002) 142.
14. S. Goriely, M. Samyn, P.-H. Heenen, J.M. Pearson, and F. Tondeur, Phys. Rev. C 66 (2002) 024326.
15. M. Samyn, S. Goriely, and J.M. Pearson, Nucl. Phys. A725 (2003) 69.
16. S. Goriely, M. Samyn, M. Bender and J.M. Pearson, Phys. Rev. C (2003) in press.
17. G. Audi and A. H. Wapstra, private communication (2001).
18. R.D. Smith and J. Wambach, Phys. Rev. C38 (1988) 100.
19. J. Wambach (2003), Private communication.
20. J.Kopecky and M. Uhl, Phys. Rev. C41 (1990) 1941.
21. P.F. Bortignon, Nucl. Phys. A687 (2001) 329c.
22. P. Demetriou and S. Goriely, Nucl. Phys. A695 (2001) 95.
23. Reference Input Parameter Library, IAEA-Tecdoc, in press (2003).
24. Goriely S., 2001, in Tours Symposium on Nuclear Physics III, AIP Conf. Proc. 561, eds. M. Arnould et al. (New York: AIP), 53; *http://www-astro.ulb.ac.be*.

Fission barriers from a microscopic model

M. Samyn* and S. Goriely*

Institut d'Astronomie et d'Astrophysique, ULB - CP 226, 1050 Brussels, Belgium

Abstract. We calculate fission barriers with the Skyrme-Hartree-Fock-Bogoliubov plus particle number projection (SHFB+PLN) method, using a Skyrme force fitted to essentially all the nuclear mass data with the same method. The reflection asymmetry is introduced to study the lowering of the outer barrier of three selected nuclei. We discuss the feasability of performing large-scale SHFB+PLN fission-barrier calculations, i.e. for about 2000 nuclei of astrophysical interest

INTRODUCTION

Under certain hydrodynamical conditions, the r-process of nucleosynthesis may produce very neutron-rich fissioning nuclei for which no experimental data are known. The need for an accurate and reliable theoretical prediction of the various properties entering the reaction rates of relevance for the r-process has led us to consider the Skyrme-Hartree-Fock-Bogoliubov (SHFB) approach. We recently studied the impact of different Skyrme parametrizations. More specifically, various aspects of the contact pairing force [1, 2, 3] as well as the effective mass [4] have been analyzed, the Skyrme force being derived by a fit to all experimental nuclear masses of the Audi and Wapstra compilation [5, 6]. Very recently, the SHFB approach has been corrected for the particle-number symmetry breaking inherent to the model. Based on that approach, a Skyrme force has been refitted and labelled BSk8 [7].

Here we will describe the present state of our efforts to extend the HFB calculations to the prediction of fission barriers.

There have been several HFBCS and HFB calculations of the fission barriers for given selected nuclei in the past, but none of these models were so far applied to the calculation of nearly 2000 barriers required for astrophysical applications. In fact, at present the only microscopic calculation of all these barriers was performed using the ETFSI (Extended Thomas-Fermi plus Strutinsky Integral) approximation to the HF method [8]. With the much greater computer power at our disposal, we can now contemplate recalculating all these barriers with the HFB method. The calculations reported here are intended to show how the HFB method leads to a satisfactory agreement with experiment.

Since we aim at a unified treatment of all nuclear physics properties of relevance for the r-process, we adopt in this HFB+PLN barrier calculation the BSk8 force [7] that emerged from the HFB+PLN mass formula. (In the same way, the ETFSI barriers were calculated with the SkSC4 force derived from the ETFSI-1 mass fit [8]).

Our calculational procedure is described in Sect. 2, and is illustrated in Sect. 3 by the cases of the double-humped barrier of ^{240}Pu, ^{246}Cm and ^{252}Cf, where the influence of the reflection asymmetry is studied. An outlook for large-scale fission barriers calculations is finally given in Sect. 4.

THE CONSTRAINED HFB MODEL AND THE CALCULATION OF BARRIER HEIGHTS

In the particle-hole (*ph*) channel, the interaction is chosen of the Skyrme type,

$$\begin{aligned}v_{ij}^{ph} &= t_0(1+x_0P_\sigma)\delta(\mathbf{r}_{ij}) \\ &+ t_1(1+x_1P_\sigma)\frac{1}{2\hbar^2}\{p_{ij}^2\delta(\mathbf{r}_{ij})+h.c.\}+t_2(1+x_2P_\sigma)\frac{1}{\hbar^2}\mathbf{p}_{ij}\cdot\delta(\mathbf{r}_{ij})\mathbf{p}_{ij} \\ &+ \frac{1}{6}t_3(1+x_3P_\sigma)\rho^\gamma\delta(\mathbf{r}_{ij})+\frac{i}{\hbar^2}W_0(\sigma_i+\sigma_j)\cdot\mathbf{p}_{ij}\times\delta(\mathbf{r}_{ij})\mathbf{p}_{ij} \quad , \end{aligned} \quad (1)$$

while in the particle-particle (*pp*) channel, a zero-range density-independent force is adopted,

$$v_{ij}^{pp} = V_{\pi q}\delta(\mathbf{r}_{ij}) \quad , \quad (2)$$

which is assumed to act in a restricted window up to a cutoff energy. In the above expressions, $\rho \equiv \rho(\mathbf{r})$ is the local density. In the context of mass formulas, the pairing strength parameter $V_{\pi q}$ is allowed to be different for neutrons and protons, and also to be slightly stronger for an odd number of nucleons ($V_{\pi q}^-$) than for an even number ($V_{\pi q}^+$), i.e., the pairing force between neutrons, for example, depends on whether N is even or odd.

The derivation, with respect to the HFB wave function,

$$|\Psi_{HFB}\rangle = \prod_{k \leq k_{max}} \beta_k |-\rangle \quad \text{with} \quad \beta_k^+ = \sum_l (U_{lk}c_l^+ + V_{lk}c_l) \quad , \quad (3)$$

(β_k^+ is the creation operator of the kth quasi-particle and $|-\rangle$ is the vacuum with respect to quasi-particles), of the expectation value of the Hamiltonian

$$\hat{H} = \sum_{i=1}^{A} \frac{\hbar^2}{2m_i}p_i^2 + \frac{1}{2}\sum_{ij} v_{ij} \quad , \quad (4)$$

leads to the HFB equations [9]. Our method for solving these equations with the Skyrme interaction has been presented earlier [10, 1] and has now been improved and extended. The improvements concern the spurious centre-of-mass (cm) energy calculation, as well as the particle number symmetry and the parity symmetry restorations. We also consider a modified prescription for the rotational correction. These new modifications are detailed below.

The HFB ground state is not an eigenstate of the total momentum operator. Thus, although the expectation value of the momentum operator $\hat{\mathbf{P}} \equiv \sum_i \hat{\mathbf{p}}_i$ in the cm frame

⟨HFB|$\hat{\mathbf{P}}$|HFB⟩ vanishes, its dispersion ⟨HFB|$\hat{\mathbf{P}}^2$|HFB⟩ does not. Gaussian overlap approximation to exact momentum projection gives for the spurious cm energy

$$E_{cm} = \frac{1}{2MA} \langle HFB|\hat{\mathbf{P}}^2|HFB \rangle, \qquad (5)$$

which has to be subtracted from the calculated total energy. The cm correction is evaluated according to Eq. (5), doing so, however, perturbatively. That is, both the diagonal and off-diagonal terms of Eq.(5) are included only in the calculation of the converged total energy, not in the variational equation that leads to the mean field in the HFB equation.

The HFB ground state is not an eigenstate of the particle number operator. The average number of nucleons is imposed using a Lagrange parameter, the Fermi energy, but the HFB wave function of the nucleus under investigation includes components of wave functions of neighbouring nuclei. To recover the exact number of nucleons, projection techniques are used. The projection after variation (PAV) is easier to implement and numericaly faster than the variation after projection (VAP) of the HFB wave function, and for this reason applied to our calculations. However, to avoid the known discontinuity in the pairing energy, the Lipkin-Nogami formalism, corresponding to an approximate projection before variation, is included in the iterative process. The method is therefore labelled HFB+PLN.

The deformed HFB wave function also breaks the rotational invariance. Projection techniques should be used to restore the exact angular momentum of the wave function, and calculate all observable with the projected wave function. However, such prescription is computer time consuming, and the cranking approximation is used to evaluate the rotational correction energy given by

$$E_{rot}^{crank} = \frac{\hbar^2}{2\mathscr{I}_{crank}} \langle \hat{J}^2 \rangle, \qquad (6)$$

where $\hat{J}$ is the angular momentum operator and $\mathscr{I}_{crank}$ the Inglis-Belyaev cranking moment of inertia [9]. To avoid the spherical divergence, but also to reproduce accurately the mass of slightly deformed nuclei, it is necessary to introduce a damping of E_{rot}^{crank} that can be parametrized by

$$E_{rot} = 0.6 E_{rot}^{crank} \operatorname{th}(4.7\beta_2) \quad, \qquad (7)$$

where the reduced quadrupole moment $\beta_l = \frac{\sqrt{(2l+1)\pi}}{3Ar_0^l} Q_l$ with $r_0 = 1.2 A^{1/3}$. This prescription is equivalent to introduce correlations beyond the mean-field of the form

$$E_{rot}^{crank}\left(1 - 0.6\operatorname{th}(4.7\beta_2)\right) \quad. \qquad (8)$$

The impact of such a rotational correction on the potential energy surface properties will be presented in [11].

Since the reflection asymmetry is included in our HFB approach, the parity symmetry of the wave function is broken. To describe left-right asymmetric shapes coherently, we

TABLE 1. Parameters of the BSk8 Skyrme force

t_0 (MeV.fm^3)	-2035.5245	x_0	0.773828
t_1 (MeV.fm^5)	398.82080	x_1	-0.822006
t_2 (MeV.fm^5)	-196.00319	x_2	-0.389640
t_3 (MeV.fm$^{3(1+\gamma)}$)	12433.359	x_3	0.130933
W_0 (MeV.fm^5)	147.80967	γ	1/4
$V_{\pi n}^+$ (MeV.fm^3)	-314.015	$V_{\pi n}^-$ (id.)	-329.780
$V_{\pi p}^+$ (MeV.fm^3)	-293.019	$V_{\pi p}^-$ (id.)	-309.924
Cutoff (MeV)	$\varepsilon_F \pm 17$		

restore the parity symmetry on the basis of the generator coordinate method, extensively described in [12]; details of our procedure are explained in [11].

Finally, with such improvements, the Skyrme parameters are determined by a fit to all the 2135 ($Z \geq 8$) nuclear masses of the 2001 Audi and Wapstra compilation [6]. The resulting parameters are given in Table 1, and are refered to as the BSk8 Skyrme force, and lead to a final rms error of 660 keV.

The HF states are expanded on an axially deformed oscillator basis. To closely follow the fission path, it is convenient to relate the deformation parameters of the basis to the so-called (c, h, α) parametrization used in the ETFSI calculations [8], defined as the elongation, the necking and the left-right asymmetry parameters, respectively. It allows for the definition of a reference surface that more or less coincides with the actual surface of the fissioning nucleus.

All information needed to calculate the fission probability and the fission fragments distribution are obtained from the analysis of the multi-dimensional energy surface in the deformation space. The only way to obtain such a surface is to constrain the HF calculation to *every* possible deformation of the nucleus. We do not constrain on multipoles of higher order than four, as they are optimized by the HF self-consistency and believed not to play a significant role. The multipole moments that must be constrained (we use the method of the quadratic constraint) are thus the quadrupole, the octupole and the hexadecapole moments, all being calculated in the frame of the centre of mass. A proper description of a system at very large deformation would require a two-centre oscillator basis [13], so that our model is not able to properly describe the fission process beyond scission.

The determination of barrier heights is relatively simple if there are just two deformation parameters, e.g, (c,h): with the total energy E of the given nucleus calculated at a sufficient number of deformations one just makes a contour plot of E in the (c,h) plane. However, for more degrees of freedom, in particular admitting a left-right asymmetry, the fission path must be determined in the 3-dimensional space spanned by the variables (c,h,α). An ingenious solution to this problem is provided by the "flooding model" of Tondeur [14]. In two dimensions we imagine water being slowly poured into the energy surface, and observe its depth, measured at the lowest point, i.e., at the ground state, as it spills over the various barriers. The virtue of this method is that its algorithm can be easily generalized to an arbitrary number of dimensions: see [8] for a detailed account. We stress that this model is applicable whether the energy E at each deformation is calculated in HF or ETFSI.

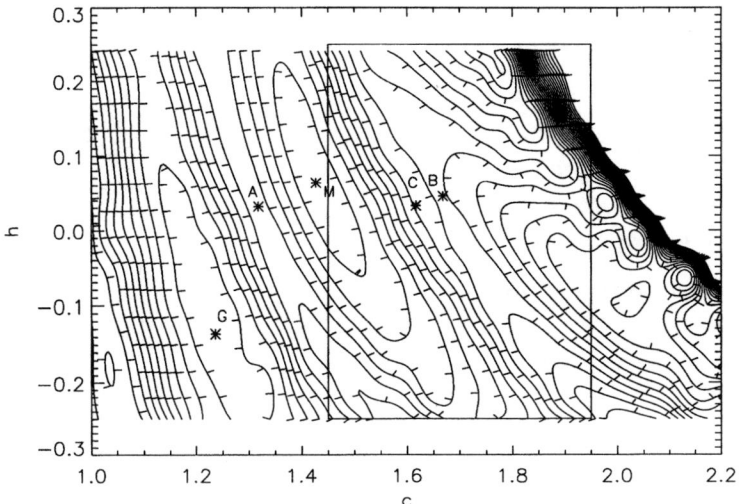

FIGURE 1. Contour plot of the left-right symmetric energy surface of ^{240}Pu in the $(c,h)_{\alpha=0}$ plane. The contour lines are spaced by 1 MeV. Small tick marks along each contour point in the downhill direction. Letters G,A,M,B,C refer to the ground-state ($\beta_2 = 0.276$), the inner saddle-point ($\beta_2 = 0.545$), the isomeric state ($\beta_2 = 0.834$), the left-right symmetric ($\beta_2 = 1.515$) and asymmetric outer saddle-point ($\beta_2 = 1.267, \beta_3 = -0.199$), respectively. The square centred on B shows the limits in the (c,h) plane taken to calculate the local 3D energy surface with its saddle-point C. The upper right zone of the panel corresponds to the fission valley

In all cases we begin with a first HFB calculation of the energy surface, assuming left-right symmetry, $\alpha = 0$. If the corresponding ETFSI calculation [8] indicates that a particular barrier in this surface is asymmetric, we calculate the HFB energy surface over the (c,h) plane in the vicinity of the concerned saddle point for each of the four values of $\alpha = n.\delta\alpha$, with $n = 1, 2, 3, 4$ and a suitable value of $\delta\alpha$ (note that α is never negative). When dynamical properties are calculated [15], it is essential to extend the asymmetric calculation at high enough deformations, in principle to an energy below the ground-state. The grid over the (c,h) plane corresponds to $\delta c = \delta h = 0.05$. Before applying the flooding model, every surface is interpolated in the c and h directions using the cubic spline method, and with respect to α using Lagrange interpolation.

FIVE TEST CASES: ^{230}TH, ^{235}U, ^{240}PU, ^{246}CM, AND ^{252}CF

We consider the ^{240}Pu case in some detail in order to illustrate our general procedure, showing the energy surface for $\alpha = 0$ in Fig. 1. The calculated (measured) energy of the ground-state (G) is -1812.61 (-1812.67 [6]) MeV, which is in agreement with the quality of the fit of BSk8 to the measured masses. The inner barrier height, measured to be of 5.8 ± 0.2 MeV [16], is predicted to be 5.9 MeV. The first shape isomeric state (M) is at 1.914 MeV above the ground-state (experimentally, about 2.25 ± 0.2 MeV [17]). The

TABLE 2. PES properties [MeV] predicted by the HFB+PLN(BSk8) model; the influence of reflection asymmetry is given for the outer barrier; experimental data are given between brackets (the experimental uncertainty of the energy above ground state G of the shape isomer, M, and of the saddle points A and C, is of the order of ± 0.3 MeV (in some cases, values from different references differ by more than .5 MeV)). The upperscript letters refer to references: [a] Ref. [6]; [b] Ref. [18]; [c] Ref. [16]; [d] Ref. [19]; [e] Ref. [17]

	G	A	M	B	C
^{230}Th	-1754.68 (-1754.41)[a]	4.4 (5.4)[b]	2.58,5.53 (2.25,5.55)[b]	11.71	7.25,9.21 (5.75,6.45)[b]
^{235}U	-1783.67 (-1783.10)[a]	5.1 (5.2)[c]	2.33 (2.5)[d]	10.78	6.56 (6.0)[c]
^{240}Pu	-1812.61 (-1812.67)[a]	5.9 (5.8)[c]	1.91 (2.25)[e]	9.6	5.9 (5.45)[c]
^{246}Cm	-1846.86 (-1846.97)[a]	6.0 (6.0)[c]	1.71 (-)	7.5	5.3 (4.8)[c]
^{252}Cf	-1880.41 (-1880.37)[a]	6.7 (5.3)[c]	1.06 (-)	6.2	4.4 (3.5)[c]

outer symmetric barrier B (at 9.6 MeV) has been recalculated by the method explained before, including the third dimension α within a local variation of c and h as shown by the square centred on B. The resulting asymmetric saddle-point (C) gives an outer barrier height of 5.9 MeV, while it is measured to be 5.45 ± 0.2 MeV [17]. The effect of the left-right asymmetry on the outer barrier is to lower it by 3.7 MeV.

This procedure has been applied for four other cases: the results are summarized in Table 2 and compared to experimental data.

The sensitivity of barriers to various aspects of the Skyrme interaction and to the rotational correction is studied in [11] in great detail.

We see from Table 2 that ^{230}Th presents a triple-humped barrier (when the reflection asymmetry is taken into account), as found experimentally [18]. In spite of this qualitative agreement with experiment, the intermediate and outer barriers are overestimated, whereas the inner barrier is underestimated; these features are thought to be related to an interplay of shell effects and pairing effects, as explained in [11]. However, the isomeric state energy agrees with [18]. For $Z \geq 92$, the agreement with experimental data is within the theoretical and experimental uncertainties.

A word of caution should be given about the comparison between the calculated inner barrier heights and the experimental data. Triaxiality has been shown to lower the ETFSI(SkSC4) inner barrier height of ^{233}Th, ^{236}U, ^{240}Pu, ^{244}Cm and the ^{252}Cf by about .4, .3, .8, .9 and 1.3 MeV, respectively [20]. This effect is however force dependent and expected to be smaller for the strong BSk8 pairing force, in a similar way as the pairing affects the asymmetric outer barrier [11].

OUTLOOK AND CONCLUSIONS

The present HFB+PLN calculation shows that microscopic models can compete with more phenomenological highly parametrized models in the reproduction of experimental data. The large-scale HFB calculation of the symmetric and asymmetric barriers for all the thousands nuclei of relevance for the r-process has now become feasible, although it still faces some technical difficulties, that are partly discussed in [11]. A coherent and accurate determination of nuclear masses and fission barriers within one unique mean field approach definitely represents one of the most challenging issues for nuclear astrophysics applications in the future. Work is in progress.

ACKNOWLEDGMENTS

M.S. and S.G. are FNRS Research Fellow and Associate, respectively. We thank J.M. Pearson, P.-H. Heenen, M. Bender and P. Demetriou for stimulating and clarifying discussions.

REFERENCES

1. M. Samyn, S. Goriely, P.-H. Heenen, J. M. Pearson and F. Tondeur, *Nucl. Phys.* **A700**, 142 (2002).
2. S. Goriely, M. Samyn, P.-H. Heenen, J. M. Pearson and F. Tondeur, *Phys. Rev.* **C66**, 024326 (2002); (http://www-astro.ulb.ac.be).
3. M. Samyn, S. Goriely and J.M. Pearson, Nucl. Phys. **A725** (2003) 69.
4. S. Goriely, M. Samyn, M. Bender and J.M. Pearson, Phys. Rev. **C** (2003), in press.
5. G. Audi and A. H. Wapstra, Nucl. Phys. **A595**, 409 (1995).
6. A. H. Wapstra and G. Audi, private communication (2001).
7. M. Samyn, S. Goriely, M. Bender and J.M. Pearson, (2003) in preparation.
8. A. Mamdouh, J. M. Pearson, M. Rayet and F. Tondeur, *Nucl. Phys.* **A644**, 389 (1998); *Nucl. Phys.* **A679**, 337 (2001); (http://www-astro.ulb.ac.be).
9. P. Ring and P. Schuck, *The nuclear many-body problem*, Springer, New York (1980).
10. F. Tondeur, S. Goriely, J.M. Pearson and M. Onsi, *Phys. Rev.* **C62**, 024308 (2000)
11. M. Samyn, S. Goriely, (2003) in preparation.
12. P. Bonche, J. Dobaczewski, H. Flocard, P.-H. Heenen and J. Meyer, Nucl. Phys. **A510** (1990) 466.
13. J.F. Berger and D. Gogny, *Nucl. Phys.* **A333**, 302 (1980).
14. F. Tondeur, private communication.
15. P. Demetriou, this conference.
16. *Reference Input Parameter Library – 2*, IAEA-TecDoc (2003), in press; also available at http://www-nds.iaea.org.
17. M. Hunyadi *et al.*, Phys. Lett. B 505 (2001) 27.
18. J. Blons, B. Fabbro, C. Mazur, D. Paya and M. Ribrag, Nucl. Phys. **A477**, 231 (1988).
19. S. Bjørnholm and J. E. Lynn, Rev. Mod. Phys. **52** (1980) 725.
20. A. K. Dutta, J. M. Pearson, and F. Tondeur, *Phys. Rev.* **C61**, 054303 (2000).

Weak interaction processes in stars

I.N. Borzov

Institut d'Astronomie et d'Astrophysique, Université Libre de Bruxelles,
CP226, Bvd. du Triomphe, Belgium

Abstract. Major astrophysical applications involve a huge number of exotic nuclei. An important effort has been devoted during the last decades to the measurements of the masses and β-decay rates of very neutron-rich nuclei at RIB facilities. However, most of them cannot be synthesised in terrestrial laboratories and only theoretical predictions can fill the gap. We concentrate mainly on the β-decay rates needed for the stellar r-process modeling and for performing the RIB experiments. The continuum QRPA approach based on the self-consistent ground state description in the framework of the density functional theory is briefly described. The model for the large-scale calculations of total β-decay half-lives accounts for the Gamow-Teller and first-forbidden transitions. Due to the shell configuration effect, the first-forbidden decays have a strong impact on the total half-lives of the r-process relevant nuclei at N=126, Z=60-70. The performance of existing global models for the nuclides near the r-process paths at N=126 is critically analyzed and confronted with the recent RIB experiments in the region "east" of ^{208}Pb. The possible role of neutrino nucleosynthesis is exemplified by the production of rare isotope ^{138}La via the neutrino process.

INTRODUCTION

The quality of the nucleosynthesis models depends crucially on the choice of the astrophysical site and of the nuclear input. A reliable prediction of the nuclear properties for thousands of experimentally unknown nuclei is a challenge for microscopic theory. The rates of the weak interaction are among the essential ingredients to be known at best. Phenomenological nuclear models fitted to experimental data close to the β-stability line usually allow for a crude extrapolation to high isospin values. The adequate treatment of the β-strength function demands for a self-consistent approach. At least, the same ground state description should be employed as the one used to calculate the nuclear masses [1]. This is essential in order to ensure the internal consistency of both the nuclear structure and nucleosynthesis models.

The total β-decay half-life is a quite peculiar quantity to deal with. Indeed, extrapolations of this quantity to the regions far from stability can hardly be considered. Close to the β-stability line the microscopic calculations of the β-strength function are obviously not accurate enough. At the nucleon drip-lines, the dynamics of loosely bound quantum systems should be properly taken into account. The best way to test the predicted half-lives is to confront them with the experimental ones for the short-lived nuclei.

Experiments using a new generation of the radioactive ion beam facilities are crucial in validating the existing theories. Great attention has been paid to the nuclei near the closed shells at $Z = 28, N = 50$; $Z = 50, N = 82$ and $Z = 82, N = 126$ providing an important benchmarks for theoretical models. The high precision data for short-lived Ni isotopes near ^{78}Ni [2], $^{121-129}$Ag isotopes [3] and $^{133-137}$Sn [4] have been measured at

RILIS (CERN), which benefits from the high isotopic and isobaric selectivity reached by a laser ion source and by mass spectroscopy, respectively. The current experimental studies in the neutron-rich and neutron-deficient lead regions [5] are very important for studying the nuclear structure at extreme neutron-proton asymmetry, as well as for elucidating the weak interaction and β–decay theories.

The neutron-rich nuclei near the closed-shells at $N = 50$, 82 and 126 provide a clear case for application of the QRPA-type approach, for a comparison of different methods and for experimental checks. First, they are not extreme drip-line systems ($S_n \approx 2.0 - 3.0$ MeV $> \Delta$, where Δ is the pairing potential), therefore an application of the mean-field approach should not yield too large errors. Second, most of these nuclei are spherical, hence their β-decay half-lives are very sensitive to nuclear structure effects. Third, the nuclei with $Z \approx 28$ near ^{78}Ni and $Z \approx 50$ near ^{132}Sn, as well as the ones near $Z \approx 60–80$ at $N=126$ undergo high-energy Gamow-Teller (GT) and/or first-forbidden (FF) β-decays. In these conditions, the spherical, 1p1h-QRPA model accounting for the allowed and FF transitions have enough accuracy in predicting the β-decay half-lives.

THEORETICAL FRAMEWORK

The ultimate model of the β–decay should be based on the self-consistent mean-field and pairing potential and on the universal effective NN-interaction are to be derived from a single nuclear energy-density functional (DF). The phenomenological DF can be constructed in the local density approximation (see i.e. [6] and Refs. therein). This could gives a possibility to describe *ab initio* the properties of nuclear ground and excited states. In practice, some restrictions exist in applications of the fully self-consistent approach. The spin-dependent (time-odd) component of the DF mostly responsible for the β-decay properties, and its scalar (time-even) part which characterizes the ground state are interconnected through the density dependence of the effective NN-interaction. Hence, the parameters of the spin-isospin part of the DF are constrained by the values of the scalar parameters. In Landau limit, the spin-isospin parameter g' derived from the available Skyrme DFs turns out to be much lower than its empirical value [7]. The satisfactory spin-dependent DFs are lacked for spin-unsaturated nuclei [6, 8].

On the other hand, the ground state properties are rather insensitive to the spin and spin-isospin dependent components of the DF (except for the spin-orbit term). Thus, the scalar and spin-isospin components of the DF can be decoupled and the effective NN-interactions in the scalar and spin-isospin channels can be introduced independently [9]. The DF+QRPA approach developed in such an approximation may gain an advantage of using the well-founded spin-isospin effective NN-interaction of the finite Fermi system theory (FFS) [10] augmented by the medium-renormalized π and ρ-exchange terms. When treated within the full single-particle basis framework of the continuum-QRPA, this interaction turns to be universal (A-independent). That is of prime importance for the large-scale applications. In the DF+CQRPA approach, the Landau-Migdal constant is very close to the empirical value, thereby ensuring a reliable description of the GT strength function in the whole energy scale including the continuum.

We have followed the approach to the large-scale calculations of the allowed Gamow-Teller (GT) and first forbidden (FF) decays developed in [11]. The specific features of the model are as follows:

1) The ground states properties are treated in terms of the Fayans phenomenological DF [6] consisting of a normal and a pairing part. The DF3 version of this functional [9] contains the two-body spin-orbit and velocity dependent effective NN-interactions important for the full consistency. The isovector spin-orbit force ensures a correct description of the single-particle levels near the "magic-cross" at ^{132}Sn [12]. In [7, 11] we have studied aslo a possibility of the ground state description given by the Skyrme SkSC17 [7] and MSk7 forces [13].

2) The pairing energy density depends on the anomalous nucleon density v as $\varepsilon_{pair} = \frac{1}{2} v F^\xi v$. Here F^ξ(T=1) is the effective NN-interaction in the particle–particle (pp) channel chosen in the density-dependent form as $F^\xi(\vec{r}_{12}) = -4N_0^{-1} f^\xi(x) \delta(\vec{r}_{12})$ where N_0=150 MeV fm^3 is the inverse half-density of states at the Fermi surface in equilibrium nuclear matter, $x = (\rho_p + \rho_n)/2\rho_0$, where the $\rho_{p,(n)}$ are the proton (neutron) densities; the $f^\xi(x)$ is expressed in a Skyrme-like form [6]. A local cutoff treatment of the pairing energy density [6] helps to avoid the problem of the choice of the cutoff energy. An efficient pairing regularization procedure has been suggested recently in [14].

3) For the excited states, the continuum QRPA (CQRPA) equations of the finite Fermi system theory [10] are solved with an exact treatment of the particle-hole (ph) continuum, pairing and effective NN-interaction in the ph and pp channels [15]. Thus, the SO(8) quasispin symmetry of the QRPA problem is preserved.

4) The method to include the ph continuum for the $\Delta T = 1$ excitations of superfluid nuclei [15] is similar to the one for $\Delta T = 0$ excitations [16]. It is based on exact treatment of the pairing in "valence λ-space" ($\mu^\tau - \xi < \varepsilon_\lambda < \mu^\tau + \xi$), where μ^τ are the neutron and proton chemical potentials. Far from the Fermi surface, the ph propagator is the same as in the system with no pairing. It is calculated via the Green functions constructed in the r-space which allows the exact inclusion of the ph-continuum.

5) The universal medium renormalization (beyond the QRPA-type correlations) of the spin-isospin fields is taken into account via the quasiparticle local charge operators $\hat{e}_{qi} = e_q[V_0^{JLS}]$; the so-called "quenching factor" is $Q = e_{qs}[\sigma \tau]^2 = (g_A/G_A)^2$. The smaller Q, the less strength contained in the low-energy part ($\omega < \varepsilon_F$) of the spin-isospin response, and therefore the longer the β–decay half-lives.

6) In the β-decay studies, the effective NN-interaction should preserve the appropriate ballance of the short-range and finite-range components, as well as of repulsive and attractive parts. In the ph channel it is chosen in a $\delta + \pi + \rho$ form. The one-π and one-ρ exchange terms modified by the nuclear medium are important in describing the magnetic properties of nuclei and the nuclear spin-isospin responses. Their competition determines a degree of "softness" of the pionic modes in nuclei that directly affects the β–decay half-lives. [1] The effective T=0 NN-interaction in the particle-particle channel

---

[1] So far, two basic sets of FFS parameters have been used: (A) $Q = 0.81$, $g\prime$=0.9-1.0 derived from the magnetic moments, which allows for a moderately soft π-modes [17]; (B) $Q = 0.64$, $g\prime=1.0$-1.1 derived from observed GT and M1 strength distributions [18, 19]. A strong quenching excludes the existence of

is assumed to have a form similar to a like-particle pairing. It can not be neglected, as this would destroy the SO(8) symmetry of the QRPA equations and cause unrealistic odd-even staggering of total β–decay half-lives. The CQRPA equations of the FFS allows a reasonable description of nuclear spin-isospin modes in the region far from the instability point in the pp channel [15].

7) Total β–decay half-life is calculated through the corresponding β–decay strength functions $S_{j\pi}(\omega)$, where ω is the decay transition energy. [2] The strength functions are found solving the CQRPA equations for the following driving operators $\vec{\sigma}$; γ_5, $[\vec{\sigma}\vec{r}]^{(J=0)}$; $\alpha, \vec{r}, [\vec{\sigma}\vec{r}]^{(J=1)}$; $[\vec{\sigma}\vec{r}]^{(J=2)}$. Thus, the relativistic vector operator α, and axial charge operator γ_5 should be included alongside with the space-like operators. An efficient approximation used in global calculations of total half-lives is to replace α, γ_5, by the space-dependent fields. The exact non-relativistic relation for the matrix element of the time-like operator $<\alpha> = \xi/\lambda_e \cdot \Lambda_1 \cdot <i\vec{r}>$ can be applied which reflects the conservation of the nuclear vector current (CVC). In a fully self-consistent approach a precise cancellation of all the terms except the averaged Coulomb potential takes place, thus the translation factor Λ_1 reads $\xi\Lambda_1 = \omega_{if} + \bar{u}_C$. Due to the PCAC, no analogous exact relation exist for the axial charge operator γ_5 and its space-like counterpart $\vec{\sigma}\cdot\vec{r}$. On the basis of the chiral symmetry and soft-pion theorem it has been shown that the $<\gamma_5>$ vertex is amplified in the nuclear medium due to the axial-exchange currents and the effective NN-interactions [22]. The self-consistent FFS sum rule approach [18] is used to approximate the operator γ_5 by the space-like operator $\vec{\sigma}\cdot\vec{r}$ taking into account the medium corrections which are mainly due to the spin-orbit and velocity dependent interactions [11]. With the resulting set of the space-dependent external fields, the large-scale calculations of the β-decay half-lives are feasible.

THE RESULTS

β–decay in the Z =28, N=50 region.

According to available experimental [2] and predicted [7] decay schemes, the nuclei at Z=28, N=50 shell sequence undergo high–energy GT decays which are built mainly on the simple shell-model configurations $\nu 1f_{7/2,5/2} \to \pi 1f_{7/2,5/2}$ and $\nu 1g_{9/2} \to \pi 1g_{9/2}$. The first-forbidden decays are due mainly to the $\nu 1g_{9/2} \to \pi 1f_{7/2}$ and $\nu 1f_{5/2} \to \pi 1d_{5/2}$ transitions with the energies close to that of the GT decays. Thus, the first-forbidden transitions may be of little effect for nuclei near Z=28, N=50. It is of importance that the

---

soft π-modes. The recent analysis of (p,n) reaction spectra at E_p=295 MeV [20] and excitation energies up to E_x <50 MeV gives some evidences for $Q = 0.93 \pm 0.05$ i.e. of a lower quenching than previously deduced in [18] from the old (p,n) data at E_x <30 MeV [21]. The experimental uncertainties remain still large, and both sets will be applied in the calculations of the β–decay half-lives.

[2] Though the excitation energies in the daughter nuclei are used for the experimental decay schemes, the adequate variable within the RPA-type approaches is clearly the β–decay transition energy ω relative to the parent nucleus ground state [1].

TABLE 1. Total β-decay half-lives for Ni isotopic chain

A	Ref.	72	73	74	75	76	77	78	79	80
	[11]	1770	1940	1160	770	600	350	240	123	109
$T_{1/2}$	[28]	7026	1306	1203	460	430	207	224	37	72
(ms)	[27]	3450	627	563	211	117	83	40	33	10
	[26]	1948	717	382	239	112	114	54	35	14
	[2]	1570±50	840±30	900±200	700±400	440±400	-	-	-	-

high–energy GT transitions exist both at Z<28 and at Z=28-29. Even when the $\pi 1f_{7/2}$ orbital is completely blocked still many high-energy GT transitions remain open. Indeed, we have found very small impact of the forbidden decays on the total half-lives in Z=28, N=50 region [11]. In the Table our predictions for Ni isotopes are compared with the other calculations and recent experimental data taken from [2].

β–decay in the Z <50 and Z$\geq$50, N=82 regions.

Unlike Z=28, N≈50 region, the only high-energy GT transition in the Z<50, N≈82 region is built on the $v1g_{7/2} \to \pi 1g_{9/2}$ configuration. The high-energy first-forbidden decay is mainly due to the $v1h_{11/2} \to \pi 1g_{9/2}$ configuration. Our results (Fig.1a) show a moderate (up to a factor of two) impact of the FF decay in Z<50, N=82 region [11]. In contrary, for the nuclides in the ^{132}Sn region with Z=50-51 and $N \approx 82$, the high-energy GT and FF transitions mentioned above are no longer possible, as the $1\pi g_{9/2}$–orbital is blocked. The higher energy FF transitions related to the $v2f_{7/2} \to \pi 1g_{7/2}$, $vf_{7/2} \to \pi 2d_{5/2}$ configurations dominate the total half-life. For Z$\geq$52 and N≈82 nuclei, the unblocking of the $1\pi g_{9/2}$–orbital starts due to pairing but the FF decays still dominate. It can be well exemplified by the $^{133-137}$Sn nuclei (Fig.2) studied at RILIS, CERN [4]. Within the Q_β–window of 9.0 MeV, An experimental decay scheme for ^{135}Sn [4] shows the low energy GT transitions at 4-5 MeV and high energy FF transitions at 7-9 MeV. As seen from Fig.1b, the GT calculations by [23] overestimate the experimental half-lives for $^{133-137}$Sn [4]. An agreement of our calculations [11] with the experimental data shows that the FF decay of these nuclei is dominated by Coulomb mechanism.

β–decay vs (v_e, e^-)–capture in N=126 region.

Currently favored r-proces scenarios demand much shorter r-process time scales than the one of the canonical model obtained in the allowed transition approximation. Though the "thumb rule" is that the β^-–decay half-lives become longer as closed (sub)shells are approached, the $T_{1/2}$s are sensitive to the specific N,Z-shell sequence. It is the interplay of the low-energy GT and high-energy FF transitions that may result in hindering the β–half-lives due to a phase-space amplification of the high-energy FF decays. In the

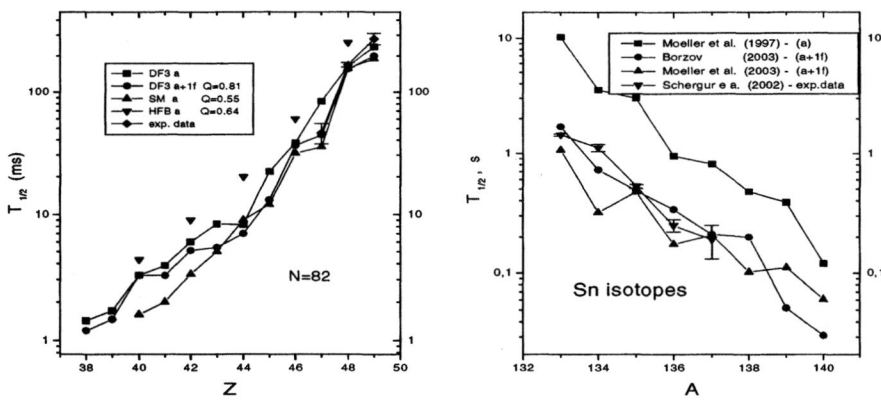

FIGURE 1. Left: a) the total half-lives for N=82 isotones; right: b) the same for Sn isotopes.

previous microscopic large-scale calculations a validity of the allowed transition approximation has not been analyzed. Within the proposed microscopic approach both the GT and FF transitions are treated self-consistently and on the same footing. In the important r-process region near Z≈60-80 and N=126, the role of the FF decays is decisive. These nuclei undergo high-energy first-forbidden decays related to the $\nu 1i_{13/2} \to \pi 1h_{11/2}$ configuration. At the same time, the unperturbed β–decay energy of the main GT decay configuration $\nu 1i_{13/2} \to \pi 1i_{11/2}$ is low (about 1 MeV).

Our calculated half-lives are displayed in Fig.2a together with the shell–model (SM) calculation [24] performed in the GT approximation. We see that the inclusion of the FF transitions made in [11] results in noticeably shorter half-lives in the N=126 region. The deviation with the GT approximation amounts to typically a factor 5 to 10, and is more pronounced for heavier nuclei approaching the closed proton shell at Z=82. Note that the shell–model half-lives [24] (obtained for the GT decay with the SM-quenching factor $(\frac{g_A}{G_A})^2=0.55$) would be shorter if the first-forbidden decays were included.

Also shown in Fig.2a are our calculated inverse (ν_e, e^-)-rates at the neutrino temperature $T_\nu=4$ MeV [7] (corrected for the forbidden transitions) and the RPA calculations by [25] for $T_\nu=8$ MeV (the neutrino luminosity is $L_\nu \approx 10^{52}$ erg s^{-1}, the distance from the center of the neutron star is R=100km). One observes from Fig.2b that even at $T_\nu=8$ MeV (with complete neutrino conversion $\nu_e \rightleftharpoons \nu_{\mu,\tau}$ in effect [25]), the GT+FF β–decays dominate over charged-current electron neutrino captures in this region of the r-process path (for the accepted values of L_ν, R and T).

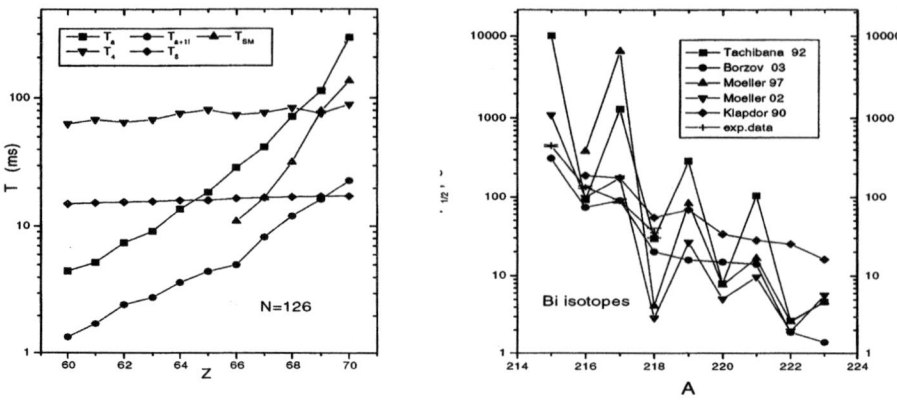

FIGURE 2. Left: a) the total half-lives for N=126 isotones; right: b) the same for Bi isotopes.

β–decay in the region "east" of ^{208}Pb

At the time being, the predictions in Z=60–70, N=126 region can not be verified experimentally. The current experiments [5] concern mostly the nuclei "east" of ^{208}Pb. In the region close to ^{208}Pb, the strong first-forbidden β–decays are well known experimentally. The main problem the models encounter lies in a near cancellation of the leading matrix elements for J=0,1 transitions. In Fig.2b we compare the measured half-lives [5] with the predictions from the existent global approaches [26, 27, 23, 11, 28]. The total half-lives for the GT decay calculated within the FRDM+RPA model [23] overestimate the experimental ones by orders of magnitude. The spurious odd-even staggering amounting the orders of magnitude in [23] are due to the omission of the T=0 pn-effective interaction which leads to the violation of the QRPA O(8)-symmetry.

The results of statistical model (gt) [26] with a parametric description of the FF transitions, also vary significantly from the odd-A to even nuclei (Fig.2b). The results from the "micro-statistical" (FRDM-RPA-gt) model [28] (in which the "gross-theory" calculations of the FF decay is combined with FRDM+RPA description of the GT decay) are closer to the experimental data. However, obtained strong renormalization of the half-lives (for example in ^{217}Bi) is hard to explain. It may well come from the inconsistency in the microscopic and "gross-theory" inputs within this hybrid models.

In the QRPA calculations by [27], the strengths of the separable ph and pp NN-interactions are fitted to the experimentally known half-lives for each isotopic chain. As the odd-even behavior of the half-lives from [27] is realistic (the T=0 paring has been included), such a procedure may give a sound local extrapolation. The FF transitions has not been included, and the calculations by [27] overestimate the experimental half-lives in the region of Bi isotopes (Fig.2b).

The results of the GT+FF calculation [11] show a fairly regular behavior with some underestimation of the experimental total half-lives (Fig.2b). The calculation may still be oversimplified in the specific region "east" of ^{208}Pb, especially if the $\Delta J=0$ transitions dominate in the decay schemes. This is mainly due to the neglect of the velocity-dependent terms in the effective NN-interaction and to the use of the Coulomb (ξ) approximation. However, the results are in a qualitative agreement with available experimental data on total half-lives.

Neutrino process and ^{138}La production

The odd-odd neutron-deficient heavy nuclides ^{138}La and isomeric ^{180}Ta^m are among the rarest solar system species (no information exists for other locations), with ^{138}La/^{139}La $\approx 10^{-3}$ and ^{180}Ta^m/^{181}Ta $\approx 10^{-4}$. In spite of their scarcity, their origin has long been a puzzle. As initially claimed by [29] and confirmed by [30] ^{180}Ta^m appears to be a natural product of the p-process in the O/Ne-rich layers of Type II supernovae (SNII). In contrast, ^{138}La is underproduced in all thermonuclear p-process calculations.

In view of the low ^{138}La abundance, it has been attempted to explain its production by non-thermonuclear processes involving neutrino-induced transmutations [31, 32]. This mechanism was predicted by Woosley et al. (1990) to be able to overproduce the solar ^{138}La/^{139}La ratio by a factor of about 50. In view of the qualitative nature of the evaluation, the prediction needs to be verified in the network calculations.

In the [33], the model for a $Z_\odot$ 8 $M_\odot$ helium star (main sequence mass of about 25 $M_\odot$) already considered by [30] has been used. As P-Process Layers (PPLs), 25 O/Ne-rich zones has been taken with explosion temperatures peaking in the $(1.7-3.3)\times 10^9$ K range. The p-process reaction network includes some 2500 stable and neutron-deficient nuclides with $Z \leq 84$. The n-, p- and α-capture reactions on all nuclei are considered, as well as their (γ,n), (γ,p) and (γ, α) photodisintegrations. The nuclear reaction rates are available in the Brussels Nuclear Astrophysics Library http://www-astro.ulb.ac.be). Finally, the experimentally-based neutron capture rates of [34] are used. The analysis [33] clearly demonstrated that the ^{138}La problem can not be solved within the framework of thermonuclear model assuming the existing uncertainties of the nuclear physics input. The main sources of uncertainties lie in the nuclear level densities, γ-ray strength functions and neutron optical potentials. Related errors in the capture rate predictions can amount to about a factor of 2 for ^{138}La(n,γ)^{139}La at the temperature of relevance (2.4 10^9 K) for the ^{138}La synthesis. In view of scarcer experimental information, larger uncertainties affect the ^{137}La(n,γ)^{138}La, ^{138}La(n,γ)^{139}La and ^{139}La(γ,n)^{138}La photodisintegration rates. An experimental study of the above-mentioned key rates is no longer out of reach and is eagerly awaited. Experimental determination of the ^{138}La and ^{139}La photodisintegration rates could really put the theoretical conclusions on a safe footing.

In order to examine the impact of the neutrino interactions on the p-nuclide yields, and in particular on ^{138}La, the same reaction network used above has been augmented with all neutrino-induced reactions (charged and neutral currents) up to Kr (all rates are from [36]), as well as all charged-current neutrino and antineutrino scatterings up to Po (these rates are from [7]). The neutral-current scattering off nuclei heavier than Kr are

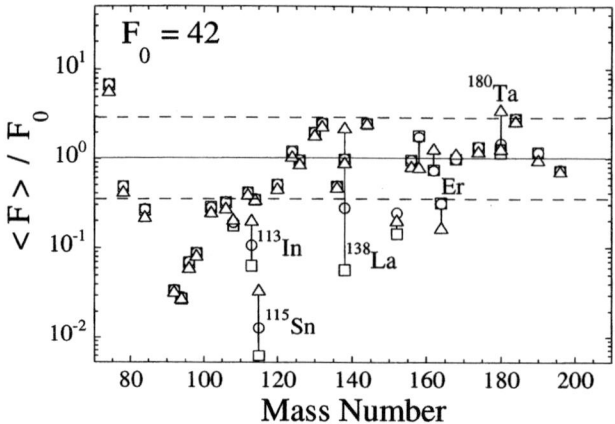

FIGURE 3. Comparison of the p-process abundances obtained with and without (squares) neutrino interactions (with standard MOST rates). Two sets of luminosities $L_\nu[10^{51}$ ergs s$^{-1}]$ =(3, 4, 16) (circles) and (30, 40, 160) for ($\nu_e, \bar{\nu}_e, \nu_x$)(triangles) are adopted for the calculations with neutrinos.

considered only for the Ba, La and Ce isotopes. The neutrino and antineutrino scattering cross sections are averaged over supernova (anti)neutrino spectra that are approximated by zero-degeneracy Fermi-Dirac distributions corresponding to typical temperatures $T_\nu = 8, 5$ and 4 MeV for the ν_x (where x stands for μ and τ (anti)neutrinos), $\bar{\nu}_e$ and ν_e, respectively [35]. The allowed transitions for the charged-current (anti)neutrino scatterings are treated in terms of the β-strength function calculated within the cQRPA approximation based on the ETFSI ground state description [7]. As the average $\nu_e(\bar{\nu}_e)$ energies are relatively low, for a conservative estimate one can use an approximation of allowed transitions. In particular, the resulting averaged ν_e-capture cross section by ^{138}Ba is estimated to be 7.5 10^{-41} cm^2 at $T_{\nu_e} = 4$ MeV. The neutral current ν-scattering contribution is also estimated in a rough way by assuming an approximate scaling of the $(\nu_x, \nu_x\prime)$ cross section with mass number.

Fig.3 shows the impact of the neutrino interactions on the p-nuclide production for the two sets of typical luminosities [35] $L_\nu[10^{51}$ ergs s$^{-1}] = $ (3, 4, 16) and (30, 40, 160) for ($\nu_e, \bar{\nu}_e$ and ν_x). These two L_ν combinations lead to increases of the ^{138}La production with respect to the case without neutrinos by factors of 4.8 and 36, respectively. This enhanced synthesis originates entirely from the ν_e-capture by ^{138}Ba, the neutral current scattering on ^{139}La being found to have a negligible impact. Thus, despite relatively similar cross sections, the ^{138}Ba ν_e-capture is found particularly efficient due to the large initial ^{138}Ba abundance (about 10 times the ^{139}La abundance). For the high luminosity set, the ν_e-captures also enhance the production of ^{113}In, ^{115}Sn, ^{162}Er and ^{180}Ta. These conclusions have been independently supported by [37]. Thus, despite the numerous uncertainties still affecting supernova models and the neutrino physics in supernovae (spectra, luminosity, temperature, oscillation, interaction cross sections, ...), ν_e-captures appear so far to be the most efficient production mechanism of the solar ^{138}La.

ACKNOWLEDGMENTS

The author have a pleasure to thank Profs. M. Arnould, P. VanDuppen and M. Huyse for their interest in this work, Dr. S. Goriely for performing the astrophysical network calculations, and the OSTC, Belgium for the support within the PAI Program IAP P5/07 "Exotic nuclei for Nuclear Physics and Astrophysics".

REFERENCES

1. I.N. Borzov, S. Goriely, *Phys. Part. Nucl.* **63**, 1375 (2003).
2. S. Franchoo et al. *Phys. Rev. Lett.* **81**, 3100 (1998).
3. V.N. Fedoseev et al. *Zeit.Phys. A* **353**, 9 (1995).
4. J. Shergur et al. *Phys. Rev. C* **65**, 034313 (2002).
5. H. DeWitte, A.N. Andreev, I.N.Borzov et al. *Phys. Rev. C* (2003, in print).
6. S.A. Fayans et al, *Nucl. Phys.* **676**, 49 (2000).
7. I.N. Borzov, S. Goriely, *Phys. Rev. C* **62**, 035501 (2000).
8. M .Bender, J. Dobaczewski, J. Engel, W. Nazarewicz *Phys. Rev. C* **65**, 054322 (2002).
9. I.N. Borzov, S.A. Fayans, E. Kromer, D. Zawischa. *Zeit.Phys. A* **355**, 117 (1996).
10. A.B. Migdal *Theory of finite Fermi systems and atomic nuclei properties* (Russ.original 2nd ed.), Nauka, Moscow, 1983.
11. I.N. Borzov *Phys. Rev. C* **67**, 025802 (2003).
12. K.M.Mezilev et al., *Phys. Scripta T* **56**, 227 (1995).
13. S. Goriely et al. *At.Data Nucl.Data Tables* **77**, 311 (2001).
14. A. Bulgac *Phys. Rev. C* **65**, R051305 (2002).
15. I.N. Borzov, E.L. Trykov *Sov.J. Nucl. Phys* **52**, 33 (1990).
16. A.P. Platonov, E.E. Saperstein *Nucl. Phys.* **486**, 63 (1988).
17. I.N. Borzov et al. *Sov. J. Part. Nucl.* **12**, 848 (1981).
18. N.I. Pyatov and S.A. Fayans. *Sov. J. Part. Nucl.* **14**, 401 (1983).
19. I.N. Borzov, S.V. Tolokonnikov,S.A. Fayans *Sov.J. Nucl. Phys* **40**, 732 (1984).
20. T. Wakasa et al. *Phys. Rev. C* **55**, 2909 (1999).
21. C. Gaarde *Nucl. Phys.* **396**, 127c (1983).
22. K. Kubodera et al., *Phys. Rev. Lett.* **40**, 755 (1987).
23. P. Möller et al. ADNDT 66, 26434 (1997).
24. G. Martinez-Pinedo., *Nucl. Phys.* **668**, 357c (2000).
25. A.Hektor et al., *Phys. Rev. C* **61**, 055803 (2000).
26. T. Tachibana et al. Prog. Theor. Phys. **84**, 641 (2000).
27. M. Hirsch, et al. *At.Data Nucl.Data Tables* **51**, 244 (1992).
28. P. Möller et al. *Phys. Rev. C* **67**, 26434 (2003).
29. N. Prantzos,M. Hashimoto,M. Rayet M, M. Arnould *Astron. & Astroph.*, **238**, 455 (1990).
30. M.Rayet et al., *Astron. & Astroph.*, **298**, 517 (1990).
31. G.V. Domogatskii, D.K. Nadyozhin *Ap.Sp.Sci.*, **70**, 33 (1977).
32. S.E. Woosley, D.H. Hartmann, R.D. Hoffman, W.C. Haxton, *Astroph. J.*, **356**, 272 (199).
33. S.Goriely, M.Arnould, I.Borzov, M.Rayet., *Astro. & Astroph.* **375**, L35 (2001).
34. Z.Y.Bao, H.Beer, F. Käppeler et al., *At. Data Nucl. Data Tables*, **76**, 70 (2000).
35. G.M.Fuller, B.S.Meyer, *Astroph. J.,* **453**, 792 (1995).
36. R.D.Hoffman, S.E.Woosley, *http://ie.lbl.gov/astro.html*(1992).
37. A.Heger et.al. *astro-ph 0307546v1* (2003).

Nuclear Data for Low-Energy Astrophysics and Other applications — An Addendum

Kohji Takahashi

Gesellschaft für Schwerionenforschung mbH, Planckstraße 1, 64291 Darmstadt, Germany[1]

Abstract. We briefly review the short contributions and discussions called forth in a "round table" time-slot following the formal presentations on nuclear data needs. The subjects covered are: microscopic model investigations of nuclear fissions, experimental and theoretical aspects of fixing masses and half-lives of nuclei far off stability, and measurements of photo-disintegration cross sections needed for nucleosynthesis studies.

PREFACE

The idea of "round-table discussions" on nuclear astrophysics was introduced to Tours Symposium three years ago to add an extra flavor to this series of joint symposia of nuclear and astrophysicists. One learns from its summary [1], though heavily edited it might be, that the session was held among a handful and most exclusively astrophysics-oriented participants. This time, in a sharp contrast, most of the "pure" nuclear physicists registered to the symposium remained to attend this (and the preceding) nuclear astrophysics sessions (despite, one may add, that the conference excursion and banquet had already been held the day before... or for that?). That was indeed a pleasant surprise to me, but it also obliged me as a moderator to find quickly a common denominator among the interests of the attendants. Did I succeed in so doing ? It depends...

SHORT CONTRIBUTIONS AND DISCUSSIONS

Here is a very brief summary of what was discussed, although I take the liberty to sprinkle it with some of my more benign comments.

Fissions

The first and the only scheduled short contribution was presented by P. "Vivian" Demetriou [2]. This work aims at deriving theoretically spontaneous fission life-times and neutron-induced fission cross sections in comparison with the experimental data.

---

[1] E-mail : K.Takahashi@gsi.de

In the first instance, a microscopic model of Skyrme-HFB type [3] is used to compute fission potential-energy surfaces, with which the dynamical fission paths and thus lifetimes are derived. Secondly, an attempt is made to estimate neutron-induced fission cross-sections from a model with its physical ingredients obtained as microscopically as possible.

Along with β-delayed fissions, neutron-induced fissions of very neutron-rich nuclei may play an important role in the r-process of terminating the flow (the "r-process path") towards the heaviest mass region. Judging from a comparison presented in [2] of the calculated cross sections and the existing data, one cannot but invoke further (and perhaps dramatic) improvements of theoretical treatments before any reliable predictions could be made.

Masses and half-lives: SPIRAL II, GSI and RIKEN projects

A session on Physics of Exotic Nuclei had been devoted to the "present status and future plans of radioactive beam facilities", and in particular to the comprehensive reviews by three invited speakers representing GSI [4], SPIRAL II [5] and RIKEN [6]. Those projects enjoy keen attentions also of (nuclear)astrophysicists because, to say the least, of the possibilities of measuring the masses and half-lives of very neutron-rich nuclei, which play major roles in the r-process. Gottfried Münzenberg kindly agreed to extract just essential parts that were associated with mass and life-time measurements and discuss the strengths and limitations *in that regard* of the three facilities.

There are two basic method to produce radioactive beams. One is referred to as ISOL (Isotope Separator On Line), which is applied at SPIRAL II, and the other is the so-called "in-flight", which is in use both at RIKEN and GSI. (See e.g. [7] for general ideas of radioactive beam technologies and facilities. As for the most recent developments at facilities other than the above three, see e.g. [8] particularly for those in the North America.)

Tables I and II, as prepared by G. Münzenberg, compare some characteristics of the mass and half-life measurements at those laboratories. Although the contents are largely self-explanatory, some of the comments he attached there may be worth repeating.

The SPIRAL II has a very good beam quality and its strength relates to decay spectroscopy. That it does not allow for refractory elements is an disadvantage. The RIKEN upgrade deals most favorably with light ions. The beam energy of $E/u < 400$ MeV (as opposed to E/u as high as 1 GeV at GSI) is not high enough to fully strip heavy elements, leading to the charge ambiguity.

TABLE 1. Masses and Half-lives(I)

mass range	production	LABs
light ($A \leq 100$)	fragmentation, fission	SpII/GSI/RIKEN
medium ($A \leq 150$)	fission	SpII/GSI/RIKEN
heavy ($A > 150$)	fragmentation	SpII/GSI

TABLE 2. Masses and Half-lives (II)

Method	Precision	Efficiency	single-atom ?	$T_{1/2}$-limit	$T_{1/2}$?	LABs
Indirect:						
Q_{decay}	≤ 100 keV	50 %	no	ms(ISOL)	yes	SpII (GSI/RIKEN)
Direct:						
ESR-TOF	< 100 keV	5 %	yes	μs	no(?)	GSI
ESR-e$^-$cool	< 30 keV	5 %	yes	s	yes	GSI
Trap	< 3 keV	10^{-3}	yes	ms	no	SpII GSI/RIKEN

The real strength at GSI has to do with its use of the Experimental Storage Ring (ESR). One of the virtues is that many nuclei can be measured simultaneously. The ESR with e$^-$ cooling makes it possible to study the decays of bare, H-, He-, and Li-like ions, enabling to unravel interesting phenomena. The so-called bound-state β-decays of bare ions of some near-stable nuclear species have already been detected (including ^{187}Re^{+75} with the half-life of 33 yr as opposed to 42 Gyr of the neutral ^{187}Re. See e.g. [9] for the astrophysical meaning of this finding). Another phenomenon of interest concerns the total (or partial) suppressions of orbital electron captures, a mere example being ^{54}Mn^{+25}, which with its expectedly very slow β^--decays is a potential clock for the secondary cosmic rays [10]. (Recall that the measured half-life of the neutral ^{54}Mn is as short as 312 days.)

As shown in Table II, the combination of the ESR and the TOF technique allows mass measurements of very-short lived isotopes. It would be worth challenging to figure out how to measure decay half-lives of the order of milliseconds to microseconds.

Masses and half-lives: A theoretical consideration

Experimental explorations of masses and half-lives (and perhaps other quantities) of nuclei far off stability will certainly help revealing more about the structural peculiarities of those "exotic nuclei". How about their astrophysical applications? Are they very important for r-process studies? For one reason or another, everybody (well, almost everybody) involved in the game would reply unhesitatingly in the affirmative. But, as in many other cases of "astrophysical importance", it is not at all trivial how to *quantify* the importance to the satisfaction of experimental planners.

Yuko Motizuki volunteered to introduce a brave(?) attempt [11] to set those minimum accuracies in mass (and perhaps half-life) determinations *in the near future* which would be required for the acquired data to be meaningfully used to improve theoretical models by readjusting, for instance, parameter values. To begin with, the use is made of the observation that, despite equally good fit to the existing data, two mass models based on utterly different philosophies [12, 13] predict quite different values for the masses of very neutron-rich nuclei along a likely r-process path.

Any r-process calculations are burdened with the lack of those and various other kinds of data on very neutron-rich nuclei. One also has to remember that the very quest for "the" astrophysical r-process scenario(s) remains unrequited.

Photodisintegrations

The Konan University, the very organizer of this symposium, has recently launched quite an innovative project aiming at measuring astrophysically-interesting photo-disintegration cross-sections with various γ-ray sources [14]. Representing a younger generation, Kaoru Y. Hara agreed to present, literally impromptu, the results on the $d(\gamma,n)p$ cross-sections in relation to the Big Bang nucleosynthesis [15], which indicated a successful reduction of the uncertainty in the inverse $p(n,\gamma)d$ reaction at relevant energies (cf. [16]). Photo-disintegration cross-section data on many nuclides, and in particular the "p-nuclei" ([17] for a review), are expected by the next Tours Symposium.

ACKNOWLEDGMENTS

My special thanks go to G. Münzenberg for having allowed me to reproduce here Tables I and II and the accompanying comments. The responsibility for blame with regard to the text must be placed fully on me. Useful communications with P. Demetriou, Y. Motizuki, T. Motobayashi, C. Scheidenberger, and H. Utsunomiya are acknowledged with thanks.

REFERENCES

1. Arnould M. & Takahashi K., 2001, *Tours Symposium on Nuclear Physics IV*, AIP Conf. Proc. 561, 76
2. Demetriou P., Samyn M. & Goriely S., this volume
3. Samyn M. & Goriely S., this volume
4. Weick H., this volume
5. Villari A., this volume
6. Motobayashi T., this volume
7. Bennett R., Van Duppen P., Geissel H., Heyde K., Jonson B., Kester O., Körner G.-E., Mittig W., Mueller A. C., Münzenberg G., Ravn H. L., Riisager K., Schrieder G., Schotter A., Vaagen J. S. & Vervier J., 2000, *Radioactive Nuclear Beam Facilities*, NuPECC Report
8. Schotter A., in *Future Astronuclear Physics*, EDP Sciences, to appear
9. Takahashi K., 1998, *Tours Symposium on Nuclear Physics III*, AIP Conf. Proc. 425, 616
10. Binns W. R., this volume
11. Motizuki Y., Tachibana T. & Goriely S., in *Future Astronuclear Physics*, EDP Sciences, to appear
12. Koura H., this volume and references therein
13. Goriely S., this volume and references therein
14. Utsunomiya H., this volume
15. Hara K. Y., Utsunomiya H., Goko S., Akimune H., Yamagata T., Ohta M., Toyokawa H., Kudo K., Uritani A., Shibata Y., Lui Y.-W. & Ohgaki H., this volume
16. Tomyo A., Nagai Y., Suzuki T. S., Kikuchi T., Shima T., Kii T. & Igashira M., 2003, Nucl. Phys. A718, 401c
17. Arnould M. & Goriely S., preprint

Cross section measurements of capture reactions relevant to the p process: Status and perspectives

Sotirios V. Harissopulos

Institute of Nuclear Physics, NCSR "Demokritos", 153.10 Aghia Paraskevi, Athens, Greece

Abstract. A systematic study aiming at establishing an extended cross-section database of mainly proton capture reactions of nuclei in the Se-Sb region is presented. Such a database is required to perform a validity test of the Hauser-Feshbach (HF) calculations in the medium-mass region that will enable to investigate the impact of nuclear physics uncertainties on p-nuclei abundance calculations. The results of this systematic work as well as a comparison of all existing cross-section data with the predictions of the HF theory are presented. Several aspects of all the experiments performed so far including a recent in-beam cross section measurement of the ^{92}Mo$(\alpha,\gamma)^{96}$Ru reaction are outlined. The question of whether the experimental information obtained so far is sufficient to draw final conclusions is discussed.

1. INTRODUCTION

The origin of the so-called p nuclei is one of the most challenging puzzles to be solved by any model of heavy-element nucleosynthesis. The term p nuclei refers to 35 stable proton-rich nuclei that ly on the "neutron-deficient" side of the stability valley, between ^{74}Se and ^{196}Hg. These nuclei cannot be synthesized by the so-called s and/or r processes [1], i.e. by neutron capture reactions, as is the case of all the other nuclei heavier than iron. Instead, the production of the p nuclei requires a special mechanism, named p process that has recently been reviewed in [2]. Although various p-process calculations have been successful in reproducing the abundances of most of the p nuclei within a factor of 3 (see e.g. [2] and references therein), this is not the case in the low mass region (A≤120) where significant discrepancies have yet to be resolved.

The relevance of the present work to the p-process modelling results from the fact that the Hauser-Feshbach (HF) theory [3] is strongly involved in p-nuclei abundance calculations: The reproduction of these abundances requires extended network calculations involving more than 20000 nuclear reactions with about 2000 nuclei Such a network is shown in Fig. 1, where the p nuclei are also displayed. Since it is hardly possible to measure the cross sections of all these reactions, the HF theory has to be extensively used to calculate the cross sections for most of the reactions involved in the network. Hence, the reliability of the p-process abundance calculations depends on that of the HF theory. For this reason, the reliability of the HF predictions has to be checked independently of any improvements in the modelling of the p process. The HF calculations depend predominantly on nuclear properties such as nuclear masses, nuclear level densities (NLDs), the nucleon-nucleus and α-nucleus optical model potentials (OMPs), and finally γ-ray strength functions. A sensitive check of the uncertainties related to these properties can be performed by comparisons with experimental cross section data over a wide mass renge. Due to scarce experimental cross section data, an intensive experimental effort has been devoted during the last five years to the study mainly of proton-capture reactions at energies well below the Coulom barrier that are relevant to the *p* process. These studies aim at establishing an extended cross-section database for capture reactions that will enable us to derive global input parameters for HF calculations as well as to investigate the impact of nuclear physics uncertainties on the *p*-nuclei abundance calculations.

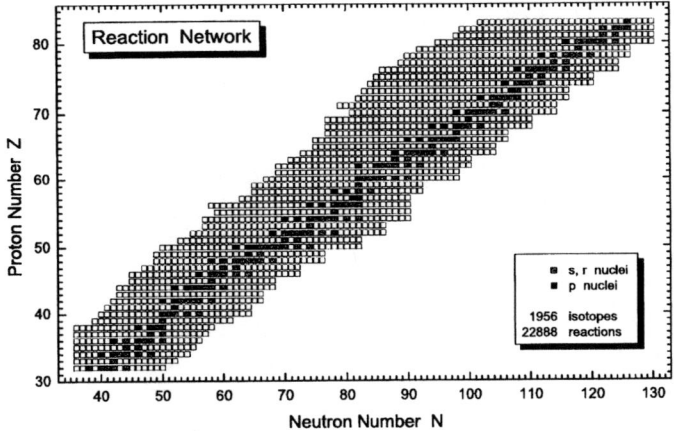

FIGURE 1. A typical reaction network used in the *p*-nuclei abundance calculations. The stable isotopes are indicated with black or grey boxes. The former correspond to the *p* nuclei whereas the latter to the *s* and *r* isotopes ([4]).

2. EXPERIMENTAL PROCEDURES

Cross section measurements for astrophysical applications are carried out at a certain range of beam energies defined by the so-called Gamow window [5], which is calculated according to the temperatures of the stellar sites where the *p* process occurs. All *p*-process models agree that *p* nuclei are produced at stellar temperatures ranging from 1.8 to 3.3 10^9 Kelvin. These temparature limits correspond to laboratory energies that range from ≈ 1.3 to 5 MeV for proton capture reactions in the Se-Sn mass region. Hence, measurements have to be carried out at energies well below half the value of the corresponding Coulomb barrier which is 7.9 and 10 MeV for protons incident on Se and Sn, respectively. At such energies, the cross sections are expected to be very small and their measurement is, therefore, a challenge for experimentalists. Indeed, the HF theory predicts a proton capture cross section of ≈ 0.1 μb and 1 mb at 1.3 and 5 MeV incident energy, respectively. The situation for the case of (α, γ) reactions in the medium mass region is even worse: Here, the corresponding Gamow windows extend from ≈ 5 to 11 MeV. However, the cross sections are expected to be at least 10 times smaller than those predicted for the (p,γ) reactions.

Due to these difficulties, measurements have to be carried out using very efficient setups, combined preferably with intense proton or α beams. The energy region where cross sections have to be measured can in principle, be covered by low-energy *tandems* or *single-ended* accelerators. These machines are capable of delivering very stable and intense proton beams. In addition, the *single-ended* accelerators can generally provide intense α beams. However, most of the latter type of accelerators available worldwide cannot completely cover the relevant energy region. Low energy cyclotrons can be therefore complementarily used, mainly for α-capture cross-section measurements.

In order to determine the cross section of a capture reaction leading to the formation of a compound nucleus, the absolute number of the produced compound nuclei has to be determined. This can be done by either measuring the absolute number of photons emitted by the reaction or by measuring the absolute activity from the subsequent decay of the produced nuclei. The latter method obviously presupposes that the produced nuclei are unstable, which, however, is not always the case. The two procedures mentioned above form the two distinctly different methods of measuring cross sections of capture reactions, namely, the *in-beam* method and the *activation* technique, respectively. The common feature of the methods is that they lead to an overall accuracy of $\approx$5-10% in the resulting cross sections. Experimental details on recent activation measurements can be found in e.g. refs. [6, 7]. Hence, in the following we shall focus on the *in-beam* method only.

In the in-beam method, the total cross section σ_T is derived from the reaction yield Y_0, i.e. the absolute number of γ rays emitted by the reaction in 4π per beam-particle. Y_0 can be determined either from the angular distributions of all γ-transitions feeding the ground state of the produced nucleus or from angle-integrated γ-fluxes measured by means of a 4π "summing" detector (4π calorimeter). The total cross section is then given by

$$\sigma_T = (A/N_A \xi) Y_0 \quad (1)$$

where A is the atomic weight of the target in amu, N_A is the Avogadro number, and ξ is the target thickness in μg/cm^2.

Angular distribution measurements can basically be performed by means of a high-purity Ge detector (HPGe) placed on a goniometric table that rotates around the target. γ-singles spectra are usually taken at 5 at least angles with respect to the beam. The number of the incoming proton-beam particles is obtained from the charge accumulated at each angle. This is determined by using a current integrator, i.e. the current of the beam is properly measured and further integrated over the measurement time. This procedure is repeated at different beam energies with a proper energy step, so that finally the relevant Gamow window is covered. In this way, the intensities of all the relevant γ transitions are determined at all angles and all beam energies. These intensities are first normalized to the corresponding charge and then they are corrected for the absolute efficiency of the detector used. Hence, at each beam energy one obtains for each γ transition an angular distribution, i.e. the quantity $Y(\theta)$=(photons/charge-unit/4π). The angular distribution is then fitted by a sum of Legendre polynomials $P_k(\theta)$ given by $W(\theta) = A_0(1 + \sum_k \alpha_k P_k(\theta))$. The coefficients A_0 and α_k that are determined by the fitting procedure are energy dependent. The maximum value of index k, with $k \geq 2$, depends on the multipolarity of the γ transition in consideration. In this way one obtains at each beam energy N different A_0 coefficients in *photons/mC*, one for every γ transition feeding the ground state of the produced compound nucleus. The total yield Y_0 is then taken from

$$Y_0 = \sum_i^N A_0^i \quad (2)$$

and is inserted in Eq. 1 to give the total cross section σ_T.

The determination of cross sections from γ angular distribution measurements can obviously be a time-consuming task. The use of an array of HPGe detectors rather than just one detector may considerably reduce the required time. Such an array is shown in Fig. 2. It was used by the Nuclear Astro-Group of NCSR "Demokritos", Athens, in collaboration with the Institut für Strahlenphysik of the University of Stuttgart, Germany to study the reactions listed in Table 1. The Stuttgart setup consisted of 4 large volume HPGe detectors with a relative efficiency of $\varepsilon_r \approx 100$ % each. They were all shielded with BGO crystals for Compton background supression. The detectors were placed on a rotating table at distances between 10 and 20 cm from the target. By rotating the table by 15° γ-single spectra were measured at eight angles with respect to the beam direction. A typical γ-singles spectrum measured with the Stuttgart setup for the ^{89}Y(p,γ)^{90}Zr reaction and some typical γ-angular distributions of this reaction are shown in Fig. 3.

TABLE 1. List of the in-beam cross section measurements carried out in the present work. E_p is the corresponding energy range covered. ΔG is the astrophysically relevant region (Gamow window). Details on the targets used are given in the last 3 columns.

Reaction	E_p (MeV)	ΔG (MeV)	Target	Enrichment (%)	ξ (μg/cm^2)
^{78}Se(p,γ)^{79}Br	1.5–3.5	1.27–3.87	metallic	97.80	85
^{80}Se(p,γ)^{81}Br	1.5–3.5	1.27–3.87	metallic	99.82	106, 132
^{86}Sr(p,γ)^{87}Y	2.0–3.6	1.39–4.13	^{86}SrCO$_3$	96.89	103
^{87}Sr(p,γ)^{88}Y	2.0–3.6	1.39–4.13	^{87}SrCO$_3$	91.50	82
^{88}Sr(p,γ)^{89}Y	2.0–3.6	1.39–4.13	^{86}Sr(NO$_3$)$_2$	99.84	168
^{89}Y(p,γ)^{90}Zr	1.4–4.9	1.42–4.20	metallic	100 (nat.)	97,436
^{93}Nb(p,γ)^{94}Mo	1.4–4.9	1.48–4.32	Nb$_2$O$_5$, metallic	100 (nat.)	106,371
^{103}Rh(p,γ)^{104}Pd	2.0–3.6	1.60–4.56	metallic	100 (nat.)	171
^{113}In(p,γ)^{114}Sn	2.0–3.5	1.72–4.79	metallic	4.3 (nat.)	10,37
^{115}In(p,γ)^{116}Sn	2.0–3.5	1.72–4.79	metallic	95.7 (nat)	209
^{116}Sn(p,γ)^{117}Sb	2.0–3.5	1.74–4.85	^{116}SnO$_2$	95.74	237
^{118}Sn(p,γ)^{119}Sb	2.0–3.5	1.74–4.85	metallic	97.06	180
^{121}Sb(p,γ)^{122}Te	2.4–3.4	1.77–4.91	metallic	57.2 (nat.)	178
^{123}Sb(p,γ)^{124}Te	2.4–3.4	1.77–4.91	metallic	42.8 (nat.)	134

The use of high-efficiency detector arrays with BGO shields in in-beam experiments enables us to measure very low capture cross sections, often smaller than 1μb. It also shortens the measuring time significantly. However, the data analysis itself is a time-consuming procedure, as the number of γ transitions to be taken into account remains the same and very often numerous γ-ray spectra have to be analyzed. For this reason, it is very useful to carry out "angle-integrated" measurements which deliver the total reaction yield Y_0 directly. This can be achieved by using a sufficiently large NaI(Tl) detector covering a solid angle of almost 4π around the target, shown in Fig. 4. Such a detector is installed in the Institute of Nuclear Physics of NCSR "Demokritos", Athens.

The detector consists of two scintillation crystals of cylindrical shape. Each crystal is 6 inches long and 12 inches wide. It is further segmented in 4 equal parts. The detector has an axial throughhole (Ø 35 mm) as well as a radial throughhole (Ø 83 mm). In the latter hole, two extra NaI(Tl) detectors are inserted. Hence, γ rays emitted from the target are detected over a solid angle of $\approx 96\%$ of 4π. The energy signals from the photomultipliers are summed to obtain the final signal which is then guided to the data acquisition system.

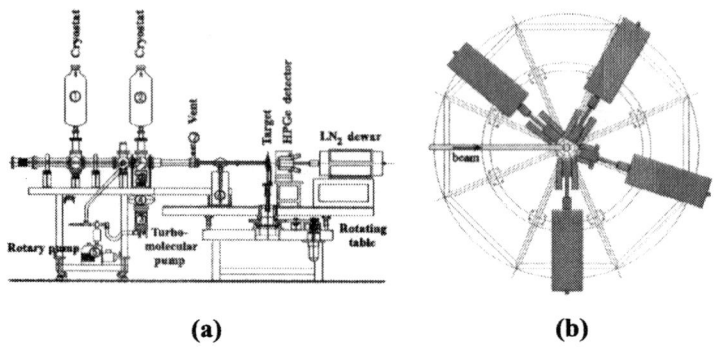

FIGURE 2. The Stuttgart setup for γ angular distribution measurements. a) Side view of the beam line and its components. b) Top view of the HPGe-detector array as placed on motor-driven rotating table.

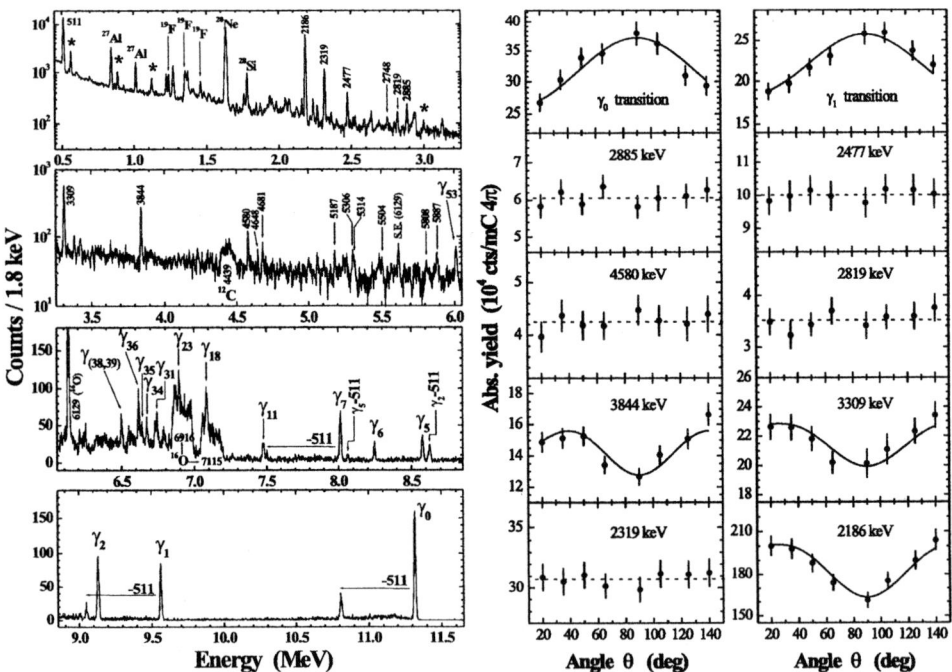

FIGURE 3. LEFT: In beam γ-singles spectrum of the ^{89}Y(p,γ)^{90}Zr reaction measured at E_p=3 MeV and at $\theta = 90°$ with the setup of Fig. 2. The primary γ rays observed are labeled with γ_x, where x indicate the xth excited state of the produced ^{90}Zr nucleus as given by Firestone *et al.*[8]. The γ rays observed feeding the ground state are labeled with their energies. RIGHT: Typical γ-angular distributions.

The main advantage in using a 4π summing detector is that the response of the NaI(Tl) crystal leads predominantly to a single peak, called *sum peak*, at the sum of the energies of the cascading transitions, i.e. at $E_\gamma = Q + E_{cm}$, where Q is the Q-value of the reaction and E_{cm} is the center-of-mass projectile energy. The area under the sum peak is determined by a simple integration and is further corrected for the corresponding efficiency ε. The efficiency ε can be independently determined experimentally by using calibrated γ-ray sources and/or by yield measurements of certain resonant nuclear reactions. Often, however, ε has to be determined by Monte-Carlo simulations.

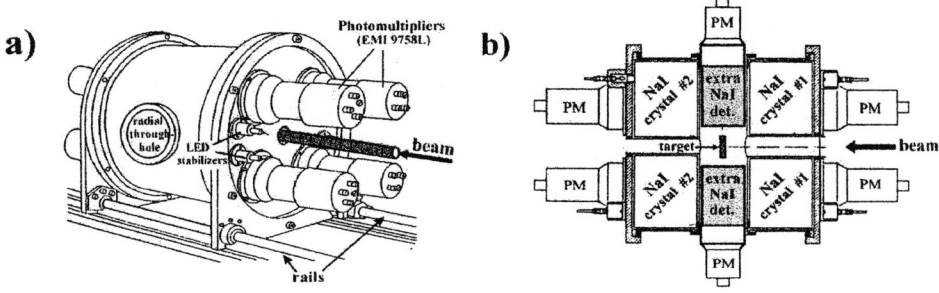

FIGURE 4. The 4π NaI(Tl) summing detector installed at "Demokritos", Athens.

Due to the 4π geometry covered by the summing crystal additional corrections for angular distribution effects are not necessary. The "ideal" picture, however, of having just one peak is not achieved in "real" experiments where the spectra also include the Compton continuum, since high-energy photons are not fully absorbed in the NaI(Tl) crystal. In the case of segmented NaI crystals, like the one shown in Fig. 4, even fewer photons are fully absorbed and the summation is imperfect. A typical γ spectrum measured with this summing detector is shown in Fig. 5a. The measured γ rays result in a continuum-like distribution rather then a single (sum) peak. The integration of the

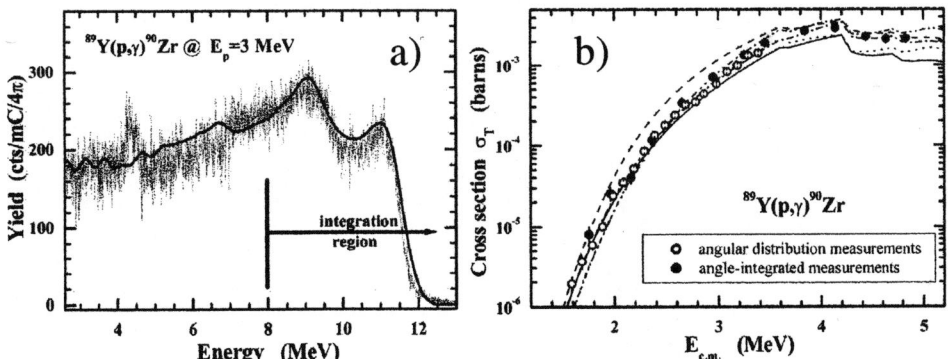

FIGURE 5. a) γ-singles spectrum taken with the 4π NaI summing detector installed at "Demokritos", Athens. The solid curve indicates the GEANT-simulated spectrum. b) Cross sections resulting from angle-integrated measurements (solid circles) and γ-angular distribution measurements (open circles). The curves indicate the corresponding predictions of the Hauser-Feshbach theory (see in chapter 3).

γ events included in this spectrum yields the angle integrated γ photon flux F. The total reaction yield Y_0 is then obtained by

$$Y_0 = F/Q\varepsilon \,, \qquad (3)$$

where Q is the corresponding accumulated charge and ε the absolute efficiency of the detector. Y_0 is inserted in Eq. 1 to give the cross section. In our case, an integration window from 8 to 14 MeV was used to determine the photon flux F at different beam energies. This was chosen so as to avoid possible contributions from contaminating reactions ocurring mainly in the backing of the target, like e.g. the ^{19}F(p,$\alpha\gamma$)^{16}O reaction yielding photons up to 7.2 MeV. The corresponding efficiency ε was derived from Monte-Carlo simulations. In our investigations, the cross section of the ^{89}Y(p,γ)^{90}Zr reaction was measured by both methods; by angular-distribution measurements as well as by angle-integrated γ-flux measurements. A comparison of the cross sections resulting from both techniques is given in Fig. 5b. As can be seen, the results from both methods are completely compatible.

3. HAUSER-FESHBACH CROSS-SECTION CALCULATIONS

As was mentioned in the introduction, all cross section measurements of proton capture reactions performed so far at low energies, aim at checking the nuclear input of the HF calculations in order to obtain global input parameters for astrophysical applications. To date, the amount of the existing (p,γ) cross section data allows for a comparison with HF calculations over a sufficiently wide mass range, which is shown in Fig. 6. In the HF theory [3], the cross section for the decay of a compound nucleus depends on a) the transmission coefficients for particle and photon emission, b) the nuclear level densities of the compound and residual nuclei in the different decay channels, and c) on the Q values of the involved nuclei.

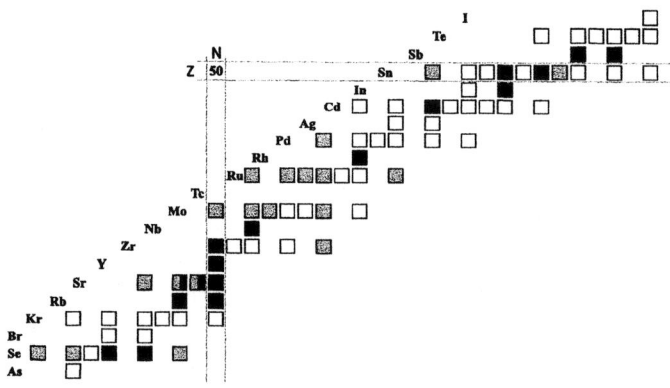

FIGURE 6. Stable isotopes used so far as targets in cross section measurements of (p,γ) reactions at energies relevant to the p process. Black boxes correspond to in-beam measurements carried out by "Demokritos" and IfS-Stuttgart (see also in [9, 10, 11, 12]) except ^{90}Zr ([13]). The grey isotopes were used in activation measurements ([6, 7, 14, 15, 16, 17]) in which natural targets were mainly used.

For the nuclear level densities (NLDs), there exist global phenomenological models, based on the Fermi Gas model description of the excited nucleus, and also microscopic models based on single-particle spectra associated with realistic effective potentials. The main advantage of the latter models is that they take into account the discrete structure of the nucleus and treat shell, pairing and deformation effects consistently, whereas the former models consider these effects by means of empirical corrections. For the transmission coefficients for particle emission there exist purely phenomenological global optical model potentials and microscopic ones. The latter are obtained from nuclear matter calculations using realistic effective interactions. For photon emission, assuming that dipole transitions dominate in the γ emission channel, the electric- and magnetic-dipole (GDR) strength functions are commonly described by Lorentz-type functions. The GDR energies and widths are obtained from experimental data where they exist, otherwise they are determined by appropriate global parametrizations. Recently, microscopic calculations of the GDR strength functions have also been performed using a Hartree-Fock-QRPA model [18].

In order to derive global input parameters for HF calculations we took into account all existing cross section data of capture reactions on the targets displayed in Fig. 6. The calculations of the HF cross sections were carried out using the statistical model code MOST [19, 20]. All available experimental data on nuclear masses, deformation, spectra of low-lying states, and GDR energies and widths were taken into account. The nuclear masses were taken from the compilation of Audi and Wapstra [21] and the ground state properties (matter density, single-particle level scheme) from the microscopic Hartree-Fock-BCS model of [22]. At the low incident energies studied in the current experiment, the most dominant decay channels are those of photon and neutron emission, the α-particle emission channel becoming important at energies above ≈ 7 MeV. On the other hand, the effect of using different Lorentz-type or microscopic strength functions for the photon transmission coefficient turns out to be negligible. Therefore, in all the following calculations we adopt the same α-Nucleus potential of Demetriou *et al.* [23] and E1 strength functions given by the hybrid model of Goriely [24]. The M1 strength functions are parametrized according to [25], with the energies and widths chosen according to the latest recommendations of [26]. While the HF cross sections do not seem to be sensitive to the above-mentioned nuclear ingredients, they show a strong dependence on the NLDs and nucleon OMPs. We therefore, use different combinations of NLDs and OMPs in order to investigate the range of uncertainties they give rise to.

In particular, we use the OMPs of a) Koning and Delaroche [27], b) Jeukenne *et al.* [28] and c) Bauge *et al.* [29]. The first potential is purely phenomenological whereas the last two are based on microscopic infinite nuclear matter calculations applied with the local density approximation. We adopt two NLD formulae, namely the purely macroscopic densities of Thielemann *et al.* [30] and the statistical microscopic ones of Demetriou and Goriely [31].

Five different combinations of these OMPs and NLDs have been considered in the HF calculations. The resulting cross sections for the ^{89}Y(p,γ)^{90}Zr reaction are compared with the data in Fig. 5b). The solid, dashed and dotted curves correspond to the same microscopic NLDs, namely those of Demetriou and Goriely [31], and the OMPs of Jeukenne *et al.* [28], Bauge *et al.* [29], and Koning and Delaroche [27], respectively. The dot-dashed and dot-dot-dashed curves are obtained from the combinations of the macroscopic NLD of Thielemann et al. [30] with the microscopic OMP of Jeukenne et al. [28] and the phenomenological OMP of Koning and Delaroche [27], respectively. Overall, similar results are obtained for all existing data.

CONCLUSIONS

Based on the comparison of all the existing (p,γ) cross section data one can conclude that: The HF theory agrees with the experiments within a factor 2-3. The predictions are more sensitive to the optical model potentials (OMPs) rather than to nuclear level densities (NLDs). Moreover, assuming the microscopic NLDs of Demetriou and Goriely [31] and varying the OMPs, we observe that the OMP of Bauge et al. [29] overestimates the data systematically, whereas the OMP of Jeukenne et al. [28] either reproduces the data satisfactorily or underestimate them by at most a factor of ≈ 1.3. Finally, the OMP of Koning and Delaroche [27], shows no systematic behaviour, i.e. in some cases there is agreement in others disagreement of at most 20-40%. The most important conclusion of astrophysical relevance, however, is that uncertainties of the above mentioned magnitude in the nuclear data input in nucleosynthesis calculations, are of minor importance compared to the huge uncertainties (often exceeding orders of magnitudes) involved in the "pure astrophysics" modelling. These findings however have to be further confirmed in the case of α-particle capture reactions for which hardly any data exist in the mass region in consideration. Indeed, there are only three experimental papers ([32, 33, 34]) reporting on cross sections of (α, γ) reactions at energies relevant to p process above the lightest p-nucleus ^{74}Se. For this reason a new research project was recently initiated by the Nuclear Astro-Group of "Demokritos" in collaboration with the University of Bochum aiming at measuring (α, γ)-reaction cross sections by using the 4π summing detector (a 12inch×12inch NaI *mono*crystal) installed at Bochum. The first measurements carried out recently were on the ^{92}Mo$(\alpha,\gamma)^{96}$Ru as well as the ^{92}Zr$(\alpha,\gamma)^{96}$Mo reaction. A typical γ-spectrum of the former reaction is shown in Fig. 7. The data analysis is in progress.

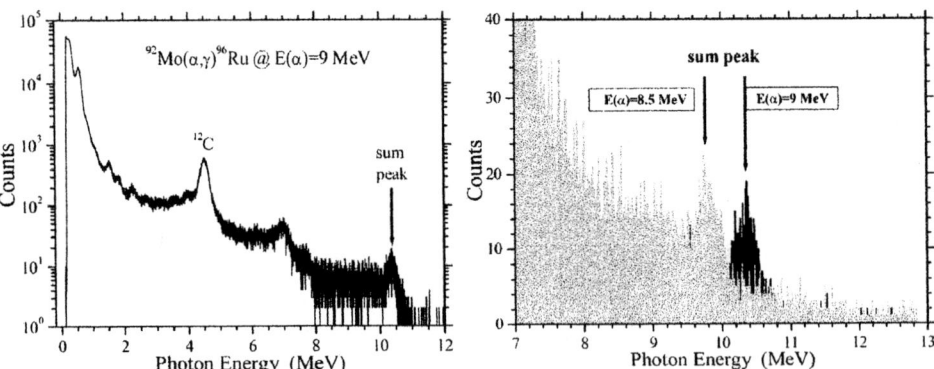

FIGURE 7. LEFT: Gamma-spectrum of the ^{92}Mo$(\alpha,\gamma)^{96}$Ru reaction measured at E$_\alpha$=9 MeV with the 4π calorimeter installed at the University of Bochum. RIGHT: Display of the sum peaks observed at E$_\alpha$=8.5 MeV (grey spectrum) and at E$_\alpha$=9 MeV (black peak).

ACKNOWLEDGMENTS

This work was mainly supported by NATO (Contract CRG961086), the German Academic Exchange Service DAAD, and the Greek-Hungarian bilateral programme (GSRT/Demokritos/E797). The measurements listed in Table 1 would not have been realized without the efforts of all the members of the "Demokritos" Nuclear Astro-Group. Many thanks to Dr. P. Demetriou (NCSR "Demokritos" and IAA, ULB, Brussels) for the numerous HF calculations, the excellent collaboration on the theoretical aspects of the present project, and her decisive contribution to its sucess. During the Stuttgart runs the support by Dr. W. Hammer and his Group was valuable. The collaboration with the ATOMKI group of Dr. E. Somorjai has been quite fruitful. The interaction with Dr. S. Goriely (IAA, ULB, Brussels) on issues of the p process has been particularily illuminating.

REFERENCES

1. Wallerstein, G., et al., Rev. Mod. Phys. **69**, 995 (1997).
2. Arnould, M., and Goriely, S., Phys. Rep. 384, 1 (2003).
3. Hauser, W., and Feshbach, H., Phys. Rev. **87**, 366 (1952).
4. Rayet, M., *private communication*, 2000.
5. Rolfs, C. E., and Rodney, W. S., *Cauldrons in the Cosmos*, (The University of Chicago Press, Chicago, 1988).
6. Sauter, T., and Käppeler, F., Phys. Rev. C**55**, 3127 (1997).
7. Gyürky, Gy., et al., Phys. Rev. C **64**, 065803 (2001)
8. Firestone, R. B., et al., *Table of Isotopes*, 8th edition (Wiley-Interscience, New York, 1996).
9. Harissopulos, S., et al., Phys. Rev. C**64**, 055804 (2001)
10. Harissopulos, S., et al., Nucl. Phys. A**688**, 421 (2001).
11. Harissopulos, S., et al., Nucl. Phys. A**719**, 115 (2003).
12. Galanopoulos, S., et al., Phys. Rev. C. **67**, 015801 (2003).
13. Laird, C. E., et al. Phys. Rev. C**35**, 1265 (1987).
14. Bork, J., et al., Phys. Rev. C**58**, 524 (1998).
15. Chloupek, R. F., et al., Nucl. Phys. A**652**, 391 (1999).
16. Özkan, N., et al., Nucl. Phys. A**710**, 469 (2002).
17. Gyürky, Gy., et al., Phys. Rev. C *in press*, (2003)
18. Goriely, S., and Khan E., Nucl. Phys. A**706**, 217 (2002)
19. Demetriou, P., *private communication*, (2003).
20. Goriely, S., in *Nuclei in the Cosmos V*, edited by N. Prantzos and S. Harissopulos, (Edition Frontières, Paris, 1998), p. 314 (see also http://www-astro.ulb.ac.be).
21. Audi, G., and Wapstra, A. H., Nucl. Phys. A**595**, 409 (1995).
22. Goriely, S., Tondeur, F., and Pearson, J. M., At. Data. Nucl. Data Tables **77**, 311 (2001).
23. Demetriou, P., Grama, C., and Goriely, S., Nucl. Phys. A**707**, 141 (2002).
24. Goriely, S., Phys. Lett. B**436**, 10 (1998).
25. Kopecky, J., and Chrien, R. E., Nucl. Phys. A**468**, 285 (1987).
26. Reference Input Parameter Library, IAEA-Tecdoc-1034 (1998), (see also http://iaeand.iaea.or.at/ripl).
27. Koning, A., and Delaroche, J. P., Nucl. Phys. A**713**, 231 (2003).
28. Jeukenne, J. P., Lejeune, A., and Mahaux, C., Phys. Rev. C**16**, 80 (1977).
29. Bauge, E., Delaroche, J. P., and Girod, M., Phys. Rev. C**63**, 024607 (2001).
30. Thielemann, F.-K, Arnould, M., and Truran, J. W., in *Advances in Nuclear Astrophysics*, edited by E. Vangioni-Flam, J. Audouze, M. Cassé, J.-P. Chieze, and J. Tran Thanh Van, (Editions Frontières, Gif-sur-Yvette, 1986), p.525.
31. Demetriou, P., and Goriely, S., Nucl. Phys. A**695**, 95 (2001).
32. Hahn, R. L., Phys. Rev. 137, 1491 (1965).
33. Somorjai, E., et al., Astron. Astrophys. 333, 1112 (1998).
34. Rapp, W., et al., Phys. Rev. C **66**, 015803 (2002).

Experiments with radioactive beams in nuclear astrophysics : evolutions and perspectives

Pierre Leleux

Université catholique de Louvain, Louvain-la-Neuve, Belgium

Abstract. The evolution of experiments with radioactive beams in nuclear astrophysics is reviewed, regarding the techniques (beams and detectors), the astrophysical consequences, the status of indirect methods. Perspectives are outlined.

INTRODUCTION

Nuclear astrophysics with radioactive beams started in the 1980's when people realised that : i) it was possible to produce such beams with decent intensity and purity in nuclear physics laboratories and ii) some reactions induced by such beams were of interest to astrophysicists developing models of explosive environments.

Radioactive nuclides had been ignored in the first era of nuclear astrophysics, in which nuclear reactions of interest in quiet stars (like our sun) had been measured. Indeed in quiet stars, the low temperature and density favor strongly the "decay" channel with respect to the "reaction" channel.

The situation is quite different in explosive environments (novae, supernovae, X-ray bursts). Here, hydrodynamic H- or He-burning is active, implying the occurrence of many reactions of radioactive nuclides with H or He. In the so-called hot-CNO cycles, involving nuclei of mass $M \leq 20$, which is the main subject of this contribution, radioactive nuclides have short lifetimes prohibiting their use as targets, whence the realization of the "inverse kinematics" strategy, in which a radioactive beam interacts with a light H- or He-target.

Very briefly, the main characteristics of these reactions are the following :

- The temperature and density of the astrophysical environment are large, and so is the Gamow energy, implying that cross sections are bigger than the ones in the quiet stars.
- Radioactive beams are secondary beams, which are necessarily of low intensity.
- Working with radioactive beams means that a beam-induced background will affect the detectors.

Detailed reference to previous work in this field can be found in the proceedings of the "Nuclei in the Cosmos" conferences [1].

EVOLUTION OF TECHNIQUES

The main ingredients needed to perform the above experiments are the beams and the detectors. Progress regarding both items will be considered successively.

Beams

It is generally accepted by now that the most suited method for beam production in our field is the ISOL - or the two-accelerator-method. Let us remind that the first accelerator is used for the production of radioactive atoms, which are ionized in a source and subsequently selected and finally accelerated to the requested astrophysical energy by the second accelerator. If the second accelerator is a cyclotron, it can fulfill both functions of selection and acceleration. The radioactive beam intensity I_{RB} is obtained from the formula :

$$I_{RB} = I_{SB} \cdot \varepsilon_R \cdot \varepsilon_{Ex} \cdot \varepsilon_I \cdot \varepsilon_S \cdot \varepsilon_A$$

where I_{SB} is the intensity of the stable beam accelerated in the first device ;
ε_R is the efficiency for production of radioactive atoms from the stable beam-induced nuclear reaction ;
ε_{Ex} is the efficiency for extraction of radioactive atoms from the production target ;
ε_I is the ionization efficiency in the ion source ;
ε_S is the efficiency for selecting the radioactive species of interest ;
ε_A is the acceleration efficiency in the second device.

Very often in the past, expected values of beam intensity in planned facilities were obtained by multiplying "best values" derived from stand-alone operation of individual parts (production target, source, separator, post-accelerator), leading to unrealistic numbers (i.e. much too large). The production target, which has to withstand a large dissipated power, is a very delicate part of the installation, but it should be stressed that bottlenecks often occur in the coupling of successive elements. For example, when increasing the primary beam intensity, not only the number of radioactive atoms produced, but also the amount of gas extracted from the production target to the ion source are increased, which could lead to a decrease of the ionization efficiency of the latter. This may explain why only a modest increase of the radioactive beams intensity was noticed over the last decade ; presently beams up to a few 10^9 particles per second on target are available. More spectacular progress was observed in two other issues, i.e. the reliability of the beams and the diversity of the available beams. With regard to the former, periods of several days with a constant accelerated intensity are now commonly obtained for a given beam. Regarding the second point, Table 1 lists the different radioactive beams produced versus time in the Cyclotron Research Center (CRC), Louvain-la-Neuve.

TABLE 1. Radioactive beams available at CRC vs time.

1988	^{13}N
1992	^{13}N, ^{19}Ne
1996	^{6}He, ^{11}C, ^{13}N, ^{18}Ne, ^{19}Ne, ^{35}Ar
2000	^{6}He, ^{7}Be, ^{11}C, ^{13}N, ^{15}O, ^{18}F, ^{19}Ne, ^{35}Ar

Detection methods

In hot CNO-cycles at moderate temperature, nuclear reactions are organized in chains starting from ^{12}C and ^{16}O seed nuclei, e.g. ^{12}C(p,γ)^{13}N(p,γ)^{14}O(β^+)^{14}N(p,γ)^{15}O or ^{16}O(p,γ)^{17}F(p,γ)^{18}Ne(β^+)^{18}F(p,α)^{15}O. The subsequent chain ^{15}O(α,γ)^{19}Ne(p,γ)^{20}Na leads to an escape from CNO cycles to rp-process. At higher temperatures, other reactions like ^{14}O(α,p) or ^{18}Ne(α,p) become important. From the experimentalist point of view, nuclear reactions of interest are divided in two classes, the ones in which the light particle in the exit channel is a γ-ray (the radiative capture reactions) and the ones in which a charge light particle (α-particle or proton) is emitted. Detection methods for both classes will be examined successively.

Radiative capture reactions

In these reactions, performed in inverse kinematics, three detection schemes exist : prompt γ-rays, radioactive decays of final nuclei, or final nuclei themselves, are detected. The three were experienced in previous measurements. I have selected one measurement in each class to quantify the evolution in detection efficiency.

In a first experiment, ^{13}N(p,γ)^{14}O, performed in 1990 [2] in Louvain-la-Neuve, prompt γ-rays were detected in a single large volume HPGe detector, with a global efficiency (intrinsic times geometrical), of about 0.1 %. Another reaction, ^{19}Ne(p,γ)^{20}Na was studied in the mid-nineties in a series of experiments [3] : both the positrons from ^{20}Na decays to ^{20}Ne and the α-particles from excited states in ^{20}Ne to ^{16}O were detected, with efficiencies of 1-2 %. Finally, a third reaction ^{21}Na(p,γ)^{22}Mg was measured in TRIUMF-ISAC last year [4], in a recoil separator, DRAGON, that allowed the detection of both the final nuclei ^{22}Mg and the prompt γ-rays, with a total efficiency of about 20 %. The resonance strengths effectively measured in the three experiments, i.e. 3.3 ± 0.7 eV, < 21 meV, and $1.03 \pm 0.16 \pm 0.14$ meV respectively are witnessing the tremendous increase in the detection efficiency.

(p,α) or (α,p) reactions

In this domain, a new type of detector, the Silicon Strip Detector (SSD), which appeared in the mid-nineties, made life much easier to experimentalists. This detector, of which the first used was better known as LEDA or Louvain-Edinburgh Detector Array [5], covered typically 10 % of the total solid angle accessible to outgoing protons or α-particles. In a 300 μm-thick LEDA, the typical energy resolution for 5.5 MeV α-particles is 23 keV. Originally, particle identification required the combination of two informations, the energy deposited in the strip, and the time-of-flight of the particle, obtained from the RF of the postaccelerator. Subsequently, thin SSD's were developed which permitted a classical ΔE-E procedure for particle identification. Precise angular distributions of the outgoing particles were obtained from the granularity of the SSD's. It should be noticed that SSD's were not only used in (α,p) or (p,α) reactions of astrophysical interest, but also in the measurement of (p,p) elastic scattering that was performed in conjunction with every (p,γ) measurement.

EVOLUTION OF THE ASTROPHYSICAL CONSEQUENCES

For a given reaction, nuclear physicists measure cross sections, or resonance strengths from which a reaction rate $<\sigma v>$ is calculated [6]. Reaction rates are incorporated as ingredients in astrophysical models of explosive events. A priori the impact of a particular reaction can be estimated by changing arbitrarily its rate by a large factor and looking at consequences in terms of e.g. ejecta from the explosion. To my knowledge, this procedure was not followed very often ; people were indeed very pragmatic, and they first measured reactions which appeared feasible regarding the availability of a radioactive beam of sufficient intensity, and the existence of a detection system of sufficient efficiency.

The reaction rate can be used to deduce the lifetime (τ) of a radioactive nuclide in the presence of H or He :

$$\tau = \frac{1}{N. <\sigma v>}$$

where N is the density of H or He in the astrophysical environment, and
   $<\sigma v>$ the reaction rate.

This lifetime can be compared to the natural lifetime of the radioactive nuclide, τ_{β^+}. The (density, temperature) (ρ,T) plane is divided in two regions, one in which the reaction dominates, another one in which β-decay dominates. Both regions are separated by the line where both times are equal. Knowing the (ρ,T) conditions in the astrophysical environment, one can infer which situation prevail.

In recent years, more attention was paid to the field of γ-ray astronomy, as a consequence of the nice results obtained by the SIGMA and CGRO mission [7] and the approaching launch of the INTEGRAL mission [8]. The above mentioned competition between reaction and β-decay can in fact be used to deduce the amount of 511 keV γ-rays that would be emitted from an explosive environment ; knowing the sensitivity of the instrument in space, the maximal distance from which such an event could be observed was deduced.

One should emphasize, that the "astrophysical consequences" are extremely model-dependent. For example, the fact that the "reaction" channel would surpass the "β-decay" channel depends strongly on the time the radioactive nuclide will spend in the neighbourhood of H or He after its formation : if radioactive nucleides are expelled very soon after outburst, the "reaction" channel would be severely inhibited.

THE EVOLUTION OF INDIRECT METHODS

It was clear from the beginning that some reactions of astrophysical interest would be very difficult to measure directly : the predicted cross section and the expected beam intensity resulted in a much too low expected counting rate. Other measurement aiming at obtaining the same result without measuring the difficult cross section were proposed, the so-called indirect measurements or indirect methods.

The first indirect method to be applied was the Coulomb dissociation or Coulomb break-up [9] : instead of measuring a difficult (p,γ) radiactive capture reaction, the inverse reaction (γ,p) was measured. Moreover, γ-rays are not real, but they are provided by the Coulomb field of a heavy target, like Pb. This method does not exempt people from producing a radioactive beam (but the method is quite different from the ISOL method, as the required beam energy is much larger), but the gain in cross section is very important, as the detailed balance theorem is applied.

Other indirect methods were subsequently developed. Two of them are quoted hereafter :

- The Trojan Horse Method [10] : instead of measuring the X + B $\rightarrow$ c + d reaction of interest at low energy, the A + B $\rightarrow$ S + c + d is measured at high energy, where A = X + S. The kinematical conditions in the exit channel are such that S was a spectator in the collision. In simple words, one can say that the large velocity of A helps to surpass the Coulomb repulsion of B, and that X is brought surreptitiously nearby B.
- The Asymptotic Normalization Coefficient (ANC) Method : instead of measuring the A(p,γ)B reaction, i.e. an electromagnetic interaction with a low cross section, the A(d,n)B or the A(^{3}He,d)B reaction is measured, i.e. a strong interaction with a cross section larger by three or four orders of magnitude. At low energy, both reactions proceed at large distances, through the tail of the wave functions. This method is relevant to (p,γ) reactions dominated by direct capture [11].

A general comment about the three indirect methods quoted above is the following : the positive factor - a large cross section - is accompanied by negative factors - some theoretical corrections to apply before extracting the final result. In my opinion, direct measurements should be performed in all cases where they are feasible.

Finally, another "indirect method" has existed long before radioactive beams became available : cross sections are dominated by isolated resonances ; levels of interest can be produced by transfer reactions induced by stable beams on stable targets, and their decay properties can be studied by conventional spectroscopic methods. A recent example of such work is the production, in the ^{21}Ne(p,t)^{19}Ne reaction, of the level in ^{19}Ne that is of interest in the ^{15}O(α,γ) reaction [12] : α-decays from this level were observed, allowing the authors to deduce an upper limit to the ratio Γ_α/Γ of the α-width to the total width for this level.

PERSPECTIVES

This field has witnessed tremendous progress in somewhat more than a decade. A bright future is still in sight, in two directions :

- New facilities partly dedicated to nuclear astrophysics started operation in the recent years in Europe, Japan and in North America. Sophisticated detection set-up were installed as well, which strongly increased the efficiency, as mentioned above. Measurements of reactions not yet studied ($^{15}O(\alpha,\gamma)$?), and extension to lower c.m. energies of previously measured reactions are expected.
- Extensive studies have started in the same three parts of the world for implanting big facilities aiming at the production of radioactive beams up to very far from stability, in particular to the neutron-rich side. The most interesting point here is the future availability of new beams, very exotic, for measurements of static properties (masses, Q_β, ...) of interest to the r- and s-processes.

CONCLUSION

After more than a decade of work aiming at the measurement of reactions of interest in explosive astrophysical environments, it is not obvious to point out well-defined "astrophysical consequences" that would have been drawn from such data. It appears indeed that such consequences are very much model-dependent.

Though the main motivation of the above measurements was nuclear astrophysics, my impression is that the first impact was to nuclear physics : every time a (p,γ) reaction was measured, the (p,p) elastic scattering was obtained as well and my feeling is that the latter data (resonance energies, total width of levels) were more debated in the nuclear physics community than were the former one (the (p,γ) reaction rate) in the astrophysics community. The reason for that situation is probably that nuclear physics uncertainties in explosive events' models are presently second-order effects, astrophysicist being in fact tackling more basic problems, like the treatment of convection, the type of explosion, the number of dimensions in models, ...

In my opinion, we nuclear physicists involved in this field should not be deceived by this seeming lack of interest. We should keep in mind that the same situation prevailed for some decades in a neighbouring field, the nuclear reactions relevant in quiet stellar environments like our sun : a well-defined solar standard model has existed many years after first measurements of such reactions had been performed. I am confident that soon or later, the same evolution will occur in the explosive environments ; as soon as a "nova standard model" or a "supernova standard model" will exist, nuclear physics uncertainties will come back to the forefront of the astrophysicists' concerns.

ACKNOWLEDGMENTS

The author is a Research Director of the National Fund for Scientific Research, Brussels.

REFERENCES

1. Proceedings of the 7th Int. Symp. on Nuclei in the Cosmos, 2002, Ed. by S. Kubono, T. Teranishi, T. Kazino, K. Nomoto and I. Tanihata, 2003 Nucl. Phys. A718, 1c-758c
2. Decrock P. et al., 1991, Phys. Rev. Lett. 67, 808-811
3. Vancraeynest G. et al., 1998, Phys. Rev. C57, 2711-2723
4. Bishop S. et al., 2003, Phys. Rev. Lett. 90, 162501
5. Davinson T. et al., 2000, Nucl. Instr. Meth. Phys. Res. A454, 350-358
6. Rolfs C. and Rodney W.S., 1988, Cauldrons in the Cosmos, The University of Chicago Press
7. The Universe in Gamma Rays. V. Schönfelder (Ed.), Springer-Verlag, 2001
8. Winkler C., in Exploring the Gamma-Ray Universe, Fourth INTEGRAL Workshop, Alicante, Spain (2000) ; ESA-SP 459, A. Gimino, V. Reglero, and C. Winkler eds, 2001, pp. 471-478
9. Baur G. and Rebel H., 1994, J. Phys. G : Nucl. Part. Phys. 20, 1-33
10. Spitaleri C. et al., 2003, Nucl. Phys. A719, 99c-106c
11. Kroha V. et al., 2003, Nucl. Phys. A719, 119c-122c
12. Davids B. et al., 2003, Phys. Rev. C67, 012801

Photoreaction Cross Section Measurements for Astrophysics

H. Utsunomiya*, S. Goko*, K.Y. Hara*, H. Akimune*, T. Yamagata*, M. Ohta*, H. Ohgaki[†], H. Toyokawa**, T. Hayakawa[‡], T. Shizuma[‡], Y.-W. Lui[§] and P. Mohr[¶]

*Department of Physics, Konan University, 8-9-1 Okamoto, Higashinada, Kobe 658-8501, Japan
[†]Institute of Advanced Energy, Kyoto University, Gokanosho, Uji, Kyoto 611-0011, Japan
**National Institute of Advanced Industrial Science and Technology (AIST), 1-1-1 Umezono, Tsukuba, Ibaraki 305-8568, Japan
[‡]Advanced Photon Research Center, Japan Atomic Energy Research Institute, Tokai, Ibaraki 319-1195, Japan
[§]Cyclotron Institute, Texas A & M University, College Station, Texas 77843, USA
[¶]Strahlentherapie, Diakoniekrankenhaus Schwäbisch Hall, D-74523 Schwäbisch Hall, Germany

Abstract. γ-ray beams at synchrotron radiation facilities will give us unprecedented opportunities to study photonuclear reactions of astrophysical interest. We report our photonuclear reaction studies and give a prospect of a fertile research field to be pioneered in the next 5 years.

INTRODUCTION

There are three major real-photon sources that have been used in nuclear physics: radioactive isotopes, bremsstrahlung, and positron annihilation in flight [1, 2]. Recently, one witnesses emergence of a new generation of γ-ray sources at synchrotron radiation facilities that are devoted to material science. An idea of producing γ rays in laboratory by interactions between laser photons and relativistic electrons was born in 1963 [3, 4]. The idea was first put into reality for practical use at Frascati in 1980's [5]. However, its use for astronuclear physics had been ignored until recently.

Photonuclear reaction data of astrophysical relevance are (γ,γ), (γ,n), (γ,p), and (γ,α) cross sections near particle thresholds without/with Coulomb barrier effects in exit channels. As shown in the literature [6], photoneutron reaction rates are determined by the product of a Planck spectrum of black-body radiation and photoneutron cross sections. As a result, (γ,n) cross sections immediately above thresholds are of astrophysical significance. Although a nuclear physics database for electric giant dipole resonance was constructed after major contributions at Lawrence Livermore National Laboratory and Saclay [7], the data near thresholds lack sufficient accuracy for astrophysical use. (γ,p) and (γ,α) reactions are difficult to measure because of the Coulomb barrier effects. (γ,γ) cross sections below particle thresholds should never be forgotten for photoreactions on nuclei thermally equilibrated in excited states [8].

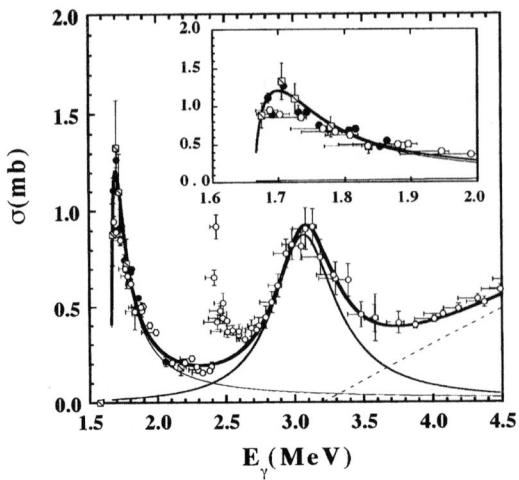

FIGURE 1. Photodisintegration cross sections for ^{9}Be$(\gamma,n)\alpha\alpha$ [10, 11].

γ BEAMS AT AIST

Real photon beams in the MeV region have been developed at the National Institute of Advanced Industrial Science and Technology (AIST) in *head-on* collisions of laser photons with relativistic electrons stored in an accumulator ring TERAS [9]. The laser Compton scattering (LCS) plays a role of *photon accelerator*, producing quasi-monochromatic γ beams in the energy range of 1 - 40 MeV. They are bremsstrahlung-free unlike the positron annihilation in flight, and 100 % linearly (circularly) polarized. Because of the monochromaticy, the AIST-LCS beam is best suited to excitation function measurements of photoreaction cross sections with enriched target materials. In addition, photo-activation of natural foils by the AIST-LCS beam can be done for nuclei whose isotopic abundance is sufficiently large.

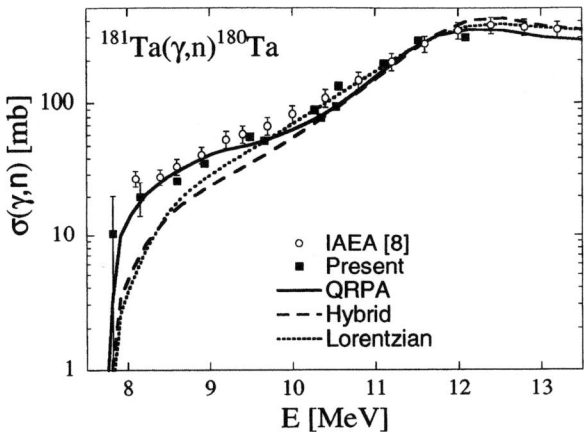

FIGURE 2. Photodisintegration cross sections for ^{181}Ta$(\gamma,n)^{180}$Ta [12].

PHOTOREACTION CROSS SECTION MEASUREMENTS AT AIST

The AIST-LCS γ beam with a rather limited intensity (10^{4-5} photons/sec) was so far used to measure cross sections of ^{9}Be$(\gamma,n)\alpha\alpha$ for supernovae nucleosynthesis [10, 11], ^{181}Ta$(\gamma,n)^{180}$Ta [12] for the p-process nucleosynthesis, and D(γ,n)p [13] for big bang nucleosynthesis. Further the latest photoneutron cross section measurements included ^{186}W, ^{187}Re, and ^{188}Os for the s-process nucleosynthesis and cosmochronometer, and ^{93}Nb and ^{139}La for the p-process nucleosynthesis.

Figure 1 shows photonuclear data for ^{9}Be [11]. Large differences were found in electromagnetic quantities ($B(E1)\downarrow$ and $\Gamma_\gamma(M1)$) for ^{9}Be between the present real-photon measurement and the electron scattering (virtual-photon) measurement. The present experiment is followed by microscopic-model studies of a ^{9}Be nucleus [14, 15, 16]. The reaction rate of $\alpha\alpha(n,\gamma)^9Be$ was evaluated from the photoneutron cross section and compared with those of the CF88 [17] and the NACRE [18] compilations [11]. A factor of two difference was found to CF88 over the temperature range $T_9 = 0.1$ - 10, while reasonable agreement with NACRE was obtained, putting aside details in the treatment of individual states in ^{9}Be.

Figure 2 shows photonuclear data for ^{181}Ta [12]. The data for ^{181}Ta necessitated a microscopic understanding of threshold behavior of photoneutron cross sections [12], showing the advantage of a QRPA calculation over a conventional Lorentzian-model or a hybrid-model analysis. The deduced overproduction factor shows that the nature's rarest isotope and the only naturally occurring isomer ^{180}Ta can be a p-process nuclide, leaving the neutrino process as a most competing process. In the context of the p-process nucleosynthesis, one should challenge to photo-destruction cross section measurements for ^{180}Ta. Also needed are determination of partial photoneutron cross sections leading to the 9^- isomeric state in ^{180}Ta for a dynamical p-process calculation without an explicit assumption of thermalization during type-II supernovae explosion.

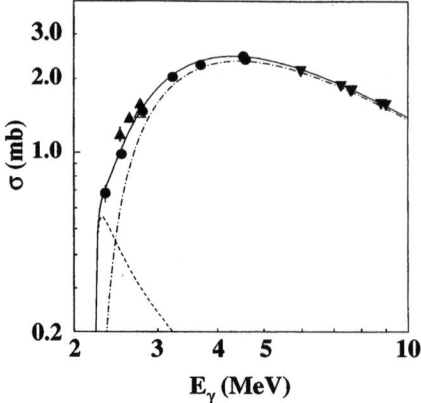

FIGURE 3. Photodisintegration cross sections for D(γ,n)p [13]. The filled circles stand for the present result. The filled triangles are from [48], the open triangle from [49], and the filled reversed triangles from [50].

Figure 3 shows photonuclear data for deuterium [13]. The photodisintegration data for deuterium were converted to cross sections of the capture reaction, p(n,γ)D [13]. A least-squares fit to the present cross section combined with the existing capture data [19, 20] was obtained. It is to be pointed out that the capture cross section was evaluated only theoretically for more than 30 years since the FCZI compilation in 1967 [21]. Furthermore, details of the latest evaluation by Hale *et al.* compiled in ENDF/B-VI [22] can no longer be traced. The present work has provided an experimental confirmation of the theoretical evaluation, offering the foundation for recent discussions of the baryon density based on observations of primeval deuterium-to-hydrogen ratios in metal-poor hydrogen clouds at high-redshifts toward quasers [23, 24, 25]. One may be entering a new era of big bang nucleosynthesis upon an independent confirmation of the baryon density by the observation of temperature anisotropies of the cosmic microwave background (CMB) radiation by the Wilkinson Microwave Anisotropy Probe (WMAP) satellite [26].

γ BEAMS AT SPRING-8

Currently two γ sources are under development at the 8 GeV Super-Photon ring (SPring-8) in Japan: one γ source is based on inverse Compton scattering of far-infrared laser photons with $\lambda = 118.8$ μm from 8-GeV electrons; and the other one is high-energy radiation from a 10 tesla superconducting wiggler (SCW) [27]. The former is rather white with the maximum energy ~ 10 MeV, while the latter is characterized by exponential tails extended to several MeV. It is expected that their intensities are 10^{7-8} photons/sec/MeV near neutron thresholds, ~ 8 MeV.

Figure 4 shows energy spectra of the SCW radiations calculated following the prescription of [28, 29]. It is remarkable that the exponential tails of the radiation are equiv-

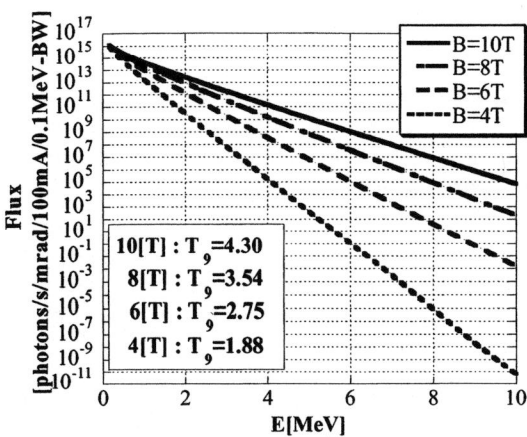

FIGURE 4. Synchrotron radiation from a 10 tesla superconducting wiggler at SPring-8 (calculations).

alent to temperatures of a few to several billions of kelvin (T_9). In other words, the SCW radiation *mimics* the Planck spectrum of black-body radiations generated during supernova explosions. In a promising astrophysical site for the p-process, the O/Ne layers of massive stars during their pre-supernovae phase [30, 31, 32] or during their explosions as type-II supernovae [33, 34, 31, 32], temperatures of the blackbody radiation lie in the range of T_9 = 1.7 - 3.3. This temperature range can be accessed by the SCW at lower magnetic fields. The intensity of the high-energy tail of the SCW radiation strongly depends on the magnetic field. Under this condition, it is noted that the SCW radiation can be used to directly determine photonuclear reaction rates by activation technique without such manipulation [35] as superposition of several Bremsstrahlung with different end-point energies.

Experimentally, the higher the intensity the more feasible the measurement. The intensity at the maximum 10 tesla field is indeed attractive. The threshold behavior of the photoneutron cross section can be parametrized by

$$\sigma(E) = \sigma_0[(E_\gamma - S_n)/S_n]^p, \quad (1)$$

where E_γ is γ-ray energy and S_n is neutron separation energy. When a neutron with an orbital angular momentum ℓ is emitted in photodisintegration, p is given by $\ell + 1/2$. Generally different values of ℓ contribute, depending on the spin and the parity of the residual nucleus in photodisintegration. This is in good contrast with the fact that s-wave neutrons are preferentially captured at low energies in the neutron capture process. σ_0 and p are parameters to be determined experimentally. Based on Eq.(1), a few measurements of the photoreaction rate with the SCW radiation at highest magnetic fields enable us to determine these experimental parameters. This method is more practical and promising in the p-process study, opening up a variety of applications. We have a long list of photoreactions to be studied by photoactivation technique though not listed here.

PROSPECT OF PHOTONUCLEAR REACTION STUDIES

A variety of photonuclear reactions can be studied at the synchrotron radiation facilities in Japan in the context of big bang nucleosynthesis, neutron poison, p-process nucleosynthesis, s-process branching, and cosmochronometer. They include photo-destruction of the nature's rarest isotope, ^{180}Ta[0.012%]$(\gamma,n)^{179}$Ta[$T_{1/2}=1.82$ y]. Preparing enriched ^{180}Ta material for photonuclear reactions is out of question. This reaction can be studied by photoactivation of natural tantalum foils with a high-intensity γ source at SPring-8 (see [36] for details). On the other hand, *more light* of the order of 10^{11} photons/sec is awaited by another p-process nuclide ^{138}La[0.09%] whose photo-destruction leads to ^{137}La[$T_{1/2}=6\times 10^4$ y].

Nuclear waste known as long-lived fission products (LLFPs) is nothing but the s-process branch-point nuclei whose neutron capture can compete with β decay, depending upon such physical parameters as temperature and neutron flux of a promising s-process site, AGB stars [37]. Nuclear transmutation by Accelerator-driven Systems aims to convert LLFPs with a halflife of the order of 10^5 - 10^6 years to stable or short-lived nuclei [38]. However, cross sections for capture of thermal, epithermal, and keV neutrons by radioactive LLFPs are difficult to measure. Basic nuclear data for ^{79}Se, ^{93}Zr, and ^{107}Pd can be evaluated by photodisintegration of stable nuclei (^{80}Se, ^{94}Zr, and ^{108}Pd). Such evaluation was done in astrophysics with bremsstrahlung for an s-process branch-point nucleus ^{185}W [39]. Recently, photodisintegration of ^{186}W was also investigated with the AIST-LCS γ beams.

The detection of cosmic-ray events above 10^{20} eV has raised a great excitement [40, 41, 42]. If they were protons, it may be due to violation of the Lorentz invariance that generates energies above the Greisen-Zatsepin-Kuz'min (GZK) cutoff imposed by photo-pion production in the intergalactic photon field [43]. A propagation problem was also addressed for iron nuclei originated from an extragalactic point source [44], where photodisintegration of nuclei occurs in interactions with the CMB radiation and intergalactic infrared background radiation. It was shown that a cluster structure appears in the energy spectrum of the cosmic-ray events [45]. A ground-based Pierre Auger Observatory [46] and an Extreme Universe Space Observatory (EUSO) to be installed on the International Space Station [47] may reveal the composition of ultrahigh-energy cosmic rays.

We are at the entrance of unexplored field of photonuclear reactions of astrophysical interest. In addition to reactions induced by charged particles and neutrons, photonuclear reactions can constitute an astrophysics database with devoted efforts by international collaborations.

ACKNOWLEDGMENTS

This work was supported in part by the Japan Private School Promotion Foundation and by the Japan Society for the Promotion of Science.

REFERENCES

1. R.L. Bramblett, J.T. Cladwell, G.F. Auchampaugh, and S.C. Fultz, Phys. Rev. **129**, 2723 (1963).
2. R. Bergère, H. Beil, and A. Veyssière, Nucl. Phys. **A121**, 463 (1968).
3. R.H. Milburn, Phys. Rev. Lett. **10**, 75 (1963).
4. F.R. Arutyunian and V.A. Tumanian, Phys. Lett. **4**, 176 (1963).
5. L. Federici et al., Nuovo Cimento **59 B**, 247 (1963).
6. P. Mohr, K. Vogt, M. Babilon, J. Enders, T. Hartmann, C. Hutter, T. Rauscher, S. Volz, and A. Zilges, Phys. Lett. **B488**, 127 (2000).
7. S.S. Dietrich and B.L. Berman, At. Data Nucl. Data Tables **38**, 199 (1988).
8. M. Arnould and S. Goriely, Phys. Rep. **384**, 1 (2003).
9. H. Ohgaki, S. Sugiyama, T. Yamazaki, T. Mikado, M. Chiwaki, K. Yamada, R. Suzuki, T. Noguchi, and T. Tomimasu, IEEE Trans. Nucl. Sci. **38**, 386 (1991).
10. H. Utsunomiya, Y. Yonezawa, H. Akimune, T. Yamagata, M. Ohta, M. Fujishiro, H. Toyokawa, and H. Ohgaki, Phys. Rev. **C63**, 018801 (2001).
11. K. Sumiyoshi, H. Utsunomiya, S. Goko, and T. Kajino, Nucl. Phys. **A709**, 467 (2002).
12. H. Utsunomiya, H. Akimune, S. Goko, M. Ohta, H. Ueda, T. Yamagata, K. Yamasaki, H. Ohgaki, H. Toyokawa, Y.-W. Lui, T. Hayakawa, T. Shizuma, E. Khan, and S. Goriely, Phys. Rev. **C67**, 015807 (2003).
13. K.Y. Hara, H. Utsunomiya, S. Goko, H. Akimune, T. Yamagata, M. Ohta, H. Toyokawa, K. Kudo, A. Uritani, Y. Shibata, Y.-W. Lui, and H. Ohgaki, Phys. Rev. **D68**, 072001 (2003).
14. P. Descouvemont, Eur. Phys. J. **A12**, 413 (2000).
15. N. Itagaki and K. Hagino, Phys. Rev. **C66**, 057301 (2002).
16. K. Arai, P. Descouvemont, D. Baye, and W.N. Catford, Phys. Rev. **C68**, 014310 (2003).
17. G.R. Caughlan and W.A. Fowler, At. Data Nucl. Data Tables **40**, 283 (1988).
18. C. Angulo et al., Nucl. Phys. **A656**, 3 (1999).
19. T.S. Suzuki et al., Astrophys. J. **439**, L59 (1995).
20. Y. Nagai et al., Phys. Rev. **C56**, 3173 (1997).
21. W.A. Fowler, G.R. Caughlan, and B.A. Zimmerman, Ann. Rev. Astr. Ap. **5**, 525 (1967).
22. G.M. Hale, D.C. Dodder, E.R. Siciliano, and W.B. Wilson, Los Alamos National Laboratory, ENDF/B-VI evaluation, Mat #125, Rev. 1, 1991.
23. C.B. Netterfeld et al., Astrophys. J. **571**, 604 (2002).
24. D. Stompor et al., Astrophys. J. **561**, L7 (2001).
25. C. Pryke et al., Astrophys. J. **568**, 46 (2002).
26. D.N. Spergel et al., Astrphys. J. Suppl. **148**, 175S (2003).
27. A. Ando et al., J. Synchrotron **3**, 201 (1996); ibid. **5**, 360 (1998).
28. J.D.Jackson, "Classical Electrodynamics", (Wisley, 2nd ed.), Section 14.6.
29. V.O.Kostroun, Nucl. Instr. Meth. **172**, 371 (1980).
30. M. Arnould, Astron. Astrophys. **46**, 117 (1976).
31. M. Rayet, M. Arnould, M. Hashimoto, N. Prantzos, and K. Nomoto, Astron. Astrophys. **298**, 517 (1995).
32. T. Rauscher, A. Heger, R.D. Hoffman, and S.E. Woosley, Astrophys. J. **576**, 323 (2002).
33. S.E. Woosley and W.M. Howard, Astrophys. J. Suppl. **36**, 285 (1978).
34. M. Arnould, M. Rayet, and M. Hashimoto, in *Unstable Nuclei in Astrophysics*, edited by S. Kubono and T. Kajino (World Scientific, Singapore, 1992), p. 23.
35. K. Vogt, P. Mohr, M. Babilon, J. Enders, T. Hartmann, C. Hutter, T. Rauscher, S. Volz, and A. Zilges, Phys. Rev. **C63**, 055802 (2001).
36. S. Goko, this symposium.
37. M. Busso, R. Gallino, and G.J. Wasserburg, Annu. Rev. Astron. Astrophys. **37**, 239 (1999).
38. Accelerator-driven Systems (ADS) and Fast Reactors (FR) in Advanced Nuclear Fuel Cycles, A Comprehensive Study, NEA#03109, ISBN: 92-64-18482-1 (2002).
39. K. Sonnabend, A. Mengoni, P. Mohr, T. Rauscher, K. Vogt, and A. Zilges, Nucl. Phys. **A718**, 533c (2003); ibid. **719**, 123c (2003).
40. N. Hayashi et al., Phys. Rev. Lett. **73**, 3491 (1994).
41. D.G. Bird et al., Astrophys. J. **441**, 144 (1995).
42. M. Takeda et al., Phys. Rev. Lett. **81**, 1163 (1998).

43. H. Sato and T. Tati, Prog. Theor. Phys. **42**, 1788 (1972).
44. F.W. Stecker and M.H. Salamon, Astrophys. J. **512**, 521 (1999).
45. T. Yamamoto *et al.*, Astropart. Phys. **20**, 405 (2003); this symposium.
46. http://www.auger.org/
47. M. Teshima, this symposium; http://www.euso-mission.org/
48. G.R. Bishop *et al.*, Phys. Rev. **80**, 211 (1950).
49. R. Moreh, T.J. Kennett, and W.V. Prestwich, Phys. Rev. **C39**, 1247 (1989).
50. Y. Birenbaum, S. Kahane, and R. Moreh, Phys. Rev. **C32**, 1825 (1985).

Low-energy radioactive-ion beam separator at CNS and resonance scattering experiments

T. Teranishi[*], S. Kubono[†], J.J. He[†], M. Notani[†], T. Fukuchi[†],
S. Michimasa[†], S. Shimoura[†], S. Nishimura[**], M. Nishimura[**],
Y. Yanagisawa[**], M. Kurokawa[**], Y. Wakabayashi[‡], N. Hokoiwa[‡],
Y. Gono[‡], T. Morikawa[‡], A. Odahara[§], H. Ishiyama[¶], Y.X. Watanabe[¶],
T. Hashimoto[¶], T. Ishikawa[¶], M.H. Tanaka[¶], H. Miyatake[¶], J.Y. Moon[∥],
J.H. Lee[∥], J.C. Kim[∥], C.S. Lee[∥], V. Guimarães[††], R.F. Lihitenthaler[††],
H. Baba[‡‡], A. Saito[‡‡], K. Sato[§§], T. Kawamura[§§], S. Kato[§§], H. Iwasaki[¶¶],
K. Ue[¶¶], Y. Satou[***] and Zs. Fülöp[†††]

[*]*Center for Nuclear Study (CNS), University of Tokyo, RIKEN campus, 2-1 Hirosawa, Wako, Saitama 351-0198, Japan*
[†]*Center for Nuclear Study (CNS), University of Tokyo, Japan*
[**]*The Institute of Physical and Chemical Research (RIKEN), Japan*
[‡]*Department of Physics, Kyushu University, Japan*
[§]*Nishinippon Institute of Technology, Japan*
[¶]*IPNS, KEK, Japan*
[∥]*Chung-Ang University, Korea*
[††]*Department of Nuclear Physics, São Paulo University, Brazil*
[‡‡]*Department of Physics, Rikkyo University, Japan*
[§§]*Department of Physics, Yamagata University*
[¶¶]*Department of Physics, University of Tokyo*
[***]*Tokyo Institute of Technology, Japan*
[†††]*Institute of Nuclear Research of the Hungarian Academy of Sciences (ATOMKI), Hungary*

Abstract. A low-energy radioactive-ion beam separator based on an in-flight technique was developed at CNS, University of Tokyo and was installed at RIKEN. Using the separator, experiments of the elastic resonance scattering of ^{11}C+p, ^{12}N+p, and ^{23}Mg+p were performed to study resonance levels in ^{12}N, ^{13}O, and ^{24}Al, respectively. New information on spin-parities and widths for these levels may help understand stellar (p,γ) reaction rates.

INTRODUCTION

With the development of ISOL based facilities, it has been realized that low-energy radioactive-ion (RI) beams below 10 MeV/nucleon are useful to study nuclear-reaction properties for nuclear astrophysics. As an alternative to ISOL, we have developed an in-flight technique dedicated to low-energy RI beams. In this technique, a low-energy heavy-ion primary beam of below 10 MeV/nucleon and a light-ion target of 0.1–1 mg/cm^2 are used for an RI production reaction in inverse kinematics. Secondary ions, emitted out from the thin target, fly toward forward angles in the laboratory frame with energies close to the primary beam energy. After the target, a magnetic separator purifies and focuses the secondary ions to utilize as an RI beam. In the ISOL technique, development and operation of target ion sources are not simple since one needs to deal with chemical properties of ions to be extracted. On the other hand, development and operation of the in-flight separator are technically easier.

The in-flight separator, called CNS Radioactive-Ion Beam separator (CRIB) [1, 2], was constructed in RIKEN Accelerator Research Facility (RARF) under cooperation of RARF and Center for Nuclear Study (CNS), University of Tokyo. As the first step for nuclear studies using the separator, we are performing experiments of elastic resonance scattering with low-energy RI beams and proton targets to investigate resonance levels in the RI+p systems.

LOW-ENERGY RADIOACTIVE-ION BEAM SEPARATOR

Primary heavy-ion beams for RI production are supplied by an AVF cyclotron with $K = 70$ at RARF. With ECR ion sources attached to the cyclotron, heavy-ion beams are available with reasonably high intensities. For example, a ^{14}N beam was produced with 500 pnA on target at an energy of 8.4 MeV/nucleon. Production targets are usually gases of light elements. Hydrogen and CH_4 gases can be used for (p,n) reactions in inverse kinematics. Deuterium and ^{3}He gases can be used, for example, for (d,p) and (^{3}He,n) reactions in inverse kinematics, respectively. An RI production rate of 10^6 particles/sec can be achieved with a primary-beam intensity of 500 pnA, a cross section of 10 mb and a target thickness of 0.1 mg/cm^2. The actual RI beam intensity depends also on the kinematic acceptance of the separator.

The CRIB separator consists of two dipole magnets and four quadrupole magnets. There is a momentum dispersive focal plane (F1) between the two dipole magnets and an achromatic focal plane (F2) after the second dipole magnet. Secondary particles are selected by the magnetic rigidity using a slit at F1. The maximum aperture of the slit corresponds to a momentum acceptance of $\pm 7.5\%$. The acceptance of separator in solid angle is about 5.6 msr. For secondary particles by (p,n) reactions in inverse kinematics, a transmission efficiency from the target point to F2 is usually more than 50% because of the large momentum and solid-angle acceptances. An energy degrader can be inserted at F1 to have an additional particle separation power by means of energy loss.

A primary beam is swept out by the first dipole magnet and stopped in a beam dump set inside the magnet. However, halo components of the primary beam cause scattering at inner walls of beam pipes and the magnets. Scattered ions of the same nuclide as the primary beam often reach F2 and become a major contaminant in the secondary beam. Therefore, the purity of an RI beam at F2 depends on the RI intensity and the rate of scattered contaminant. For (p,n) reactions, purities of about 90% can be easily obtained. For other reactions with smaller cross sections or smaller acceptances, the purities often go down to 10% or less. To improve the purity further, a Wien filter system is now under construction and will be attached to F2. The Wien filter system has a velocity separation section of 1.5-m long with a maximum magnetic field of 3 kGauss and a maximum electric field of 50 kV/cm.

EXPERIMENTS OF ELASTIC RESONANCE SCATTERING

With low-energy RI beams of ^{11}C, ^{12}N and ^{23}Mg produced by CRIB, experiments for the elastic resonance scattering of ^{11}C+p, ^{12}N+p, ^{23}Mg+p were performed, respectively. To test the experimental technique, contaminant particles of ^{24}Mg stable nucleus, which came together with the ^{23}Mg beam, were used to observe known resonance levels in the ^{24}Mg+p scattering. The experimental technique and results of these experiments are discussed below.

Experimental Technique

An elastic scattering experiment was performed in inverse kinematics with an RI beam and a $(CH_2)_n$ target (as a proton target). To measure the excitation function $(d\sigma/d\Omega(E))$ efficiently, a thick-target technique [3, 4] was used. The thickness of the target was chosen to be a little larger than the stopping range of the beam. Utilizing energy-loss process of the beam in the target, a wide range of center-of-mass energy (E_{CM}) was scanned without changing the beam energy before the target. While the beam particles were completely stopped in the target, most of the recoil protons went out from the target with small energy losses. The recoil protons were detected by two sets of silicon detectors at laboratory angles of $\theta_{LAB} = 0°$ and $17°$. The E_{CM} and center-of-mass angle (θ_{CM}) were determined by measuring the energy and angle of proton. At $\theta_{LAB} = 0°$, the proton energy is roughly four times of E_{CM}. Because of this kinematic factor, it is rather easy to achieve an E_{CM} resolution of 30 keV (FWHM) by using a silicon detector.

The secondary beams produced by CRIB often have contaminants. Therefore, it is necessary to identify beam particles on an event-by-event basis. For this purpose, two parallel-plate avalanche counters (PPACs) [5] were set for beam particles at the upstream side of the $(CH_2)_n$ target. The PPACs gave information of timing and two-dimensional hit position. Time-of-flight between the two PPACs was used to identify beam nuclides. The beam incident angle and reaction position on the target were determined by the hit positions at the two PPACs.

A carbon target was also used to check the background contribution to the proton spectrum from C atoms in the $(CH_2)_n$ target. The proton spectrum with the C target was subtracted from that with the $(CH_2)_n$ target to deduce the excitation function of proton elastic scattering.

In the excitation function, a resonance level can be identified as an interference pattern of potential scattering and resonance scattering. The energy, width (Γ) and spin-parity value (J^π) of the resonance level may be deduced from an R-matrix analysis for the interference pattern. The angle of $\theta_{LAB} = 0°$ corresponds to $\theta_{CM} = 180°$, where the Coulomb-potential scattering amplitude is minimum. Therefore, angles around $\theta_{LAB} = 0°$ are suitable for observing resonance contributions.

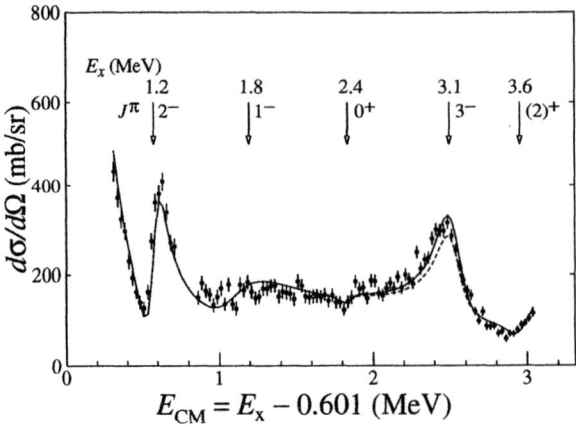

FIGURE 1. Excitation function of the ^{11}C+p elastic scattering at $\theta_{CM} \sim 180°$. The solid curve represents a result of R-matrix based analysis.

^{11}C+p and ^{12}N+p

The measurement of ^{11}C+p and ^{12}N+p aimed at studying resonance levels in ^{12}N and ^{13}O, respectively, above the proton threshold energies. Some of these levels may play important roles in the hydrogen burning processes of ^{11}C(p,γ)^{12}N and ^{12}N(p,γ)^{13}O at high temperatures in metal-deficient massive stars [6, 7].

The ^{11}C and ^{12}N beam particles were produced simultaneously with a ^{10}B primary beam of 7.8 MeV/nucleon and a ^{3}He gas target of 0.25 mg/cm^2. The production reactions were ^{3}He(^{10}B,^{11}C)np for ^{11}C and ^{3}He(^{10}B,^{12}N)n for ^{12}N. The secondary beam had an intensity of about 10^5 particles/sec and consisted of ^{11}C (15%), ^{12}N (3%), and other contaminants, mainly ^{10}B. The energies of ^{12}N and ^{11}C were 3.9 and 3.4 MeV/nucleon, respectively.

Figure 1 shows the result of excitation function of ^{11}C+p, covering $E_x = 0.9$–3.7 MeV in ^{12}N. There are several known levels in this region. The spectrum was fitted with an R-matrix based resonance formula [8] as indicated by the solid line in Fig. 1. The energies and widths of the levels determined by the present experiment are consistent with the known values. A peak seen in the spectrum at $E_{CM} \sim 2.5$ MeV corresponds to the 3.13-MeV level, whose J^π value was not clearly determined before. From the shape and width of the peak, the level was attributed to a d-wave resonance, which indicates the negative parity for the level. The height of the peak depends on J and is consistent with $J = 3$. Therefore, we have made an assignment of $J^\pi = 3^-$ for the level at $E_x = 3.13$ MeV in ^{12}N. From the J^π assignment, it can be concluded that the 3.13-MeV level does not contribute to the (p,γ) reaction rate so much, even at high temperatures, since the level decays by M2 or E3 transitions to the 1^+ ground state in ^{12}N, which is the unique bound state in ^{12}N.

The excitation function of ^{12}N+p is under analysis to investigate levels in ^{13}O. There are only three known excited levels in ^{13}O and most of their J^π and Γ values have not

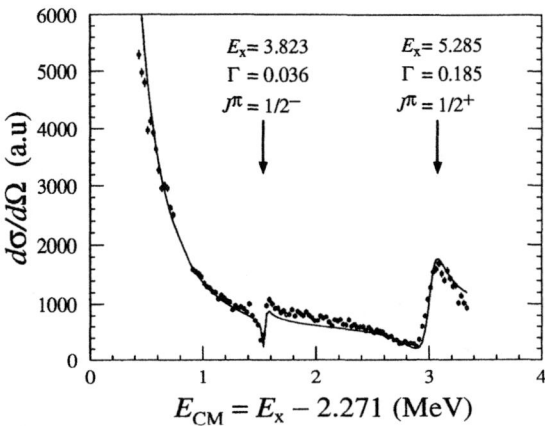

FIGURE 2. Preliminary result of the excitation function for the elastic scattering of protons on ^{24}Mg stable nucleus at $\theta_{CM} \sim 180°$. The solid curve represents a tentative result of an R-matrix fit.

been determined yet [9]. In a preliminary spectrum, a peak was observed at $E_x \sim 2.7$ MeV in ^{13}O, corresponding to one of the known levels. From the shape and width of the peak, the level might be attributed to an s-wave resonance.

^{23}Mg+p and ^{24}Mg+p

Levels in ^{24}Al above the proton threshold of $E_x = 1.871$ MeV may play important roles in the astrophysical reaction of ^{23}Mg$(p,\gamma)^{24}$Al, which is one of the breakout paths from the Ne-Na cycle [10]. Values of J^π are not precisely determined for many of these levels. There are almost no experimental data of Γ for resonance levels in ^{24}Al [11]. The ^{23}Mg+p experiment aimed at deducing new information on J^π and Γ for these levels from the excitation function of ^{23}Mg+p.

The RI beam of ^{23}Mg was produced by the ^{24}Mg$(d,t)^{23}$Mg reaction in inverse kinematics with a ^{24}Mg primary beam of 7.5 MeV/nucleon and a deuterium gas target of 0.33 mg/cm^2. The energy and intensity of ^{23}Mg were 3.5 MeV/nucleon and 3.2×10^4 particles/sec, respectively. The purity of ^{23}Mg in the secondary beam was 12%. A major contaminant was ^{24}Mg, which was also used to observe known resonance levels of ^{24}Mg+p.

The ^{24}Mg+p data, taken simultaneously with the ^{23}Mg+p data, is useful to check the energy calibration of the silicon detectors and the analysis procedures. Figure 2 shows the result of ^{24}Mg+p excitation function, covering $E_x = 2.7$–5.6 MeV in ^{25}Al. The spectrum clearly shows resonance shapes due to two known levels in ^{25}Al at $E_x = 3.823$ and 5.285 MeV with $\Gamma = 0.036$ and 0.185 MeV, respectively. The solid curve in Fig 2. represents a result of R-matrix analysis and agrees well with the experimental data. There are many other known levels in the present spectral range. However, their contributions to the spectrum were not clear because most of their widths are known to be less than 10 keV [11], which is smaller than the energy resolution of about 30 keV (FWHM).

In a preliminary spectrum for ^{23}Mg+p, several peaks were observed. Most of them seem to correspond to known levels in ^{24}Al. Analysis is in progress for these peaks to restrict J^{π} values and to determine Γ values.

SUMMARY

We have produced so far low-energy RI beams of about ten nuclides at energies around 5 MeV/nucleon. The data of ^{11}C+p demonstrated usefulness of low-energy RI beams by the in-flight technique for studying resonance levels in unstable nuclei. The experimental technique for elastic resonance scattering was also confirmed by the data of known resonance levels in ^{24}Mg+p. Measurement and analysis for other RI+p systems are in progress or being planned. These beams may be also useful to study resonance reactions of astrophysical interests, such as the (p,α) and (α,p) reactions. The present work was partially supported by Grant-in-Aid for Science Research from the Japan Ministry of Education, Culture, Sports, and Technology under the contract numbers 13440071 and 14740156, and also by the Joint Research Program under the Korea-Japan Basic Science Promotion Program (KOSEF 2002-JR015).

REFERENCES

1. T. Teranishi et al., *Nucl. Phys.*, **A718**, 207c–213c (2003).
2. S. Kubono et al., *Eur. Phys. J.*, **A13**, 217–220 (2002).
3. K.P. Artemov et al., *Sov. J. Nucl. Phys.*, **52**, 408–411 (1990).
4. S. Kubono, *Nucl. Phys.*, **A693**, 221–248 (2001).
5. H. Kumagai et al., *Nucl. Instr. and Meth.*, **A470**, 562–570 (2001).
6. M. Wiescher et al., *Astrophys. J.*, **343**, 352–364 (1989).
7. R. Mitalas, *Astrophys. J.*, **290**, 273–275 (1985).
8. T. Teranishi et al., *Phys. Lett.*, **B556**, 27–32 (2003).
9. F. Ajzenberg-Selove, *Nucl. Phys.*, **A523**, 1–196 (1991).
10. S. Kubono et al., *Nucl. Phys.*, **A588**, 521–536 (1995).
11. P.M. Endt, *Nucl. Phys.*, **A633**, 1–220 (1998).

Study of Astrophysical (α, n) and (p, n) Reactions on Light Neutron-Rich Nuclei by Means of Low-Energy RNB

Hironobu Ishiyama[*], Hiroari Miyatake[*], Masa-Hiko Tanaka[*],
Yutaka Watanabe[*], Nobuharu Yoshikawa[*], Sunchan Jeong[*],
Yoshitaka Matsuyama[*], Yoshihide Fuchi[*], Ichiro Katayama[*],
Toru Nomura[*], Takashi Hashimoto [ᶺ], Tomoko Ishikawa [ᶺ], Kouji Nakai [ᶺ],
Suranjan K. Das , Pranab K. Saha , Tomokazu Fukuda ,
Katsuhisa Nishio , Shinichi Mitsuoka , Hiroshi Ikezoe ,
Makoto Matsuda , Shinichi Ichikawa , Takeshi Furukawa ,
Hideaki Izumi , Tadashi Shimoda , Yutaka Mizoi ⸸, Mariko Terasawa

Institute of Particle and Nuclear Studies, High Energy Accelerator Research organization (KEK), Tsukuba, Ibaraki, 305-0801 Japan
ᶺDepartment of Physics, Tokyo University of Science, Noda, Chiba, 278-8510 Japan
Osaka Electro-Communication University, Neyagawa, Osaka, 572-8530 Japan
Japan Atomic Energy Research Institute (JAERI), Tokai, Ibaraki, 319-1195 Japan
Department of Physics, Osaka University, Toyonaka, Osaka, 560-0043
⸸The Institute of Physical and Chemical Research (RIKEN), Wako, Saitama, 351-0198 Japan
Center for Nuclear Study, University of Tokyo, Wako, Saitama, 351-0198 Japan

Abstract. A systematic study of astrophysical reaction rates of (α, n) and (p, n) reactions on light neutron-rich nuclei by using low-energy radioactive nuclear beam is in progress at the tandem facility of Japan Atomic Energy Research Institute. Exclusive measurements of ^{8}Li(α, n)^{11}B and ^{16}N(α, n)^{19}F reaction cross sections have been performed successfully. Their excitation functions together with the experimental method are presented.

INTRODUCTION

Where the rapid process takes place has been a long-standing puzzle in the history of theoretical studies of element synthesis in the universe. One of the most probable sites for the r-process often discussed today is the so-called " hot bubble" in the supernova explosion. That is, the r-process is considered to occur in the region between the surface of a pre-neutron star and the outward-moving shock wave during the explosion. The nuclear statistical equilibrium favors abundant free neutrons and alpha particles in this region as long as the relevant temperature is high. When we follow the paper of Ref. [1], even seed nuclei for the r-process can be produced in this region in the early α-capture process at around $T_9 = 3$. When the temperature and density become lower and charged-particle induced reactions almost cease, the usual r-process starts from such seed nuclei produced and a large number of free neutrons. Therefore, nuclear reactions such as (α, n) on light neutron-rich nuclei play an important role as the r-process starting point [1].

However, there is little experimental data on cross sections of reactions on light neutron-rich nuclei. We are therefore proceeding an experimental project to measure systematically cross sections of astrophysical interest for (α, n) and (p, n) reactions on ^{6}He, ^{8}Li, ^{10}Be, ^{12}B, ^{16}N, and ^{20}F using radioactive nuclear beams (RNB) [2, 3] at the tandem facility of Japan Atomic Energy Research Institute (JAERI). Direct measurements of ^{8}Li(α, n)^{11}B, ^{16}N(α, n)^{19}F and ^{16}N(p, n)^{16}O reaction rates have been already carried out and their analysis is in progress.

The present method of production of low-energy neutron-rich RNB, some characteristics of the detection system as well as the resultant excitation functions of ^{8}Li(α, n)^{11}B and ^{16}N(α, n)^{19}F reactions are presented in this paper.

RNB PRODUCTION

Because the beam energy available is relatively low, we have decided to produce low-energy neutron-rich beam by using transfer reactions on light targets. In this case, it is important to avoid impurities originating from the primary beam particles [4]. We have utilized a recoil mass separator (RMS) existing at the tandem facility [5, 6], which consists of two electric dipoles and a magnetic dipole as shown in FIG. 1. The RNB can be separated from the primary beam using the difference of the magnetic and electric rigidities.

The d(^{18}O, ^{16}N)α reaction was chosen for production of ^{16}N-beam. The production target is D$_2$ gas of 1 atmospheric pressure contained in a 5 cm long chamber separated from the vacuum region with Havar foils of proper thickness. The typical intensity of the primary ^{18}O^{6+}-beam of 73 MeV was about 300 enA on the target. The ^{16}N-beam energy became 32 MeV at the exit of the target. The intensity of the secondary beam was then 4.7 kpps at the focal plane of RMS. The contamination of ^{18}O particles in the ^{16}N-beam measured with a ΔE-E telescope at the RMS focal plane is about 1.5%.

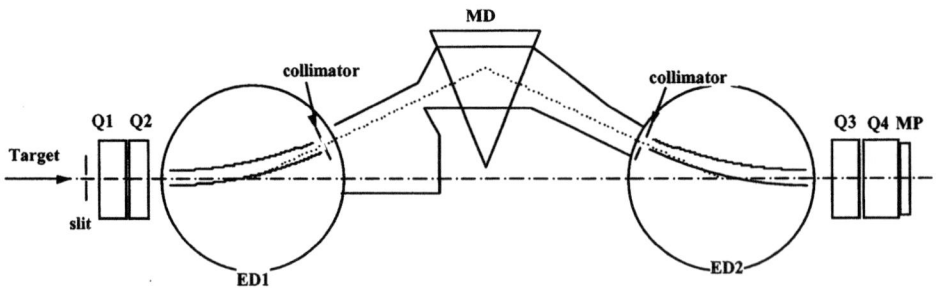

FIGURE 1. Ion optical configuration of JAERI-RMS. Q, ED, MD, and MP stand for magnetic quadrupole, electric dipole, magnetic dipole, and magnetic multi-pole, respectively.

TABLE 1. The measured yield and purity of RNBs.

RNB	production reaction	E [MeV]	Yield (100enA^{-1})	contamination
^{6}He	d(^{7}Li, ^{6}He)^{3}He	10	5.9 ~10^3	t = 42%
^{8}Li	^{9}Be(^{7}Li, ^{8}Li)^{8}Be	14.6	4.8 ~10^4	^{6}He = 1%
^{12}B	d(^{11}B, ^{12}B)p	20	2.0 ~10^4	^{12}C, ^{11}B < 10 %
^{16}N	d(^{18}O, ^{16}N)α	32	1.6 ~10^3	^{18}O = 1.5%

If we define the beam suppression factor as the ratio of the number of ^{18}O particles contained in ^{16}N beam to that in the primary beam, it is 2.2 ~10^{-10}. It should be noted, however, that the ^{16}N-beam consists of the ground state ($J^\pi = 2^-$) and the isomeric state ($E_x = 120$ keV, $J^\pi = 0^-$, $t_{1/2} = 5.25$ μs). The isomer ratio has been measured to be 35 } 1 %.

The ^{8}Li-beam has been produced via ^{9}Be(^{7}Li, ^{8}Li) reaction. A 42μm ^{9}Be foil was set at the target position. In order to make the energy resolution of the resultant ^{8}Li-beam better, the target foil was tilted at 40 with respect to the beam axis. The maximum intensity of the primary ^{7}Li-beam was about 200 enA, its initial energy being 24 MeV. The intensity of ^{8}Li is 4.8 ~10^3 pps /10 enA ^{7}Li beam at the focal plane of RMS. Its energy and resolution were 14.6 MeV and 5%, respectively. Although a little amount of ^{6}He particles are mixed in the secondary ^{8}Li–beam obtained, no ^{7}Li impurities have been observed. The purity of ^{8}Li-RNB was 99%.

Table 1 lists RNB so far produced. Concerning the ^{6}He-beam case, the main contaminant was triton, of which the intensity was the same order of magnitude as that of ^{6}He, because their velocity was nearly the same as that of ^{6}He particles. However, it is no serious problem under the present detector system described later, because tritons can be easily distinguished from ^{6}He.

DETECTOR SYSTEM

The present detector system is schematically shown in FIG. 2. It consists of a beam pick up detector system, a "multi-sampling and tracking proportional chamber" (MSTPC) [7] placed at the focal-plane of RMS, and a neutron detector array, in which the first one is composed of a multi-channel plate (MCP) and a parallel plate avalanche counter (PPAC). The absolute energy of RNB is determined by time-of-flight (TOF) information between MCP and PPAC. Then, the RNB is injected into the MSTPC filled with gas of He+CO_2 or CH_4 which works as counter gas and gas target.

FIG. 3 shows the cross-sectional view of the MSTPC. The MSTPC can measure a three-dimensional track of a charged particle and the energy loss along its trajectory. Electrons produced by the incoming charged particle in the drift space are drifted toward the proportional region consisting of segmented cathode, 24 pad cells, and anode wires.

The vertical position of a trajectory can be determined by drift time of electrons. The horizontal position is determined from the comparison of signals from the right-hand and left-hand sides of a pad cell divided into 2 electrodes. The position along the beam direction can be determined by the position of the segmented cathode. Of course, the energy loss of each pad can be determined.

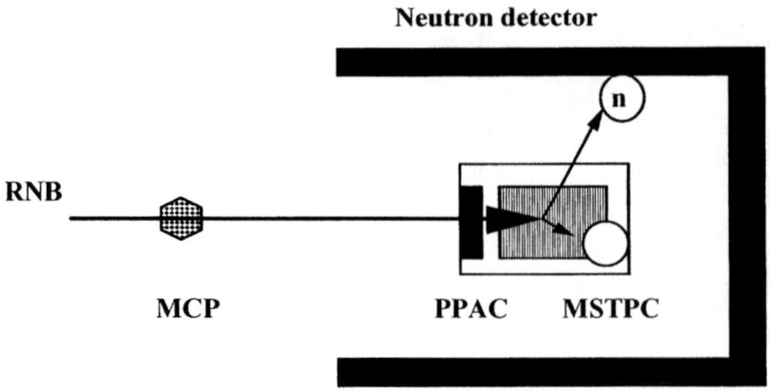

FIGURE 2. Schematic view of the experimental set-up. The RNB provided from JAERI-RMS is directly injected into the MSTPC filled with He +CO_2 or CH_4 gas, which acts as counter gas and as gas target. A neutron detector array consisting of 28 plastic scintillators is placed around the MSTPC.

When a nuclear reaction takes place inside the MSTPC, the energy loss (dE/dx) changes largely mainly due to the change of the relevant atomic numbers. Therefore, where the reaction occurs can be determined by detecting the dE/dx change, and the beam energy at the reaction point can be evaluated using the range and energy loss relation.

A neutron detector array to detect neutrons emitted from nuclear reactions is placed to surround the MSTPC as much as possible. It consists of 28 pieces of BC408 plastic sintillators, covering 31.2% solid angle of 4 π. The absolute energy of a neutron is obtained by TOF measurement between the PPAC and the plastic scintillator. The vertical position of the neutron is determined by the position of a plastic sintillator itself, and the horizontal position is evaluated by TOF of right-hand side and left-hand

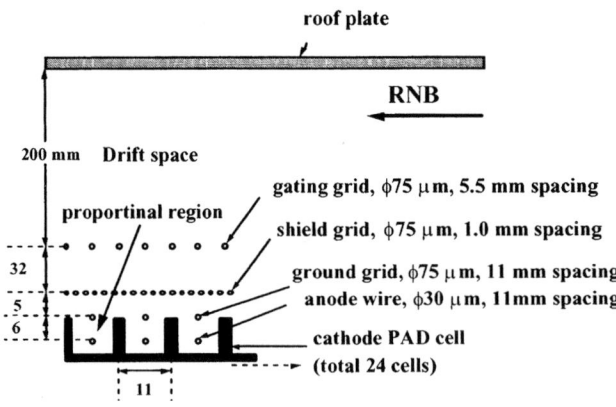

FIGURE 3. Schematic cross-sectional view of the MSTPC (not scaled).

side signals from a plastic scintillator. The typical efficiency for a 5 MeV neutron measured by using a ^{252}Cf-fission source is about 40%.

The MSTPC should work under relative high injection rate ($\sim 10^4$ pps) of low-energy RNB particles. However, large pulse height defects in energy loss signals from the MSTPC were first observed under high injection rate larger than 10^3 pps. Because these defects are considered to come from space charge gain limitation at the anode wires due to numerous ions generated by multiplication processes, a gating grid [8] was installed between the drift region and the proportional region inside the MSTPC as shown in FIG. 3. Alternate wires of the gating grid are connected with two separate voltage sources. The transparency of the gating grid plane can be controlled by changing the electric potential difference between the alternative wires. The gating grid plane is normally kept opaque so that electrons cannot reach anode wires. When an external signal generated by the neutron counters triggers the gating grid, it becomes transparent during some limited time so that the drift electrons can reach anode wires.

A performance test of the gating grid using sample beam of ^{14}N ions (E = 2 MeV/u) was made at the tandem accelerator center of University of Tsukuba. In this case, the gating grid was triggered by a signal from SSD installed at the end of the MSTPC. The trigger rate was reduced properly by a down scalar and was fixed at 20 cps.

The results are shown in FIG. 4. All pulse heights of energy loss signals from cathode pads are normalized by those obtained under the injection rate of 20 cps. Without operation of the gating grid, the pulse height defect became about 15 % around the center of the MSTPC at the injection rate of 3.6 kpps as shown in FIG. 4 (a). This value is consistent with a value of 9 % estimated using the equation in Ref. [9]. Under operation of the gating grid with injection rate less than 5.6 kpps, the pulse

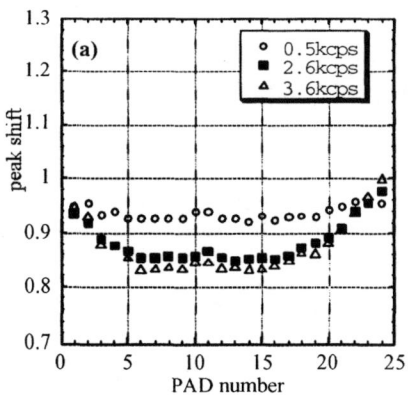
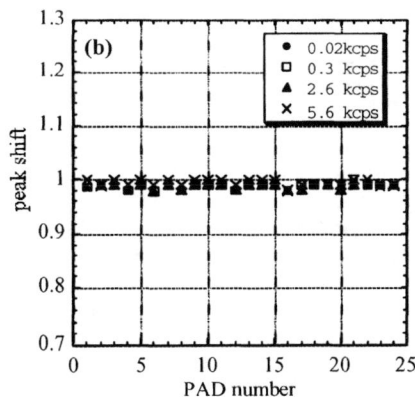

FIGURE 4. Relative pulse height of signal from each cathode pad with various injection rates from 0.02 k pps to 5.6 k pps without operation of the gating grid (a), and with operation of the gating grid (b). The horizontal axis shows the number of cathode pads counted from the upper stream of the beam.

height defect is suppressed successfully within about 2 %. This value is small enough for the present experiment.

EXPERIMENTS

All the experiments to measure cross sections of (α, n) and (p, n) reactions have been carried out at the JAERI tandem facility.

RNB injection rates into the MSTPC for measurements of ^{16}N(α, n) and ^{8}Li(α, n) were 2 ~10^3 pps and 5 ~10^3 pps, respectively. These rates were restricted by the upper limit of the present data acquisition system. The trigger for data acquisition was generated by the coincident signal between the PPAC and one of the plastic scintillators. The maximum trigger rate was 20 cps, which was mainly due to accidental coincidences between the injected RNB and background signals of plastic scintillators.

The MSTPC was filled with He + CO_2 (10 %) gas. Gas pressure for measurements of ^{16}N(α, n) and ^{8}Li(α, n) cross sections were 129 Torr and 220 Torr, respectively. Some typical parameters used for the ^{16}N(α, n) measurement are tabulated in Table. 2. Those for the measurement of the ^{8}Li(α, n) reaction are given in ref. [10].

Because the accidental rate mentioned above is far higher than the true event rate, it is necessary to distinguish true events from accidental. It is possible to select true events by comparison with the energy loss between a certain pad and the next pad. The threshold of energy loss difference between the two pads was set at 150 keV for the selection of the ^{16}N(α, n) reaction events by considering the energy loss of the emitted ^{19}F. In the case of the ^{8}Li(α, n) reaction, it was set at 40 keV by considering the energy loss of the emitted ^{11}B. The reaction event thus selected was checked with its kinematical condition by using all information on its reaction energy, scattering angle and energy of ejected nuclei and neutron. The selected typical event of the ^{16}N(α, n) ^{19}F reaction is shown in FIG. 5.

TABLE 2. Some typical parameters for the measurement of the ^{16}N(α, n) reaction.

parameter	value
electrode of MCP	1.5 μm Mylar, evaporated Au
entrance window of PPAC	2.0 μm Mylar, 40 mm (diameter)
electrode of PPAC	1.5 μm Mylar, evaporated Al
exit window of PPAC	3.5 μm Mylar, 40 mm (diameter)
gas in PPAC	iso-butane, 6 Torr
length between MCP and PPAC	900 mm
gas in MSTPC	He + CO_2 (10 %), 129 Torr
roof plate voltage of MSTPC	-1170 V
gating grid voltage (V_0) of MSTPC	-199 V
gating grid voltage (}ΔV) of MSTPC	40 V
anode wire voltage of MSTPC	790.3 V
length between MSTPC and plastic scintillators	900 mm
size of a typical plastic scintillator	50 ~150 ~1500 mm

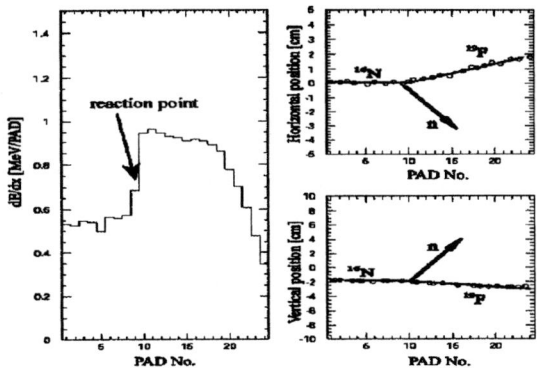

FIGURE 5. Selected typical event of the ^{16}N(α, n) reaction. The left-hand side of the figure shows the dE/dx spectrum. The horizontal axis shows the relative length from the beam injection point inside the MSTPC given by the pad number. One pad corresponds to 11mm. The right-hand shows horizontal and vertical projections of the 3-dimensional particle trajectory. A neutron is detected at the direction marked by arrow.

RESULTS

The measured excitation function of the ^{16}N(α, n) ^{19}F reaction is shown in FIG. 6. Black circles indicate the measured cross sections and the curve shows the theoretical estimation by Fowler [11]. Arrows indicate the Gamow energy region at $T_9 = 2$ and 3.

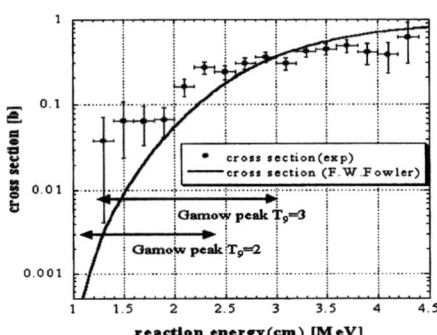

FIGURE 6. Measured excitation function of ^{16}N(α, n)^{19}F reaction. The horizontal axis is the center-of-mass energy and the vertical one is the cross section given in unit of barns.

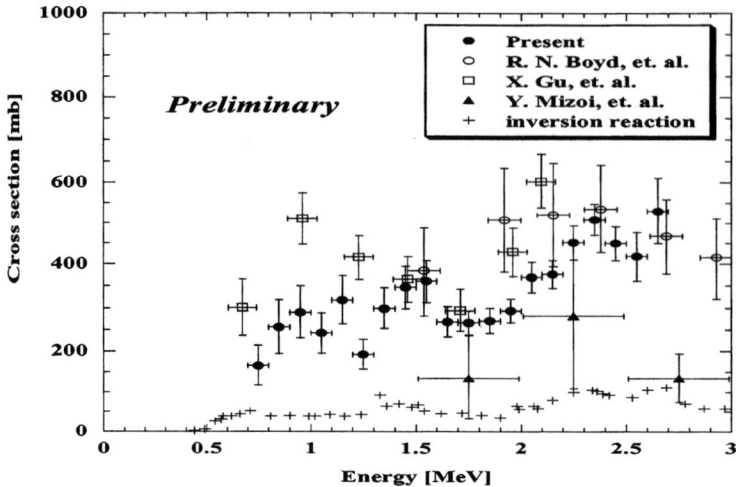

FIGURE 7. The measured excitation function of ^{8}Li$(\alpha, n)^{11}$B reaction (black circles). The horizontal axis is the center-of-mass energy and the vertical one is the cross section in unit of mb.

The excitation function of ^{16}N$(\alpha, n)^{19}$F reaction was determined successfully in the energy region of E_{cm} = 1.5 – 4.0 MeV. It is to be noted that the present measured cross sections around the energy region of T_9 = 3 are a few times larger than the theoretical estimation.

FIG. 7 shows the measured excitation function of the ^{8}Li$(\alpha, n)^{11}$B reaction. It is still preliminary, because the analysis is not yet complete in various aspects. The given errors are only statistical.

Black circles indicate our experimental cross sections and the other symbols show previous experimental data. Open circles and squares indicate the data by Boyd et al. [12] and Gu et al. [13] based on inclusive measurements without neutron detection. Black triangles indicate the data by Mizoi et al. [14] based on the exclusive measurement using a detector system similar to that in the present work. Black crosses show cross sections leading to the ground state only estimated from the inversion reaction [15].

It is to be noted that the present result has good statistics compared with the previous data. Further analysis including the excitation function in the energy region from E_{cm} = 0.8 MeV to 2.7 MeV is still in progress.

In addition, it is possible to obtain branching ratios of neutron decay channels from the compound ^{12}B in our experiment. Excited states of residual ^{11}B are open up to the 8th state (Ex = 8.559 MeV, J^π = 3/2$^-$) in our experimental condition. The excitation-energy spectrum of ^{11}B is shown in FIG. 8. The ground and the first excited states are not separated clearly. But the other higher excited states can be seen in this spectrum and positions of measured peaks look consistent with those expected. Its branching ratios will be determined in further analysis.

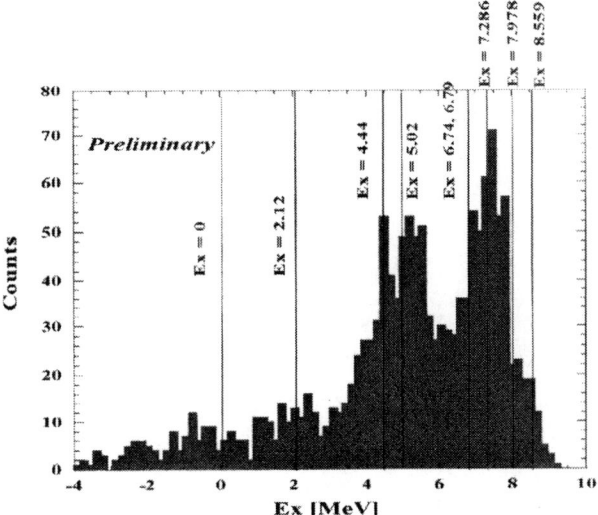

FIGURE 8. Excitation-energy spectrum of ^{11}B states. The horizontal axis shows the excitation energy in unit of MeV and the vertical axis shows the number of raw events summed over the energy region from E_{cm} = 0.8 MeV to 2.7 MeV. Each line indicates excitation energy of an ^{11}B state.

SUMMARY

Exclusive measurements of reaction rates of (α, n) reactions on light neutron-rich nuclei are in progress using low-energy radioactive nuclear beam. The excitation function of the ^{16}N(α, n) reaction has been measured in the energy region of E_{cm} = 1.5 – 4.0 MeV corresponding to the Gamow energy at T_9 = 2 – 6. In the energy region of T_9 = 3, the experimental cross sections obtained are a few times larger than the theoretical estimation. Preliminary results of the ^{8}Li(α, n) reaction cross sections measured in the energy region of E_{cm} = 0.8 – 2.7 MeV are reported. Further measurement of the same reaction rates below E_{cm} = 0.8 MeV will be carried out soon. Measurements of ^{6}He(α, n), ^{10}Be(α, n) and (p, n), ^{12}B(α, n) and (p, n), ^{20}F(α, n) and (p, n) reactions are also being planned.

ACKNOWLEDGMENTS

The authors wish to thank Prof. Y. Tagishi and Dr. T. Komatsubara at University of Tsukuba for their helpful support in the performance test of the MSTPC. We also thank the staff members of the JAERI tandem facility for their kind operation of the tandem accelerator.

REFERENCES

1. Terasawa, M. et. al., *Nucl. Phys.* **A688**, 581c(2001).
2. Ishiyama, H. et. al., *Nucl. Phys.* **A718**, 481c-483c (2003).
3. Ishikawa, T. et. al., *Nucl. Phys.* **A718**, 484c-486c (2003).
4. Becchetti, F. D. et. al., *Nucl. Instrum. Meth. Phys. Res.* **B56/57**, 554 (1991).
5. Ikezoe, H. et. al., *Nucl. Instrum. Meth. Phys. Res.* **A376**, 470 (1996).
6. Kuzumaki, T. et. al., *Nucl. Instrum. Meth. Phys. Res.* **A437**, 107 (1999).
7. Mizoi, Y. et. al., *Nucl. Instrum. Meth. Phys. Res.* **A431**, 112 (1999).
8. Nemethy, P. et. al, *Nucl. Instrum. Meth. Phys. Res.* **A212**, 73(1983).
9. Hendricks, R. W., *Rev. Sci. Instrum.* **40**, 1216(1969).
10. Hashimoto, H. et. al., to be submitted in *Nucl. Instrum. Meth.*
11. Fowler, W. A., *Astrophy. J. Supp.* **91**,201 (1964).
12. Boyd, R.N. et. al., *Phys. Rev. Lett.* **68**, 1283 (1992).
13. Gu, X. et. al., *Phys. Lett.* **B343**, 31 (1995).
14. Mizoi, Y. et. al., *Phys. Rev.* **C62**, 065801 (2000).
15. Paradellis, T. et. al., *Z. Phys.* **A337**, 211 (1990).

The (n,γ) cross sections of short-living s-process branching points

K. Sonnabend*, A. Mengoni†, P. Mohr*, T. Rauscher**, K. Vogt* and A. Zilges*

*Institut für Kernphysik, Technische Universität Darmstadt, Germany
†CERN, Geneva, Switzerland
**Institut für Physik, Universität Basel, Switzerland

Abstract. An experimental method to determine the (n,γ) cross section of short-living s-process branching points using data of the inverse (γ,n) reaction is presented. The method was used to observe the branching point nucleus ^{95}Zr because the elemental abundance patterns corresponding to this branching point cannot be reproduced by full stellar models and a possible error source is the neutron capture cross section of ^{95}Zr. The analysis of the experiment is still under progress, we will outline the current status in this manuscript.

1. INTRODUCTION

The nucleosynthesis of the elements heavier than iron is today mainly explained by three processes: s-, r-, and p-process. s- and r-process are based on neutron-capture reactions with adjacent β-decays while the p-process is governed by photodisintegration reactions such as (γ,n), (γ,p), or (γ,α).

The distinction between s- and r-process is due to the different neutron densities n_n involved. During s-process nucleosynthesis n_n is in the order of 10^8 neutrons/cm^3 while typical r-process sites deal with $n_n > 10^{20}$ neutrons/cm^3 [1]. Hence, the neutron capture rates $\lambda_{(n,\gamma)}$ are larger than typical β-decay rates λ_β during r-process nucleosynthesis ($\lambda_{(n,\gamma)} \gg \lambda_\beta$, r: rapid). In the s-process the situation is the other way round ($\lambda_{(n,\gamma)} \ll \lambda_\beta$, s: slow) and the involved nuclei are close to the valley of β-stability.

However even in a s-prozess scenario, if the half-life $T_{1/2}$ is long enough and the neutron capture cross section high enough another neutron capture might take place and the path "branches" out. Therefore, these nuclides are called branching points of the s-process. The branching ratio – i.e. how often each of the paths is taken – determines the corresponding elemental abundance patterns. Using a model for s-process nucleosynthesis and knowing precisely the nuclear physics input (half-life $T_{1/2}$ and Maxwellian averaged capture cross section (MACS) at typical s-process temperatures $kT = 30$ keV) it is possible to determine the astrophysical parameters temperature T and neutron density n_n.

In the so-called classical approach temperature T and neutron density n_n are considered to be constant. Thus, three different components produced with different astrophysical parameters are needed to reproduce the observed abundances of s-only nuclei: the weak component corresponds to mass numbers $A < 90$ while the strong component only describes the termination of the s-process path at lead and bismuth. In between these borders the isotopes belong to the so-called main s-process [2]. The abundances of s-only isotopes that are not affected by a branching can be reproduced with a mean square deviation of about 3% using this simple model [3].

A more realistic approach is a full stellar model, e.g. the AGB star model described in Ref. [4]. These models need very precise nuclear physics input data to reproduce the s-only abundances. Especially the MACS of several branching points hamper the reliability of the predictions: due to the lack of experimental data the theoretical predictions of the MACS sometimes show a broad spread (see [5]).

A direct measurement of the MACS of the branching points is only possible if their half-lives are in the order of years (e.g. ^{147}Pm with $T_{1/2} = 2.62$ yr [6]). In the case of short-living branching points with half-lives of about a dozen days or even less, adequate samples are not available. Thus, direct measurements are possible.

To solve this problem we have investigated an experimental method using the data of the inverse (γ,n) reaction as described in Section 2. Section 3 presents ^{95}Zr focusing on astrophysical aspects as well as on the constraints of our experimental method for this nuclide. The present status of the analysis is summarized in Section 3.3. In Section 4 we describe the next step in the data evaluation.

2. EXPERIMENTAL METHOD

The aim of our experimental method is to constrain the theoretical predictions of the MACS of short-living s-process branching nuclei. Short-living under s-process conditions denotes half-lives of about a dozen days or less, i.e. direct measurements are not possible due to the lack of adequate samples. Therefore, our method is based on data of the inverse (γ,n) reaction.

The first step is to measure the so-called energy-integrated cross section I_σ using the photoactivation technique:

$$I_\sigma = \int_{S_n}^{E_{max}} \sigma(E) \cdot N_\gamma(E, E_{max}) \cdot dE \qquad (1)$$

with $\sigma(E)$ being the (γ,n) cross section and $N_\gamma(E, E_{max})$ the experimental photon flux.

We use the monoenergetic electron beam provided by the S–DALINAC [7] to produce a continuous bremsstrahlung spectrum by fully stopping the beam in a thick rotating copper target. The spectrum ranges from $E = 0$ to E_{e^-}, so that E_{max} of Eq. (1) equals E_e. Due to the location of our setup directly behind the injector of the S–DALINAC, our maximum energy is limited to about $E_{e^-} = 10$ MeV presently [8]. While the shape of $N_\gamma(E, E_{max})$ is well known the absolute value has to be determined by a relative measurement with gold as a calibration standard.

After irradiating the targets the yield Y of the activation is measured by counting the γ-rays emitted in the decay of the produced radioactive nuclei with a HPGe-detector that is well-shielded against background by several layers of lead. The yield Y is proportional to the energy-integrated cross section I_σ with the constant of proportionality being explained in detail in Ref. [9].

Because the (γ,n) cross section of ^{197}Au is known to behave like

$$\sigma_{Au}(E) = 146.2 \text{ mb} \cdot \left(\frac{E - S_n}{S_n}\right)^{0.545} \tag{2}$$

near threshold [10], a predicted energy dependence of the target's cross section $\sigma_{target}(E)$ can be normalized to the observed ratio $I_\sigma^{target}/I_\sigma^{Au}$ by a correction factor f for each measured energy E_{max}. In doing so, either a parametrization or a theoretical calculation of $\sigma_{target}(E)$ can be used. If f depends on E_{max} the shape of $\sigma_{target}(E)$ is not accurately described (compare Eq. 1) and f can not be used for the further procedure without modifications.

Although the photoactivation technique is very sensitive, several limitations concerning the activation reaction $^A X(\gamma,n)^{A-1}X$ and the following decay of ^{A-1}X have to be taken into account. To test the energy dependece of the cross section $\sigma_{target}(E)$ it is necessary to measure at different energies E_{max}. Because our present setup is limited to 10 MeV, isotopes with neutron thresholds $S_n > 9$ MeV cannot be checked.

If one wants to use naturally composed targets, the fraction of the isotope of interest must not be too small and the activation of other isotopes should not produce too much background radiation when the decay of ^{A-1}X is observed. To reach a good peak-to-background ratio in the measured spectra only a few γ-rays with high branchings should be emitted in the decay of ^{A-1}X. Furthermore, the analyzed γ-decays should not be part of a multi-level decay and their energies E_γ should not allow self-absorption in the target to avoid additional uncertainties in the analysis.

If the normalization of a theoretical prediction of $\sigma_{target}(E)$ yields a constant factor f, we use f to correct the MACS predicted in the same model with equal parameters. This second step of using the correction factor for the inverse reaction is based on the detailed balance assumption. In our case, scaling the MACS with a constant factor f equals scaling the γ-ray strength function. For further discussion see [11].

Although the corrected MACS is still based on a theoretical prediction, its value is constrained by experimental knowledge of the inverse reaction due to the determination of f. Thus, it is more reliable than a value purely based on theory.

3. THE BRANCHING POINT ^{95}ZR

3.1. Position on the s-process path

The branching point ^{95}Zr is located near the borderline between the weak and the main component of the s-process [2]. Therefore, the prediction of the elemental abundance patterns corresponding to this branching is a crucial test for the validity of a full stellar model described in [4].

Several problems concering the prediction of the zirconium abundance patterns measured in SiC grains are reported in Ref. [12]. The uncertainty in the predicted MACS of

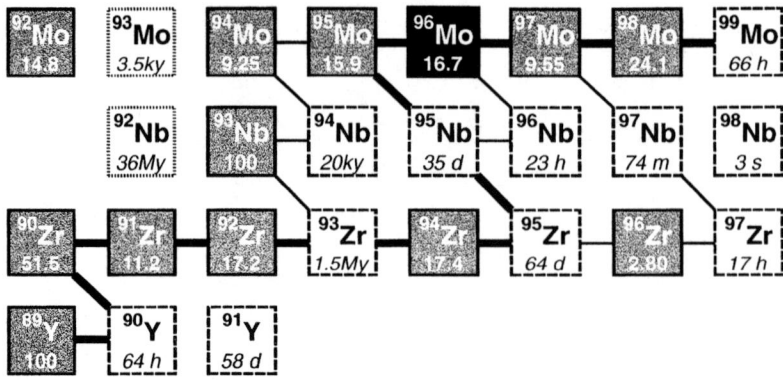

FIGURE 1. The s-process path around the branching point ^{95}Zr. The main flow of the s-process is marked with thick lines, thinner lines indicate branches. Grey shaded boxes illustrate stable isotopes with the numbers at the bottom of the boxes being their natural abundances. The black box of ^{96}Mo stresses that we have an s-only nucleus. The white boxes display β-instable nuclei: dashed boundaries are used for β^--decay, dotted lines are used to indicate β^+- or ε-instable isotopes. In this cases the numbers at the bottom of the boxes are the half-lives of the ground-states.

^{95}Zr is mentioned as one of the possible error sources due to the wide range of predicted values from 23 mb to 126 mb at $kT = 30$ keV (see [13], [14], [15], and [16]). Another reason for the deviation between the measured and predicted abundance patterns might be the uncertainty in the half-life $T_{1/2}$ of ^{95}Zr at s-process temperatures. However, as confirmed by a recent measurement [17] the first excited level is at $E = 954$ keV and thus, is not significantly populated at $kT = 30$ keV. Hence, the half-life of ^{95}Zr does not depend on temperature under s-process conditions and can be omitted as an error source because of its small error: $T_{1/2}(^{95}\text{Zr}) = (64.032 \pm 0.006)$ d [18].

3.2. Constraints of the experiment

To use our experimental method the constraints described in Section 2 have to be fullfilled for the activation reaction ^{96}Zr$(\gamma,n)^{95}$Zr. The threshold of this reaction is $S_n = 7854$ keV, which can be easily reached at our setup providing energies up to 10 MeV. The abundance of ^{96}Zr in a naturally composed target is 2.8% and thus, high enough for the activation method.

The properties of the decay of ^{95}Zr are summarized in Fig. 2. The characteristics claimed in Section 2 are achieved for $E_\gamma = 724.2$ keV and $E_\gamma = 756.7$ keV.

Due to the half-life of ^{95}Zr of about 64 days it is mandatory to observe the decay for several hours to reach good statistics. During this time the β^--decay of the daughter nucleus ^{95}Nb also starts to take place. The assignment of the measured γ-rays to the decays of ^{95}Zr or ^{95}Nb can either be done by a determination of the observed half-life or by comparing their different behaviour in time. As shown in Fig. 3 the ratio of the

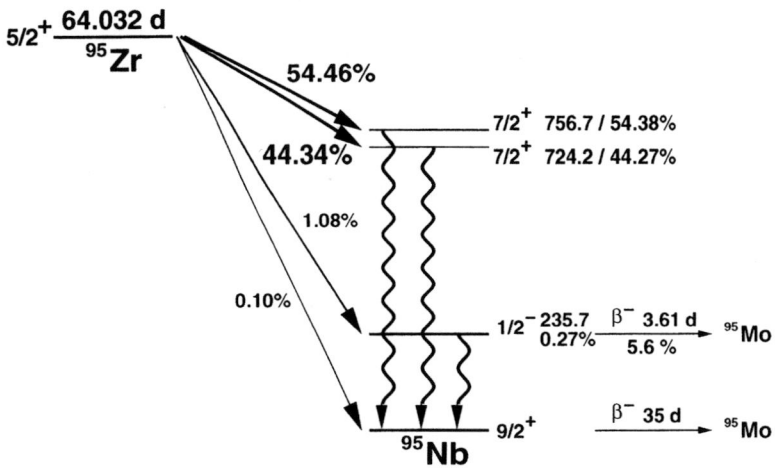

FIGURE 2. Section of the decay scheme of ^{95}Zr based on [18]

two peaks at 724.2 and 756.7 keV remains unchanged if the measurement takes place different times after activation. Thus, both lines correspond to the same decay and are identified to be the most prominent lines of the decay of ^{95}Zr. In contrast, the peak at 765.8 keV becomes more and more dominant with time. This behaviour is consistent

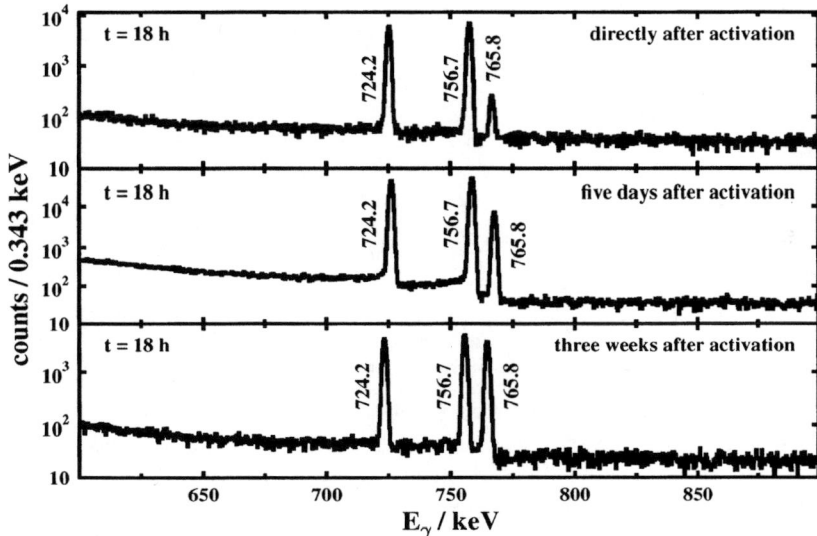

FIGURE 3. Typical decay spectra of ^{95}Zr each measured for $t = 18$ h. From the top to the bottom panel the time between activation and start of the measurement is changing as indicated on the top right. For further explanation see text.

with the one expected of a line corresponding to the decay of the daughter nucleus ^{95}Nb. Actually, the energy $E_\gamma = 765.8$ keV is exactly the energy supposed for the most prominent line of this decay.

3.3. Current status of analysis

The photoactivation of zirconium has been done at six different energies E_{max} ranging from 8325 to 9900 keV. The lower limit is due to the separation energy of the used calibration standard: $S_n(Au) = 8071$ keV. In a first approach, we used a parametrization that describes the energy dependence of the (γ,n) cross section near threshold:

$$\sigma(E) = \sigma_0 \cdot \left(\frac{E - S_n}{S_n}\right)^p \qquad (3)$$

In this case, the exponent p changes the shape of the energy dependence, whereas the factor σ_0 is responsible for the absolute value of $\sigma(E)$. Therefore, σ_0 corresponds to the correction factor f of Section 2. Due to the fact that this parametrization is derived from the time reversal symmetry of (n,γ) and (γ,n) reactions, the exponent p is connected to the wave-character of the emitted neutron: an exponent of $p = 0.5 + l$ corresponds to an l-wave decay. Thus, exponents less than 0.5 cannot be explained in this basic approach.

However, if a least-mean-square-fit of our data to this parametrization is carried out using σ_0 and p as free parameters, we obtain $p = 0.339$ and $\sigma_0 = 31.81$ mb as shown in Fig. 4. The single results $\sigma_0(E_{max})$ fluctuate around the mean value, thus, the overall energy dependence of $\sigma(E)$ seems to be described accurately.

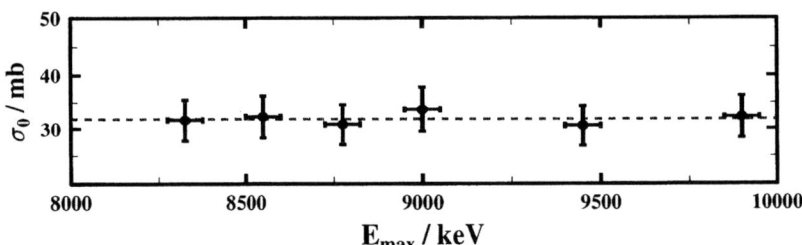

FIGURE 4. Result of a least-mean-square-fit of Eq. (3) to the measured data at different energies E_{max}. The dots are the results for σ_0 derived with an exponent of $p = 0.339$ while the dashed line is the mean value $\sigma_0 = 31.81$ mb. The horizontal error bars are due to uncertainties in the beam energy.

The result $p < 0.5$ can be explained by assuming a resonance on top of the smooth behaviour described by Eq. (3). We measure the energy integrated cross section I_σ and compare it with the result of Eq. (1) using $\sigma(E)$ of Eq. (3). Thus, we average the measured (γ,n) cross section $\sigma(E)$ over the energy interval ranging from S_n to E_{max}. A resonance in this energy region would yield the measured bigger I_σ.

The smaller p, the steeper is the rise of the cross section at the threshold energy S_n. Hence, the calculated I_σ becomes bigger and suits the measured value. Thus, we accept the derived values for p and σ_0 as a description of the mean energy dependence of the (γ,n) cross section: $\sigma(E) = 31.81$ mb $\cdot ((E - S_n)/S_n)^{0.339}$.

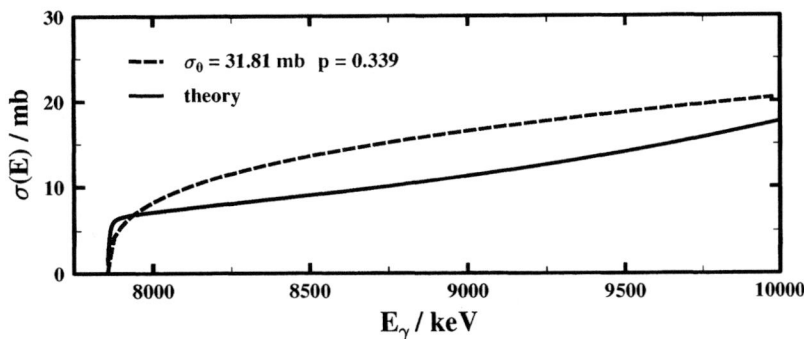

FIGURE 5. Comparison of the energy dependence of the cross section $\sigma(E)$ for the reaction ^{96}Zr(γ,n). The solid line is a theoretical prediction based on Hauser-Feshbach calculations. The dashed line is the result of a least-mean-square-fit of the parametrization described in Eq. (3) to our data.

Fig. 5 shows the mean $\sigma(E)$ and a theoretical prediction of the (γ,n) cross section $\sigma_{theo}(E)$ based on Hauser-Feshbach calculations [19, 20]. If the shapes of the two energy dependences are compared, a significant deviation is obvious: the theoretical prediction rises very steeply close to the threshold, but its slope becomes flatter 50 keV above threshold. In contrast, our parametrization is not so steep near threshold but more continuously rising.

Although the parametrization of $\sigma(E)$ describes only a mean energy dependence of the cross section, the different behaviour illustrates that the energy dependence of the theoretical prediction does not suit the data as well. Obviously, no resonances are predicted by theory. A correction factor f, derived as explained in Section 2, is only a mean correction. f can compensate the absence of resonances in theory. Furthermore, f is a function of E_{max}. Therefore, the MACS of ^{95}Zr predicted by this theory cannot simply be corrected by this mean value of f but the (n,γ) cross section must be corrected by $f(E_\gamma)$ before calculating the MACS. $f(E_\gamma)$ can be calculated weighting $f(E_{max})$ by $\sigma(E) \cdot N_\gamma(E, E_{max})$. For this purpose, it is necessary to measure the (γ,n) cross section as close as possible to the reaction threshold to locate the position of the resonances.

So far, the experiment was limited by the threshold of the calibration standard. The lowest measured energy was $E_{max} = 8325$ keV i.e. still about 500 keV above the threshold S_n of ^{96}Zr. In order to measure as close as possible near the target's threshold the neutron separation energy of a new calibration standard has to be lower.

In addition to a low neutron separation energy the calibration standard has to fullfill other characteristics: despite the general features mentioned in Section 2, the half-life $T_{1/2}$ of the produced radioactive isotope has to be short so that the counting can be done in short times with good statistics.

We decided to use ^{187}Re with a neutron separation energy of only $S_n = 7359$ keV and a natural abundance of 62.6%. The half-life of the produced ^{186}Re is $T_{1/2} = 3.72$ d and allows short measuring times although the branching of the two dominant γ-decays is quite low: $E_\gamma = 137.2$ keV with 9.42% and $E_\gamma = 122.6$ keV with 0.6%. Self-absorption in the rhenium target limits the amount of material, however, based on experience of former experiments [11] very thin rhenium targets with a thickness of $d = 50\ \mu$m should provide enough activity after a typical activation.

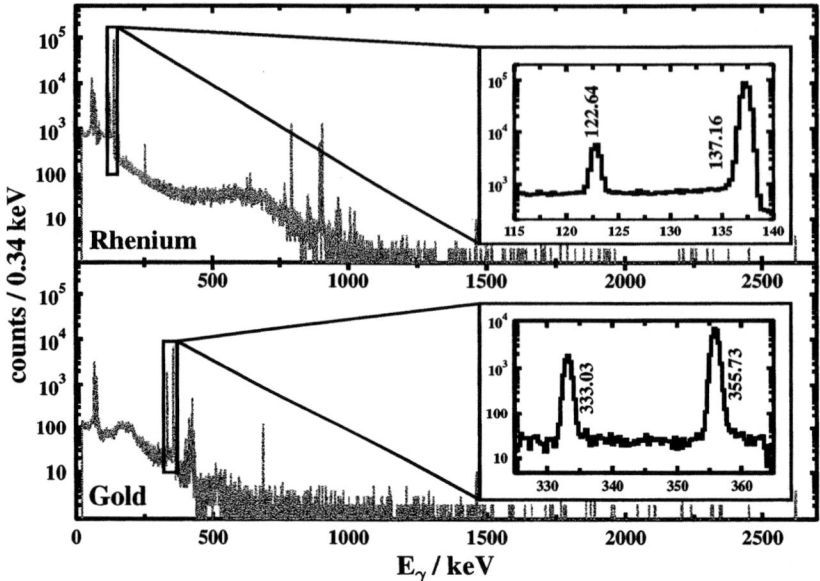

FIGURE 6. Comparison of the decay spectra of the calibration standards gold and rhenium. The inlays show the γ peaks that are used for the calibration. The peak-to-background ratio is in the same order of magnitude for both standards.

^{187}Re was measured in May 2003 and the analysis is still under progress. Fig. 6 compares the spectra of rhenium and gold measured after an activation at $E_{max} = 8325$ keV. The upper panel shows the spectrum of rhenium measured for three hours directly after the activation. The peaks between 750 and 1000 keV correspond to the decay of ^{184}Re that is produced by ^{185}Re(γ,n). Due to these lines the two peaks of the decay of ^{186}Re at $E_\gamma = 137.2$ keV and $E_\gamma = 122.6$ keV are located above a high background. However, as shown in the inlay the peak-to-background ratio of both lines is good enough to analyze the peak volumina with a low statistical error. For comparison the inlay in the lower panel shows the two most prominent lines of a gold spectrum also measured for three hours. The peak-to-background ratio is in the same order of magnitude.

4. SUMMARY AND OUTLOOK

The precise knowledge of the (n,γ) cross section of the s-process branching point ^{95}Zr is a crucial input parameter to test the reliability of the AGB star model. Due to the very wide range of the theoretically predicted values a measurement observing the inverse reaction with the photoactivation technique was carried out at the S–DALINAC in June 2002.

During the analysis of the data several problems occured: The energy dependence of the (γ,n) cross section of ^{96}Zr extracted from the data differs significantly from the theoretical prediction. In our opinion, the deviation is due to resonances that are not included in the used theory. Therefore, the experimental method described in Section 2 determines an energy dependent correction factor $f(E_\gamma)$ that cannot be used to correct the predicted MACS of ^{95}Zr without further studies.

Hence, the experiment was extended in May 2003 by measuring the (γ,n) cross section of ^{96}Zr closer to the reaction threshold. Therefore, a new calibration standard with lower neutron separation energy was measured during the same beam-time. The analysis of the new standard ^{187}Re is still under progress, preliminary results like the spectrum shown in Fig. 6 are encouraging that it will work as well as our first standard ^{197}Au.

Once a description of the (γ,n) cross section of ^{187}Re near threshold is available the new data point of ^{96}Zr taken at $E_{max} = 8100$ keV has to be added to the existing analysis. Therewith, the discrepancy between the experimental result and the theoretical prediction will hopefully disappear or at least become explainable. The outcome of the whole analysis that is the experimentally confirmed MACS of ^{95}Zr can then be used in the AGB star model and might solve the problems in explaining the observed zirconium abundance patterns in SiC grains.

In future, we will take more data points at different energies E_{max} in order to reduce difficulties due to resonances on top of the cross section. If the distance between two data points ΔE_{max} is smaller, the determination of the location of a resonance is possible with higher precision.

ACKNOWLEDGMENTS

The authors would like to thank R. Gallino (Torino, Italy) and F. Käppeler (Karlsruhe, Germany) for fruitful discussions concerning s-process modelling. This work was supported by the Deutsche Forschungsgemeinschaft (contracts Zi 510/2-2 and SFB 634) and Swiss NSF (grants 2124-055832.98, 2000-061822.00, 2024-067428.01).

REFERENCES

1. Beer, H., *Encyc. Astron. Astroph.* **2**, 1866 (2000).
2. Käppeler, F., *Prog. Part. Nucl. Phys.* **43**, 419 (1999).
3. Käppeler, F., Gallino, R., Busso, M., Picchio, G., and Raiteri, C., *Astrophys. J.* **354**, 630 (1990).
4. Lugaro, M., Herwig, F., Lattanzio, J., Gallino, R., and Straniero, O., *Astrophys. J.* **586**, 1305 (2003).
5. Bao, Z. Y., Beer, H., Käppeler, F., Voss, F., Wisshak, K., and Rauscher, T., *At. Data Nucl. Data Tables* **76**, 70 (2000).
6. Reifarth, R., Arlandini, C., Heil, M., Käppeler, F., Sedyshev, P., Mengoni, A., Herman, M., Rauscher, T., and Gallino, R., *Astrophys. J.* **582**, 1251 (2003).
7. Richter, A., *Prog. Part. Nucl. Phys.* **44**, 3 (2000).
8. Mohr, P., Enders, J., Hartmann, T., Kaiser, H., Schiesser, D., Schmitt, S., Volz, S., Wissel, F., and Zilges, A., *Nucl. Instr. and Meth. A* **423**, 480 (1999).
9. Vogt, K., Mohr, P., Babilon, M., Enders, J., Hartmann, T., Hutter, C., Rauscher, T., Volz, S., and Zilges, A., *Phys. Rev. C* **63**, 055802 (2001).

10. Vogt, K., Mohr, P., Babilon, M., Bayer, W., Galaviz, D., Hartmann, T., Hutter, C., Rauscher, T., Sonnabend, K., Volz, S., and Zilges, A., *Nucl. Phys.* **A707**, 241 (2002).
11. Sonnabend, K., Mohr, P., Vogt, K., Zilges, . A., Mengoni, A., Rauscher, T., Beer, H., Käppeler, F., and Gallino, R., *Astrophys. J.* **583**, 506 (2003).
12. Lugaro, M., Davis, A., Gallino, R., Pellin, M., Straniero, O., and Käppeler, F., *Astrophys. J.* **593**, 486 (2003).
13. Holmes, J., Woosley, S., Fowler, W., and Zimmermann, B., *At. Data Nucl. Data Tables* **18**, 305 (1976).
14. Toukan, K., and Käppeler, F., *Astrophys. J.* **348**, 357 (1990).
15. Rauscher, T., and Thielemann, F.-K., *At. Data Nucl. Data Tables* **75**, 1 (2000).
16. Goriely, S., Nuclear astrophysics data base, http://www.astro.ulb.ac.be/Nucdata/ (2003).
17. Sonnabend, K., Mohr, P., Zilges, A., Hertenberger, R., Wirth, H.-F., Graw, G., and Faestermann, T., *Phys. Rev. C* **68**, 048802 (2003).
18. NNDC Online Data Service, ENSDF database, http://www.nndc.bnl.gov/nndc/ensdf/ (2003).
19. Wolfenstein, L., *Phys. Rev.* **82**, 690 (1951).
20. Hauser, W., and Feshbach, H., *Phys. Rev.* **87**, 366 (1952).

Microscopic nuclear equation of state with three-body forces and neutron star structure

U. Lombardo*, G. F. Burgio[†], H.-J. Schulze[†], W. Zuo** and X. R. Zhou[‡]

*INFN-LNS and Dipartimento di Fisica, Via S. Sofia 62, I-95123 Catania, Italy
[†]INFN Sezione di Catania, Via S. Sofia 64, I-95123 Catania, Italy
**Institute of Modern Physics, CAS Lanzhou 730000, China
[‡]Department of Physics, Tsinghua University, 100084 Beijing, China

Abstract.
The equation of state (EOS) of nuclear matter is discussed within the Brueckner-Bethe-Goldstone approach. First the energy per particle E/A is calculated in the Brueckner-Hartree-Fock limit with the Argonne v_{18} potential, using the continuous choice as auxiliary potential. Then, the contribution of three-body clusters is determined by solving the Bethe-Faddeev equation, and the equivalence with the same calculations based on the standard choice as auxiliary potential, is demonstrated. In spite of reaching a quite good convergence of the hole-line expansion, the resulting EOS does not fit the empirical saturation density ($\rho_0 = 0.17$ fm^{-3}). To this end, three-body forces (TBF) are introduced. A first class of microscopic TBF comprises effects due to $N\bar{N}$ virtual excitations via σ and ω-meson exchanges (the main relativistic correction to Brueckner theory), the 2π-exchange, and the virtual excitation of the lowest nucleonic resonance $N^*(1440)$. We compare with a phenomenological TBF, involving two parameters adjusted on the saturation density and energy. Next, using microscopic or phenomenological TBF, the symmetry energy of nuclear matter is computed, allowing to determine the EOS of beta-stable and charge neutral matter, and the properties of neutron stars, in particular the mass-radius curve.

1. PRESENT STATUS OF THE BRUECKNER THEORY

Over the last two decades the increasing interest for the equation of state of nuclear matter has stimulated a great deal of theoretical activity. Phenomenological and microscopic models of the EOS have been developed along parallel lines with complementary roles. The former models include nonrelativistic mean field theory based on Skyrme interactions [1] and relativistic mean field theory based on meson-exchange interactions (Walecka model) [2]. Both of them fit the parameters of the interaction in order to reproduce the empirical saturation properties of nuclear matter extracted from the nuclear mass table. The latter ones include nonrelativistic Brueckner-Hartree-Fock (BHF) theory [3] and its relativistic counterpart, the Dirac-Brueckner (DB) theory [4], the nonrelativistic variational approach also corrected by relativistic effects [5], and more recently the chiral perturbation theory [6]. In these approaches the parameters of the interaction are fixed by the experimental nucleon-nucleon and/or nucleon-meson scattering data.

For states of nuclear matter with high density and high isospin asymmetry the experimental constraints on the EOS are rather scarse and indirect. Different approaches lead to theoretical predictions for the nuclear matter properties, which are quite different from each other if not contradictory. The interest for the latter properties lies, to a large extent, in the study of astrophysical objects, i.e., supernovae and neutron stars. In particular, the structure of a neutron star is very sensitive to the compressibility and the symmetry energy. The neutron star mass, measured in binary systems, has been proposed as a constraint for the EOS of nuclear matter [7].

One of the most advanced approaches to the EOS of nuclear matter is the Brueckner theory. In the recent years, it has made a rapid progress in several aspects. The convergence of the Brueckner-Bethe-Goldstone (BBG) expansion has been firmly established [8, 9]. Important relativistic effects have been incorporated by including into the interaction the virtual nucleon-antinucleon excitations, and the relationship with the DB approach has been numerically clarified [10]. Finally, the addition of microscopic three-body forces based on nucleon excitations via pion exchange and heavy meson fields, permitted to improve to a large extent the agreement with the empirical saturation properties [11].

In the present paper the EOS of nuclear matter is discussed within the Brueckner theory with special emphasis to the role played by the TBF. The predictions of the microscopic TBF are discussed in comparison with those from phenomenological TBF. The EOS is used as input to solve the Tolman-Oppenheimer-Volkov (TOV) equation for determining the structure of a neutron star.

2. CONVERGENCE OF THE HOLE-LINE EXPANSION

The nonrelativistic BBG expansion of the nuclear matter energy shift E/A can be cast as a power series in terms of the number of hole lines contained in the corresponding diagrams, which amounts to a density power expansion [3]. The two hole-line truncation is named the Brueckner-Hartree-Fock (BHF) approximation. At this order the energy shift D_2 is very much affected by the choice of the auxiliary potential, as shown in Fig. 1 (solid lines), where the numerical results are reported for symmetric nuclear matter with the gap and the continuous choice.

But, adding the three-hole line contributions D_3, the resulting EOS is almost insensitive to the choice of the auxiliary potential, as also shown in the same figure [8]. A fine convergence is also found in the case of neutron matter [9]. In spite of the satisfactory convergence, the saturation density misses the empirical value extracted from the nuclear mass tables. This confirms the belief that the concept of a many nucleon system interacting with only a two-body force (2BF) is not adequate to describe nuclear matter, especially at high density.

3. RELATIVISTIC CORRECTIONS

Before the possible effects of TBF are examined, one should introduce the relativistic corrections in the preceding nonrelativistic BHF predictions. How to identify the most important relativistic corrections? One has to look at the Dirac-Brueckner approach, which is the relativistic counterpart of the Brueckner theory. Comparing the EOS predicted by the two theories (left plot of Fig. 2, dot-dashed and short-dashed curves, re-

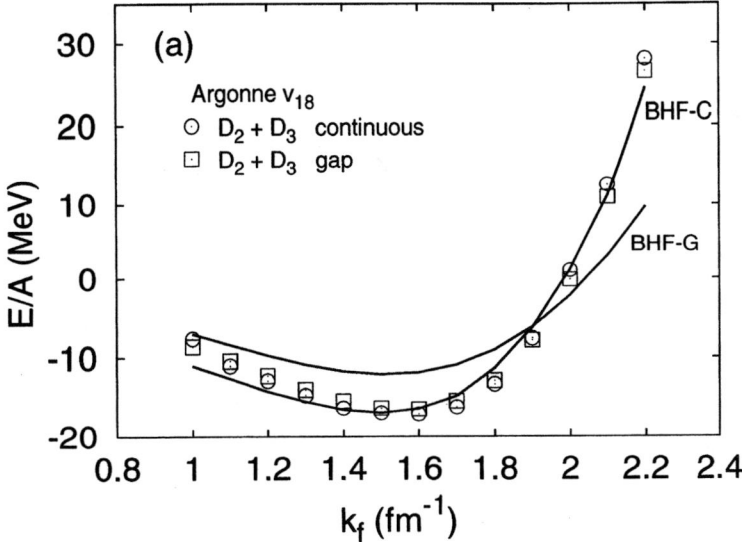

FIGURE 1. Comparison of BHF two hole-line (lines) and three hole-line (markers) results for symmetric nuclear matter, using continuous and gap choice for the single-particle potentials.

spectively) it seems that the relativistic effect is quite large, increasing rapidly with density, with the consequence that the saturation density is pushed back to the empirical value.

The main difference between DB and BHF theory is that the nucleons, instead of propagating as plane waves, propagate as spinors in a mean field with a scalar component U_S and a vector component U_V, self-consistently determined together with the G-matrix. The self-energy, also calculated self-consistently with the G-matrix, can be expanded in terms of the free spinors (corresponding to the expansion in powers of the scalar field):

$$\Sigma(p) = U_V + \sqrt{p^2 + (M+U_S)^2} \approx U_V + p_0 + \frac{MU_S}{p_0} + \frac{p^2}{p_0^3}U_S^2 + \ldots \quad (1)$$

The second-order term has been interpreted as due to the interaction between two particles with the virtual excitation of a nucleon-antinucleon pair [12]. This interaction is a TBF with the exchange of the scalar field, as illustrated by the diagram (d) of Fig. 3. Actually this diagram represents a class of TBF with the exchange of light and heavy mesons. Adding to the nonrelativistic EOS the contribution of only the σ-meson exchange (identified with the scalar field), one obtains a result very close to the DB prediction, as shown in Fig. 2 (left plot, solid line).

However, the other terms, in particular that due to the exchange of the vector (ω) meson, must be added and it turns out that they do not cancel each other. The overall effect of the $N\bar{N}$ terms is shown in Fig. 2 (right plot) along with other effects discussed later, and exhibits saturation at too small density. This and the other signatures above mentioned call for the evaluation of the additional effects due to true TBF.

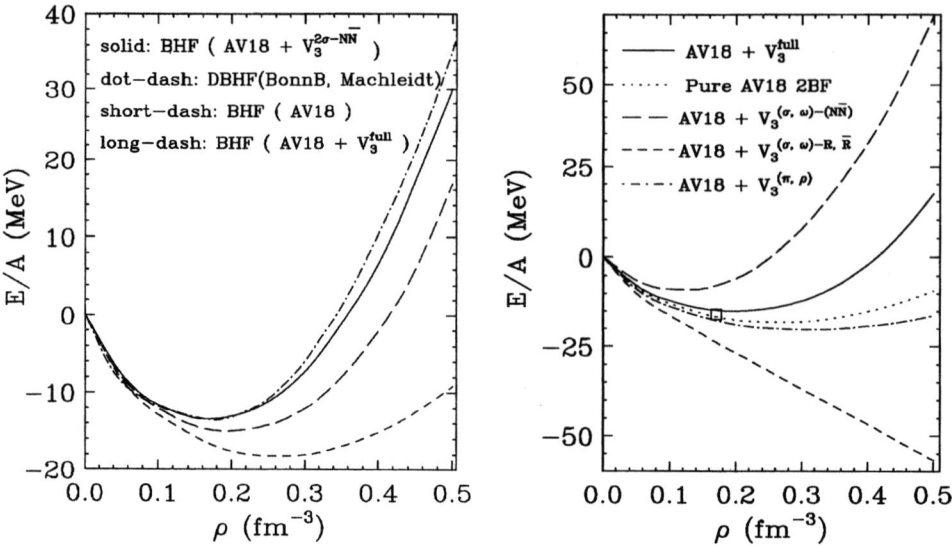

FIGURE 2. Different TBF contributions to the binding energy of symmetric nuclear matter, and comparison with DB results. See text for details.

4. THREE-BODY FORCES

Since long it is well known that two-body forces are not enough to explain some nuclear properties, and TBF have to be introduced. Typical examples are: the binding energy of light nuclei, the spin dynamics of nucleon-deuteron scattering, and the saturation point of nuclear matter. Phenomenological and microscopic TBF have been widely used to describe the above mentioned properties.

In the framework of the Brueckner theory a rigorous treatment of TBF would first ask for the solution of the Bethe-Faddeev equation, describing the dynamics of three bodies embedded in the nuclear matter. In practice a much simpler approach is employed, namely the TBF is reduced to an effective, density dependent, two-body force by averaging over the third nucleon in the medium, taking account of the nucleon-nucleon correlations by means of the BHF defect function g_{ij},

$$\langle 12|V(\rho)|1'2'\rangle = \sum_{33'} \Psi^*_{123} \langle 123|V|1'2'3'\rangle \Psi_{1'2'3'}. \tag{2}$$

Here $\Psi_{123} = \phi_3(1-g_{12})(1-g_{13})$ and ϕ_3 is the free wave function of the third particle. This effective two-body force is added to the bare two-body force and recalculated at each step of the iterative procedure.

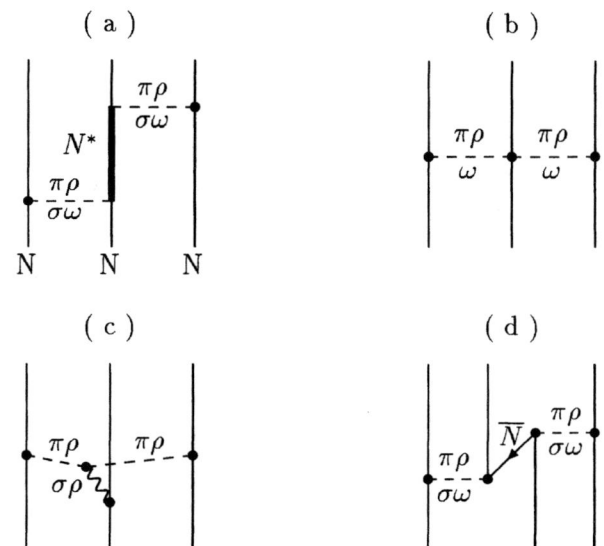

FIGURE 3. Various diagrams contributing to the microscopic TBF.

4.1. Microscopic TBF

The microscopic TBF [10, 11] has the great advantage to exhibit explicitly how the interaction mechanisms prevailing in different density domains work. The TBF we are going to discuss are based on the meson-exchange mechanisms accompanied by the excitation of nucleonic resonances.

They are represented by the diagrams plotted in Fig. 3. Besides the TBF arising from the excitation of $N\bar{N}$ via heavy fields already discussed in the preceding section [included in the diagram (d)], a first class of TBF contains the excitation of $N\bar{N}$ and the isobar Δ resonance excitations via the exchange of light mesons, namely π and ρ. A second class involves the non-isobar lowest nucleon excitation, the $N^*(1440)$, excited by the heavy meson (σ and ω) exchanges. The two classes are included in the diagrams (d) and (a), respectively. Diagrams (b) and (c) are included only for completeness and anyhow their minor role is discussed elsewhere [10].

The TBF with N^* plays a very crucial role in that it largely cancels the strongly repulsive effect of the TBF with a $N\bar{N}$ intermediate state excited from one side by σ-meson exchange and from the other side by ω-meson exchange. The processes with light meson exchange are incorporated in the so-called Tucson-Melbourne potential based on the off-shell extension of the $N\pi$ scattering T-matrix [13]. The effect of this contribution is negligible in comparison with relativistic terms and the other TBF.

The various contributions to the nuclear EOS are presented in Fig. 2. The global effect of TBF results in a remarkable improvement of the saturation properties of nuclear matter. Compared with the BHF prediction with only two-body forces, the saturation energy is shifted from -18 to -15 MeV, the saturation density from 0.26 to 0.19 fm^{-3}, and the compression modulus from 230 to 210 MeV. The spin and isospin properties with TBF exhibit also quite satisfactory behaviour [14].

4.2. Phenomenological TBF

A second class of TBF that are widely used in the literature, in particular for variational calculations of finite nuclei and nuclear matter [5], are the phenomenological Urbana TBF [15]. We remind that the Urbana IX TBF model contains a two-pion exchange potential $V_{ijk}^{2\pi}$ supplemented by a phenomenological repulsive term V_{ijk}^{R},

$$V_{ijk} = V_{ijk}^{2\pi} + V_{ijk}^{R}, \qquad (3)$$

where

$$V_{ijk}^{2\pi} = A \sum_{\text{cyc}} \left[\{X_{ij}, X_{jk}\} \{\tau_i \cdot \tau_j, \tau_j \cdot \tau_k\} + \frac{1}{4} [X_{ij}, X_{jk}] [\tau_i \cdot \tau_j, \tau_j \cdot \tau_k] \right], \qquad (4)$$

$$V_{ijk}^{R} = U \sum_{\text{cyc}} T^2(m_\pi r_{ij}) T^2(m_\pi r_{jk}). \qquad (5)$$

The two-pion exchange operator X_{ij} is given by

$$X_{ij} = Y(m_\pi r_{ij}) \sigma_i \cdot \sigma_j + T(m_\pi r_{ij}) S_{ij}, \qquad (6)$$

where σ and τ are the Pauli spin and isospin operators, and $S_{ij} = 3(\sigma_i \cdot \hat{r}_{ij})(\sigma_j \cdot \hat{r}_{ij}) - \sigma_i \cdot \sigma_j$ is the tensor operator. Y and T are the Yukawa and tensor functions, respectively, associated to the one-pion exchange [15].

After reducing this TBF to an effective, density dependent, two-body force by the averaging procedure described earlier, the resulting effective two-nucleon potential assumes a simple structure,

$$\overline{V}_{ij}^{\text{pheno}}(r) = (\tau_i \cdot \tau_j) \left[(\sigma_i \cdot \sigma_j) V_C^{2\pi}(r) + S_{ij}(\hat{r}) V_T^{2\pi}(r) \right] + V^R(r), \qquad (7)$$

containing central and tensor two-pion exchange components as well as a central repulsive contribution. For comparison, the averaged microscopic TBF involves five different components:

$$\overline{V}_{ij}^{\text{micro}}(r) = (\tau_i \cdot \tau_j)(\sigma_i \cdot \sigma_j) V_C^{\tau\sigma}(r) + (\sigma_i \cdot \sigma_j) V_C^{\sigma}(r) + V_C(r)$$
$$+ S_{ij}(\hat{r}) \left[(\tau_i \cdot \tau_j) V_T^{\tau}(r) + V_T(r) \right]. \qquad (8)$$

In the variational approach the two parameters A and U are determined by fitting the triton binding energy together with the saturation density of nuclear matter (yielding however too little attraction, $E/A \approx -12$ MeV, in the latter case [5]). In the

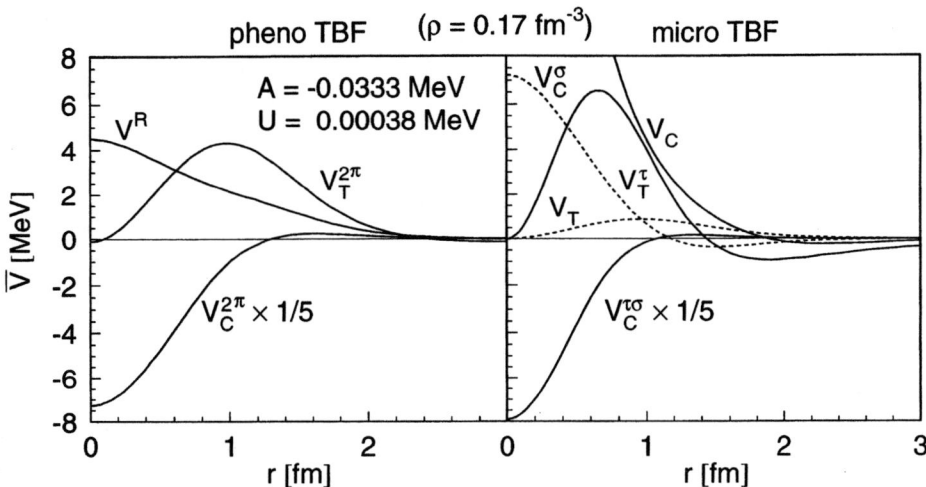

FIGURE 4. Comparison of the different components of averaged phenomenological and microscopic TBF, Eqs. (7) and (8).

BHF calculations they are instead chosen to reproduce the empirical saturation density together with the binding energy of nuclear matter. The resulting parameter values are $A = -0.0293$ MeV and $U = 0.0048$ MeV in the variational Urbana IX model, whereas for the optimal BHF+TBF calculations we require $A = -0.0333$ MeV and $U = 0.00038$ MeV, yielding a saturation point at $k_F \approx 1.36$ fm^{-1}, $E/A \approx -15.5$ MeV, and an incompressibility $K \approx 210$ MeV.

These values of A and U have been obtained by using the Argonne v_{18} two-body force [16] both in the BHF and in the variational many-body theories. However, the required repulsive component ($\sim U$) is much weaker in the BHF approach, consistent with the observation that in the variational calculations usually heavier nuclei as well as nuclear matter are underbound. Indeed, less repulsive TBF became available recently [17] in order to address this problem.

5. EOS OF NUCLEAR MATTER FROM DIFFERENT TBF

In Fig. 4 we compare the different components $V_C^{2\pi}, V_T^{2\pi}, V^R$, Eq. (7) and $V_C^{\tau\sigma}, V_C^{\sigma}, V_C, V_T^{\tau}, V_T$, Eq. (8), of the averaged phenomenological and microscopic TBF potentials in symmetric matter at normal density. One notes that the attractive components $V_C^{2\pi}, V_T^{2\pi}$ and $V_C^{\tau\sigma}, V_T^{\tau}$ roughly correspond to each other, whereas the repulsive part (V^R vs. V_C) is larger for the microscopic TBF. With the choice of parameters A and U given above, one would therefore expect a more repulsive behaviour of the microscopic TBF, which is indeed confirmed in the following.

Let us now confront the EOS predicted by the phenomenological TBF and the microscopic one. In both cases the BHF approximation has been adopted with same two-body

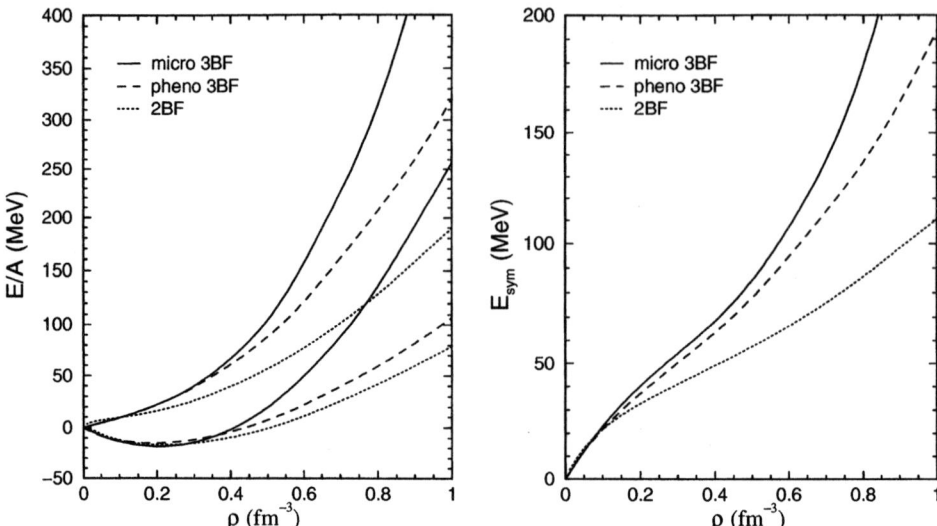

FIGURE 5. Left plot: Binding energy per nucleon of symmetric nuclear matter (lower curves using a given linestyle) and pure neutron matter (upper curves), employing different TBF. Right plot: Corresponding symmetry energy of nuclear matter.

force (Argonne v_{18}). In the left panel of Fig. 5 we display the equation of state both for symmetric matter (lower curves) and pure neutron matter (upper curves). We show results obtained for several cases, i.e., i) only two-body forces are included (dotted lines), ii) TBF implemented within the phenomenological Urbana IX model (dashed lines), and iii) TBF treated within the microscopic meson-exchange approach (solid lines). We notice that the EOS for symmetric matter with TBF reproduces the correct nuclear matter saturation point. Moreover, the incompressibility turns out to be compatible with the values extracted from phenomenology, i.e., $K \approx 210$ MeV. Up to a density of $\rho \approx 0.4$ fm^{-3} the microscopic and phenomenological TBF are in fair agreement, whereas at higher density the microscopic TBF turn out to be more repulsive.

In the right panel of Fig. 5 we display the symmetry energy as a function of the nucleon density ρ for different choices of the TBF. We observe results in agreement with the characteristics of the EOS shown in the left panel. In fact, the stiffest equation of state, i.e., the one calculated with the microscopic TBF, yields larger symmetry energies compared to the ones obtained with the Urbana phenomenological TBF. Moreover, the symmetry energy calculated (with or without TBF) at the saturation point yields a value $E_{\text{sym}} \approx 30$ MeV, compatible with nuclear phenomenology.

6. NEUTRON STAR STRUCTURE

In order to study the effects of different TBF on neutron star structure, we have to calculate the composition and the EOS of cold, catalyzed matter. We require that the

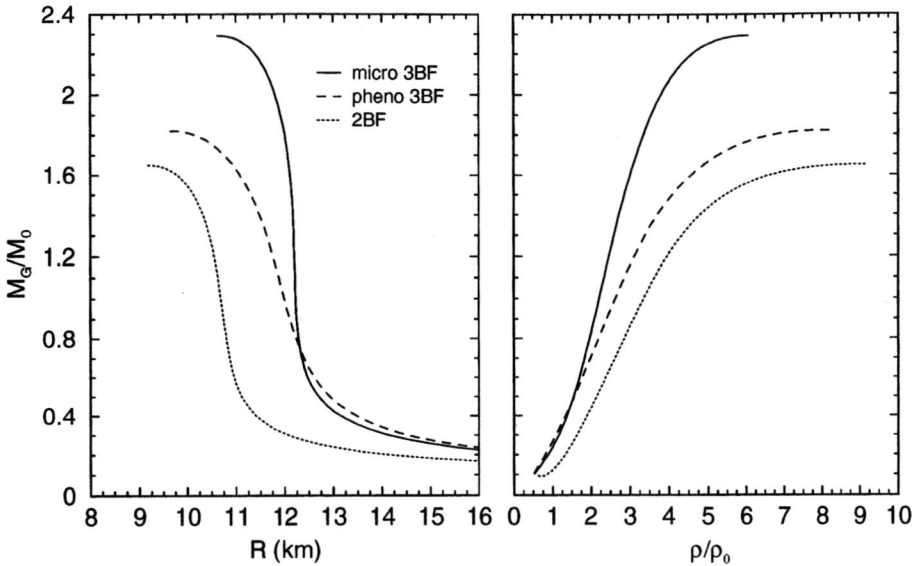

FIGURE 6. The neutron star gravitational mass (in units of solar mass $M_\odot$) is displayed vs. the radius (left panel) and the central baryon density ρ_c, normalized with respect to the value at saturation ρ_0 (right panel).

neutron star contains charge neutral matter consisting of neutrons, protons, and leptons in beta equilibrium. No transitions to other phases are considered in this work. Following standard procedures [18, 19], we compute the proton fraction and the equation of state for charge neutral and beta-stable matter, using the various TBF discussed above. In order to calculate the mass-radius relation, one has to solve the well-known Tolman-Oppenheimer-Volkov equations [18], with the newly constructed EOS for the charge neutral and beta-stable case as input.

The results are shown in Fig. 6. We notice that the EOS calculated with the microscopic TBF produces the largest gravitational masses, with the maximum mass of the order of 2.3 $M_\odot$, whereas the phenomenological TBF yields a maximum mass of about 1.8 $M_\odot$. In the latter case, neutron stars are characterized by smaller radii and larger central densities, i.e., the Urbana TBF produce more compact stellar objects. For completeness, we also show a sequence of stellar configurations obtained using only two-body forces. In this case the maximum mass is slightly above 1.6 $M_\odot$, with a radius of 9 km and a central density equal to 9 times the saturation value.

These results should however be taken with a grain of salt, since it is well known that the inclusion of hyperons [20] or quark matter [21] may strongly affect the structure of the star, in particular reducing substantially the maximum mass.

REFERENCES

1. P. Bonche, E. Chabanat, P. Haensel, J. Meyer, and R. Schaeffer, Nucl. Phys. **A 643**, 441 (1998).
2. B. D. Serot and J. D. Walecka, Adv. Nucl. Phys. **16**, 1 (1986).
3. M. Baldo, *The many body theory of the nuclear equation of state* in Nuclear Methods and the Nuclear Equation of State, 1999, Ed. M. Baldo, World Scientific, Singapore.
4. R. Machleidt, Adv. Nucl. Phys. **19**, 189 (1989) and references quoted therein.
5. A. Akmal and V. R. Pandharipande, Phys. Rev. **C56**, 2261 (1997); A. Akmal, V. R. Pandharipande, and D. G. Ravenhall, Phys. Rev. **C58**, 1804 (1998); J. Morales, V. R. Pandharipande, and D. G. Ravenhall, Phys. Rev. **C66**, 054308 (2002).
6. N. Kaiser, S. Fritsch, and W. Weise, Nucl. Phys. **A 697**, 255 (2002).
7. N. K. Glendenning, Nucl. Phys. **A 493**, 521 (1989); *Compact Stars, Nuclear Physics, Particle Physics, and General Relativity*, 2nd ed., 2000, Springer-Verlag, New York.
8. H. Q. Song, M. Baldo, G. Giansiracusa, and U. Lombardo, Phys. Rev. Lett. **81**, 1584 (1998).
9. M. Baldo, A. Fiasconaro, H. Q. Song, G. Giansiracusa, and U. Lombardo, Phys. Rev. **C65**, 017303 (2002).
10. P. Grangé, A. Lejeune, M. Martzolff, and J.-F. Mathiot, Phys. Rev. **C40**, 1040 (1989).
11. A. Lejeune, U. Lombardo, and W. Zuo, Phys. Lett. **B477**, 45 (2000); W. Zuo, A. Lejeune, U. Lombardo, and J.-F. Mathiot, Nucl. Phys. **A 706**, 418 (2002); Eur. Phys. Journ. **A 14**, 469 (2002).
12. G. E. Brown et al., Comm. Nucl. Part. Phys. **17**, 39 (1987).
13. S. A. Coon, M. D. Scadron, P. C. McNamee, B. R. Barrett, D. W. E. Blatt, and B. H. J. McKellar, Nucl. Phys. **A 317**, 242 (1979).
14. W. Zuo, Caiwan Shen, and U. Lombardo, Phys. Rev. **C67**, 037301 (2003).
15. B. S. Pudliner, V. R. Pandharipande, J. Carlson, and R. B. Wiringa, Phys. Rev. Lett. **74**, 4396 (1995).
16. R. B. Wiringa, V. G. J. Stoks, and R. Schiavilla, Phys. Rev. **C51**, 38 (1995).
17. S. C. Pieper, V. R. Pandharipande, R. B. Wiringa, and J. Carlson, Phys. Rev. **C64**, 014001 (2001).
18. S. Shapiro and S. A. Teukolsky, *Black Holes, White Dwarfs, and Neutron Stars*, 1983, ed. John Wiley & Sons, New York.
19. M. Baldo, I. Bombaci, and G. F. Burgio, Astron. Astroph. **328**, 274 (1997).
20. M. Baldo, G. F. Burgio, and H.-J. Schulze, Phys. Rev. **C58**, 3688 (1998); **C61**, 055801 (2000).
21. G. F. Burgio, M. Baldo, P. K. Sahu, and H.-J. Schulze, Phys. Rev. **C66**, 025802 (2002); M. Baldo, M. Buballa, G. F. Burgio, F. Neumann, M. Oertel, and H.-J. Schulze, Phys. Lett. **B562**, 153 (2003).

UNEXPECTED GOINGS-ON IN THE STRUCTURE OF A NEUTRON STAR CRUST

Aurel Bulgac*, Paul-Henri Heenen†, Piotr Magierski**, Andreas Wirzba‡ and Yongle Yu*

*Department of Physics, University of Washington, Seattle WA 98195-1560, USA
†Service de Physique Nucléaire Th'eorique, Univerite Libre de Bruxelles,B 1050, Brussels, Belgium
**Faculty of Physics, Warsaw University of Technology, ul. Koszykowa 75, 00-662, Warsaw, Poland
‡Helmholtz-Institut für Strahlen- und Kernphysik, Universität Bonn,D-53115 Bonn, Germany

Abstract.
We present a brief account of two phenomena taking place in a neutron star crust: the Fermionic Casimir effect and the major density depletion of the cores of the superfluid neutron vortices.

FERMIONIC CASIMIR EFFECT AND NEUTRON STAR CRUST

At a depth of about 500 m or so below the surface of a neutron crust the nuclear matter (which consists mostly of neutrons plus a small percentage of protons and electrons in β-equilibrium) organize themselves in some exotic inhomogeneous solid phase [1]. As a matter of fact, neutron star crusts seem to be just about the only other places in the entire Universe, apart from planets, where one can find condensed matter, in particular a solid phase [2]. Moving from the neutron star surface inward, one finds at first a Coulomb crystal lattice of nuclei immersed in a very low density neutron gas and even lower density electron gas. With increasing depth, the density and pressure increase, the nuclei get closer to each other and start evolving into some unusual elongated nuclei, which eventually become rods. These nuclear rods evolve gradually into plates, their place being taken later by tubes and bubbles (dubbed "inside out" nuclei) just before the average density becomes almost equal to the nuclear saturation density and the entire mixture of neutrons, protons and electrons become an homogeneous phase. The properties of this part of the neutron star have been the subject of a lot of studies, see Refs. [1, 3, 4, 5, 6, 7, 8, 9, 10, 11, 12, 13, 14, 15, 16, 17] and other references therein. Most of these approaches however have missed a rather subtle and apparently important physical phenomenon, the fermionic counterpart of the Casimir interaction in such a medium [10, 11, 12, 13, 14].

In order to quickly explain the main physics ideas behind this new phenomenon, let us consider an over-simplified model of the neutron star crust. One can ask the rather innocuous question: "What is the ground state energy of an infinite homogeneous Fermi sea of noninteracting neutral particles with two hard spheres of radii a, separated by a distance r?" The naive and somewhat startling answer that perhaps one can place the two hard spheres almost anywhere with respect to each other and that the energy of the system will not be affected if one were to move the hard spheres around. The "theoretical argument" which can lead to such a conclusion is based on the same type of argumentation, which was used in Refs.bbp,pra,dhm,dha and allowed these authors to establish that by going deeper and deeper into the interior of the neutron star one finds a well defined sequence of "exotic" nuclear shapes. This traditional argumentation is based essentially on liquid drop model, which includes the volume, surface, Coulomb contributions to the ground state energy only. This is basically "classical thinking." For a person using "quantum reasoning" instead, the fact that the ground state of such a system in infinitely degenerate (corresponding to an arbitrary relative arrangement of the two hard spheres) will find such an answer most likely wrong. An indeed, a careful analysis of the problem reveals the fact that indeed a system of two hard spheres, immersed in an infinite Fermi see of noninteracting particles at zero temperature has a well defined ground state. The correct answer, namely that the "interaction energy" of the two hard spheres is

$$E_C \approx -\frac{\hbar^2 k_F^2}{m} \frac{a^2}{2\pi r(r-2a)} j_1[2k_F(r-2a)],$$

where $j_1(x)$ is the spherical Bessel function and k_F is the Fermi momentum is somewhat even more surprising. "Why would this "interaction energy" be a non-monotonic function of the hard sphere separation r?" and, moreover, "How does interaction really emerges here, where one starts with such a simple system of non-interacting particles?" As one soon "discovers" the "culprit" is the wave character of the Quantum Mechanics really. Fermions even at zero temperature do not stop moving and the space is really "filled" with an infinite number of de Broglie's waves. These waves reflect from the two hard spheres and as in the case of any wind musical instrument, for some frequencies one would have a favorable wave interference while for other frequencies there will not such a favorable interference. In an infinite Fermi sea there is an infinite number of waves with all frequencies ranging from zero to the Fermi frequency. If one carefully adds up the effects of all these waves one readily arrives at the result above [10, 11]. Things cat a little bit more complicated when one adds more hard spheres, as then one naturally discovers that besides the "natural" two-body interactions there are genuine three- and four- and many-body interactions among these spheres. Moreover, there is absolutely no reason why not consider other type of objects, which could be immersed in this Fermi sea, like "logs" and "boards" and in principle almost anything else. Surprisingly all these combinations of various objects in various arrangements can be analyzed rather easily. What is surprising however is the fact that the characteristic interaction energy between such objects is of the same order as the energy differences between various phases in a neutron star crust [10, 12, 13, 14] and when taken into account this fermionic Casimir energy can in "ruin perfect crystalline structures" found in all previous studies. These conclusions have been backed by more sophisticated fully microscopic calculations of the nuclear matter in a neutron star crust [8, 9].

Instead of describing in more detail results which have been published already, we shall instead draw the attention of our readers here to another element which was overlooked in studies of the neutron star crust, and which is apparently going to influence a great deal of properties. In order to analyze the thermal and electric conductivities of the crust, which are important for understanding of the thermal evolution of neutron stars one has to go beyond the static approximation. The "nuclei" which are immersed in the neutron fluid, which indeed is a superfluid, can and do move. As with boats on a lake, when they start moving they make waves and one has to include the dynamics of the surrounding superfluid in any analysis. We shall limit ourselves here to quoting a single result, namely the kinetic energy of two penetrable spheres immersed in a superfluid at velocities below the critical velocity for the loss of superfluidity. One then finds [18] that the kinetic energy of two such spheres becomes:

$$T = \frac{1}{2}(M_1^{ren}u_1^2 + M_2^{ren}u_2^2) + 4\pi m \rho_{out}\left(\frac{1-\gamma}{2\gamma+1}\right)^2 \left(\frac{R_1R_2}{r}\right)^3 \left[\vec{u}_1 \cdot \vec{u}_2 - \frac{3}{r^2}(\vec{u}_1 \cdot \vec{r})(\vec{u}_2 \cdot \vec{r})\right]$$

where the renormalized masses of nuclei have the form:

$$M_i^{ren} = \frac{4}{3}m\rho_{in}\pi R_i^3 \frac{(1-\gamma)^2}{2\gamma+1} = M_i \frac{(1-\gamma)^2}{2\gamma+1},$$

where $\vec{u}_i$ are the velocities of the two nuclei, $i = 1,2$ and M_i and R_i denote the nuclear bare mass and radii of the $i-th$ nucleus, $\gamma = \rho_{in}/\rho_{out}$ and $\rho_{in,out}$ are the densities inside and outside the two nuclei. The somewhat unexpected cross term appearing above shows that the existence of mere motion of the two objects in a perfect fluid can lead to a velocity-dependent interaction, which decays with the separation as slows as the static Casimir Fermionic energy, namely as $1/r^3$. Further analysis shows that this velocity dependent-interaction is important as well when considering dynamical properties of neutron star crust.

THE SPATIAL STRUCTURE OF A VORTEX IN LOW-DENSITY SUPERFLUID NEUTRON MATTER

There is a long held belief that vortices in Fermi systems do not show any appreciable normal density variations and that only the anomalous density vanishes along the vortex axis, similarly to the behavior of the density (which is the order parameter) in Bose systems [19, 20, 21]. Thus it came as somewhat of a surprise the fact that in Fermi systems one can have a spatial structure of a vortex with a significant normal density depletion along the vortex axis [22, 23, 24]. What happens in low density superfluid neutron matter for example is the following. The magnitude of the pairing gap becomes comparable with the Fermi energy,

The possibility that the value of the superfluid gap can attain large values was raised more than two decades ago in connection with the BCS $\to$ BEC crossover [28, 29]. One can imagine that one can increase the strength of the two–particle interaction in such a manner that at some point a real two–bound state forms, and in that case $a \to -\infty$. By continuing to increase the strength of the two–particle interaction, the scattering length becomes positive and starts decreasing. A dilute system of fermions, when $\rho r_0^3 \ll 1$ (here r_0 is the interaction radius), will thus undergo a transition from a weakly coupled BCS system, when $a < 0$ and $a = \mathcal{O}(r_0)$, to a BEC system of tightly bound Fermion pairs, when $a > 0$ and $a = \mathcal{O}(r_0)$ again. In the weakly coupled BCS limit the size of the Cooper pair is given by the so called coherence length $\xi \propto \frac{\hbar^2 k_F}{m\Delta}$, which is much larger than the inter-particle separation $\approx \lambda_F = 2\pi/k_F$. In the opposite limit, when $k_F a \ll 1$ and $a > 0$, and when tightly bound pairs/dimers of size a are formed, the dimers are widely separated from one another. Surprisingly, these dimers also repel each other with an estimated scattering length $\approx 0.6...2a$ [26, 27] and thus the BEC phase is also (meta)stable. The bulk of the theoretical analysis in the intermediate region where $k_F|a| > 1$ was based on the BCS formalism [26, 28, 29, 30] and thus is highly questionable. Even the simplest polarization corrections have not been included into this type of analysis so far. In particular, it is well known that in the low density region, where $a < 0$ and $k_F|a| \ll 1$ the polarization corrections to the BCS equations lead to a noticeable reduction of the gap [25]. Only a truly *ab initio* calculation could really describe the structure of a many Fermion system with $k_F|a| \gg 1$. In the limit $a = \pm\infty$, when the two–body bound state has exactly zero energy, and if $k_F r_0 \ll 1$, one can expect that the energy per particle of the system is proportional to $\varepsilon_F = \hbar^2 k_F^2/2m$, as it was recently confirmed by the variational calculations of Refs. [31, 32]. The normal density at the vortex core is lowered, while the pairing field vanishes at the vortex axis as expected. In hindsight this result could have been expected. Large values of the pairing field correspond to the formation of atom pairs/dimers of relatively small sizes. When these dimers are relatively strongly bound and when they are also widely separated from one another, they undergo a Bose–Einstein condensation. For a vortex state in a 100% BEC system the density at the vortex axis vanishes identically. Therefore, by increasing the strength of the two–particle interaction, the Fermion system simply approaches more and more an ideal BEC system, for which a density depletion of the vortex core is expected.

Almost thirty years ago Anderson and Itoh [2] put forward the idea that vortices should appear in neutron stars and that they can also get pinned to the solid crust. They argued that the "star–quakes," observable on Earth as pulsar "glitches," apparently are caused by the vortex de–pinning in neutron star crust. This idea and its various implications have been examined by numerous authors, see Refs. [20, 23] and further references therein, but a general consensus does not seem to have emerged so far.

The profile of a vortex in neutron matter is typically determined using a Ginzburg–Landau equation, which is expected to give mostly a qualitative picture and its accuracy is difficult to estimate. Surprisingly, prior to Ref. [22] there exists only one microscopic calculation of a vortex in low density neutron matter [21]. The existence of a strong density depletion in the vortex core is going to affect appreciably the energetics of a neutron star crust. One can obtain a gross estimate of the pinning energy of a vortex on a nucleus as $E^V_{pin} = [\varepsilon(\rho_{out})\rho_{out} - \varepsilon(\rho_{in})\rho_{in}]V$, where $\varepsilon(\rho)$ is the energy per particle at density ρ, ρ_{in} and ρ_{out} are the densities inside and outside the vortex core and V is the volume of the nucleus. Naturally, this simple formula does not take into account a number of factors, in particular surface effects and the changes in the velocity profile and the pairing field. These last contributions were accounted for (with some variations) in the past [2, 20]. However, if the density inside the vortex core and outside differ significantly one expects E^V_{pin} to be the dominant contribution. In the low density region, where $\varepsilon(\rho_{out})\rho_{out}/\varepsilon(\rho_{in})\rho_{in}$ is largest, one expects a particularly large anti–pinning effect ($E^V_{pin} > 0$). The energy per unit length of a simple vortex is expected to be significantly lowered when compared with previous estimates [2, 20] by $\approx [\varepsilon(\rho_{out})\rho_{out} - \varepsilon(\rho_{in})\rho_{in}]\pi R^2$, where R is an approximate core radius.

REFERENCES

1. G. Baym, H.A. Bethe, C.J. Pethick, Nucl. Phys. **A175** 225 (1971).
2. P.W. Anderson and N. Itoh, Nature, **256**, 25 (1975).
3. J.W. Negele and D. Vautherin, Nucl. Phys. **A207**, 298(1973).
4. P. Bonche and D. Vautherin, Nucl. Phys. **A372** 496.(1981); Astron. Astrophys. **112** 268 (1982).
5. C.J. Pethick and D.G. Ravenhall, Annu. Rev. Nucl. Part. Sci. **45** 429 (1995).
6. F. Douchin, P. Haensel, J. Meyer, Nucl. Phys. **A 665** 419 (2000).
7. F. Douchin, P. Haensel, Phys. Lett. **B485** 107 (2000).
8. P. Magierski, P.-H. Heenen, Phys. Rev. **C65** 045804 (2002).
9. P. Magierski, A. Bulgac, P.-H. Heenen, Int. J. Mod. Phys. **A17** 1059 (2002).
10. A. Bulgac, P. Magierski, Nucl. Phys. **A683** 695 (2001); Erratum: Nucl. Phys. **A703** 892 (2002).
11. A. Bulgac and A. Wirzba, Phys. Rev Lett. **87**, 120404 (2001).
12. A. Bulgac, P. Magierski, Phys. Scripta **T90** 150 (2001).
13. A. Bulgac, P. Magierski, Acta Phys. Pol. **B32** 1099 (2001).
14. P. Magierski, A. Bulgac, Acta Phys. Pol. **B32** 2713 (2001).
15. G. Watanabe, K. Sato, K. Yasuoka, T. Ebisuzaki, Phys. Rev. C 66, 012801 (2002); Phys. Rev. C 68, 035806 (2003).
16. C. J. Pethick and A. Y. Potekhin, Phys. Lett. **B 427**, 7 (1998).
17. P.B. Jones, Phys. Rev. Lett. **83**, 3589 (1999).
18. P. Magierski and A. Bulgac, preprint *Nuclear Hydrodynamics in the Inner Crust of Neutron Stars*.
19. P.G. de Gennes, *Superconductivity of Metals and Alloys*, Addison–Wesley, Reading MA, (1998); F. Gygi and M. Schlüter, Phys. Rev. B **43**, 7609 (1991); P.I. Soininen *et al.*, Phys. Rev. B **50**, 13883 (1994); N. Hayashi, *et al.*, Phys. Rev. Lett. **80**, 2921 (1998); M. Franz and Z. Tešanović, Phys. Rev. Lett. **80**, 4763 (1998); N. Nygaard, *et al.*, Phys. Rev. Lett. **90**, 210402 (2003).
20. P.W. Anderson, M.A. Alpar, D. Pines and J. Shaham, Phil. Mag. **45**, 227 (1982); M.A. Alpar, P.W. Anderson, D. Pines and J. Shaham, Ap. J. **278**, 791 (1984); M.A. Alpar, K.S. Cheng and D. Pines, Ap. J. **346**, 823 (1989); R.I. Epstein and G. Baym, Ap. J. **328**, 680 (1988); R.I. Epstein and G. Baym, Ap. J. **387**, 276 (1992); B. Link, R.I. Epstein and G. Baym, Ap. J. **403**, 285 (1993); P.B. Jones, Phys. Rev. Lett. **79**, 792 (1997). *ibid* **81**, 4560 (1998); *ibid* Mon. Not. R. Astron. Soc. **257**, 501 (1992); *ibid* Mon. Not. R. Astron. Soc. **296**, 217 (1998); P.M. Pizzochero, L. Viverit and R.A. Broglia, Phys. Rev. Lett. **79**, 3347 (1997).
21. F.V. De Blasio and Ø. Elgarøy, Phys. Rev. Lett. **82**, 1815 (1999); Ø. Elgarøy and F.V. De Blasio, A&A, **370**, 939 (2001).
22. Y. Yu and A. Bulgac. Phys. Rev. Lett. **90**, 161101 (2003).
23. P. Donati and P.M. Pizzochero, Phys. Rev. Lett. **90**, 211101 (2003).
24. A. Bulgac and Y. Yu, cond-mat/0303235, Phys. Rev. Lett. **91**, in press (2003)
25. L.P. Gorkov and T.K. Melik–Barkhudarov, Zh. Eksp. Teor. Fiz. **40**, 1452 (1961) [Sov. Phys. JETP **13**, 1018 (1961)]; H. Heiselberg, *et al.*, Phys. Rev. Lett. **85**, 2418 (2000).
26. M. Randeria, in *Bose–Einstein Condensation*, eds. A. Griffin, *et al.*, Cambridge Univ. Press (1995), pp 355–392.
27. P. Pieri and G.C. Strinati, Phys. Rev. B **61**, 15370 (2000) and cond-mat/0307421; D.S. Petrov, C. Salomon, G.V. Shlyapnikov, cond-mat/0309010.
28. A.J. Leggett, in *Modern Trends in the Theory of Condesed Matter*, eds. A. Pekalski and R. Przystawa, Springer–Verlag, Berlin, 1980; J. Phys. (Paris) Colloq. **41**, C7–19 (1980).
29. P. Nozières and S. Schmitt–Rink, J. Low Temp. Phys. **59**, 195 (1985).
30. C.A.R. Sá de Mello, *et al.*, Phys. Rev. Lett. **71**, 3202 (1993); J.R. Engelbrecht, *et al.*, Phys. Rev. B **55**, 15153 (1997).
31. J. Carlson, *et al.*, nucl-th/0302041.
32. J. Carlson, *et al.*, Phys. Rev. Lett. **91**, 050401 (2003).

Nucleosynthesis in Supernovae and the Early Universe

T. Kajino*†, T. Sasaqui*†, M. Orito**, K. Otsuki*‡, G. J. Mathews§, S. Honda*, W. Aoki* and S. Chiba¶

*National Astronomical Observatory and Graduate University for Advanced Studies,
Mitaka, Tokyo 181-8588, Japan
†Department of Astronomy, Graduate School of Science, University of Tokyo,
Bunkyo-ku, Tokyo 113-0033, Japan
**Research Laboratory for Nuclear Reactors, Tokyo Institute of Technology,
Meguro-ku, Tokyo, 152-8550, Japan
‡Department of Physics and Center for Astrophysics, University of Notre Dame,
Notre Dame, IN 46556-5670, U.S.A.
§ Department of Physics and Center for Astrophysics, University of Notre Dame,
Notre Dame, IN 46556-5670, U.S.A.
¶Advanced Science Research Center, Japan Atomic Energy Research Institute, Tokai, Naka, Ibaraki
319-1195, Japan

Abstract.
Universal baryon-density parameter Ω_B inferred from the Big-Bang nucleosynthesis and cosmic microwave background fluctuations disagrees with each other at the 1σ level. We propose new cosmological models for the Big-Bang nucleosynthesis to try to recover the concordance of Ω_B. Nuclear processes in the Big-Bang nucleosynthesis are similar to those in the r-process nucleosynthesis in supernovae. Even in the nucleosynthesis of heavy elements, entropy and density in the neutrino-driven winds of Type II supernovae are so high that nuclear statistical equilibrium favors production of abundant light nuclei. In such explosive circumstances many radioactive light-to-intermediate mass nuclei as well as heavy mass nuclei play the significant roles.

INTRODUCTION

Recent data of cosmic microwave background (CMB) anisotropy [1] have indicated profound implication in cosmology and nucleosynthesis models: The cosmic expansion seems accelerating for dark energy, $\Omega_\Lambda = 0.69 \pm 0.03$, and needs cold dark matter, $\Omega_{DM} = 0.27 \pm 0.02$. A tiny fraction, $\Omega_B = 0.044^{+0.03}_{-0.02}$, consists of known particles and nuclei. Although inferred values are subject to several prior assumptins of flat cosmology and scalar fluctuations of the CMB, the biggest challenge is to understand the true nature of dark matter and dark energy.

We have recently proposed a new theoretical model [2] that a massive cold dark matter particle is likely to disappear when it is quantized in a Randall-Sundrum noncompact higher dimensional AdS_5 spacetime. We looked for cosmological evidence for this new paradigm and found that the model is consistent with all data at the 95% C.L., satisfying presently available all sorts of observational constraints from the redshift-luminosity realation of Type Ia supernovae, the mass-to-light ratios of galaxy clusters, and the CMB power spectrum. In this paradigm there appears a new term in generalized Friedmann equation which is called dark radiation. Upper and lower limits on the dark radiation can be deduced from the Big-Bang nucleosynthesis (BBN) [3].

We have also studied dark energy in terms of a dynamical attractor-like solution for the evolution of a quintessence scalar field [4]. A generic feature of our solution is the possibility of significant energy density in the scalar field during the radiation dominated epoch. This model can affect the primordial BBN and the epoch of photon decoupling, i.e. CMB. Therefore, the BBN should make an important constraint on the quintessence model of dark energy.

There exists a potential conflict among various Ω_B values inferred from different analyses of the CMB anisotropy, the BBN, the Lyman-α forests, the light-to-mass ratios of galaxy clusters, and the X-ray gas fractions of rich clusters. Recent focus is the different Ω_B value deduced from the CMB ($\Omega_b h^2 \approx 0.0224 \pm 0.0009$) and the BBN ($\Omega_B h^2 \leq 0.017$), where theoretical upper limit arises from the severe constraints on the primordial ^{4}He and ^{7}Li abundances [5]. h is the present value of the Hubble constant in units of 100 km sec^{-1} Mpc^{-1}. It is very critical to look for the concordance of Ω_B in view of testing the theoretical models for Ω_Λ and Ω_{DM}.

From the nuclear physics viewpoints, nuclear processes in the early stage of the r-process in supernova explosions are very similar to those of the BBN [5, 6]. Even in heavy element production, entropy and density in the neutrino-driven winds are so high that the NSE favors abundant light nuclei. In such an initial explosive condition, many radioactive light-to-intermediate mass nuclei as well as heavy mass nuclei play the significant roles in the production of r-process elements [7, 8, 9]. They are also critical for the production of actinides which are used for cosmochronometers [10].

BIG-BANG NUCLEOSYNTHESIS

Lepton Asymmetric Model

The standard BBN model takes account of finite baryon asymmetry (B $\neq$ 0), while the lepton asymetry is assumed to be zero (L = 0). It has been proposed recently [11, 12] that models based upon the Affleck-Dine scenario of baryogenesis might generate naturally lepton number asymmetry which is seven to ten orders of magnitude larger than the baryon number asymmetry. L $\neq$ 0 realizes when the universal neutrino chemical potential μ_v is not equal to zero.

Non-zero lepton numbers affect most strongly the BBN in the following three ways [13, 14, 15]: First, neutrinos and anti-neutrinos drop out of thermal equilibrium with the background thermal plasma when the weak reaction rate becomes slower than the universal expansion rate. L $\neq$ 0 leads to an earlier decoupling, and the ratio of the neutrino to photon temperatures, T_v/T_γ is reduced. This makes universal expansion slower. Secondly, neutrino degeneracy increases the expansion rate and increases the ^{4}He production. Thirdly, the equilibrium n/p ratio is affected by the electron neutrino chemical potential, n/p = $\exp\{-(\Delta M/T_{n\leftrightarrow p}) - \mu_{v_e}/T_{v_e}\}$, where ΔM is the neutron-proton mass difference and $T_{n\leftrightarrow p}$ is the freeze-out temperature for the relevant weak

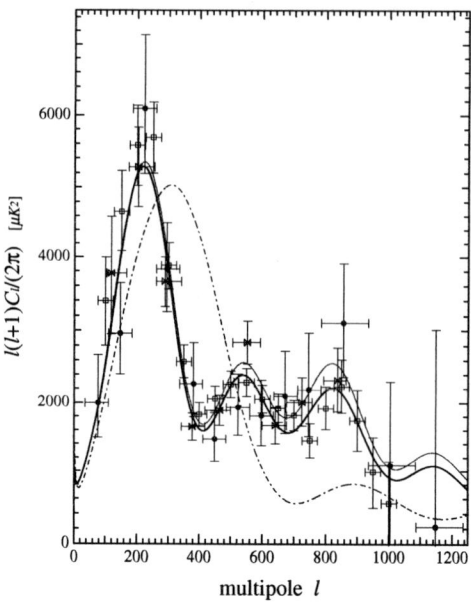

FIGURE 1. Fits to the power spectrum of fluctuations in the CMB from BOOMERANG [18], MAXIMA-1 [19], and DASI [20]. The solid line shows the best neutrino-degenerate fit [17]. The dotted line shows a best non-degenerate model. For illustration, the dot-dashed line also shows the large-degeneracy.

reactions. These effects either increase or decrease production of ^{4}He and other light elements, depending upon the sign of μ_ν. For these mechanisms neutrino degeneracy can even allow baryonic densities up to $\Omega_B h^2 = 1$ for positive universal lepton number [15].

Neutrino degeneracy dramatically alters the CMB power spectrum [16]. We limit our consideration to flat $\Omega_{tot} = \Omega_{DM} + \Omega_B + \Omega_\Lambda = 1$ cosmological models with ionization parameter $\tau = 0$. We then obtained that optimum neutrino-degenerate model for which $(\Omega_B h^2, \mu_{\nu_{\mu,\tau}}/T_{\nu_{\mu,\tau}}, \mu_{\nu_e}/T_{\nu_e}, \Omega_\Lambda, h, n) = (0.021, 1.0, 0.09, 0.74, 0.74, 0.93)$, where n is tilt of primordial fluctuation, implies a nearly perfect fit, as shown in Figure 1 [15, 17]. We can conclude from our present studies of BBN and CMB that the recent CMB data seem to require a slight increase in the baryonic contribution to the closure density, $\Omega_B h^2 = 0.021$, as allowed in our neutrino-degenerate models [15, 17].

Baryon Inhomogeneous Model

Baryon inhomogeneous BBN models [21, 22, 23] have an advantage to allow wider range of $\Omega_B h^2$ values, i.e. $0.01 \leq \Omega_B h^2 \leq 0.05$ which well cover the constraint from recent CMB data. Note that a slight lepton asymmetry is needed even in the inhomogeneous BBN in order to satisfy the light element abundance constraints [14].

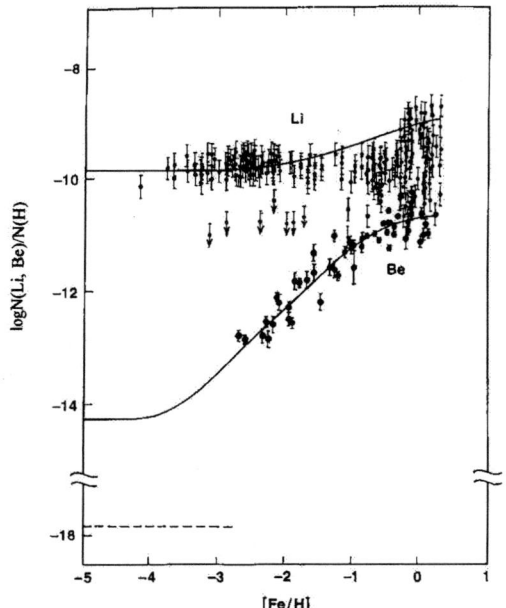

FIGURE 2. Observed Li and Be abundances in various metallicity stars. Metallicity is defined as metal abundance relative to H, relative to the solar abundance; $[Fe/H] = log(Fe/H) - log(Fe/H)_\odot$. Solid curves are the calculated results in Galactic chemical evolution model, and the plateaus at low metallicity region ([Fe/H]$\leq$-2 for Li, and [Fe/H]$\leq$-3.5 for Be) are the predicted primordial abundances in the baryon inhomogeneous BBN model (solid) [23, 28] and the standard homogeneous BBN model (dashed) [24].

The nucleosynthesis occurs in an environment of proton-neutron segregated inhomogeneous distribution due to most likely the first order osmological QCD phase transition and subsequent proton and neutron diffusion in a different matter from each other. Therefore, the radioactive nuclear reactions play the significant roles as well as stable and meta-stable nuclear reactions [24, 25]. Several new reaction chains are identified to be important:

^{4}He$(^3$H$,\gamma)^7$Li$(n,\gamma)^8$Li$(\alpha,n)^{11}$B$(n,\gamma)^{12}$B$(\beta\ v)^{12}$C$(n,\gamma)^{13}$C$(n,\gamma)^{14}$C...,

^{7}Li$(n,\gamma)^8$Li$(n,\gamma)^9$Li$(\beta\ v)^9$Be$(n,\gamma)^{10}$Be$(n,\gamma)^{11}$Be$(\beta\ v)^{11}$B...,

^{4}He$(^3$He$,\gamma)^7$Be$(n,p)^7$Li$(\alpha,\gamma)^{11}$B...

^{7}Li$(^3$H$,n)^9$Be$(^3$H$,n)^{11}$B... or ^{7}Li$(^3$He$,p)^9$Be$(^3$He$,p)^{11}$B...

The first two reaction chains and the third chain play the key roles in the production of heavy isotopes [21, 23] in the neutron-rich and proton-rich zones, respectively.

We also found that the fourth reaction chains are extremely important for the production of ^{9}Be and ^{11}B [26]. With these reactions being included in the network, the ^{9}Be and ^{11}B abundances are predicted to enhance by three or four orders of magnitude from the standard homogeneous BBN value (dashed line). In Figure 2, plateau (solid line) at low metallicity region [Fe/H] $\leq$ -3.5 is the predicted primordial ^{9}Be abundance in the inhomogeneous BBN model [27, 28]. It is highly desirable to look for the plateau in Be and B abundances to test the baryon inhomogeneous model.

R-PROCESS NUCLEOSYNTHESIS IN SUPERNOVAE

Observation

Massive stars culminate their evolution by supernova explosions that are presumed to be most viable candidate sites for r-process nucleosynthesis. Appropriate physical condition for a successful r-process, however, has not been uniquely identified until recently because we did not have clear astronomical data which show pure r-process products from a single supernova nucleosynthesis.

Recent measurements using high-dispersion spectrographs with large Telescopes or the Hubble Space Telescope have made it possible to detect minute amounts of r-process elements in faint metal-deficient ([Fe/H] $\leq$ -2.5) stars. Stellar metallicity, [Fe/H] = log[N(Fe)/N(H)] - log[N(Fe)/N(H)]$_\odot$, obeys an approximate time-metallicity relation $t/10^{10}$yr $\sim 10^{[Fe/H]}$. Massive stars with $10 M_\odot \leq M$ evolve in short life time $\sim 10^7$ yr and eject material into the interstellar medium when they explode as supervovae. Therefore, the first generation supernovae are presumed to be a unique source of the r-process elements discoveded in metal-deficient stars with [Fe/H] $\leq$ -2.5.

Incidentally, large dispersion in heavy neutron-capture elements has been observed [29, 30, 31, 32, 33] in halo stars, as displayed in Figure 3. This is interpreted as a natural consequence of inhomogeneous mixing of supernova ejecta with interstellar medium. Namely, metal-deficient halo stars had formed from a single supernova ejecta before they were well mixed with surrounding primeval gas that has almost no metal. Honda et al. [31, 32] have recently discovered observationally that the Be/Eu abundance ratios fit almost perfectly the pure r-process ratio as shown in Figure 4, even should each abundance Ba or Eu scatter very largerly from star to star by more than 2-3 dex (Figures 3).

Theoretical Problems

Hot neutron stars just born in the gravitational core collapse SNeII release most of their energy as neutrinos during the Kelvin-Helmholtz cooling phase. An intense flux of neutrinos heat the material near the neutron star surface and drive matter outflow called neutrino-driven winds.

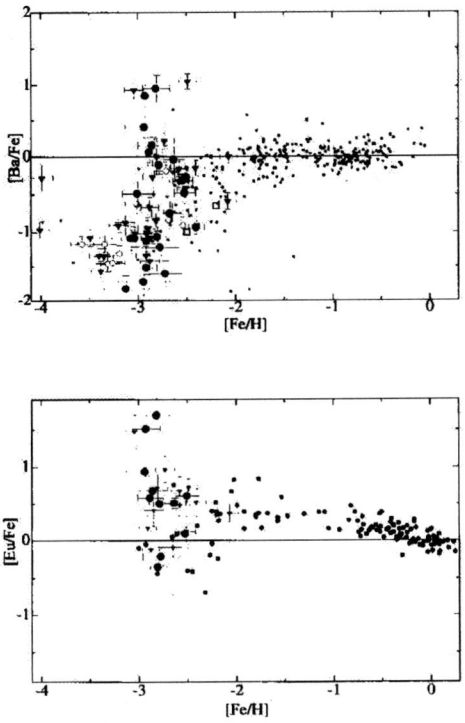

FIGURE 3. [Ba/Fe] and [Eu/Fe] as a function of metallicity [Fe/H]. Large closed circles for -3.2 ≤ [Fe/H] ≤ -2.3 are the observed data by using High Dispersion Spectrograph equipped with SUBARU Telescope [31, 32], and small dots are from the previous observation.

Although Woosley et al. [34] demonstrated a profound possibility that the r-process could occur in these winds, several difficulties were subsequently identified. First, independent non-relativistic numerical supernova models [35, 36] have difficulty producing the required entropy, S/k ~ 400 (high entropy problem). Second, even should the entropy so high, the effects of neutrino absorption $v_e + n \rightarrow p + e^-$ and $v_e + A(Z,N) \rightarrow A(Z+1, N-1) + e^-$ may decrease the neutron fraction during the nucleosynthesis process. As a result, a deficiency of free neutrons prohibits the r-process [37] (neutrino interaction problem).

Solution

In order to resolve the high entropy problem, we have studied [7, 38] neutrino-driven winds in a Schwarzschild geometry under reasonable assumption of spherical steady-state flow. We obtained two important resuls. First, the general relativistic effects make

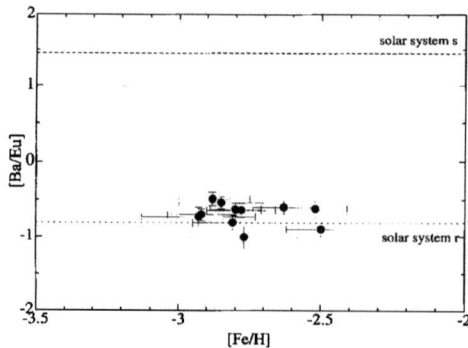

FIGURE 4. [Ba/Eu] as a function of metallicity [Fe/H]. Data are from the observation by using High Dispersion Spectrograph equipped with SUBARU Telescope [31, 32]. Expected ratios from pure r- and s-process abundances are respectively labeled as "solar system r" and "solar system s".

expansion dynamic time scale τ_{dyn}, which is defined as the duration time of the α-process when the temprature drops from T $\approx$ 0.5 MeV to 0.5/e MeV, much shorter and entropy per baryon, S/k, larger by about 40 % from the Newtonian case. Second, the neutrino-driven winds from massive and compact proto-neutron stars produce successfully the r-process elements. The critical condition is the short expansion time $\tau_{dyn} \sim 10$ msec even at medium entropy S/k $\approx$ 130 [7].

Since the initial nuclear composition of the relativistic plasma consists of neutrons and protons, α-burning begins when the temperature cools below T $\sim$ 0.5 MeV. The ^{4}He$(\alpha\alpha, \gamma)^{12}$C reaction is too slow at this temperature, and hence an alternative nuclear reaction chain ^{4}He$(\alpha n, \gamma)^9$Be$(\alpha, n)^{12}$C triggers explosive α-burning to produce seed elements with A $\sim$ 100. Therefore, time scale for nuclear reactions is regulated by the ^{4}He$(\alpha n, \gamma)^9$Be reaction, $\tau_N \equiv (\rho_b^2 Y_\alpha^2 Y_n \lambda(\alpha\alpha n \rightarrow ^9Be))^{-1} \sim 100$ ms. Since the condition $\tau_{dyn} \approx 10ms \ll \tau_N \approx 100ms$, is satisfied, fewer seed nuclei are produced during the α-process with plenty of free neutrons left over at the time when the r-process begins. The high neutron-to-seed ratio, $n/s \sim 100$, leads to appreciable production of r-process elements, producing the 2nd (A $\sim$ 130) and 3rd (A $\sim$ 195) abundance peaks and the hill of rare-earth elements (A $\sim$ 165) in Figure 5.

Collision time scale for neutrino-nucleus interactions, $v_e + n \rightarrow p + e^-$ and $v_e + A(Z,N) \rightarrow A(Z+1, N-1) + e^-$, takes another key to resolve the neutrino interaction problem. It is given by $\tau_v \approx 201 \times L_{v,51}^{-1} \times (\varepsilon_v/\text{MeV})(r/100\text{km})^2 (<\sigma_v> /10^{-41}\text{cm}^2)^{-1}$ msec, where $L_{i,51}$ is the individual neutrino or antineutrino luminosity in units of 10^{51} ergs/s, $\varepsilon_i = <E_i^2>/<E_i>$ in MeV ($i = v_e, \bar{v}_e,$ etc.), and $\langle\sigma_v\rangle$ is the cross section averaged over the neutrino energy spectrum. At the α-burning site r $\approx$ 100 km for $L_{v,51} \approx 10$, $\varepsilon_{v_e} = 12$ MeV, and $\langle\sigma_v\rangle \approx 10^{-41}\text{cm}^2$, τ_{v_e}(r=100 km) turns out to be $\approx$ 240 ms. This is larger than the expansion dynamic time scale,

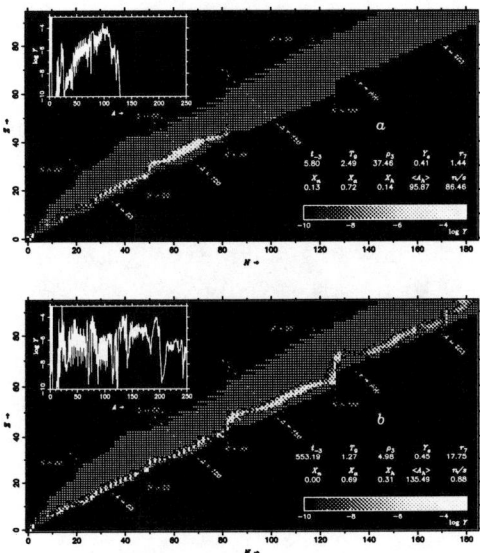

FIGURE 5. Nuclear reaction flow patterns in Z-N plane, and abundances in the insets at the time; (a) $t \approx 18ms$, and (b) $t \approx 568ms$. Time zero refers to the time when the neutrino-driven wind leaves off the surface of proto-neutron star. Highlight abundance scales to $\log Y$, where Y is the number fraction of each nucleus. This result is from [40].

$\tau_{dyn} \approx 10 msec \ll \tau_{\nu_e}(r = 100km) \approx 240 msec$, showing that ν_e's interact with neutrons so slowly in such rapidly expanding neutrino-driven winds that the neutron fraction is not affected strongly. Note that the opposite condition $\tau_{dyn} > \tau_{\nu_e}$ applies to slowly expanding winds as adopted in the previous studies [34, 35, 39] which produce only the r-process elements up to the 2nd abundance peak [37].

Relevant Nuclear Reactions

We exploited a fully implicit single network code, which includes over 3000 isotopes and 11000 nuclear and particle processes among them, for the whole processes of NSE - α-process - r-process [7, 38]. Previous r-process calculations had complexity that the seed abundance distribution at $T_9 = 2.5$ was first calculated in a smaller network code for the α-process of the light-to-intermediate mass nuclei, and then the r-process nucleosynthesis calculation was extensively carried out by using another network code [34, 41]. The white dots in Figure 5 show the nuclei included in our network code.

We found [8] that even light neutron-rich nuclei play the significant roles in the r-process. At early epoch of the wind expansion, $t \leq$ a few dozens msec (Figure 5(a)), both temperature and density are so high that the charged particles interact with one another to proceed α-process nucleosynthesis around the β-stability line, which is triggered by ^{4}He$(\alpha n, \gamma)^9$Be$(\alpha, n)^{12}$C [41, 42, 43]. At relatively later epoch when the temperature drops below T = 0.5/e MeV even after the α-rich freeze out, new reaction paths open [8]:

^{4}He$(\alpha n,\gamma)^9$Be$(n,\gamma)^{10}$Be$(\alpha,\gamma)^{14}$C$(n,\gamma)^{15}$C,

^{3}H$(\alpha,\gamma)^7$Li$(n,\gamma)^8$Li$(\alpha,n)^{11}$B$(n,\gamma)^{12}$B$(n,\gamma)^{13}$B$(n,\gamma)^{14}$B$(n,\gamma)^{15}$B$(e^-\nu)^{15}$C.

These new reaction flow paths also take appreciable flux of baryon number and continuously produce seed nuclei. A side flow ^{7}Li$(\alpha,\gamma)^{11}$B$(p,2\alpha)^4$He is important, too. As such, the classical r-process flow, (n,γ) followed by beta decay, has already started from the light neutron-rich nuclei. This is a very different and new result from the previous picture that the r-process starts from only intermediate-mass seed nuclei $A \approx 100$.

There are several branching points for the (n,γ) and (α,n) reactions. They are at ^{18}C, ^{24}O, ^{36}Mg, etc. Sensitivity study of the r-process yields to these unmeasured reaction cross sections has recently been carried out theoretically [9].

Cosmochronology

Cosmological model analysis of the WMAP-CMB data infers the cosmic expansion time $t_U \approx 13.7 \pm 0.2$ Gyr. With the success of such observation, it is highly desired to construct chronometers for the estimate of various stellar objects such as the clouds, stars, galaxies and galaxy clusters for the studies of cosmic and Galactic chemical evolution.

Thorium (Th) and Uranium (U), which are the typical r-process elements and have half-lives of 14 Gyr (Th) and 4.5Gyr (^{238}U), are ideal chronometers. These long-lived radioactive nuclei have recently been found in very metal-deficient halo stars, as discussed in the previous section [29, 30, 32]. These elements are presumed to be the r-process products from a single primeval (population III) supernova. Nevertheless the observed abundance pattern in these metal-deficient stars is very similar to the solar, which is called "universality". Thus in the previous Th/Eu chronometer studies the universality was assumed for the initial abundances of actinides.

We found evidence that the observed universality for $56 < Z < 75$ elements does not imply a unique astrophysical site for the r-process [10]. Neither does it imply a universality of abundances of nuclei lighter or heavier than this range. In particular, we showed that a variety of astrophysical r-process models can be constructed which reproduce the same observed universal abundance pattern for $56 < Z < 75$ nuclei, yet have vastly different abundances for $Z \geq 75$ and $Z \leq 56$ nuclei. This introduces an uncertainty into the use of the Th/Eu chronometer as a means to estimate the ages of the metal-deficient stars.

We do find that the U/Th ratio is a robust chronometer. This is because the initial production ratio of U to Th is almost independent of the astrophysical nucleosynthesis environment, yet remaining the largest uncertainties in the U/Th initial production ratio due to the input nuclear physics models. Further tight collaboration among nuclear physics, astrophysics, and astronomy is highly desirable.

This work has been supported in part by Grants-in-Aid for Scientific Research

(12047233, 13640313, 14540271) and for Specially Promoted Reseach (13002001) of the Ministry of Education, Science, Sports and Culture of Japan, and also by the U.S. Department of Energy under Nuclear Theory Grant DE-FG02-95-ER40934.

REFERENCES

1. D. Spergel, et al. (WMAP Collaboration), Astrophys. J. Suppl. **148** (2003), 175.
2. K. Ichiki, P. M. Garnavich, T. Kajino, G. J. Mathes, & M. Yahiro, Phys. Rev. D. **68** (2003), 083518.
3. K. Ichiki, M. Yahiro, T. Kajino, M. Orito, & G. J. Mathews, Phys. Rev. D. **66** (2002), 043521.
4. N. Yahiro, G. J. Mathews, K. Ichiki, T. Kajino, & M. Orito, Phys. Rev. D. **65** (2002), 063502.
5. T. Kajino, K. Otsuki, & M. Orito, Prog. Theor. Phys. Suppl. **146** (2002), 247.
6. T. Kajino, S. Wanajo, & G. J. Mathews, Nucl. Phys. A. **704** (2002), 175c.
7. K. Otsuki, H. Tagoshi, T. Kajino, & S. Wanajo, Astrophys. J. **533** (2000), 424.
8. M. Terasawa, K. Sumiyoshi, T. Kajino, G. J. Mathews, & I. Tanihata, Astrophys. J. **562** (2001), 470.
9. T. Sasaqui, K. Otsuki, T. Kajino, G. J. Mathews, & T. Nakamura, Astrophys. J. (2004), submitted.
10. K. Otsuki, G. J. Mathews, & T. Kajino, New Astronomy 8 (2003) 767.
11. A. Casas, W. Y. Cheng, & G. Gelmini, Nucl. Phys. B. **538** (1999), 297.
12. I. Affleck, & M. Dine, Nucl. Phys. B. **249** (1985), 361.
13. H. Kang, & G. Steigman Nucl. Phys. B. **372** (1992), 494.
14. T. Kajino, & M. Orito, Nucl. Phys. A. **629** (1998), 538c.
15. M. Orito, T. Kajino, G. J. Mathews, & R. N. Boyd, Nucl. Phys. A. **688** (2001), 17c.
16. W. K. Kinney, & A. Riotto, Phys. Rev. Lett. **83** (1999), 3366.
17. M. Orito, T. Kajino, G. J. Mathews, & Y. Wang, Phys. Rev. D. **65** (2002), 123504.
18. C. B. Netterfield, et al. (BOOMERANG Collaboration), Astrophys. J. **571** (2002), 604.
19. S. Hanany, et al. (MAXIMA-1 Collaboration), Astrophys. J. **545** (2000), L5.
20. C. Pryke, et al. (DASI Collaboration), Astrophys. J. **568** (2002), 46.
21. J. H. Applegate, C. J. Hogan, & R. J. Scherrer, Phys. Rev. D. **35** (1987), 1151.
22. C. R. Alcock, G. M. Fuller, & G. J. Mathews, Astrophys. J. **320** (1987), 439.
23. T. Kajino, G. J. Mathews, & G. M. Fuller, Astrophys. J. **364** (1987), 7.
24. M. S. Smith, L. H. Kawano, & R. A. Malaney, Astrophys. J. Suppl. **85** (1993), 219.
25. A. Coc, et al., (2003) (astro-ph/0309480, astro-ph/0401008).
26. R. N. Boyd, & T. Kajino, Astrophys. J. **336** (1989), L55.
27. T. Kajino, & R. N. Boyd, Astrophys. J. **359** (1990), 267.
28. M. Orito, T. Kajino, & G. J. Mathews, Astrophys. J. **488** (1997), 515.
29. C. Sneden, A. McWilliam, G. W. Preston, J. J. Cowan, D. L. Burris, & B. J. Armosky, Astrophys. J. **467** (1996), 819.
30. R. Cayrel, et al. Nature. **409** (2001), 681.
31. S. Honda, W. Aoki, H. Ando, & T. Kajino, Nucl. Phys. A. **718** (2003), 674c.
32. S. Honda, et al. (Subaru/HDS Collaboration), Astrophys. J. (2004), in press.
33. Ryan, S. G., Norris, J. E., & Beers, T. C., Astrophys. J. **471** (1996), 254.
34. S. E. Woosley, J. R. Wilson, G. J. Mathews, R. D. Hoffman, & B. S. Meyer, Astrophys. J. **433** (1994), 229.
35. J. Witti, H.-Th. Janka, & K. Takahashi, Astron.&Astrophys. **286** (1994), 842.
36. Y. Z. Qian, & S. E. Woosley, Astrophys. J. **471** (1996), 331.
37. B. S. Meyer, Astrophys. J. **449** (1995), L55.
38. S. Wanajo, T. Kajino, G. J. Mathews, & K. Otsuki, Astrophys. J. **554** (2001), 578.
39. B. S. Meyer, G. J. Mathews, W. M. Howard, S. E. Woosley, & R. D. Hoffman, Astrophys. J. **399** (1992), 656.
40. T. Kajino, S. Wanajo, & G. J. Mathews, Nucl. Phys. A. **704** (2002), 165c.
41. S. E. Woosley, & R. D. Hoffman, Astrophys. J. **395** (1992), 202.
42. H. Utsunomiya, Y. Yonezawa, H. Akimune, T. Yamagata, M. Ohta, M. Fujishiro, H. Toyokawa, & H. Ohgaki, Phys. Rev. C. **63** (2001), 018801.
43. K. Sumiyoshi, H. Utsunomiya, S. Goko, & T. Kajino, Nucl. Phys. A. **709** (2002), 467.

POSTER SESSIONS

Neutron evaporation as a probe for dynamical effects in heavy ion fusion reactions

Ajay Kumar[1], A. Kumar[1], G. Singh[1], Hardev Singh[1], R.P. Singh[2], Rakesh Kumar[2], K.S. Golda[2], S.K. Datta[2] and I.M. Govil[1]

[1] *Department of Physics, Panjab University, Chandigarh-160014, India*
[2] *Nuclear Science Centre, New Delhi-110067, India*

Abstract: The compound nucleus ^{76}Kr* was populated at the excitation energy of 75 MeV and angular momentum of 39 $\hbar$ in fusion reactions with two complementary, mass-symmetric (^{31}P+^{45}Sc) and mass asymmetric (^{12}C+^{64}Zn) entrance channels. The neutron evaporation spectra were measured and compared with the predictions of the statistical model calculations. The results for the mass-asymmetric reaction are found to be consistent with the predictions of the statistical model calculations. However, for the mass-symmetric reaction (^{31}P+^{45}Sc), the experimental spectra are found to be harder than the theoretical neutron spectra. The dynamical model calculations of Feldmeier show that the formation time for the compound nucleus for the symmetric system is relatively larger as compared to the asymmetric system.

INTRODUCTION

Recently some papers have claimed [1, 2] that experimental neutron evaporation spectra from heavy-ion fusion reactions at higher excitation energies and angular momenta, are no longer consistent with the predictions of statistical model. Specifically, it has been observed that in such cases measured neutrons have been characterised as having more average energies than predicted [1, 3]. This is interpreted as neutron emission from the temperature equilibrated intermediate dinuclear complex during the time of its evolution towards compound nucleus formation. In order to confirm this study further we employed two systems viz mass symmetric (^{31}P+^{45}Sc) and a mass asymmetric channel (^{12}C+^{64}Zn) leading to the same compound nucleus ^{76}Kr*.

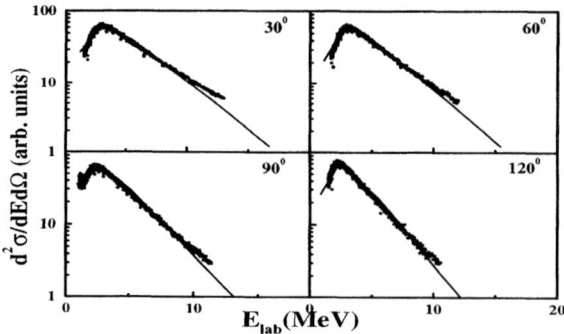

Figure 1: Comparison of the experimental neutron spectra (dots) with statistical model (solid line) using $r_0 = 1.25$ and $a = A/8$ for the asymmetric reaction $^{12}C+^{64}Zn$ with $l_{max} = 39\hbar$ and $E^* = 75$ MeV at $E_{lab} = 85$ MeV.

EXPERIMENT AND ANALYSIS

The data was obtained using 15 UD Pelletron at Nuclear Science Centre (NSC), New Delhi, India. The self-supporting isotopically enriched (99.93 %) targets of 1mg/cm^2 thickness were used in both cases. The experiment was done using the 1.5 M diameter stainless steel general purpose scattering chamber [GPSC] available at NSC. Neutron detectors having liquid scintillator cells of BC501 of 12.5 cm diameter and thicknesses of 12.5 cm were used at an angle of 30^0, 60^0, 90^0 and 120^0 respectively with respect to the beam direction and were placed at a distance of 1 M from the target. The γ-n pulse-shape discrimination was employed to reduce the γ background. The neutron energy was determined by the time of flight technique. The pulse from the neutron detectors was used as the start while the stop pulse to the TAC was provided by the pulsed beam. The time of flight spectra thus obtained were converted into neutron energy, utilizing the neutron detection efficiency code MODEFF [4]. The statistical computer code CASCADE [5] was used to perform theoretical calculations, which assumes the reaction to occur in two steps. First the formation of compound nucleus and second the statistical decay of the equilibrated system. The model HICOL developed by Feldmeier [6] gives a realistic microscopic picture of nucleus nucleus collision based on the concept of one body dissipation , where the coupling between

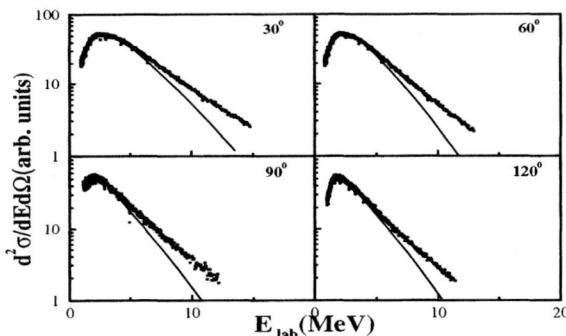

Figure 2: Comparison of the experimental neutron spectra (dots) with statistical model (solid line) using a = A/8 and r_0 = 1.25 for the symmetric reaction $^{31}P+^{45}Sc$ with l_{max} = 39$\hbar$ and E* = 70 MeV at E_{lab} = 112 MeV.

the intrinsic and collective degrees of freedom is treated in a microscopic picture of particle exchange to obtain the friction and fusion tensor.

RESULTS AND DISCUSSION

The neutron spectra of the composite system $^{76}Kr^*$ formed through the asymmetric reaction ($^{12}C+^{64}Zn$) at maximum angular momentum 39 $\hbar$ and excitation energy of 75 MeV are shown in fig.1. These are in good agreement with the statistical model calculations using the normal level density parameter **a** = A/8 MeV^{-1}, the rotating liquid drop model moment of inertia, and the optical model transmission coefficients for the respective inverse absorption channels.

The neutron emission for the mass-symmetric ($^{31}P+^{45}Sc$) system at different angles for same angular momentum (39 $\hbar$) and for the same excitation energy (75 MeV), are shown in fig.2 and fig.3 respectively. As is clear these are not be explained in terms of statistical model predictions with the normal parameters as discussed above for the asymmetric system. In particular, the higher energy part of the neutron spectra is harder than the statistical model calculations. The slope of the high energy part of the neutron spectra is very sensitive to the small level density parameter **a**. In order to verify quantitatively the experimental trends, the statistical model calculation was

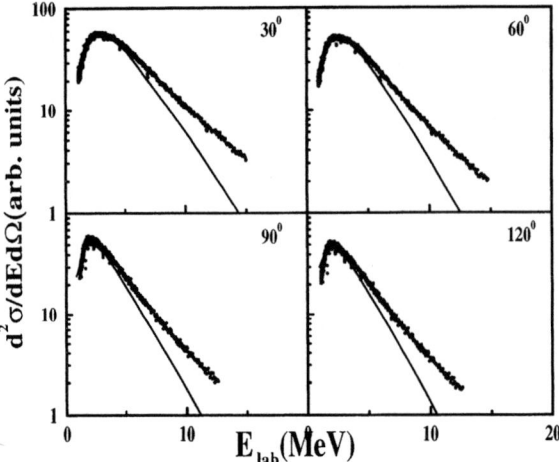

Figure 3: Comparison of the experimental neutron spectra (dots) with statistical model (solid line) using a = A/8 and r_0 = 1.25 for the symmetric reaction $^{31}P+^{45}Sc$ with l_{max} = 43$\hbar$ and E* = 70 MeV at E_{lab} = 120 MeV.

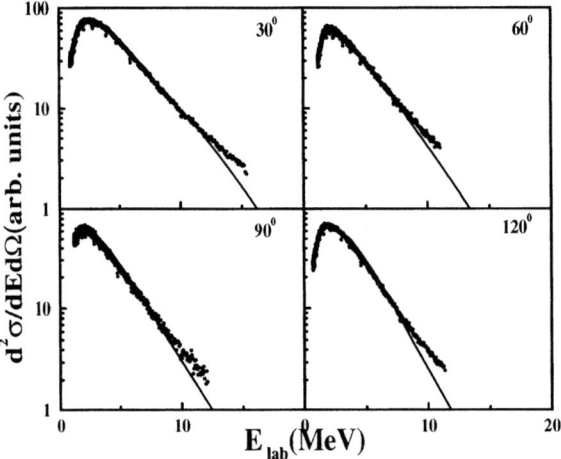

Figure 4: Comparison of the experimental neutron spectra (dots) with statistical model (solid line) using a = A/10 and r_0 = 1.25 for the symmetric reaction $^{31}P+^{45}Sc$ with l_{max} = 39$\hbar$ and E* = 70 MeV at E_{lab} = 112 MeV.

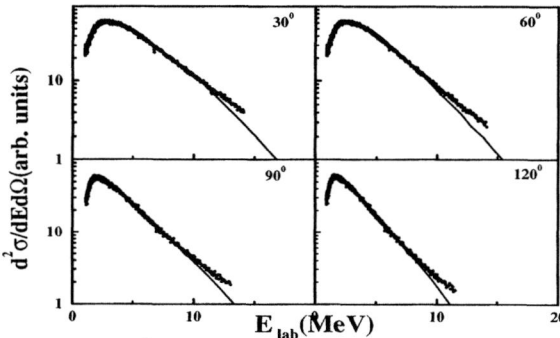

Figure 5: Comparison of the experimental neutron spectra (dots) with statistical model (solid line) using a = A/10 and r_0 = 1.25 for the symmetric reaction $^{31}P+^{45}Sc$ with l_{max} = 43$\hbar$ and E* = 75 MeV at E_{lab} = 120 MeV.

performed by varying the value of level density parameter **a** = A/10 MeV^{-1}. Results of these calculations, are shown in figs.4 and 5 respectively for the two cases. The lower value of level density parameter (**a** = A/10 MeV^{-1}) manifiest an effective higher nuclear temperature for the neutron evaporation in case of the mass-symmetric system. The observed weaker excitation energy dependence of the level density of the nuclei for the neutron evaporation indicate the dynamical effects. Therefore, to understand this behaviour, the dynamical model calculations HICOL [6] were done. The evolution of the dynamical fusion process as depicted in fig.6 for l = 30 $\hbar$, indicates that for the symmetric system ($^{31}P+^{45}Sc$) the compound nucleus formation time(37x10^{-22} sec) is larger than the formation time(29x10^{-22} sec) for the asymmetric system($^{12}C+^{64}Zn$). The possible emission of neutrons during the retarded fusion process gives rise to this observed hardening of the neutron spectra for the symmetric entrance channel.

SUMMARY

Neutron energy spectra of the asymmetric system ($^{12}C+^{64}Zn$) at different angles are well described by the statistical model predictions using the normal systematic value of small level density parameter **a** = A/8 MeV^{-1}. However, in the case of the symmetric system ($^{31}P+^{45}Sc$), the statistical model interpretation of the data requires **a** = A/10 MeV^{-1}. The harder neutron spectra

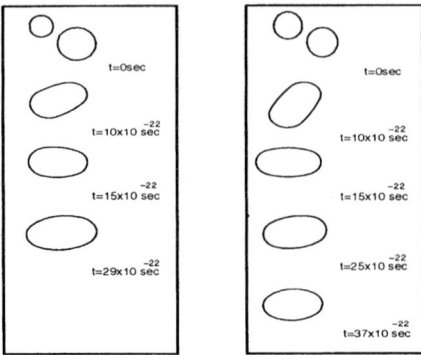

Figure 6: Time evolution of the reactions for an angular momentum of 30 $\hbar$ (a) $^{12}C+^{64}Zn$ and (b) $^{31}P+^{45}Sc$.

manifests a higher effective nuclear temperature of the nuclei involved in neutron evaporation. It is attributed to the enhanced formation time of the compound system and the evaporation of the neutrons from the intermediate thermally equilibrated dinuclear complex during the dynamical evolution of the system in case of the symmetric entrance channel.

ACKNOWLEDGMENTS

The authors are thankful to the Accelerator crew of Nuclear Science Centre, New Delhi, for providing a high quality beam during the course of this experiment. Author also acknowledge the financial support for the reseach work from CSIR, New Delhi, India.

References

[1] J. Kasagi, B. Remington, A. Galonsky, F. Hass, J.J. Kolata, L. Satkowiak, M. Xapsos, R. Racca and F.W. Prosser, Phys. Rev. **C 31**, 858 (1985).

[2] J.L. Wile et al., Phys. Rev. **C 47**, 2135 (1993).

[3] A. Saxena, A. Chatterjee, R.K. Choudhry, S.S. Kapoor, and D.M. Nadkarni, Phys. Rev. **C 49**, 932 (1994).

[4] R.A. Cecil, B.D. Anderson, and R. Madey, Nucl. Instrum. Methods **161**, 439 (1979).

[5] F. Puhlhofer, Nucl. Phys. **A 280**, 267 (1997).

[6] H. Feldmeier, Rep. Prog. Phys. **50**, 915(1987).

What Can We Learn About the Fission Process from the Energy Dependence of the Induced Fission Times Obtained by the Crystal Blocking Technique?

Vadim A. Drozdov, Dmitri O. Eremenko, Olga V. Fotina,
Sergey Yu. Platonov, and Oleg A. Yuminov

*D.V. Skobeltsyn Institute of Nuclear Physics, M.V. Lomonosov Moscow State University,
Vorobyevy gory, 119992 Moscow, Russia*

Abstract. The unified energy dependence of the induced fission lifetimes obtained by the experimental crystal blocking technique for heavy nuclei with Z = 91 - 94 in the range of initial excitation energy from 5 to 250 MeV was analyzed. It was demonstrated that for the excitation energies of investigated nuclei up to 60 – 70 MeV the fission lifetimes can be described in the framework of the statistical theory of nuclear reactions taking into account the double-humped structure of the fission barrier and the lifetimes of both classes of excited nuclear states realized in the first and second potential wells. However, for the excitation energies above 70 MeV it is needed to consider the dynamical effects in the fission channel. Analysis of the experimental data in this energy range allows us to extract information on the magnitude of nuclear viscosity.

INTRODUCTION

The nuclear reaction time is an important characteristic directly associated with both the structural features of the interacting nuclei and the decay mechanism. As it was shown in [1] the double-humped structure of the fission barrier can have a strong influence on the time characteristics of decay of excited heavy nuclei. Moreover, it is needed to stress the unique sensitivity of such observable as the induced fission lifetime to the dynamical properties of the fissioning nuclear system, i.e. magnitude of nuclear matter viscosity and so on [2].

EXPERIMENTAL DATA

Here we present the results of analysis of a large set of experimental data on the induced fission lifetimes - τ_f for heavy nuclei. The data was obtained by the experimental crystal blocking technique, which is a direct way to measure the nuclear reaction time [3, 4]. We analyze τ_f for the ^{232}Th + p, d, ^{3}He, α reactions obtained by us at the cyclotron U-120 of the Institute of Nuclear Physics, Moscow State University, at the beam energies in the range from 4 to 7.8 MeV/nucleon [5]. The experimental lifetimes range from 10^{-17} to 10^{-14} s, depending on the projectile energy. Our own τ_f

data for the ^{28}Si + ^{nat}Pt reaction at the silicon beam energies in the range from 140 to 170 MeV [6] are also analyzed. The last measurements were done with the Tandem-XTU accelerator of the LNL Laboratories (Padova, Italy) and provided experimental lifetimes ranging from 10^{-18} to 10^{-17} s. Our own data have been compared with τ_f for the U-like nuclei produced in the ^{238}U + ^{28}Si reaction at the U-beam energy of 24 MeV/nucleon, obtained at GANIL [7]. Due to the very large value of the transferred momentum in the last reaction induced in reverse kinematics, τ_f ranges from 10^{-17} to 3×10^{-19} s for excitation energies in the 10 - 250 MeV interval. Hence we have the unified global energy dependence of τ_f for heavy nuclei with Z = 91 - 94 in the range of the initial excitation energy from 5 to 250 MeV (see Fig. 1). Obviously, the properties of fissioning nuclear systems are essentially varied through this wide energy range.

Anyway, the existence of this unified dependence constructed from the experimental data obtained for different nuclear reactions (investigated both in direct and reverse kinematics) demonstrates the exceptional ability of the crystal blocking technique for the nuclear lifetime measurements.

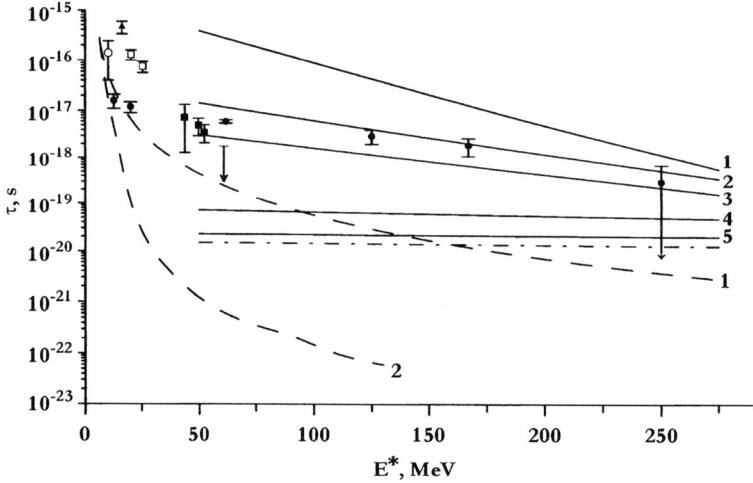

FIGURE 1. The induced fission lifetimes for nuclei with Z = 91 - 94 vs. initial excitation energy. Points are experimental data: open circle, triangle and squares are the data from the ^{232}Th(p, xnf), ^{232}Th(d, xnf), and ^{232}Th(α, xnf) reactions [5], black squares are the data from the ^{28}Si + ^{nat}Pt reaction [6], black circles are the data from the ^{238}U + ^{28}Si reaction [7]. The dashed lines are the results of the statistical calculations with allowance for (1) and with neglect (2) of the contributions of the fissioning nuclei produced in the neutron evaporation cascade. The solid curves are the results of the dynamical calculations obtained for $\beta = 5 \times 10^{21}$ s^{-1}, and $J = 20\ \hbar$ (1); $30\ \hbar$ (2); $40\ \hbar$ (3); $60\ \hbar$ (4); and $80\ \hbar$ (5). The dash-dotted curve represents the time for the fissioning system moving from the saddle to the scission for $\beta = 5 \times 10^{21}$ s^{-1} and $J = 40\ \hbar$.

ANALISYS OF THE EXPERIMENTAL DATA

As one can see from Fig. 1, the experimental fission lifetimes are much longer than that expected from the standard statistical calculations of the lifetimes for initial compound nuclei formed in the investigated reactions. Calculations in the framework of the rotating liquid drop model, taking into account only lifetimes of excited states under equilibrium deformation, underestimate the experimental data by approximately five-orders of magnitude at excitation energies about 100 MeV.

The emission of neutrons from the fissioning nuclear systems leads to the cooling of nuclei before fission and as a result to increasing the mean decay time in the fission channel. Considerations of all possible fission chances during the development of the neutron-emission cascade (each one weighted with its probabilities of occurrence - $\tau_f = \Sigma\, \tau_{fi} \times \omega_i$, where ω_i is the relative weight of the fission fragments from i-th chance, and τ_{fi} is the corresponding decay time in the fission channel) improve the fit of the experimental data. But the large difference between theory and experiment (approximately three-orders of magnitude) still remains.

Assuming the double-humped fission barrier model with allowance for the lifetimes of the both classes of excited nuclear states realized in the first and second potential wells [1] makes it possible to improve essentially the fit of τ_f in the excitation energy range below 60 - 70 MeV. The reason is that the existence of an additional time delay in the fission channel (connected with the lifetime of the second-well states) leads to a noticeable increasing of τ_f.

Investigations of τ_f in the energy range up to 20 – 30 MeV allow us to obtain information on the statistical, statical (shape symmetry) and dynamical properties of the concrete fissioning nuclei at anomalously large deformations. The unknown earlier information on the energy dependences of level densities in the second potential well [8], the fission barrier parameters (depths of the second potential well and the heights of the inner fission barrier) [9], and the types of the nuclear shape symmetry in the second potential well [5] for a large set of the Pa, U, Np and Pu isotopes was obtained.

But for higher initial excitation energies the influence of this additional time delay diminishes and finally disappears at energies above 50 - 60 MeV due to the damping of shell effects with increasing of nuclear temperature [6]. In the energy range 30 – 60 MeV the double-humped structure of the fission barrier for heavy nuclei tends to transform into a single-humped one and only one class of excited nuclear states under equilibrium deformation survives. It means that experimental τ_f values for this transitional energies contain information on the temperature behaviour of the nuclear shell structure. In [10] the concrete form (Fermi type) of the energy dependence of the shell correction was extracted for the neutron-deficit uranium nuclei from the experimental τ_f values: $F(T) = [1 + \exp(\,(T - T_0)/d\,)]^{-1}$, where parameters $d = 0.2$ MeV, $T_0 = 1.75$ MeV and T is a nuclear temperature.

For the initial excitation energy above 50 MeV the dynamical aspects of the nuclear fission process begin to play an important role. Investigations of the experimental τ_f values for such energies allow one to extract information on the magnitude and mechanism of nuclear viscosity [11].

Analysis of the obtained energy dependence of the induced fission lifetimes at energies above 50 MeV was performed in the framework of the dynamical approach based on the set of stochastic Langevin equations. In the one-dimensional case, it can be represented as

$$\frac{dr}{d\tau} = \frac{p}{m}, \qquad (1)$$

$$\frac{dp}{d\tau} = -\frac{p^2}{2}\frac{d}{dr}\left(\frac{1}{m}\right) - \frac{dV}{dr} - \beta p + f(\tau), \qquad (2)$$

where r and p are the collective coordinate and the corresponding momentum, respectively. For the r we used the distance between the centers of mass of the formed fission fragments. In Eq. (2) $f(\tau)$ is a random delta-correlated force -

$$<f(\tau)> = 0; \qquad <f(\tau_1)f(\tau_2)> = 2D\delta(\tau_1 - \tau_2), \qquad (3)$$

where D is expressed through the Einstein relation in terms of the nuclear temperature and the coefficient of nuclear friction as $D = t \times \beta \times m$, being related to the damping coefficient - β. The inertial parameter m was calculated in the incompressible fluid approximation [12]. In the proposed approach the damping coefficient β was used as an adjustable parameter. The nuclear temperature is defined as $t = (E_{int}/a)^{1/2}$ with $E_{int} = E^* - p^2/(2m) - V(r,J)$, where E^* is the total excitation energy, and a is the level-density parameter. In our analysis the level-density parameter was chosen in the form of $a = A/10$. The potential energy $V(r,J)$ was calculated within the rotating liquid-drop model with Myers-Swiatecki parameters by using the procedure proposed in [13]. The initial values of p were generated for each trajectory under the assumption of the normal momentum distribution at r corresponding to the equilibrium deformation:

$$F(p) = \frac{1}{\sqrt{2\pi mt}} \exp\left(-\frac{p^2}{2mt}\right). \qquad (4)$$

Equations (1) – (2) were solved in the framework of the Euler difference scheme. The emission of light particles (neutrons, protons and α-particles) was simulated within the method that usually used to calculate the multiplicity of pre-scission light particles in the framework of an approach based on the Langevin equations (see, for example [14]).

The induced fission lifetimes were calculated by the following relation:

$$<\tau_f> = \frac{1}{N_f}\sum_{i=1}^{N_f} \tau_{fi}, \qquad (5)$$

where N_f is the number of Langevin samples, which have fissioned, and τ_{fi} is the fission lifetime for the i-th Langevin sample. Another calculation details were described in [14].

Because of lack of information on the angular momentum J of the U-like fissioning nuclei formed in the ^{238}U + ^{28}Si reaction, we treat this value as a free parameter. So, in our calculations we used two adjustable parameters: β and J. The best description of the investigated energy dependence of τ_f was achieved for $\beta = 5 \times 10^{21}$ s^{-1} and $J = 30$ - 40 $\hbar$, see Fig. 1. These J values are consistent with the peripherical mechanism of the reaction under study. At higher values of $J = 60 - 80$ $\hbar$ we obtained τ_f values very close to the time spent by the fissioning system between the saddle and scission points.

This is a result of the fission barrier disappearance at so high J values. In addition, decreasing of J leads to the increasing of the τ_f values due to the cooling of the fissioning nucleus because of the light particle emission before the saddle.

In addition, we also tested the obtained β and J values using the experimental data on the total neutron multiplicity for the fissioning U-like nuclei formed in the ^{238}U + ^{28}Si reaction (see Fig. 2).

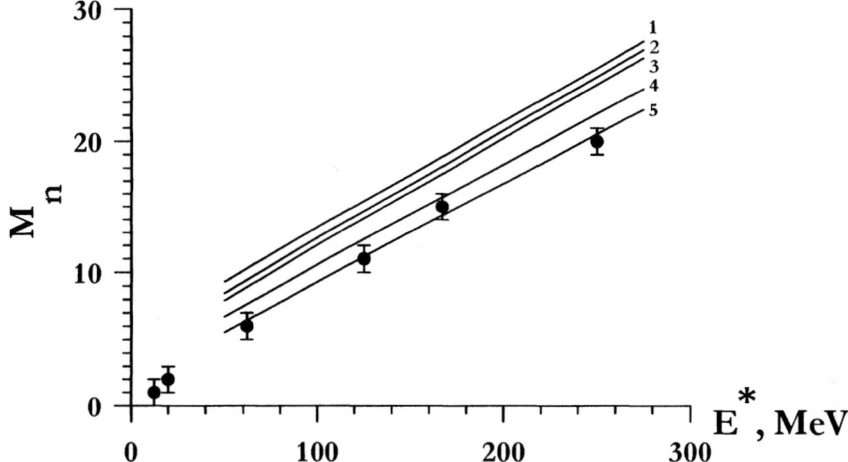

FIGURE 2. The total neutron multiplicity vs. initial excitation energy of the fissioning U-like nuclei produced in the ^{238}U + ^{28}Si reaction. Points are experimental data from [7]. Curves are the calculation results obtained for $\beta = 5 \times 10^{21}$ s^{-1}, and $J = 20$ $\hbar$ (1); 30 $\hbar$ (2); 40 $\hbar$ (3); 60 $\hbar$ (4); and 80 $\hbar$ (5).

Our calculated values of the pre-scission neutron multiplicity were summed with the data on the post-scission neutron multiplicity for the fissioning nuclei with $Z = 91$ - 95 from systematics [2]. As one can see from Fig. 2, our calculation results for $J = 30 - 40$ $\hbar$ slightly overestimate the experimental data. One of the probable reasons of this difference can be in the fact that in the systematics [2] data was obtained for complete fusion-fission reactions, while our calculations were done for the concrete discrete initial values of J.

In any case, the obtained value of the damping coefficient $\beta = 5 \times 10^{21}$ s^{-1} falls into the range of estimations of other authors [2] and corresponds to the conception of the "overdamped" collective nuclear motion.

CONCLUSION

In summary, we analyzed the unified global energy dependence of the induced fission lifetimes for heavy nuclei with $Z = 91 - 94$ in the excitation energy range from 5 to 250 MeV. The analysis allowed us to separate the excitation energy ranges in which the essentially different properties of the fissioning nuclear system can be investigated effectivelly.

In conclusion, it is necessary to stress the unique sensitivity of the induced fission lifetime to the properties of the fissioning nuclear system, i.e. magnitude of nuclear matter viscosity, structure of the excited strongly deformed nuclear states and so on. In the present paper it was discussed the experimental fission lifetime data obtained by the crystal blocking technique only. Investigations performed by this method permit us to obtain information about the energy dependence of shell effects, fission barrier parameters and level densities. These data can be very important for understanding of the induced fission process and also to plane the new experiments directed on the synthesis of superheavy elements. However, at present a lot of different experimental methods to measure the ultra-short nuclear lifetimes in the range from 10^{-14} to 10^{-21} s were developed. These methods allow one to investigate nuclear fission dynamics in the different excitation energy ranges in which the different mechanisms of excited nucleus decay are dominating. Obviously, it should be very fruitful to use the different experimental technique (for example, the method based on the measurements of the multiplicity of pre-scission neutrons and light charged particles, K-shell ionization method, "GDR clock" etc. [2]) to obtain complementary experimental information.

ACKNOWLEDGMENTS

This work was supported in part by the Russian Foundation for Basic Research (grant No. 02-02-17077-a) and by the State Program "Russian Universities" (grant No. UR.02.03.014).

REFERENCES

1. O. A. Yuminov, S. Yu. Platonov, O. V. Fotina, D. O. Eremenko, F. Malaguti, G. Giardina and A. Lamberto, *J. Phys.* **G 21**, 1243-1254 (1995).
2. D. Hilsher and H. Rossner, *Ann. Phys. (Paris)* **17**, 471-504 (1992).
3. A. F. Tulinov, *Dokl. Akad. Nauk SSSR* **162**, 546-548 (1965).
4. D. S. Gemmel and R. E. Holland, *Phys. Rev. Lett.* **14**, 945-948 (1965).
5. D. O. Eremenko, S. Yu. Platonov, O. V. Fotina and O. A. Yuminov, *Phys. Atomic Nuclei* **61**, 695-716 (1998).
6. O. A. Yuminov, S. Yu. Platonov, D. O. Eremenko, O. V. Fotina, E. Fuschini, F. Malaguti, G. Giardina, R. Ruggeri, R. Sturiale, A. Moroni, E. Fioretto, R. A. Ricci, L. Vannucci and G. Vannini, *Nucl. Instr. and Meth.* **B 164-165**, 960-964 (2000).
7. F. Goldenbaum, M. Morjean, J. Galin, E. Lienard, B. Lott, Y. Périer, M. Chevallier, D. Dauvergne, R. Kirsch, J. C. Poizat, J. Rémillieux, C. Cohen, A. L'Hoir, G. Prevot, D. Schmaus, J. Dural, M. Toulemonde and D. Jacquet, *Phys. Rev. Lett.* **82**, 5012-5015 (1999).
8. S. Yu. Platonov, O. V. Fotina and O. A. Yuminov, *Nucl. Phys.* **A 503**, 461-472 (1989).
9. V. A. Drozdov, D. O. Eremenko, O. V. Fotina, S. Yu. Platonov, A. F. Tulinov and O. A. Yuminov, *Nucl. Instr. and Meth.* **B 164-165**, 965-967 (2000).
10. O. A. Yuminov, S. Yu. Platonov, D. O. Eremenko, O. V. Fotina, E. Fuschini, F. Malaguti, G. Vannini, G. Giardina, G. Fazio, A. Lamberto, A. Taccone, A. Moroni, E. Fioretto, R. A. Ricci, L. Vannucci and R. Palamara, *J. of the Phys. Soc. of Japan* **70**, 689-695 (2001).
11. V. A. Drozdov, D. O. Eremenko, O. V. Fotina, S. Yu. Platonov and O. A. Yuminov, *Phys. Atomic Nuclei* **66**, 1628-1630 (2003).
12. K. T. R. Davies, A. J. Sierk and J. R. Nix, *Phys. Rev.* **C 13**, 2385-2405 (1976).
13. J. P. Leston, *Phys. Rev.* **C 51**, 580-585 (1995).
14. V. A. Drozdov, D. O. Eremenko, S. Yu. Platonov, O. V. Fotina and O. A. Yuminov, *Phys. Atomic Nuclei* **64**, 179-185 (2001).

Energy Dependence of the Shell Corrections Obtained from Analysis of Fission Fragment

Vadim A. Drozdov, Dmitri O. Eremenko, Sergey Yu. Platonov, Olga V. Fotina, Oleg A. Yuminov

D. V. Skobeltsyn Institute of Nuclear Physics, Moscow State University, 119992 Moscow, Russia

Abstract. Influence of the shell correction damping with increasing of excitation energy on the fission fragment angular anisotropies are considered. Experimental data on the fission fragment angular anisotropies for the compound nucleus produced in the ^{4}He + ^{238}U reactions is analyzed in the framework of the model of the transitional state in the saddle point and the statistical theory of nuclear reactions. Information about the energy dependence of the shell corrections is obtained from this analysis.

INTRODUCTION

Shell corrections to the potential energy play an important role in understanding the ground state masses of nuclei [1]. The modulation of nuclear energy surface by the shell corrections leads to the double-humped structure of fission barriers. In one's part, this extremely changes theoretical description of the decay of nuclei and made possible an understanding experimental data at low excitation energies. Usually information about shell corrections to the potential energy was obtained from analysis of the experimental data on evaporation residue cross sections, mass distributions of fission fragments, fission probabilities etc. [2].

It is well known that an increase of excitation energy results in damping of the shell corrections [3, 4]. Information about the energy dependence of the shell corrections is very important for description of fission process and synthesis of super-heavy elements [5]. It means that obtaining any information on the energy dependence of the shell corrections on the base of analysis of various experimental data is an actual problem. It should be noted that angular distributions of fission fragments are sensitive to the structure of fission barriers and may be used to obtain such dependence. With this purpose in the present paper, we analyzed angular distributions of fission fragments produced in the ^{4}He + ^{238}U reactions. This analysis was performed with consideration of contributions from all nuclei of the neutron-emission cascade.

ANALYSIS OF EXPERIMENTAL DATA ON FISSION FRAGMENT ANISOTROPIES

Usually fission fragment angular distributions are analyzed with help of the transition state models [6-8]. In the case of light projectiles and low excitation energy of compound nuclei (less than fission barriers), the states in the saddle point of the fission barrier are choused as a transition states [9]. In the framework of the transition state model in the saddle point angular distribution of fission fragments are given by

$$W(\theta) = \sum_{J=0}^{\infty} \sigma_J \sum_{K=-J}^{J} \frac{0.5(2J+1)\left|d^J_{M=0,K}(\theta)\right|^2 \rho(K)}{\sum_{K=-J}^{J} \rho(K)} \quad (1)$$

where σ_J is the partial fission cross-section, $d^J_{M,K}(\theta)$ is the θ-dependent part of the symmetric top wave function, $\rho(K)$ is

$$\rho(K) = \exp\left\{-\frac{\hbar^2 K^2}{2T}\left(\frac{1}{\Im_{II}} - \frac{1}{\Im_{\perp}}\right)\right\}, \quad \rho(K) \leq J$$
$$\rho(K) = 0, \quad \rho(K) > J \quad (2)$$

where T, $\Im_{II}$ and $\Im_{\perp}$ is the nuclear temperature, the parallel and perpendicular inertia moment's of the nucleus, respectively.

Influence of the double-humped structure of the fission barrier on the fission fragment angular distribution was considered in [10]. However, in this work the shell effect damping with increase of the nuclear temperature was neglected. Obviously, that decrease of the shell corrections with the excitation energy of the nucleus leads to change of heights of the inner and outer fission barriers of the double-humped fission barrier. Finally, the double-humped fission barrier transforms into a single-humped one. As was mentioned above the fission fragment angular distributions depend on the characteristics of the excited nucleus at the top of the fission barriers. It means that the shell effect damping with increase of the nuclear temperature has to affect on the fission fragment angular distribution. Taking into account the decrease of the shell corrections with the excitation energy of the nucleus in the characteristic points of the fission barrier (the inner and outer fission barriers and the second well) we analyzed the fission fragment angular anisotropies in the $\alpha + {}^{238}$U reaction in the energy range 23 MeV $\leq E_\alpha \leq$ 100 MeV.

In the energy range (which we consider in this paper), the double humped barrier transforms into liquid drop barrier with increasing of the excitation energy. Obviously, that at low excitation energies the fission fragment angular distributions will be determined by the characteristics of the excited nucleus at the top of the second fission barrier. At higher energies, the fission barriers become liquid drop and the top of the liquid drop barrier will determine the angular distributions. In the intermediate case, there is no saying what point in the deformation space (the first barrier or the second barrier) determines of the fission fragment angular distributions. Apparently, at these energies the transition states may be realized somewhere between the first and the second saddle points. On the other hand, in our work we are trying to use only reliable information on the parameters of the double humped fission barriers. It should be noted that in the literature such information is presented only for the characteristic points of the double-humped fission barrier (shell corrections at the inner and outer fission barriers and the second well) at low excitation energies [2]. In this situation in order to calculate the angular distributions by the unified way for the energy range under consideration, we assume that the observable fragment angular distributions are a superposition of the second barrier contribution and the second well one. In addition, we use the fact that for Pu nuclei the second well coincides with the top of the liquid barrier in the deformation space [2]. Finally, the next relation calculated the angular distribution:

$$W(\theta) = \frac{\sum_i P_i \sigma_i W_{S.B.i}(\theta)}{\sum_i \sigma_i} + \frac{\sum_i (1-P_i) \sigma_i W_{S.W.i}(\theta)}{\sum_i \sigma_i} \quad (3)$$

where σ_i is the partial fission cross-section, P_i is the probability of populating the second well of the i-s nucleus of the neutron-emission cascade, $W_{S.B.i}(\theta)$ is the contribution to the angular distribution of the second barrier and $W_{S.W.i}(\theta)$ is the angular distribution connected with the second well. As is seen from equation (3) when $P_i \to 0$ then $W(\theta) \to W_{S.W.i}(\theta)$. This case corresponds the so high excitation energies when the fission barrier becomes a liquid drop one and the angular distributions are determined by the saddle point of the liquid drop barrier. In the equation (3) $W_{S.B.i}(\theta)$ and $W_{S.W.i}(\theta)$ are given as:

$$W_i(\theta) = \sum_{J=0}^{\infty} \frac{\sigma_{i,J}}{\sigma_i} \sum_{K=-J}^{J} \frac{0.5(2J+1)\left|d_{M=0,K}^J(\theta)\right|^2 \rho_i(K)}{\sum_{K=-J}^{J} \rho_i(K)} \quad (4)$$

where J is the total angular momentum of the nuclear system.

$$\rho_i(K) = \exp\left\{-\frac{\hbar^2 K^2}{2T}\left(\frac{1}{\Im_{\parallel i}} - \frac{1}{\Im_{\perp i}}\right)\right\} \quad (5)$$

T_i, $\Im_{\parallel i}$, $\Im_{\perp i}$ is the nuclear temperature, the parallel and perpendicular inertia moment's of the i-s nucleus of the neutron-emission cascade respectively. These values are calculated at the second barrier and at the second well.

The fission cross sections and spin distributions of the initial compound nucleus and the daughter nuclei were calculated at each step of the cascade using the modified GFOT code [11]. In this code, decay of the nuclei is considered within the statistical theory of nuclear reactions. Parameters of the double-humped fission barriers for nuclei of the neutron-emission cascade were taken from [2]. The level density was calculated within the framework of Ignatyuk's phenomenological model with inclusion of the collective excitation contributions, the correlation effects of the superconductions type and the shell effects [12].

The probability of populating the second well was calculated as

$$P_i = 1 - \exp\left(-\frac{B_{fIIi}}{T_{IIi}}\right) \quad (6)$$

where $B_{fIIi} = V_{S.B.i} -. V_{S.W.i} + (\delta W_{S.B.i} - \delta W_{S.W.i})F.$ is the height of outer fission barrier relative to the second well depth of the i-s nucleus of the neutron-emission cascade, T_{IIi} is the nuclear temperature at the second well, $V_{S.B.i}$ and $V_{S.W.i}$ are the liquid drop potential energies at the deformations of the second well and the second barrier respectively, $\delta W_{S.B.i}$ and $\delta W_{S.W.i}$ are the corresponding shell corrections. To extract the shell correction values in the characteristic points of the fission barrier (at the first and second saddle points and in the second potential well), we took the difference between the liquid drop part of the fission barrier and the real parameters of the double-humped barrier [2]. F is the function determining the dependence of the shell correction on the temperature. About the type of this function sees below.

Comparison of the calculation of fission fragment angular anisotropies at the second barrier and at the liquid drop saddle point with the experimental data is presented on Fig.1. It is seen that at low energies the experimental data is reproduced by the calculation with the double humped barrier ($P_i \approx 1$) and at the highest energies the data is reproduced by the calculation with the liquid drop barrier ($P_i \approx 0$). Nevertheless, for intermediate energy range (which we consider in this paper) experimental data is not reproduced well by both the calculation with the double humped barrier and the calculation with the liquid drop barrier. In the present work for description of the experimental data on the fission fragment angular anisotropies, we used the damping function F of the Fermi type:

$$F(T) = \frac{1}{1 + \exp((T - T_0)/d)} \qquad (7)$$

where $d=0.2$ MeV is the rate of washing out of the shell corrections with the temperature [13] and T_0 was treated as an adjustment parameter. The best description of the experimental data is achieved with use of the value $T_0 = 1.15$ MeV (see fig.1).

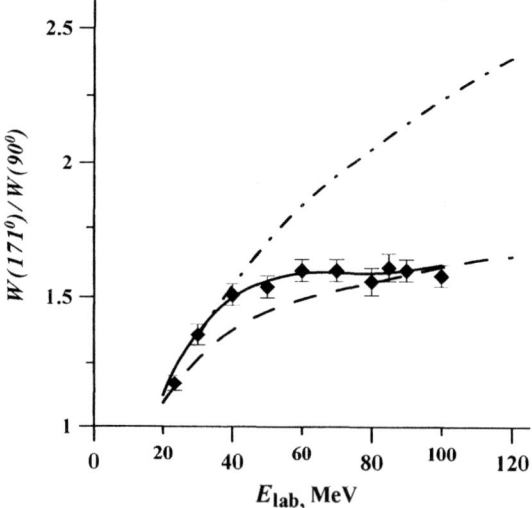

FIGURE 1 The fission fragment angular anisotropies in the $\alpha+238U$ reaction. Points are the experimental data [14]. The lines are the results of the calculations with the double humped barrier (dashed-dotted), with the liquid drop barrier (dashed) and in the framework formalism of this paper (solid).

Fig.2 presents our temperature dependence of the shell corrections in comparison with other temperature dependences.

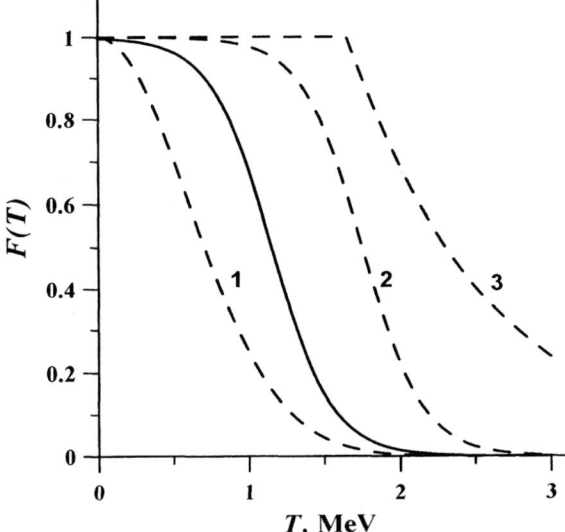

FIGURE 2 Different types of the shell corrections damping function. 1 – [15], 2 – [3], 3 – [16] and solid lines – damping function obtained from the best description of the experimental data on the fission fragment anisotropies.

Our results are compared with the data of [3], where the damping function was obtained from the measurement of the delay times in the fission channel. It is also compared with data of [16], where a semi-empirical analysis was done for temperature and spin dependence of the shell corrections and with the data of [15] that suggest a much faster decrease and from a smaller starting energy than our results. As is seen from fig.2 the data from these works gives different dependences of the damping function with increasing temperature of the nucleus. Here it should be noted that these damping functions were obtained from analysis of the various experimental data (neutron resonance, time delay in the fission channel, evaporation residue cross sections and angular anisotropies of the fission fragments). Therefore, the difference between the functions of fig. 2 probably points to the fact that the shell correction damping is critically sensitive to the nucleus deformation. In this situation in order to reproduce all available experimental data in the frames of the universal way, it is necessary to take into account also the deformation dependence of the damping function. Therefore, further inquiry of the law of the shell effect damping is an actual problem of the modern nuclear physics.

CONCLUSION

It is shown that the damping of the shell corrections with the nuclear temperature can explain the behavior of the fission fragment angular anisotropies in the cases of decay of heavy nuclei. In the analyses of the experimental data on the fission fragment angular anisotropies for the $\alpha + {}^{238}U$ reaction in the energy range 23 MeV $\leq E\alpha \leq$ 100

MeV information about the energy dependence of the shell corrections was obtained. An important property of the shell structure is that its influence on nuclear processes diminishes in highly excited nuclei and finally disappears at a certain temperature. In this work we show that fission barriers keep their double-humped structure up to the temperature 1.5 MeV. This is contradict with early works [12, 15], but close to recent works [3,17,18].

ACKNOWLEDGMENTS

This work was supported in part by the Russian Foundation for Basic Research (grant No 02-02-17077) and the State Program "Russian Universities" (grant No UR.02.03.014).

REFERENCES

1. V. M. Strutinsky, *Nucl. Phys.* **A95**, 420 (1967).
2. S. Bjornholm and J. E. Lynn, *Rev. of Mod. Phys.* **52**, 725 (1980).
3. O. A. Yuminov et al., *Nucl. Instr. Meth.* **B164-165**, 963 (2000).
4. V. M. Strutinsky, *Nucl. Phys.*, **A502**, 67 (1989).
5. G. Giardina et al., in *Proc. Int. Conf. on Nucl. Phys. "Nuclear Shells-50 years"*, Dubna, Russia, 1999, edited by Yu.Ts.Oganessian and R.Kalpakchieva, World Scientific Publishing Co. Pte. Ltd, Singapore, 2000, p.244.
6. R. Vandenbosch and J. R. Huizenga, *Nuclear Fission*, New York, Academic Press, 1973, p.1.
7. L. C. Vaz. and J. M. Alexander, *Phys. Rep.* **97**, 1 (1983).
8. I. Halpern and V. M. Strutinsky, in *Proc. Int. Conf. Peaceful uses of Atomic energy*, Geneva, V.15, 1958, p.408.
9. S. Kailas, *Phys. Rep.* **284**, 381 (1997).
10. D.O. Eremenko et al., *J. Phys. G: Nucl. Part. Phys.* **22**, 1077 (1996).
11. O. V. Grusha et al., *Izv. Akad. Nauk SSSR (ser. Fiz.)* **51**, 2055 (1987).
12. A. V. Ignatyuk, G. N. Smirenkin, A. S. Tishin, *Sov. J. Nucl. Phys.* **21**, 255, (1975).
13. G. Hansen, A. S. Jensen, *Nucl. Phys.*, **A406**, 236 (1983).
14. S. S. Kapoor et al., *Phys. Rev.* **149**, 965 (1966).
15. C. C. Sahm et al., *Nucl. Phys.* **A441**, 316 (1985).
16. A. D'Arrigo et al., *J. Phys.* **G20**, 365 (1994).
17. G.G. Chubarian et al., *Phys. Rev. Lett.* **87(5)**, 052701 (2001).
18. I.V. Pokrovski et al., *Phys. Rev.* **C60**, 041304 (1999).

Kaon Condensation and the Non-Uniform Nuclear Matter

Toshiki Maruyama*, Toshitaka Tatsumi[†], Dmitri N. Voskresensky**, Tomonori Tanigawa[‡]* and Satoshi Chiba*

*Advanced Science Research Center, Japan Atomic Energy Research Institute, Tokai, Ibaraki 319-1195, Japan
[†]Department of Physics, Kyoto University, Kyoto 606-8502, Japan
**Moscow Institute for Physics and Engineering, Kashirskoe sh. 31, Moscow 115409, Russia
[‡]Japan Society for the Promotion of Science, Tokyo 102-8471, Japan

Abstract.
Non-uniform structures of nuclear matter are studied in a wide density-range. Using the density functional theory with a relativistic mean-field model, we examine non-uniform structures at sub-nuclear densities (nuclear "pastas") and at high densities, where kaon condensate is expected. We try to give a unified view about the change of the matter structure as density increases, carefully taking into account the Coulomb screening effects from the viewpoint of first-order phase transition.

INTRODUCTION

There have been discussed various phase transitions in nuclear matter, like liquid-gas or neutron-drip phase transition, meson condensations, hadron-quark deconfinement transition, etc. In most cases they exhibit the first-order phase transitions. In the first-order phase transitions with more than one chemical potential, the structured mixed phase may be expected by way of the Gibbs conditions for the phase equilibrium [1].

At sub-nuclear densities, exotic nuclear shapes, called nuclear "pastas", are expected: with the increase of density, the matter structure is expected to change from "droplet" to "rod", "slab", "tube", "bubble" then to uniform. The existence of such "pasta" phases, instead of the crystalline lattice of nuclei, would affect several important processes in the supernova explosion by modifying the hydrodynamic properties and the neutrino opacity in supernova matter. It is also expected to influence the glitch of neutron stars via the change of the equation of state of the crust matter. Our first aim then is to study the nuclear "pasta" structure by means of a mean-field model which includes the Coulomb interaction in a fully consistent way.

At higher densities where kaon condensation may occur, it has been suggested that the structured mixed phase appears as a result of the first-order phase transition. If this is the case, we can expect the matter structure similar to the "pasta" phases [2]. In the first-order phase transitions with more than one chemical potential, the Maxwell construction (*coexisting separate phases* with local charge neutrality) does not necessarily fulfill the Gibbs conditions for the phase equilibrium,

$$T^{\mathrm{I}} = T^{\mathrm{II}}, \quad P^{\mathrm{I}} = P^{\mathrm{II}}, \quad \mu_B^{\mathrm{I}} = \mu_B^{\mathrm{II}}, \quad \mu_e^{\mathrm{I}} = \mu_e^{\mathrm{II}}; \tag{1}$$

the electron (charge) chemical potential takes different values between two phases, $\mu_e^I \neq \mu_e^{II}$, but we shall see that it means nothing but the difference in the electron number between two phases. When we naively apply the Gibbs conditions to these cases, we expect the *structured mixed phase* in a wide density range, where charge density as well as baryon density are no more uniform [1].

However, it has been suggested in recent papers that the Maxwell construction may still have a physical meaning by taking the hadron-quark matter transition as an example: the density region of the structured mixed phase is largely limited by the *Coulomb screening effect*, and results based on the Gibbs conditions become very close to the Maxwell construction curve [3]; note that if the Coulomb potential is properly included, it can give, combining with the charge chemical potential,

$$\rho_e = -(\mu_e - V_{\text{Coul}})^3/3\pi^2 \tag{2}$$

for the electron charge density in a gauge invariant way. We see later that our calculation includes the Coulomb potential consistently with other equations of motion. The second aim of this paper is to clarify the Coulomb screening effect on the structure of matter in the first-order phase transitions.

DENSITY FUNCTIONAL THEORY WITH RELATIVISTIC MEAN FIELD MODEL

We use density functional theory (DFT) with a relativistic mean field (RMF) model [4] in our study. The Coulomb potential is consistently included in the equations of motion. With this framework we can satisfy the Gibbs conditions in a proper way.

We start from the simple thermodynamic potential[2]:

$$\Omega = \Omega_B + \Omega_M + \Omega_K + \Omega_e, \tag{3}$$

$$\Omega_B = \int d^3r \left[\sum_{i=p,n} \left(\frac{2}{(2\pi)^3} \int_0^{k_{Fi}} d^3k \sqrt{m_B^{*2}+k^2} - \rho_i v_i \right) \right], \tag{4}$$

$$\Omega_M = \int d^3r \left[\frac{(\nabla\sigma)^2}{2} + \frac{m_\sigma^2 \sigma^2}{2} + U(\sigma) - \frac{(\nabla\omega_0)^2}{2} - \frac{m_\omega^2 \omega_0^2}{2} - \frac{(\nabla\rho_0)^2}{2} - \frac{m_\rho^2 \rho_0^2}{2} \right], \tag{5}$$

$$\Omega_K = \int d^3r \left[-\frac{f_K^2 \theta^2}{2} \left[-m_K^{*2} + (\mu_K - V_{\text{Coul}} + g_{\omega K}\omega_0 + g_{\rho K}\rho_0)^2 \right] + \frac{f_K^2(\nabla\theta)^2}{2} \right], \tag{6}$$

$$\Omega_e = \int d^3r \left[-\frac{1}{8\pi e^2} (\nabla V_{\text{Coul}})^2 - \frac{(V_{\text{Coul}} - \mu_e)^4}{12\pi^2} \right], \tag{7}$$

where σ, ω_0, ρ_0 are meson fields, $\mu_K = \mu_e$, $v_p = \mu_B - \mu_e + V_{\text{Coul}} - g_{\omega N}\omega_0 - g_{\rho N}\rho_0$, $v_n = \mu_B - g_{\omega N}\omega_0 + g_{\rho N}\rho_0$, $m_B^* = m_B - g_{\sigma N}\sigma$, $m_K^* = m_K - g_{\sigma K}\sigma$, and the kaon field $K = f_K \theta/\sqrt{2}$ (f_K: Kaon decay constant).[1] The parameters are chosen to reproduce the

---

[1] We here consider a linearized KN Lagrangian for simplicity, which is not chiral-symmetric.

saturation properties of nuclear matter. From $\frac{\delta\Omega}{\delta\phi_i(\mathbf{r})} = 0$ ($\phi_i = \sigma, \rho_0, \omega_0, \theta$) or $\frac{\delta\Omega}{\delta\rho_i(\mathbf{r})} = 0$ ($i = n, p, e$), we get the equations of motion for fields as

$$-\nabla^2\sigma + m_\sigma^2\sigma = -\frac{dU}{d\sigma} + g_{\sigma N}(\rho_n^{(s)} + \rho_p^{(s)}) - 2g_{\sigma K}m_K f_K^2\theta^2, \tag{8}$$

$$-\nabla^2\omega_0 + m_\omega^2\omega_0 = g_{\omega N}(\rho_p + \rho_n) + f_K^2 g_{\omega K}\theta^2(\mu_K - V_{Coul} + g_{\omega K}\omega_0 + g_{\rho K}\rho_0), \tag{9}$$

$$-\nabla^2\rho_0 + m_\rho^2\rho_0 = g_{\rho N}(\rho_p - \rho_n) + f_K^2 g_{\rho K}\theta^2(\mu_K - V_{Coul} + g_{\omega K}\omega_0 + g_{\rho K}\rho_0), \tag{10}$$

$$\nabla^2\theta = \left[m_K^{*2} - (\mu_K - V_{Coul} + g_{\omega K}\omega_0 + g_{\rho K}\rho_0)^2\right]\theta, \tag{11}$$

$$\nabla^2 V_{Coul} = 4\pi e^2\rho_{ch} \quad \text{(charge density } \rho_{ch} = \rho_p + \rho_e + \rho_K\text{)}, \tag{12}$$

$$\mu_p = \mu_B - \mu_e = \sqrt{k_{Fp}^2 + m_B^{*2}} + g_{\omega N}\omega_0 + g_{\rho N}\rho_0 - V_{Coul}, \tag{13}$$

$$\mu_n = \mu_B = \sqrt{k_{Fn}^2 + m_B^{*2}} + g_{\omega N}\omega_0 - g_{\rho N}\rho_0. \tag{14}$$

Note that the Poisson equation (12) is a highly nonlinear equation for V_{Coul}, since ρ_{ch} in RHS includes it in a complicated way.

To solve the above coupled equations, we use the Wigner-Seitz cell approximation: the space is divided into equivalent cells with spherical shape (cylindrical (slab) shape in two (one) dimensional calculation). Each cell is charge-neutral and all the physical quantities in a cell are smoothly connected to those of the neighbor cell (zero gradient at the boundary). The cell is divided into grid points ($N_{grid} \approx 100$) and the differential equations for fields are solved by a relaxation method with constraints of given baryon number and charge neutrality.

PROPERTY OF FINITE NUCLEI

Before applying our model to nuclear matter, we check how it can describe finite nuclei. In this calculation, electron density is put to be zero and the boundary condition or the charge-neutrality condition is not imposed. However, the spherical-symmetry approximation is kept. In Fig. 1 (left panel) we show the density profiles of some typical nuclei. To get a better fit, we may need to include a surface term etc. Shell effects (see the drop at the center in ^{16}O case) cannot be described by such a mean-field approach. By imposing the beta-equilibrium on the system, the most stable proton ratio can be obtained for a given mass number. Figure 1 (right panel) shows the mass-number dependence of the binding energy per nucleon and the proton-ratio. We can see that the bulk properties of finite systems (density, binding energy and proton ratio) are sufficiently reproduced.

Here we should note that we should adjust the sigma mass to be slightly smaller than that popularly used, i.e. 400 MeV to get such a good fit. If we use the popular value of $m_\sigma \approx 500$ MeV finite nuclei are overbound by about 3 MeV per nucleon. The sigma mass (or the omega mass) should be important for finite nuclei, i.e. non-uniform systems, since the meson mass is relevant to the interaction range and consequently affects, e.g., the nuclear surface tension.

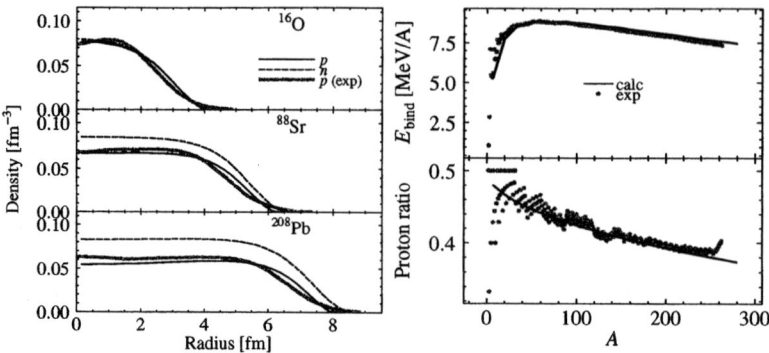

FIGURE 1. Left: the density profiles of typical nuclei. The proton densities (solid curves) are compared with the experiment. Right: binding energy per nucleon and proton ratio of finite nuclei.

NUCLEAR "PASTA" AT SUB-NUCLEAR DENSITY

In the density region where nuclei are about to melt into uniform nuclear matter, it is expected that the energetically favorable mixed phase, which consists of a nucleon liquid and a nucleon gas, possesses interesting structures, such as rod-like and slab-like nuclei and rod-like and spherical bubbles, etc. These exotic structures are referred to as nuclear "pastas". The existence of the "pasta" phases instead of the crystalline lattice of nuclei would affect the supernova explosion or glitch phenomenon of neutron stars. Due to these importance and the curiosity, the "pasta" structure has been studied by several models. It is widely accepted that the appearance of the "pasta" structure is due to the balance of the Coulomb energy and the surface tension. However, the electron density has been always treated as an uniform background in the usual treatments. Here we study the nuclear "pasta" structure with our model which consistently treats the Coulomb potential and the electron distribution. Particularly we focus on symmetric nuclear matter (relevant to the supernova matter in the initial collapsing stage) where the electron density is comparable to the baryon density.

Figure 2 (left) shows some typical profiles of symmetric nuclear matter structure obtained with our model. The nuclear "pasta" is well described. One should note the non-uniform electron distribution. The phase diagram of matter structure is shown in Fig. 2 (middle). The size of the cell $R_{\rm cell}$ is optimized with precision of 1 fm, and the lowest energy solutions are chosen. We see, in the figure, that there appears no spherical hole configuration; this depends on the effective interaction used in the calculation.

To see the Coulomb screening effect, there are two possible ways: one is to solve equations of motion for fields neglecting the Coulomb potential $V_{\rm Coul}$ (afterward, the Coulomb energy is added to the total energy), and the other is only to discard $V_{\rm Coul}$ in RHS of the Poisson equation, consequently the electron distribution becomes uniform. The first one should be standard and very clear in its definition, while it is less meaningful in our model, where the matter structure is not assumed: without the Coulomb repulsion between protons the nuclear matter would always form a bulk droplet, independent of the cell size. In the second way, on the other hand, protons interact with each

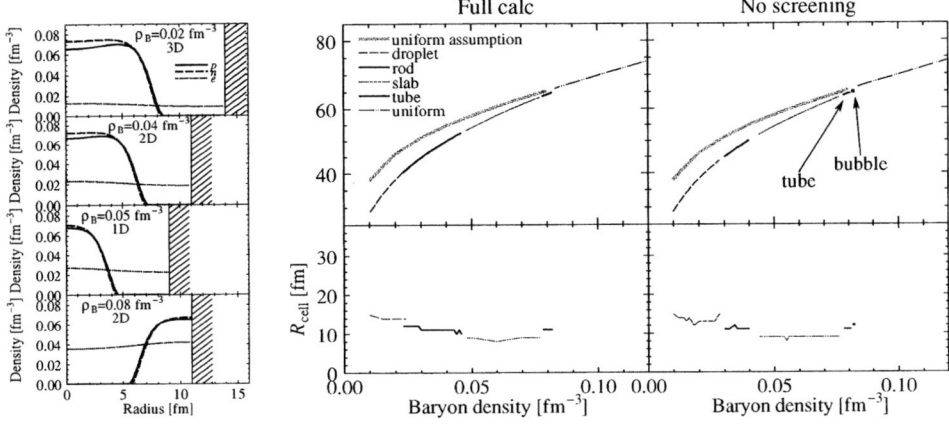

FIGURE 2. Left: examples of density profiles (droplet, rod, slab, and tube). Middle: binding energy per nucleon and the cell size of symmetric nuclear matter. Right: same as middle with uniform electron distribution.

other and may form non-uniform structure through the balance of the nuclear surface tension and the Coulomb interaction in a uniform electron background. In the next section, however, we shall see the first way is more suitable since the electron field and the kaon field are treated on an equal footing and uniform kaon distribution is meaningless.

We show in the right panel of Fig. 2 the results without the Coulomb screening (the second way), i.e. calculation with uniform electron background. The region of each structure (droplet, rod, etc.) is different from that of full calculation. Especially the "bubble" (spherical hole) appears in this case. However, such appearance of structure and its region is dependent on the very subtle energy difference, consequently on the effective interaction. The Coulomb screening effect on the bulk EOS (energy per baryon) is not so large.

KAON CONDENSATION IN HIGH-DENSITY MATTER

Next we explore the high-density nuclear matter in beta-equilibrium. This matter corresponds to the inner core of a neutron star. If the Glendenning's claim is correct, the structured mixed phase develops in a wide density range from well below to well above the critical density of the first-order kaon condensation. Then nuclear matter should exhibit the similar structure change to the nuclear "pasta" phases: the kaonic droplet, the hole, and the uniform kaonic matter. In fact we observe such structures in our calculation (see Fig. 3). Note that the above result is only for three dimensional calculation; we considered only spherical configurations for the Wigner-Seitz cell. The "complex" configuration in the diagram means not a simple droplet or a hole structure but something like a shell shape or mixture of droplet and hole. So we may not expect such configuration to be realized, when two or one dimensional structure is taken into account.

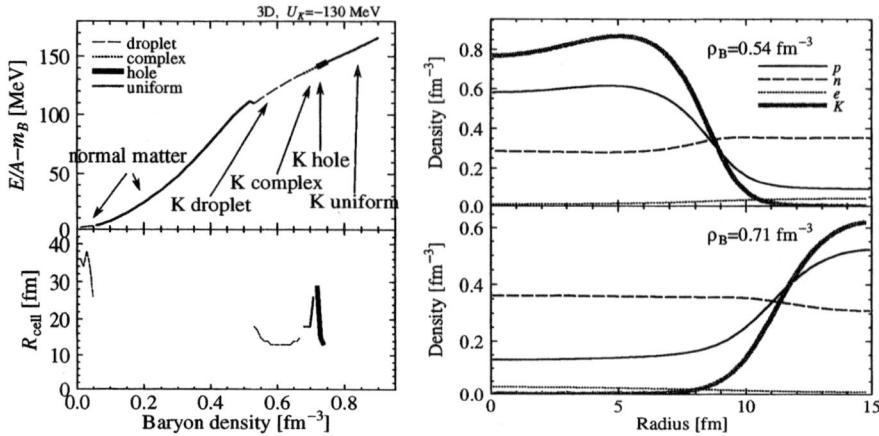

FIGURE 3. Left: binding energy per nucleon and the cell size of nuclear matter in beta equilibrium. Right: density profiles of kaonic matter. Droplet (upper panel) and hole (lower panel) configuration.

To demonstrate the Coulomb screening effects on the kaonic matter, we compare in Fig. 4 the phase diagrams in the μ_B-μ_e plane with and without the Coulomb interaction. In this calculation the Coulomb potential V_{Coul} is discarded in determining the density profile and the Coulomb energy calculated by this density profile is taken into account in the total energy. In Ref. [5], two cases, the Gibbs conditions and the Maxwell construction, are discussed. The case of the Gibbs conditions may lead to the structured mixed phase, while the Maxwell construction case to the phase separation of two bulk matters with local charge neutrality. Though we cannot definitely say now, the curve without the Coulomb interaction is similar to the one given by the Gibbs conditions and the curve with the Coulomb interaction to the one given by the Maxwell construction. If we look at the density profile, the local charge neutrality is more achieved in the case with the Coulomb interaction. These results suggest that the Maxwell construction is effectively meaningful due to the Coulomb screening.

SUMMARY AND CONCLUDING REMARKS

We have discussed how nuclear matter structure changes during the first-order phase transitions. We took nuclear "pastas" and the structured mixed phase during the course of the kaon condensation as two examples.

Using a self-consistent framework based on DFT and RMF, we took into account the Coulomb interaction in a proper way. We have seen how the self-consistent inclusion of the Coulomb interaction changes the phase diagram. It becomes more remarkable in the case of the kaon condensation; the density range of the structured mixed phase is largely limited and thereby the phase diagram becomes similar to that given by the Maxwell construction. The density profiles there also suggest the phase separation of two bulk matter. On the other hand, it brings about rather little effect on the nuclear "pastas". This

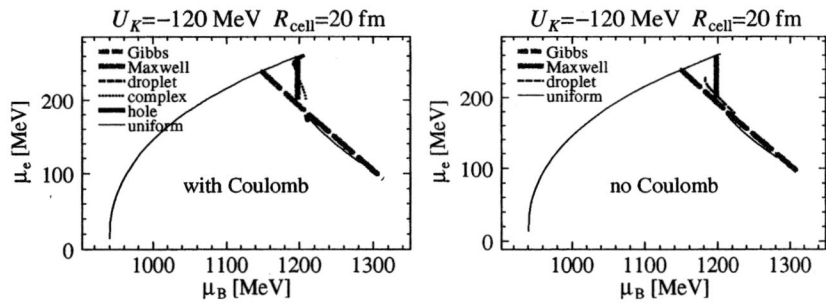

FIGURE 4. Phase diagram in the chemical potential plane. The cell radius is fixed to 20 fm. Left: full calculation. Right: Coulomb potential is discarded to determine the matter structure. Curves by Gibbs conditions and Maxwell construction are drawn in both panels. For simplicity both cases are calculated without the Coulomb interaction.

is because the electron density is rather low and the Debye screening length is rather large compared with the size of the structure or the cell. Although the importance of such a treatment has been demonstrated for the quark-hadron matter transition[3], one of our new findings here is that we could figure out the peculiar role of the screening effect without introducing an "artificial" input for the surface tension; remember that we need to introduce a sharp boundary and its surface tension by hand in discussing the quark-hadron mixed phase. By using present results we can extract the surface tension numerically. Then we can discuss the present subjects again in a similar way to the previous studies, and may confirm them.

We have shown that our model can well reproduce the bulk properties of spherical nuclei. However, we should take into account the derivative terms for the densities to describe the surface region of the density profile in more realistic ways. This inclusion should be important not only quantitatively but also in the context of the structured mixed phase mentioned above.

We used a simple model to describe kaon condensation here. In return for it, we lost some interesting features related to chiral symmetry; actually it has been known that non-linearity of the kaon field causes a serious difficulty in satisfying the Gibbs conditions[6]. Then it would be interesting to see whether we have a consistent prescription without chiral models when the Coulomb interaction is properly taken into account.

REFERENCES

1. N. K. Glendenning, Phys. Rev. D **46**, 1274 (1992).
2. T. Norsen and S. Reddy, Phys. Rev. C **63**, 65804 (2001).
3. D. N. Voskresensky, M. Yasuhira and T. Tatsumi, Phys. Lett. B **541**, 93 (2002); Nucl. Phys. **A723**, 291 (2003).
4. *Density Functional Theory*, ed. E. K. U. Gross and R. M. Dreizler, Plenum Press (1995).
5. N. K. Glendenning and J. Schaffner-Bielich, Phys. Rev. C **60**, 25803 (1999).
6. J. A. Pons, S. Reddy, P. J. Ellis, M. Prakash, J. M. Lattimer, Phys. Rev. C **62**, 035803 (2000); M. Yasuhira and T. Tatsumi, Nucl. Phys. **A690**, 769 (2001).

Production of Zero-Energy Radioactive Nuclear Beams through Extraction from the Liquid-Vapour Interface of Superfluid Helium

N. Takahashi[*][†], W. X. Huang[**], P. Dendooven[‡], K. Gloos[§], J. P. Pekola[¶] and J. Äystö[∥]

[*] *Osaka Gakuin University, Kishibe-Minami 2-36-1, Suita/Osaka, 564-8511 Japan*
[†] *Osaka City Science Museum, Nakanoshima 4-2-1, Kita-ku, Osaka, 530-0005 Japan*
[**] *Institute of Modern Physics, Chinese Academy of Sciences, Lanzhou 730000, PRC*
[‡] *KVI, Zernikelaan 25, NL-9747 AA Groningen, The Netherlands*
[§] *Oersted Laboratory, University of Copenhagen, Universitetsparken 5, DK-2100 Copenhagen, Denmark*
[¶] *Low Temperature Laboratory, Helsinki University of Technology, P.O.Box 1000, FIN-02015 HUT, Finland*
[∥] *Department of Physics, University of Jyväskylä, P.O. Box 35, FIN-40351 Jyväskylä, Finland*

Abstract. A new approach has been investigated to create an ultra-cold radioactive beam from high-energy ions. A ^{223}Ra alpha-decay recoil source has been used to produce radioactive ions in superfluid helium. The alpha spectra demonstrate that the recoiling ^{219}Rn ions have been extracted out of liquid helium. This first observation of the extraction of heavy positive ions across the superfluid helium surface has been possible thanks to the high sensitivity of radioactive ion detection. An efficiency of 36 % has been obtained for the ion extraction out of liquid helium.

INTRODUCTION

Precision spectroscopic and reaction studies with exotic nuclei involve low energy beams, typically of a few tens of keV and of energy spread of the order of 1 eV. The next generation Radioactive Ion Beam (RIB) facilities considered of high priority around the world aim at such ion beams. Exotic nuclei emerge as a high-energy beam from nuclear reactions with unavoidably large emittance and energy spread.

In this contribution the first results are presented on a new approach for creating an ultra cold ions. Superfluid liquid helium is used here as a stopping medium for energetic beam. After thermalization in superfluid helium, positive ions form owing to electrostriction "snowballs"; clusters of helium atoms around positive ions [1, 2, 3]. The formation and the fast transport of snowballs in liquid helium have been demonstrated earlier at Osaka [4, 5, 6, 7]. We concentrate on the extraction of snowballs/ions from the liquid helium into the vapour phase from where they can be injected into vacuum for further handling and/or post-acceleration. The results in combination with those from Osaka provide a new approach for producing ultracold radioactive ion beams for nuclear physics research. The method developed in this work also has applications in the studies of impurity ions and atoms in superfluid helium.

EXPERIMENTS

The experimental setup is shown schematically in Fig. 1. A ^{223}Ra ($T_{1/2}$=11.4 d) alpha source was placed at the bottom of an experimental cell with an inner diameter of 62 mm and a height of 105 mm. The decay chain consists of ^{219}Rn ($T_{1/2}$=3.96 s), ^{215}Po (1.78 ms), ^{211}Pb (36.1 min), ^{211}Bi (2.17 min), ^{207}Tl (4.77 min), and ^{207}Pb (stable). The ^{223}Ra alpha-decay products, ^{219}Rn, recoiling out of the source with an energy of approximately 100 keV, are stopped within 1 μm of liquid helium from the source and provide the source of thermalized positive ions. A surface-barrier silicon detector was mounted at the top of the cell to detect the alpha-decay of nuclei. Four ring electrodes were installed in between to provide an electric field to guide the snowballs/ions from the source onto a thin aluminium foil in front of the detector. The experimental cell was placed inside a helium cryostat and the lowest temperature attained was 1.2 K. The temperature of the experimental cell was measured with a 68 Ω Matsushita carbon resistor. The alpha energies were calibrated at a temperature of 1.22 K in the empty cell and with the aluminium foil by using the known alpha lines from the ^{223}Ra decay chain. The absolute strength of the ^{223}Ra source was also measured; typically 10^4 Bq.

The ^{223}Ra source was covered by 5 mm of liquid helium. This prevented alphas from the source to reach the detector. Without electric field no alphas were observed. Alphas appeared after correct voltages were applied to the electrodes. Based on the ion trajectory calculations with the SIMION code [8], the applied voltages were optimized by maximizing the count rate of ^{219}Rn alpha decay on the aluminium foil. The charge of the snowballs/ions was confirmed to be positive from the fact that no alphas were observed when lower voltages were applied on the source and bottom electrode than on the electrode 4, thus creating an electrostatic barrier for positive charges. Alpha spectra measured at different temperatures are shown in Fig. 2. The identification of the peaks is based on the measured energy and the calculated energy loss between the place of decay and the detector by using the SRIM code [9]. The peaks marked as Rn, Po and Bi are due to ^{219}Rn, ^{215}Po and ^{211}Bi alpha decays on the aluminuim foil in front of the detector; the

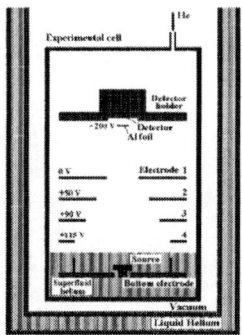

FIGURE 1. Experimental setup. The inner diameter of the experimental cell is 62 mm and the height is 105 mm. The ^{223}Ra source is covered by superfluid helium. The typical voltages on the electrodes are shown, except those on the source and bottom electrode which were varied during the measurement.

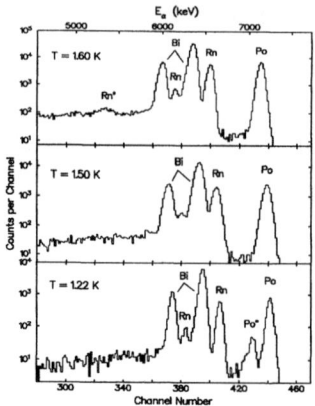

FIGURE 2. Alpha spectra with voltages on the electrodes at 1.22, 1.50 and 1.60 K. The peaks marked as "Rn", "Po" and "Bi" are from ^{219}Rn, ^{215}Po and ^{211}Bi nuclei on the aluminium foil in front of the detector. The peaks "Rn*" and "Po*" are from ^{219}Rn and ^{215}Po nuclei at the surface of liquid helium.

peaks Rn* and Po* are due to ^{219}Rn and ^{215}Po decaying at the surface of liquid helium. Other peaks from the surface are masked by the intense peaks from the decays on the aluminuim foil or the background. The vapour pressure above the liquid helium rises drastically with increasing temperature. The observed alpha spectra demonstrate that the ^{219}Rn ions have been extracted out of the liquid helium and collected on the aluminium foil in front of the detector by the static electric field. The extraction of positive ions from the superfluid helium surface has been observed for the first time, thanks to the high sensitivity of radioactivity detection.

RESULTS AND DISCUSSION

The overall transport efficiency is the ratio of the numbers of ions observed on the aluminium foil to those produced in the source and determined from the intensity of the ^{219}Rn peak. The results of a series of measurements on the overall efficiency against temperature and electric field are shown in Fig. 3. In these measurements, only the voltages on the source and bottom electrode were changed, thus basically changing the electric field in the lower part of the cell. The overall efficiency for ^{219}Rn is due to four factors: snowball formation, transport in the liquid, ion extraction and transport in the vapour.

The results of the measurements done with +450 V source voltage at 1.22, 1.34 and 1.60 K are given in Table 1. The best statistics was obtained at 1.60 K. At this temperature the overall efficiency of $(7.2\pm0.5)\times 10^{-4}$ was obtained, which includes the statistical and systematic errors. In order to obtain the extraction efficiency from the liquid surface we used the SIMION code to deduce the transport efficiency through the vapour phase and was deduced to be $38\pm5\%$. From the alpha peak of ^{219}Rn decaying

TABLE 1. Efficiencies (in %) at 1.22, 1.34 and 1.60 K measured with +450 V on the source and bottom electrodes. This corresponds to an electric field strength at the surface of 200 V/cm.

T (K)	1.22	1.34	1.60
Overall	0.029	0.032	0.072
Snowball formation	8.6	5.2	0.8
Snowball transport in liquid	100	100	100
Ion extraction out of liquid	0.9	1.6	23
Ion transport in vapour	38	38	38

on the surface of the liquid helium, the efficiency for a snowball to form and reach the surface, but not to be extracted, was deduced to be $(6.4\pm0.9)\times10^{-3}$. The efficiency for ion extraction out of liquid helium is then 23 ± 4%. In all our experimental situations, the transport time of snowballs in liquid helium is much smaller than the snowball neutralization time as mentioned below, we therefore conclude that the efficiency of snowball transportation inside the liquid is virtually 100%. This gives an efficiency for snowball formation of about 0.8% at 1.60 K.

At 1.22 and 1.34 K, our analyses show that the extraction efficiency of ^{215}Po out of the surface is significantly small. Because of its very short half-life, ^{215}Po decays essentially in the same place as its ^{219}Rn mother nucleus and almost no ^{215}Po from the source reaches the surface before its decay. Thus the Po* peak gives the same information as the Rn* peak. This equivalence is used for the measurements at the lower temperatures, where the Rn* peak is masked by the peaks from ^{211}Bi decay. Fig. 4 shows the deduced extraction efficiencies out of the liquid surface and snowball formation probabilities at 1.22 and 1.34 K against electric field strength on the liquid surface. The data show that the snowball formation probability and extraction efficiency behave in opposite ways with respect to temperature and electric field. Free electrons created by alpha-decay in the vicinity of the source can neutralize positive ions, thus preventing the creation of a snowball. Positive ions and free electrons move faster at lower temperatures (because of an increase in mobility [3, 1]) and stronger electric fields. For lower temperatures and

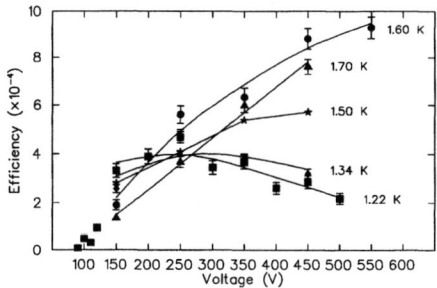

FIGURE 3. Overall efficiencies at different temperatures as a function of the source and bottom electrode voltage. The voltages applied on the other electrodes were fixed at the values shown in Fig. 1.

larger electric fields, positive ions and free electrons are separated faster, resulting in an increase in snowball formation efficiency. The fact that a higher temperature eases the release of positive ions from the surface was earlier also observed for negative charges [10, 11]. Understanding the decrease of the extraction efficiency with electric field requires further study. The interplay of the two opposite dependencies described above explains qualitatively the curves shown in Fig. 3: the overall efficiency first rises and then drops with increasing electric field. This is clearly seen at lower temperatures. For higher temperatures, the maximum efficiency is larger and can be reached at higher electric fields.

The time distribution for a snowball/ion to travel from the source to the aluminium foil was studied at a temperature of 1.50 K with an electric field strength of 85 V/cm. The ^{223}Ra recoil source can be "switched off" by putting it at a lower voltage than the bottom electrode. Moreover, the ion transport can be blocked by raising the voltage on one of the ring electrodes in order to create an electrostatic barrier. Measurements were done in which the source and transport were pulsed. A fit to the data gives a minimum delay time of less than a few milliseconds and a release time of 90±10 ms. The observed release is due to at least two processes: snowball neutralization at the surface and transport across the surface. This observation shows that ions/snowballs are, indeed, trapped on the surface prior to their extraction. This confirms earlier observations that ions can be trapped at the surface by a lateral holding potential [12].

Positive and negative ions in liquid helium near a free surface experience an attractive potential which tends to inhibit evaporation of ions across the surface. Assuming a Brownian distribution of ions, the extraction efficiency through a free surface is then proportional to $\exp(-E_b/kT)$, where k is Boltzmann constant, E_b the magnitude of the barrier. This barrier has been measured for negative charges (electrons) [10, 11], but, due to a lack of sensitivity, never for positive ions. Our experimental extraction efficiencies allow us to determine this barrier to be $E_b/k=19.4\pm4.5$ K. In this case, the electric field strength at the surface was 200 V/cm.

An attempt was made to optimize two factors: a large electric field to enhance snowball formation and a high temperature to enhance extraction from the surface. For this,

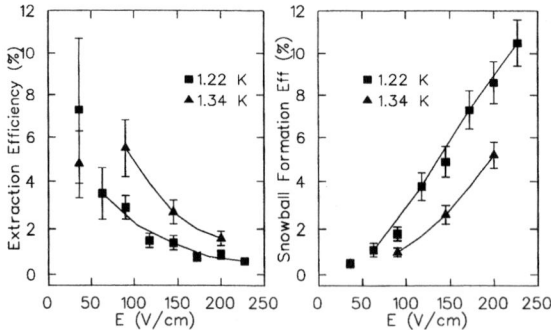

FIGURE 4. Extraction efficiency out of the liquid surface (left panel) and snowball formation efficiency at 1.22 K (squares) and 1.34 K (triangles) plotted against electric field strength on the liquid surface. The experimental conditions are identical with those in Fig. 3. The lines are to guide the eye.

we applied a pulsed second sound wave created from a circular heater around the radioactive source. Second sound is the heat transmitted in a wave-like manner rather than by diffusion and was observed in superfluid liquid helium [13]. It is expected to enhance the extraction through the free liquid surface by surpassing the barrier. The current pulses that excited the heater were 2 to 50 ms wide and were repeated every 50 to 500 ms. The increase of the overall efficiency was 10 to 30% depending on heater power.

CONCLUSION

A new method using liquid helium as stopping medium to transform a high-energy beam into ultra cold ions has been proposed and tested with an alpha-recoil source. Extraction of positive ions across the superfluid helium surface has been observed for the first time. An efficiency of $23\pm4\%$ was obtained at 1.60 K. This promises the method to be used at the next generation RIB facilities. Since a probability for snowball formation in an experiment at Osaka [5] has been found to be $20\pm10\%$ at 1.43 K, we expect a better overall performance for high-energy ions. We propose this highly sensitive method to be applied as a new tool in the study of superfluid helium properties, where the traditional electric current or charge measurements have limited sensitivity.

ACKNOWLEDGMENTS

This work has been supported by the EU Ion Catcher RTD Project HPRI-2001-50022, the Academy of Finland under the Finnish Centre of Excellence Programme 2000-2005 (Project No. 44875, Nuclear and Condensed Matter Physics Programme at JYFL) and in part by the Grants in Aid of Scientific Research of the Ministry of Education, Tokyo.

REFERENCES

1. Meyer L. and Reif F., Phys. Rev., 110, (1958), 279.
2. Atkins K. R., Phys. Rev., 116, (1959), 1339.
3. Johnson W. W. and Glaberson W. I., Phys. Rev. Lett., 29, (1972), 214.
4. Takahashi N., Shimoda T., Fujita Y., Itahashi T. and Miyatake H., Z. Phys. B, 98, (1995), 347.
5. Shimoda T., Miyatake H., Mitsuoka S., Mizoi Y., Kobayashi H., Sasaki M., Shirakura T., Ueno H., Izumi H., Asahi K., Murakami T., Morinobu S. and Takahashi N., Nucl. Phys. A, 588, (1995), 235c.
6. Takahashi N., Shimoda T., Miyatake H., Mitsuoka S., Mizoi Y., Kobayashi H., Sasaki M., Shirakura T., Ueno H., Asahi K. and Morinobu S., Hyp. Int., 97/98, (1996), 469.
7. Takahashi N., Shigematsu T., Shimizu S., Horie K., Hirayama Y., Izumi H. and Shimoda T., Physica B, 284-288, (2000), 89.
8. Dahl D. A., Delmore J. E. and Appelhans A. D., Review of Scientific Instruments, 61, (1990), 607.
9. Ziegler J. F., Computer code SRIM, IBM-Research, New York, (2000).
10. Bruschi L., Maraviglia B. and Moss F. E., Phys. Rev. Lett., 17, (1966), 682.
11. Schoepe W. and Probst C., Phys. Lett. A, 31, (1970), 490.
12. Vinen W. F., Z. Phys. B, 98, (1995), 299.
13. Tilley D. R. and Tilley J., Superfluidity and Superconductivity, Adam Hilger Ltd, Bristol,(1986).

Photon-induced Reactions in Stars and in the Laboratory: A Critical Comparison

Peter Mohr

Strahlentherapie, Diakoniekrankenhaus, D-74523 Schwäbisch Hall, Germany

Abstract. Photon-induced reactions during the astrophysical p- (or γ-) process occur at typical temperatures of $1.8 \leq T_9 \leq 3.3$. Experimental data of (γ,n), (γ,p), or (γ,α) reactions – if available in the relevant energy region – cannot be used directly to measure astrophysical (γ,n), (γ,p), or (γ,α) reaction rates because of the thermal excitation of target nuclei at these high temperatures. Usually, statistical model calculations are used to predict photon-induced reaction rates. The relations between experimental reaction cross sections, theoretical predictions, and astrophysical reaction rates will be critically discussed.

I INTRODUCTION

The nucleosynthesis of heavy neutron-deficient nuclei, so-called p-nuclei, proceeds mainly via a series of photon-induced (γ,n), (γ,p), and (γ,α) reactions in the thermal photon bath of an explosive astrophysical event. Type II supernovae are good candidates to provide the required astrophysical environment (e.g., temperatures of $1.8 \leq T_9 \leq 3.3$) [1–4].

Calculations of the astrophysical reaction rates and cross sections are based on the statistical model; input parameters for photon-induced reactions are γ-ray strength functions, optical potentials, and level densities. Recent results are summarized in [5–7].

Experimental data for photon-induced cross sections in the astrophysically relevant energy region have been obtained using two different techniques. Monochromatic photons from Compton backscattering of a Laser beam were used by [8], and a quasi-thermal photon spectrum was obtained by a superposition of bremsstrahlung spectra [9–11].

In this paper the astrophysically relevant energy window for (γ,n), (γ,p), and (γ,α) reactions [12,13] will be analyzed taking into account that the target nuclei may be thermally excited at the typical temperatures of $1.8 \leq T_9 \leq 3.3$. A critical comparison between experimental data in the laboratory and data for thermally excited nuclei in stars will be given, and relevant input parameters for the statistical model will be clearly defined.

II GAMOW WINDOW FOR (γ,n) AND (γ,α) REACTIONS

For simplicity, the following discussion will be restricted to (γ,n) and (γ,α) reactions. (γ,p) reactions play only a minor role in the reaction network for p-process nucleosynthesis. Additionally, most of the arguments given for (γ,α) reactions are valid for (γ,p) reactions, too.

The nucleus ^{148}Gd and the reactions ^{148}Gd(γ,n)^{147}Gd and ^{148}Gd(γ,α)^{144}Sm will be chosen as an example because the nucleosynthesis path of the p-process shows a branching point between (γ,n) and (γ,α) reactions which defines the production ratio between ^{146}Sm and ^{144}Sm. This ratio may be used as a chronometer for the p-process [1,4,14] because it can be measured at the time of the formation of the solar system from correlations between the ^{144}Sm abundance and isotopic anomalies in ^{142,144}Nd in meteorites [15].

The astrophysical reaction rate $\lambda(T)$ of a photon-induced reaction is given by

$$\lambda(T) = \int_0^\infty c\, n_\gamma(E,T)\, \sigma_{(\gamma,x)}(E)\, dE \qquad (1)$$

with the speed of light c and the cross section of the γ-induced reaction $\sigma_{(\gamma,x)}(E)$. The thermal photon density $n_\gamma(E,T)$ is given by the Planck distribution

$$n_\gamma(E,T) = \left(\frac{1}{\pi}\right)^2 \left(\frac{1}{\hbar c}\right)^3 \frac{E^2}{\exp(E/kT) - 1} \qquad (2)$$

where $n_\gamma(E,T)$ is the number of γ-rays at energy E per unit of volume and energy interval. The integrand in Eq. (1) is defined by the product of the cross section which increases with energy and the photon density which decreases exponentially with energy. This leads to a well-defined energy window which is astrophysically relevant (the so-called Gamow window). A comparison of typical Gamow windows for (γ,n) and (γ,α) reactions for target nuclei in their ground states is given in [13].

A ^{148}Gd(γ,n)^{147}Gd

The Gamow window for (γ,n) reactions is located close above the neutron threshold. The maximum of the integrand in Eq. (1) is located at $E_0^n \approx S_n + kT/2 \approx$ 9200 keV for $T_9 = 2.5$ where S_n is the neutron separation energy $S_n(^{148}\text{Gd}) =$ 8984 keV. The typical width of this window is about 1 MeV. Therefore, the astrophysically relevant window for the excitation energy E_x is located between S_n and $S_n + 1$ MeV (see Fig. 1). The position of the Gamow window for (γ,n) reactions depends only weakly on the temperature.

If the nucleus ^{148}Gd is in its 0^+ ground state, the dominating $E1$ transitions lead to 1^- states in the Gamow window (left part of Fig. 1, gray shaded area). These 1^- states may decay by neutron emission to low-lying states in ^{147}Gd with $E_x(^{147}\text{Gd}) < 1\,\text{MeV}$. Note that there is no Coulomb barrier for neutrons, and because of the small centrifugal barrier transitions to states with low J^π in ^{147}Gd are preferred. The cross section for this process can be measured in the laboratory. A statistical model prediction of this cross section requires the $E1$ γ-ray strength function around the energy E_0^n for the excitation process. Neutron and α optical potentials, the γ-ray strength function at $E < E_0^n$, and the level density of the residual nuclei above experimentally known levels are required for the calculation of the possible decays of excited ^{148}Gd by neutron, α, or γ emission.

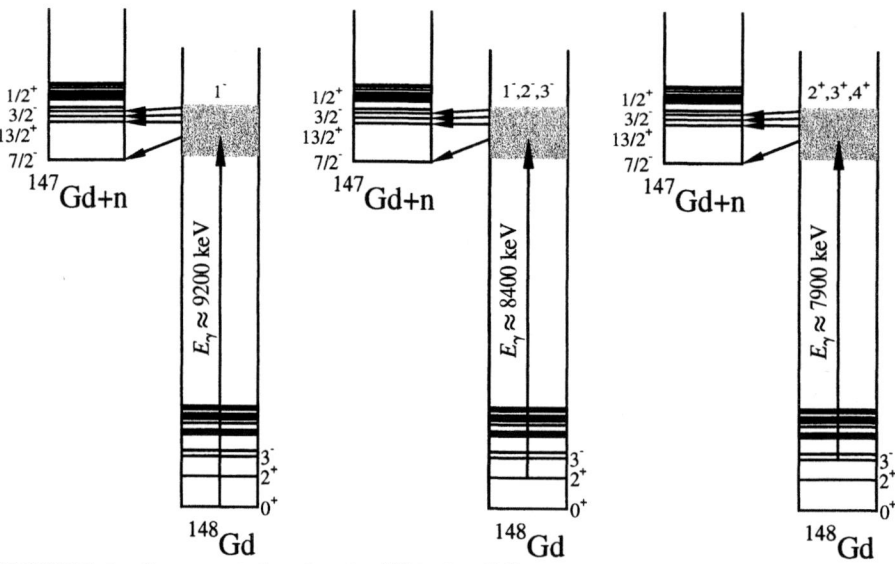

FIGURE 1. Gamow window for the ^{148}Gd$(\gamma,n)^{147}$Gd reaction for the ground state of ^{148}Gd (0^+, $0\,\text{keV}$; left) and the first excited states (2^+, $784\,\text{keV}$; middle; and 3^-, $1273\,\text{keV}$; right). Discussion see Sect. II A. All level data are from [16].

The situation changes if the nucleus ^{148}Gd is not in the ground state, but thermally excited to its low-lying levels. For simplicity, the discussion is restricted to the first two levels at $E_x = 784\,\text{keV}$ (2^+) and $1273\,\text{keV}$ (3^-). A significant contribution of these levels is already obtained at temperatures $kT < E_x$ because the ratio of population n_x/n_0 is given by the Boltzmann factor $\exp(-E_x/kT)$ and by the statistical weight of the spins

$$\frac{n_x}{n_0} = \frac{2J_x + 1}{2J_0 + 1} \exp(-E_x/kT) = (2J_x + 1)\exp(-E_x/kT) \qquad (3)$$

for even-even nuclei with $J_0^\pi = 0^+$. Assuming a similar energy dependence of the (γ,n) cross section of the excited state, one finds again a Gamow window close above the threshold at excitation energies around E_0^n. However, the required photon energy for a (γ,n) reaction is reduced by the excitation energy of the populated low-lying state: $E_\gamma = E_0^n - E_x(2^+) \approx 8400$ keV and $E_\gamma = E_0^n - E_x(3^-) \approx 7900$ keV. Starting from the 2^+ (3^-) state, $E1$ transitions may populate states with $J^\pi = 1^-, 2^-, 3^-$ ($J^\pi = 2^+, 3^+, 4^+$) as shown in Fig. 1, middle and right. These states may decay by neutron emission to low-lying states in ^{147}Gd, again preferring final states with small spin differences. This process cannot be measured in the laboratory. A statistical model calculation for these processes starting from the thermally excited 2^+ (3^-) state requires the $E1$ γ-ray strength function around the energy $E_\gamma = E_0^n - E_x(2^+) \approx 8400$ keV resp. $E_\gamma = E_0^n - E_x(3^-) \approx 7900$ keV for the excitation. For the decay the same ingredients as in the previous case are required.

The important results for the ^{148}Gd(γ,n)^{147}Gd reaction are that (i) excitation energies around $E_0^n \approx 9200$ keV are the relevant region independent of the thermal excitation of ^{148}Gd, and (ii) the $E1$ γ-ray strength function has to be known at the energy $E_0^n \approx 9200$ keV for ^{148}Gd in the ground state and at lower energies $E_0^n - E_x$ for ^{148}Gd in thermally excited states. Note that the $E1$ γ-ray strength function at energies $E_0^n - E_x < S_n$ cannot be measured by (γ,n) reactions because this strength is located below threshold! A similar phenomenon of important γ-ray strength below threshold has been found for neutron capture cross sections relevant for the r-process [17]. Usually, one extrapolates the $E1$ γ-ray strength function from the giant dipole resonance (GDR) to lower energies, and one assumes, following the Brink-Axel hypothesis [18–20], that a similar GDR and $E1$ γ-ray strength distribution can be found above each excited state. Such extrapolations of the γ-ray strength function towards lower energies are extensively discussed in [21].

B ^{148}Gd(γ,α)^{144}Sm

The position of the Gamow window for (γ,α) reactions is mainly defined by the Coulomb barrier. The maximum of the integrand in Eq. (1) for (γ,α) reactions is shifted by the α separation energy S_α compared to the (α,γ) reaction:

$$E_0^\alpha = 1.22 \, (Z_P^2 \, Z_T^2 \, A_{\rm red} \, T_6^2)^{1/3} \text{ keV} + S_\alpha \qquad (4)$$

The Gamow window for (γ,α) reactions is much broader compared to the (γ,n) reaction, and because many heavy neutron-deficient nuclei are α unbound ($S_\alpha < 0$) the energy E_0^α is often smaller than E_0^n. The position of the Gamow window depends sensitively on the temperature T. For $T_9 = 2.5$ one finds the Gamow window at $E_0^\alpha = 5520$ keV (with $S_\alpha = -3271$ keV) and with a width of about 3180 keV. $T_9 = 2.0$ (3.0) leads to $E_0^\alpha = 4300$ keV (6660 keV).

If the nucleus ^{148}Gd is in its 0^+ ground state, the dominating $E1$ transitions lead to 1^- states in the Gamow window (left part of Fig. 2, gray shaded area). These 1^- states may decay by α emission to low-lying states in ^{144}Sm. Because of the Coulomb barrier for α particles, the decay to the ground state of ^{144}Sm will be preferred; transitions to excited states in ^{144}Sm are suppressed because of the reduced tunneling probability; they are shown as dashed lines in Fig. 2. The cross section for this process can be measured in the laboratory. A statistical model prediction of this cross section requires the $E1$ γ-ray strength function around the energy E_0^α for the excitation. The α optical potential, the γ-ray strength function at $E < E_0^\alpha$, and the level density of the residual nuclei above experimentally known levels are required for the calculation of the possible decays of excited ^{148}Gd by α or γ emission; the neutron channel is not open at the low excitation energies.

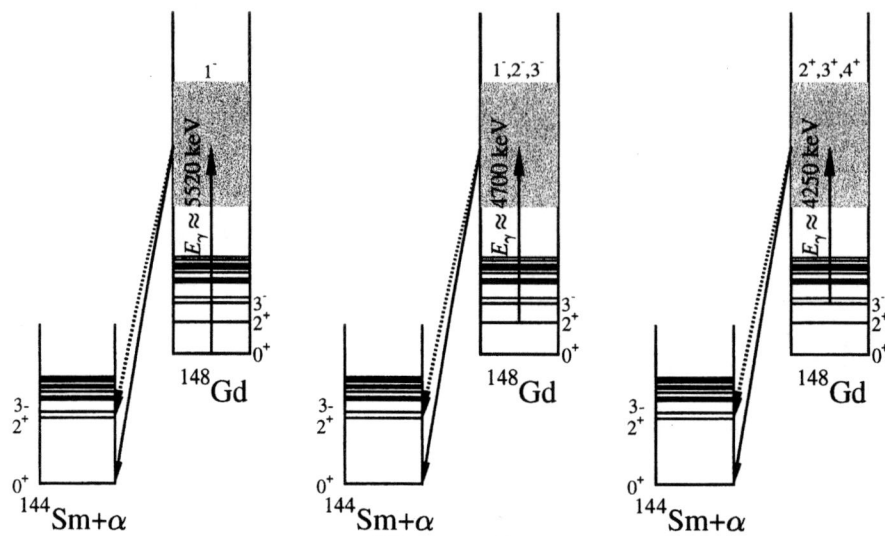

FIGURE 2. Gamow window for the ^{148}Gd$(\gamma,\alpha)^{144}$Sm reaction for the ground state of ^{148}Gd (0^+; 0 keV; left) and the first excited states (2^+; 784 keV; middle; and 3^-; 1273 keV; right) at a temperature of $T_9 = 2.5$. Further discussion see Sect. II B. All level data are from [16].

Again, the situation changes if the nucleus ^{148}Gd is not in the ground state, but thermally excited to its low-lying levels. Assuming a similar energy dependence of the (γ,α) cross section of the excited state, one finds again a Gamow window at excitation energies around $E_0^\alpha \approx 5520$ keV. However, the required photon energy for a (γ,α) reaction is reduced by the excitation energy of the populated low-lying state: $E_\gamma = E_0^\alpha - E_x(2^+) \approx 4700$ keV and $E_\gamma = E_0^\alpha - E_x(3^-) \approx 4250$ keV. Starting from the 2^+ (3^-) state, $E1$ transitions may populate states with $J^\pi = 1^-, 2^-, 3^-$ ($J^\pi = 2^+, 3^+, 4^+$) as shown in Fig. 2, middle and right. These states may decay by α emission to low-lying states in ^{144}Sm, again preferring the ground state of ^{144}Sm because of the Coulomb barrier (with the exception of the unnatural parity states with $J^\pi = 2^-$ and 3^+). This process cannot be measured in the laboratory. A statistical model calculation for these processes starting from the thermally excited 2^+ (3^-) state requires the $E1$ γ-ray strength function around the energy $E_\gamma = E_0^\alpha - E_x(2^+) \approx 4700$ keV resp. $E_\gamma = E_0^\alpha - E_x(3^-) \approx 4250$ keV for the excitation. For the decay the same ingredients as in the previous case are required.

The important results for the ^{148}Gd$(\gamma,\alpha)^{144}$Sm reaction at $T_9 = 2.5$ are that (*i*) the excitation energies around $E_0^\alpha \approx 5520\,\mathrm{keV}$ are the relevant region independent of the thermal excitation of ^{148}Gd (but E_0^α itself depends sensitively on the temperature!), and (*ii*) the $E1$ γ-ray strength function has to be known at the energy E_0^α for ^{148}Gd in the ground state and at lower energies $E_0^\alpha - E_x$ for ^{148}Gd in thermally excited states. Note that the $E1$ γ-ray strength function has now to be known at relatively low energies.

C The ratio $(\gamma,n)/(\gamma,\alpha)$

As stated above, the branching ratio between (γ,n) and (γ,α) reactions defines the nucleosynthesis in the *p*-process. For a reliable prediction of branchings between (γ,n) and (γ,α) reactions from statistical model calculations, various ingredients have to be known accurately. Besides the optical potentials for neutrons and α particles and the level densities, $E1$ γ-ray strength functions have to be known at different energies for the (γ,n) and (γ,α) reactions. Therefore a precise knowledge of the energy dependence of the γ-ray strength function at low energies is essential for the prediction of the ratio of $(\gamma,n)/(\gamma,\alpha)$ cross sections. It is highly desirable to check all ingredients of the model calculations by experimental data including the Brink-Axel hypothesis [18–20] where a partial breakdown was discussed in [21].

III COMPARISON WITH EXPERIMENTAL DATA

It is found that discrepancies between different statistical model calculations are mainly caused by different extrapolations of the $E1$ γ-ray strength function and by different α optical potentials whereas various parametrizations of level densities and neutron optical potentials lead to almost identical predictions for the cross section. The study of a global α optical potential has been described elsewhere (see Refs. [22–26]).

The $E1$ γ-ray strength function was determined for many nuclei from photoabsorption data around the GDR [27]. However, the astrophysically relevant energy region for (γ,n) and (γ,α) reactions is located at significantly lower energies. (γ,n) data close above the neutron threshold [8–11] help to restrict the $E1$ γ-ray strength function at energies around E_0^n. Experimental data with monochromatic photons [8] should be preferred because such data can be directly compared to theoretical predictions. The method of the quasi-thermal photon spectrum using a superposition of bremsstrahlung spectra [9] provides averaged cross sections which cannot be compared to theoretical predictions directly.

A standard technique to measure $E1$ γ-ray strength functions at low energies is photon scattering [28]. Bremsstrahlung experiments with unpolarized photons have no sufficient sensitivity to distinguish between $E1$ and $M1$ transitions [29] and are not well-suited for the precise determination of the $E1$ γ-ray strength function and its energy dependence. Off-axis bremssstrahlung may provide a limited photon polarization. However, the best solution would be photon scattering experiments using 100 % polarized photons from Laser-Compton scattering. Especially the SPring-8 facility with a high electron energy of several GeV and a long Laser wavelength of several μm provides an almost white spectrum with photon energies of several MeV, huge intensities, and almost 100 % polarization [30]. Alternatively, γ-ray spectra in neutron capture reactions have been used to extract the γ-ray strength function at low energies [31].

A special problem has to be mentioned. The γ-ray strength function is continuous above the neutron threshold but the measured $E1$ strength below neutron threshold consists of discrete levels. A direct comparison remains difficult. In the case of (γ,α) reactions the Gamow window is typically found at energies below the neutron threshold. Therefore, the relevant $E1$ strength is again concentrated in discrete levels, and the (γ,α) reaction rate may depend sensitively on the excitation energies of the corresponding levels.

Experimental data for the inverse capture reactions may provide further insight into the photodisintegration reaction rates. The reaction ^{144}Sm$(\alpha,\gamma)^{148}$Gd was measured close above the astrophysically relevant energies [32]. Under laboratory conditions the nucleus ^{144}Sm is in its ground state, and the (α,γ) reaction populates many levels in ^{148}Gd. These experimental conditions for the (α,γ) reaction are very close to the (γ,α) reaction under stellar conditions. Here many levels of ^{148}Gd are thermally populated, and because of tunneling probabilities through the Coulomb barrier mainly the ^{144}Sm ground state is populated in the (γ,α) reaction (see Fig. 2). However, laboratory conditions for the (γ,α) reaction with ^{148}Gd in its ground state differ significantly from stellar conditions for the (γ,α) reaction. Therefore, a measurement of the (α,γ) reaction provides the best test for statistical model predictions of astrophysical (γ,α) reaction rates. This argument does not hold for (γ,n) reactions: under stellar conditions excited states in the target and residual nucleus have to be taken into account (see Fig. 1) whereas in laboratory (γ,n) [(n,γ)] experiments the target [residual] nucleus is in its ground state.

IV CONCLUSIONS

Reaction networks for the nucleosynthesis of heavy nuclei require a huge number of reaction cross sections and reaction rates at high temperatures which are calculated using the statistical model. Experimental data are rare in the astrophysically relevant energy region; additionally, astrophysical reaction rates cannot be derived directly from experimental data in the laboratory because of the thermal excitation of target nuclei under stellar conditions. However, experimental data can provide systematic input parameters for the statistical model calculations. Improved γ-ray strength functions and a global α-nucleus potential are needed.

Although present photodisintegration experiments at astrophysically relevant energies (e.g., [8–11]) can provide valuable information for the theoretical prediction of reaction rates, there are limitations to the extent of information because only few relevant transitions are tested experimentally. Especially for the prediction of astrophysical (γ,α) reaction rates, a measurement of the inverse (α,γ) cross section seems to be a better test for the ingredients of the statistical model than a measurement of the (γ,α) reaction.

The nice idea of producing a quasi-thermal photon spectrum in the laboratory [9–11] is unfortunately in many cases not really useful because the measured laboratory reaction rates may differ significantly from the reaction rates at typical stellar conditions with temperatures of about $1.8 \leq T_9 \leq 3.3$ [10]. The relevant ingredients of statistical model calculations, namely the $E1$ γ-ray strength function, can be extracted with improved precision and reliability from experiments with monochromatic photons [8].

Discussions with T. Rauscher, H. Utsunomiya, and A. Zilges are gratefully acknowledged.

REFERENCES

1. M. Arnould and S. Goriely, Phys. Rep., submitted (2003).
2. T. Rauscher et al., Astroph. J. **576**, 323 (2002).
3. D. L. Lambert, Astron. Astrophys. Rev. **3**, 201 (1992).
4. S. E. Woosley and W. M. Howard, Astroph. J. Suppl. **36**, 285 (1978).
5. T. Rauscher and F.-K. Thielemann, At. Data Nucl. Data Tables **75**, 1 (2000).
6. T. Rauscher and F.-K. Thielemann, At. Data Nucl. Data Tables **79**, 47 (2001).
7. S. Goriely, Nucl. Phys. **A718**, 287 (2003).
8. H. Utsunomiya et al., Phys. Rev. **67**, 015807 (2003).
9. P. Mohr et al., Phys. Lett. B **488**, 127 (2000).
10. K. Vogt et al., Phys. Rev. C **63**, 055802 (2001).
11. K. Vogt et al., Nucl. Phys. **A707**, 241 (2002).
12. P. Mohr et al., Nucl. Phys. **A688**, 82c (2003).
13. P. Mohr et al., Nucl. Phys. **A719**, 90c (2003).
14. T. Rauscher, F.-K. Thielemann, H. Oberhummer, Astroph. J. Lett. **451**, L37 (1995).
15. A. Prinzhofer et al., Astroph. J. **344**, L81 (1989).
16. http://www.nndc.bnl.gov/nndc/ensdf/, revision of 30-May-2003.
17. S. Goriely, Phys. Lett. B **436**, 10 (1998).
18. P. Axel, Phys. Rev. **126**, 671 (1962).
19. N. Rosenzweig, Nucl. Phys. **A118**, 650 (1968).
20. G. A. Bartholomew et al., Adv. Nucl. Phys. **7**, 229 (1973).
21. J. Kopecky and M. Uhl, Phys. Rev. C **41**, 1941 (1990).
22. U. Atzrott et al., Phys. Rev. **53**, 1336 (1996).
23. P. Mohr et al., Phys. Rev. **55**, 1523 (1997).
24. P. Mohr, Phys. Rev. **61**, 045802 (2000).
25. Zs. Fülöp et al., Phys. Rev. **64**, 065805 (2001).
26. P. Demetriou, C. Grama, and S. Goriely, Nucl. Phys. **A707**, 253 (2002).
27. S. S. Dietrich and B. L. Berman, At. Data Nucl. Data Tables **38**, 199 (1988).
28. U. Kneissl, H. H. Pitz, and A. Zilges, Prog. Part. Nucl. Phys. **37**, 349 (1996).
29. A. Zilges et al., Phys. Lett. B **542**, 43 (2002).
30. H. Utsunomiya, private communication (2003).
31. M. Igashira et al., Nucl. Phys. **A457**, 301 (1986).
32. E. Somorjai et al., Astron. Astrophys. **333**, 1112 (1998).

Microscopic Calculations of Spontaneous Fission Life-times and Neutron-Induced Fission Cross Sections

P. Demetriou, M. Samyn, S. Goriely

Institut d'Astronomie et d'Astrophysiqe, Université Libre de Bruxelles, CP-226, Campus de la Plaine, Bd. du Triomphe, B-1050 Brussels, Belgium

Abstract. Fission potential-energy surfaces are calculated from a microscopic Hartree-Fock-Bogoliubov description of the deformed nucleus, fully constrained in the three deformation coordinates (c, h, α). The dynamical fission paths along the multidimensional symmetric and asymmetric deformation space are determined by applying the classical least action principle. The resulting dynamical fission barriers and spontaneous fission half-lives are compared with the experimental values for ^{240}Pu. Furthermore, microscopic calculations of fission barriers and nuclear level densities are used to obtain neutron-induced fission cross sections. The results for the actinides are compared with existing experimental data in the low energy region ≈ 100 keV relevant to the r-process nucleosynthesis.

INTRODUCTION

Nuclear fission could play a crucial role in the r-process nucleosynthesis (for a general review see Ref. [1]). In astrophysical environments where the neutron densities are sufficiently large to produce fissile nuclei, neutron-induced and beta-delayed fission may strongly influence the abundances in the lower mass region through the re-cycling of the r-process material, while spontaneous fission will affect the final abundance pattern, particularly the production of long-lived radiocosmochronometers Th and U. Of course all of these fission processes involve extremely neutron-rich nuclei that are unable to be measured in the laboratory. It is therefore of paramount importance to make reliable predictions of the relevant beta-delayed and neutron-induced fission rates, as well as the spontaneous fission half-lives, of all these unknown nuclides, starting from relatively close to the stability line and going out towards the drip line. In this respect, an attempt has been made to treat all aspects of fission on a microscopic basis, using a Skyrme-Hartree-Fock-Bogoliubov approach for the calculation of masses, fission barriers and fission level densities. In this work we present the results for the spontaneous fission properties of ^{240}Pu obtained with a Skyrme-Hartree-Fock-Bogoliubov method [2] in a multidimensional deformation space. The fission paths and half-lives are calculated in a dynamical approach [3, 4, 5]. In addition, we compare neutron-induced cross sections obtained with barrier heights and nuclear level densities from microscopic models [6, 7] with existing experimental data for nuclei in the actinide region.

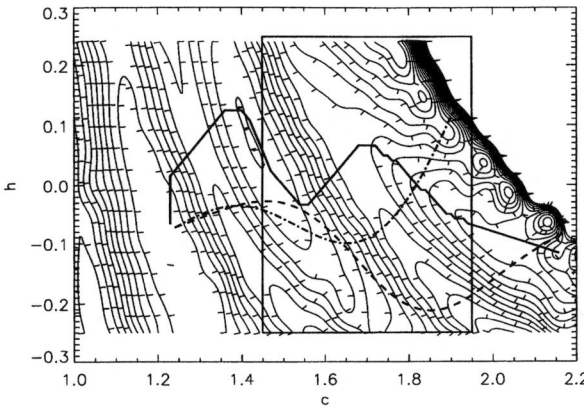

FIGURE 1. The potential energy landscape in the two-dimensional deformation space (c, h) for $\alpha = 0$. The solid and dashed lines are the symmetric dynamical and static paths, respectively. The dot-dashed line is the dynamical path obtained with $\alpha \neq 0$.

SPONTANEOUS FISSION HALF-LIVES

The potential energy surface (PES) of the nucleus is calculated with the Hartree-Fock-Bogoliubov (HFB) approach with the appropriate restoration of broken symmetries. The calculations are constrained with respect to the multipole moments (Q, O, H). The details of the PES calculations can be found in Ref. [2] so in the following we limit ourselves to the method of calculation of the fission properties. Assuming axially symmetrical deformations, the (Q, O, H) deformation space is equivalent to the so-called (c, h, α) parametrization introduced in Ref. [8]. With this parametrization all deformed shapes of a given nucleus, including total break-up into two separated fragments, can be generated continuously from a spherical configuration. With the HFB potential energy calculated at a sufficient number of deformations (c, h, α), we obtain the PES (shown in Fig. 1 for $\alpha = 0$). The 'static' barrier heights are determined by the *flooding method* described in Ref. [6]. For symmetric fission ($\alpha = 0$) we obtain $B_{inner} = 5.9$ MeV and $B_{outer} = 9.6$ MeV, while for asymmetric fission ($\alpha \neq 0$) B_{inner} is the same and $B_{outer} = 5.9$ MeV. The experimental values are $B_{inner} = 5.8$ MeV and $B_{outer} = 5.45$ MeV [9]. The asymmetry degree of fission results in reducing the outer barrier height considerably, hence bringing it closer to the experimental value.

TABLE 1. Fission barriers B (inner, outer), action integrals (S) and half-lives T_{sf} for ^{240}Pu.

	B_{inner} (MeV)	B_{outer} (MeV)	S	T_{sf} (yr)
dynamic $\alpha = 0$	6.2	12.4	49.7	1.9×10^{15}
dynamic $\alpha \neq 0$	6.2	9.5	48.1	7.8×10^{13}

Apart from the PES, another important fission quantity is the inertia tensor $B_{q_i q_j}$ that describes the inertia of the nucleus with respect to changes in the deformation $(q_i, q_j$

are the deformation parameters specifying the multidimension deformation space). It is calculated in the cranking approximation applied to the Skyrme-ETF+BCS states as prescribed in Ref. [10]. In the case of the (c, h, α) deformation space there are six components of the inertia tensor B_{cc}, B_{ch}, B_{hh}, $B_{c\alpha}$, $B_{h\alpha}$, $B_{\alpha\alpha}$. The inertia tensor is much more sensitive to the internal single-particle structure of the nucleus than the potential energy, thus it varies strongly with the deformation coordinates.

The spontaneous fission half-life T_{sf} is given by the formula

$$T_{sf} = \frac{\ln 2}{n} \frac{1}{P}, \qquad (1)$$

where n is the number of assaults of the nucleus on the fission barrier in unit time and P is the probability of penetration through the barrier for a given assault. The number of assaults is given by the frequency $\omega/2\pi$ of the vibration in the fission degree of freedom and thus by the zero-point vibration energy of the nucleus in the fission degree of freedom $E_{zp} = 0.5\hbar\omega$.

The probability P is calculated in the one-dimensional semi-classical (WKB) approximation

$$P = [1 + \exp(2S(L_{min}))]^{-1}, \qquad (2)$$

where the action integral $S(L)$ along a one-dimensional trajectory L in a multidimensional deformation space is

$$S(L) = \int_{s_1}^{s_2} \sqrt{\frac{2}{\hbar^2}(V(s) - E) B_L(s)} ds. \qquad (3)$$

$V(s)$ is the potential energy, $B_L(s)$ is the effective inertia along the trajectory L and E is the energy of the fissioning nucleus. The parameter s determines the position of a point on the trajectory and s_1, s_2 are the classical turning points given by $V(s_1) = V(s_2) = E$. The effective inertia $B(s)$ along the trajectory L is given by

$$B_L(s) = \sum_{ij} B_{q_i q_j}(s) \frac{dq_i}{ds} \frac{dq_j}{ds}, \qquad (4)$$

where $B_{q_i q_j}$ are the components of the inertia tensor and q_i, q_j the deformation coordinates (c, h, α) mentioned above.

In a dynamical calculation, the half-lives T_{sf} are determined by searching for the path that would minimize the action integral (3). This method has been applied extensively to the study of fission half-lives of actinides [8, 11] and superheavy elements [3, 4, 5] with considerable success. However, in all these previous calculations the potential energy surface and inertia parameters have been obtained in the macroscopic-microscopic approach. This is the first time a dynamical calculation of fission half-lives is performed on an entirely microscopic and fully constrained potential energy surface. The minimization of the action integral (3) is performed by means of a variational calculation following the prescription of Ref.[3]. In this method, smooth paths deviating from the straight line connecting the end points s_1, s_2, are used as trial trajectories. This turns out to be a rapidly converging problem with respect to the variation of the trial paths. The entry point s_1 is taken to be the point of minimal energy (first minimum) before the entrance into the barrier, $E_{s_1} = V_{min}(c_0, h_0, \alpha = 0) + E_{zp}$, where the zero-point energy is $E_{zp} = 0.3$ MeV [8]. The exit point corresponds to the same potential energy but lies behind the barrier.

As a first step, the dynamical calculations are performed for purely symmetric fission, i.e. $\alpha = 0$. The dynamical fission path is shown in Fig. 1. The static path obtained by the method of steepest descent is also plotted in the same figure for comparison. It crosses the static barriers through the saddle-points whereas the dynamical path prefers to go through higher barriers. The outer barrier height in particular is larger than the static one by 2.8 MeV as shown in Fig. 2. This is because the dynamical fission path depends on the effective inertia, as well as on the potential energy. A smaller effective inertia along a certain path would lead to a smaller action integral and half-life so the path would be preferred despite its crossing larger potential-energy barriers. The half-life obtained for this path is given in Table 1 and is rather large compared with the experimental half-life of $\approx 1.6 \times 10^{11}$ ys. When left-right asymmetry ($\alpha \neq 0$) is included in the calculations (for details see [2]), then the resulting dynamical outer barrier height decreases ($B_{outer} = 9.5$ MeV) with respect to the symmetric one ($B_{outer} = 12.4$ MeV), just as in the static calculations (Figs. 1 and 2). This is expected because left-right asymmetry tends to reduce the potential energy at increasing deformations with respect to the ground-state energy. However, the asymmetric dynamical barrier is still rather high compared to the corresponding static one, due to the effect of the inertia tensor described above for the symmetric case. It should be noted that the third narrow barrier appearing in Figs. 1 and 2 at $c \approx 1.9$ along the asymmetric path, is not a physical barrier but rather a result of the fact that the asymmetric calculations are confined in the deformation space outlined by the box in Fig. 1. Because of the limited deformation space, the dynamical calculations can only consider paths with turning points at $c \leq 1.95$, leaving out all possible paths going downhill in the PES with $c \geq 1.95$ like the symmetric paths (Fig. 1). For this reason, the estimate of the half-life obtained for this dynamical path is only a first approximation and should not be taken as a final result of the asymmetric dynamical calculations. Further work considering an extended deformation space in $(c, h, \alpha \neq 0)$ is in progress.

NEUTRON-INDUCED FISSION CROSS SECTIONS

Neutron-induced fission cross sections can be of importance in the r-process nucleosynthesis calculations, particularly in environments of extremely high neutron densities leading to the production of superheavy fissile nuclei. The relevant quantities in the nucleosynthesis calculations are the Maxwellian-averaged cross sections (MACS) at temperatures around 1.5×10^9 K. The fission cross sections that contribute to these MACS are in the energy range of ≈ 100 keV. It is therefore necessary to predict neutron-induced fission cross sections at the above-mentioned energies, for extremely neutron-rich nuclei extending from the stability line out towards the drip-line. Neutron-induced fission cross sections for excitation energies near or above the barrier heights, are normally calcu-

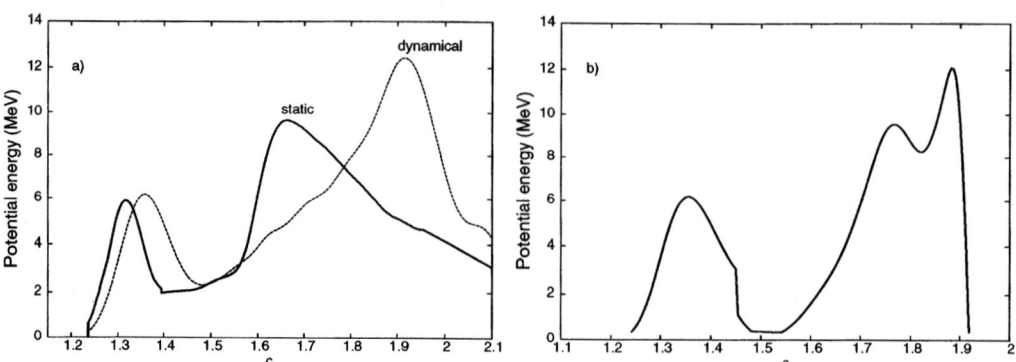

FIGURE 2. a) Shape of the potential-energy barriers along the static path (solid) and dynamical fission path with $\alpha = 0$ (dashed). b) Shape of the potential-energy barriers calculated along the dynamical fission path when asymmetry ($\alpha \neq 0$) is considered.

lated with the statistical Hauser-Feshbach (HF) theory of compound nucleus reactions. For excitation energies below the highest of the barriers, the non-negligible *subbarrier* effects are taken into account by means of the picket-fence model[12]. Important ingredients in the HF calculations are the transmission coefficients for particle/photon emission and fission, and the nuclear level densities (NLDs) at ground-state and saddle-point deformation. The fission transmission coefficients are based on the barrier heights of a large-scale calculation using the microscopic ETFSI method[6]. The widths of the barriers are taken from the systematics of Ref.[13] for even-even, odd-even and odd-A nuclei. The NLDs are obtained from microscopic statistical calculations[7]. The NLDs at saddle-point deformation are calculated with the same microscopic model on the basis of the single-particle level scheme determined at the saddle-point deformation with constrained Hartree-Fock-BCS calculations (see Ref.[9]), but without taking into account the damping of collective effects. Transition states built on top of the fission barriers are taken into account where they are known[14]. To obtain an estimate of the reliability of this choice of input, the results for the actinide nuclei are compared with experimental data. The results are shown in Fig. 3 for 18 actinides for which experimental neutron-induced fission cross sections exist in the energy range from 50 to 200 keV. An estimate of the mean deviation of the theoretical calculations from experiment is given by the root-mean-square deviation $f_{rms}=8.3$. A similar accuracy of $f_{rms}=11.1$ is found when using the barriers and NLD systematics of Ref.[14] also shown in Fig. 3.

CONCLUSIONS

Dynamical calculations of fission half-lives have been performed for ^{240}Pu based on the potential energy surfaces obtained from a fully constrained microscopic approach. The effective inertia is found to be crucial in determining the fission path and leads to dynamical barriers that are higher than the static ones. The outer barrier heights are found to decrease when the asymmetry degree of freedom is included.

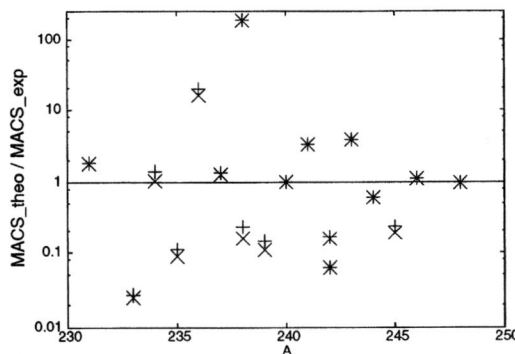

FIGURE 3. Ratio of theoretical Maxwellian-averaged cross sections over the corresponding experimental ones at $T=1.5\times 10^9$ K. The plus signs are obtained with the microscopic input and the crosses with the systematics of [14].

Global microscopic predictions of barrier heights and nuclear level densities are used in neutron-induced fission calculations giving rise to discrepancies of the order of $f_{rms}=8.3$ in the experimentally well-explored actinide region. Improvements, in particular in the NLD predictions are foreseen before performing large-scale calculations of neutron-induced cross sections and estimating the impact of neutron-induced fission on the r-process nucleosynthesis.

ACKNOWLEDGMENTS

P.D. holds a European TMR "Marie Curie" fellowship at ULB. M.S. and S.G. are FNRS Fellow and Associate, respectively.

REFERENCES

1. J.J. Cowan, F.-K. Thielemann and J.W. Truran, *Phys. Rep.* **208**, 267 (1991).
2. M. Samyn and S. Goriely, in these Proceedings.
3. A. Baran, K. Pomorski, A. Lukasiak and A. Sobiczewski, *Nucl. Phys.* **A361**, 81 (1981).
4. Z. Patyk, J. Skalski, A. Sobiczewski and S. Cwiok, *Nucl. Phys.* **A502**, 591c (1989).
5. R. Smolanczuk, J. Skalski and A. Sobiczewski, *Phys. Rev.* **C52**, 1871 (1995).
6. A. Mamdouh, J.M. Pearson, M. Rayet and F. Tondeur, *Nucl. Phys.* **A644**, 389 (1998); *Nucl. Phys.* **A679**, 337 (2001).
7. P. Demetriou and S. Goriely, *Nucl. Phys.* **A695**, 95 (2001).
8. M. Brack, J. Damgaard, A.S. Jensen, H.C. Pauli, V.M. Strutinsky and C.Y. Wong, *Rev. Mod. Phys* **44**, 320 (1972).
9. *Reference Input Parameter Library* RIPL2 (2003) IAEA-TecDoc in press. (www-nds.iaea.org)
10. C.M. Reiss, PhD Thesis, Friedrich-Alexander-Universität Erlangen-Nürnberg, 2000.
11. T. Ledeberger and H.-C. Pauli, *Nucl. Phys.* **A207**, 1 (1973).
12. J.E. Lynn and B.B. Back, *J.Phys.* **A7**, 395 (1974).
13. S. Bjornholm and J.E. Lynn, *Rev. Mod. Phys.* **52**, 725 (1980).
14. S. Maslov, in *Reference Input Parameter Library* RIPL1 (2001). (www-nds.iaea.org)

Measurement of Photo-destruction Cross Sections for ^{180}Ta^m at SPring-8

Shinji Goko*[†] and Hiroaki Utsunomiya*

*Department Physics, Konan University, Okamoto 8-9-1, Higashinada, Kobe 658-8501, Japan
[†] Photonics Research Institute, National Institute of Advanced Industrial Science and Technology (AIST) Umezono 1-1-1, Tsukuba, Ibaraki 305-8568, Japan*

Abstract. In the p-process production of ^{180}Ta^m, not only photo-production cross sections of ^{181}Ta(γ,n)^{180}Ta but also photo-destruction cross sections of ^{180}Ta^m(γ,n)^{179}Ta are needed. We propose to measure the destruction cross section with high-energy synchrotron radiation generated from a 10 tesla super-conducting wiggler at SPring-8.

THE P-PROCESS

There are 35 stable nuclides classified as the p-process nuclei [1]. They are found on the proton-rich side of the valley of nuclear stability against β decay in a nuclear chart. ^{180}Ta^m is an odd-odd nucleus, the nature's rarest isotope and only naturally occurring isomer. It is not produced by neutron capture because it is bypassed by the s-process and shielded from β^- decay after the r-process by a stable ^{180}Hf. It is also shielded from β^+ decay by a stable ^{180}W.

In the p-process, ^{180}Ta^m may be produced by photodisintegration of pre-existing ^{181}Ta in O/Ne-rich layers of massive stars during their pre-supernovae phase [2,3,4] or during their explosions as type-II supernovae [3,4,5,6]. We have measured the photo-production cross section with quasi-monochromatic γ-ray beams from laser Compton scattering at AIST [7].

In the same thermal photon bath, ^{180}Ta^m undergoes photo-destruction resulting in ^{179}Ta. The lower neutron separation energy ($S_n = 6.645$ MeV) of the odd-odd nucleus significantly increases the number of photons in the photon bath available for photodisintegration compared with that for ^{181}Ta with $S_n = 7.578$ MeV. More importantly photoneutron cross sections are unknown for ^{180}Ta^m.

The photoreaction rate (λ) is given by

$$\lambda(T) = c\int_0^\infty n_\gamma(E,T)\sigma_\gamma(E)dE, \qquad (1)$$

where $n_\gamma(E,T)$ is the number of photons per unit volume in the energy interval $E \sim E + dE$ in the black-body radiation at temperature T, $\sigma_\gamma(E)$ is the photoneutron cross section. The Planck distribution $n_\gamma(E,T)$ is given by

$$n_\gamma(E,T) = \left(\frac{1}{\pi}\right)^2 \left(\frac{1}{\eta c}\right)^3 \frac{E^2}{\exp(E/kT)-1}. \qquad (2)$$

The product of $n_\gamma(E,T)$ and $\sigma_\gamma(E)$ in Eq. (1) defines a narrow energy window for photodisintegration immediately above neutron threshold [8].

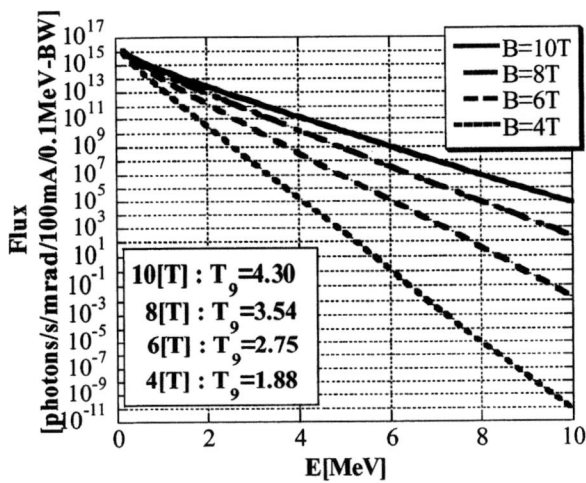

FIGURE 1. Energy spectra of synchrotron radiations from a 10 tesla super-conducting wiggler at SPring-8 as a function of the magnetic field (calculations).

PHOTO-DESTRUCTION MEASUREMENT AT SPRING-8

Figure 1 shows calculated energy spectra of the synchrotron radiation from a 10 tesla super-conducting wiggler (SCW) installed at SPring-8 [9]. The energy spectra are characterized by exponential tails with slopes depending on the magnetic field (B) of the SCW. One can find equivalent temperatures of the spectra by fitting these tails by a functional form of $E^2 \exp(-E/kT)$ in Eq. (2). It is remarkable that equivalent temperatures are a few to several billions of Kelvin, including temperatures of so-called p-process layers, $T_9 = 1.7 - 3.3$ [10]. The number of photons strongly depends on B; it is 10^{7-9} photons/sec/MeV at photon energies (E) of several MeV at the maximum magnetic field, $B = 10$ tesla. The SCW radiation is ideal for the p-process study as discussed in [11].

Photo-destruction cross sections of ^{180}Ta$^m(\gamma,n)^{179}$Ta can be measured with the intense SCW radiation by photo-activation of natural tantalum foils. Irradiation of natural foils with the SCW radiation produces ^{180}Ta(gs) with the halflife ($T_{1/2}$) 8.15 h and ^{179}Ta with $T_{1/2} = 1.82$ y through photodisintegration of ^{181}Ta[99.988%] and ^{180}Ta^m[0.012%], respectively. The amount of ^{180}Ta(gs) to be produced is by far larger than that of ^{179}Ta because of the natural abundances of parent nuclei. ^{180}Ta(gs) is also produced in ^{180}Ta^m(γ,γ')^{180}Ta(gs) [12]. Both ^{180}Ta(gs) and ^{179}Ta decay by electron capture so that Hf X rays are emitted. Photodisintegration of target impurity ^{182}W emits Ta X rays whose energies are close to those of Hf X rays. Experimental parameters are summarized in Table 1.

TABLE 1. Experimental parameters

Radioactivity	^{179}Ta	^{180}Ta(gs)	^{181}W
Production reaction (threshold energy)	^{180}Ta^m(γ,n)^{179}Ta (6.645MeV)	^{181}Ta(γ,n)^{180}Ta(gs) (7.577MeV)	^{182}W(γ,n)^{181}W (8.064MeV)
Halflife ($T_{1/2}$)	1.82 y	8.152 h	121.2 d
Natural Abundance	0.012%	99.988%	26.5%
Energies of KX rays	55.8 keV($K_{\alpha1}$) 54.6 keV($K_{\alpha2}$)	55.8 keV($K_{\alpha1}$) 54.6 KeV($K_{\alpha2}$)	57.5 keV($K_{\alpha1}$) 56.3 keV($K_{\alpha2}$)

The production of radioactive nuclei is expressed by

$$\frac{dN(E,t)}{dt} = \rho_t \sigma(E) I_\gamma(E) - \lambda N(E,t), \quad (3)$$

where $N(E,t)$ is the number of daughter nuclei, ρ_t is the areal density of target nuclei, $\sigma(E)$ is the photoneutron cross section, $I_\gamma(E)$ is the photon flux, and λ is the decay constant of daughter nuclei. The total number of daughter nuclei produced after irradiation time t is written by

$$N(t) = \frac{\rho_t}{\lambda}(1 - \frac{1}{e^{\lambda t}}) \int_0^\infty \sigma(E) I_\gamma(E) dE. \quad (4)$$

Figure 2 shows the initial amounts of daughter nuclei produced after irradiation of ten natural tantalum foils of each 100μm thickness during 100 hours and their decays as a function of time. The same threshold behavior of photoneutron cross sections as that for ^{181}Ta [7] was assumed for ^{180}Ta^m and ^{182}W. A catalog value of the maximum amount of target impurity (^{182}W) was used in the production of ^{181}W. The huge amount of ^{180}Ta(gs) decays quickly in such a away that the radioactivity (λN) of ^{179}Ta is equal to that of ^{180}Ta(gs) approximately after 6-days cooling. At about 10 days after irradiation, the KX-ray emission rate is 141 counts/hour (cph) for ^{179}Ta, 53 cph for ^{180}Ta(gs), and 0.03 cph for ^{181}W. Thus, these KX rays can be detected after sufficient cooling time with high-efficiency germanium detectors in a low-background environment.

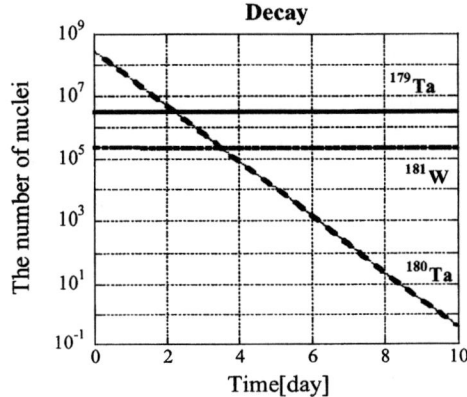

FIGURE 2. The number of radioactive nuclei (^{179}Ta, ^{180}Ta(gs)) produced by irradiation of ten natural tantalum foils of each 100μm thickness with the SCW radiation during 100 hours and their changes with time. Photo-destruction of ^{180}Ta^m initially produces a small number of ^{179}Ta, but survives with the long halflife 1.82 years. Photo-disintegration of possible target impurity ^{182}W produces a certain amount of ^{181}W, but their KX rays can be separated from those of ^{179}Ta.

PROSPECT

A proposal of constructing a dedicated beam line to use the SCW radiation for nuclear physics, astronuclear physics, and atomic physics is under preparation as of December 2003 and will be submitted in early 2004.

Photo-activation experiments will be the first-phase experiments to be carried out inside the vault of the SPring-8. Because of the nature of the SCW radiation mentioned above, potentially SPring-8 can be a center of the p-process study.

ACKNOWLEDGMENTS

This work is supported in part by the Japan Private School Promotion Foundation and by the Japan Society for the Promotion of Science.

REFERENCES

1. D.L. Lambert, *Astron. Astrophys. Rev.* **3**, 201 (1992)
2. M. Arnould, *Astron. Astrophys.* **46**, 117 (1976)
3. M. Rayet et al., *Astron. Astrophys.* **298**, 517 (1995)
4. T. Tauscher et al., *Astrophus.J.* **576**, 323 (2002)
5. S.E. Woosley and W.H. Howard, *Astrophus.J. Suppl.* **36**, 285 (1978)
6. M. Rayet, N. Prantzos, M. Arnould, *Astron. Astrophys.* **227**, 271 (1990)

7. H. Utsunomiya et al., *Phys.Rev.* **C67**, 015807 (2003).
8. P. Mohr et al., *Phys. Lett. B* **488**, 127 (2000).
9. K. Soutome, private communications.
10. M. Arnould and S. Goriely, Phys. Rep. **382**, 1 (2003).
11. H.Utsunomiya et al., this symposium.
12. D. Belic et al., *Phys. Rev. Lett.* **83**, 5242 (1999).

Photodisintegration of Deuterium in the Precision Era of Big Bang Nucleosynthesis

K.Y. Hara*, H. Utsunomiya*, H. Akimune*, S. Goko*, K. Kudo[†],
Y.-W. Lui**, H. Ohgaki[‡], M. Ohta*, Y. Shibata[†], H. Toyokawa[†], A. Uritani[†]
and T. Yamagata*

*Department of Physics, Konan University, Kobe 658-8501, Japan
[†]National Institute of Advanced Industrial Science and Technology (AIST), 1-1-1 Umezono,
Tsukuba, Ibaraki 305-8568, Japan
**Cyclotron Institute, Texas A & M University, College Station, Texas 77843, USA
[‡]Institute of Advanced Energy, Kyoto University, Gokanosho, Uji, Kyoto 611-0011, Japan

Abstract. Photodisintegration cross sections were measured for deuterium with Laser Compton scattering γ beams at E_γ = 2.3 - 4.6 MeV. The experimental evaluation of $R(E) = N_a \sigma v$ for the p(n,γ)D reaction was performed in the energy region relevant to big bang nucleosynthesis (BBN) by using the present data combined with the preceding data. The result confirms the past theoretical evdaluation with 6% uncertainty. The reaction rate for the p(n,γ)D reaction is presented for the BBN in the precision era.

INTRODUCTION

Since the discovery of the cosmic microwave background (CMB) in 1965 [1], big bang nucleosynthesis (BBN) has been developed based on the primeval abundances of four light elements D, ^{3}He, ^{4}He, and ^{7}Li [2-9]. BBN may be entering a precision era in view of the latest observations of deuterium primeval abundances for quaser systems [10-13] and temperature anisotropies of CMB by WMAP [14]. The primeval abundance of deuterium may play a role of a *cosmic baryometer* in the precision era [15-18].

Recently, a Monte Carlo method of directly incorporating nuclear inputs in the standard BBN calculations dramatically reduced uncertainties in the calculated abundances by as large as a factor of three [16, 19]. Among nuclear inputs for twelve key reactions in the standard BBN, however, only the one for p(n,γ)D is very scarce. Capture data for D are available only at four energies relevant to the BBN [20, 21] though a large collection of photodisintegration data is available above 5 MeV [22-28]. In the energy region of the BBN, the cross section is small and starts deviating from the 1/v law for the M1 capture due to the contribution of the E1 capture. The scarcity of data in this transitional energy region makes a theoretical evaluation of the cross section mandatory. Although the cross section is available in the ENDF/B-VI data library [29], it is said that details of the theoretical evaluation are not possible to trace [16].

Experimental cross sections for deuterium with sufficient accuracy are desired in the precision era. In this work, photodisintegration cross sections for deuterium were measured at 7 energies near threshold. We discuss the dependence of the p(n,γ)D reaction

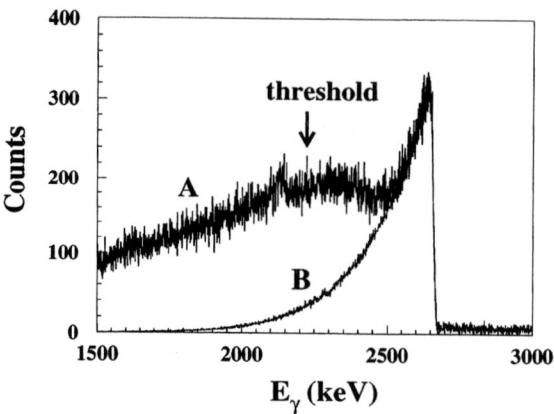

FIGURE 1. Response of a 120% Ge detector to the LCS γ rays (A) and an energy distribution of the LCS γ beam determined by a Monte Carlo analysis of the Ge response with the code EGS4 (B).

cross section on the energy relevant to the BBN in comparison with theoretical evaluations. We provide the reaction rate for the p(n,γ)D reaction.

EXPERIMENTAL PROCEDURE

A new γ-ray source, Laser Compton scattering (LCS) γ beam, was developed at the National Institute of Advanced Industrial Science and Technology (AIST) [30]. The LCS γ rays are produced in head-on collisions of Nd:YLF Q-switch laser photons with relativistic electron in the accumulator ring TERAS. Quasi-monochromatic γ beams, which were collimated into 2 mm in diameter with a 20 cm Pb block, were used to irradiate heavy water.

Energy spectra of the LCS γ rays were measured with a 120% Ge detector and analyzed with a Monte Carlo code EGS4 [31] to determine the tail profile of the LCS beam. An energy spectra of the LCS γ rays that best reproduced the Ge response (A) is shown (B) in Fig. 1. The fraction of LCS γ rays above 2.22 MeV was responsible for photodisintegration.

The total number of γ rays was determined from responses of a large volume (8 in. in diameter and 12 in. in thickness) NaI(Tl) detector to multi photons per pulse of the 1 kHz LCS beam and to single photons of the DC beam. The uncertainty in the total flux arose from nonlinearity in the response of our beam monitoring system to the pulsed multi photons. In view of the statistical analysis of pile-up spectra [32], we assigned 3% uncertainty to the γ flux.

The neutron detector consists of 16 ^{3}He proportional counters (EURISYS MEASURES 96NH45) embedded in a polyethylene moderator. Eight counters were placed in a concentric ring at 7 cm from the beam axis; the other eight at 10 cm. The neutron detection efficiency was measured with a neutron source of ^{252}Cf whose uncertainty in

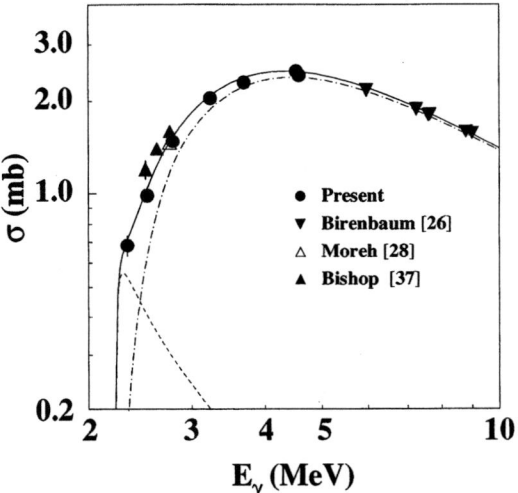

FIGURE 2. Photodisintegration cross sections for deuterium. The JENDL evaluations are shown by the dashed line for the M1 cross section, by the dot-dashed line for the E1 cross section, and by the solid line for the sum, respectively

the absolute neutron emission rate is 5%. The results for the ^{252}Cf source were well reproduced by Monte Calro simulations with the MCNP code [33]. The efficiencies for monoenergetic neutrons were calculated with the same code and were used in the data analysis.

RESULTS AND DISCUSSION

Figure 2 shows photodisintegration cross sections for deuterium as a function of the average γ-ray energy. The data analysis method and the numerical values of the cross section were given in our recent paper [34, 35]. The photonuclear data [26, 28] compiled in the IAEA document [36] are shown in Fig. 2. The datum of Moreh et al. [28] is consistent with our data, whereas the data of Bishop et al. [37] are not. The solid line is the JENDL evaluation [38, 39] which is the sum of the E1 (the dot-dashed line) and the M1 (the dashed line) cross sections. The JENDL evaluation is based on the M1 cross section of Segre [40] and the E1 cross section of the simplified Marshall-Guth model [41] below 10 MeV and that of Partovi above 10 MeV [42].

The systematic uncertainty of the cross section has three source: the neutron emission rate of the ^{252}Cf source (5%), the total flux of the LCS γ rays (3%), and the angular distribution of neutrons (2%). The overall systematic uncertainty is 6.2% after adding three sources in quadrature.

The present data were converted to capture cross sections with the detailed balance theorem. Figure 3 shows $R(E) = N_a \sigma v$ for the p(n,γ)D reaction as a function of the

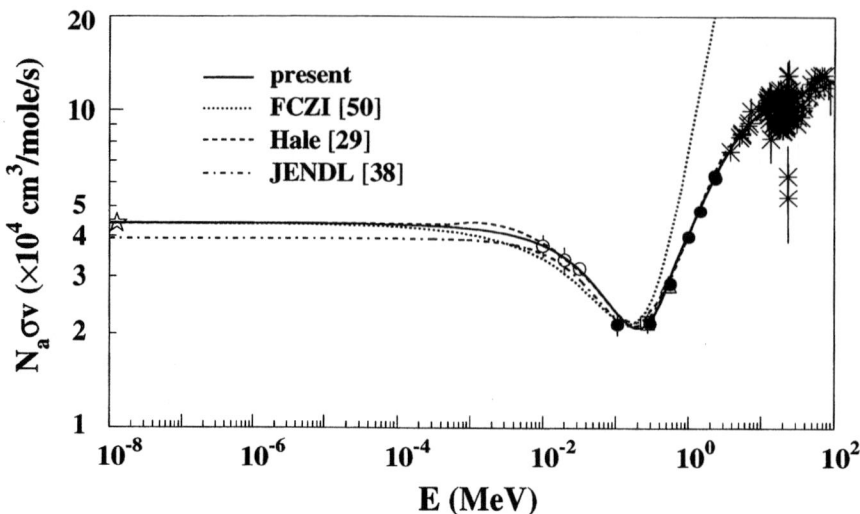

FIGURE 3. $R(E)=N_a\sigma v$ for the p(n,γ)D reaction as a function of the c.m. energy. Keys for the data are solid circles (present); open star [49]; open circles [20]; open square [21]; open triangle [28]. Only statistical uncertainties are shown for the present data. The high-energy data are from Refs. [43, 44]. The dotted line, the dashed line, and dot-dashed line stand for the theoretical evaluations of FCZI [50], Hale [29], and the JENDL [38], respectively. The solid line shows the best fit to the data connects to the JENDL evaluation at 1 MeV.

center of mass energy E, where N_a is the Avogadro's number, σ is the capture cross section, and v is the c.m. velocity. High-energy capture data [43-48] are also shown in the figure. A least squares fit was performed to the present data combined with the preceding data. The preceding data included the latest thermal neutron capture datum [49], the capture data [20, 21], and the photodisintegration datum [28]. The data of Ref. [37] were not included in the fit. The same polynomial expansion formula as that [Eq. (19), m=5] in Ref. [8] was used. The solid line shows the best fit to the data which is connected to the JENDL evaluation [38] at 1 MeV. The χ^2 value of the best fit was 0.61. The error involved in the fit was estimated to be 6%, which is dominated by the systematic uncertainty of the present measurement. For comparison, the theoretical evaluations of Fowler, Caughlan, and Zimmerman (FCZI) [50], Hale *et al.*, and the JENDL are shown by the dotted line, the dashed line, and the dot-dashed line, respectively.

The reaction rate, which is the thermal average of the present $R(E)$ funcion over the Maxwell-Boltzman velocity distribution, was calculated at temperature $0.01 < T_9 < 100$. The numerical intergration was made from 0.1 keV to 100 MeV. An analytical formula, the best fit to numerical values of the reaction rate, is expressed by

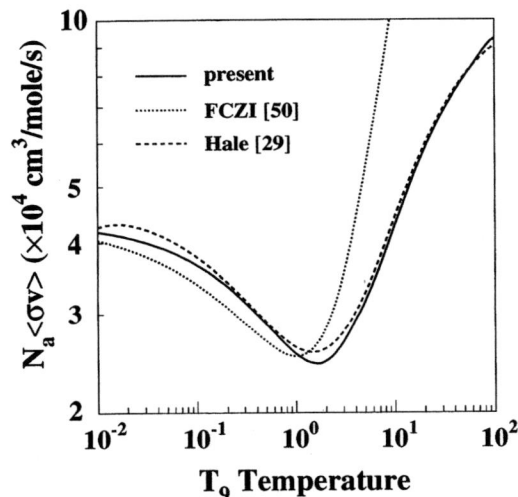

FIGURE 4. Reaction rate for the p(n,γ)D reaction. The solid line shows the present evaluation based on experimantal data. The theoretical evaluation of FCZI [50] and Hale et al. [29] for the reaction rate were shown by the dotted line and the dashed line, respectively.

$$N_a \langle \sigma v \rangle = \begin{cases} 4.41 \times 10^4 (1 - 0.489 T_9^{1/2} - 0.268 T_9 + 0.491 T_9^{3/2} - 0.182 T_9^2 - 0.0217 T_9^{5/2}) \\ \qquad\qquad\qquad\qquad\qquad\qquad\qquad\qquad\qquad\text{for } 0.01 < T_9 \leq 4 \\ 3.71 \times 10^4 (1 - 0.691 T_9^{1/2} + 0.429 T_9 - 0.0813 T_9^{3/2} - 0.00687 T_9^2 - 0.000219 T_9^{5/2}) \\ \qquad\qquad\qquad\qquad\qquad\qquad\qquad\qquad\qquad\text{for } 4 < T_9 < 100, \end{cases}$$

where the reaction rate is given in the unit of cm^3 $mole^{-1}$ sec^{-1}. The resultant reaction rate is shown by the solid line in Fig. 4 in comparison with the theoretical evaluation of FCZI (dotted line) and Hale et al. (dashed line). It is consistent with the thoretical evaluation of Hale et al. to within 4.5%.

CONCLUSIONS

Photodisintegration cross sections for deuterium were measured at 7 energies near threshold with the LCS γ beams at AIST. These cross sections resolve the scarcity of data relevant to BBN. The present data combined with the preceding data provide an experimantal fundation for the p(n,γ)D reaction cross section which has been evaluated only theoretically for more than three decades since the FCZI. The present $R(E)$ evaluated with 6% uncertainty confirms those theoretical evaluations made in the past. The reaction rate for the p(n,γ)D reaction is presented in the analytical form.

REFERENCES

1. A.A. Penzias and R.W. Wilson, *Astrophys. J.* **142**, 420 (1965).
2. P.J.E. Peebles, *Phys. Rev. Lett.* **16**, 410 (1966); *Astrophys. J.* **146**, 542 (1966).
3. R.V. Wagoner, W.A. Fowler, and F. Hoyle, *Astrophys. J.* **148**, 3 (1967).
4. H. Sato, *Prog. Theor. Phys.* **38**, 1083 (1967).
5. H. Reeves, J. Audouze, W.A. Fowler, and D. Schramm, *Astrophys. J.* **179**, 909 (1973).
6. L.M. Krauss and P. Romanelli, *Astrophys. J.* **358**, 47 (1990).
7. T.P. Walker, G. Steigman, D.N. Schramm, K.A. Olive, and H.-S. Kang, *Astrophys. J.* **376**, 51 (1991).
8. M.S. Smith, L.H. Kawano, and R.A. Malaney, *Astrophys. J. Suppl. Ser.* **85**, 219 (1993).
9. C.J. Copi, D.N. Schramm, and M.S. Turner, *Science* **267**, 192 (1995).
10. S. Burles and D. Tytler, *Astrophys. J.* **499**, 699 (1998).
11. S. Burles and D. Tytler, *Astrophys. J.* **507**, 732 (1998).
12. D. Kirkman, D. Tytler, S. Burles, D. Lubin, and J.M. O'Meara, *Astrophys. J.* **529**, 655 (2000).
13. J.M. O'Meara, D. Tytler, D. Kirkman, N. Suzuki, and J.X. Provhaska, *Astrophys. J.* **552**, 718 (2001).
14. See, http://map.gsfc.nasa.gov/
15. D.N. Schramm and M.S. Turner, *Rev. Mod. Phys.* **70**, 303 (1998).
16. K.M. Nollett and S. Burles, *Phys. Rev D* **61**, 123505 (2000).
17. K.A. Olive, G. Steigman, and T.P. Walker, *Phys. Rep.* **333-334**, 389 (2000).
18. S. Burles, K.M. Nollett, and M.S. Turner, *Astrophys. J. Lett.* **552**, L1 (2001).
19. S. Burles, K.M. Nollett, J.W. Truran, and M.S. Turner, *Phys. Rev. Lett.* **82**, 4176 (1999).
20. T.S. Suzuki *et al.*, *Astrophys. J. Lett.* **439**, L59 (1995).
21. Y. Nagai *et al.*, *Phys. Rev. C* **56**, 3173 (1997).
22. J. Ahrens *et al.*, *Phys. Lett.* **56B**, 49 (1974).
23. J.E.E. Baglin, R.W. Carr, E.J. Bentz, C.-P. Wu, *Nucl. Phys.* **A201**, 593 (1973).
24. R. Bernabei *et al.*, *Phys. Rev. Lett.* **57**, 1542 (1986).
25. R. Bernabei *et al.*, *Phys. Rev. C* **38**, 1990 (1988).
26. Y. Birenbaum, S. Kahane, and R. Moreh, *Phys. Rev. C* **32**, 1825 (1985).
27. J.A. Galey, *Phys. Rev.* **117**, 763 (1960).
28. R. Moreh, T.J. Kennett, and W.V. Prestwich, *Phys. Rev. C* **39**, 1247 (1989).
29. G.M. Hale, D.C. Dodder, E.R. Siciliano, and W.B. Wilson, Los Alamos National Laboratory, ENDF/B-VI evaluation, Mat No. 125, Rev. 1, 1991.
30. H. Ohgaki *et al.*, *IEEE Trans. Nucl. Sci.* **38**, 386 (1991).
31. W.R. Nelson, H. Hirayama and W.O. Roger, "The EGS4 Code Systems" SLAC-Report No. 265, 1985.
32. H. Toyokawa *et al.*, *IEEE Trans. Nucl. Sci.* **47**, 1954 (2000).
33. J.F. Briesmeister, computer code MCNP, Version 4C (Los Alamos National Labratry, Los Alamos, 2000).
34. H. Utsunomiya *et al.*, *Phys. Rev. C* **67**, 015807 (2003).
35. K.Y. Hara *et al.*, *Phys. Rev. D* **68**, 072001 (2003).
36. Photonuclear data for applications, "Cross Sections and Spectra", IAEA Report No. 1178, 2000.
37. G.R. Bishop *et al.*, *Phys. Rev.* **80**, 211 (1950).
38. T. Murata, Technical Report No. JAERI-M 94-019, 1994.
39. N. Kishida, T. Murata, T. Asami, K. Maki, and T. Fukahori, *J. Nucl. Sci. Technol.* **2**, 56 (2002).
40. E. Segre, *Nuclei and Particles* (Benjamin/Cummings, Mento Park, CA, 1977), p. 496.
41. J.F. Marshall and E. Guth, *Phys. Rev.* **78**, 738 (1950).
42. F. Partovi, *Ann. Phys.* (N.Y.) **27**, 79 (1964).
43. M. Bosman *et al.*, *Phys. Lett. B* **82**, 212 (1979).
44. T. Stiehler *et al.*, *Phys. Lett. B* **151**, 185 (1985).
45. M. Cerineo, K. Ilakovac, I. Šlaus, and P. Tomaš, *Phys. Rev.* **124**, 1947 (1961).
46. C. Dupont, P. Leleux, P. Lipnik, P. Macq, and A. Ninane, *Nucl. Phys.* **A445**, 13 (1985).
47. P. Michel, K. Moeller, J. Moesner, and G. Schmidt, *J. Phys. G* **15**, 1025 (1989).
48. P. Wauters *et al.*, *Few-Body Systems* **8**, 1 (1990).
49. S.F. Mughbghab, M. Divadeenam, and N.E. Holden, *Neutron Cross Sections, Vol. 1, Neutron Resonance Parameters and Thermal Cross Sections*, Part A, Z = 1-60 (Academic, New York, 1981).
50. W.A. Fowler, G.R. Caughlan, and B.A. Zimmerman, *Annu. Rev. Astron. Astrophys.* **5**, 525 (1967).

On the excitation energy for maximum cold fusion reactions in superheavy mass region

A. Fukushima, T. Wada, and M. Ohta, Y. Aritomo*

Department of Physics, Konan University, Okamoto, Kobe 658-8501, Japan
*Flerov Laboratory of Nuclear Reactions, JINR, Dubna, Moscow region, 141980 Russia
*Department of Physics, University of Tokyo, Tokyo 113-0033, Japan

Abstract. We have analyzed cold fusion reactions, especially, (HI(Heavy-Ion), xn) reactions with ^{208}Pb target series. The ratio a_f/a_n used in the statistical model is determined by fitting experimental data of Γ_n/Γ_{total} for Z=102 and the extrapolative use is made for Z>103. Theoretical predictions for excitation functions in ^{208}Pb(^{50}Ti, ^{51}V, ^{54}Cr, ^{55}Mn, ^{58}Fe, ^{59}Co, ^{65}Cu, ^{70}Zn, xn) reactions are compared with experiments. We found that peak positions of (HI, xn) reactions except for the sub-barrier fusion enhancement effect stay almost constant when we increase the Z-number of compound nucleus from Z=104 to Z=110.

1. INTRODUCTION

In the researches on synthesis cross sections for superheavy elements, the incident projectile energies or corresponding excitation energies of compound nuclei for which maximum yields of (HI, xn) reactions are observed are of great interest experimentally. Today, in many laboratories, either experiments or intensive preparatory work are in progress (RIKEN, GANIL, Berkeley, Darmstadt, and Dubna).

Theoretical support for these elaborate experiments is very important for setting a combination of fusing nuclei and their collision energy, and for an evaluation of cross sections and identification of evaporation residues.

Fusion dynamics change remarkably with increasing Z-number of compound nuclei. On the other hand, formation cross sections for compound nuclei decrease very rapidly in increasing order of the Z-number. The dynamics of so-called quasi-fission process causes mainly decrease in the formation probabilities, and plays decisive roles in determining survival probabilities of the compound nuclei in the processes of their cooling.

In this paper, the formation probabilities are evaluated by a phenomenological formula [1] and the survival probabilities by the statistical model. The aim of this work is to investigate where the maximum cross section appears in the cold fusion reaction forming the compound nuclei with Z=104 to 110. The phenomenological formula mentioned above does not contain a sub-barrier enhancement. However, we can see where the cross section becomes optimum by the subtle balance between the position of the entrance Coulomb barrier and the energy dependence of the survival probabilities. The comparison of our simple prediction with the experimental data also seems to derive useful information for the sub-barrier fusion enhancement by comparing the prediction with the data.

2. FORMATION PROBABILITY

The evaporation residue cross sections are given by the formula

$$\sigma_{ER} = \frac{\pi\eta^2}{2\mu E_{cm}} \sum_\lambda (2\lambda+1) P_{for} P_{sur}. \tag{1}$$

Here P_{for} and P_{sur} denote formation and survival probabilities for the total angular momentum λ, respectively, and are independent of each other except conserved quantities, such as the total angular momentum, etc.. E_{cm} is the incident energy in the center-of-mass system.

The logarithm of P_{for}, the probability of forming a compound nucleus, can be parameterized by [1]

$$\log_{10} P_{for}(Z,\alpha,E^*) = -\frac{3.2+\left[(E^*-E_B^*+5)^{-\frac{1}{2}}+0.085\right](Z-100)}{1+\exp[(\alpha-0.55)/d]}, \tag{2}$$

where Z and E^* are the atomic number and the excitation energy of the compound nucleus, and $\alpha = |A_1 - A_2|/(A_1 + A_2)$. A_1 and A_2 are the mass number of target and projectile nuclei, respectively. The excitation energy corresponding to the Bass potential barrier is denoted by E_B^* and $d = 0.05$. Note that this formula is valid for $Z > 102$, because the function in Equation (2) is derived under the assumption that the liquid-drop part of the fission barrier is less than 1 MeV. The guiding principle of the derivation of this functional form is based on the probability of overcoming a potential barrier ΔV by thermal diffusion connected with a heat bath of temperature T [1]:

$$P_{for} \approx \exp(-\Delta V/T) h(Z,\alpha). \tag{3}$$

Here, we assume that, in order to form the compound nucleus, the colliding partners have to overcome an extra barrier ΔV along a fusion path after contacting each other; ΔV is mainly a function of Z and α. The function $h(Z,\alpha)$ indicates the probability that the system reaches the compound nucleus after overcoming the barrier ΔV. If the colliding partners can not overcome the barrier ΔV, they separate into two fragments through the quasi-fission process. The function $h(Z,\alpha)$ should be investigated by means of the exact calculation for the dynamical evolution of the system. But now, we consider that this function is effectively included in Equation (2) in the process of data fitting. The dependence on the excitation energy in Equation (2) comes from the temperature T according to the relation $E^* = aT^2$, where a is the level density parameter.

The potential difference ΔV between the contact point and the ridge of the potential energy surface located slightly inside of the contact point is investigated by the two-center parameterization [2]. We have obtained the following relation [1]:

$$\Delta V(Z,\alpha) \propto \frac{Z-100}{1+\exp[(\alpha-0.55)/0.05]}. \quad (4)$$

The substitution of the relation (4) into Equation (3) leads to Equation (2) by introducing some constant factors to reproduce the experimental data.

3. SURVIVAL PROBABILITY

The survival probability is calculated by the traditional statistical model [3]. We expect to obtain the probability of finding evaporation residue nuclei in competition with the fission process

$$P_{sur} = \prod_{i=1}^{N} \Gamma_n^{(i)}/(\Gamma_n^{(i)} + \Gamma_f^{(i)}) \quad (5)$$

where N is the number of emitted neutrons, and $\Gamma_n^{(i)}$ and $\Gamma_f^{(i)}$ are the decay widths of neutron evaporation and fission before the i-th neutron emission, respectively. The essential quantity for evaluating the survival probability is Γ_n/Γ_f at any excitation energy of compound nucleus. We can write the probability in the form [3]

$$\frac{\Gamma_n}{\Gamma_f} = \frac{k_{coll}(g.s.)}{k_{coll}(saddle)k_{Kramars}} A_0 \exp\left[2\sqrt{a_n E_n^*} - 2\sqrt{a_f E_f^*}\right] \quad (6)$$

with $E_n^* = E_{int}^* - B_n$, $E_f^* = E_{int}^* - B_f$, $A_0 = 4A^{2/3} a_f E_n^*/K_0 a_n \left[2\sqrt{a_f E_f^*} - 1\right]$, and $K_0 \equiv \eta^2/2\mu r_0^2$ where E_{int}^* presents the intrinsic excitation energy of compound nucleus, B_n the neutron separation energy, B_f the fission barrier height, μ the reduced mass and r_0 the radius parameter. The factors $k_{coll}(g.s.)$ and $k_{coll}(saddle)$ are the collective enhancement factors for the level density in the ground state and saddle shape, respectively, and the Kramars factor [4] is expressed as $k_{Kramars} = (\eta\omega_1/\sqrt{E_{int}^*})(\sqrt{1+x^2}-x)$ with $x \equiv \gamma/2\omega_0$; ω_0 and ω_1 are curvature of potential energy surface at ground state configuration and saddle shape, γ is the strength of one-body friction. B_f and a_n depend on the excitation energy as follows

$$B_f = B_f^{LDM} - (E_{rot}^{sad}(\lambda) - E_{rot}^{gr}(\lambda) - \delta_{shell}(T=0)), \quad (7)$$

$$a_n = \tilde{a}_n \left\{1 + \frac{\delta_{shell}(T=0)}{E_n^*}\left[1-\exp\left(-\frac{E_n^*}{E_D}\right)\right]\right\}. \quad (8)$$

$\tilde{a}_n$ is given by Töke and Swiatecki [5]. $E_{rot}^{sad}(\lambda)$ and $E_{rot}^{gr}(\lambda)$ are the rotational energy of the saddle shape nucleus with angular momentum λ and that of the ground state, respectively. δ_{shell} is the shell correction energy at the ground state given by Möller et al. [6]. B_f^{LDM} is the macroscopic part of the fission barrier in the liquid drop model. The parameter a_f also depends on E_f^*. We derive an empirical relation between a_n and a_f so as to reproduce the experimental data as will be discerned in the next section. E_D means the shell dumping energy [5].

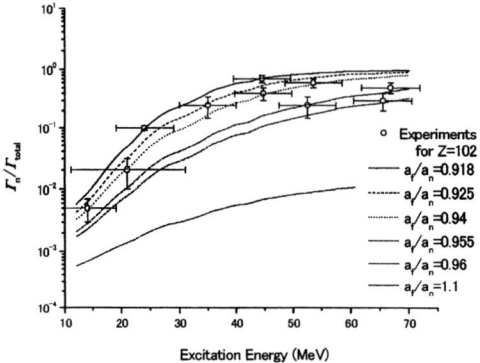

FIGURE 1. Comparison of the ratio $\langle\Gamma_n/(\Gamma_n+\Gamma_f)\rangle$'s for Z=102 with the theoretical curve for various value of a_f/a_n. In the calculations, the shell dumping factor E_D =20MeV, and friction parameter x =5.0.

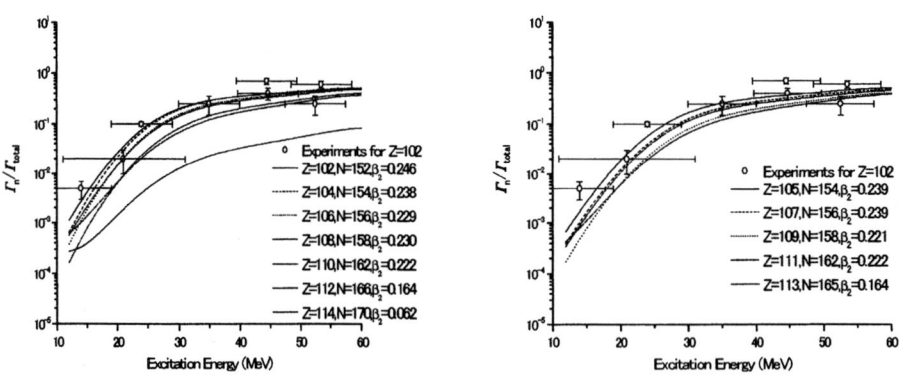

FIGURE 2. Z-dependence of excitation curves of $\langle\Gamma_n/(\Gamma_n+\Gamma_f)\rangle$ for Z>104. Left: The excitation curves of each $\langle\Gamma_n/(\Gamma_n+\Gamma_f)\rangle$s for even-Z nuclei. Right: Those for odd-Z nuclei.

4. RESULTS AND DISCUSSIONS

When the fission barrier height is determined by taking account of the shell correction energy which is evaluated by using an appropriate mass table (for example, the mass table by Möller et al. [6]), one of ambiguous parts for estimating P_{sur} arises from the ratio a_f/a_n in the framework of the statistical model.

We determined the ratio a_f/a_n by fitting the experimental data $\Gamma_n/(\Gamma_n+\Gamma_f)$ [7] which is presented in Fig.1. The experimental values of $\Gamma_n/(\Gamma_n+\Gamma_f)$ for Z=102 nucleus at various excitation energy are corresponding to the value of a_f/a_n from 0.918 to 1.1 as shown Fig.1. By fitting the experimental data, we can find a_f/a_n as a function of the excitation energy. By using the E^*-dependent a_f/a_n, the $\Gamma_n/(\Gamma_n+\Gamma_f)$ for the compound nuclei with Z>103 are calculated and shown in Fig.2. We can observe the variation due to the shell correction energies and to the β_2-deformations of the compound nuclei.

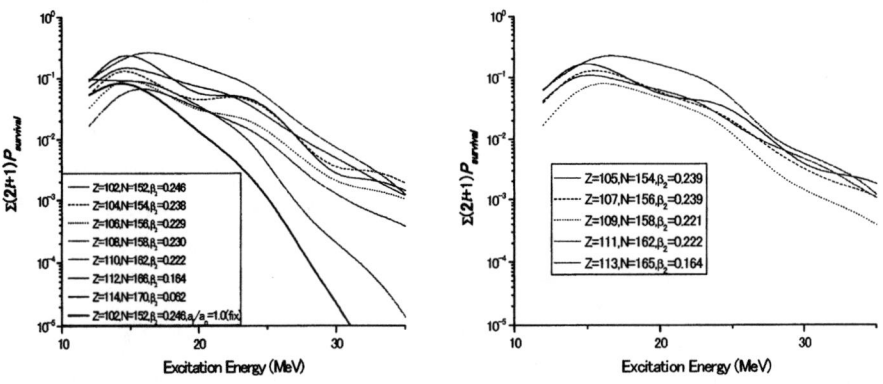

FIGURE 3. Excitation energy dependence of survival probabilities for Z>104. Left: The excitation curves of each $\sum(2\lambda+1)P_{sur}$ s for even-Z nuclei. Right: Those for odd-Z nuclei.

When we know $\Gamma_n/(\Gamma_n+\Gamma_f)$ at each E^*, we can calculate the survival probabilities by Equation (5). And by combining them with formation probabilities given by Equation (2), we can obtain the excitation functions for (HI, xn) reaction.

Excitation functions for Z>104 given by Equation (1) are shown in Fig.4. The theoretical curve is multiplied by 1/10 in order to fit the experimental data.

Our discussion is summarized as follows: In the present calculation in which the sub-barrier fusion enhancement is not taken into account, the structure around $E^* = 15$ MeV is not explained clearly. For Z>109 where the Bass barrier become less than about 15 MeV, the peak position of the excitation function stays almost constant, around 20 MeV of excitation energy, in this cold fusion reaction series. There seems no effect of the sub-barrier fusion enhancement to be found for Z>109. As shown in Fig. 2, the value of $\Gamma_n/(\Gamma_n+\Gamma_f)$ at 20 MeV is larger by one order or more than that at

15 MeV where we expect the 1n channel to dominate. Therefore, in the contribution from the flat region of the survival probabilities shown in Fig.3, the excitation functions appear to have maxima around 20 MeV. The essential point is that the value of $\Gamma_n/(\Gamma_n+\Gamma_f)$ is sensitive to the level density parameter ratio a_f/a_n.

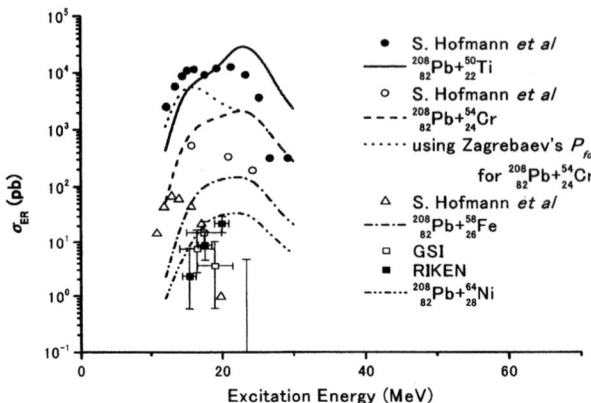

FIGURE 4. Excitation functions for $Z>104$.

We are now proceeding to investigate the ratio a_f/a_n for low excitation energy region where the shell effect is still important and also to study how the level density parameter depends on the saddle point deformation.

REFERENCES

1. M. Ohta and Y. Aritomo, *J. Nucl. Sci. and Technology, Supl.* **2**, 586 (2002); *Phys. of Atomic Nuclei,* **66**, 1026 (2003).
2. K. Sato, A. Iwamoto, K. Harada, et al., *Z. Phys.* **A288**, 387 (1978).
3. R. Vandenbosh and J. R. Huizenger, *Nuclear Fission* (Academic, New York, 1973), p.233.
4. H. A. Kramers, *Physica* (Utrecht) **7**, 284 (1940).
5. A. V. Ignatyk, G.N. Smirenkin, and A. S. Tishin, *Sov. J. Nucl.Phys.* **21** 255(1975).
6. J. Töke and W. J. Swiatecki, *Nucl. Phys.* **A372**, 141 (1981).
7. P. Möller, J. R. Nix, W. D. Myers and W. J. Swiatecki, *Atomic Mass Nucl. Data Table* **59**, 185 (1995).
8. T. Sikeland et al., *Phys. Rev.* **169**, 1000 (1968); *Phys. Rev.* **172**, 1232 (1968); A. N. Andreyev et al., *Z. Phys.* **A345**, 389 (1993); G. N. Flerov et al., *Nucl. Phys.* **A160**, 181 (1970).
9. V. I. Zagrebaev, Y. Aritomo, M. G. Itkis, Yu. Ts. Oganessian, and M. Ohta, *Phys. Rev.* **C65**, 014607 (2001).
10. S. Hofmann, *Rep. Prog. Phys.* **61**, 639-689 (1998).
11. K. Morita et al., *RIKEN Accel. Prog. Rep.* **35** (2002).

New ANC measurement of the astrophysical S_{17}-factor from d(^{7}Be,^{8}B)n reaction.

J.J. Das[*], V.M. Datar[†], P. Sugathan[*], N. Madhavan[*], P.V. Madhusudhana Rao[***], A. Navin[‡], A. Jhingan[*], T. Varughese[*], S. Nath[*], A. Ray[§], S. Barua[¶], R. Singh[‖], S.K. Dhiman[††], R. Shyam[‡‡], R.G. Kulkarni[§§], A.K. Sinha[¶¶] and D.L. Sastry[***]

[*]*Nuclear Science Centre, Post Box 10502, New Delhi- 110067, India.*
[†]*Nuclear Physics Division, Bhabha Atomic Research Centre, Trombay, Mumbai-400085, India.*
[**]*Department of Nuclear Physics, Andhra University, Visakhapatnam-530003, India.*
[‡]*Nuclear Physics Division, Bhabha Atomic Research Centre,Trombay, Mumbai-400085, India.*
[§]*Variable Energy Cyclotron Centre, 1/AF Bidhan Nagar, Kolkata -700064, India*
[¶]*Department of Physics, Gauhati University, Jalukabari-Guwahati -787014, India*
[‖]*Department of Physics and Astro-physics, Delhi University, New Delhi- 10007, India*
[††]*Department of Physics, Himachal Pradesh University, Shimla-171005, India.*
[‡‡]*Saha Institute of Nuclear Physics,1/AF Bidhan Nagar. Kolkata - 700064, India*
[§§]*Department of Physics, Saurashtra University, Rajkot - 360005, India*
[¶¶]*Inter University Centre for DAE Facilities, Kolkata Centre, Bidhan Nagar, Kolkata - 700092, India*
[***]*Department of Nuclear Physics, Andhra University, Visakhapatnam - 530003, India.*

Recent measurements[1, 2] of total ^{8}B solar neutrino flux comprising neutrinos of different flavors account for all the missing high energy ^{8}B solar neutrinos and agree well with Standard Solar Model (SSM)[3] calculations. Since only electron neutrinos should be produced at the center of the sun, so the observations of neutrinos of different flavors provide evidence in the support of neutrino flavor oscillation. However at present, the uncertainty on the measured ^{8}B solar neutrino flux is over 10% and that on the predicted flux from SSM calculations is also over 10%. So it is important to improve upon the accuracy of both the measured and predicted ^{8}B solar neutrino fluxes for a precise comparison between them. Such comparison should be important for better understanding of both neutrino and solar physics. The most uncertain parameter in the calculation of ^{8}B solar neutrino flux using SSM is ^{7}Be(p,γ)^{8}B reaction rate at solar Gammow energy. This reaction rate is usually represented by $S_{17}(0)$ factor and so high precision measurement of $S_{17}(0)$ factor is required.

There have been a number of measurements of ^{7}Be(p,γ)^{8}B reaction cross section which are limited only to higher energies i.e., $E_{cm} > 100$ keV. The recent direct capture precision measurements yielded S-factor which are clustered around 21.3 eV b and 18.5 eV b [4]. Apparent disagreement between different measurements were attributed to uncertainties in target thickness and extrapolation to lower energy. However, as these measurements do not agree within the quoted errors in these measurements alternative experiments involving different systematic are desirable [5] for accurate estimation of $S_{17}(0)$. Alternate methods attempted so far were (1) Coulomb dissociation of ^{8}B [6] and the most recent measurement of this kind reported the $S_{17}(0)$ value of 18.6 eV b [7]. (2) Asymptotic normalization coefficient (ANC)[8] is the method adopted by us and would be discussed in some details in this paper. In this method, ANC corresponding to the overlap integral of the ^{7}Be and ^{8}B ground states are obtained from the proton pickup reaction (^{7}Be,^{8}B) reaction at low energies where the proton transfer is expected to be peripheral at forward angles.

In recent studies (^{7}Be,^{8}B) reaction was used to extract $S_{17}(0)$ by different groups by measuring the cross section for ^{10}B(^{7}Be,^{8}B)^{9}Be at 84MeV and ^{14}N(^{7}Be,^{8}B)^{13}C at 85MeV[10]. The extracted $S_{17}(0)$ from these measurements are 17.8 ± 2.8 and 16.6 ±1.9 eV b respectively.

An experimental program was undertaken at Nuclear Science Centre, New Delhi, to extract the $S_{17}(0)$-factor by employing the ANC method from the proton transfer reaction d(^{7}Be,^{8}B)n at further low energy with better statistics. Thus this measurement should be able to provide a more accurate measurement. The ^{7}Be beam was produced by employing a new ion optical configuration[16] for the recoil mass separator HIRA[14] with added new hardware. Primary beam of ^{7}Li from the 15UD Pelletron was used in inverse kinematics on a hydrogen target to produce the ^{7}Be radioactive beam. A large area polyethylene $(CH_2)_n$ foil, of 20 μm thickness, was used as hydrogen target. Target foil was mounted on a rotary and linear motion drive[15] to prevent the beam induced damage of the foil at high beam currents. This device ensures the helical path of the beam irradiation profile on the target, so that the effective target area is as large as ~3500 mm^2, so that the irradiation beam spot on the target is in continuously moved over the target which enable the use of the target for long runs without damage as the heat generated by the beam is dissipated through the target area.

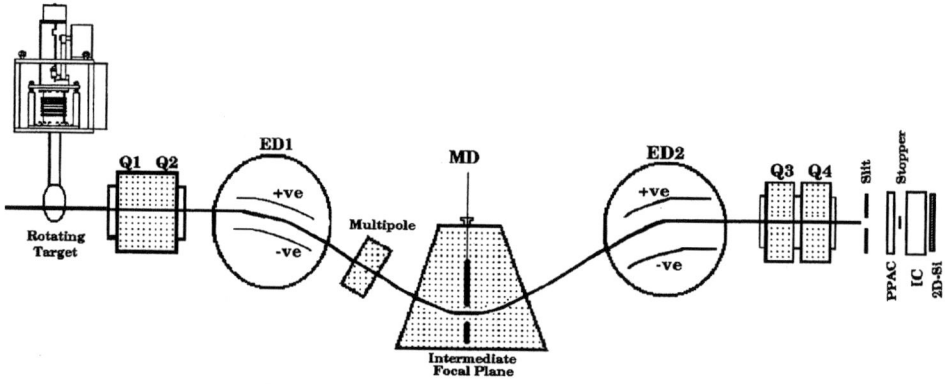

FIGURE 1. Schematic of the ^{7}Be radioactive beam production and separation with HIRA and detection system at the focal plane.

The ^{7}Be radioactive beam, produced in the inverse kinematic reaction ^{7}Li(p,n)^{7}Be, is separated from the primary beam in-flight using HIRA. Since the first dispersive element is an electric dipole, HIRA provides a very good separation between the primary and secondary beam energy and hence trajectory. A special ion-optical mode was used for HIRA to achieve a point-to-point focusing of the secondary beam at two different positions, one at the center of dipole magnet and the other at secondary target position. A momentum filter aperture is used at the first focal point to block the primary beam and transmit the pure ^{7}Be beam through HIRA. With this a very good beam quality of better than 99.9% purity, beam spot size $\approx$3 mm and an angular divergence of $\pm 1°$ was achieved. Such a good quality beam is ideally required for the precision angular distribution measurement of the reaction products. A schematic of the ^{7}Be production, separation and detection system is shown in Figure 1.

Present measurement was performed at a ^{7}Be beam energy of 21.0 (pm0.5) MeV on a deuterated polyethylene (CD$_2$)$_n$ target foil of thickness $\sim$ 1 mg/cm^2. The ^{7}Be beam was produced with the 25.0 MeV primary ^{7}Li pulsed beam from the Pelletron with a repetition rate of 4 MHz. The primary beam current was restricted to maintain the ^{7}Be intensity below 3 kHz. The detection system include a two dimensional(2D) position sensitive MWPC followed by a gas ΔE and 50$\times$50 i mm^2 2D-position sensitive Silicon detector. The X-Y position profile of the ^{7}Be was monitored with the MWPC after the secondary target. High intensity flux of uninteracted beam particles were stopped with a 4 mm diameter tantalum stopper which blocked almost 90% of the beam component from reaching the main detectors. The angular coverage of the stopper was limited to about 1° in laboratory system. Isometric projection of the position and intensity profile with and without beam stopper was shown in figure 2. It is known that the rates of pile-up events in DE detector increases with data rate, which can make particle identification ambiguous. It is specially because the ^{8}B data events are limited to 1 for 300,000 ^{7}Be flux, so it is crucial to maintain the low count rate in the detectors to avoid the pile-up event contamination in the ^{8}B data. Thus by using the stopper enabled the use of more beam flux without compromising the data quality and was useful in collecting more data to minimize the statistical errors. This detection system provided particle identification and the angular distribution of the reaction products.

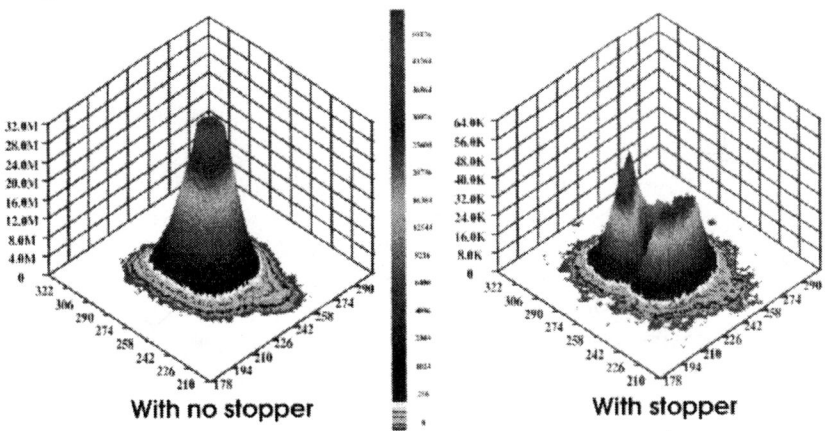

FIGURE 2. Position and intensity profile of the ^{7}Be beam on Silicon detector with and without the stopper for the amount of beam dose.

To check the beam stability with time a surface barrier detector was inserted beam position periodically which gave the normalization. With this beam stability was found to be less than 5%[?]. Detection system was calibrated in-situ particle identification. This was achieved by using the primary beams of ^{7}Li and ^{12}C. The HIRA was rotated to $2°$ to minimize the primary beam flux on the Si-detector. With this a clear separation of Li, Be and C particle groups was obtained. For energy loss calculations through target, different foils and gaseous media, the program SRIM [18] was used. An alternate particle identification with MZ^2 dependance, in figure 3 was also obtained in the offline analysis using E, ΔE and the effective thickness of the ΔE detector with the help of a particle identifier (PID) algorithm [19].

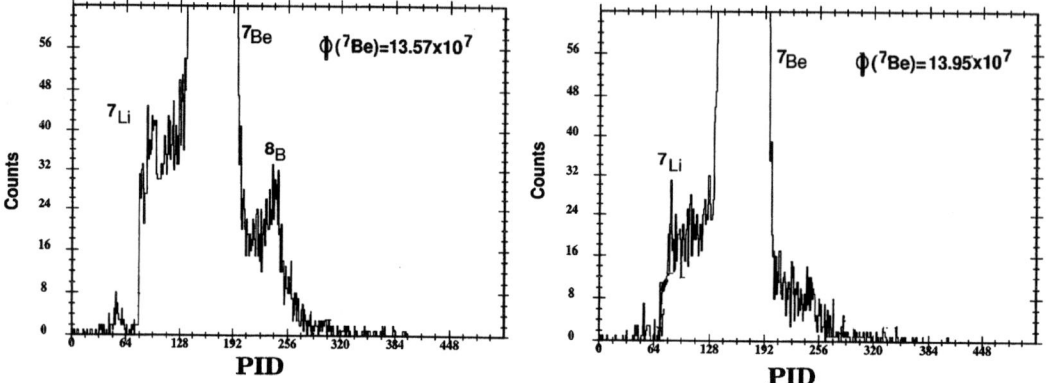

FIGURE 3. Particle Identification (PID) spectra from CH_2 and CD_2 target for incident ^{7}Be flux of 13.9 x 10^7 particles is shown.

The measured cross section of $^2H(^7Be,^8B)n$ direct transfer reaction has been analyzed within the finite range distorted wave Born approximation (FRDWBA). The astrophysical $S_{17}(0)$ has been extracted by employing the ANC method [8]. In Fig. 4, we present the extracted results for astrophysical S-factor by using three different set of optical potential parameters. The optical potential parameters denoted as OM1 and OM2 in Fig. 4 are set1 and set2 respectively of Liu [12]. The OM3, in Fig. 4 shows the results of S_{17}-factor using deuteron optical parameters from elastic scattering on ^{6}Li [20] at the corrected energy ($E_{cm} = 4.5$ MeV) for d+^{7}Be state and final state optical parameters are taken from the set1 of Liu [12]. The deuteron (n - p) state overlap function has been obtained from the deuteron wave function for both s-wave and p-wave states, corresponding to the Reid soft potential. ^8B single particle wave function has been calculated with Woods Saxon potential [21]. Proper account of compound nuclear contribution has also been accounted for using the Housher Feshbach code [22].

Recently the elastic scattering angular distribution in the $^2H(^7Be,^7Be)^2H$ system has been measured at 20.25 MeV using the same CDsub2 target using kinematics coincidence using a pair of annular detectors [17]. Through this measurement it would be possible to further constrain the optical model parameters in the entrance channel.

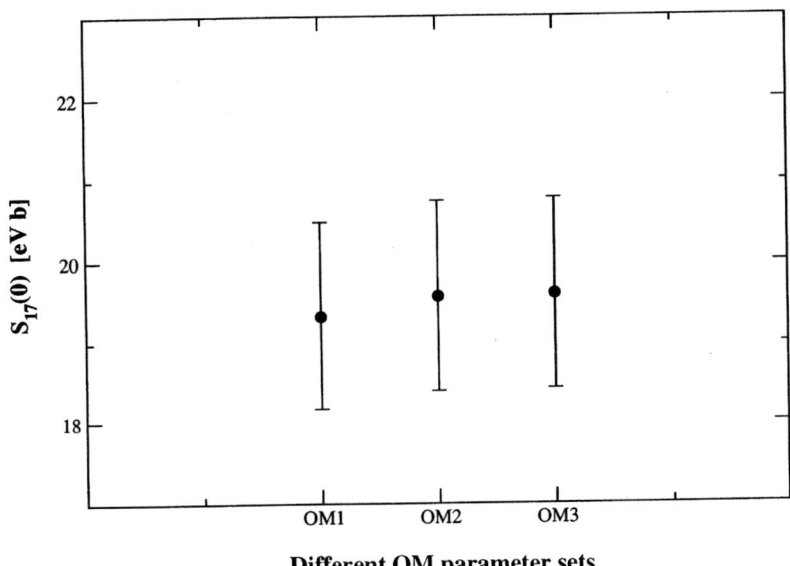

FIGURE 4. extracted $S_{17}(0)$ factor from our data using different sets of optical model parameters for entrance and exit channel)

ACKNOWLEDGMENTS

This work was supported by the Department of Science and Technology, Govt. of India under grant no. SP/S2/K-26/1997.

REFERENCES

1. S. Fukuda et al., Phys. Rev. Lett. **86**, 5651(2002) and *ibid* Phys. Rev. Lett. **86**,5656(2002).
2. Q .R. Ahmad et al., Phys. Rev. Lett. **89**, 011301 (2002); S.N. Ahmed et al., http://arXiv.org/nucl-ex/0309004. K. Eguchi et al., Phys. Rev. Lett. **90**, 21802 (2003).
3. J.N. Bahcall and M. H. Pinsonneault, Rev. Mod. Phys. **64**, 885 (1992), *ibid* Astrophys. J. **555**, 990 (2001).
4. F. Hammache et al., Phys. Rev. Lett. **86**, 3985 (2001); A.R. Junghans et al., Phys. Rev. Lett. **88**, 041101 (2002); L.T. Baby et al., Phys. Rev. Lett. **90**, 022501 (2003) and references therein.
5. E.G. Adelberger et al., Rev. Mod. Phys. **70**, 1265 (1998).
6. T. Motobayashi, Nucl. Phys. **A718**, 101c (2003) and references therein:
7. F. Schumann et al., Phys. Rev. Lett. **90**, 232501 (2003).
8. H.M. Xu, C. A. Gagliardi, R. E. Tribble, A M Mukhamedzhanov, and N.K. Timofeyuk, Phys. Rev. Lett. **73**, 2027 (1994).
9. C.A. Gagliardi et al., Phys. Rev. C **59**, 1149 (1999).

10. A. Azhari et al., Phys. Rev. Lett. **82**, 3960 (1999); A. Azhari et al., Phys. Rev. C **60**, 055803 (1999).
11. J.C. Fernandes, R. Crespo, F.M. Nunes, and I.J. Thompson, Phys. Rev. C **59**, 2865 (1999).
12. W. Liu et al., Phys. Rev. Lett. **77**, 611 (1996).
13. C.A. Gagliardi, A.M. Mukhamedzhanov, R.E. Tribble, and H.M. Xu, Phys. Rev. Lett. **80**, 421 (1998).
14. A.K. Sinha et al., Nucl. Instr. Meth., **A339**, 543 (1994).
15. T. Varughese et al., Nucl. Instr. Meth. (2003) (*in press*).
16. J.J. Das et al., J.Phys. G **24**, 1371 (1998).
17. A. Jhingan et al., *Proc. DAE-BRNS Symp. on Nucl. Phys.* **45B**, 416 (2002).
18. J.F. Zeigler, www.srim.org
19. F.S. Goulding, in *Treatise on Heavy Ion Science*, Vol.7 (Plenum Press, New York, 1985), ed. D.A. Bromley.
20. H. G. Bingham et al., Nucl. Phys. **A173**, 265 (1971).
21. F.C. Barker, Aust. J. Phys. **33**, 177 (1980).
22. M.A. Eswaran, et al., Phys. Rev. C **39**, 1856 (1989). **A716**, 211 (2003).

List Of Participants

Akimune, Hidetoshi
Department of Physics, Konan Univ.
8-9-1 Okamoto Higashinada-ku
Kobe   658-8501
Japan

akimune@konan-u.ac.jp

Aoyama, Shigeyoshi
Kitami Institute of Technology
Koentyo 165, Kitami 090-8507
Japan

aoyama@mail.kitami-it.ac.jp

Arai, Koji
Department of Physics, Univ. of Surrey
Guildford, Surrey GU2 7XH
United Kingdom

k.arai@surrey.ac.uk

Aritomo, Yoshihiro
Department of Physics, Univ. of Tokyo
Hongo, Bunkyo-ku, Tokyo 113-0023
Japan

aritomo@nt.phys.s.u-tokyo.ac.jp

Arnould, Marcel
Institut d'Astronomie et d'Astrophysique
Univ. Libre de Bruxelles
Campus Plaine - CP 226, Bld du Triomphe 50,
Brussels B-1050
Belgium

marnould@astro.ulb.ac.be

Asano, Tomomasa
Department of Physics, Konan Univ.
8-9-1 Okamoto Higashinada-ku
Kobe 658-8501
Japan

dn121001@center.konan-u.ac.jp

Bauge, Eric
CEA/Bruyeres-le-Chatel
CEA/DIF/DPTA/SPN,
BP 12, Bruyeres-le-Chatel F-91680
France

eric.bauge@cea.fr

Binns, Walter
Department of Physics
Washington Univ in St. Louis
CB1105, 1 Brookings Drive,
St. Louis, MO 63130
USA

wrb@wuphys.wustl.edu

Blank, Bertram
CEN Bordeaux-Gradignan
Le Haut-Vigneau, Gradignan F-33175
France

blank@cenbg.in2p3.fr

Bonaccorso, Angela
INFN, Sezione di Pisa
via F. Buonarroti 2, Pisa I-56127
Italy

bonaccorso@pi.infn.it

Borzov, Ivan N.
IPPE Obninsk
CP226 Bvd.du Triomphe IAA ULB Brussels
B-1050
Belgium

iborzov@astro.ulb.ac.be

Bulgac, Aurel
Department of Physics, Univ. of Washington
Seattle, WA 98195-1560
USA

bulgac@phys.washington.edu

Catford, Wilton
Department of Physics, Univ. of Surrey
Guildford, Surrey GU2 7XH
United Kingdom

W.Catford@surrey.ac.uk

De Oliveira Santos, Francis
GANIL/CNRS
Bd Henri Becquerel - BP 55027 - CAEN
CEDEX 5 F-14076
France

oliveira@ganil.fr

Denisov, Vitali
GSI Darmstadt
Planckstrasse 1, Darmstadt D-64291
Germany

v.denisov@gsi.de

Fukushima, Akira
Department of Physics, Konan Univ.
8-9-1 Okamoto Higashinada-ku
Kobe 658-8501
Japan

dn921001@center.konan-u.ac.jp

Goriely, Stephane
Institut d'Astronomie et d'Astrophysique
Univ. Libre de Bruxelles
Campus Plaine - CP 226, Bld du Triomphe 50,
Brussels B-1050
Belgium

sgoriely@astro.ulb.ac.be

Greiner, Walter
Institut fuer Theoretical Physics
Johann Wolfgang Goethe-Univ.
Robert-Mayer-Str. 8-10, Frankfurt am Main
D-60054
Germany

greiner@th.physik.uni-frankfurt.de

Coc, Alain
CSNSM
Bat. 104, Orsay Campus F-91405
France

coc@csnsm.in2p3.fr

Demetriou, Paraskevi
Institut d'Astronomie et d'Astrophysique
Univ. Libre de Bruxelles
Campus Plaine - CP 226, Bld du Triomphe 50,
Brussels B-1050
Belgium

vdemetri@astro.ulb.ac.be

Drozdov, Vadim
Skobeltsyn Institute of Nuclear Physics
Moscow State Univ.
Moscow 119992
Russia

drozdov@p6-lnr.sinp.msu.ru

Goko, Shinji
Department of Physics, Konan Univ.
8-9-1 Okamoto Higashinada-ku
Kobe 658-0072
Japan

gokou@konan-u.ac.jp

Goutte, Dominique
GANIL
Bd Henri Becquerel - BP 55027 - CAEN
CEDEX 5 F-14076
France

goutte@ganil.fr

Gupta, Mohini
Manipal Academy of Higher Education
Oceana, 214 Marine Drive, Bombay 400 020
India

nuclear@rolta.net

Hagino, Kouichi
Yukawa Institute for Theoretical Physics
Kyoto Univ.
Oiwake-cho, Kitashirakawa, Kyoto 606-8502
Japan

hagino@yukawa.kyoto-u.ac.jp

Harissopulos, Sotirios
NCSR "Demokritos"
Institute of Nuclear Physics
153.10 Aghia Paraskevi, Athens
Greece

sharisop@inp.demokritos.gr

Hofmann, Sigurd
GSI Darmstadt
Planckstrasse 1, Darmstadt D-64291
Germany

s.hofmann@gsi.de

Ishiyama, Hironobu
KEK (Institute of Particle and Nuclear Studies)
1-1 Oho, Tsukuba-shi, Ibaraki 305-0801
Japan

hironobu.ishiyama@kek.jp

Iwamoto, Akira
Japan Atomic Energy Research Institute
Shirakata Shirane 2-4, Tokai, Naka,
Ibaraki 319-1195
Japan

iwamoto@hadron01.tokai.jaeri.go.jp

Junghans, Arnd
Forschungszentrum Rossendorf
Postfach 510119, Dresden D-01314
Germany

A.Junghans@fz-rossendorf.de

Hanappe, Francis
PNTPM Univ. Libre de Bruxelles
CP229, Avenue F.D. Roosevelt, 50, Bruxelles
B-1050
Belgium

fhanappe@ulb.ac.be

Hashizume, Kazuaki
Department of Physics, Konan Univ.
8-9-1 Okamoto Higashinada-ku Kobe
658-8501
Japan

dn321002@center.konan-u.ac.jp

Ikezoe, Hiroshi
Japan Atomic Energy Research Institute
Shirakata Shirane 2-4, Tokai, Naka, Ibaraki
319-1195
Japan

ikezoe@popsvr.tokai.jaeri.go.jp

Ivanyuk, Fedor
GSI Darmstadt
Planckstrasse 1, Darmstadt D-64291
Germany

F.Ivanyuk@theory.gsi.de

Janas, Zenon
Institute of Experimental Physics
Univ. of Warsaw
ul. Hoza 69, Warsaw PL 00-681
Poland

janas@mimuw.edu.pl

Kajino, Taka
National Astronomical Observatory
2-21-1 Osawa, Mitaka,
Tokyo 181-8588
Japan

kajino@nao.ac.jp

Karataglidis, Steven
CEA/Bruyeres-le-Chatel
CEA/DIF/DPTA/SPN, B.P. 12,
Bruyeres-le-Chatel F-91680
France

steven.karataglidis@cea.fr

Kawai, Shingo
Department of Physics, Konan Univ.
8-9-1 Okamoto Higashinada-ku Kobe
658-8501
Japan

mn221006@center.konan-u.ac.jp

Kiener, Juergen
CSNSM
Bat. 104, Orsay Campus F-91405
France

kiener@csnsm.in2p3.fr

Knodlseder, Jurgen
C.E.S.R.
9, avenue du Colonel-Roche, Toulouse F-31028
France

knodlseder@cesr.fr

Kozouline, Edouard
Flerov Laboratory of Nuclear Reactions JINR
Joliot Curie 6, Dubna, Moscow region 141980
Russia

kozulin@jinr.ru

Leleux, Pierre
Univ. of Louvain
Chemin du Cyclotron 2, Louvain-la-Neuve
B-1348
Belgium

leleux@fynu.ucl.ac.be

Kato, Kiyoshi
Department of Physics, Hokkaido Univ.
Kita-8 Nishi-10, Sapporo 060-0810
Japan

kato@nucl.sci.hokudai.ac.jp

Khan, Elias
Institut de Physique Nucleaire
Orsay cedex F-91406
France

khan@ipno.in2p3.fr

Kinoshita, Maki
Department of Physics, Konan Univ.
8-9-1 Okamoto Higashinada-ku
Kobe 658-8501
Japan

mn321004@center.konan-u.ac.jp

Koura, Hiroyuki
RIKEN
Hirosawa 2-1, Wako, Saitama 351-0198
Japan

koura@aoni.waseda.jp

Kumar, Ajay
Cyclotron Laboratory, Physics Department
Panjab Univ.
CHANDIGARH 160014
India

atyagi44@indiatimes.com

Lewitowicz, Marek
GANIL
Bd Henri Becquerel - BP 55027 - CAEN
CEDEX 5 F-14076
France

lewitowicz@ganil.fr

Leya, Ingo
Inst. for Isotope Geology and Mineral Resources ETH Zurich
Sonneggstrass 5, Zurich CH-8092
Switzerland

Leya@erdw.ethz.ch

Marques, Miguel
LPC Caen
6 Bd du Marechal Juin CAEN cedex F-14050
France

marques@caelav.in2p3.fr

Mazzocco, Marco
Dipartimento di Fisica "Galileo Galilei" & INFN Univ. degli Studi di Padova
via Marzolo 8, Padova I-35131
Italy

marco.mazzocco@pd.infn.it

Morita, Kosuke
Cyclotron Center RIKEN
Hirosawa 2-1, Wako, Saitama 351-0198
Japan

morita@rarfaxp.riken.go.jp

Motobayashi, Tohru
Heavy Ion Nuclear Physics Laboratory RIKEN
Hirosawa 2-1, Wako, Saitama 351-0198
Japan

motobaya@riken.jp

Muenzenberg, Gottfried
GSI Darmstadt
Planckstrasse 1, Darmstadt D-64291
Germany

g.muenzenberg@gsi.de

Lombardo, Umberto
Dept of Physics/INFN-LNS Catania
via S. Sofia 44, Catania I-95100
Italy

lombardo@lns.infn.it

Maruyama, Toshiki
Japan Atomic Energy Research Institute
Shirakata Shirane 2-4, Tokai, Naka, Ibaraki 319-1195
Japan

maru@hadron02.tokai.jaeri.go.jp

Möller, Peter
Los Alamos National Laboratory
T-16 MS B243 87545
USA

moller@moller.lanl.gov

Motizuki, Yuko
RIKEN
Hirosawa 2-1, Wako, Saitama 351-0198
Japan

motizuki@riken.jp

Mueller, Alex C.
CNRS/IN2P3/IPN Orsay/Accelerator Division
IPN ORSAY cedex F-91406
France

mueller@ipno.in2p3.fr

Nagai, Yasuki
Research Center for Nuclear Physics
Osaka Univ.
Mihogaoka 10-1, Ibaraki, Osaka 567-0047
Japan

nagai@rcnp.osaka-u.ac.jp

Nishikawa, Yoshihisa
Department of Physics, Konan Univ.
8-9-1 Okamoto Higashinada-ku Kobe
658-8501
Japan

mn221011@center.konan-u.ac.jp

Ohta, Masahisa
Departmenft of Physics, Konan Univ.
8-9-1 Okamoto Higashinada-ku
Kobe 658-8501
Japan

masaota@konan-u.ac.jp

Peter, Jean
LPC/ENSI Caen
6 Bd du Marechal Juin CAEN cedex F-14050
France

jpeter@in2p3.fr

Romoli, Mauro
INFN - Sezione di Napoli
Compl. Univ. Monte S. Angelo - Via Cintia - Napoli I-80126
Italy

mauro.romoli@na.infn.it

Schoenfelder, Volker
Max Planck Institute for Extraterrestrial Physics
P.o.box 1312 D-85741
Germany

vos@mpe.mpg.de

Sobiczewski, Adam
Soltan Institute for Nuclear Studies
ul. Hoza 69, Warsaw PL 00-681
Poland

sobicz@fuw.edu.pl

Nomura, Toru
KEK (Institute of Particle and Nuclear Studies)
1-1 Oho, Tsukuba-shi, Ibaraki 305-0801
Japan

toru.nomura@kek.jp

Osaka, Koji
Department of Physics, Konan Univ.
8-9-1 Okamoto Higashinada-ku
Kobe 658-8501
Japan

mn221005@center.konan-u.ac.jp

Rao, Pallem Venkata Madhusudhana
Nuclear Science Centre, New Delhi
Post Box No. 10502, New Delhi 110 067
India

madhu@nsc.ernet.in

Samyn, Mathieu
Institut d'Astronomie et d'Astrophysique
Univ. Libre de Bruxelles
Campus Plaine - CP 226, Bld du Triomphe 50,
Brussels B-1050
Belgium

msamyn@astro.ulb.ac.be

Signorini, Cosimo
Dipartimento di Fisica "Galileo Galilei" &
INFN Univ. degli Studi di Padova
via Marzolo 8, Padova I-35131
Italy

signorini@pd.infn.it

Sonnabend, Kerstin
Institut fuer Kernphysik Technische Univ. Darmstadt
Schlossgartenstr. 9, Darmstadt D-64289
Germany

kerstin@ikp.tu-darmstadt.de

Tachibana, Takahiro
Senior High School of Waseda University
3-31-1 Kamishakujii, Nerima-ku, Tokyo
177-0044
Japan

ttachi@waseda.jp

Takahashi, Noriaki
Osaka Gakuin Univ.
Kishibe-Minami 2-36-1, Suita
Osaka 564-8511
Japan

ntakahas@utc.osaka-gu.ac.jp

Teshima, Masahiro
Max-Planck-Institute for Physics
Foehringer Ring 6, 80805 Munchen D-80805
Germany

mteshima@mppmu.mpg.de

Toyoshima, Yuka
Department of Physics Konan Univ.
8-9-1 Okamoto Higashinada-ku
Kobe 658-8501
Japan

mn321009@center.konan-u.ac.jp

Villari, Antonio
GANIL
Bd Henri Becquerel - BP 55027 - CAEN
CEDEX 5 F-14076
France

villari@ganil.fr

Wada, Takahiro
Department of Physics, Konan Univ.
8-9-1 Okamoto Higashinada-ku
Kobe 658-8501
Japan

wada@konan-u.ac.jp

Takahashi, Kohji
Schwalbenweg 23, Leimen D-69181
Germany

K.Takahashi@gsi.de

Teranishi, Takashi
Center for Nuclear Study
Univ. of Tokyo/RIKEN Campus
Hirosawa 2-1, Wako, Saitama 351-0198
Japan

teranisi@cns.s.u-tokyo.ac.jp

Tinyakov, Peter
Institute of Theoretical Physics
Univ. of Lausanne
BSP-Dorigny CH-1015
Switzerland

peter.tinyakov@cern.ch

Utsunomiya, Hiroaki
Department of Physics, Konan Univ.
8-9-1 Okamoto Higashinada-ku
Kobe 658-8501
Japan

hiro@konan-u.ac.jp

von Oertzen, Wolfram
Hahn Meitner Institut Berlin
Glienickerstr.100, Berlin D-14109
Germany

oertzen@hmi.de

Weick, Helmut
GSI Darmstadt
Planckstrasse 1, Darmstadt D-64291
Germany

h.weick@gsi.de

Wieleczko, Jean-Pierre
GANIL
Bd Henri Becquerel - BP 55027 - CAEN
CEDEX 5 F-14076
France

wieleczko@ganil.fr

Y. Hara, Kaoru
Department of Physics, Konan Univ.
8-9-1 Okamoto Higashinada-ku
Kobe 658-8501
Japan

kaoru@konan-u.ac.jp

Yamamoto, Tokonatsu
Center for Cosmological, Physics Univ. of Chicago
5640 S. Ellis Ave, Chicago, IL 60637
USA

yamamoto@cfcp.uchicago.edu

Wolski, Roman
Flerov Laboratory of Nuclear Reactions JINR
Joliot Curie 6, Dubna, Moscow region 141980
Russia

wolski@lnr.jinr.ru

Yamagata, Tamio
Department of Physics, Konan Univ.
8-9-1 Okamoto Higashinada-ku
Kobe 658-8501
Japan

yamagata@center.konan-u.ac.jp

Zagrebaev, Valeri
Flerov Laboratory of Nuclear Reactions JINR
Joliot Curie 6, Dubna, Moscow region 141980
Russia

valeri.zagrebaev@jinr.ru

AUTHOR INDEX

A

Adahchour, A., 341
Akimune, H., 253, 261, 439, 551
Angulo, C., 341
Aoki, W., 488
Arai, K., 283
Aritomo, Y., 139, 147, 557
Asano, T., 111
Äystö, J., 526

B

Baba, H., 447
Barbui, M., 273
Barua, S., 563
Batist, L., 176
Bauge, E., 385
Blank, B., 159
Blazhev, A., 176
Bogatchev, A., 139
Bonaccorso, A., 218
Bonetti, R., 202
Borzov, I. N., 408
Bouchat, V., 139
Brüchle, W., 176
Brondi, A., 273
Bulgac, A., 483
Burgio, G. F., 473

C

Caballero, L., 185
Catford, W. N., 185
Chapman, R., 185
Chiba, S., 488, 519
Cinausero, M., 273
Coc, A., 341

D

Das, J. J., 563
Das, S. K., 453
Dasgupta, M., 82
Datar, V. M., 563
Datta, S. K., 501
De Francesco, A., 202
Demetriou, P., 540
Dendooven, P., 526
Denisov, V. Y., 92
de Oliveira Santos, F., 195
De Rosa, A., 202
Descouvemont, P., 341
Dhiman, S. K., 563
Di Pietro, M., 202
Döring, J., 176
Dorvaux, O., 139
Drozdov, V. A., 130, 507, 513

E

Eremenko, D. O., 130, 507, 513

F

Fabris, D., 273
Faestermann, T., 176
Fioretto, E., 273
Fortunato, L., 273
Fotina, O. V., 130, 507, 513
Fuchi, Y., 453
Fujiwara, M., 253, 261
Fukuchi, T., 447
Fukuda, N., 212
Fukuda, T., 212, 453
Fukushima, A., 557
Fülöp, Z., 447
Furukawa, T., 453
Fushimi, K., 253, 261

G

Gierlik, M., 176
Giot, L., 301
Glodariu, T., 202
Gloos, K., 526
Goko, S., 439, 546, 551
Golda, K. S., 501

Gono, Y., 447
Goriely, G., 401
Goriely, S., 369, 375, 395, 540
Górska, M., 176
Goto, S., 13
Govil, I. M., 501
Grawe, H., 176
Greene, J., 202
Greenfield, M. B., 261
Greiner, W., 3
Guglielmetti, A., 202
Guimar, V., 447

Ivanyuk, F., 120
Iwamoto, A., 49
Iwasaki, H., 447

J

Janas, Z., 176
Jeong, S.-C., 73, 453
Jhingan, A., 563
Jiang, C. L., 202
Jungclaus, A., 176
Junghans, A. R., 102

H

Haba, H., 13
Hagino, K., 82
Hanappe, F., 139, 147
Hara, K., 253, 261
Hara, K. Y., 253, 261, 439, 551
Harissopulos, S. V., 176, 422
Hashimoto, K., 261
Hashimoto, T., 447, 453
Hayakawa, T., 439
He, J. J., 447
Heenen, P.-H., 483
Heinz, A., 202
Heinz, A. M., 102
Henderson, D., 202
Hinde, D. J., 82
Hofmann, S., 21
Hokoiwa, N., 447
Honda, S., 488
Huang, W. X., 526

K

Kaji, D., 13
Kajino, T., 488
Kanungo, R., 13
Karny, M., 176
Katayama, I., 453
Kato, K., 293
Kato, S., 447
Katori, K., 13
Kavatsyuk, M., 176
Kavatsyuk, O., 176
Kawamura, T., 447
Kawase, K., 253, 261
Khan, E., 395
Kim, J. C., 447
Kinnard, V., 139
Kinoshita, M., 261
Kirchner, R., 176
Koura, H., 13, 60
Kubono, S., 447
Kudo, H., 13
Kudo, K., 551
Kulkarni, R. G., 563
Kumagai, S., 369
Kumar, A., 501
Kumar, Aj., 501
Kumar, R., 501
Kurgalin, S. D., 301
Kurokawa, M., 447

I

Ichihara, K., 253, 261
Ichikawa, S., 453
Ichikawa, T., 49, 111
Idegichi, E., 13
Ignatyuk, A. V., 102
Ikeda, K., 293
Ikezoe, H., 73, 453
Inglima, G., 202
Ishihara, M., 212
Ishikawa, T., 447, 453
Ishiyama, H., 447, 453

L

Labiche, M., 185
La Commara, M., 176, 202
La Rana, G., 273

Lee, C. S., 447
Lee, J. H., 447
Leleux, P., 432
Lemmon, R. C., 185
Leya, I., 331
Liang, J. F., 202
Lihitenthaler, R. F., 447
Lin, C.-J., 73
Lombardo, U., 473
Lui, Y.-W., 439, 551
Lunardon, M., 273

M

Madhaven, N., 563
Madhusudhana Rao, P. V., 563
Magierski, P., 483
Marqués Morano, F. M., 169
Martin, B., 202
Maruyama, T., 519
Masone, V., 202
Materna, T., 139, 147
Mathews, G. J., 488
Matsuda, M., 453
Matsui, Y., 253, 261
Matsuyama, Y., 453
Mazzocchi, C., 176
Mazzocco, M., 202, 212, 273
Mengoni, A., 463
Michimasa, S., 447
Mitsuoka, S., 73, 453
Miyatake, H., 447, 453
Mizoi, Y., 212, 453
Mohr, P., 463, 532
Mohr, Y.-W., 439
Möller, P., 49
Moon, J. Y., 447
Moore, E. F., 202
Morikawa, T., 447
Morimoto, K., 13
Morita, K., 13
Moro, R., 273
Motizuki, Y., 369
Motobayashi, T., 245
Mueller, A. C., 313
Mukha, I., 176
Muntian, I., 41
Myo, M., 293

N

Nagai, Y., 323
Nakahara, H., 111
Nakai, K., 453
Nakanishi, K., 253, 261
Nakayama, S., 253, 261
Nath, S., 563
Navin, A., 563
Newton, J. O., 82
Nishimura, M., 447
Nishimura, S., 447
Nishio, K., 73, 453
Nomura, T., 453
Notani, M., 447

O

Odahara, A., 447
Ohgaki, H., 439, 551
Ohnishi, T., 13
Ohta, M., 111, 139, 147, 439, 551, 557
Ohtsuki, T., 82
Ordine, A., 273
Orito, M., 488
Otsuki, K., 488
Ozawa, A., 13

P

Parascandolo, P., 202
Pardo, R. C., 202
Parkhomenko, O., 41
Pekola, J. P., 526
Peter, J. C., 13
Pierroutsakou, D., 202, 212
Platonov, S. Y., 130, 507, 513
Plettner, C., 176
Płochocki, A., 176
Prete, G., 273
Prokhorova, E., 139

R

Rauscher, T., 463
Ray, A., 563
Rehm, K. E., 202

Rizzi, V., 273
Roeckl, E., 176
Romoli, M., 176, 202
Roussel-Chomaz, P., 301
Rowley, N., 82
Rusek, K., 301

S

Saha, P. K., 453
Saito, A., 447
Sakurai, H., 212
Samyn, M., 395, 401, 540
Sandoli, M., 202
Sasaqui, T., 488
Sastry, D. L., 563
Sato, K., 447
Satou, K.-i, 73
Satou, Y., 447
Schädel, M., 176
Schmidt, K.-H., 102
Schmitt, C., 139
Schönfelder, V., 361
Schulze, H.-J., 473
Schwengner, R., 176
Scopel, P., 202, 273
Shibata, Y., 551
Shimoda, T., 453
Shimoura, S., 447
Shiokawa, A., 253
Shizuma, T., 439
Shyam, R., 563
Sierk, A. J., 49
Signorini, C., 202, 212, 273
Singh, G., 501
Singh, H., 501
Singh, R., 563
Singh, R. P., 501
Sinha, A. K., 563
Siwek-Wilczynska, K., 139
Sobiczewski, A., 41
Sonnabend, K., 463
Soramel, F., 202, 212, 273
Stroe, L., 202, 212, 273
Stuttgé, L., 139, 147
Suda, T., 13
Sueki, K., 13
Sugathan, P., 563

T

Takahashi, K., 418
Takahashi, N., 526
Tanaka, M., 253, 261
Tanaka, M. H., 447, 453
Tanigawa, T., 519
Tanihata, I., 13
Tatsumi, T., 519
Tchuvil'sky, Y. M., 301
Ter-Akopian, G. M., 301
Teranishi, T., 447
Terasawa, M., 453
Teshima, M., 351
Thompson, I. J., 273
Timis, C. N., 185
Tokanai, F., 13
Toyokawa, H., 439, 551
Trotta, M., 273
Tsuruta, K., 73

U

Ue, K., 447
Uritani, A., 551
Utsunomiya, H., 253, 261, 439, 546, 551

V

Vangioni-Flam, E., 341
Vardaci, E., 202, 273
Varughese, T., 563
Viesti, G., 273
Villari, A. C. C., 234
Vitturi, A., 273
Vogt, K., 463
Voskresensky, D. N., 519

W

Wada, T., 111, 557
Wakabayashi, Y., 447
Watanabe, Y. X., 212, 447, 453
Weick, H., 228
Wirzba, A., 483
Wolski, R., 301
Wuosmaa, A., 202

X

Xu, H., 13

Y

Yamagata, T., 253, 261, 439
Yamaji, S., 111
Yamamoto, T., 351
Yanagisawa, Y., 447
Yeremin, A. V., 13
Yoneda, A., 13
Yoshida, A., 13, 212
Yoshikawa, N., 453

Yosoi, M., 253, 261
Yu, Y., 483
Yuminov, O. A., 130, 507, 513

Z

Zagrebaev, V. I., 31
Zhao, Y.-L., 13
Zheng, T., 13
Zhou, X. R., 473
Zilges, A., 463
Zuo, W., 473
Żylicz, J., 176